Methods in Enzymology

Volume 381
OXYGEN SENSING

METHODS IN ENZYMOLOGY

EDITORS-IN-CHIEF

John N. Abelson   Melvin I. Simon

DIVISION OF BIOLOGY
CALIFORNIA INSTITUTE OF TECHNOLOGY
PASADENA, CALIFORNIA

FOUNDING EDITORS

Sidney P. Colowick and Nathan O. Kaplan

Methods in Enzymology

Volume 381

Oxygen Sensing

EDITED BY

Chandan K. Sen

THE OHIO STATE UNIVERSITY MEDICAL CENTER
COLUMBUS, OHIO

Gregg L. Semenza

JOHNS HOPKINS HOSPITAL
BALTIMORE, MARYLAND

ELSEVIER
ACADEMIC
PRESS

AMSTERDAM • BOSTON • HEIDELBERG • LONDON
NEW YORK • OXFORD • PARIS • SAN DIEGO
SAN FRANCISCO • SINGAPORE • SYDNEY • TOKYO

Academic Press is an imprint of Elsevier

Elsevier Academic Press
525 B Street, Suite 1900, San Diego, California 92101-4495, USA
84 Theobald's Road, London WC1X 8RR, UK

This book is printed on acid-free paper. ∞

For all information on all Academic Press publications
visit our Web site at www.academicpress.com

ISBN: 0-12-182785-2

PRINTED IN THE UNITED STATES OF AMERICA
04 05 06 07 08 9 8 7 6 5 4 3 2 1

Table of Contents

Section I. Lung and the Airways

Section II. Cardiovascular and Blood

Section III. Ion Channels

Section IV. Tumor Biology

Section V. Metabolism

Section VI. Nervous System

Section VII. General

Section VIII. Wound Healing

Section IX. Unicellular Systems

Section X. Physical Detection of Oxygen

Contributors to Volume 381

Article numbers are in parentheses and following the names of contributors.
Affiliations listed are current.

STEPHANIE ACHEBACH (41), *Institut fur Mikrobiologie und Weinforschung der, University of Mainz, Mainz 55099, Germany*

HELMUT ACKER (32), *Facharzt fuer Physiologie, Max-Planck-Intitut fuer moleculare Physiologie, Dortmund D-44227, Germany*

MARIA TERESA AGAPITO (3), *Departamento de Bioqumíca y Biología Molecular y Fisiologia, Facultad de Medicina, Universidad de Valladolid, Valladolid 47005, Spain*

MANSOOR AHMAAD (10), *Department of Physiology, New York Medical College, Valhalla, New York 10595*

JORGE E. ALBINA (34), *Division of Surgical Research, Department of Surgery, Brown Medical School, Providence, Rhode Island 02903*

SHIGETOSHI AONO (40), *Center for Integrative Bioscience, Okazaki National Research Institutes, Okazaki 444-8585, Japan*

STEPHEN G. BALL (20), *Institute for Cardiovascular Research, University of Leeds, Leeds LS2 9JT, United Kingdom*

ROBERT BARTLETT (37), *National Healing Corporation, Lexington, South Carolina 29072*

HORST D. BECKER (36), *UCSF, Wound Healing Laboratory, San Francisco, California 94143-0522*

VALERIE BERGDALL (37), *University Laboratory Animal Research, The Ohio State University Medical Center, Columbus, Ohio 43210*

MYRIAM BERNAUDIN (27), *UMR-CNRS 6551, CYCERON, BD Henri Becquerel, BP5229, F-14074, France*

KEITH BUCKLER (17), *Laboratory of Physiology, University of Oxford, Oxford OX1 3PT, United Kingdom*

DONALD G. BUERK (43), *Department of Physiology and Bioengineering, University of Pennsylvania, Philadelphia, Pennsylvania 19104-6085*

ARTURO J. CARDOUNEL (11, 28), *Laboratory of Molecular Medicine, Departments of Surgery and Molecular and Cellular Biochemistry, Davis Heart & Lung Research Institute, The Ohio State University Medical Center, Columbus, Ohio 43210*

MARILIA CASCALHO (16), *Transplantation Biology, Departments of Surgery and Immunology, Mayo Clinic, Rochester, Minnesota 55905*

JIANZHU CHEN (16), *Department of Biology, Center for Cancer Research, Massachusetts Institute of Technology, Cambridge, Massachusetts 02139*

SEAN P. COLGAN (25), *Center for Experimental Therapeutics and Reperfusion Injury, Brigham and Women's Hospital, Harvard Medical School, Boston, Massachusetts 02115*

KATRINA M. COMERFORD (25, 33), *Department of Medicine, Conway Institute, Belfield, Dublin, Ireland*

ERNEST CUTZ (2), *Division of Pathology, Department of Pediatric Laboratory Medicine, The Hospital for Sick Children, Toronto, Ontario M5G1X8, Canada*

PAMELA DAVID (42), *Department of Molecular, Cellular, and Developmental Biology, University of Colorado, Boulder, Colorado 80309-0347*

LOWELL DAVIS (15), *Department of OB/GYN, Oregon Health and Science University, Portland, Oregon 97201-3098*

MONIQUE DENAVIT-SAUBIE (29), *UPR 2216 Neurobiologie Genetique et Integrative, CNRS Institut de Neurobiologie Alfred Fessard, Yvette, France*

JOSÉ DEL CAMPO (39), *Department of Chemistry and Biochemistry, Los Angeles, California 90095-1569*

REINHARD DIRMEIER (38, 42), *Department of Molecular, Cellular, and Developmental Biology, University of Colorado, Boulder, Colorado 80309-0347*

ATHENA DODD (38, 42), *Department of Molecular, Cellular, and Developmental Biology, University of Colorado, Boulder, Colorado 80309-0347*

MARCELLA ENGLE (38), *Department of Molecular, Cellular, and Developmental Biology, University of Colorado, Boulder, Colorado 80309-0347*

JOHN R. FALCK (9), *Department of Biochemistry, University of Texas Southwestern Medical Centre, Dallas, Texas 75390-9038*

IAN M. FEARON (20), *Department of Biology, McMaster University, Hamilton, Ontario L8S 4K1, Canada*

TSUNEO K. FERGUSON (30), *Department of Genome Science, University of Cincinnati College of Medicine, Cincinnati, Ohio 45267-0505*

JEFFERSON C. FRISBEE (9), *Department of Physiology, Medical College of Wisconsin, Milwaukee, Wisconsin 53226-1408*

MEGAN C. FROST (45), *Department of Chemistry, University of Michigan, Ann Arbor, Michigan 48109*

X. W. FU (2), *Division of Pathology, Department of Pediatric Laboratory Medicine, The Hospital for Sick Children, Toronto, Ontario M5G1X8, Canada*

PETER W. GAGE (19), *Membrane Biology Program, John Curtin School of Medical Research, Australian National University, Canberra 2601, Australia*

MARK W. GERACI (5), *Pulmonary Hypertension Center, Division of Pulmonary Sciences and Critical Care Medicine, University of Colorado Health Sciences Center, Denver, Colorado*

Q. PERVEEN GHANI (36), *UCSF, Wound Healing Laboratory, San Francisco, California 94143-0522*

AMATO J. GIACCIA (26), *Division of Radiation and Cancer Biology, Department of Radiation Oncology, Stanford University, Stanford, California 94305*

CONSTANCIO GONZALEZ (3), *Departamento de Bioquímica y Biología Molecular y Fisiologia, Facultad de Medicina, Universidad de Valladolid, Valladolid 47005, Spain*

GAYLE M. GORDILLO (37), *Laboratory of Molecular Medicine, Department of Surgery, The Ohio State University Medical Center, Columbus, Ohio 43210*

JEFF GORMAN (31), *Institute for Molecular Bioscience, Queensland Bioscience Precinct, University of Queensland, St. Lucia, Queensland 4072, Australia*

SACHIN A. GUPTE (10), *Department of Physiology, New York Medical College, Valhalla, New York 10595*

ANNA K. M. HAMMARSTROM (19), *Membrane Biology Program, John Curtin School of Medical Research, Australian National University, Canberra 2601, Australia*

MICHAEL HOCKEL (23), *Department of Obstetrics and Gynecology, University of Leipzig, Leipzig 04103, Germany*

ROGER HOHIMER (15), *Department of OB/ GYN, Oregon Health and Science University, Portland, Oregon 97201-3098*

ERIC HONORE (17), *Institut de Pharmacologie, Moleculaire et Cellulaire, Valbonne, France*

HARRIET W. HOPF (35), *Department of Surgery, UCSF, Wound Healing Laboratory, San Francisco, California 94143-0522*

CHRISTINE HUCKSTORF (32), *Institut fur Physiologie, Medizinische Fakultät, Universitat Rostock, Rostock D-18057, Germany*

LISA M. HUMPHREY (35), *Department of Anesthesia and Perioperative Care, University of California, San Francisco, San Francisco, California 94143-0522*

THOMAS K. HUNT (35, 36), *Department of Surgery, UCSF, Wound Healing Laboratory, San Francisco, California 94143-0522*

ZAMIR HUSSAIN (36), *Department of Preventative and Restorative Dental Sciences, University of California, San Francisco, San Francisco, California 94143*

GOVINDASAMY ILANGOVAN (48), *Biomedial EPR Spectroscopy & Imaging Center, The Ohio State University, Columbus, Ohio 43210*

DAVID ILES (18), *School of Medicine, University of Leeds, Leeds LS2 9JT, United Kingdom*

KAIKOBAD IRANI (12), *Division of Cardiology, Department of Medicine, The Johns Hopkins University School of Medicine, Baltimore, Maryland 21205*

MIRCEA IVAN (22), *Dana-Farber Cancer Institute, Brigham and Women's Hospital, Harvard Medical School, Boston, Massachusetts 02115*

JAROSLAV JELINEK (14), *Department of Medicine, Hematology, and Oncology, Baylor College of Medicine, Houston, Texas 77030*

BRENDAN T. JONES (16), *Department of Biology, Center for Cancer Research, Massachusetts Institute of Technology, Cambridge, Massachusetts 02139*

WILLIAM G. KAELIN, JR. (22), *Dana-Farber Cancer Institute, Harvard Medical School, Howard Hughes Medical Institute, Boston, Massachusetts 02115*

PAWEL M. KAMINSKI (10), *Department of Physiology, New York Medical College, Valhalla, New York 10595*

PAUL J. KEMP (1, 18), *School of Biomedical Sciences, University of Leeds, Leeds LS2 9JT, United Kingdom*

SAVITA KHANNA (8, 11, 13), *Laboratory of Molecular Medicine, Departments of Surgery and Molecular and Cellular Biochemistry, Davis Heart & Lung Research Institute, The Ohio State University Medical Center, Columbus, Ohio 43210*

THOMAS KIETZMANN (24), *Institut fur Biochemie und Molekulare Zellbiologie, Universitat Gottingen, Gottingen D-37073, Germany*

WILLIAM KIM (22), *Dana-Farber Cancer Institute, Harvard Medical School, Boston, Massachusetts 02115*

TEIZO KITAGAWA (40), *Center for Integrative Bioscience, Okazaki National Research Institutes, Okazaki, Japan 444-8585*

HIDEFUMI KOJIMA (16), *Division of Immunology, Institute for Medical Science, Dokkyo University School of Medicine, Tochigi 321-093, Japan*

GANESH K. KUMAR (6), *Department of Physiology and Biophysics, School of Medicine, Case Western Reserve University, Cleveland, Ohio 44106*

PERIANNAN KUPPUSAMY (48), *Center for Biomedical EPR and Spectroscopy and Imaging, Davis Heart & Lung Research Insitute, The Ohio State University, Columbus, Ohio 43210*

DAVID LANDO (31), *Wellcome Trust/ Cancer Research United Kingdom Institute, Cambridge CB2 1QR, United Kingdom*

MARTIN O. LEONARD (33), *The Conway Institute for Biomolecular and Biomedical Research, University College Dublin, Belfield, Dublin 4, Ireland*

JULIAN LOMBARD (9), *Department of Physiology, Medical College of Wisconsin, Milwaukee, Wisconsin 53226-1408*

NORMA MASSON (21), *The Henry Wellcome Building of Genomic Medicine, University of Oxford, Oxford OX3 7BN, United Kingdom*

ARNULF MAYER (23), *Institute of Physiology and Pathophysiology, University of Mainz, Mainz D-55099, Germany*

SHARON MCGRATH (5), *Division of Cardiopulmonary Pathology, Department of Pathology, The Johns Hopkins University School of Medicine, Baltimore, Maryland 21205*

SABEEHA MERCHANT (39), *Department of Chemistry and Biochemistry, Los Angeles, California 90095-1569*

MARK E. MEYERHOFF (45), *Department of Chemistry, University of Michigan, Ann Arbor, Michigan 48109*

DAVID E. MILLHORN (30), *Department of Genome Science, University of Cincinnati College of Medicine, Cincinnati, Ohio 45267-0505*

JUNG-HYUN MIN (22), *Howard Hughes Medical Institute, Memorial Sloan-Kettering Cancer Center, Harvard Medical School, Boston, Massachusetts 02115*

MARIE-PIERRE MORIN-SURUN (29), *UPR 2216 Neurobiologie Genetique et Integrative, CNRS Institut de Neurobiologie Alfred Fessard, Yvette, France*

HIROSHI NAKAJIMA (40), *Center for Integrative Bioscience, Okazaki National Research Institutes, Okazaki, Japan 444-8585*

ANA OBESO (3), *Departamento de Bioquímica y Biología Molecular y Fisiologia, Facultad de Medicina, Universidad de Valladolid, Valladolid 47005, Spain*

KRISTIN O'BRIEN (38, 42), *Department of Molecular, Cellular, and Developmental Biology, University of Colorado, Boulder, Colorado 80309-0347*

TAKEHIRO OHTA (40), *Center for Integrative Bioscience, Okazaki National Research Institutes, Okazaki, Japan 444-8585*

SUSAN O'REILLY (33), *The Conway Institute for Biomolecular and Biomedical Research, University College Dublin, Belfield, Dublin 4, Ireland*

JEFFERY L. OVERHOLT (6), *Department of Physiology and Biophysics, School of Medicine, Case Western Reserve University, Cleveland, Ohio 44106*

MICHITAKA OZAKI (11), *Division of Cardiology, Department of Medicine, Johns Hopkins University School of Medicine, Baltimore, Maryland 21205*

PAUL PANTANO (47), *Department of Chemistry, University of Texas at Dallas, Richardson, Texas 75083*

DMITRI PAPKOVSKY (46), *Biochemistry Department, University College Cork, Cork, Ireland*

OLIVIER PASCUAL (29), *Department of Neuroscience, University of Pennsylvania School of Medicine, Philadelphia, Pennsylvania 19104*

CHRIS PEERS (1, 18, 20), *School of Medicine, Institute for Cardiovascular Research, University of Leeds, Leeds LS2 9JT, United Kingdom*

DANIEL J. PEET (31), *Department of Molecular Biosciences (Biochemistry), Centre for Molecular Genetics of Development, University of Adelaide, Adelaide, South Australia, Australia*

YING-JIE PENG (6), *Department of Physiology and Biophysics, School of Medicine, Case Western Reserve University, Cleveland, Ohio 44106*

JEAN-MARC PEQUIGNOT (29), *UPR 2216 Neurobiologie Genetique et Integrative, CNRS Institut de Neurobiologie Alfred Fessard, Yvette, France*

ROBERT O. POYTON (38, 42), *Department of Molecular, Cellular, and Developmental Biology, University of Colorado, Boulder, Colorado 80309-0347*

NANDURI R. PRABHAKAR (6), *Department of Physiology and Biophysics, School of Medicine, Case Western Reserve University, Cleveland, Ohio 44106*

JOSEF PRCHAL (14), *Department of Medicine, Hematology, and Oncology, Baylor College of Medicine, Houston, Texas 77030*

JEANETTE M. QUINN (39), *Stratagene, La Jolla, California 92037*

PETER J. RATCLIFFE (21), *The Henry Wellcome Building of Genomic Medicine, University of Oxford, Oxford OX3 7BN, United Kingdom*

JONATHAN S. REICHNER (34), *Division of Surgical Research, Department of Surgery, Brown Medical School, Providence, Rhode Island 02903*

MARK D. ROLLINS (35), *Department of Anesthesia and Perioperative Care, University of California, San Francisco, San Francisco, California 94143-0522*

RICHARD J. ROMAN (9), *Department of Physiology, Medical College of Wisconsin, Milwaukee, Wisconsin, 53226*

JEAN-CHRISTOPHE ROUX (29), *Laboratoire de physiologie des regulations energetique, Cellulaires et Moleculaires, Lyon cedex 08 69373, France*

SASHWATI ROY (8, 13), *Laboratory of Molecular Medicine, Departments of Surgery and Molecular and Cellular Biochemistry, Davis Heart & Lung Research Institute, The Ohio State University Medical Center, Columbus, Ohio 43210*

GLORIA SANZ-ALFAYATE (3), *Departamento de Bioquímica y Biología Molecular y Fisiologia, Facultad de Medicina, Universidad de Valladolid, Valladolid 47005, Spain*

HEINRICH SAUER (32), *Zentrum Physiologie und Pathophysiologie, Institut fur Neurophysiologie, Robert-Kosch Str. 39, Koln D-50931, Germany*

HEINZ SCHEUENSTUHL (35), *Department of Surgery, UCSF, Wound Healing Laboratory, San Francisco, California 94143-0522*

RICHARD SCHLANGER (37), *Department of Surgery, The Ohio State University Medical Center, Columbus, Ohio 43210*

RUTH A. SCHMITZ (41), *Institute for Microbiology and Genetics, University of Göttingen, Göttingen 37077, Germany*

ARNOLD SCHWARTZ (20), *Institute of Molecular Pharmacology and Biophysics, University of Cincinnati, Cincinnati, Ohio 45267-0828*

JASON L. SCRAGG (20), *Institute for Cardiovascular Research, University of Leeds, Leeds LS2 9JT, United Kingdom*

GREGG L. SEMENZA (7), *Department of Pediatrics, Medicine, and Radiation Oncology, Institute of Genetic Medicine, The Johns Hopkins University School of Medicine, Baltimore, Maryland 21287-3914*

CHANDAN K. SEN (8, 11, 13, 28), *Laboratory of Molecular Medicine, Departments of Surgery and Molecular and Cellular Biochemistry, Davis Heart & Lung Research Institute, The Ohio State University Medical Center, Columbus, Ohio 43210*

KAREN A. SETA (30), *Department of Genome Science, University of Cincinnati College of Medicine, Cincinnati, Ohio 45267-0505*

FRANK R. SHARP (27), *Department of Neurology and Neuroscience Program, University of Cincinnati, Cincinnati, Ohio 45267*

A. DEAN SHERRY (47), *Department of Chemistry, University of Texas at Dallas, Richardson, Texas 75083*

LARRISSA A. SHIMODA (7), *Division of Pulmonary and Critical Care Medicine, Department of Medicine, The Johns Hopkins University School of Medicine, Baltimore, Maryland 21224*

RICHARD SHLANGER (37), *509 Davis Heart & Lung Research Institute, The Ohio State University Medical Center, Columbus, Ohio 43210*

MICHAIL V. SITKOVSKY (16), *Biochemistry and Immunopharmacology Section, Laboratory of Immunology, National Institute of Allergy and Infectious Diseases, National Institutes of Health, Bethesda, Maryland 20892*

VLADIMIR A. SNETKOV (4), *Department of Asthma, Allergy, and Respiratory Science, Guy's King's and St. Thomas School of Medicine, King's College London, London SE1 9RT, United Kingdom*

CHRISTOPHE SOULAGE (29), *Laboratoire de physiologie des regulations energetique, Cellulaires et Moleculaires, Lyon cedex 08 69373, France*

ERICK SPEARS (38), *Department of Molecular, Cellular, and Developmental Biology, University of Colorado, Boulder, Colorado 80309-0347*

TINO STRELLER (32), *Facharzt fuer Physiologie, Max-Planck-Intitut fuer moleculare Physiologie, Dortmund D-44227, Germany*

LILLIAN M. SWIERSZ (26), *Department of Gynecology and Obstetrics, Stanford University, Stanford, California 94305*

CORMAC T. TAYLOR (33), *The Conway Institute for Biomolecular and Biomedical Research, University College Dublin, Belfield, Dublin 4, Ireland*

ARUN K. TEWARI (13), *Department of Physiology and Cell Biology, Davis Heart & Lung Research Institute, The Ohio State University Medical Center, Columbus, Ohio 43210*

RUBIN M. TUDER (5), *Division of Cardiopulmonary Pathology, Department of Pathology, The Johns Hopkins University School of Medicine, Baltimore, Maryland 21205*

GOTTFRIED UNDEN (41), *Institut fur Mikrobiologie und Weinforschung der, University of Mainz, Mainz 55099, Germany*

GYULA VARADI (20), *Institute of Molecular Pharmacology and Biophysics, Denver, Colorado 80206, University of Cincinnati, Cincinnati, Ohio 45267-0828*

PETER VAUPEL (23), *Institute of Physiology and Pathophysiology, University of Mainz, Mainz D-55099, Germany*

NORBERT F. VOELKEL (5), *Division of Pulmonary and Critical Care Sciences, Department of Medicine, Health Sciences Center, University of Colorado, Denver, Colorado 80206*

SILVIA WAGNER (36), *UCSF, Wound Healing Laboratory, San Francisco, California 94143-0522*

WILLIAM A. WALLACE (11, 28, 37), *Laboratory of Molecular Medicine, Departments of Surgery and Molecular and Cellular Biochemistry, Davis Heart & Lung Research Institute, The Ohio State University Medical Center, Columbus, Ohio 43210*

JEREMY P. T. WARD (4), *Department of Asthma, Allergy, and Respiratory Science, Guy's King's and St. Thomas School of Medicine, King's College London, London SE1 9RT, United Kingdom*

MARIA WARTERNBERG (32), *Zentrum Physiologie und Pathophysiologie, Institut fuer Neurophysiologie, Universitat Koeln, Koeln D-50931, Germany*

XIAO WEN FU (2), *Department of Pediatric Laboratory Medicine, The Research Institute, The Hospital for Sick Children, Toronto, Ontario M5G1X8, Canada*

JUDITH M. WEST (35), *Department of Surgery, UCSF, Wound Healing Laboratory, San Francisco, California 94143-0522*

DEAN A. WHELAN (31), *Monash Institute for Reproduction and Development, Monash Medical Centre, Monash University, Clayton, Victoria, Australia*

MURRAY L. WHITELAW (31), *Department of Molecular Biosciences (Biochemistry), Centre for Molecular Genetics of Development, University of Adelaide, Adelaide, South Australia, Australia*

DAVID F. WILSON (44), *Department of Biochemistry and Biophysics, University of Pennsylvania, Philadelphia, Pennsylvania 19104-6059*

MICHAEL S. WOLIN (10), *Department of Physiology, New York Medical College, Valhalla, New York 10595*

HAIEFENG YANG (22), *Dana-Farber Cancer Institute, Harvard Medical School, Boston, Massachusetts 02115*

HERMAN YEGER (2), *Division of Pathology, Department of Pediatric Laboratory Medicine, The Hospital for Sick Children, Toronto, Ontario M5G1X8, Canada*

ZHONG YUN (26), *Division of Radiation and Cancer Biology, Department of Radiation Oncology, Stanford University, Stanford, California 94305*

YADONG ZHAO (47), *Department of Chemistry, University of Texas at Dallas, Richardson, Texas 75083*

PIYU ZHAO (47), *Department of Chemistry, University of Texas at Dallas, Richardson, Texas 75083*

JAY L. ZWEIER (48), *Center for Biomedical EPR Spectroscopy and Imaging, Davis Heart & Lung Research Insitute, The Ohio State University, Columbus, Ohio 43210*

Preface

Cellular O_2 concentrations are maintained within a narrow range (perceived as "normoxia") due to the risk of oxidative damage from excess O_2 (hyperoxia), and of metabolic demise from insufficient O_2 (hypoxia). Oxygen partial pressure (pO_2) ranges from 90 to below 3 Torr in mammalian organs under normoxic conditions with arterial pO_2 of about 100 Torr or $\sim14\%$ O_2. Thus, "normoxia" for cells is a variable that is dependent on the specific localization of the cell in organs and functional status of the specific tissue in question. O_2-sensing signaling responses are required to adjust to physiological or pathophysiological variations in pO_2. Whereas acute responses often entail changes in the activity of pre-existing proteins, chronic responses invariably involve oxygen-sensitive changes in signal transduction and gene expression. There is a vast body of literature addressing the biological mechanisms underlying the sensing and management of hypoxia. For example, a hypoxic microenvironment serves as a cue to drive angiogenesis with the objective to correct the compromised state of oxygenation at the affected site.

Recent studies have identified that the exposure of biological cells to a state of oxygenation that is in excess of the pO_2 to which the cells are adjusted results in perceived hyperoxia. Perceived hyperoxia is characterized by cellular responses that may range from growth arrest, extracellular matrix reorganization, and differentiation to oxidative injury, depending on the extent and duration of hyperoxic insult. On one hand, perceived hyperoxia seems to be a significant event in ischemia-reoxygenation biology including post-reoxygenation tissue remodeling. On the other hand, it leads to the notion that biological cells are not "blind" to a change of O_2 concentration from $<10\%$ (*in vivo* normoxia) to $\sim20\%$ (interpreted as *in vitro* normoxia). Indeed, recent evidence establishes that simple exposure of primary cells, isolated from an organ, to room air significantly influences the transcriptome profile as well as cellular functionality. These developments lay the rationale to revisit the relevance of studying cell biology under culture conditions involving room air ambience.

The ability of cells to sense and respond to changes in oxygenation underlies a multitude of developmental, physiological, and pathological processes. This volume of *Methods in Enzymology* provides a comprehensive compendium of experimental approaches to the study of oxygen sensing in 48 chapters that are written by leaders in their fields. The excellent editorial assistance of Dr. Savita Khanna and the outstanding contribution of authors are gratefully acknowledged. We hope that this volume will contribute to the further development of this fundamentally important field of biomedical research.

METHODS IN ENZYMOLOGY

VOLUME 72. Lipids (Part D)
Edited by JOHN M. LOWENSTEIN

VOLUME 73. Immunochemical Techniques (Part B)
Edited by JOHN J. LANGONE AND HELEN VAN VUNAKIS

VOLUME 74. Immunochemical Techniques (Part C)
Edited by JOHN J. LANGONE AND HELEN VAN VUNAKIS

VOLUME 75. Cumulative Subject Index Volumes XXXI, XXXII, XXXIV–LX
Edited by EDWARD A. DENNIS AND MARTHA G. DENNIS

VOLUME 76. Hemoglobins
Edited by ERALDO ANTONINI, LUIGI ROSSI-BERNARDI, AND EMILIA CHIANCONE

VOLUME 77. Detoxication and Drug Metabolism
Edited by WILLIAM B. JAKOBY

VOLUME 78. Interferons (Part A)
Edited by SIDNEY PESTKA

VOLUME 79. Interferons (Part B)
Edited by SIDNEY PESTKA

VOLUME 80. Proteolytic Enzymes (Part C)
Edited by LASZLO LORAND

VOLUME 81. Biomembranes (Part H: Visual Pigments and Purple Membranes, I)
Edited by LESTER PACKER

VOLUME 82. Structural and Contractile Proteins (Part A: Extracellular Matrix)
Edited by LEON W. CUNNINGHAM AND DIXIE W. FREDERIKSEN

VOLUME 83. Complex Carbohydrates (Part D)
Edited by VICTOR GINSBURG

VOLUME 84. Immunochemical Techniques (Part D: Selected Immunoassays)
Edited by JOHN J. LANGONE AND HELEN VAN VUNAKIS

VOLUME 85. Structural and Contractile Proteins (Part B: The Contractile
Apparatus and the Cytoskeleton)
Edited by DIXIE W. FREDERIKSEN AND LEON W. CUNNINGHAM

VOLUME 86. Prostaglandins and Arachidonate Metabolites
Edited by WILLIAM E. M. LANDS AND WILLIAM L. SMITH

VOLUME 87. Enzyme Kinetics and Mechanism (Part C: Intermediates,
Stereo-chemistry, and Rate Studies)
Edited by DANIEL L. PURICH

VOLUME 88. Biomembranes (Part I: Visual Pigments and Purple Membranes, II)
Edited by LESTER PACKER

VOLUME 89. Carbohydrate Metabolism (Part D)
Edited by WILLIS A. WOOD

VOLUME 226. Metallobiochemistry (Part C: Spectroscopic and
Physical Methods for Probing Metal Ion Environments in Metalloenzymes
and Metalloproteins)
Edited by JAMES F. RIORDAN AND BERT L. VALLEE

VOLUME 227. Metallobiochemistry (Part D: Physical and Spectroscopic
Methods for Probing Metal Ion Environments in Metalloproteins)
Edited by JAMES F. RIORDAN AND BERT L. VALLEE

VOLUME 228. Aqueous Two-Phase Systems
Edited by HARRY WALTER AND GÖTE JOHANSSON

VOLUME 229. Cumulative Subject Index Volumes 195–198, 200–227

VOLUME 230. Guide to Techniques in Glycobiology
Edited by WILLIAM J. LENNARZ AND GERALD W. HART

VOLUME 231. Hemoglobins (Part B: Biochemical and Analytical Methods)
Edited by JOHANNES EVERSE, KIM D. VANDEGRIFF, AND ROBERT M. WINSLOW

VOLUME 232. Hemoglobins (Part C: Biophysical Methods)
Edited by JOHANNES EVERSE, KIM D. VANDEGRIFF, AND ROBERT M. WINSLOW

VOLUME 233. Oxygen Radicals in Biological Systems (Part C)
Edited by LESTER PACKER

VOLUME 234. Oxygen Radicals in Biological Systems (Part D)
Edited by LESTER PACKER

VOLUME 235. Bacterial Pathogenesis (Part A: Identification and Regulation of
Virulence Factors)
Edited by VIRGINIA L. CLARK AND PATRIK M. BAVOIL

VOLUME 236. Bacterial Pathogenesis (Part B: Integration of Pathogenic
Bacteria with Host Cells)
Edited by VIRGINIA L. CLARK AND PATRIK M. BAVOIL

VOLUME 237. Heterotrimeric G Proteins
Edited by RAVI IYENGAR

VOLUME 238. Heterotrimeric G-Protein Effectors
Edited by RAVI IYENGAR

VOLUME 239. Nuclear Magnetic Resonance (Part C)
Edited by THOMAS L. JAMES AND NORMAN J. OPPENHEIMER

VOLUME 240. Numerical Computer Methods (Part B)
Edited by MICHAEL L. JOHNSON AND LUDWIG BRAND

VOLUME 241. Retroviral Proteases
Edited by LAWRENCE C. KUO AND JULES A. SHAFER

VOLUME 242. Neoglycoconjugates (Part A)
Edited by Y. C. LEE AND REIKO T. LEE

VOLUME 243. Inorganic Microbial Sulfur Metabolism
Edited by HARRY D. PECK, JR., AND JEAN LEGALL

Section I

Lung and the Airways

[1] Assessment of Oxygen Sensing by Model Airway and Arterial Chemoreceptors

By PAUL J. KEMP and CHRIS PEERS

Introduction

1988 was a landmark year in the field of O_2 sensing. In that year, the first publication appeared that placed ion channels in the spotlight as key molecules involved in O_2 chemoreception by the longest-studied O_2 sensing organ, the carotid body. For decades, researchers had understood the importance of the carotid body as a sensory organ that provides key afferent information concerning blood gas status to the central nervous system and that this information initiates cardiorespiratory reflexes designed to optimize O_2 collection and delivery. However, the cellular mechanisms by which the chemosensory elements of the carotid body type I cells responded to hypoxia, hypercapnia, or acidosis to increase afferent carotid sinus nerve discharge remained elusive. The demonstration of Lopez-Barneo et al.[1] that acute hypoxia could reversibly inhibit K^+ channels in type I cells immediately provided a candidate mechanism for type I cell chemoreception: selective K^+ channel inhibition sets in motion a series of events that lead to neurotransmitter release via membrane depolarization, increased excitability, activation of voltage-gated Ca^{2+} channels, and increased Ca^{2+} influx.

In the intervening years, the role of ion channels in O_2 chemoreception has received much support not only in the carotid body, but also in a variety of other O_2-sensitive tissues.[2,3] In the pulmonary vasculature, the involvement of K^+ channels in hypoxic vasoconstriction is well supported,[4] but is by no means uncontested.[5] Less controversially, hypoxic vasodilation in the systemic vasculature appears to be dependent on the ability of hypoxia to inhibit selectively voltage-gated L-type Ca^{2+} channels.[6] Access to vascular smooth muscle cells is relatively straightforward, given the abundance of available tissue and the ease with which a pure cell population can be harvested, and so research in this tissue has progressed steadily.

[1] J. Lopez-Barneo, J. R. Lopez-Lopez, J. Urena, and C. Gonzalez, *Science* **241**, 580 (1988).
[2] C. Peers, *Trends Pharmacol. Sci.* **18**, 405 (1997).
[3] J. Lopez-Barneo, R. Pardal, and P. Ortega-Saenz, *Annu. Rev. Physiol.* **63**, 259 (2001).
[4] E. K. Weir and S. L. Archer, *FASEB J.* **9**, 183 (1995).
[5] J. P. Ward and P. I. Aaronson, *Respir. Physiol.* **115**, 261 (1999).
[6] A. Franco-Obregon, J. Urena, and J. Lopez-Barneo, *Proc. Natl. Acad. Sci. USA* **92**, 4715 (1995).

In contrast, the carotid body is a very small, inaccessible organ and this has limited research considerably. However, such limitations are minimal when compared to the studies to date of another important O_2 sensing tissue, the neuroepithelial body (NEB) of the lung.[7] NEBs are discrete clusters of cells, located primarily at airway bifurcations, and may serve as airway chemoreceptors. While information concerning the importance of their role as physiological O_2 sensors is relatively limited, their location, innervation, and morphology suggest strongly that they can respond to hypoxia in a manner comparable to that of the carotid body, potentially initiating afferent nerve activity and regulating local blood flow via the release of vasoactive amines and peptides.[8] Their paucity, especially in the adult lung, has clearly impeded progress toward understanding both their role in O_2 homeostasis and the cellular mechanisms underlying their ability to sense hypoxia. Technical obstacles notwithstanding, a number of elegant studies have demonstrated clearly that they possess O_2-sensitive K^+ channels, hypoxic inhibition of which leads to cell depolarization, and Ca^{2+}-dependent neurotransmitter release.[9–11]

Despite these significant advances in our understanding of O_2 sensing by specific chemosensory tissues, many questions remain unanswered. Of primary interest is the molecular nature of the O_2 sensor itself and the identity of the specific ion channels that are selectively modulated by hypoxia. Various groups are currently focused on these questions, and the importance of molecular biological techniques, in combination with functional studies, such as patch-clamp recording and amperometry, is becoming increasingly apparent. This presents a technological barrier to further progress, particularly as some chemosensory tissues are in low abundance and extremely inaccessible, making molecular approaches highly challenging. One means by which such problems are circumvented is to employ model O_2 sensing systems, namely immortalized cell lines. These share many of the features of chemosensory tissues, but can be grown in continuous cell culture in order that molecular biological and postgenomic technologies can be applied more easily. To date, two specific cell lines have received much attention: the PC12 pheochromocytoma, derived from rat adrenal chromaffin tissue and originally considered representative of type

[7] E. Cutz and A. Jackson, *Respir. Physiol.* **115,** 201 (1999).

[8] D. Adriaensen, I. Brouns, J. Van Genechten, and J. P. Timmermans, *Anat. Rec.* **270,** 25 (2003).

[9] X. W. Fu, D. Wang, C. Nurse, M. C. Dinauer, and E. Cutz, *Proc. Natl. Acad. Sci. USA* **97,** 4374 (2000).

[10] X. W. Fu, C. A. Nurse, V. Wong, and E. Cutz, *J. Physiol.* **539,** 503 (2002).

[11] X. W. Fu, C. A. Nurse, Y. T. Wang, and E. Cutz, *J. Physiol.* **514,** 139 (1999).

I carotid body cells and/or immature adrenal chromaffin cells,[12,13] and H146 cells, a clonal line of a small cell lung carcinoma, believed to share a common cell lineage with native NEB cells.[14,15] The validation and use of these model cell lines are convenient for many reasons. They can be grown in bulk for molecular/proteomic studies and their adaptation to periods of prolonged (chronic) hypoxia, which causes marked remodeling of ion channel expression, can be controlled carefully. The progress that has been made with these model cell lines, which includes some surprising and potentially useful observation, is the focus of this article.

PC12 Cells: Sensors of Acute Hypoxia

Derived from a rat adrenal medulla tumor (or pheochromocytoma), the PC12 cell has made an enormous impact in the fields of neuroscience and cell biology and continues to be a widely used cell line. It has been used extensively as a model system for studying, among other topics, molecular mechanisms underlying fundamental cellular processes such as exocytosis and differentiation.[12] Zhu and colleagues[13] were among the first to demonstrate that PC12 cells also shared many common features with type I carotid body cells. First, and most obviously, they secrete the catecholamines dopamine and noradrenaline (as well as other transmitters) and express a comparable array of voltage-gated ion channels. Most importantly, however, Zhu et al.[13] and Conforti et al.[14] demonstrated that they possessed a K$^+$ current that could be reversibly inhibited by acute hypoxia. Furthermore, inhibition of this channel led to membrane depolarization that was of sufficient magnitude to activate voltage-gated Ca^{2+} channels. This immediately suggested that PC12 cells may represent a good model for O$_2$ sensing by type I carotid body cells. When undifferentiated, these cells appear near spherical and, when stimulated, release large quantities of catecholamines in a quantal manner, apparently from most of the plasma membrane surface. This feature means that they are highly useful for direct monitoring of exocytosis by measuring membrane capacitance changes.[16] In addition to their amenability for electrophysiological study, they have also proved useful for direct measurement of quantal catecholamine release from

[12] A. S. Tischler and L. A. Greene, *Nature* **258**, 341 (1975).
[13] W. H. Zhu, L. Conforti, M. F. Czyzyk-Krzeska, and D. E. Millhorn, *Am. J. Physiol.* **271**, C658 (1996).
[14] D. Wang, C. Youngson, V. Wong, H. Yeger, M. C. Dinauer, E. Vega-Saenez de Miera, B. Rudy, and E. Cutz. *Proc. Natl. Acad. Sci. USA* **93**, 13182 (1996).
[15] I. O'Kelly, C. Peers, and P. J. Kemp, *Am. J. Physiol.* **275**, L709 (1998).
[16] H. Kasai, H. Takagi, Y. Ninomiya, T. Kishimoto, K. Ito, A. Yoshida, T. Yoshioka, and Y. Miyashita, *J. Physiol.* **494**, 53 (1996).

individual cells in real time, using the noninvasive technique of amperometry.[17] This latter technique can only be used when transmitter species are oxidized easily and, fortunately, the predominant transmitters released from PC12 cells are dopamine and noradrenaline, both of which are susceptible to oxidation.

PC12 Cell Culture

PC12 cells are obtained from the American Type Tissue Collection (Manassas, VA) and cultured in suspension in Roswell Park Memorial Institute (RPMI) 1640 medium (containing L-glutamine) supplemented with 20% fetal calf serum and 1% penicillin/streptomycin (all from GIBCO BRL, Paisley, Strathclyde, UK). Cells are incubated at 37° in a humidified atmosphere of 5% CO_2/95% air, passaged every 7 days (1:7) but only used for a maximum of 20 passages. Cells are transferred to smaller (25 cm^2) flasks in 10 ml of medium, to which is added 1 μM dexamethasone (Sigma Aldrich, Poole, Dorset, UK; from a stock solution of 1 mM in ultrapure water), and are cultured for a further 72–96 h prior to use in experiments. This period of exposure to dexamethasone enriches catecholamine stores and so increases the signal-to-noise ratio, which we find significantly aids amperometric detection.[18]

Amperometry

On each experimental day, aliquots of PC12 cells are plated onto poly-L-lysine-coated (1 mg/ml^{-1} in ultrapure water; Sigma Aldrich) coverslips and allowed to adhere for approximately 1 h. Fragments of coverslip are then transferred to a recording chamber (volume ca. 80 μl), which is mounted on the stage of an inverted microscope (Olympus CK2 or Nikon TMD) and perfused continually under gravity (flow rate 1–2 ml/min^{-1}) with a solution of composition (in mM) of NaCl 135, KCl 5, $MgSO_4$ 1.2, $CaCl_2$ 2.5, HEPES 5, and glucose 10 (pH 7.4, osmolarity adjusted to 300 mOsm with sucrose using a freezing point osmometer, Camlab Ltd., Cambridge, UK). Whenever "Ca^{2+}-free" solutions are employed, $CaCl_2$ is omitted and 1 mM EGTA is added instead. All experiments are conveniently conducted at room temperature (21–24°). Drugs of interest are applied via the perfusate, except in the case of specific, irreversible toxins [such as ω-conotoxin GVIA toxin (ω-CgTx) and ω-agatoxin GIVA (ω-Aga-VIA) Alomone Labs, Jerusalem, Israel] where the effects of these agents are investigated after preincubation of cells for at least 10 min. It

[17] S. C. Taylor and C. Peers, *Biochem. Biophys. Res. Commun.* **148**, 13 (1998).
[18] A. S. Tischler, R. L. Perlman, G. M. Morse, and B. E. Sheard, *J. Neurochem.* **40**, 364 (1983).

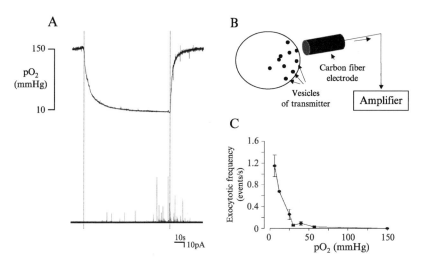

FIG. 1. Monitoring hypoxia-evoked exocytosis from single cells by amperometry. (A) (Upper trace) Recording of perfusate pO_2 before, during, and after application of solution preequilibated and bubbled continually with N_2. Solution exchange occurred at the point indicated by dashed lines. (Lower trace) Representative secretory response of a PC12 cell to the hypoxic challenge. (B) Position of a carbon fiber microelectrode tip adjacent to a PC12 cell in order to monitor exocytosis amperometrically. (C) Plot of mean ($\pm$ SEM) exocytotic frequency versus pO_2 level determined in PC12 cells (n >5 for each point). Adapted from Taylor and Peers.[17]

should be remembered that experiments investigating the effects of photo-labile drugs, such as the L-type Ca^{2+} channel antagonist nifedipine (Sigma Aldrich), must always be conducted at low-light intensity and be made regularly from a stock solution. To sustain the experimental life of such compounds, we find it convenient to cover the reservoir and perfusion lines with kitchen-grade aluminum foil (Morrisons, Bradford, West Yorks, UK). For studies investigating modulation of secretion by O_2, solutions are made hypoxic by bubbling with N_2 for at least 30 min. Normoxic solutions are equilibrated in room air. These maneuvers produce no change in solution pH or bath temperature. For maximum removal of O_2 from solutions, all tubing must be Tygon (BDH, Atherstone, Berkshire, UK), which is flexible and almost completely gas impermeant. It is important in experiments employing decreased O_2 levels that pO_2 is measured directly; this is particularly convenient during amperometric recordings, as the same carbon fibers (see later) serve as O_2 electrodes when their polarity is reversed[19] (see Figs. 1A and 2A for examples of such recordings). Although a number

[19] M. H. Mojet, E. Mills, and M. R. Duchen, *J. Physiol.* **504,** 175 (1997).

of devices are available specifically for use with carbon fiber micro-electrodes, any electrophysiology laboratory can conveniently use this technology by employing a standard patch-clamp amplifier that has an extended voltage range, such as the Axopatch 200B. Using such equipment, we connect the carbon fiber electrode to the headstage via a mercury bridge and polarize to +800 mV (Fig. 1B). Our carbon fiber electrodes of choice have a diameter of 5 μm and are now available prefabricated (proCFE, Axon Instruments, Foster City, CA). Some investigators, however, choose to fabricate their own electrodes, and a useful description of this process can be found in Schulte and Chow.[20] Electrodes are positioned adjacent to individual PC12 cells using a Narishige MX-2 micromanipulator. Amperometric currents are filtered at 1 kHz and digitized at 2 kHz before being stored on a computer. All acquisition is performed using a Digidata 1200 interface and Fetchex software from the pClamp 6.0.3 suite (Axon Instruments).

This noninvasive, high-resolution approach allows quantal catecholamine secretion to be monitored in real time, and the release of transmitters from single vesicles in response to a stimulus (e.g., hypoxia; Fig. 1A) appears as discrete spike-like current events (see also Wightman *et al.*[21]). The availability of Minian 16 software (Jaejin Software, Columbia, NY), which was originally designed to quantify synaptic neurotransmission, has aided the quantification of release considerably and its use is highly recommended. This software is particularly helpful as it allows visual inspection of each spike event, which ensures that artifacts (due, for example, to solution switches) can be rejected from analyses.

The use of amperometry has allowed us to validate the PC12 cell as a model secretory line by screening rapidly with known stimuli of the native carotid body. Thus, for example, both graded hypoxia (Figs. 1A and 1C) and pH (Figs. 2A and 2B) evoke exocytosis. Furthermore, the PC12 model has allowed rapid dissection of the role of specific Ca^{2+} channels in stimulus–secretion coupling. With either stimulus, secretory responses are fully dependent on extracellular Ca^{2+} and can be inhibited by selective blockade of voltage-gated Ca^{2+} channels. Thus, for example, ω-conotoxin GVIA inhibits the exocytosis seen in response to hypoxia, suggesting that N-type Ca^{2+} channels are the primary Ca^{2+} influx route coupling hypoxia to catecholamine release.[22] Similarly, using PC12 cells has allowed us to

[20] A. Schulte and R. H. Chow, *Anal. Chem.* **68,** 3054 (1996).
[21] R. M. Wightman, J. A. Jankowski, R. T. Kennedy, K. T. Kawagoe, T. J. Schroeder, D. J. Leszczyszyn, J. A. Near, E. J. Diliberto, Jr., and O. H. Viveros, *Proc. Natl. Acad. Sci. USA* **88,** 10754 (1991).
[22] S. C. Taylor and C. Peers, *J. Neurochem.* **73,** 874 (1999).

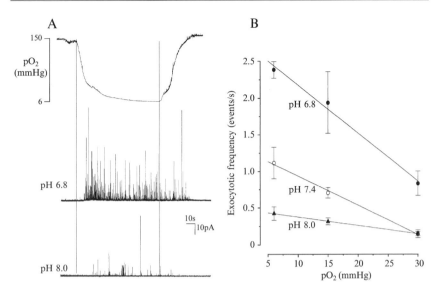

Fig. 2. Multiplicative interactions of chemostimuli to evoke exocytosis from PC12 cells. (A) (Upper trace) Recording of perfusate PO$_2$ before, during, and after application of solution preequilibrated and bubbled continually with N$_2$. (Middle trace) Secretory response of a representative PC12 cell to the hypoxic challenge at pH 6.8. (Lower trace) Secretory response of another PC12 cell to the hypoxic challenge at pH 8.0. Scale bars apply to both traces. (B) Plot of exocytotic frequency (determined over a 1-min period, 90 s after switching to hypoxic solution) as a function of final bath PO$_2$ at three different pH$_o$ levels as indicated. Each point shows the mean (with vertical SEM bars, $n = 6$–18 cells in each case) exocytotic frequency determined from experiments such as those shown in A. Adapted from Taylor et al.[23]

investigate a long-standing but poorly understood phenomenon in the carotid body, that is, the basis of the multiplicative interaction of hypoxia and hypercapnia/acidity. The observation that these stimuli together can excite afferent chemosensory fibers of the carotid body to a greater degree than the sum of the individual stimuli was reported as early as the 1970s,[24] yet the cellular basis of this effect remained unknown. Indeed, it took some 25 years before it was firmly established that an individual type I cell could respond to both stimuli separately with a rise of $[Ca^{2+}]_i$.[25] However, when both stimuli are applied simultaneously, rises of $[Ca^{2+}]_i$ of type I cells are

[23] S. C. Taylor, M. L. Roberts, and C. Peers, J. Physiol. **519,** 765 (1999).

[24] S. Lahiri, E. Mulligan, T. Nishino, A. Mokashi, and R. O. Davies, J. Appl. Physiol. **50,** 580 (1981).

[25] L. L. Dasso, K. J. Buckler, and R. D. Vaughan-Jones, Am. J. Physiol. Lung Cell Mol. Physiol. **279,** L36 (2000).

not greater than additive. The PC12 model has allowed us to investigate this apparent paradox and to suggest that multiplicative secretory responses to hypoxia and acidity (Fig. 2C) are generated by the differential recruitment of discrete secretion-linked classes of Ca^{2+} channels.[23] This raises the possibility that greater than additive hypoxic and hypercapnic interactions can occur at the level of individual type I cells rather than via mechanisms that involve multicellular responses, for example, recruitment of additional type I cells that may preferentially sense CO_2 rather than O_2 levels. Such an idea has yet to be tested in native type I cells primarily because when isolated enzymatically, their secretory responses to hypoxia are weak. This is possibly due to their being hyperpolarized artifactually when dispersed.[26] However, with introduction of the carotid body thin slice preparation,[27] stimulus interactions may soon be investigated directly.

The experiments described earlier highlight the usefulness of PC12 cells as model arterial chemoreceptors; however, as with most model systems, results obtained from such cells cannot be assumed to be true for the tissue they are purported to represent. It is important to understand the limitations of model systems and, if understanding the cellular basis of arterial chemoreception is the primary aim, results obtained should act merely to inform the design of appropriate experiments within the technically demanding native tissues. Indeed, there are important examples of discrepancies between PC12 and type I carotid body cells that are worthy of note. The first is the molecular identity of the specific O_2-sensitive K^+ channel. In PC12 cells, there is strong evidence that Kv1.2 is specifically inhibited by acute hypoxia, leading to cell depolarization.[28] In contrast, at least four separate K^+ channel types have been shown to be inhibited by hypoxia in type I carotid body cells, and species variation is apparent. Thus, in the rat, both a high-conductance, Ca^{2+}-dependent (maxi-K) channel[29] and an acid-sensitive, background K^+ channel (possibly TASK-1[30]) have been identified. In the rabbit, both an inactivating, voltage-gated K^+ channel (most likely of the Kv4.3 family[31]) and a HERG-like K^+ current[32] have been shown to be O_2 sensitive. None of these four K^+ currents resembles the O_2-sensitive K^+ current in PC12 cells. The second notable difference

[26] E. Carpenter, C. J. Hatton, and C. Peers, *J. Physiol.* **523,** 719 (2000).

[27] R. Pardal, U. Ludewig, J. Garcia-Hirschfeld, and J. Lopez-Barneo, *Proc. Natl. Acad. Sci. USA* **97,** 2361 (2000).

[28] L. Conforti, I. Bodi, J. W. Nisbet, and D. E. Millhorn, *J. Physiol.* **524,** 783 (2000).

[29] C. N. Wyatt and C. Peers, *J. Physiol.* **483,** 559 (1995).

[30] K. J. Buckler, B. A. Williams, and E. Honore, *J. Physiol.* **525,** 135 (2000).

[31] M. T. Perez-Garcia, J. R. Lopez-Lopez, and C. Gonzalez, *J. Gen. Physiol.* **113,** 897 (1999).

[32] J. L. Overholt, E. Ficker, T. Yang, H. Shams, G. R. Bright, and N. R. Prabhakar, *J. Neurophysiol.* **83,** 1150 (2000).

between PC12 cells and type I carotid body cells is the molecular identity of the voltage-gated Ca^{2+} channels mediating stimulus-evoked secretion. Much evidence indicates that L-type Ca^{2+} channels play a primary role in mediating Ca^{2+} influx during carotid body stimulation.[33] However, in PC12 cells, N-type (and, to a lesser extent, P/Q-type) channels are the major routes of Ca^{2+} entry during exposure to physiological stimuli.[26,34] These two discrepancies highlight the limitations of PC12 cells as a model system for studying acute arterial chemoreception, but the two tissue types are likely to share common, O_2-sensitive intracellular signaling pathways. Given their abundance and ease of maintenance, PC12 cells may yet prove more useful as a model system for studying cellular responses to chronic hypoxia at the molecular level.

PC12 Cells: A System for Studying Remodeling of Cell Function by Chronic Hypoxia

The usefulness of cell lines to monitor molecular events associated with prolonged hypoxia has been exploited well over the last decade. Czyzyk-Krzeska and colleagues first reported that chronic hypoxia could enhance tyrosine hydroxylase (TH) expression in type I carotid body cells,[35] but turned to the PC12 cell to investigate this phenomenon in depth.[36] These workers found that TH mRNA was stabilized in hypoxia via formation of a ribonucleoprotein complex, involving a binding sequence in the 3′-untranslated region of TH mRNA, termed the hypoxia-inducible protein-binding site.[37] This is necessary for constitutive and hypoxia-regulated stability of TH mRNA. This represents one example of the way in which prolonged hypoxia causes remodeling of cellular function, and other such changes have also been documented utilizing PC12 cells. For example, hypoxic modulation of the expression and functional coupling of adenosine A_{2A} and dopamine D2 receptors has been characterized in depth,[38,39] along with alterations in glutamate metabolism and transport[40] and, using PC12 cells that have been differentiated into a

[33] C. Gonzalez, L. Almaraz, A. Obeso, and R. Rigual, *Physiol. Rev.* **74,** 829 (1994).
[34] S. C. Taylor and C. Peers, *Biochem. Biophys. Res. Commun.* **248,** 13 (1988).
[35] M. F. Czyzyk-Krzeska, D. A. Bayliss, E. E. Lawson, and D. E. Millhorn, *J. Neurochem.* **58,** 1538 (1992).
[36] M. F. Czyzyk-Krzeska, B. A. Furnari, E. E. Lawson, and D. E. Millhorn, *J. Biol. Chem.* **269,** 760 (1994).
[37] M. F. Czyzyk-Krzeska and J. E. Beresh, *J. Biol. Chem.* **271,** 3293 (1996).
[38] S. Kobayashi and D. E. Millhorn, *J. Biol. Chem.* **274,** 20358 (1999).
[39] W. H. Zhu, L. Conforti, and D. E. Millhorn, *Am. J. Physiol.* **42,** C1143 (1997).
[40] S. Kobayashi and D. E. Millhorn, *J. Neurochem.* **76,** 1935 (2001).

neuronal phenotype, N-methyl-D-asparatate receptors.[41] These latter studies are of particular importance in providing preliminary indications of the
adaptive reponses that central neurons may undergo in order to establish
protective mechanisms in the face of prolonged hypoxia. It is clear from
the studies published to date (arising primarily from the laboratories of
Millhorn and Czyzyk-Krzeska) that numerous, complex interacting second
messenger pathways mediate such responses to hypoxia. Given the marked
heterogeneity of such preparations, such studies would originally have
been extremely difficult using central nervous system tissue. However,
now that these observations have been reported in PC12 cells, several
exciting avenues of future research have been opened to investigate
the protective (and deleterious) adaptive responses of central neurons to
prolonged hypoxia.

As discussed in the previous section, PC12 cells have been valuable for
monitoring exocytosis. We also use this technique to investigate the effects
of prolonged hypoxia on depolarization-evoked catecholamine secretion.
In cells that are cultured in normoxia (pO_2 ca. 143 mm Hg), secretion is to-
tally dependent on extracellular Ca^{2+} (Fig. 3A) and is fully blocked by
Cd^{2+} (Fig. 3B), a nonselective inhibitor of voltage-gated Ca^{2+} channels. Secretory responses in PC12 cells cultured in chronic hypoxia (i.e., cells cultured for 24 h at a pO_2 ca. 78 mm Hg) are enhanced but remain fully
dependent on extracellular Ca^{2+} (Fig. 3C). However, following such a
period of chronic hypoxia, a significant fraction of the secretory response
is now resistant to the effects of Cd^{2+} (Fig. 3D). Pharmacological investigation of this Cd^{2+}-resistant Ca^{2+} influx pathway, which is induced by chronic
hypoxia, shows that it can be suppressed by La^{3+} or Zn^{2+} and, more importantly, by the addition of Congo red to the culture medium during exposure to hypoxia.[42] Congo red is an agent that prevents the formation of
protein aggregates and is known in particular to prevent aggregation of the
amyloid β peptides (AβPs) of Alzheimer's disease (AD).[43] Furthermore,
amyloid peptides can form Ca^{2+}-permeable, Zn^{2+}-sensitive pores in both
native membranes and artificial bilayers.[44] This raises the surprising possibility that chronic hypoxia increases the production of AβPs, which then
aggregate to form Ca^{2+}-permeable pores in the plasma membrane. Such
an idea is supported by our immunofluorescent studies (utilizing a monoclonal anti-AβP antibody), which show marked enhancement of plasma
membrane-associated AβP immunoreactivity.[42] In addition, exposure of

[41] S. Kobayashi and D. E. Millhorn, *Neuroscience* **101**, 1153 (2000).

[42] S. C. Taylor, T. F. Batten, and C. Peers, *J. Biol. Chem.* **274**, 31217 (1999).

[43] D. Burdick, B. Soreghan, M. Kwon, J. Kosmoski, M. Knauer, A. Henschen, J. Yates,
C. Cotman, and C. Glabe, *J. Biol. Chem.* **267**, 546 (1992).

[44] N. Arispe, H. B. Pollard, and E. Rojas, *Proc. Natl. Acad. Sci. USA* **93**, 1710 (1996).

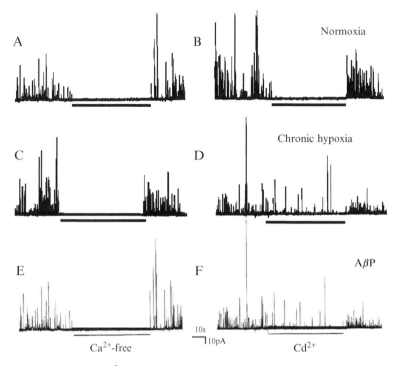

FIG. 3. Induction of Cd²⁺-resistant exocytosis by chronic hypoxia and amyloid peptide. Amperometric detection of secretion evoked in control (A, B) and chronically hypoxic (C, D) PC12 cells and in cells exposed to 100 nM AβP$_{(1-40)}$ (E, F) in response to 50 mM K⁺ (applied 20 s before the commencement of each trace). For the periods indicated by horizontal bars, cells were exposed to either Ca²⁺-free perfusate (A, C, E) or 200 μM Cd²⁺ in the presence of 2.5 mM Ca²⁺ (B, D, F) in the continued presence of 50 mM K⁺. Scale bars apply to all traces. Note that in chronically hypoxic and AβP$_{(1-40)}$-treated cells, 200 μM Cd²⁺ failed to inhibit secretion completely (D, F). Adapted from Taylor et al.[42]

normoxic cells to AβPs causes the same Ca²⁺-dependent enhancement of secretion (Fig. 3E), including a significant fraction that is resistant to blockade by Cd²⁺ (Fig. 3F). In other words, chronic hypoxia is mimicked by AβP. These unexpected findings may be pertinent to specific clinical situations: the incidence of dementias (of which AD is the most common form) is well known to increase in individuals following the prolonged periods of hypoxia that can arise from numerous cardiorespiratory diseases or following a stroke.[45,46] With the lucidity of hindsight, it is remarkable

[45] J. T. Moroney, E. Bagiella, D. W. Desmond, M. C. Paik, Y. Stern, and T. K. Tatemichi, Ann. N. Y. Acad. Sci. **826,** 433 (1997).

[46] J. T. Moroney, E. Bagiella, D. W. Desmond, M. C. Paik, Y. Stern, and T. K. Tatemichi, Stroke **27,** 1283 (1996).

that PC12 cells might represent a more useful model for modeling the molecular mechanisms underlying hypoxia-associated observations in neurodegeneration than acute hypoxic regulation *per se* studying hypoxia-associated neurodegeneration. This suggestion is fully supported by recent observations in brain tissue, which produced results remarkably similar to our original findings in PC12 cells.

H146 Cells: Model Airway Chemoreceptor Cells

While the physiological role of airway NEBs has yet to be fully established, their abundance in the neonatal lung and the observation that they undergo hyperplasia in sudden infant death syndrome[47] and apnea of prematurity[48] strongly implicate their involvement in these clinical conditions. Furthermore, as described earlier, their innervation and anatomical location (in contact with the lung airway lumen, yet also in close proximity to blood vessels) predict a likely important physiological role in O_2 homeostasis.[8] Even more so than in the carotid body, the study of NEBs and the potential mechanisms by which they sense changes in local O_2 levels has been hampered by their low abundance and inaccessibility. Indeed, the only cellular physiological studies to date have emerged from the collaborative efforts of the laboratories of Cutz and Nurse.[9–11,49] These pioneering reports (based on studies employing both isolated NEB cells and recordings from NEBs in a novel lung slice preparation; see Cutz *et al.*[49a]) have shown that hypoxia causes reversible inhibition of NEB cell K^+ currents, leading to cell depolarization and release of serotonin.[9,10] Furthermore, use of a knockout mouse lacking gp91[phox]—a major subunit of NADPH oxidase—has indicated that NEB cells use this enzyme as an O_2 sensor.[9] When O_2 is abundant, this complex generates reactive oxygen species from O_2 (using NADPH as an electron donor), which are then likely to regulate K^+ channel activity through the formation of H_2O_2.

Such progress in understanding the mechanisms underlying O_2 sensing by NEBs has been remarkable given the practical difficulties associated with accessing these cells, but is likely to be limited when attention turns to the molecular identification of specific cellular mechanisms underlying chemoreception by NEBs. For this reason, we have turned to H146 cells as a model system for studying airway chemoreception. H146 cells represent one of a number of cell lines of human small cell lung carcinoma

[47] D. G. Perrin, T. J. McDonald, and E. Cutz, *Pediatr. Pathol.* **11,** 431 (1991).

[48] E. Cutz, T. K. Ma, D. G. Perrin, A. M. Moore, and L. E. Becker, *Am. J. Respir. Crit. Care Med.* **155,** 358 (1997).

[49] C. Youngson, C. Nurse, H. Yeger, and E. Cutz, *Nature* **365,** 153 (1993).

[49a] E. Cutz, X. W. Fu, and H. Yeger, *Methods Enzymol.* **381** [2] 2003 (this volume).

origin believed to be derived from the same precursor pool as NEBs and to share numerous commonalities.[50] Thus, they possess an array of voltage-gated ion channels expected of excitable cells[15,51] and also synthesize, store, and release numerous transmitters, including the vasoactive amine serotonin.[8]

Culture of H146 Cells

The small cell lung carcinoma cell line H146 can be purchased from American Tissue Type Cell Collection (Manassas, VA) and is maintained in suspension culture in RPMI 1640 medium (containing L-glutamine) supplemented with 10% fetal calf serum, 2% sodium pyruvate, and 2% penicillin/streptomycin (all from GIBCO BRL) at 37° in a humidified atmosphere of 5% CO$_2$/95% air. Medium is changed every 2 days, and cells are passaged every 6–7 days by splitting in the ratio 1:5. Cells are used between passage 1 and 10.

Patch Clamp

For patch-clamp studies, the standard pipette solution is K$^+$ rich and contains (in mM) 10 NaCl, 117 KCl, 2 MgCl$_2$, 10 HEPES, 11 EGTA, 1 CaCl$_2$, 2 Na$_2$ATP; pH 7.2 with KOH; free [Ca^{2+}] = 27 nM. The standard bath solution is Na$^+$ rich and contains (in mM) 135 NaCl, 5 KCl, 1.2 MgCl$_2$, 5 HEPES, 2.5 CaCl$_2$, 10 D-glucose; pH 7.4 with NaOH. All tubing is gas impermeant (see earlier discussion). Cells are allowed to adhere at 37° for at least 1 h to poly-L-lysine-coated glass coverslips before being placed in a perfusion chamber mounted on the stage of an Oympus CK40 inverted microscope. Patch pipettes are manufactured from standard-walled borosilicate glass capillary tubing (Clarke Electromedical Instruments) on a two-stage Narishige PP-830 pipette puller (Narishige Scientific Instrument Laboratory, Kasuya, Tokyo, Japan), are heat polished on a Narishige microforge, and have measured tip resistances of 5–8 MΩ. A resistive feedback voltage clamp is achieved using an Axopatch 200B amplifier (Axon Instruments). Voltage protocols are generated, and currents are recorded using pClamp 8 software employing a Digidata 1320 A/D converter (Axon Instruments). Data are filtered (4-pole Bessel) at 1 kHz and digitized at 5 kHz. Following successful transition to the whole cell recording mode,[52] capacitance transients are compensated for

[50] A. F. Gazdar, L. J. Helman, M. A. Israel, E. K. Russell, R. I. Linnoila, J. L. Mulshine, H. M. Schuller, and J. G. Park, *Cancer Res.* **48,** 4078 (1988).
[51] J. J. Pancrazio, M. P. Viglione, I. A. Tabbara, and Y. I. Kim, *Cancer Res.* **49,** 5901 (1989).
[52] O. P. Hamill, A. Marty, B. Sakmann, and F. J. Sigworth, *Pflug. Arch.* **391,** 85 (1981).

and measured. To evoke ionic currents in H146 cells, a ramp-step voltage-clamp protocol is usefully employed (from -100 to $+60$ mV from a holding potential of -70 mV followed by sequential steps to 0 mV and $+60$ mV for 200 ms at 0.1 Hz). The magnitude of the outward currents, for construction of the time course graphs, is measured at 0 mV from the ramp or the step.

We have validated the use of H146 cells as model airway chemoreceptors by characterizing their ion currents using patch-clamp and, most importantly, demonstrating a K^+ current that is reversibly inhibited by acute hypoxia (Fig. 4A[15]). This current strongly influences H146 cell membrane potential (Fig. 4B), and we find an excellent correlation between O_2-sensitive channel activity and membrane depolarization during graded hypoxia (Figs. 4C and 4D). Furthermore, the validity of this cell line as a model is evidenced by our demonstration that the upstream O_2 sensor is, like the native NEB cell, NADPH oxidase.[53,54]

Based on the premise that H146 cells resemble NEB cells strongly in respect to the O_2 sensor, we have been in the unique position of determining the molecular identity of a NADPH oxidase-linked, O_2-sensitive K^+ channel[55] and have characterized the potential involvement of mitochondrial reactive oxygen species (ROS) in the cellular responses to acute hypoxia.[56]

Generating Cells without Functional Mitochondria

Mitochondrial DNA (mtDNA)-depleted cell lines (ρ^0) are generated from wild-type cultured H146 by incubation for >3 weeks in RPMI growth media supplemented with 10% fetal bovine serum containing ethidium bromide (25 $ngml^{-1}$), pyruvate (1 mM), and uridine (50 μgml^{-1}). The wild-type, parental H146 cells are maintained for the same time period in normal culture medium. The ρ^0 cells are selected on the day of experimentation by incubation with rotenone (1 μgml^{-1}) and antimycin A (1 μgml^{-1}) for 2 h. To compare the mtDNA content between control and ρ^0 H146 cells, total cellular DNA can be extracted using the DNeasy tissue kit (Qiagen, Crawley, West Sussex, UK), and the yield, purity, and integrity of the DNA are verified by spectrophotometry at 260/280 nm. Equal amounts of wild-type and ρ^0 DNA are subjected to polymerase chain reaction (PCR) amplification in a volume of 100 μl containing 1 μl of Taq

[53] I. O'Kelly, A. Lewis, C. Peers, and P. J. Kemp, *J. Biol. Chem.* **275**, 7684 (2000).

[54] P. J. Kemp, G. J. Searle, M. E. Hartness, A. Lewis, P. Miller, S. Williams, P. Wootton, D. Adriaensen, and C. Peers, *Anat. Rec.* **270**, 41 (2003).

[55] M. E. Hartness, A. Lewis, G. J. Searle, I. O'Kelly, C. Peers, and P. J. Kemp, *J. Biol. Chem.* **276**, 26499 (2001).

[56] G. J. Searle, M. E. Hartness, R. Hoareau, C. Peers, and P. J. Kemp, *Biochem. Biophys. Res. Commun.* **291**, 332 (2002).

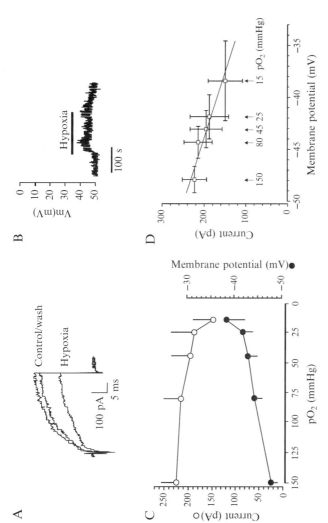

Fig. 4. Resting membrane potential and K⁺ current responses of H146 cells to graded hypoxia. (A) Typical voltage clamp record showing outward K⁺ currents evoked during a step depolarization from −70 to 0 mV for 50 ms before (control), during (hypoxia), and after (wash) perfusion with hypoxic solution at *ca.* 40 mm Hg. (B) Typical current clamp record (I = 0 pA) of resting membrane potential before, during, and after perfusion with hypoxic solution at *ca.* 40 mm Hg. Current recording taken from the same cell as the voltage recording shown in A. The period of hypoxic perfusate is shown by the horizontal bar above the trace. (C) Mean outward K⁺ currents (left axis, ○) and membrane potentials (right axis, ●) recorded during graded depression in pO₂ from 150 to 15 mm Hg. Currents were recorded during 50-ms step depolarizations from a holding potential of −70 to 0 mV; *n* = 7 cells. Membrane potential was measured in current clamp with I = 0 pA; *n* = 7 cells. (D) Mean current amplitude versus mean membrane potential at five separate pO₂ values. R = 0.96, *n* = 7. Adapted from O'Kelly *et al.*[53]

DNA polymerase (Promega, Southampton, Hants., UK) and oligonucleo-tide primer pairs that are sequence specific for human mtDNA: upstream 5′-CCT AGG GAT AAC AGC GCA AT-3′ and downstream 5′-TAG AAG AGC GAT GGT GAG AG-3′ (630-bp product) and β-actin: upstream 5′-TGG CCG GGA CCT GAC TGA CTA C-3′ and downstream 5′-CGT GGC CAT CTC TTG CTC GAA G-3′ (150-bp product). PCR conditions include an initial denaturing step for 5 min followed by denaturing for 30 s at 94°, annealing for 30 s at 60°, and extending for 40 s at 72° using a Hybaid Express thermal cycler (Hybaid, Ashford, Middlesex, UK). The progress of the PCR reaction is monitored by taking aliquots after 20, 22, 24, 26, 28, 30, 32, and 34 cycles. These aliquots are analyzed by 1% agarose gel electrophoresis, and product concentration is determined by ultraviolet densitometry of ethidium bromide-stained gels. Using this method, selective loss of mitochondrial DNA can be verified (as exemplified in Fig. 5A). This genetic manipulation results in no diminution of hypoxic K^+ channel inhibition (Fig. 5B), which shows that mitochondria-derived ROS, although a potential modulatory route in other O_2 sensing systems, such as pulmonary arterial smooth muscle,[57,58] do not contribute to the hypoxic response in the H146 model, a notion that supports other data in native NEB cells.[59]

RT-PCR Screening for mRNA Encoding Human K_{2P} Channels

One significant advantage that immortalized cells offer over their native counterparts is the ability to perform long-term molecular abrogation protocols in order to identify specific components in a signal transduction pathway by loss of function. The initial step in identification of O_2-sensitive K^+ channels of any system is to determine which genes are expressed. For H146 cells, our previous pharmacological characterization led us to suspect that the channel of interest was one of the members of the K_{2P} family.[60] To screen for mRNA of such channels, total RNA is extracted from pelleted H146 cells using the RNeasy mini kit (Qiagen) and treated with RQ-1 RNase-free DNase (1 Uμg^{-1} RNA; Promega, Southampton, Hants., UK) to remove genomic DNA contamination before reextraction using the RNeasy mini kit. The yield, purity, and integrity of the RNA are verified by spectrophotometry at 260/280 nm, followed by electrophoresis on 1%

[57] R. M. Leach, H. M. Hill, V. A. Snetkov, T. P. Robertson, and J. P. Ward, *J. Physiol.* **536**, 211 (2001).

[58] G. B. Waypa, N. S. Chandel, and P. T. Schumacker, *Circ. Res.* **88**, 1259 (2001).

[59] D. Wang, C. Youngson, V. Wong, H. Yeger, M. C. Dinauer, E. Vega-Saenz de Miera, B. Rudy, and E. Cutz, *Proc. Natl. Acad. Sci. USA* **93**, 13182 (1996).

[60] I. O'Kelly, R. H. Stephens, C. Peers, and P. J. Kemp, *Am. J. Physiol.* **276**, L96 (1999).

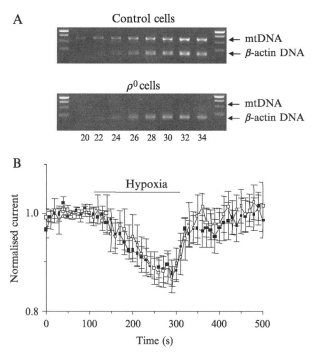

FIG. 5. Acute O_2 sensing in H146 cells lacking functional mitochondria (ρ^0 cells). (A) Exemplar semiquantitative RT-PCR reactions of control and ρ^0 cells using primer pairs directed against mitochondrial DNA (mtDNA; upper band) and β-actin DNA (lower band). The right lane in both gels shows DNA size markers of 150, 300, 500, and 1000 bp. Numbers at the bottom of the lower gel indicate the number of cycles and apply to both gels. (B) Mean time courses of hypoxia-evoked inhibition of K^+ currents recorded from control (solid symbols) and ρ^0 cells (open symbols). K^+ currents were measured at 0 mV. The period of perfusion with a hypoxic (30 mm Hg) solution is indicated by the horizontal bar. Adapted from Searle et al.[56]

agarose, and high-quality RNA can then be stored in an aqueous solution at $-80°$. Reverse transcription is performed on 1-μg aliquots of RNA using the Reverse Transcription System A3500 (Promega, Southampton, Hants., UK), comprising avian myeloblastosis virus reverse transcriptase and oligo dT(15) primers (42°, 15 min). The resulting cDNA is amplified by PCR using a panel of oligonucleotide primer pairs designed against the published sequences of the human homologues of TASK1 (Genbank accession number AF006823), TASK2 (AF084830), TASK3 (AC007869), TREK1 (AF004711), TREK2 (XM012342), TWIK1 (U90065), TWIK2 (AF117708), and TRAAK, (XM006543). We have had success using the

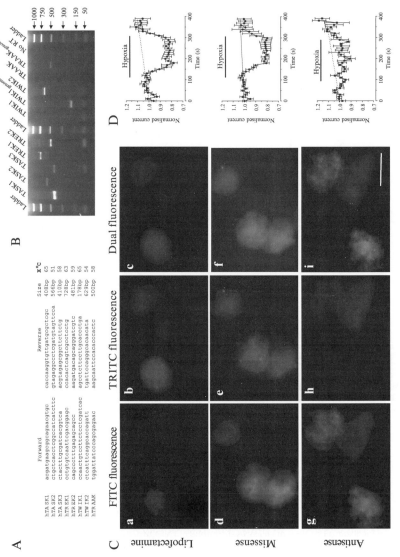

Fig. 6. (*continued*)

primer pairs defined in Fig. 6A. Amplification of 1 μl cDNA (equivalent to 160 ng of reverse-transcribed RNA) is performed in a volume of 50 μl containing 1 μl of Advantage-GC2 polymerase (Promega) under optimized conditions. The hot-start PCR protocol is $94°/1$ min, $X°/1$ min, and $72°/1$ min for 35 cycles with a final extension period of 10 min at $72°$. Optimized annealing temperatures $(X°)$ are gene specific and are shown in Fig. 6A. Products are separated on 2% agarose gels and visualized with ethidium bromide/ultraviolet transillumination. Sequencing of all amplicons is carried out by dye terminator PCR with an ABI PRISM-automated sequencer. Using this approach, we can identify mRNA from TASK1, 2, and 3; TREK1 and 2; and TWIK2 (Fig. 6B). It is noteworthy that two members of the family yield negative results, TWIK1 and TRAAK. That these were not false negatives is reinforced by the positive controls employing genomic DNA from the same cells (Fig. 6B). This validation of PCR primers and protocols is essential when negative results are found. In addition, the use of high-fidelity DNase before the reverse transcription stage in order to remove potential contamination from genomic DNA excludes the possibility of false positives.

Loss of Function Employing Antisense Oligodeoxynucleotides

Based on a combination of RT-PCR screening and pharmacological profiling, we recommend moving to the use of antisense oligodeoxynucleotides in an attempt to downregulate transcription/translation of candidate

---

FIG. 6. RT-PCR screening for tandem P-domain K⁺ channel mRNA and loss-of-function experiments in H146 cells. (A) Primer pair, optimal annealing temperatures, and predicted amplicon size employed for RT-PCR screening. (B) A 2% agarose gel showing products amplified from reverse-transcribed H146 mRNA employing the specific primer pairs complementary to the K⁺ channels indicated above each lane. Also shown is a reaction in which mRNA was not reverse transcribed (No RT) and three DNA ladders running at the base pair number indicated to the right. For TWIK1 and TRAAK, reactions employing genomic DNA are also shown as a positive control for the efficiency of those reactions. The forward and reverse primer sequences employed to amplify each specific channel mRNA, the optimized annealing temperature (in $°$C), and the expected product size for each reaction are shown in A. (C) Exemplar high-power images from lipofectamine-only (a–c), missense (d–f), or antisense oligodeoxynucleotide (g–i)-treated H146 cells. Column 1 (a, d, and g) shows FITC localization of the probe, where applicable. The second column (b, e, and h) shows TRITC localization of the anti-hTASK-1 antibody. The third column (c, f, and i) shows dual-fluorescence images and demonstrates that only antisense transfection resulted in specific hTASK-1 protein knock down. The scale bar represents 10 μm and applies to all panels. (D) Mean time courses of the effect of acute hypoxia (15–25 mm Hg, applied for the period indicated by the horizontal bar) on cells treated only with lipofectamine (A; $n = 11$), cells treated with missense oligodeoxynucleotide (B; $n = 5$), and cells treated with antisense oligodeoxynucleotide (C; $n = 12$). Adapted from Hartness *et al.*[55] (See color insert.)

K_{2P} channels. In H146 cells, we have employed this technology successfully to downregulate TASK1/3. The specific considerations and methodology are defined. H146 cells are transfected with (1) lipofectamine only; (2) 5'-FITC-labeled, phosphothioate-modified antisense (sequence 5'-cgttctgccgcttcatcg-3'; Genosys Biotechnologies, Pampisford, Cambs, UK) directed across the translation start site of human TASK1 and TASK3, or (3) 5'-FITC-labeled, phosphothioate-modified missense (sequence 5'-gccgtctatcttcgcgct-3'; Genosys Biotechnologies) that consist of the same bases as employed in the antisense probe but in "random" order. Odds of success are shortened considerably when the antisense deoxynucleotide probes are designed to span the *atg* start codon. FITC labeling of the probes allows us to follow transfection progress and efficiency by fluorescence microscopy, and phosphorothiolation increases the probe half-life within the transfected cells.

Cells are seeded in six-well plates at a density of *ca.* 2×10^6 cells per well in 0.8 ml of serum-free RPMI 1640 medium. The oligodeoxynucleotides are diluted in 0.1 ml of serum-free RPMI 1640 medium and mixed with 0.1 ml of 6% (v/v) lipofectamine (Gibco LifeTechnologies) in serum-free RPMI 1640 medium. The resulting oligodeoxynucleotide and cationic lipid mixture is incubated at room temperature for 30 min to allow the formation of DNA/liposome complexes, which are then added (0.2 ml) to the cell suspension, mixed gently, and incubated in a humidified atmosphere of 5% CO_2/95% air at 37° for 4 h. Following incubation, 4 ml of complete RPMI 1640 medium is added to each well. Cells are then cultured as normal for up to 5 days. The concentrations and time course of transfection are followed by measuring cellular/nuclear FITC fluorescence incorporation and, in our case, are optimal at 1 μM and for 4–5 days, respectively.

Thus, at optimal time following lipofection, cells are split into two batches and used in parallel in functional (patch clamp; see earlier discussion) and protein abundance (immunocytochemical) assays. Ideally, protein knockdown should be followed using specific antibodies in immunocytochemistry. H146 cells (*ca.* 5×10^5 cells) are cytospun onto glass poly-L-lysine-coated microscope slides for 5 min at 1200 rpm. Cells are heat fixed by placing the slides on a hot plate for 10 s and are then fixed in 10% neutral-buffered formalin for 5 min at 37°. The remaining procedures are carried out at room temperature. Cells are permeabilized by incubating with 0.5% Triton X-100 PBS for 15 min, refixed in 10% formalin for 5 min, washed in PBS for 10 min, and then incubated with blocking solution [10% (v/v) fetal calf serum, 0.1% (w/v) bovine serum albumin and 0.01% (w/v) NaN_3 in phosphate-buffered saline (PBS)] for 1 h. Cells are then incubated overnight in channel-specific antibodies (1:500 dilution of rabbit anti-hTASK in our case; Alomone Laboratories, Jerusalem, Israel) in blocking solution. Following antibody incubation, cells are washed three times in PBS and incubated in TRITC-labeled secondary antirabbit antiserum for 3 h. Using a TRITC-labeled second layer allows us to distinguish between protein expression and incorporation of the FITC probe (Fig. 6C). Finally, cells are washed a further three times in PBS prior to mounting and viewing. The specificity of immunoreactivity is confirmed by an antigen preadsorption step using the peptide to which the antibody was originally raised. This is important, and our preadsorption step employs excess peptide (1.2 μg per 1 ml of diluted antibody) and is carried out for 1 h at room temperature.

Using this approach, incorporation of antisense and missense probes is commonly more than 80%, indicating high transfection efficiency. Furthermore, only transfection with the antisense causes significant reduction in protein immunofluorescence (Fig. 6C). That this protein knockdown is reflected in functional abrogation of hypoxia-dependent depression of K$^+$ currents is indicated in Fig. 6D, demonstrating that this approach is able to identify either TASK1 or TASK3 as the O$_2$-sensitive channel in this model cell line.

H146 Cells: A Potential System for Studying Remodeling by Chronic Hypoxia

Remodeling of O$_2$ sensing by chronic hypoxia in NEBs has received little, if any, attention to date. However, any possible effects are worthy of future study given that NEBs can undergo hyperplasia in certain hypoxic disease states (see earlier discussion), as does the carotid body during chronic hypoxia. The H146 model presents a unique opportunity to study such phenomena under precisely controlled conditions in a system well defined in terms of its ion channel profile. Using microfluorimetric recordings of [Ca^{2+}]$_i$, we demonstrated that chronic hypoxia causes a suppression of depolarization-evoked Ca^{2+} influx into H146 cells (Fig. 7A,B), an effect associated with marked, differential alterations in the functional expression profile of voltage-gated Ca^{2+} channels (Fig. 7C,D). Whether such effects occur in NEB cells remains to be determined, but such dramatic changes would doubtless alter fundamental processes in these cells (such as stimulus-evoked transmitter release) that may be associated with dysfunction in hypoxic disease states.

Concluding Remarks

The present article serves to highlight two specific cell lines—PC12 and H146—as useful model systems for studying arterial and airway chemoreception, respectively. Both cell lines respond to acute hypoxia in a manner

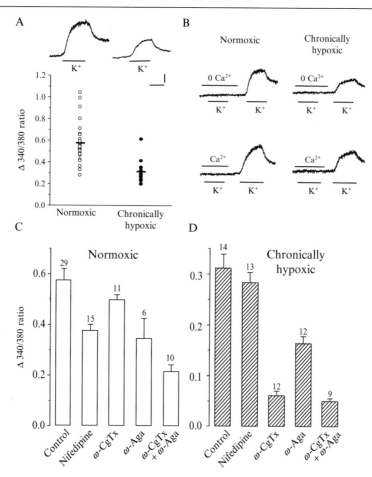

FIG. 7. Remodeling of calcium channels in H146 cells treated with chronic hypoxia. (A) Example recordings of $[Ca^{2+}]_i$ rises in response to cell depolarization evoked by the application of 50 mM K^+ (K^+, applied for the periods indicated by the horizontal bars). Recordings were made from normoxic (top left) and chronically hypoxic (top right) cells. Also plotted (bottom) are peak rises of $[Ca^{2+}]_i$ evoked by 50 mM K^+ in all normoxic cells (o) and all chronically hypoxic cells (●). Each point represents the individual responses of 29 control recordings and 14 recordings from chronically hypoxic cells. Horizontal bars represent the mean response for each group. (B) Typical effects of Ca^{2+}-free perfusate on K^+-evoked rises of $[Ca^{2+}]_i$ in normoxic (top left) and chronically hypoxic (top right) cells. Also shown are typical effects of 200 μM Cd^{2+} on K^+-evoked rises of $[Ca^{2+}]_i$ in normoxic (lower left) and chronically hypoxic (lower right) cells. For all examples, the first application of 50 mM K^+ was during perfusion with Ca^{2+}-free or Cd^{2+}-containing perfusate, and the second application was following reperfusion with normal bath solution (as indicated). Scale bars: vertical, 0.2 ratio units; and horizontal, 30 s. These apply to all traces shown in A and B. (C and D) Mean rises of $[Ca^{2+}]_i$ recorded in normoxic cells (open bars, C) and chronically hypoxic cells (hatched bars, D)

that compares well with their native counterparts and so their usefulness in studying the cellular basis of such O$_2$ sensing is well founded. However, it must be remembered that information gleaned may be limited, as notable differences exist in the native systems and their immortal counterparts. This is particularly true for PC12 cells as model carotid body type I cells where the apparent differences cannot merely be attributed to species variation. For H146 cells, the limits of their use as models of native NEB cells have yet to be firmly established, as no cell physiological data are currently available for human NEBs. These two cell lines, being readily amenable to molecular intervention, may prove more useful in studying the mechanisms underlying adaptation to chronic hypoxia. Whether such information (which is already substantial in the case of PC12 cells) can be applied to the relevant native tissues remains to be fully determined. However, information gathered to date may also be utilized as a starting point for investigating physiological and pathological consequences of prolonged hypoxia in other tissues such as the brain. Given the present technological difficulties that limit advances in more complex and inaccessible tissues, PC12 and H146 cells are likely to remain valuable cell types in the study of cellular responses of O$_2$-sensitive cells for the foreseeable future.

Acknowledgments

The authors' contributions to the work discussed were supported by the Wellcome Trust, the British Heart Foundation, the Alzheimer's Society, and the Medical Research Council. We are also grateful for the contributions of past and present graduate students and postdoctoral workers, with all of whom it has been a pleasure to work.

---

in response to the application of perfusate containing 50 mM K$^+$. Recordings were made in the absence of Ca^{2+} channel blockers (control) or in the presence of 2 μM nifedipine, after preincubation with 1 μM ω-conotoxin GVIA or 200 nM ω-agatoxin GIVA, or both toxins, as indicated. Each bar represents the mean response (with vertical SEM bar) taken from the number of recordings shown above each bar. Adapted from Colebrooke et al.[61]

[61] R. L. Colebrooke, I. F. Smith, P. J. Kemp, and C. Peers, *Neurosci. Lett.* **318,** 69 (2002).

[2] Methods to Study Neuroepithelial Bodies as Airway Oxygen Sensors

By E. Cutz, X. W. Fu, and H. Yeger

Introduction

Pulmonary neuroepithelial bodies (NEB) are composed of innervated clusters of amine (serotonin, 5-HT) and peptide producing cells widely distributed within the airway mucosa of human and animal lungs.[1,2] Although earlier morphological studies suggested that NEB may represent hypoxia-sensitive airway chemoreceptors,[3] it has been only recently that strong evidence for a role of NEB as airway sensors has been obtained.[4] The major challenge for the studies of NEB function has been their relatively small number and widespread distribution within a sponge-like lung parenchyma. To overcome these obstacles, we have used two general approaches: (a) isolation and culture of NEB and (b) the use of fresh lung slice preparations.

Initially we developed a method for the isolation and culture of NEB from near-term rabbit fetal lungs.[5] This approach took advantage of an observation that NEB at these stages are more frequent compared with adult lung. In addition, near-term fetal as well as neonatal lungs could be microdissected readily to yield bronchial trees, which, when performed carefully, retained several generations of branches containing a significant fraction of the NEB, as they are concentrated along the main airways and at airway bifurcations.

A critical step, permitting electrophysiological studies of NEB cell membrane properties, has been our finding that a supravital dye, neutral red, selectively stained NEB cells in a living state both in culture[4] and in fresh lung slice preparations.[6] Our studies on O_2 sensing by NEB using these methods have been reviewed elsewhere.[7,8] This article provides

[1] E. Cutz, "Cellular and Molecular Biology of Airway Chemoreceptors," Landes Bioscience and Chapman Hall, Austin, TX, 1997.

[2] A. Van Lommel, T. Bolle, W. Fannes, and J. M. Lauweryns, *Arch. Histol. Cytol.* **62,** 1 (1999).

[3] J. M. Lauweryns and M. Cockleare, *Z. Zellforsch.* **193,** 373 (1978).

[4] C. Youngson, C. A. Nurse, H. Yeger, and E. Cutz, *Nature* **365,** 153 (1993).

[5] E. Cutz, V. Speirs, H. Yeger, C. Newman, D. Wang, and D. G. Perrin, *Anat. Rec.* **236,** 41 (1993).

[6] X. W. Fu, C. A. Nurse, Y. T. Wang, and E. Cutz, *J. Physiol.* **514,** 139 (1999).

[7] E. Cutz and A. Jackson, *Respir. Physiol.* **115,** 201 (1999).

[8] E. Cutz, X. W. Fu, H. Yeger, C. Peers, and P. J. Kemp, *in* "Oxygen Sensing: Responses and Adaptation to Hypoxia" (S. Lahiri, G. Semenza, and N. Prabhakar, eds.), p.567. Dekker, New York, 2003.

detailed protocols for isolation and culture of NEB, as well as the use of fresh lung slice preparation for electrophysiological recordings and amperometric detection of 5-HT release from NEB.

In Vitro Models for Isolation and Culture of NEB

Protocol for the Isolation and Culture of NEB

The following protocol details procedures for the isolation and culture of NEB from rabbit lungs but is also applicable to other species (Fig. 1).

Step 1. Fetal rabbit lung (New Zealand White, E26–E28) and postnatal days 1–2 lungs are removed surgically with trachea intact and transferred to a petri dish on ice. Culture medium, alpha medium of Minimal Essential

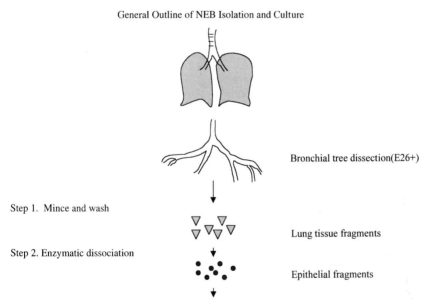

General Outline of NEB Isolation and Culture

Bronchial tree dissection(E26+)

Step 1. Mince and wash

Lung tissue fragments

Step 2. Enzymatic dissociation

Epithelial fragments

Step 3. Culture on LabTeks, glass coverslips, and plastics

Fig. 1. Schema depicting the general procedure for isolation and culture of NEB from rabbit fetal/neonatal lungs. Lungs of fetuses, E26+, are microdissected to produce bronchial trees. The trees are minced into small fragments that are digested enzymatically with 0.25% collagenase to yield epithelial microfragments containing epithelial, mesenchymal, and NEB components. The microfragments are plated in culture medium on a variety of culture substrata, including LabTek chambers, glass coverslips, tissue culture dishes, multiwell plates, and flasks. NEBs are subsequently identified by neutral red staining if used for electrophysiological studies.

Medium (αMEM [GIBCO-BRL]) containing antibiotics (penicillin/ streptomycin) and antimycotic (amphotericin B), is used to wet the tissue during the microdissection of bronchial trees. Lungs are placed on a dental wax sheet or on tissue paper placed into a 100-mm petri dish and are viewed under a dissecting microscope. Using fine tweezers and needles and beginning at the hilar region, the lung parenchyma is teased away gently from the bronchi and more distal bronchioles in a progressive manner working toward the lung apex. It should be possible to tease out at least four generations of branches so that the net result is a tree-like structure (Fig. 2A). On a fixed whole mount sample of bronchial tree immunostained for 5-HT (marker of NEB), the distribution and relatively high density of NEB during the fetal/neonatal stage becomes evident (Fig. 2B). During dissection the tissue is kept submerged in the culture medium and the dissected bronchial tree is placed into cold medium with antibiotics/antimycotic until ready for enzyme digestion. Fetal lungs of <E24–E26 may still be used to obtain bronchial trees, but the soft tissue precludes microdissection of more than two branches. Therefore, these younger lungs are minced finely in the carrier solution after grouping on

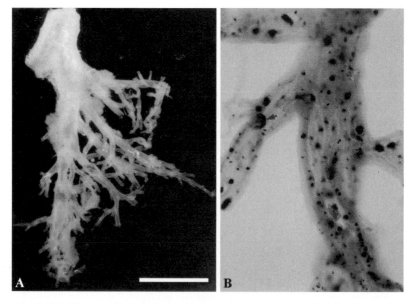

Fig. 2. (A) Dissected bronchial tree from one lung of a 4-day-old rabbit suspended in culture medium. Up to four generations of airway branches are retained (bar: 2 mm). (B) Whole mount preparation of dissected bronchial tree from fetal (E26) lung fixed in neutral-buffered formalin and immunostained for 5-HT. Numerous NEBs (dark foci) are distributed along the length of airway and at airway branch points (arrow) (bar: 200 μm).

top of a nylon slab and using a No. 10 blade cross cutting the tissue until approximately 1-mm^3 cubes are produced (Fig. 1). The mince is then scooped into a 50-ml polypropylene tube (Falcon) and washed two times with the carrier medium with brief low-speed centrifugation to separate viable tissue and remove damaged cell debris. To prepare bronchial trees for digestion, a similar mince step is performed, but here only ~1-mm rings need to be cut. Overmincing of the finer bronchioles may damage too many cells and reduce yields.

Step 2. The objective of the enzymatic digestion step is to produce small epithelial fragments while digesting away the bulk of the collagenous network of the lung. Although a number of different collagenases, including dispase, can be used, type IV collagenase works well in most instances. A 0.25% solution of collagenase type IV (Sigma) is made in αMEM plus antibiotics and added to the minced tissues in the 50-ml tube in a ratio of 10-ml enzyme solution/ml of mince. A few grains of DNase I can be added to degrade the DNA released from damaged cells if it is noted that the fragments are trapped in a viscous aggregate. The tube is then rocked at 37° on a rocking platform for 1 h after which the contents are triturated gently with a 5- or 10-ml pipette until well dispersed into a cloudy suspension of small fragments invisible to the naked eye. The appearance and size of the fragments can be assessed further microscopically.

Note: If digestion appears to be slow, due to batch enzyme differences, incubation is continued for another 30 min at 37°. The digest is triturated once more and stopped once the appropriate suspension of microfragments has been generated. One volume of medium containing 10% fetal bovine serum is added to the digest, pipetted gently several times, allowed to stand for a few minutes, and then filtered through a square of presterilized nitex nylon filter (70 μm) inserted in the form of a cone shape into a fresh 50-ml tube. This filtration step removes large fragments, including cartilage and reaggregated collagen fibers, leaving behind lung tissue epithelial microfragments in the filtrate. The filtrate is centrifuged at 500 rpm for 2–3 min, transferred to a 15-ml conical tube, and washed/centrifuged repeatedly an additional three times. The resulting cleaned fragments are now ready for culture (Fig. 3A).

Step 3. Microfragments containing airway epithelium with NEB (Fig. 3B) and a small component of adjacent mesenchymal cells are resuspended in growth culture medium, αMEM with supplements,[5] and maintained at 37° in 5% CO$_2$. Cultures are set up in a ratio of medium to cells that yields approximately 30% confluency after overnight attachment. This can be determined after two to three trials and can be varied if a more concentrated seeding is required for electrophysiology. Culture vessels that have been employed successfully include tissue culture flasks, multiwell

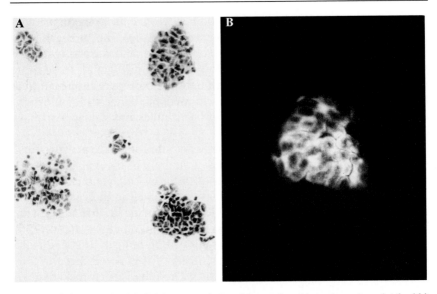

FIG. 3. (A) Purified epithelial fragments isolated from minced lung tissue from fetal rabbit (E26) (hematoxylin & eosin stain, ×120). (B) Isolated epithelial fragment containing NEB with diffuse cytoplasmic formaldehyde-induced fluorescence due to 5-HT content in the same preparation as in a formaldehyde-induced fluorescence method (×550).

dishes, individual 35-mm dishes and glass coverslips (autoclaved or flame sterilized after dipping in alcohol), and LabTek chambers (NUNC). When using glass, the surfaces are precoated with type 1 collagen (rat tail, 1 mg/ml in dH$_2$O; air dried and neutralized/washed prior to seeding with medium) or 0.1% poly-L-lysine/dH$_2$O (air dried and neutralized/washed with medium). Once cells have attached, cultures receive a medium change every 2–3 days. Reducing the serum concentration to <2% (as low as 0.5%) is effective in slowing down the overgrowth of fibroblasts.

Step 4. Primary cultures of lung tissue are incubated with 0.02 mg/ml neutral red in medium or phosphate-buffered saline (PBS) for 15–30 min and mounted in an inverted microscope for the determination of dye uptake. NEB take up neutral red more rapidly than surrounding cells due to their acidic content of 5-HT in secretory granules (Fig. 4) In fact, 5-HT levels and neutral red positivity can increase after 2–3 days of culture. It would appear that enzymatic tissue dissociation enhances the degranulation of NEB cells. *In vitro* culture restores endogenous 5-HT content to levels that can exceed those seen in native tissue.

Our methodological approach for the isolation and culture of NEB has been applied to different animal species and works well with mouse,

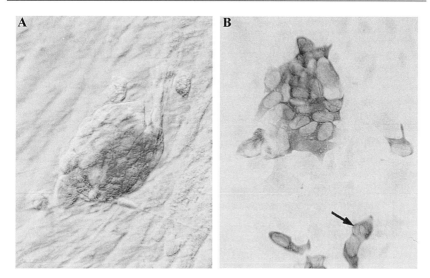

FIG. 4. (A) Vital staining of a 3-day culture of NEB isolated from rabbit fetal lung (E26) with neutral red visualized under Hoffman modulating contrast optics. The culture was incubated with 0.02 mg/ml neutral red for 30 min at 37°. Positive selective staining of the NEB cell cytoplasm (red) contrasts with other lung cells, which are unstained (neutral red, ×550). (B) NEB in a 3-day culture fixed in neutral-buffered formalin and immunostained for 5-HT. NEB forms a closely packed cluster of 5-HT-immunoreactive cells (dark cytoplasm) and negative nuclei. Few isolated 5-HT-immunoreactive NEB cells are seen outside of the cell cluster (arrow) (immunoperoxidase method for 5-HT, ×550). (See color insert.)

hamster, and rat lungs with only small protocol adjustments necessary as dictated by species variations, gestational age, and the actual time required for enzymatic dissociation based on the collagen content of the lung tissue. In all instances, dissociated lung cells are subsequently seeded either on tissue culture plastics or on type 1 collagen-coated glass surfaces (LabTek chamber glass slides), which are inherently less conducive to good cell attachment when used uncoated. The aforementioned protocols for the isolation and culture of NEB have been used successfully in studies of hypoxia-induced amine release,[5] electrophysiological characterization of membrane currents, and O$_2$ sensing mechanism,[4] as well as immunohistochemical and molecular identification of neuroendocrine markers and various functional proteins including O$_2$ sensing protein (NADPH oxidase),[4,9]

[9] D. Wang, C. Youngson, V. Wong, H. Yeger, M. Dinauer, E. Vega-Saenz de Miera, B. Rudy, and E. Cutz, *Proc. Natl. Acad. Sci. USA* **93,** 13182 (1996).

cystic fibrosis transmembrane regulator (CFTR),[10] and serotonin-3 receptor.[11] These culture preparations were also found suitable for RT-PCR analysis using either NEB cell clusters or even single NEB cells recovered by means of micropipette removal.[10]

Fresh Lung Slice Preparation for Patch-Clamp Recordings of O_2-Sensitive K^+ Current and Amperometric Measurement of 5-HT Release

Pulmonary NEB are functionally analogous to the type I or glomus cells of the carotid body and behave as airway chemoreceptors. In chemoreceptor cells, hypoxia inhibition of K^+ channels induces cell depolarization and neurotransmitter release.[12] We developed a fresh neonatal rabbit lung slice model to characterize the electrophysiological properties of "intact" NEBs with patch-clamp, whole cell recording.[6]

While the method for the isolation and culture of NEB is highly reproducible, it is time-consuming and may induce potential artifacts related to enzymatic treatment and cell culture. Advantages of the lung slice model include (1) patch-clamp recordings are made directly from NEB in their "natural" environment; (2) this preparation may be suitable for investigation of synaptic connections between NEB cells and their innervation; and (3) the technique is easily applied to different animal species.

Lung Slice Preparation. Neonatal New Zealand white rabbits,[6] Syrian gold hamster,[11] and black C57/BL/6L mice were used between 1 and 10 days of age. All the experiments are carried out with the approval of the local ethics committee and in accordance with the Institutional Guidelines for Animal Care. The animals are killed by an intraperitoneal euthanyl (pentobarbital sodium, 100 mg/kg) injection. The lung tissue is cut to around 2- to 3-mm^2 size fragments and subsequently embedded in 2% agarose (FWC Bioproducts, Rockland, ME). Sectioning is performed with tissue immersed in ice-cold Krebs solution that has the following composition (in mM): 140 NaCl, 3 KCl, 1.8 CaCl$_2$, 1 MgCl$_2$, 10 HEPES, and 5 glucose at pH 7.3 adjusted with NaOH. Transverse lung slices (200–300 μm) are cut with a Vibratome (Ted Pella, Redding, CA) (Fig. 5A).

Note: The present protocol uses 2% low-melting temperature agarose to embed the fragments of lung for sectioning. The temperature of agarose should not exceed 37 ± 0.5° as a higher temperature may damage NEB cells with failure of recordings.

---

[10] H. Yeger, J. Pan, X. W. Fu, C. Bear, and E. Cutz, *Am. J. Physiol. Lung Cell. Mol. Physiol.* **281,** L713 (2001).

[11] X. W. Fu, D. Wang, J, Pan, S. M. Farragher, V. Wong, and E. Cutz, *Am. J. Physiol. Lung Cell. Mol. Physiol.* **281,** L931 (2001).

[12] J. Lopez-Barneo, R. Pardal, and P. Ortega-Saenz, *Annu. Rev. Physiol.* **63,** 259 (2001).

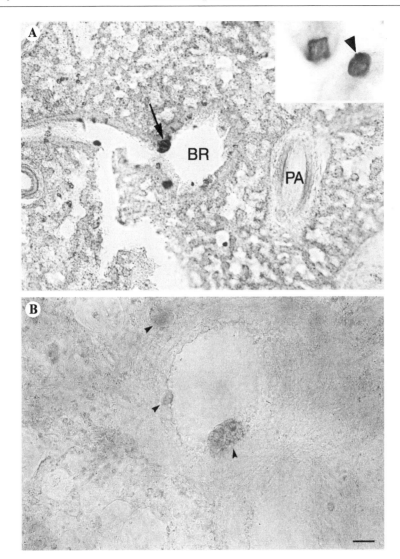

FIG. 5. (A) Low magnification view of a neonatal rabbit lung slice (~200 μm thick) fixed in neutral-buffered formalin and immunostained for 5-HT. A small bronchiole (BR) with a NEB cell cluster strongly positive for 5-HT is located at airway bifurcation (arrow). (Immunoperoxidase method for 5-HT; bar: 100 μm.) PA, pulmonary artery. (Inset) Higher magnification of airway epithelium with two 5-HT-immunoreactive NEBs facing each other with their apical surfaces exposed to airway lumen (arrowhead) (immunoperoxidase method for 5-HT; bar: 5 μm). From Fu et al.[13] (B) Staining of NEB cells. Arrowheads show neutral red staining of NEB cells in a fresh slice (~200 μm thick) of neonatal rabbit lung (2 days old) (bar: 5 μm). From Fu et al.[6] (See color insert.)

Identification of NEB by Neutral Red. The slices are incubated with L-15 medium (Leibovitz's, GIBCO BRL, Concord, Canada) in tissue culture dish at room temperature. To identify NEB in lung slices, we adopt the method from NEB in culture.[4] In lung slices incubated with neutral red (0.02 mg/ml^{-1} for 15–20 min at 37°), NEB appear as reddish-pink cell clusters within the airway epithelium, which remain unstained (Fig. 5B). If in doubt, verify that neutral red-stained cells are NEB cells using immunostaining with a specific antibody against 5-HT, a marker for NEB using conventional immunohistochemical methods on fixed tissue.[4,5]

Note: In mouse lungs, NEB staining with neutral red is less intense (light pink) compared to rabbit or hamster lungs. This is due to a lower endogenous content of 5-HT in NEB cells of the mouse lung. To increase 5-HT content and subsequently neutral red staining, preincubate lung slices with 5-hydroxytryptophane or L-DOPA (50–100 μg/ml) in Krebs solution for 30 min at 37°.

Solutions for Electrophysiology. The external solution contains (in mM) 130 NaCl, 3 KCl, 2.5 CaCl$_2$, 1 MgCl$_2$, 10 NaHCO$_3$, 10 HEPES, and 10 glucose (pH 7.35–7.4). When K$^+$ currents are recorded, the internal recording solution contains (in mM) 30 KCl, 100 potassium gluconate, 1 MgCl$_2$, 4 Mg-ATP, 5 EGTA, and 10 HEPES. The pH of the solution is adjusted to 7.25 with KOH. For inward currents, the internal recording solution contains (in mM): 130 CsCl, 5 TEACl, 2 MgCl$_2$, and 10 HEPES; pH adjusted to 7.2 with CsOH.

Electrophysiology Recordings. Whole cell recordings are performed as described previously.[4,6] The recording electrodes are advanced with a hydraulic manipulator (ONM-1, Narishge, Japan) to NEB cells at a 45° angle under visual guidance ($\times$400 magnification). The seal resistance is typically 1–2 GΩ. The hypoxic solution is prepared by bubbling 95% N$_2$ into the reservoir, which feeds the perfusion chamber via low-gas permeability tubing. Samples of perfusion medium from the recording chamber are taken for measurements of PO$_2$ levels using a PO$_2$ electrode (1610 pH/blood Gas Analyser, Instrumentation Laboratory, Lexington, MA). The PO$_2$ of the control normoxic perfusion medium in the recording chamber is 145–155 mm Hg. The level of PO$_2$ of the hypoxic solution in the recording chamber varies between 15 and 20 mm Hg.

Data Analysis. An Axopatch 200B (Axon Instruments, Foster City, CA) amplifier is used for recording in the whole cell voltage-clamp mode. Data are filtered at 5 kHz. Voltage commands and data acquisition are done using pCLAMP6 software and Digidata 1200B interface (Axon Instruments). The leak current is subtracted from all current records using

[13] X. W. Fu, C. A. Nurse, V. Wong, and E. Cutz, *J. Physiol.* **539,** 503 (2002).

pCLAMP software (*P*/4 subtraction protocol). The resting membrane potential is measured soon after rupture of the cell membrane. The resting input resistance (R_m) is estimated from liner current–voltage relationships over the range −80 to −50 mV in the voltage-clamp mode. The cell capacitance (C_m) is determined from the transient current response during a small hyperpolarizing voltage step. The time constant (τ_m) is calculated from the input resistance and capacitance; $\tau_m = R_m \times C_m$.[6]

Results. Hypoxia suppressed roughly equally both $I_{K(Ca)}$ and $I_{K(v)}$ components of K$^+$ outward current, and no further inhibition of K$^+$ current was seen with either K$^+$ channel blockers, TEA, or 4-AP plus hypoxia in NEBs. The ion channel-based mechanisms of NEB cell transduction of PO$_2$ center on the role played by O$_2$-sensitive K$^+$ channels located in the plasma membrane.[12] Low PO$_2$ inhibits a voltage-dependent, delayed rectifier type of K$^+$ channel (Fig. 6) that is not inhibited by cellular levels of ATP. This decreased channel activity is primarily dependent on a decrease in K$^+$ channel densities. In NEB cells, three types of K$^+$ current have been described: (1) voltage-gated Ca^{2+}-independent [$I_{K(V)}$], delayed rectifier current; (2) Ca^{2+}-dependent K$^+$ current [$I_{K(Ca)}$]; and (3) TEA-insensitive K$^+$ current.[4,6] Of functional relevance is that this decrease in K$^+$ channel activity is presumed to depolarize the NEB cells; this depolarization is manifested by an increase in excitability of NEB cells and an increase in action potentials traveling along the vagal afferent fibers to the central nervous system (CNS). NEB cells express the oxygen-binding protein NADPH oxidase on their plasma membrane. Hypoxia affects the function of the oxidase via a decrease in the availability of substrate oxygen, resulting in a reduced production of oxygen-reactive intermediates, including H$_2$O$_2$. The H$_2$O$_2$-sensitive channel Kv 3.3a is coexpressed with the membrane components of the NADPH oxidase (gp91^{phox} and p22^{phox}) in NEBs of rabbit and neonatal human lungs.[9] Exposure of NEBs to hydrogen peroxide results in an increase in K$^+$ channel amplitude.[6,9,14] The important role of NADPH oxidase as the oxygen sensor in NEB cells was confirmed in a gp91^{phox} knockout mouse model.[14]

Amperometric Detection of 5-HT Release from NEB Cells

Electrochemical methods based on the oxidation of reduction of specific transmitters enable exquisitely sensitive measurements of secretion from single cells. Advantages of the electrochemical approach compared to the capacitance method are that (1) exocytosis can be monitored without

---

[14] X. W. Fu, D. Wang, C. A. Nurse, M. C. Dinauer, and E. Cutz, *Proc. Natl. Acad. Sci. USA* **97**, 4374 (2000).

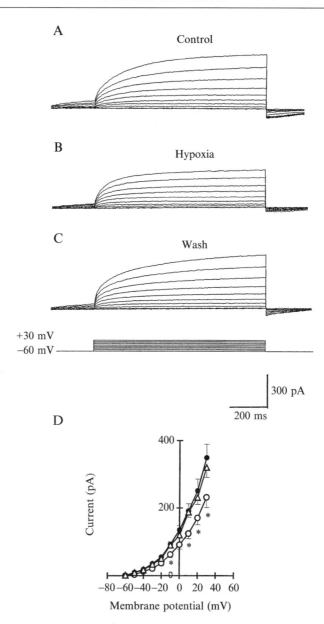

A   Control

B   Hypoxia

C   Wash

+30 mV
−60 mV

300 pA

200 ms

D

Current (pA)

400

200

0

−80 −60 −40 −20 0   20 40 60

Membrane potential (mV)

FIG. 6. Effect of hypoxia on K^+ current in neonatal rabbit NEB cells. (A) Outward K^+ current evoked by depolarizing steps from −60 to 30 mV in control normoxic Krebs solution. (B) Outward current evoked by same voltage steps as in A was reduced by hypoxia. (C) Washout of the hypoxic solution caused a recovery of the outward K^+ current. (D) I–V

interference from overlapping endocytosis; (2) the released products can be monitored directly (not a model-dependent parameter such as electrical capacitance); (3) voltage-clamp control is not necessary and, therefore, there are fewer restrictions on the cell shape; and (4) the method is "non-invasive" in the sense that the cell is not subject to the dialysis of cytosolic components that occurs with (whole cell) patch-clamp measurements.[15] Many secreted products are readily oxidizable; among these, the most intensively studied are norepinephrine, epinephrine, dopamine, and 5-HT.[15]

The predominant amine in mammalian and human NEB cells is serotonin, 5-HT, as demonstrated by immunohistochemistry and HPLC measurement.[1,5] Earlier *in vivo* and *in vitro* studies have provided indirect evidence for 5-HT release from NEB cells in reponse to hypoxia challenge.[3,5] Although the precise function of 5-HT in the lung is not known, the postulated local effects include bronchoconstriction, modulation of vasomotor tone, and/or growth factor-like properties. In addition, this amine is also postulated to act as a neurotransmitter of hypoxia signaling.[8] To study hypoxia-induced 5-HT release from NEB cells directly, we have adapted the use of amperometric technique to our lung slice preparation.[13]

Solutions for Amperometric Detection of 5-HT Release

All recordings are made from submerged lung slices at room temperature (see earlier discussion). The perfusing Krebs solution has the following composition (mM): 130 NaCl, 3 KCl, 2.5 CaCl$_2$, 1 MgCl$_2$, 10 NaHCO$_3$, 5 HEPES, and 10 glucose at pH 7.35–7.4. The recording carbon fiber microelectrode is backfilled with 3 M KCl as the internal solution and calibrated by adding known amounts of 5-HT to the medium.

Equipment for Amperometric Detection of 5-HT

Carbon fiber microelectrodes (ProCFE, Dagan Corporation, 5–8 μm diameter) used in our study consist of polypropylene-insulated carbon fiber designed for the detection of oxidizable analytes. Detectable neurotransmitters include catecholamines (adrenaline and epinephrine), noradrenaline

[15] R. H. Chow and L. von Ruden, in "Single-Channel Recording" (B. Sakmann and E. Neher, eds.), p. 245. Plenum Press, New York, 1995.

relationships for the current in the control solution (●) and in the hypoxic solution (○) are plotted together with recovery K$^+$ current (△). Holding potential was −60 mV. Asterisks indicate significant differences from control ($P < 0.05$). Data represent means ± SEM for a sample of 10 cells. From Fu *et al.*[6]

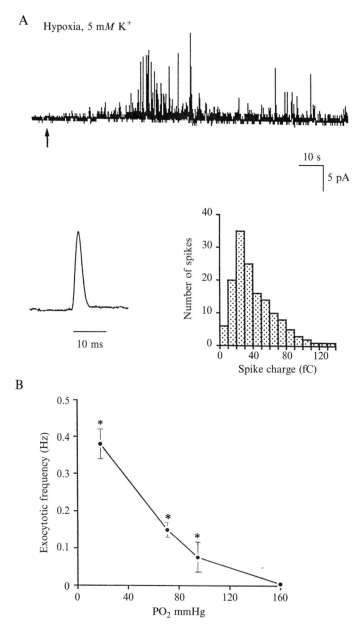

Fig. 7. Amperometric recording of hypoxia-induced 5-HT release from a NEB cell in a neonatal rabbit lung slice preparation. (A) Example of exocytosis with 5-HT release induced by perfusion with a hypoxia solution ($PO_2 \sim 18$ mm Hg). At the point indicated by the arrow,

(norepinephrine and dopamine), and indolamines (serotonin and 5-HT). The ProCFE microelectrode oxidizes these compounds within a voltage range of 600–900 mV. The microelectrodes are positioned adjacent to NEBs identified with neutral red (see earlier discussion) and are polarized to 700 mV. At this voltage, 5-HT is the main substance contained in NEB cell vesicles capable of being oxidized; therefore, the amperometric signal mostly represents 5-HT release. The positive voltage is given by an Axopatch 200B amplifier (Axon Instruments). All potential is reported with respect to an Ag–AgCl reference electrode. Currents are recorded using an Axopatch 200B amplifier (with extended voltage range), filtered at 1 kHz, kHz, and digitized at 2 kHz before storage on a personal computer. All data acquisitions are performed using a Digidata 1200B interface and Fetchex software from pCLAMP 6.04 (Axon Instruments).

Data Analysis

Exocytosis is expressed as the frequency of quantal events determined with the aid of the Mini Analysis Program (Synaptosoft Inc., Leonia, NJ) by counting the number of secretory events over a 60-s period, beginning 15 s after switching to the test solution. Quantal secretion is detected as transient upward current deflections, each arising from oxidation of the released contents from a single vesicle containing 5-HT. The integral of the transient oxidation current can be directly related to the calculated number of applied molecules by Faraday's law: $Q = Mez$, where M is the number of molecules, e is the elementary charge 1.6×10^{-19} coulombs, and z is the number of moles of electrons transferred per mole of 5-HT reacted.[16]

Results

Under normoxia (PO₂ ∼ 155 mm Hg; 1 mm Hg ≈ 133 Pa), most NEB cells did not exhibit detectable secretory activity; however, hypoxia elicited a dose-dependent (PO₂ range; 95–18 mm Hg) vesicular amine release (Fig. 7). The integral of the oxidation current increased linearly with the amount of applied transmitter and correlated well with the prediction of

[16] D. Bruns and J. Reinhard, *Nature* **377,** 62 (1995).

the perfusate was changed to a hypoxia solution. Large spike-like exocytic events are shown at an expanded time base (bottom left). Also shown is frequency distribution of the charge of secretory events evoked by hypoxia in six different cells (bottom right). (B) Graded response to hypoxia with four different pO₂ levels on a secretory response from NEB cells. Each point represents mean values of between 8 and 18 cells (mean ± SEM). From Fu *et al.*[13]

four electrons transferred per oxidized 5-HT molecule, agreeing with the mechanism proposed for the dominant electrochemical oxidation reaction.[16] In NEB cells, the average ($\pm$ SEM) quantal charge of secretory events during hypoxia was 33.1 ± 2.4 fC [$n = 157$ events from six cells, range 2.3–183 fC (Fig. 7)]. This value was obtained from the time integral of selected spikes with the fast-rising phase and slow decay typical of secretory events occurring at the membrane facing the amperometric electrode. Assuming that one 5-HT molecule contributes an average of four electrons, it is estimated that a single synaptic vesicle or quantum in NEB cells releases an average of $13{,}000 \pm 971$ ($n = 157$; six cells) molecules of 5-HT.[13] High extracellular K^+ (50 mM) induced a secretory response similar to that elicited by severe hypoxia. Exocytosis was stimulated in normoxic NEB cells after exposure to tetraethylammonium (20 mM) or 4-aminopyridine (2 mM). Hypoxia-induced secretion was abolished by the nonspecific Ca^{2+} channel blocker, Cd^{2+} (100 μM). Secretion was also largely inhibited by the L-type Ca^{2+} channel blocker, nifedipine (2 μM), but not by the N-type Ca^{2+} channel blocker, ω-conotoxin GVIA (1 μM). The 5-HT$_3$ receptor blocker, ICS 205 930, also inhibited secretion from NEB cells under hypoxia. These results suggest that hypoxia stimulates 5-HT secretion from intact NEBs via the inhibition of K^+ channels and calcium entry through L-type Ca^{2+} channels, as well as by positive feedback activation of 5-HT$_3$ autoreceptors.[13]

[3] Role of Glutathione Redox State in Oxygen Sensing by Carotid Body Chemoreceptor Cells

By Constancio Gonzalez, Gloria Sanz-Alfayate, Ana Obeso, and Maria Teresa Agapito

Introduction

This article first presents some basic structural traits of the carotid body (CB) arterial chemoreceptors to understand the relationship between the arterial blood PO$_2$ and the activation of chemoreceptor cells, which are the O$_2$ sensing structures of the CB. Some considerations in relation to the intensity of CB blood flow and O$_2$ consumption of the organ would allow us to define the threshold for the detection of the hypoxic stimulus, which would lead us to the cardinal theme of the article, namely whether at the PO$_2$ levels detected by the CB there alterations in the genesis of reactive oxygen species (ROS). An alteration in the rate of ROS production

METHODS IN ENZYMOLOGY, VOL. 381

would impinge on the glutathione system [reduced glutathione (GSH) and oxidized glutathione (GSSG)], causing modifications in the GSH/GSSG ratio that are detected by direct measurement; the GSH/GSSG system represents the quantitatively most important mechanism to dispose ROS and to maintain the overall redox status or redox environment in mammalian cells.[1] The relationship between GSH/GSSG and oxygen chemoreception is approached from two different points of view. We will measure GSH/GSSG levels and calculate the redox environment of the cells and correlation with the activity of chemoreceptor cells in normoxia and in hypoxia. We will also present data on pharmacological manipulation of the redox environment of the cells, as assessed by GSH/GSSG quotients, and possible correlations with the level of activity of chemoreceptor cells. The possible mechanisms of coupling between ROS and the GSH/GSSG system to the cellular effector machineries have been reviewed.[2,3]

Structure, Blood Flow, Oxygen Consumption, and PO_2 of Carotid
   Body Tissue

Carotid bodies are small paired organs, round to pear shaped, located in the proximity of the carotid artery bifurcation, which weigh ≈ 500 μg in the cat and ≈ 50 μg in the rat; in humans it is estimated that each CB weighs ≈ 1 and 2 mg. The CB receives sensory innervation via the carotid sinus nerve (CSN), a branch of the IX cranial nerve, by its cephalic pole. The parenchymatous cells of CB are organized in cluster-separated connective tissue converging on the surface of the organ to form the CB capsule[4] (Fig. 1). The size of the clusters varies considerably, some have 5–8 cells and others up to 20–30 cells; in any case, chemoreceptor cells, located toward the center of the clusters, exceed glial-like sustentacular cells by a factor of 3–5. In the connective tissue of the organ there is a dense net of capillaries with tortuous trajectories and variable diameters (8 to 20 μm). This ample vascularization of the CB constitutes the most prominent feature in a section of a well-perfused CB, being 25–33% of the surface of the section occupied by the capillary lumens[5] (Fig. 1). The endothelium is thin,

[1] F. Q. Schaffer and G. R. Buettner, *Free Radic. Biol. Med.* **30,** 1191 (2001).
[2] V. J. Thannickal and B. L. Fanburg, *Am. J. Physiol. Lung Cell. Mol. Physiol.* **279,** L1005 (2000).
[3] C. Gonzalez, M. T. Agapito, A. Rocher, G. Sanz-Alfayate, and A. Obeso, *in* "Oxygen Sensing: Responses and Adaptation to Hypoxia" (S. Lahiri *et al.,* eds.), p. 489. Dekker, New York, 2003.
[4] A. Verna, *in* "The Carotid Body Chemoreceptors" (C. Gonzalez, ed.), p. 1. Springer Verlag, New York, 1997.
[5] C. Gonzalez, L. Almaraz, A. Obeso, and C. Gonzalez, *Physiol. Rev.* **74,** 829 (1994).

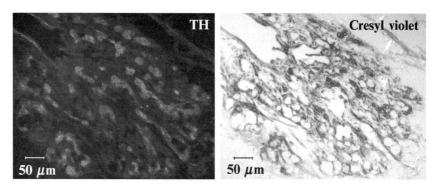

FIG. 1. Histological sections near the equator of a rat carotid body. The section was first immunostained using a monoclonal antibody against tyrosine hydroxylase (TH) and a secondary antibody labeled with fluorescein isothiocyanate to label the TH-containing chemoreceptor cells that are grouped in clusters. The section was later counterstained with cresyl violet to show the great density of capillaries in the CB tissue. (See color insert.)

sometimes fenestrated, and partially surrounded by pericyte processes. The capillaries resume into venules that emerge from the organ to form a dense venous plexus in the surface of the CB. This rich vascularization determines that the distance from the center of most chemoreceptor cells to capillaries is between 10 and 20 μm with a median distance of 15.77 μm from the center of CB cell clusters to the nearest capillary in the adult cat CB.[6] Paralleling this ample vascularization, the CB is the organ with the highest blood flow of the organism, 1417 ml/min/100 g tissue, although with discrepancies it is estimated that the O_2 consumption in basal conditions is $\approx$1.3 ml/min/100 g tissue at a perfusing pressure and PO_2 in the perfusates (whether blood or saline) of 100 mm Hg.[7,8] These basic data on the vascularization, blood flow, and O_2 consumption show that the blood supply and distribution inside the CB have important physiological roles beyond the nutritional requirement. The aforementioned data and *in vitro* studies with the saline-superfused preparation of the CB show that the organ can survive, sense, and transduce O_2 and CO_2 levels just with O_2 dissolved at a PO_2 of 100 mm Hg. However, despite the apparent excess of blood flow in normal CB, there is a further increase

[6] D. W. Lübbers, L. Teckhaus, and E. Seidl, *in* "Chemoreception in the Carotid Body" (H. Acker *et al.*, eds.), p. 62. Springer Verlag, Berlin, 1977.

[7] W. J. Whalen and P. Nair, *in* "Physiology of the Peripheral Arterial Chemoreceptors" (H. Acker and R. G. O'Regan, eds.), p. 117. Elsevier, Amsterdam, 1983.

[8] A. Obeso, A. Rocher, B. Herreros, and C. Gonzalez, *in* "The Carotid Body Chemoreceptors" (C. Gonzalez, ed.), p. 31. Springer Verlag, New York, 1997.

in the vascularization of the CB with increased diameters and neoformation of the capillaries in situations of chronic hypoxia.[4] This increase in capillarity reduces the mean distance from capillaries to the border of cell clusters from 3 to 1.8 μm.[4]

What is the purpose of this high blood flow? The high blood flow and the increase in vascularization that occurs during chronic hypoxia (acute hypoxia produces an important vasodilatation) guarantee that the arteriovenous difference of O_2 is very small and, therefore, that CB tissue PO_2 is the optimal of the organism at any arterial PO_2. Although there are inconsistencies in the actual tissue PO_2 in the CB,[8] Lübbers et al.[6] calculated that only about 4% of the PO_2 values in the CB would be below 40 mm Hg (in fact, the PO_2 should be higher because Lübbers et al.[6] used data for the O_2 consumption that were four to five times higher than those obtained in more recent studies). In line with those calculations, Whalen and Nair (reviewed in Refs. 7 and 8) found that mean CB tissue PO_2 is around 65 mm Hg when perfusing the CB with blood equilibrated at normal PO_2 (above 85 mm Hg). Perfused with blood at normal PO_2, these authors found tissue PO_2 values below 40 mm Hg in some studies and, when perfused with blood at PO_2 in the range of 30–49 mm Hg, the mean CB tissue PO_2 was 20 mm Hg with very few values below 5 mm Hg. Using air-equilibrated saline solutions to perfuse, they found normal tissue PO_2, validating the empirical observation that the CB functions normally in the saline-superfused preparation. Whalen and Nair[7] found that the hypoxic threshold for CSN discharge is a CB tissue PO_2 oscillating between 50 and 65 mm Hg, the P_{50} for discharges oscillates between 10 and 32 mm Hg, and the peak CSN discharge is reached at tissues PO_2 of 3–5 mm Hg. These values correspond to arterial PO_2 of 70–75 mmHg (threshold), 40 mm Hg (P_{50}), and 10 mm Hg (peak CSN discharges).[5] To conclude, and giving an answer to the question formulated at the beginning of this paragraph, we consider that the high blood flow of the CB and the neovascularization occurring in chronic hypoxia tend to show that the CB (which constitutes the origin of a regulatory loop aimed to secure the availability of O_2 to the organism) receives an adequate O_2 supply to support its activity. Activity, however, increases in parallel to the decrease in arterial PO_2.

Genesis of ROS and Tissue PO_2

From the aforementioned data we can state that the entire range of activity of chemoreceptor cells occurs at PO_2 in their near environment at about 65 mm Hg in normoxia and above 5 mm Hg in situations of extreme hypoxia hardly compatible with life. At lower PO_2, CBs are still able to function and, in fact, they continue functioning for long periods of time after the death of experimental animals (the CB has been considered the *"ultimum moriens"*). Questions to be asked include (1) is there a modification in the rate of ROS productions in the range of PO_2 where the CB is activated, (2) what is the mechanism responsible for the modification in the rate of ROS production, and (3) is the modification a cause or consequence of chemoreceptor cell activation? At the outset we must state that published experimental data related specifically to chemoreceptor cells are not available.

The first question does not have an univocal answer as evidenced by the dual hypothesis put forward by Jones *et al.*[9] referring to hypoxic pulmonary vasoconstriction: "vasoconstrictive ROS are produced under hypoxia" and "a reduced production of vasodilatory ROS occurs under hypoxia." We[3,10] have summarized information on the putative relationship between PO_2 and the rate of ROS production. The classical view[11,12] is that ROS production occurs in proportion to the available O_2, except in situations of ischemia–reperfusion when xanthine dehydrogenase might be converted into xanthine oxidase by oxidative or proteolytic processes; the new enzyme transfers electrons from the purines directly to O_2 to form $O_2^{-\bullet}$.

According to this view, the first hypothesis on O_2 sensing in chemoreceptor cells in considering the participation of ROS assumed that hypoxia decreased the rate of ROS production (the same occurred in pulmonary artery smooth muscle cells, which like CB cells are stimulated by hypoxia). A decreased ROS would increase the GSH/GSSG ratio and determine that the O_2 sensor and additional proteins involved in the generation of the response to hypoxia were in reduced form (Prot-SH). The reduction of these proteins would produce activation of the cells and generate the hypoxic responses. Semiquantitative data giving support to this hypothesis were provided by Cross *et al.*[13] using dihydrorodamine 123 fluorescence in the CB and by Archer *et al.*[14] and Weir and Archer[15] using lucigenin chemiluminescence in the lung. The decreased ROS production during hypoxia would result from a putative decrease in the flow of electrons in the mitochondrial respiratory chain and from a decreased

[9] R. D. Jones, J. T. Hancock, and A. L. Morice, *Free Radic. Biol. Med.* **29,** 416 (2000).

[10] C. Gonzalez, G. Sanz-Alfayate, M. T. Agapito, A. Gomez-Niño, A. Rocher, and A. Obeso, *Respir. Physiol. Neurobiol.* **132,** 17 (2002).

[11] B. Chance, H. Sies, and A. Boveris, *Physiol. Rev.* **59,** 527 (1979).

[12] B. Halliwell and J. MC. Gutteridge, "Free Radicals in Biology and Medicine." Oxford Univ. Press, Oxford, 1999.

[13] A. R. Cross, L. Henderson, O. T. G. Jones, M. Delpiano, M. A. Hentschel, and H. Acker, *Biochem. J.* **272,** 743 (1990).

[14] S. L. Archer, J. Huang, T. Henry, D. Peterson, and E. K. Weir, *Cir. Res.* **73,** 1100 (1993).

[15] E. K. Weir and S. L. Archer, *FASEB J.* **9,** 183 (1995).

velocity of superoxide ($O_2^{-\bullet}$) formation by a NADPH oxidase similar to the one present in phagocytes. Consistent with the classical view, it was also observed that the production of ROS in HepG-2 cells decreased with PO_2 in a very ample range of oxygen pressures.[16] More recently Archer et al.[17] found that hypoxia (36 mm Hg) decreased ROS production in endothelium-free rings of resistance rat pulmonary arteries from apparently two different sources: as hypoxia decreased the production of ROS and diphenylene iodonium (DPI; used as an inhibitor of NADPH oxidase) caused a further decrease in ROS production, ROS would be originated in mitochondria and at the level of NADPH oxidase. The same group[18] also reported that hypoxia (40 mm Hg) decreased ROS production in pulmonary artery rings, but increased ROS production in renal artery rings (measured as lucigenin chemiluminescence, but measured as 2,7-dichlorofluorescein fluorescence or by the peroxidase-based AmplexRed kit, hypoxia did not alter the production of ROS in renal artery rings). As in the previous study, rotenone (but not myxothiazol) and DPI also inhibited the production of ROS, but curiously enough, antimycin A, which blocks the respiratory chain distally to the quinone pool (the step in the respiratory chain where most ROS appear to be generated[10]), also inhibited the production of ROS in pulmonary rings but cyanide did not alter the rate of ROS production in either pulmonary or renal artery rings (in Archer et al.,[14] cyanide increased and antimycin A decreased the production of ROS in the lung).

In the opposite view, where hypoxia increased ROS production, Marshall and co-workers[19] found that an NADPH oxidase-like enzyme increased ROS production in lung tissue. More recently, the view that hypoxia increases ROS production has been marshaled by Chandel and co-workers who, in different preparations, such as cardiomyocytes,[20] pulmonary artery smooth muscle cells,[21] or alveolar epithelial cells,[22] found that hypoxias of intensities in the range of PO_2 of 20–30 mm Hg (1.5–3.0% oxygen-equilibrated solutions) produced increases in ROS levels, measured as

[16] J. Fandrey, S. Frede, and W. Jemkelmann, *Biochem. J.* **303,** 507 (1994).

[17] S. L. Archer, H. L. Reeve, E. Michelakis et al., *Proc. Natl. Acad. Sci. USA* **96,** 7944 (1999).

[18] E. D. Michelakis, V. Hampl, A. Nsair, G. Harry, A. Haromy, R. Gurtu, and S. L. Archer, *Cir. Res.* **90,** 1307 (2002).

[19] C. Marshall, A. J. Mamary, A. J. Verhoeven, and B. E. Marshall, *Am. J. Respir. Cell Mol. Biol.* **15,** 633 (1996).

[20] J. Duranteau, N. S. Chandel, A. Kulisz, Z. Shao, and P. T. Schumacker, *J. Biol. Chem.* **273,** 11619 (1998).

[21] G. B. Waypa, N. S. Chandel, and P. T. Schumacker, *Circ. Res.* **88,** 1259 (2001).

[22] L. A. Dada, N. S. Chandel, K. M. Ridge, C. Pedemonte, A. M. Bertorello, and J. I. Sznaider, *J. Clin. Invest.* **111,** 1057 (2003).

increased fluorescence due to the oxidation of 2′,7′-dichlorofluorescein, ranging from about 30 to 500–1000% above control (perfusion with 15–16% O_2). However, the origin of ROS according to Chandel and co-workers is mitochondrial, being produced mainly at the level of the quin-one pool, because inhibitors of the respiratory chain proximal to this level (exemplified by rotenone and myxothiazol) abolished the hypoxic in-crease of ROS, as well as the responses elicited by hypoxia (but see Refs. 3 and 10), whereas inhibitors distal to the quinone pool (antimycin A, cyanide, azide) augmented ROS levels and mimicked hypoxia. Many of the findings of Chandel group have been contested by other groups[23] (see Gonzalez et al.[3] for additional references). Using 2′,7′-dichlorofluorescein fluorescence computed from the cells or the tissue sections, Kummer and co-workers also found that hypoxia increased mitochondrial ROS produc-tion in PC12 cells,[24] decreased ROS in neurons of the nodose ganglion,[25] and increased them in pulmonary artery smooth muscle cells,[26] being also this increase of mitochondrial origin. With a similar method, Killilea and co-workers[27] reported a 500% increase in ROS production in pulmonary artery smooth muscle cells exposed to ≈25 mm Hg for 1 h.

To the unbiased reader, the collection of data presented in previous paragraphs must seem unintelligible. If it is difficult to accept that hypoxia acting at the mitochondrial level affects the rate of ROS production differ-ently in one cell type vs another, then it is impossible to understand that hypoxia and inhibitors of the proximal vs the distal complexes of the re-spiratory chain decrease ROS production in some laboratories, whereas in other laboratories they act conversely. Where are the pitfalls? Available literature would indicate that the main problems may relate to the methods used to detect ROS. Serious doubts on the meaning of the infor-mation obtained with 2,7-dichlorofluorescein fluorescence have been cast by many authors.[28–33] Similar criticisms have been made as to the use of

[23] N. Enomoto, N. Koshikawa, M. Gassmann, and K. Takenaga, *Biochem. Biophys. Res. Commun.* **297,** 346 (2002).
[24] W. Kummer, B. Hohler, A. Goldenberg, and B. Lange, *Adv. Exp. Med. Biol.* **475,** 371 (2000).
[25] Y. Yamammoto, M. Henrich, R. L. Snipes, and W. Kummer, *Brain Res.* **961,** 1 (2003).
[26] R. Paddenberg, B. Ishaq, A. Goldenberg, P. Faulhammer, F. Rose, N. Weissmann, R. C. Braun-Dullaeus, and W. Kummer, *Am. J. Physiol. Lung Cell. Mol. Physiol.* **284,** L710 (2003).
[27] D. W. Killilea, R. Hester, R. Balczon, P. Babal, and M. N. Gillespie, *Am. J. Physiol. Lung Cell. Mol. Physiol.* **279,** L408 (2000).
[28] C. Rota, Y. C. Fann, and R. P. Mason, *J. Biol. Chem.* **274,** 28161 (1999).
[29] C. Rota, C. F. Chignell, and R. P. Mason, *Free Radic. Biol. Med.* **27,** 873 (1999).
[30] W. Jakubowski and G. Bartosz, *Cell Biol. Int.* **24,** 757 (2000).
[31] M. j. Burkitt and P. Wardman, *Biochem. Biophys. Res. Commun.* **282,** 329 (2001).

lucigenin[34] (see Janiszewski et al.[35] for additional references). In addition to those critiques, there is solid evidence that time-dependent light-induced production of ROS may affect the findings enormously.[36–38] With the use of 2,7-dichlorofluorescein, we have experienced the pitfalls mentioned earlier. The increase in fluorescence detected in short-term cultured chemoreceptor cells was mostly dependent on the parameters of stimulation (intensity of the lamp, the time of illumination per frame, and the number of frames per minute), and we could not detect any clear signal that could unequivocally be assigned to hypoxic stimulation. Those findings recommended a turn in our experimental approach to measure GSH/GSSG as the main determinant of the redox environment of the cells.

The answer to the first question is that we do not know if hypoxia increases or decreases the rate of ROS production. Regarding the second question, that is, the mechanism responsible for the alteration in ROS production, we must make a double assumption. The initial assumption would be that hypoxia decreases the rate of ROS production both at the level of mitochondria, because ROS production parallels PO_2 at the respiratory chain,[10,39] and at the level of NADPH oxidase, because the decrease in PO_2 and the K_m of the enzyme would make the oxidase work at a lower rate and to produce less ROS.[13] Using dihydrorodamine 123 fluorescence, Cross et al.[13] found a decrease in fluorescence during hypoxia and favored the notion that the decrease in ROS levels was due to a decrease in the activity of NADPH oxidase. However, data from our laboratory showed that inhibition of this oxidase does not prevent the detection and the genesis of a normal response to hypoxia in chemoreceptor cells of the rat or rabbit.[40] However, a minor modulatory role for NADPH-derived ROS could not be excluded from our study. Our second assumption would be that hypoxia increases ROS. The authors concluded that the weight of the literature supports the notion that mitochondria cannot be the source of those increased levels of ROS production. In the ranges of PO_2 (and probably even at

[32] S. I. Liochev and I. Fridovich, J. Biol. Chem. **276,** 35253 (2001).

[33] J. L. Brubacher and N. C. Bols, J. Immunol. Methods **251,** 81 (2001).

[34] S. I. Liochev and I. Fridovich, Arch. Biochem. Biophys. **337,** 115 (1997).

[35] M. Janiszewski, H. P. Souza, X. Liu, M. A. Pedro, J. L. Zweier, and F. R. Laurindo, Free Radic. Biol. Med. **32,** 446 (2002).

[36] P. E. Hockberger, T. A. Skimina, V. E. Centonze, C. Levin, S. Chu, S. Dadras, J. K. Reddy, and J. G. White, Proc. Natl. Acad. Sci. USA **96,** 6255 (1999).

[37] M. Afzal, S. Matsugo, M. Sasai, B. Xu, K. Aoyama, and T. Takeuchi, Biochem. Biophys. Res. Commun. **304,** 619 (2003).

[38] P. Bilski, A. G. Belanger, and C. F. Chignell, Free Radic. Biol. Med. **33,** 938 (2002).

[39] L. E. Costa, S. Llesuy, and A. Boveris, Am. J. Physiol. **264,** C1395 (1993).

[40] A. Obeso, A. Gomez-Niño, and C. Gonzalez, Am. J. Physiol. **276,** C593 (1999).

lower PO_2) where the CB works, there is no limitation in the availability of O_2 to accept the electrons flowing through the respiratory chain due to the great affinity for O_2 of cytochrome oxidase.[38-43] As stated explicitly by several authors, mitochondria with the physiologically available substrates for oxidation do not release measurable levels of ROS (even less than the 1–3% of the consumed O_2 as classically suggested[11]) unless the respiratory chain is inhibited to build up reduced forms of the initial mitochondrial complexes and unless there is O_2 available. The most plausible explanation of the observed increased rate of ROS production by mitochondria would be the result of unspecific interactions between the dyes used to measure ROS and mitochondria.[41,43] The proposal made by Staniek and Nohl[41] that mitochondrial respiration seems not to be required as permanent sources of ROS for physiological activities, such as cell signaling, gains full support from studies of several groups showing that hypoxia-inducible factor-dependent gene expression[23] (see Gonzalez et al.[3] for additional references), as well as acute membrane-linked O_2 chemoreception in a model of airway chemoreceptors,[44] takes place in cells lacking functional mitochondria (ρ° cells). It is of special relevance for this article that the observation of the first known effector in the oxygen chemoreception cascade, represented by specific K^+ currents, is inhibited in an identical percentage and with an identical time course in control and ρ° H146 cells. Staniek and Nohl[41] also proposed that it is more likely that any ROS involved in the physiological function of cell signaling is produced in specific compartments in the vicinity of the effector molecules. This microdomain-centered production of ROS would have two purposes: (1) to avoid inflicting unnecessary damage and activation of alternative pathways and (2) to be effective in reaching the target before the ROS are inactivated by the scavenging mechanisms of the cells.

Excluding mitochondria as the source of ROS at the hypoxic levels, which activate the CB chemoreceptors physiologically, where could ROS be generated if we stay with the assumption that hypoxia increases ROS production? Potential sources of ROS during hypoxia in mammalian cells include smooth endoplasmic reticulum-oxidizing enzymes that use the cytochrome P450 and b_5 electron transport chain, microsomal cyclooxygenases and cytoplasmic lipooxygenases, and the leucocyte-type NADPH-oxidase system.[10,12] Although some suggestions have been made regarding the

[41] K. Staniek and H. Nohl, *Biochim. Biophys. Acta* **1460,** 268 (2000).

[42] E. Gnaiger, G. Mendez, and S. C. Hand, *Proc. Natl. Acad. Sci. USA* **97,** 11080 (2000).

[43] J. St-Pierre, J. A. Buckingham, S. J. Roebuck, and M. D. Brand, *J. Biol. Chem.* **277,** 44780 (2002).

[44] G. J. Searle, M. E. Hartness, R. Hoareau, C. Peers, and P. J. Kemp, *Biochem. Biophys. Res. Commun.* **291,** 332 (2002).

potential role of cytochrome P450-dependent systems in hypoxic signaling, they were based on the use of inhibitors of this cytochrome that later proved to be direct inhibitors of K^+ channels. Phospholipase A_2, the arachidonic acid-releasing enzyme, is Ca^{2+} dependent and thereby the entire process of prostaglandin, thromboxane, and leukotriene synthesis by cyclooxygenases and lipooxygenases also becomes Ca^{2+} dependent. Therefore, it is possible that the primary activation of chemoreceptor cells occurs by a mechanism unrelated to ROS, leading to an increase in $[Ca^{2+}]_i$ and to an increase in formation of the lipidic second messengers and ROS, both capable of modulating the ongoing chemoreceptor cell activity. The case for prostaglandin has been proved by Gomez-Niño et al.[45,46] by showing that hypoxia increases the production of PGE_2 and that this prostaglandin inhibited, in a dose-dependent manner, the Ca^{2+} currents and the release of neurotransmitters elicited by hypoxia in chemoreceptor cells. Finally, NADPH oxidase, a multienzymatic complex responsible for the oxidative burst and the genesis of ROS used by phagocytes to destroy bacteria and fungi, has been found to be a very ubiquitous enzymatic complex with several isoforms in many tissues.[47,48] The phagocytic multienzymatic complex has two subunits (gp91*phox* and p22*phox*) located in the plasma membrane and three cytoplasmic subunits (p67*phox*, p47*phox*, and p40*phox*) that, with the concurrence of two low molecular weight GTP-binding proteins, assemble in the membrane to form the active enzymatic complex. Interestingly, several isoforms of the enzyme are activated by Ca^{2+} and, contrary to the phagocytic enzyme, they exist already as preassembled enzymes associated with the cytoskeleton, providing a means to produce ROS in particular microdomains of the cells.[49] Available data suggest that an isoform of NADPH oxidase could modulate, but not trigger, CB chemoreceptor activity, as knockouts of p47*phox* subunits exhibit a greater ventilatory and CSN response to hypoxia than wild-type animals.[50] These findings suggest that the NADPH oxidase-mediated generation of ROS is activated during hypoxia and that ROS modulate the chemoreception process negatively.

The third question formulated at the outset of this section was if the presumptive alteration of ROS during hypoxia was a cause or consequence of CB chemoreceptor activation. As mentioned in the preceding paragraphs,

[45] A. Gomez Niño, L. Almaraz, and C. Gonzalez, *J. Physiol.* **476,** 257 (1994).
[46] A. Gomez Niño, J. R. Lopez-Lopez, L. Almaraz, and C. Gonzalez, *J. Physiol.* **476,** 269 (1994).
[47] K. K. Griendling, D. Sorescu, and M. Ushio-Fukai, *Circ. Res.* **86,** 494 (2000).
[48] J. D. Lambeth, *Curr. Opin. Hematol.* **9,** 11 (2002).
[49] J. M. Li and A. M. Shah, *J. Biol. Chem.* **277,** 19952 (2002).
[50] K. A. Sanders, K. M. Sundar, L. He, B. Dinger, S. Fidone, and J. R. Hoidal, *J. Appl. Physiol.* **93,** 1357 (2002).

data of Obeso *et al.*[40] and Sanders *et al.*[50] indicate that ROS could be modulators, but not triggers, of the CB chemoreceptor response to hypoxia. In the same direction, data of Sanz-Alfayate *et al.*[51] showed that the GSH precursor and the ROS scavenger *N*-acetylcysteine did not affect normoxic nor hypoxic activity in chemoreceptor cells, implying that intracellular ROS levels, whether increased or decreased by hypoxia, do not participate in setting the response of chemoreceptor cells.

ROS and the GSH/GSSG System

Except for their role in phagocytes, ROS have classically been considered damaging by-products of the metabolism; therefore, research has been oriented to characterize the defense mechanisms against the oxidative damage produced by ROS. Along these lines have been described[3,10,12] (1) the metabolic processes producers of ROS; (2) many toxic-damaging reactions of ROS with lipids, proteins, and nucleic acids, and (3) several enzymatic systems and a group of small molecules that serve a protective function against oxidative damage *in vivo,* that is, they act as scavengers of ROS. The recognition of ROS as presumptive physiological signaling molecules has expanded the inventory of reactions produced by ROS. It is evident, however, that if ROS have to function as second messengers, a competition between scavengers and physiological targets of ROS must occur, although the compartmentalized production (and scavenging) of ROS would reduce such competition. Finally, if ROS function as second messengers, they must be disposed to terminate signaling; presumably, the same systems controlling the oxidative damage are responsible for the disposal of ROS to terminate the signal. These considerations demonstrate the need to know the metabolic pathways of ROS to understand the role of ROS in cell signaling and to make predictions on the modifications in cellular responses mediated by ROS when the redox environment of the cells[1] is altered experimentally.

ROS species derived directly from normal metabolism are basically limited to $O_2^{•-}$ (oxygen transport, respiratory chain, cytochrome P450-using enzymes, NADPH oxidase, cyclooxygenase, and lipooxygenase), H_2O_2 (in reactions catalyzed by peroxixomal oxidases and monoamino oxidase), and $NO^•$ (synthesized by several isoforms of nitric oxide synthase), plus some $OH^•$ (hydroxyl radical) generated on the surface of the organism by the homolytic action of UV light on H_2O_2 ($H_2O_2 \rightarrow 2\ OH^•$) or in any part of the organism by the radiolysis of water by high-energy radiation

[51] G. Sanz-Alfayate, A. Obeso, M. T. Agapito, and C. Gonzalez, *J. Physiol.* **537,** 209 (2001).

($H_2O \rightarrow OH^\bullet + H^\bullet$). The superoxide radical, $O_2^{-\bullet}$, is no doubt the quantitatively most relevant primary ROS capable of reacting with many molecules to generate altered structures and new secondary ROS; specific enzymatic systems (superoxide dismutases) have been developed to dispose $O_2^{-\bullet}$ by transformation into H_2O_2, which is considerably less reactive (see later). However, H_2O_2 can be decomposed by ultraviolet light in 2 $OH^\bullet$ and can react with transition metals in their reduced form (Fenton reaction) to yield $OH^\bullet + OH^- +$ oxidized metal. $OH^\bullet$ is the most harmful ROS due to its high oxidizing power (reduction potential = 1900 mV), which makes it cable of attacking and removing one electron from many molecules (e.g., lipids), transforming the attacked molecules in radicals, which in turn are capable of continuing a destructive chain reaction until some scavenger molecule is capable of reacting with the radical and stops the reaction. In many of these secondary reactions, new ROS species, including H_2O_2, can appear. The Fenton reaction in healthy biological systems is kept to a minimum: transition metals are at very low concentrations as free molecules; they are usually complexed with proteins to limit their reactivity. The potential harm of H_2O_2 explains the existence of several enzymes aimed to dispose H_2O_2. This article does not deal with $NO^\bullet$.

The physiologically more relevant protective molecules, or ROS scavengers, include vitamin E, ascorbic acid, and lipoic acid. All of them are able to react with several ROS species, including lipid peroxides, being transformed in molecular species that are either radicals with low reactivity or molecules lacking redox reactivity that may be degraded or recycled back to active scavengers by the action of enzymes systems. For example, ascorbate can react with $O_2^{-\bullet}$ or with Prot-TyrO$^\bullet$ radicals to form the poorly reactive semidehydroascorbate capable nonetheless of transferring an additional electron to another more oxidant radical, inactivating it, and becoming dehydroascorbate, which is unreactive. Dehydroascorbate can be degraded and lost, implying a depletion of ascorbate, or can be recycled enzymatically by the GSH-dependent dehydroascorbate reductase.[12] Comparable reactions occur with vitamin E, probably the most important antioxidant, physiologically speaking, and with lipoic acid. All three antioxidants can react among them and regenerate each other: for example, a deficit of vitamin E can be covered partially by vitamin C (ascorbate), and glutathione reductase and thioredoxin reductase can act on lipoic acid to reduce it back into dihydrolipoic acid.

The enzyme systems devoted to eliminate ROS include superoxide dismutases, peroxidases, and catalase. Being $O_2^{-\bullet}$ the most important ROS species produced physiologically in mitochondria and cytoplasm, there is a mitochondrial superoxide dismutase, the manganese superoxide dismutase, and a cytoplasmic enzyme, the copper-zinc superoxide dismutase.

This cytoplasmic form is located not only free in the cell cytoplasm, but in various organellae, including lysosomes, nucleus, peroxisomes, and the intermembrane space of the mitochondria. This last location would represent a mechanism of defense against a fraction of $O_2^{-\bullet}$ produced in the mitochondria and released to this space.[43] In extracellular fluids, there is a secreted form of copper-zinc dismutase with a higher molecular weight and with a tetrameric composition instead of the dimeric of the cellular enzymes. In the reaction that they catalyze, the metal (copper or manganese; zinc serves a noncatalytic enzyme-stabilizing function) acts sequentially as an acceptor and as a donor of one electron with the final result of the elimination of two molecules of $O_2^{-\bullet}$ and the genesis of a molecule of O_2 and another of H_2O_2, as shown for the manganese superoxide dismutase:

$$\text{Enz-Mn}^{3+} + O_2^{-\bullet} \rightarrow \text{Enz-Mn}^{2+} + O_2$$
$$\text{Enz-Mn}^{2+} + 2H^+ + O_2^{-\bullet} \rightarrow \text{Enz-Mn}^{3+} + H_2O_2$$

(1)

Having transformed $O_2^{-\bullet}$ into H_2O_2, which somehow represents the common final path of all ROS,[3,10] we need to know how cells prevent H_2O_2-damaging effects. Cells rely on two different enzymatic systems, namely catalases and peroxidases, with variable relative significance in different tissues. In animal cells, catalases are enzymes that are located mainly in peroxisomes, where many oxidases are also located and H_2O_2 is generated physiologically. It catalyzes the decomposition of H_2O_2 into water and molecular O_2 ($2H_2O_2 \rightarrow 2H_2O + O_2$). Catalases are tetrameric hemoproteins with the ferric hemogroup buried in a pocket like that of hemoglobin capable of limiting the accessibility of large molecules. The enzymatic reaction occurs in two steps with a similar rate constant: in the first step, the ferric ion of hemo is oxidized by H_2O_2 to Fe^{5+} (being H_2O_2 itself reduced to H_2O) and the entire catalase subunit, called compound I in this oxidized state, acquires spectral absorbance characteristics that allow measurement of the catalase-catalyzed reactions in intact cell systems. In the second step, compound I oxidizes a second H_2O_2 to $H_2O + O_2$ and ferric catalase is regenerated.[12,52] Peroxidases catalyze the decomposition of H_2O_2 by oxidizing another substrate ($SH_2 + H_2O_2 \rightarrow S + 2H_2O$). Aside from tissue-specific peroxidases,[12] glutathione peroxidase and thioredoxin peroxidase are present in essentially all mammalian cells and serve a homeostatic role in the maintenance of their redox environment. Thioredoxin, a small cysteine-rich protein molecule with several isoforms ranging from 12 to 32 kDa, is a cofactor for ribonucleotide reductase that catalyzes

[52] M. S. Wolin, *Arterioscler. Thromb. Vasc. Boil.* **20,** 1430 (2000).

the conversion of ribonucleotides to deoxyribonucleotides, regulates the activity of several transcription factors, and acts as a growth factor.[53] Additionally, it also plays an antioxidant role by being the substrate of a cycle of redox reactions catalyzed by thioredoxin peroxidase and thioredoxin reductase. The glutathione couple, GSH/GSSG, and the correspondent peroxidase and reductase constitute the main antioxidant system of cells responsible for the maintenance of a reduced intracellular environment[1] (Fig. 2). Both peroxidases react with H_2O_2, but they can also react with other peroxides, such as fatty acid hydroperoxides. Although the intimate mechanisms of reaction are different (in the reaction catalyzed by glutathione peroxidase, a selenium located in the active center participates, and thioredoxin peroxidase does not contain a catalytic selenium), both peroxidases reduce the peroxides to alcohols and two $-SH$ groups from two reduced molecules of glutathione or thioredoxin are oxidized to form a $-S-S-$ bond that, in the case of thioredoxin, can be intra- or intermolecular. The resulting oxidized forms of the two redox pairs are back reduced to restore their capacity as antioxidants by the action of the correspondent reductases that use $NADPH + H^+$ (as donors of equivalents of reduction) provided by the hexose monophosphate pathway (Fig. 2). The glutathione and thioredoxin antioxidant systems are interrelated through the correspondent $NADPH + H^+$-dependent reductases so that a thermodynamic connection exists among the three redox couples in the cells. Thus, the redox status of the cells is defined by the three interrelated redox pairs. It should be noted that despite the parallelisms between GSH and

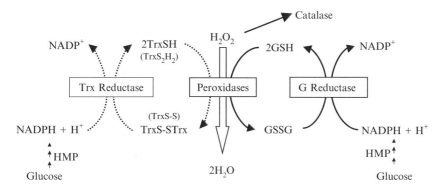

Fig. 2. Glutathione and thioredoxin peroxidase and reductase systems. TrxSH and TrxS$_2$H$_2$, reduced thioredoxin; TrxS-S and TrxS-STrx, oxidized thioredoxin with intra- and intermolecular disulfide bonds. GSH and GSSG, reduced and oxidized glutathione, respectively; HMP, hexomonophosphate pathway.

[53] G. Powis and W. R. Monfort, *Annu. Rev. Pharmacol.* **41,** 261 (2001).

thioredoxin antioxidant systems and despite the fact that they have similar reducing ability (similar standard redox potential), the limited availability of thioredoxin, (about 6 nM in plasma and 2–12 μM in tissues), in comparison to the GSH/GSSG system (about 1–5 μM in plasma and 0.5 to 10 mM in tissues), makes the thioredoxin system a relatively small significance in the overall maintenance of the reduced environment of the cells.[1]

The redox potential of any redox pair in the cells depends on its standard redox potential and on the ratio of the concentrations of the reduced and oxidized forms of the pair following the Nernst equation. However, the reducing capacity of a redox pair depends on the concentration of the reduced form of this particular redox pair. The standard redox potential of a redox pair defines the easiness or intrinsic tendency of the reduced form to donate equivalents of reduction (the more negative the standard redox potential, the greater the tendency to donate electrons), but if the concentration of the reduced form of a particular redox pair is small, its capacity to reduce is very small because it will be exhausted in a few cycles of reduction. The product of the standard redox potential times the concentration of the reduced member of a redox pair in the cells determines the "quality" (quality $\approx$ ability to maintain the cell interior in a reduced form) of the redox pair in any cell system. Then, the redox state of all the interrelated redox pairs of the cells, called *the redox environment of the cells* by Schafer and Buettner,[1] would be

$$\text{Redox environment} = E_1 \times [\text{reduced form pair } 1] + \dots$$
$$+ E_n \times [\text{reduced form pair } n] \qquad (2)$$

When the redox environment of a cell is calculated using typical concentrations of the most important redox pairs (GSH = 5 mM, E = -240 mV; NADPH = 0.1 mM, E = -370 mV; thioredoxin = 0.01 mM, E = -270 mV; dihydrolipoic acid < 0.001 mM, E = -320 mV) it becomes evident that glutathione is indeed mainly responsible for the redox environment of the cells, representing the major redox buffer. The ascorbate/dehydroascorbate pair, despite concentrations in the near millimolar range (with concentrations in densecore cromaffin granules of up to 10–20 mM), does not constitute a significant redox pair because its standard potential is near 70 mV. Therefore, aside from considerations related to possible specific situations derived from compartmentalizations of the cell, whether in terms of ROS production or elimination, measurement of the redox potential of the GSH/GSSG pair in tissues would represent a very reliable index of the redox environment of the cells: a decrease in the redox potential of glutathione in response to a given stimulus would imply that the stimulus *per se* or the molecular machinery involved in the generation of the cell response to the stimulus activates the production of ROS and the subsequent oxidation of GSH with the decrease in its redox potential. The converse would also be true. These considerations also imply that the glutathione peroxidase/glutathione reductase system, together with catalase, is responsible for the elimination of H_2O_2, which represents the common final path of all ROS in the cells. It should be recalled, however, that the congenital absence of catalase does not cause serious clinical problems,[12] indicating that the glutathione peroxidase system may suffice to eliminate H_2O_2 when produced at normal physiological rates. The necessity for catalase only becomes evident when there is an extra production of H_2O_2.

Another consideration of relevant physiological and pathophysiological significance is that all the enzyme systems involved in the disposal of ROS (superoxide dismutase, glutation and thioredoxin peroxidases and reductases, and catalase and thioredoxin itself) are up- and downregulated as a function of the rate of production of ROS.[38,54-56] Additionally, in general terms, the level of activity of these enzymes in some species or strains of animals versus others or in different developmental periods correlates directly with their resistance to the oxidative insults.[57]

Finally, the definition of a correct correlation between GSH levels measured experimentally and any given cellular function needs some more considerations. Figure 2 shows a 2:1 molar ratio in the utilization of GSH per mole of H_2O_2 removed. Knowing that cells consume O_2 at a rate of ≈1 mmol/kg/min or less, that ROS production is around 0.01–0.03 mmol/kg/min (i.e., 1–3% of the O_2 consumed), and that the concentration of GSH in the cells is >1 mmol/kg, it follows that to decrease GSH levels significantly in a scale of minutes, a large burst of ROS production would be required. This decrease would even be minimized by the NADPH-dependent cycling of GSSG back to GSH produced by glutathione reductase. In other words, the high redox buffer power of GSH tends to prevent large changes of the redox environment of the cells to preserve the functionality of the cells. This in turn implies that when stimulation of a cell system alters the "redox status" or redox environment of the cells, the stimulus causes a dramatic change in the rate of production of ROS in the cells. If in a situation of increased ROS production, glutathione reductase is unable of cycling back to GSH the extra amount of GSSG produced, we would assist a time-dependent decrease in GSH and a tendency of

[54] L. Frank, *Fed. Proc.* **44,** 2328 (1985).
[55] D. Mustacich and G. Powis, *Biochem. J.* **346,** 1 (2000).
[56] S. Hoshida, N. Yamashita, K. Otsu, and M. Hori, *J. Am. Coll. Cardiol.* **40,** 826 (2002).
[57] M. A. Hass and D. Massoro, *Am. J. Physiol.* **253,** C66 (1987).

GSSG to increase, and therefore, to a decrease of the redox environment of cell according to the equation (at 37°):

$$\text{Redox environment(mV)} \approx E_{GSH} = -240 - 30.75 \times \log[GSH]^2/[GSSG] \tag{3}$$

However, as soon as GSSG starts to increase, cells export GSSG[1]; this release or leak of GSSG aims to avoid an intracellular accumulation of high concentrations of GSSG and therefore intends to maintain an adequate GSH/GSSG ratio and a redox environment as constant as possible inside the cells. Of course the cost of this is a loss of redox buffer capacity with a net tendency to decrease in GSH. Additional information to consider for the correct evaluation of the correlation between GSH and ROS signaling is that the turnover of GSH seems to be rather slow, as it is required to inhibit glutathione synthesis for long periods of time to deplete GSH levels significantly in the cells. For example, the inhibition of glutamylcysteine synthetase (the first enzyme in glutathione biosynthesis) with buthionine sulfoximine at concentrations of 100 and 200 μM reduced in 24 h the GSH content by the same percentage ($\approx 75\%$) in lens cells in culture,[58] and at 200 μM depleted at a constant rate of about 10%/h during 5 h.[59] This in turn reflects the fact that the rate (mole/min) of ROS production by cells is small. In sum, these considerations would suggest that under physiological stimulation of any cell system it should not be surprising to see only minor changes or no measurable changes in the GSH/GSSG system and in the redox environment of the cells. Additionally, imposed experimental alterations of the GSH/GSSG system aimed at mimicking a strong stimulation should not produce very dramatic changes, otherwise we would bring the redox environment of the cells to levels only observed in pathologic situations.

The presumable existence of compartmentalized microdomains in the production of ROS induced by specific stimulus in the near vicinity of sensing and effector molecules[41] would impose a completely different set of considerations. Under this scope, it would be conceivable that the stimulus produces a dramatic increase in ROS production in a specific cell compartment, and therefore that there is a local dramatic alteration of the redox environment of the cells without a large change in the overall intracellular redox environment. In these physiological circumstances, we should expect only minor changes (or no changes) in the bulk levels of the GSH/GSSG pair or in its redox potential as measured, for example,

[58] F. Shang, M. Lu, B. Dudek, J. Reddan, and A. Taylor, *Free Radic. Biol. Med.* **34,** 521 (2003).
[59] S. Sinbandhit-Tricot, J. Cillard, M. Chevanne, I. Morel, P. Cillard, and O. Sergent, *Free Radic. Biol. Med.* **34,** 1283 (2003).

in a tissue homogenate. Additionally, experimental maneuvers aimed to alter the GSH system, and through it the redox environment of the cells to mimic physiological stimulation, would tend to be rather unspecific because they would alter the entire redox environment of the cells without the precise spatial (and temporal) coordinates of the physiological stimulus.

Experimental Methods

Measurement of Glutathione

In the present study we measured GSH and GSSG in quarters of rat diaphragm in all experimental conditions here. A previous study[51] measured GSH/GSSG in calf CB, but the experiments proved to be enormously time-consuming and inconvenient because around 40–50 mg of tissue equivalent to four to five calf CBs was needed and, in addition, the crisis of mad cows restricted access to the slaughterhouse. Selection of the rat diaphragm as the test tissue to evaluate the effects of hypoxia and other experimental treatments is due to the fact that the diaphragm is a sheet of tissue comparable in thickness to the rabbit CB (<1 mm), therefore presenting no problems for gas and drug diffusion to the interior of the tissue. Therefore, most of our experiments compare data on GSH/GSSG content in the diaphragm with data on the activity of chemoreceptor cells obtained in the rabbit CB treated identically to diaphragms. In some occasions, we also used liver and brain tissue to test the effects of specific drugs and compared findings in the diaphragm.

Removal and Incubation of Tissues. Wistar rats with a 250- to 300-g body weight are anesthetized with sodium pentobarbital (60 mg/kg; ip). After an incision in the abdomen, the entire diaphragm is removed carefully, clamping as needed to avoid bleeding. Animals are killed by an intracardiac overdose of sodium pentobarbital. All measures are taken to prevent distress in the animals. The Committee for Animal Care and Use at the University of Valladolid approved the protocols. The diaphragm is freed of blood by washing in ice-cold 100% O_2-saturated Tyrode (in mM: NaCl, 140; KCl, 5; CaCl$_2$, 2; MgCl$_2$, 1.1; glucose, 5.5; HEPES, 10; pH 7.40 with 1 N NaOH). Under a dissecting microscope, small pieces of fat, small vessels, and phrenic tendon are eliminated, and the diaphragm is cut in to four quadrants of comparable size.

The four pieces of diaphragm are transferred individually to glass scintillation vials placed in a metabolic shaker at 37° containing 10 ml of bicarbonate-buffered Tyrode (composition as before except that 24 mM NaCl is eliminated and 24 mM NaHCO$_3$ is added) that are

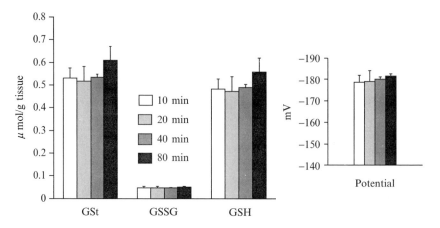

FIG. 3. Effects of duration of incubation on levels of GSt (total glutathione), GSSG, GSH, and GSH/GSSG redox potential. (See color insert.)

bubbled constantly with water-saturated 5% CO_2-containing gas mixtures. All pieces are incubated for 30 min while bubbling with 21% O_2/5% CO_2 to allow the tissues to recover from the trauma of the surgical procedures[60]; thereafter the incubating solutions are renewed every 10 min for up to 40 min. In control tissues, the renewing solution is 21% O_2/5% CO_2-equilibrated bicarbonate Tyrode, in hypoxic tissues the last 10-min incubating period the solution is equilibrated, and the vials are bubbled continuously with 7% O_2/5% CO_2. In drug-treated tissues the solutions are equilibrated and the vials are bubbled continuously with 21% O_2/5% CO_2 and contain (from 0 to 40 min) 0.050 mM p-chloromercurybenzosulfonate (PCMBS) sodium salt, 0.5 mM carmustine [1,3-bis(2-chloroethyl)-nitrosourea; BCNU], 2 mM N-acetylcysteine (NAC), or 0.2 mM diamide (DIA). Incubation time does not modify GSH or GSSG levels for up to 80 min (Fig. 3). At the end of the incubation the tissues are transferred to new vials kept at 0–4° containing 10 ml of Tyrode equilibrated with 100% O_2 for 5 min, dry blotted by touch on filter paper, weighed, and placed in Eppendorf tubes containing a solution of 5-sulfosalicylic acid (SSA) at 5% and 0.25 mM EDTA (at 4° it is stable for months) whose volume is adjusted to five times the weight of the tissue. The tissues are stored at −80° until the day of the assay or are immediately glass-to-glass homogenized at 0–4° and centrifuged in a microfuge (4°, 10 min) and the supernatant is used to measure GSH/GSSG. The assay can be performed

[60] A. Obeso, L. Almaraz, and C. Gonzalez, *Brain Res.* **371,** 25 (1986).

TABLE I

GLUTATHIONE LEVELS IN SUPERNATANTS OF HOMOGENATES OF RAT DIAPHRAGM ASSAYED
IMMEDIATELY OR AFTER STORAGE FOR 1 MONTH AT $-80°$ [a]

Supernatant conditions	Total glutathione (μM/g tissue)	GSSG (μM/g tissue)	GSH (μM/g tissue)	Gltutathione redox potential (mV)
Assayed immediately	0.533 ± 0.014 ($n = 41$)	0.044 ± 0.002 ($n = 41$)	0.489 ± 0.014 ($n = 41$)	-179.9 ± 1.1 ($n = 41$)
Assayed after frozen ($-80°$) for 1 month	0.549 ± 0.018 ($n = 23$)	0.048 ± 0.002 ($n = 23$)	0.500 ± 0.015 ($n = 23$)	-179.1 ± 0.8 ($n = 23$)

[a] The 41 samples of tissue used for the assay immediately after obtaining came from 20 different animals, whereas those used to be frozen were obtained from 11 rats. GSSG levels are expressed as GSH, i.e., the actual concentration of GSSG is half because each molecule of GSSG yields two molecules of GSH (see calculations in the text).

immediately or the supernatant can be stored at $-80°$C until assay. No differences have been observed whether the tissues or the supernatants are stored for up to 1 month at $-80°$ (Table I).

Assay for GSH and GSSG. The measurement of GSH and GSSG is made by the method of Griffith.[61] GSH reacts spontaneously with 5,5'-dithiobis-(2-nitrobenzoic acid) (DTNB) to generate GSSG and 5-thio-2-nitrobenzoic acid (TNB; peak absorbance at 412 nm). The GSSG formed is reduced enzymatically to GSH by glutathione reductase coupled to NADPH oxidation. In this cycling assay, concentrations of DTNB, NADPH, and glutathione reductase are chosen so that the rate of color formation followed with a spectrophotometer is linear with time for 2–3 min and the slope of the line relating the increase of absorbance and time (Δ absorbance/Δ time, min; $\Delta A/\Delta t$, min) is directly proportional to the concentration of total glutathione (GSH + GSSG; GSt) in the sample. The plot of the slopes of these lines as a function of the concentration of GSt in the samples is also linear for an ample range of concentrations and defines the standard curves used to determine the concentrations of GSt in the problem samples by interpolation. GSSG is measured identically, but first the GSH present in the samples is masked by derivatization with 2-vinylpyridine. The size of the sample is higher due to the usually much lower concentration of GSSG than GSH. In all instances (construction of standard curves and tissue homogenates), the assays are made by triplicate. The assay is highly specific, due to the enzymatic step, and is highly sensitive due to cycling; the sensitivity of this colorimetric method is in the range of 10^{-10} mol/assay.

[61] O. W. Griffith, *Anal. Biochem.* **106,** 207 (1980).

SOLUTIONS. The description that follows corresponds to a typical working day that includes 12–16 tissue samples to be determined by triplicate plus the standard curves for GSt and GSSG, also in triplicate. Some stock solutions can be prepared and, if stored conveniently, can be used for months. Stock solutions include

- A stock buffer solution of 125 mM sodium phosphate buffer containing 6.30 mM disodium EDTA (pH = 7.50) is maintained at 4° (1 liter).
- A stock solution of 100 mM of GSH in distilled water is aliquoted (0.1 ml) in Eppendorf tubes and maintained at −20°.
- A stock solution of 6 mM DTNB in stock buffer and maintained at −20° is aliquoted in 5-ml plastic tubes.
- 2-Vinyl pyridine (Aldrich) is stored at −20° as provided by the supplier and can be used (usually by several months) until it starts to acquire a yellow tinge (5 ml).
- Triethanolamine (Sigma) undiluted is maintained at room temperature (100 ml).

The rest of the solutions are prepared daily and include

- 4.31% SSA solution by diluting with water the 5% SSA + 0.25 mM EDTA solution used to homogenate the tissues (2 ml of SSA at 5% + 0.32 ml water).
- 0.3 mM NADPH (Sigma) in the stock buffer (100 ml).
- GSH standard solution: the 100 mM GSH stock solution (100 μl) is diluted to 0.5 mM with the 5% SSA + 0.25 mM EDTA solution to construct the standard curves for GSt.
- GSSG standard solution: the 100 mM GSH stock solution (100 μl) is diluted to 0.1 mM GSH with the 5% SSA + 0.25 mM EDTA solution to construct the standard curves for GSSG; alternatively, it can be prepared from a solution 0.05 mM GSSG in identical conditions.
- 1 unit of GSSG reductase (GR; Sigma)/5 μl in stock buffer (1 ml).

PROCEDURE. The assay of GSt in the test samples is performed as follows. For the first sample of the run, the cuvette of the spectrophotometer is placed in a bath at 37° before addition of the assay mixture (700 μl of the NADPH solution + 100 μl of DTNB solution + 195 μl of H$_2$O + 5 μl of the supernatant of the tissue homogenate). The cuvette is maintained for at least 4 min at 37° and then 5 μl of GSSG reductase solution is added, mixed rapidly by gentle vortexing, and introduced in the thermostatized (37°) reading chamber of the spectrophotometer (Hitachi U-1100/ U1100; Hitachi Scientific Instruments, Pacisa, Madrid). The recording of absorbances starts after 30 s (to reach the 37°) and proceeds for 2 min with recordings every 10 s, with the slope (Δabsorbance/min) calculated automatically (U-1100 Data Manager Software) and stored in the computer.

Additional samples are prewarmed in Eppendorf tubes, and the enzyme is added on transference of the assay mixture to the cuvette immediately prior to starting the recording of absorbances. During this warming period, the assay mixture acquires a yellowish tinge due to the reduction of a small fraction DTNB produced by the GSH present in the sample that is transformed into GSSG, but the real development of color occurs when the reductase is added and GSSG is cycled back to GSH with many cycles during the 2-min reading period. The assay of GSSG in the test samples involves an additional preparative step: to derivatize (destroy) GSH. To a 100-μl aliquot of supernatant of tissue homogenate is added, under vigorous vortexing, 2 μl of 2-vinyl pyridine, and the pH of the mixture is adjusted to between 6.8 and 7.2 with triethanolamine (4 μl, but due to its high viscosity, the precise volume can vary with experimenter) that should be added while vortexing, preferably to the side of the tube wall to avoid sharp peaks of pH locally in the solution that will destroy GSSG. If the pH (measured with narrow-range pH paper) is higher than 7.2, derivatization should be repeated. The derivatization reaction is allowed to proceed at room temperature for 1 h and then the assay is performed with an identical assay mixture used for GSt, except for the water aliquot (175 μl) and the size of the sample and enzyme solution (20 μl each).

The standard curves for GSt and GSSG are constructed with an assay procedure identical to the one just described for the test samples (except for the derivatization step, which is not required) with the assay mixtures shown in Table II. The slopes (Δabsorbance/min) obtained for each amount of glutathione are plotted as a function of glutathione amount itself; the best-fit regression line is obtained by the least-square method, and the slopes (Δabsorbance/min) of the test samples are interpolated in the correspondent standard curves. When a drug has been used in the treatment of tissues and it is suspected that it may interfere with the assay procedure (e.g., inhibiting glutathione reductase), such a possibility needs to be excluded or corrected. To exclude that possibility, a group of assays should be performed with assay mixtures containing fixed and known amounts (e.g., 1 nmol) of GSH in a fixed 1-μl standard sample as the internal standard and increasing volumes of tissue homogenates (0–4 μl) and decreasing volumes of 4.31% SSA (4–0 μl). An additional assay with just 5 μl of tissue homogenate should also be included. A comparable set of assays should be made with the 1-μl test sample (0.2 nmol) and up to 19 μl of tissue homogenate derivatized previously with 2-vinyl pyridine. If there is no interference, measurements will give additive GSH or GSSG values; however, the combination of homogenate + standard sample will yield values less than additive, and the effect will be greater with the higher

TABLE II

ASSAY MIXTURES USED TO CONSTRUCT STANDARD CURVES FOR GSt AND GSSG[a]

GSH (nmol)	NADPH solution (μl)	DTNB solution (μl)	H$_2$O (μl)	Standard solution (sample) (μl)	4.31% SSA solution (μl)	GR solution (μl)
0–0	700	100	195–175	0	5–20	5–20
0.5–0.1	700	100	195–175	1	4–19	5–20
1.0–0.2	700	100	195–175	2	3–18	5–20
1.5–0.3	700	100	195–175	3	2–17	5–20
2.0–0.4	700	100	195–175	4	1–16	5–20
?	700	100	195–175	5–20	0	5–20

[a] Note that GSH solutions are used for both the standard curve to GSt and the standard curve to GSSG in the samples (see text for explanation). A single figure means that the same volume of that component is used for both standard curves; when there are two figures, the first one is for the GSt standard curve and the second for the GSSG standard curve. In the last row, figures correspond to the components used in the assay of the test samples. SSA is added to standard curves to maintain the same concentration of SSA that is in the test samples due to the fact that SSA tends to inhibit glutathione reductase. Note that the SSA solution used to add to the assay mixture in the standard curves is 4.31%: tissue water represents $\approx$80% of tissue weight and thereby when we homogenize in a volume of 5% SSA equivalent to five times the tissue weight, we end up diluting the SSA solution with water contained in the tissue. The final volume in the supernatant would be $\approx$5.8 volumes with a reduced concentration of SSA to 4.31%: 5 vol. •5% = 5.8 vol. •x%; x% = 4.31%.

volumes of homogenate. To correct the possible interference, standard curves are constructed containing the drug at concentrations equivalent to those present in the usual 5 μl of the tissue supernatant used to measure GSt and in the 20 μl used to measure GSSG. These equivalent concentrations are determined empirically as those concentrations producing an inhibition comparable to that observed with use of the internal standard. The assay mixtures for control tissues (that have not been treated with the drug under study) should also be supplemented with the drug at the concentration used in the standard curves.

Note that the standard curve to assay GSSG can be made with standard solutions of GSH or GSSG. The reason for that is when the reading of the absorbance starts, all glutathione in the assay mixtures is in an oxidized form as GSSG due to the spontaneous reaction with DTNB. It is in the cycling process initiated by the addition of glutathione reductase where all GSSG is first transformed into GSH, which in turn reduces new molecules of DTNB generating color and becoming newly oxidized into GSSG, and so on. Therefore, it does not matter whether the construction of the standard curves is initiated with one or another form of glutathione.

However, it should be kept in mind that each molecule of GSSG would yield two molecules of GSH, and this would affect the calculations of the concentration of glutathione in the tissue.

CALCULATIONS OF TISSUE CONCENTRATION OF GSt, GSSG, AND GSH. It should be realized that GSt in the tissue equals GSH + GSSG. Because each molecule of GSSG yields two molecules of GSH and our standard curve is made against GSH concentrations, our GSt in tissues equals GSH + 0.5 GSSG concentration in the tissue. Referring to the calculations for GSSG concentrations, two different standard curves can be used. If a standard curve made with GSH is used, we are expressing the GSSG as GSH molar units, and therefore the actual GSSG molar concentration is half the one obtained by interpolation in the standard curve. If the standard curve is made with GSSG, the actual molar concentrations in the tissues are those given directly by interpolation in the standard curve. Usually, the GSSG is expressed as GSH because it facilitates the calculation of the actual GSH concentration in the test samples obtained as a difference: GSH = GSt − GSSG.

The concentration of either GSt or GSSG in the tissue is obtained according to the following equation:

$$GSt \text{ or } GSSG(\mu mol/g \text{ tissue}) = \frac{A \times [\text{tissue weight}(g) \times 5.8]/\text{assay volume of the test sample (ml)}}{\text{tissue weight}(g)} \quad (4)$$

where A is the amount obtained by interpolation in the standard curve, but expressed in micromoles. Simplifying the equation

$$GSt \text{ or } GSSG(\mu mol/g \text{ tissue}) = \frac{A \times 5.8}{\text{assay volume of the test sample (ml)}} \quad (5)$$

The GSH/GSSG redox potential is calculated as in Eq. (3).

Assessment of CB Chemoreceptor Cell Function

To assess the functionality of chemoreceptor cells of the CB, we have used an *in vitro* preparation of intact CB of rabbit. The rabbit CB weighs ≈400 μg, and our group has provided ample experimental evidence of its normal functioning *in vitro* with the ability of chemoreceptor cells to detect a great variety of stimuli (including the natural ones hypoxia and hypercapnia/low pH) and to respond with a neurosecretory response best monitored as the release of catecholamines (CA).[62] The procedure used to monitor

[62] C. Gonzalez, L. Almaraz, A. Obeso, and R. Rigual, *Trends Neurosci.* **15**, 146 (1992).

the neurosecretory response varies from one laboratory to the next. In our laboratory, two alternative methods are used: a radioisotopic method, which monitors the release of [³H]CA synthesized from the natural precursor [³H]tyrosine, which is described in some detail later, and a voltametric method that monitors continuously the release of CA from chemoreceptor cells as concentrations of endogenous CA in the extracellular space of the CB tissue. Any variation in the CA concentration associated with CB stimulation implies a response of chemoreceptor cells to such stimulus (see Rigual et al.[63] for a description of the method).

Surgical Procedures. Rabbits are anesthetized with sodium pentobarbital (40 mg/kg, iv through the lateral vein of the ear) and, after a longitudinal incision in the ventral face of the neck, the carotid arteries are dissected past the carotid bifurcation. After convenient clamping, a block of tissue containing the carotid bifurcation is removed and placed in a lucite chamber filled with ice-cold/100% O_2-equilibrated Tyrode (see earlier discussion) to dissect the CB free of surrounding connective tissue. Animals are killed by an intracardiac overdose of sodium pentobarbital. The Committee for Animal Care and Use at the University of Valladolid approved the protocols. Animals did not suffer any distress in all the experimental procedure.

Labeling of CA Stores, Synthesis of [³H-]CA from [³H-]tyrosine, and Release of [³H-]CA. The CA stores of CBs are labeled by incubating the organs in small glass vials (eight CBs/vial) and are placed in a shaker bath at 37° containing 0.5 ml of a 100% O_2-preequilibrated Tyrode solution. The incubating solution contains [3,5-³H]tyrosine (30 μM; 20 Ci/mmol), 100 μM 6-methyl-tetrahydropterine, and 1 mM ascorbic acid, cofactors for tyrosine hydroxylase and dopamine-β-hydroxylase, respectively.[64] The incubation to label [³H]CA stores lasts 2 h. After labeling the [³H]CA stores, the CB are transferred individually to glass scintillation vials, containing 4 ml of precursor-free bicarbonate-buffered Tyrode solution (see earlier discussion) and are kept in a shaker bath at 37° for the rest of the experiment. Solutions are bubbled continuously with 20% O_2/5% CO_2/ 75% N_2 saturated with water vapor. When hypoxia is applied as the stimulus, the solutions are bubbled with a hypoxic gas mixture (containing 0–21% O_2), which in these experiments was 7% O_2/5% CO_2/balance N_2. During the first hour, incubating solutions are renewed every 20 min and discarded. During the rest of the experiment, incubating solutions are collected every 10 min and are saved for ulterior analysis in their [³H]CA content. Stimulus to CBs consisted in their incubation during a 10-min period

[63] R. Rigual, L. Almaraz, C. Gonzalez, and D. F. Donnelly, *Pflug. Arch.* **439,** 463 (2000).
[64] S. Fidone and C. Gonzalez, *J. Physiol.* **333,** 69 (1982).

with the hypoxic (7% O_2-equilibrated) solution. In some experiments we also stimulated the CBs by incubation for 10 min with a solution containing 35 mM K$^+$ (balanced osmotically by the removal of NaCl). Collected solutions are acidified with glacial acetic acid to pH 3 and are maintained at 4° to prevent degradation of the [^{3}H]CA released until analysis. At the end of the experiments, CB tissues are homogenized in 0.4 N perchloric acid and centrifuged in a microfuge (4° 10 min) and the supernatant is stored for analysis in [^{3}H]CA.

Analytical Procedures. The analysis of [^{3}H]catechols present in the collected incubating solutions and the supernatants of the CB homogenates included adsorption to alumina (100 mg) at alkaline pH (by the addition, under shaking, of 3.6 ml of 2.0 M Tris buffer, pH 8.7), extensive washing of the alumina with distilled water, bulk elution of all catechols[65] with 1 ml of 1 N HCl, and liquid scintillation counting. When identification of the synthesized or released [^{3}H]catechols was needed, only 100 μl of the alumina eluates was counted and the remaining 900 μl was used for identification of the labeled [^{3}H]catechols by HPLC-ED.[66]

Quantification of the Release of [^{3}H]CA. The basal normoxic release of [^{3}H]CA in the sequential 10-min periods is obtained from the scintillation counter as dpm. It can be transformed easily in [^{3}H]CA molar units by taking into account that their specific activity is half of the precursor [^{3}H]tyrosine because the label in position 3 is lost in the first step of [^{3}H]CA synthesis. Either as dpm or as mole units it can be referred to CB or to mg of CB tissue. Alternatively, it can be expressed as a fraction of the total [^{3}H]CA synthesized by each CB. The [^{3}H]CA synthesized amounts to the [^{3}H]CA present in the tissue plus those released present in the sequentially collected incubated solutions. Basal release can also be normalized by referring as 100% the release obtained in a given collected fraction in the experiment: this allows following the normal decay of the basal release as the percentage of that obtained in the indicated period of time of the experiment. The stimulus-evoked release of [^{3}H]CA is first calculated as the dpm above basal normoxic release. It can be expressed as dpm, as mole units, or as percentage of tissue content and referred as per CB or per milligram of CB tissue; the evoked release in a experimental group of CBs can also be expressed as a percentage of that obtained in the correspondent control group of CBs. Alternative forms of expressing the release are also possible.[51]

[65] H. Weil-Malherbe, *in* "Análisis of Biogenic Amines and Their Related Enzymes" (D. Glick, ed.), p. 119. Wiley, New York, 1971.
[66] I. Vicario, R. Rigual, A. Obeso, and C. Gonzalez, *Am. J. Physiol. Cell. Physiol.* **278,** C490 (2000).

Results and Discussion

The experiments performed were aimed to disclose the relationship between the GSH/GSSG system and oxygen chemoreception. We have measured glutathione in rat diaphragm in several experimental conditions, including control or normoxic, 40-min incubation with 21% O_2/5% CO_2; hypoxic, 30 min 21% O_2/5% CO_2 + 10 min 7% O_2/5% CO_2; BCNU-treated, 40-min normoxic incubation in the presence of 0.5 mM BCNU; diamide (DIA)-treated, 40-min normoxic incubation in the presence of 0.2 mM diamide; PCMBS-treated, 40-min normoxic incubation in the presence of 50 μM PCMBS; and NAC-treated, 40-min normoxic incubation in the presence of 2 mM NAC (Fig. 4). The control levels of GSt, GSSG, and GSH, as well as the control redox potential shown in Fig. 4, correspond to the mean of all control groups and correspond to the arithmetic mean of 50 samples of tissue. Figure 4 presents this mean control value rather than the individual controls for each experimental group (8–12 pieces of tissue each) to simplify the figure, but the statistical significances shown have been calculated by comparing each experimental group with its corresponding control group. It is evident that hypoxia did not alter any of the parameters significantly. Carmustine (BCNU) and DIA decreased GSt and GSH levels significantly; in addition, DIA also increased GSSG levels significantly. Both agents, acting by completely different mechanisms (see Discussion), lowered the redox environment of the diaphragm similarly and significantly. However, NAC increased GSt and GSH and produced a very significant increase in the redox environment of the tissue. Identical results with NAC treatment were obtained in a

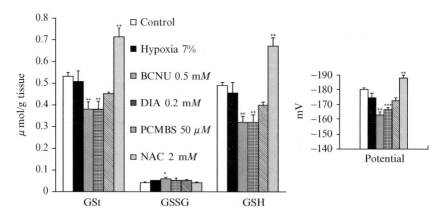

FIG. 4. Levels of GSt, GSSG, GSH, and GSH/GSSG redox potential in the experimental conditions shown. Data are means ± SEM of 8–12 individual data.

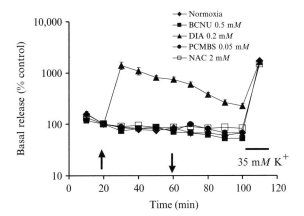

FIG. 5. Effects of several agents on the basal or normoxic release of [³H]CA by the intact rabbit CB. As indicated by arrows, drugs were applied during 40 min. At the end of the experiments, a pulse (10 min) of high external K^+ was added to test the viability of the preparations. Data are means ± SEM of 6 to 10 individual data. Standard error bars are omitted (except for DIA) for clarity.

previous study[51] with calf CB and in the present study with rat liver (not shown).

BCNU, DIA, PCMBS, and NAC were tested for their effect on the basal normoxic release of [³H]CA from rabbit CB chemoreceptor cells at the same concentrations used to measure glutathione in rat diaphragm. The findings are shown in Fig. 5. Note that the basal normoxic release of [³H]CA in the 100–120 min of duration of the experiments follows a monotonic slow decay. If the basal release for each CB in the second 10-min collection is normalized to 100%, at the end of the experiments the release in a 10-min period of incubation has decayed to 60–75% of that at the initiation of the experiment. When at the end of the experiment a depolarizing stimulus (35 mM extracellular K^+) or a hypoxic stimulus is applied, there is a burst release during application of the stimulus that continues during two or three additional 10-min periods. This apparently delayed release of [³H]CA represents the washout of [³H]CA, which, upon release, has been taken up again by all the structures of the CB and catabolized, leaving the tissue very slowly.[67] The basal release during a 10-min period in the rabbit CB represents from 0.1 to 0.4% of the [³H]CA synthesized in the 2-h period of loading with [³H]tyrosine, which is equivalent to 400–1600 dpm/CB. Among the tested agents, only DIA (0.2 mM) altered the normoxic release of [³H]CA. The release elicited by DIA at this concentration in most

[67] E. Gonzalez, R. Rigual, S. J. Fidone, and C. Gonzalez, *J. Auton. Nerv. Syst.* **18,** 249 (1987).

preparations peaks during the first 10 min of exposure to the drug and later decays slowly, despite the presence of the drug. Even after retiring the drug, the release of [^{3}H]CA slowly tends toward the normal basal release curve. BCNU (0.05 mM), which produced an alteration in the GSH/GSSG system very similar to the one produced by DIA, did not alter the basal release of [^{3}H]CA. Neither did NAC (2 mM), even though it affected the GSH/GSSG in the opposite direction than DIA and BCNU. Also, PCMBS (50 μM) did not affect the basal release of [^{3}H]CA. In sum, it would appear that the ability of DIA to activate chemoreceptor cells does not relate in a cause–effect manner with its ability to decrease the redox potential of the cells.

In other series of experiments, such as the ones shown in Figs. 6A and 6B, the same drugs were tested for their effects on the release of [^{3}H]CA elicited by hypoxia. In these experiments, one CB served as a control (Fig. 6A) and the other served as experimental (Fig. 6B). In the control CB, a hypoxic stimulus was applied and the evoked release (dpm above the horizontal dotted line in the Fig 6A) was calculated and taken as 100%. It is evident that hypoxia, which did not alter the GSH/GSSG system (see Fig. 4), produced a marked activation of chemoreceptor cells, indicating again that a correlation does not exist between the activity of chemoreceptor cells and their redox environment. The hypoxic release (10 min, 7% O$_2$) in drug-free CBs amounted to 2000–8000 dpm/CB equivalent to 0.8–4% of the amount synthesized. In the experimental CB, the drug to be tested (0.2 mM DIA in Fig. 6B) was added 40 min prior to and during the application of the hypoxic stimulus. The stimulating effect of DIA in normoxic conditions is evident, and the additional effect of the hypoxic stimulus is also clearly shown. The release of [^{3}H]CA elicited by hypoxia in the presence of DIA is calculated as in the control CB (dpm above the dotted line in Fig. 6B) and is expressed as a percentage of the hypoxic release in control CB. It was found that DIA did not alter the hypoxic response; the effects of DIA and hypoxia were additive. In fact, none of the agents tested produced a significant alteration of the release of [^{3}H]CA induced by hypoxia (Fig. 6C).

As a prerequisite to understanding this complex and conflictive field of cellular biology, this article provided ample background on the biology of ROS and enzyme systems involved in the maintenance of cell redox homeostasis. Along this line, we emphasized methodological aspects: we believe that most of the conflictive results are generated by methods that are not well contrasted. We provided a detailed description of the method of Griffith[61] to measure glutathione. This method is a very specific and sensitive assay. We also provided clear experimental guidelines to study the functionality of chemoreceptor cells through their neurosecretory response in an intact preparation of rabbit CB.

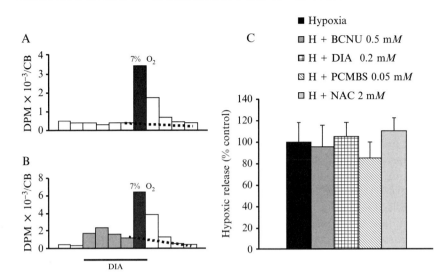

FIG. 6. Effects of the several agents on the release of [³H]CA induced by hypoxia. (A and B) Single experiments obtained in individual CBs that show the experimental protocol in control CBs subjected to hypoxic stimulation and in an experimental CB treated with diamide (0.2 mM) and subjected to the same hypoxic stimulation. The dotted lines separate normoxic release from that elicited by hypoxia. Each column represents the release in a 10-min incubating period. (C) Normalized data for the release induced by hypoxia in control conditions (hypoxia) and in the presence of the agents shown. In all cases, data are means ± SEM (n = 6 to 9).

Experimental data presented indicated that a hypoxic stimulus intense enough to activate chemoreceptor cells to produce a vigorous release of their neurotransmitters ([³H]CA) does not produce a significant modification in the redox environment of the cells, whether they be CB cells[51] or diaphragm muscle cells. This observation *per se* would indicate that a change in the general redox environment of chemoreceptor cells is not the signal triggering hypoxic activation in this stirps of cells specialized in oxygen sensing. This conclusion is reinforced by the additional observations of the present study. If an increase or a decrease in the production of ROS (and therefore the opposite modification in the GSH/GSSG redox potential) happens to be important for the activation of chemoreceptor cells, then the pharmacological manipulation GSH/GSSG redox potential of the cells in one of the directions should mimic hypoxia. Thus, manipulation in the opposite direction should inhibit the effect of hypoxia. By reducing the GSH/GSSG redox potential, DIA and BCNU would mimic a generalized increase in the production of ROS in the cells and mimic hypoxia if we assume that hypoxia increases the rate of ROS production. If this were the case, then NAC should inhibit the neurosecretory response elicited by hypoxia. The converse would be true if we assume that hypoxia decreases the production of ROS in a generalized manner. Data presented earlier do not support either possibility.

The observation that diamide and BCNU, which produce a very similar change in the redox environment of the cells, behave differently regarding their ability to activate chemoreceptor cells is intriguing. Although we do not know the reason for this difference, we propose the following explanation. In the time scale of our experiments, BCNU is a selective drug acting only intracellularly via inhibition of the glutathione reductase, thereby decreasing the concentration of GSH and the glutathione redox potential. However, DIA acts extracellularly and intracellularly. In the outside of the cells, DIA would react with –SH groups and promote the formation of disulfide bonds on the extracellular domains of the plasma membrane proteins. In the cell interior it will promote the formation of disulfide bonds among GSH molecules and then decrease the concentration of GSH and the glutathione redox potential of the cells. The ability of DIA to stimulate chemoreceptor cells would be due, according to our suggestion, to its extracellular actions. However, additional experiments are required to sustain this suggestion. It would also appear that whatever the mechanisms involved in the normoxic activation of chemoreceptor cells by DIA, those mechanisms do not participate in the response to hypoxia. Finally, PCMBS, which does not penetrate in the cells,[68] does not affect the GSH/GSSG system nor the functionality of chemoreceptor cells in normoxia or hypoxia. The ability of PCMBS to activate the action potential frequency in the CSN[69] would be produced by a direct action on the sensory nerve endings or nerve fibers.

An alternative explanation to our findings would be that the tested agents and stimulus produce different quantitative effects in the GSH/GSSG system of the rabbit CB than in the GSH/GSSG system of the rat diaphragm. We cannot exclude such a possibility, but NAC produced very similar effects in the rat diaphragm and liver (present study) and in the calf CB. Similarly, hypoxia was ineffective in altering the GSH/GSSG system in calf CB and in rat diaphragm. In addition, the drugs used in this study have been tested in many other systems[1,12] and in all of them they produce comparable effects to the ones found in the rat diaphragm.

Finally, if hypoxic stimulation triggers the production of ROS in a very restricted compartment of the cells, it is possible that we cannot detect any alteration in the GSH/GSSG system (see earlier), However, it is

[68] R. I. Fonteriz, C. Villalobos, and J. Garcia-Sancho, *Am. J. Physiol. Cell. Physiol.* **282**, C864 (2002).
[69] S. Lahiri, *Science* **212**, 1065 (1981).

difficult to accept that the redox environment of the cells is critically important for oxygen sensing (see also Searle et al.[44]) because alteration of the general redox environment of cells, which would include alteration of the putative cell compartment, does not alter the hypoxic response of CB chemoreceptor cells.

Acknowledgments

This work was supported by Spanish DGICYT grant BFI 2001–1713, by J. C. Y. L. Grant VA092/03, and by a FISS Grant to the Red Respira-SEPAR.

[4] Determination of Signaling Pathways Responsible for Hypoxic Pulmonary Vasoconstriction: Use of the Small Vessel Myograph

By JEREMY P. T. WARD and VLADIMIR A. SNETKOV

Introduction

Hypoxic pulmonary vasoconstriction (HPV) was first described as an important regulatory mechanism of the ventilation–perfusion ratio by Von Euler and Liljestrand in 1946.[1] Many studies have shown that HPV can be elicited in isolated pulmonary arteries, demonstrating that both the sensor for hypoxia and effector mechanisms reside within the artery. However, the signaling pathways underlying HPV are still poorly understood. It is becoming increasingly evident that HPV is multifactorial in origin, with mechanisms in both the vascular smooth muscle (VSM) and the endothelium[2,3]; there is, however, controversy regarding the precise identity of these mechanisms. Hypoxia sensing has been attributed to changes in redox state (e.g., NAD/NADH redox couples) and generation of reactive oxygen species (ROS) from either an NAD(P)H oxidase or the mitochondria.[4–6] The origin of the hypoxia-induced rise in VSM intracellular [Ca^{2+}] is equally controversial, with evidence for inhibition of K^+

[1] U. S. Von Euler and G. Liljestrand, *Acta Physiol. Scand.* **12,** 301 (1946).
[2] J. P. T. Ward and P. I. Aaronson, *Respir. Physiol.* **115,** 261 (1999).
[3] P. I. Aaronson, T. P. Robertson, and J. P. Ward, *Respir. Physiol. Neurobiol.* **132,** 107 (2002).
[4] R. M. Leach, H. M. Hill, V. A. Snetkov, T. P. Robertson, and J. P. Ward, *J. Physiol.* **536,** 211 (2001).
[5] G. B. Waypa, N. Chandel, and P. T. Schumacker, *Circ. Res.* **88,** 1259 (2001).
[6] E. K. Weir, Z. Hong, V. A. Porter, and H. L. Reeve, *Respir. Physiol. Neurobiol.* **132,** 121 (2002).

channels and depolarization, release of Ca^{2+} from intracellular stores, and possibly entry via nonselective and capacitative Ca^{2+} entry pathways.[7–9]

Determination of the signaling pathways involved in HPV is more straightforward when the complexity of the experimental model is limited. Isolated pulmonary artery preparations obviate the influence of mediators derived from other cells within the lung while maintaining functional integrity, in particular cell–cell interactions and specifically those of VSM and endothelium. This article describes methodologies for use of the functionally important small pulmonary arteries and techniques by which contractile function can be recorded simultaneously with estimates of key intracellular messengers using fluorescence microscopy (e.g., intracellular Ca^{2+}, NADH, ROS).

Isolated Small Pulmonary Arteries: Small Vessel Myography

The hypoxic pressor response is primarily due to the vasoconstriction of small muscular intrapulmonary arteries (IPA) with an internal diameter of 100–500 μm.[10] IPA differ in a variety of physiological and pharmacological respects from large capacitance pulmonary arteries.[11–13] Classical gut bath techniques used for rings or strips from large vessels are unsuitable for IPA, and these are best studied using a small vessel myograph, as originally designed for small systemic arteries.[14] The small vessel myograph consists of a heated stainless steel chamber with a volume of ~10 ml. The vessel is mounted (see later) onto two opposing jaws, one of which is connected to an isometric tension transducer and the other to a micrometer screw gauge to adjust the width of preparation and initial tension. Protocols are modified to take into account the low pulmonary artery pressure, as characterized previously.[12]

Equipment and Materials

> Small vessel myograph: Dual Wire Myograph system 410A (manual) or 510A (automatic), Danish Myo Technology A/S, Denmark (www.dmt.dk)

[7] S. L. Archer, E. K. Weir, H. L. Reeve, and E. Michelakis, *Adv. Exp. Med. Biol.* **475,** 219 (2000).

[8] H. L. Wilson, M. Dipp, J. M. Thomas, C. Lad, A. Galione *et al., J. Biol. Chem.* **276,** 11180 (2001).

[9] T. P. Robertson, D. E. Hague, P. I. Aaronson, and J. P. T. Ward, *J. Physiol.* **525,** 669 (2000).

[10] A. P. Fishman, *Circ. Res.* **38,** 221 (1976).

[11] R. M. Priest, D. Hucks, and J. P. Ward, *Br. J. Pharmacol.* **122,** 1375 (1997).

[12] R. M. Leach, C. H. Twort, I. R. Cameron, and J. P. T. Ward, *Clin. Sci.* **82,** 55 (1992).

[13] S. L. Archer, J. M. C. Huang, H. L. Reeve, V. Hampl, S. Tolarova *et al., Circ. Res.* **78,** 431 (1996).

[14] M. J. Mulvany and W. Halpern, *Circ. Res.* **41,** 19 (1977).

Chart recorder or Myodaq software and PC for data logging, Danish Myo Technology A/S

40-μm stainless steel wire for mounting arteries, Danish Myo Technology A/S

Binocular dissecting microscope ($\times$2–5), eyepiece graticule, and light source, World Precision Instruments Inc. (WPI), Sarasota, Florida (www.wpiinc.com)

Microforceps, Dumont #5/5c, WPI Inc.

Ophthalmic microspring scissors (Vannas), with superfine tips, WPI Inc.

Pipettors (1–200 μl)

Oxygen meter, SI782 Strathkelvin Instruments Ltd., Glasgow, UK (www.strathkelvin.com)

Mini Clark-style Oxygen electrode type 733, Diamond General Development Corp., Ann Arbor, Michigan (www.diamondgeneral.com)

Solutions

Physiological salt solution (PSS; mM): NaCl 118; NaHCO$_3$ 24; MgSO$_4$ 1; NaH$_2$PO$_4$ 0.44; glucose 5.6; CaCl$_2$ 1.8; KCl 4 (pH 7.4 with 5% CO$_2$)

KPSS (high K$^+$ PSS): NaCl 42; NaHCO$_3$ 24; MgSO$_4$ 1; NaH$_2$PO$_4$ 0.44; glucose 5.6; CaCl$_2$ 1.8; KCl 80

Drugs and Inhibitors

Compounds added to the chamber are made up daily in stock solutions; total additions during the experiment should be <1% of chamber volume.

Dissection and Mounting of Arteries

1. This description refers to rat, but similar procedures are followed for other species. Rats of $\sim$250–300 g (male Wistar in our studies) tend to provide the best preparations. Animals are killed by cervical dislocation. Unless performed carefully, cervical dislocation and/or stunning can result in damage to the pulmonary circulation and blood clots; this can substantially reduce preparation viability. Damage is avoided by terminal anesthesia (e.g., 50 mg/kg ip sodium pentobarbital), although pentobarbital may have long-term effects on VSM Ca^{2+} handling.[15]

2. The chest is opened and heart and lungs are excised rapidly and rinsed in cold PSS. One lobe of the lungs is placed into a dissecting dish

---

[15] M. Dipp, C. G. Nye, and A. M. Evans, *Am. J. Physiol.* **281**, L318 (2001).

with the pulmonary vein uppermost. Under a binocular microscope using fine forceps and scissors the lobe is dissected to reveal the bronchioles (whitish fibrous appearance). Careful dissection reveals the pulmonary artery running beneath. The artery is separated from the bronchiole and dissected free of adventitia; it is vital that the artery is subject to minimal handling and stretch. A ~2-mm length of artery is excised for mounting. Arteries of the fourth division and below provide preparations of ~100–500 μm i.d.

3. The myograph chamber is filled with PSS at room temperature. A 2- to 3-cm section of 40-μm stainless steel wire is attached to one screw of one of the myograph jaws (Fig. 1A) and the artery segment is threaded onto it. Great care needs to be taken so that the artery and endothelium are not damaged or stretched during any of these procedures. The free end of the wire is threaded across the jaw and made taut under the other screw so that the artery segment lies in the cutout of the jaw; any connective tissue is removed. Another section of wire is threaded through the lumen of the artery, and the myograph jaws are brought together before this is fastened under the screws of the other jaw (Figs. 1B and 1C). In the dual

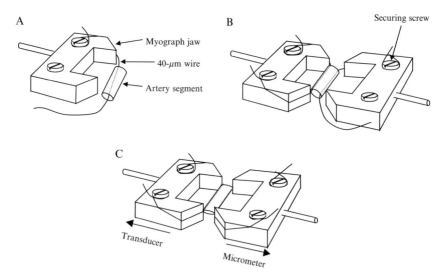

Fɪɢ. 1. Diagram of myograph jaws showing the mounting procedure. (A) Artery segment being threaded onto 40-μm stainless steel wire. (B) The wire is tightened, leaving the artery in the cutout of the jaw, and the second wire from the other jaw is threaded through the lumen. (C) Wires on both jaws must be made taut and fastened tightly under the securing screws. If there is any appreciable freedom of movement in the wires, poor results will be obtained and the artery may be damaged.

myograph, a second artery section can be mounted. The myograph chamber should be gassed (normoxia: 5% CO_2, 95% air), and the temperature brought to 37°. The preparation is allowed to equilibrate for >30 min before normalization.

Normalization Procedure

The normalization procedure determines the resting internal circumference of the artery that it would have had *in vivo* for the desired transmural pressure. In the myograph, the internal circumference is twice the distance between the jaws plus the circumference of the wire ($\sim$126 μm for 40-μm wire; Fig. 2). For IPA, transmural pressure is set to either $\sim$17 mm Hg (representing the situation *in vivo*) or 30 mm Hg, the peak of the length–tension curve.[12] The latter tends to provide more reproducible results, with no other qualitative differences. The relationship among the internal circumference (L_P), the transmural pressure (P, in kPa), and artery wall tension (T) is described by a modification of the Laplace equation: $L_P = T\pi/P$.[14]

4. The artery is stretched incrementally using the micrometer, and when tension stabilizes it is plotted against the distance between the jaws for each step; the process is then reversed (i.e., the jaws are brought closer together) and the mean plot is determined. The intersection between actual data and the theoretical plot derived from the aforementioned equation determines the appropriate jaw width (and thus L_P) for the required transmural pressure (Fig. 3). In automated myographs, the whole normalization procedure is automatic. For manual myographs, calculations are performed using computer software (e.g., Myodaq, Danish Myo Technology). The jaw width is then set to the calculated value, and the

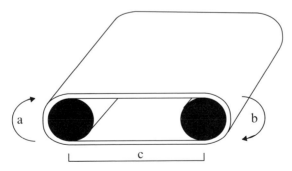

Fig. 2. Geometry of an artery after normalization to 30 mm Hg. The internal circumference at 30 mm Hg (L_{30}) equals twice the distance between the jaws (c) plus a and b, that is, the circumference of the wire itself.

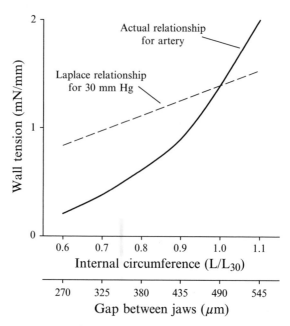

FIG. 3. A typical graph of tension plotted against the gap between myograph jaws for a 350 μm i.d. pulmonary artery. The tension measured by the myograph, corrected for unit length, is shown as the solid line. The dashed line represents data calculated from the modified Laplace equation for a transmural pressure of 30 mm Hg. The intersection point of calculated and actual data represents the distance to which the jaws should be set to normalize the internal circumference (L) of the artery for 30 mm Hg (L_{30}). The second axis shows the equivalent ratio between L and L_{30} as the artery is stretched.

preparation is allowed to equilibrate for >45 min in gassed PSS at 37°; the PSS is refreshed every 15 min.

Final Equilibration and Checking Integrity of the Endothelium

5. The artery is subjected to 3–4 KPSS challenges of 2 min each, ~6 min apart. The tension induced by the last two KPSS challenges should be identical; if not, further challenges are administered. KPSS-induced tension in rat IPA should normally be >5–10 mN/mm; IPA developing <2 mN/mm are discarded. KPSS-induced tension does not plateau for ~5–10 min; measurements are taken at exactly 2 min. Obtaining useful preparations is highly dependent on minimal artery handling and the time taken to mount.

6. Endothelium integrity must be established. Arteries are constricted with a suitable agonist; for example, 10 μM prostaglandin $F_{2\alpha}$.

α-Adrenoceptor agonists are poor vasoconstrictors in IPA.[11] Acetylcholine (ACh, 10 μM) is used as an endothelium-dependent vasorelaxant; endothelium integrity is confirmed if >50% vasorelaxation occurs (IPA are less responsive to ACh than systemic or large pulmonary arteries).

Removal of the Endothelium

7. Where the endothelium is to be removed, the lumen of the artery is rubbed gently with a human hair (male forearm is best); endothelium integrity is ascertained as described earlier. To check whether this procedure has damaged the preparation, a further KPSS challenge is performed. If the tension is less than 90% of that obtained in the last challenge before endothelium removal, the preparation is discarded.

Data Analysis

8. Tension is expressed as millinewtons per millimeter artery length, cut end to cut end, as measured under the microscope using the eyepiece graticule. Length should be 1.5–2 mm.

9. Small differences in the diameter of small arteries have large effects on force generation (cf. Laplace equation); to normalize between arteries tension is commonly expressed as a percentage of the response to a KPSS challenge.

Induction of Hypoxic Pulmonary Vasoconstriction

In IPA, hypoxia commonly induces a multiphasic response, most often a transient phase 1 superimposed on a more slowly developing phase 2 (Fig. 4A).[2,3,16] The absolute and relative size of each phase depends on the amount and type of preinduced tension (pretone) and degree of hypoxia challenge; phase 2 is particularly sensitive to the quality of the preparation and is endothelium dependent.[3] A small amount of pretone potentiates HPV,[10,12,17] although is not always required.[8]

1. Arteries are mounted on the myograph as described earlier. The Perspex lid of the chamber is sealed with silicon grease so that diffusion of O_2 from the atmosphere is limited. A port is left open for the exit of aerating gases and for the entry and removal of solutions. The O_2 electrode is placed through another port with the tip central in the chamber fluid.

---

[16] R. M. Leach, H. S. Hill, V. A. Snetkov, T. P. Robertson, and J. P. T. Ward, *J. Physiol.* **536,** 211 (2001).

[17] M. Ozaki, C. Marshall, Y. Amaki, and B. E. Marshall, *Am. J. Physiol.* **19,** L1069 (1998).

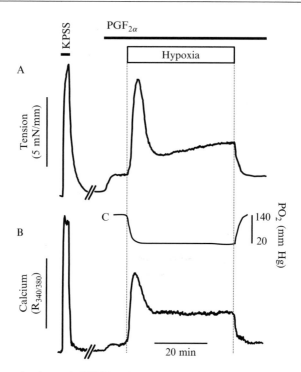

FIG. 4. Trace showing typical KPSS and HPV responses in a small ($\sim$400-μm i.d.) pulmonary artery. Trace (A) shows tension, and trace (B) shows intracellular [Ca^{2+}] as estimated by the Fura PE-3 ratio ($R_{340/380}$). Trace (C) is the output of the O_2 electrode in the myograph bath. Prostaglandin $F_{2\alpha}$ ($PGF_{2\alpha}$; 2 μM) was present during and after the hypoxic challenge, as shown by the bar.

2. A KPSS challenge is performed for the subsequent normalization of tensions.

3. Pretone is induced by addition of the vasoconstrictor. Prostaglandin $F_{2\alpha}$ (1–3 μM) is often used in IPA, but endothelin 1 or raised K^+ (20–35 mM) is also suitable (although see Robertson et al.[9]). Pretone must be matched precisely between experiments by titration of the vasoconstrictor (i.e., after the addition of any experimental drugs) and should be no more than $\sim$10–20% of the response to KPSS.[9,17]

4. Hypoxia is induced by changing the aerating gas to one containing 0–2% O_2. The PO_2 in the chamber should fall to the desired level within 1–2 min and remain stable. Artery preparations are less sensitive to hypoxia than blood-perfused lungs or intact animals,[2] and a PO_2 of <35 mm Hg is generally required for significant HPV; 20–25 mm Hg generally produces

the best response. A PO_2 of <15 mm Hg counts as severe and may have other effects.[3]

5. The hypoxic challenge should be maintained for 40–60 min; IPA should exhibit a "classical" biphasic response (Fig. 4A).

6. On restoration of normoxia, tension should fall rapidly to the prehypoxic level and remain stable. If it does not, consideration should be given to tissue viability.

7. The addition of experimental drugs and mediators should be made 15 min before the induction of pretone or, if not possible due to experimental design, any effects on pretone should be first determined. Timed control experiments should always be performed. Up to three hypoxic challenges can be performed on each preparation; these are highly reproducible if 60 min is left between each challenge.[9,16]

Estimation of Intracellular Ca^{2+} and Other Key Mediators in Intact Arteries

The use of small vessel myography coupled with fluorescence microscopy is a powerful technique that can be used to correlate contractile function directly with changes in intracellular messengers measurable by fluorescent probes or by autofluorescence. IPA ($<\sim$400 μm i.d.) are good for this purpose, as they contain small amounts of connective tissue compared to smooth muscle, and the walls are relatively thin. Although standard myographs (see earlier discussion) can be used, a specially modified version that minimizes the distance between the microscope objective and the artery preparation gives better results. The glass window in the base of the myograph is highly UV absorbent and should be as thin as possible; a glass coverslip is suitable, although a quartz window could be fitted. Methods relate to the measurement of Ca^{2+} with Fura PE-3 and reactive oxygen species with DCF, but are basically the same for any fluorescent probe, with appropriate excitation and emission filters. Details on available probes and their excitation/emission characteristics are in the Molecular Probes handbook on their web site (www.molecularprobes.com).

Equipment

Small vessel myograph: Ideally a Confocal wire myograph designed specifically for this use (Fig. 5), Danish Myo Technology A/S, Denmark (www.dmt.dk)

Inverted epifluorescence microscope (Olympus, Nikon, Zeiss) equipped with a 10 or 20× UV objective with a high UV transmission rate down to 340 nm (UAPO for Olympus, CFI S

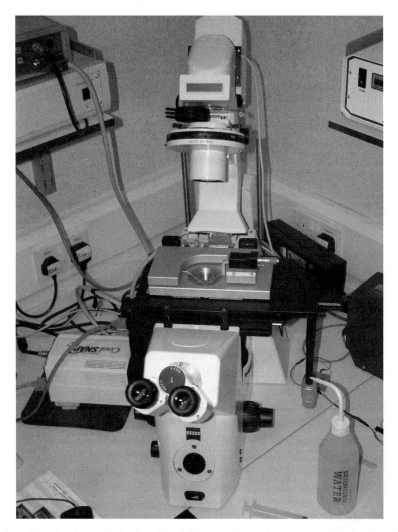

Fɪɢ. 5. Confocal myograph placed in position on a Zeiss Axiovert 200A equipped with a fluorescence-imaging system.

Fluor for Nikon, Fluar for Zeiss), a UV light source, and a mechanical stage suitable for supporting myograph; microspectrofluorimeter system, for example, Cairn Research Ltd., Faversham, UK [excitation filter wheel and appropriate excitation filters (340, 360, and 380 nm for Fura PE-3), appropriate dichroic mirror and emission filter (>500 nm), recording photomultiplier tube, controller, PC, and software] or fluorescence imaging system: Universal Imaging Co. (excitation and emission filters wheels and filters as described previously, dichroic mirrors, 12-bit CCD camera, controllers, PC, and software); otherwise as for standard myography.

Reagents and Solutions

   Ca^{2+}-sensitive fluorescent probe: Fura PE-3/AM, Molecular Probes Inc. (www.molecularprobes.com)
   ROS-sensitive probe: 5-(and-6)-chloromethyl-2',7'-dichlorodihydro-fluorescein diacetyl ester (CM-DCF/DA), Molecular Probes Inc.
   Rotenone (100 μM stock)
   Low and high substrate PSS (0 and 15 mM glucose).

Measurement of Intracellular [Ca^{2+}] and Tension in IPA

   1. The tension output from the myograph is recorded on the same apparatus as the fluorescence signal (or vice versa) so that correct time correlations can be made. Analogue inputs are available on most fluorescence imaging and photometer systems.
   2. IPA are mounted on the myograph and normalized as described earlier.
   3. Arteries are incubated with PSS containing 2 μM Fura PE-3/AM for 90 min at room temperature, followed by a further loading period of 30 min at 37° to allow intracellular hydrolytic cleavage of the dye.
   4. The preparation is washed several times with fresh PSS and stimulated repeatedly with KPSS until a stable tension response is achieved.
   5. The myograph is transferred to the microscope stage (Fig. 5), and the microscope is focused on the surface of a central area of the artery, away from the wires and cut ends. If the microscope is well focused, the contribution to the signal from the endothelium is negligible.[9,16]
   6. The artery is illuminated with UV light alternating between 340 and 380 nm by spinning the filter wheel. The intensity of emitted light >500 nm is measured using the photomultiplier or CCD camera. The ratio of intensities of emitted light corresponding to excitation by 340 and 380 nm ($R_{340/380}$) is a measure of intracellular [Ca^{2+}]. The speed at which the excitation wavelengths are alternated defines time resolution. There is a balance between the latter and sensitivity of the system; at faster speeds the amount of emitted light may not be sufficient for a reasonable signal-to-noise ratio unless the artery is well loaded with Fura PE-3 and the detection system is very sensitive. During HPV, one ratio measurement every 2–8 s is generally adequate (e.g., see Fig. 4B).

7. Calibration: Taking into account known difficulties with the calibration of intracellular $[Ca^{2+}]$ using fluorescent dyes even in single cells, we strongly advise against any attempt to calculate an absolute measurement of $[Ca^{2+}]$, as this is likely to be inaccurate and misleading. Instead, we recommend either using just $R_{340/380}$ as a qualitative indicator of intracellular $[Ca^{2+}]$ or, for comparison between arteries, expressing experimental values of $R_{340/380}$ as a percentage of that obtained from a KPSS challenge (e.g., Robertson et al.[9,18]).

8. Immediately before and after each experimental challenge (i.e., induction of hypoxia), the preparation should be challenged with KPSS both as a control and to ensure that the system is working correctly. Depolarization with KPSS causes a large increase in intracellular $[Ca^{2+}]$ and concomitant vasoconstriction (Fig. 4) almost entirely due to Ca^{2+} entry via L-type Ca^{2+} channels.[9]

Manganese Quench: Measurement of Ca^{2+} Entry

A useful adjunct to the aforementioned technique is the use of Mn^{2+} quench of Fura PE-3 to estimate Ca^{2+} entry into the cell; this is only feasible for Mn^{2+}-permeable Ca^{2+} entry pathways.[9] The technique relies on the fact that Mn^{2+} quenches Fura PE-3 fluorescence, thus the rate of decline in Fura PE-3 fluorescence reflects the rate of Mn^{2+} entry. Fluorescence is measured following excitation at 360 nm, the isosbestic point for Fura, where changes in $[Ca^{2+}]$ itself have little effect.

1. Arteries are mounted and loaded with Fura PE-3 as described previously.

2. Fluorescence is recorded with 360-nm excitation; a slow decline is generally seen representing normal loss and/or bleaching of Fura PE-3 (Fig. 6).

3. On adding Mn^{2+} (50 μM) to the PSS, the rate of decline increases, presenting Mn^{2+} entry into the cells and quench of fluorescence. The rate of decline can be adjusted to suit by varying $[Mn^{2+}]$. Removal of Ca^{2+} also increases the rate, as Ca^{2+} competes for the same entry pathway. After ~1–4 min the chamber is washed with Mn^{2+}-free PSS and the rate should decline to the resting level.

4. The artery is exposed to the experimental challenge, and Mn^{2+} is reapplied. For example, thapsigargin (100 nM to empty stores) substantially increases the rate by activation of a Mn^{2+}-permeable capacitative Ca^{2+} entry pathway (Fig. 6[9]).

5. Data are expressed in terms of total quenchable fluorescence, determined by incubating the preparation with 1 mM Mn^{2+} in the

[18] T. P. Robertson, P. I. Aaronson, and J. P. T. Ward, *Am. J. Physiol.* **268,** H301 (1995).

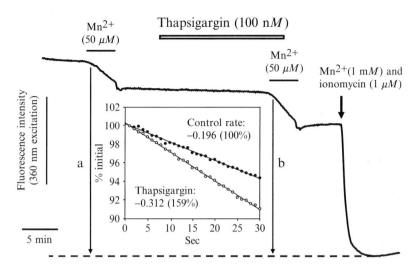

FIG. 6. Typical Mn^{2+} quench experiment in an IPA. Quench of Fura PE-3 fluorescence is observed during the addition of Mn^{2+} as shown by the bars. At the end of the experiment, Fura PE-3 fluorescence is quenched completely by the addition of a high concentration of Mn^{2+} plus ionomycin. Data are normalized by expressing the rate of quench in terms of the total quenchable fluorescence present at the start of the Mn^{2+} addition, shown by arrows labeled a and b for control and experimental runs. (Inset) Normalized data and the increase in quench rate observed after thapsigargin, indicating increased Mn^{2+} entry.

presence of 10 μM ionomycin (Ca^{2+} ionophore) at the end of the experiment and allowing the Fura PE-3 signal to decay (Fig. 6).

Known Methodical Problems with Fluorescent Dyes

"Leak" of Dye. Use of Fura-2 in intact preparations is limited by the "leak" of dye from loaded arteries, as Fura-2 is actively pumped out of the tissue by an anion transporter. The introduction of Fura PE-3, which contains twice the number of negative charges compared to Fura-2, reduces dye loss substantially, and run-down of fluorescence is not generally a problem in experiments lasting less then 4 h.

Bleaching. Most fluorescent dyes bleach under high-intensity UV illumination. To limit this, the intensity should be limited to the minimum sufficient to produce a good signal-to-noise ratio (e.g., using neutral density filters). Preparations should also only be illuminated when necessary; use the light path shutter rather than switching off the UV source.

Autofluorescence and Hypoxia. Although unloaded arteries often possess a significant degree of autofluorescence, this is often not considered to be a problem, as if necessary this can be determined after the experiment by quenching Fura PE-3 fluorescence with 1 mM Mn^{2+} in the presence of Ca^{2+} ionophores (ionomycin, A23187) and substracted from the emission signal at each wavelength. However, it has been determined that a large part of the autofluorescence is due to NAD(P)H, which has a broad excitation spectrum between 340 and 380 nm, that is, the precise band used with Fura PE-3. As NAD(P)H is not necessarily constant during experiments, especially during hypoxia[16] (see later); this correction is not always feasible or appropriate. During hypoxia changes in NAD(P)H, fluorescence will tend to overestimate the Fura PE-3 Ca^{2+} signal. The degree of overestimation will depend on how well the arteries are loaded with Fura PE-3; in well-loaded arteries, it will be small to negligible. An estimate of the error can be calculated by comparing fluorescence emission in loaded and unloaded arteries or by treatment with rotenone (100 nM), which causes a maximum increase in NAD(P)H fluorescence (i.e., complete reduction), although note that this compound is irreversible and, in some hands, is said to increase Ca^{2+} itself.[16]

Movement Artifacts. Although the preparation is nominally isometric, some movement and redistribution of smooth muscle ("bulging") occur during powerful contractions. This causes parallel changes in fluorescence intensity at all excitation wavelengths. The use of ratiometric dyes (e.g., Fura PE-3) compensates for the changes to a large extent, and indeed a powerful vasoconstriction induced by phorbol ester (Ca^{2+} independent) is not accompanied by significant changes in $R_{340/380}$. However, movement artifacts can become significant if the rate of change in tension is very fast.

Estimation of NAD(P)H in Isolated Arteries during Hypoxia

Mammalian cells exhibit autofluorescence partly due to reduced pyridine nucleotides (NADH and NADPH) and, to a lesser extent, flavin coenzymes (FAD and FMN). The former have absorption and emission maxima at 340 and 460 nm, respectively, and can potentially interfere with Fura-2 or Fura PE-3-based Ca^{2+} measurements (see earlier discussion). Estimation of NADH is commonly performed in isolated cells and other systems using fluorescence during 340-nm excitation. This is not possible in preparations that move, and we modified a ratiometric technique used in heart[19] for use in arteries mounted in the myograph.[16] This does not distinguish between NADH and NADPH; the term NAD(P)H is used to reflect this.

1. IPA are mounted on the myograph and normalized as described previously. The equipment is set up in an identical fashion to that described for Fura PE-3, with the same excitation and emission filters, with

[19] D. A. Scott, L. W. Grotyohann, J. Y. Cheung, and R. C. Scaduto, *Am. J. Physiol.* **267,** H636 (1994).

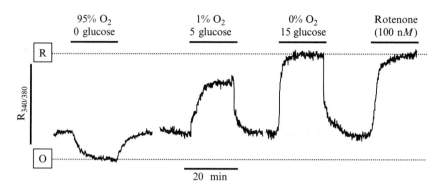

FIG. 7. Typical trace of NAD(P)H autofluorescence ($R_{340/380}$) in a small pulmonary artery mounted in a myograph. Dotted lines represent the fully oxidized state (O), obtained by rapid gassing with high O_2 and no substrate, and the fully reduced state (R), obtained by rapid gassing with zero O_2 and high substrate (15 mM glucose). Hypoxia (1% O_2; ~20 mm Hg PO_2 in this case) in the presence of a normal substrate causes a rapid increase in the NAD(P)H signal. Inhibition of mitochondrial NADH consumption by the complex I inhibitor rotenone causes full reduction.

the exception that the amplifier gain may have to be increased to account for the much reduced amount of fluorescence in the unloaded artery.

2. Experiments are performed in an identical fashion. A KPSS challenge should be performed as usual; this should have no significant effect on NAD(P)H fluorescence ratio, except for occasionally producing a small movement artifact.[16]

3. In IPA, hypoxia causes a rapid and well-defined PO_2-dependent increase in the NAD(P)H fluorescence ratio ($R_{340/380}$) (Fig. 7).

4. Calibration: Only a proportion of autofluorescence is due to NAD(P)H. Data are expressed in terms of the maximally reduced and oxidized states. These are obtained by incubating the artery in high substrate PSS while gassing rapidly with 0% O_2, and zero substrate PSS gassed with 95% O_2, 5% CO_2, respectively. Additionally, 100 nM rotenone can be used to determine the maximally reduced state, as this blocks complex 1 of the mitochondria and essentially abolishes the large majority of NADH consumption (Fig. 7).

Estimation of Reactive Oxygen Species

The role of reactive oxygen species in the signal transduction pathways underlying HPV and other oxygen-sensitive mechanisms is highly controversial.[6,16,20,21] One of the key reasons for this concerns the methods by which ROS is estimated. Several different methodologies have been used, but one of the most convenient is dichlorodihydrofluorescein (DCF) and its

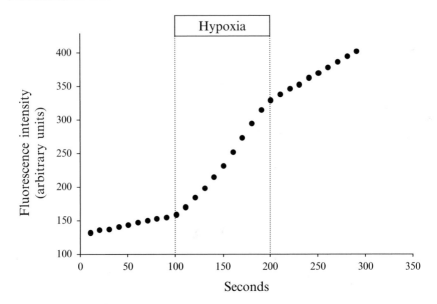

FIG. 8. Typical experiment showing the effect of hypoxia (PO$_2$ ~20 mm Hg) on DCF oxidation and fluorescence in an IPA (~400 μm i.d.). In this case the preparation was illuminated for 50 ms every 10 s.

derivatives, which are oxidized to a fluorescent form by ROS.[5,22] This is essentially irreversible and the rate of ROS production is estimated from the rate of DCF oxidation, and hence increase in fluorescence. Unfortunately, the chemistry of DCF oxidation is complicated; while it is widely used to study ROS generation, it is not clear to which individual ROS it is sensitive (see also Chandel and Schumacker[20]). Oxidation is, moreover, enhanced by light, and there is the possibility that the hydrolysis of DCF esters could itself produce ROS, thus causing autooxidation.[23] Caution should therefore be used in the interpretation of data obtained using DCF.

     1. IPA are mounted and normalized as described earlier. The equipment is set up in a similar fashion as for Fura PE-3, except that a single excitation wavelength of 490 nm is used; peak emission is around 550 nm.

     2. Arteries are loaded with 5 μM CM-DCF/DA for 30 min at 37° before being washed. The CM-DCF derivative is better retained than earlier forms.

[20] N. S. Chandel and P. T. Schumacker, *J. Appl. Physiol.* **88**, 1880 (2000).

[21] S. Archer and E. Michelakis, *News Physiol. Sci.* **17**, 131 (2002).

[22] D. W. Killilea, R. Hester, R. Balczon, P. Babal, and M. N. Gillespie, *Am. J. Physiol.* **279**, L408 (2000).

[23] C. Rota, C. F. Chignell, and R. P. Mason, *Free Radic. Biol. Med.* **27**, 873 (1999).

3. In resting arteries, a slow increase in fluorescence is observed, presumably related to background ROS generation. This is light dependent, and between experiments the light shutter should be closed to prevent the exhaustion of dye. In the presence of hypoxia, the rate of oxidation increases sharply[5,22] (Fig. 8). As the rate of oxidation is also dependent on excitation light intensity, this should be minimized by using short periods of exposure (e.g., 50 ms every 10 s) rather than continuous exposure.

4. Data are best expressed in terms of the rate of change in fluorescence calculated as a proportion of that observed under the initial control conditions in the same preparation.

Known Methodological Problems with CM-DCF

1. As discussed earlier, the validity of DCF oxidation as an indicator for ROS production is currently under scrutiny, although it is still widely used for this purpose.
2. Experiments should be kept short, as the rate of oxidation often accelerates with time, possibly due to autooxidation[23]; internal and time controls are critical.

Acknowledgment

We are grateful to the Wellcome Trust for support of our studies.

[5] Oxygen-Dependent Regulation of Pulmonary Circulation

By Rubin M. Tuder, Sharon McGrath, Norbert F. Voelkel, and Mark W. Geraci

Definition of Pulmonary Hypertension

The central function of the lung is to oxygenate the blood. To carry out this task throughout our lifetime, an extensive capillary network branches through alveolar structures to create an alveolar surface area equivalent to a tennis court, which fits approximately in a structure with a volume of 8 liters. Oxygen probably orchestrates the developmental patterning of the lung and its function in the adult life. Indeed, the lung has to rapidly adapt its perfusion to circulation patterns to keep the proper oxygenation of the blood to an optimum. Central to this regulation is the matching of pulmonary blood flow to areas of varying ventilation accordingly to match the local ventilation, that is, the precapillary pulmonary arteries can contract or dilate accordingly to match the local levels of alveolar oxygen. Pulmonary blood flow control is positional, to some degree, gravity dependent but to a large measure determined by the design and the branching pattern of the lung vascular tree. Physiologically, the pulmonary artery pressure is low, one-fifth of the systemic pressure (usually 25 mm Hg systolic).

Structurally, pulmonary arteries contain a medial muscular layer that extends from vessels 500 μm in diameter down to vessels measuring 70 to 50 μm in diameter.[1] The lack of a well-defined medial smooth muscle cell layer in this vascular segment probably contributes to the low perfusion pressure at the capillary level, a vital physiological property of the pulmonary arterial circulation to avoid pulmonary edema.[2] Contributing to low perfusion pressures are the recruitment of vessels, hormonal modulation of intracellular cyclic mucleotide levels, and the status of K^+ channels in smooth muscle cells, despite fold increases in cardiac output, such as during exercise.

If there is global lung hypoxia, the pulmonary circulation in its entirety is affected by precapillary pulmonary vasoconstriction, although there may also be some small degree of postcapillary (venous) vasoconstriction. Persistence of vasoconstriction for several days starts a process of vascular remodeling; the pulmonary artery pressure continues to increase, but usually not to the level encountered in severe pulmonary hypertension (PH).

Pulmonary hypertension is an important clinical complication in approximately 30% interstitial and other nonneoplastic lung diseases in humans.[3] The levels of PH usually are between 25 and 40 mm Hg and of clinical importance if it progressively compromises right heart function. The underlying mechanisms of PH in these conditions are related to vessel remodeling, which is in turn affected by progressive alveolar hypoxia and small vessel destruction due to inflammation and/or scarring. In contrast, a small number of patients have severe PH in whom the pulmonary artery pressures are in excess of 40 mm Hg and commonly associated with important and life-threatening right ventricular failure.[4]

Pulmonary vascular remodeling is the distinctive structural component associated with PH. Pulmonary arteries are composed of three layers, each with its characteristic cellular component. Whereas endothelial cells are the predominant cell in the intima, the migration of medial smooth muscle

---

[1] C. A. Wagenvoort and N. Wagenvoort, in "Pathology of Pulmonary Hypertension" (C. A. Wagenvoort and N. Wagenvoort, eds.), p. 1. Wiley, New York, 1977.

[2] D. M. Rodman and N. F. Voelkel, in "The Lung" (R. G. Crystal, J. B. West, E. R. Weibel, and P. J. Barnes, eds.), p. 1473. Lippincott-Raven, Philadelphia, 1997.

[3] R. M. Tuder, C. D. Cool, C. Jennings, and N. F. Voelkel, in "Interstitial Lung Disease" (S. M. Schwartz and T. E. J. King, eds.), p. 251. Decker, Hamilton, Ontario, 1998.

[4] N. F. Voelkel and R. M. Tuder, Eur. Respir. J. **14**, 1246 (1999).

cells occurs when there is a marked increase in pulmonary artery pressures, vessel injury, or local thrombosis. The vascular medium is composed almost exclusively of smooth muscle cells; alterations in media are the most dramatic alteration seen in hypoxic PH (HPH). The adventitia is composed of fibroblasts, which, like medial smooth muscle cells, can also undergo alterations in cell size and number in HPH. The relative importance and how to assess the cellular contribution to pulmonary vascular remodeling are addressed in the methodological section. Whether pulmonary vascular remodeling antedates the rise in pulmonary artery pressures or whether the converse occurs is unknown. In fact, the pathobiology of HPN probably relies on a mutual interaction of hypoxic vasoconstriction with hypoxia-dependent growth factor overexpression, downregulated vasodilating substances, vascular remodeling, and effect of shear stress. Therefore, ascribing a primary causal role to any of the aforementioned events is artificial; it follows, as will become apparent later on, that several equally important end points related to these pathobiological processes linked to PH are used to characterize experimental models of hypoxia effect on pulmonary circulation.

It is the authors' goal to relate traditional approaches to study the effect of chronic hypoxia on the lung circulation to modern and state of the art approaches to unravel pathobiologically relevant information in HPN. Furthermore, it is our intent to review the importance of studies related to HPN to the understanding of diseases of pulmonary arteries, particularly PH.

Chronic Hypoxic Models and Their Pathobiological Significance

The readers are referred to a previous review of the overall relevance of the model of hypoxic PH.[5] To what extent does the pathobiological information obtained with HPN models apply to the human disease is unknown. Excessive medial smooth muscle proliferation is not a feature of most cases of severe human PH, whereas a group of patients with this disease show tumor-like proliferation of pulmonary endothelial cells, which is not a feature of HPN.[6]

Historically, the chronic hypoxia model of PH has been investigated because of the rationale that pathobiological information obtained from the chronic hypoxic model may also explain human PH. This belief was based to a large extent on the importance given to pulmonary vascular remodeling in the installation, progression, and regression of PH. Common to pulmonary hypertensive states is pulmonary vascular remodeling, particularly medial muscular thickening and extension of the muscular layer to peripheral, usually nonmuscularized, pulmonary arteries. It has been postulated

[5] N. F. Voelkel and R. M. Tuder, *J. Clin. Invest.* **106,** 733 (2000).
[6] R. M. Tuder and A. L. Zaiman, *Am. J. Respir. Cell Mol. Biol.* **26,** 171 (2002).

that hypoxic vasoconstriction leads to increased shear stress and to shear stress-activated expression of growth factors. Furthermore, several mediators of vasoconstriction double as pro- or antiproliferative growth factors. The complexity of signaling during hypoxia is illustrated by the activation of proinflammatory signals, such as the expression of 5-lipoxygenase, platelet-activating factor, or vascular oxidative stress.

The role of inflammation in HPN has been highlighted by the decrease in pulmonary artery pressures in rats that received a platelet-activating factor inhibitor[7] or the 5-lipoxygenase inhibitor MK866 or mice lacking 5-lipoxygenase.[8]

Using gene array methodology, Hoshikawa et al.[9] compared chronically hypoxic rat and mouse lungs and found a great number of genes expressed in the rat but not in the mouse. These different gene expression responses to hypoxia may begin to explain the vast species differences examined and demonstrated during the last two decades. Hoshikawa et al.[9] also showed expression of MAP kinase in HPN and MAP kinase inhibitor that blocks HPN.

Serotonin, which has been shown to be increased in the serum of patients with human primary PH (PPH),[10] also plays a role in medial smooth muscle cell remodeling in HPN models, as mice lacking the serotonin transporter or receptor show decreased HPN when compared to wild-type controls.[11,12] Of note, a polymorphism of the 5-hydroxytryptamine transporter gene has been identified in PPH patients. This polymorphism renders cultured pulmonary artery smooth muscle cells more susceptible to the growth-promoting effects of serotonin.[13]

Prostacyclin treatment has significantly improved the survival of patients with severe PH.[14] The rationale for the use of prostacyclin in human

[7] S. Ono, J. Y. Westcott, and N. F. Voelkel, J. Appl. Physiol. **73**, 1084 (1992).

[8] N. F. Voelkel, R. M. Tuder, K. Wade, M. Hoper, R. A. Lepley, J. L. Goulet, B. H. Koller, and F. Fitzpatrick, J. Clin. Invest. **97**, 2491 (1996).

[9] Y. Hoshikawa, P. Nana-Sinkam, M. D. Moore, S. Sotto-Santiago, T. Phang, R. L. Keith, K. G. Morris, T. Kondo, R. M. Tuder, N. F. Voelkel, and M. W. Geraci, Physiol. Genomics 00081 (2002).

[10] P. Herve, J. M. Launay, M. L. Scrobohaci, F. Brenot, G. Simonneau, P. Petitpretz, P. Poubeau, J. Cerrina, P. Duroux, and L. Drouet, Am. J. Med. **99**, 249 (1995).

[11] S. Eddahibi, N. Hanoun, L. Lanfumey, K. P. Lesch, B. Raffestin, M. Hamon, and S. Adnot, J. Clin. Invest. **105**, 1555 (2000).

[12] J. M. Launay, P. Herve, K. Peoc'h, C. Tournois, J. Callebert, C. G. Nebigil, N. Etienne, L. Drouet, M. Humbert, G. Simonneau, and L. Maroteaux, Nature Med. **8**, 1129 (2002).

[13] S. Eddahibi, M. Humbert, E. Fadel, B. Raffestin, M. Darmon, F. Capron, G. Simonneau, P. Dartevelle, M. Hamon, and S. Adnot, J. Clin. Invest. **108**, 1141 (2001).

[14] D. B. Badesch, V. F. Tapson, M. D. McGoon, B. H. Brundage, L. J. Rubin, F. M. Wigley, S. Rich, R. J. Barst, P. S. Barrett, K. M. Kral, M. M. Jobsis, J. E. Loyd, S. Murali, A. Frost, R. Girgis, R. C. Bourge, D. D. Ralph, C. G. Elliott, N. S. Hill, D. Langleben, R. J. Schilz, V. V. McLaughlin, I. M. Robbins, B. M. Groves, S. Shapiro, and T. A. Medsger, Jr., Ann. Intern. Med. **132**, 425 (2000).

PH was initially developed in HPN models.[15,16] Using modern transgenic approaches, it has been demonstrated that lung overexpression of prostacyclin synthase ameliorates HPN.[17]

Novel therapeutic approaches have been equally developed in HPN models. Hypoxia upregulates endothelin expression,[18] and endothelin receptor blockade attenuates HPN.[19] These findings and the fact that endothelin expression is increased in severe PH[20] have led to the use of endothelin receptor blockers in human severe PH.[21] Statins, which may represent an exciting and novel therapy in severe PH (Taraseviciene-Stewart et al., unpublished observations),[22] have equally been shown to improve hypoxic PH.[23] Finally, proteases may play either a beneficial or a detrimental role in HPN. Vascular elastic protease inhibition with elafin also ameliorates HPN,[24] whereas a metalloprotease inhibitor worsens HPN.[25]

We have described that hypoxia leads to upregulation of vascular endothelial growth factor (VEGF) in the lung and that this upregulation was dependent on the levels of nitric oxide (NO).[26] Because VEGF leads to nitric oxide and prostacyclin production, VEGF may be part of a lung response to acute and chronic hypoxia to dampen hypoxic vasoconstriction and hypoxic pulmonary vascular remodeling.[27] Indeed, adenovirus-induced lung overexpression of VEGF improves pulmonary vascular hemodynamics in the HPN rat model.[28]

[15] J. G. Gerber, N. F. Voelkel, A. S. Nies, I. F. McMurtry, and J. T. Reeves, *J. Appl. Physiol.* **49,** 107 (1980).

[16] T. Higenbottam, D. Wheeldon, F. Wells, and J. Wallwork, *Lancet* **1,** 1046 (1984).

[17] M. W. Geraci, B. Gao, D. Shepherd, M. D. Moore, J. Y. Westcott, L. A. Alger, R. M. Tuder, and N. F. Voelkel, *J. Clin. Invest.* **103,** 1509 (1999).

[18] K. Yamashita, D. J. Discher, J. Hu, N. H. Bishopric, and K. A. Webster, *J. Biol. Chem.* **276,** 12645 (2001).

[19] R. A. Bialecki, C. S. Fisher, B. M. Abbott, H. G. Barthlow, R. G. Caccese, R. B. Stow, J. Rumsey, and W. Rumsey, *Pulm. Pharmacol. Ther.* **12,** 303 (1999).

[20] A. Giaid, M. Yanagisawa, D. Langleben, R. P. Michel, R. Levy, H. Shennib, S. Kimura, T. Masaki, W. P. Duguid, and D. J. Stewart, *N. Engl. J. Med.* **328,** 1732 (1993).

[21] L. J. Rubin, D. B. Badesch, R. J. Barst, N. Galie, C. M. Black, A. Keogh, T. Pulido, A. Frost, S. Roux, I. Leconte, M. Landzberg, G. Simonneau, and T. E. Bosentan Randomized, *N. Engl. J. Med.* **346,** 896 (2002).

[22] T. Nishimura, J. L. Faul, G. J. Berry, L. T. Vaszar, D. Qiu, R. G. Pearl, and P. N. Kao, *Am. J. Respir. Crit. Care Med.* **166,** 1403 (2002).

[23] R. E. Girgis, D. Li, X. Zhan, J. G. M. Garcia, R. M. Tuder, P. M. Hassoun, and R. A. Johns, *Am. J. Physiol. Lung Cell,* submitted for publication.

[24] S. H. Zaidi, X. M. You, S. Ciura, M. Husain, and M. Rabinovitch, *Circulation* **105,** 516 (2002).

[25] A. Vieillard-Baron, E. Frisdal, S. Eddahibi, I. Deprez, A. H. Baker, A. C. Newby, P. Berger, M. Levame, B. Raffestin, S. Adnot, and M. P. d'Ortho, *Circ. Res.* **87,** 418 (2000).

[26] R. M. Tuder, B. E. Flook, and N. F. Voelkel, *J. Clin. Invest.* **95,** 1798 (1995).

[27] R. M. Tuder, J. Allard, and N. F. Voelkel, *Circulation* **94,** 164 (1996).

[28] C. Partovian, S. Adnot, B. Raffestin, V. Louzier, M. Levame, I. M. Mavier, P. Lemarchand, and S. Eddahibi, *Am. J. Respir. Cell Mol. Biol.* **23,** 762 (2000).

Notwithstanding the issues of relevance of HPN models to the human disease, chronic hypoxia in association with VEGF receptor blockade gives rise to severe, lethal PH in rats, whereas VEGF receptor blockade alone leads to medial thickening, mild elevation of pulmonary artery pressures, and emphysema.[29,30] The role of chronic hypoxia in this model, which closely parallels the human disease,[31] has been elucidated by Gleany et al. (unpublished observations). VEGF receptor blockade in rats in which one lung has been removed, thus causing redirection of pulmonary blood flow to the remaining pulmonary circulation, developed severe PH with endothelial cell proliferation. These studies indicate that increased pulmonary vascular shear stress common to chronic hypoxia and pneumectomy plays a critical role in the development of endothelial cell proliferation and severe PH.

Oxygen Regulation of Pulmonary Circulation in the Neonate

In fetal life, the pulmonary circulation is a high resistance system. The preacinar pulmonary vessels develop by 20 weeks gestation and the intra-acinar vessels form during late fetal life and after birth. Blood is shunted away from the lungs through a patent ductus arteriosus. At birth, there is a dramatic decrease in pulmonary pressures. This process is associated with both functional and structural changes taking place within the neonatal pulmonary circulation. Factors that help initiate the drop in pulmonary vascular resistance at birth include inflation of the lungs, increase in oxygen tension and pH, and changes in the concentration of specific vasoactive substances. At birth it has been shown that metabolites of NO, a potent vasodilator, are low and that levels of plasma endothelin (ET), a potent vasoconstrictor, are high. Within the first 5 days of life the NO metabolites increase and the plasma level of ET decreases significantly, concomitant with a drop in pulmonary arterial pressure.[32] Furthermore, at birth the vessel wall lumen, within the pulmonary circulation, increases in diameter. This occurs through changes in endothelial and smooth muscle length, resulting in decreased resistance in the pulmonary vasculature. The smooth muscle cells within the pulmonary vessels have marked plasticity early in postnatal life, with differences in myofilament concentrations.[33] Smooth muscle cell

[29] Y. Kasahara, R. M. Tuder, L. Taraseviciene-Stewart, T. D. Le Cras, S. H. Abman, P. Hirth, J. Waltenberger, and N. F. Voelkel, *J. Clin. Invest.* **106,** 1311 (2000).

[30] L. Taraseviciene-Stewart, Y. Kasahara, L. Alger, P. Hirth, G. G. McMahon, J. Waltenberger, N. F. Voelkel, and R. M. Tuder, *FASEB J.* **15,** 427 (2001).

[31] R. M. Tuder, M. Chacon, L. A. Alger, J. Wang, L. Taraseviciene-Stewart, Y. Kasahara, C. D. Cool, A. E. Bishop, M. W. Geraci, G. L. Semenza, M. Yacoub, J. M. Polak, and N. F. Voelkel, *J. Pathol.* **195,** 367 (2001).

[32] A. Endo, M. Ayusawa, M. Minato, M. Takada, S. Takahashi, and K. Harada, *Pediatr. Int.* **42,** 26 (2000).

[33] S. G. Haworth, *Exp. Physiol.* **80,** 843 (1995).

proliferation is also downregulated, and changes in collagen content and type occur as the animal ages.

Chronic hypoxia in the neonate may also impair septal development in the lung, leading to alveolar hypoplasia. The mechanisms by which this occurs are not completely understood, but have been shown to be associated with dysregulation of transforming growth factor (TGF)-β.[34] Vascular endothelial growth factor receptor 2 blockade has also been shown to cause impairment of endothelial cell growth, abnormal alveolar development, and PH.[35] The role of HIF-1 (hypoxia inducible factor) in the neonatal pulmonary circulation has not been well elucidated, but induction of the HIF-1 subunits, HIF-1α and HIF-2α, has been demonstrated in lung exposed to hypoxia.[36] HIF-2α has also been shown to be increased markedly in type 2 epithelial cells exposed to hypoxia.[37] HIF-2α knockouts are deficient in surfactant and have subtle abnormalities in capillary development, resulting in fatal respiratory distress at birth.[38] HIF-1 also regulates nitric oxide synthase and endothelin-1, genes involved in pulmonary circulation vasoresponsiveness.[39]

Chronic hypoxia is used as a model for PH in the neonate. The response to chronic hypoxia, with respect to the pulmonary bed, however, differs between the neonate and the adult animal. In neonatal rat, exposure to chronic hypoxia results in decreased levels of endothelial NO and NO metabolites. In contrast, the adult animal exposed to chronic hypoxia has increased levels of NO and NO metabolites.[40] Chronic hypoxia may also impair development of the alveolar septum, leading to alveolar hypoplasia.

Methodological Approaches in HPN

As stated earlier, HPN is characterized by abnormal hemodynamics, abnormal pulmonary vascular remodeling, and altered expression of molecules.

[34] A. G. Vicencio, O. Eickelberg, M. C. Stankewich, M. Kashgarian, and G. G. Haddad, *J. Appl. Physiol.* **93**, 1123 (2002).
[35] T. D. Le Cras, N. E. Markham, R. M. Tuder, N. F. Voelkel, and S. H. Abman, *Am. J. Physiol. Lung Cell Mol. Physiol.* **283**, L555 (2002).
[36] A. Y. Yu, M. G. Frid, L. A. Shimoda, C. M. Wiener, K. Stenmark, and G. L. Semenza, *Am. J. Physiol.* **275**, L818 (1998).
[37] M. S. Wiesener, J. S. Jurgensen, C. Rosenberger, C. Scholze, J. H. Horstrup, C. Warnecke, S. Mandriota, I. Bechmann, U. A. Frei, C. W. Pugh, P. J. Ratcliffe, S. Bachmann, P. H. Maxwell, and K. U. Eckardt, *FASEB J.* **17**, 271 (2003).
[38] V. Compernolle, K. Brusselmans, T. Acker, P. Hoet, M. Tjwa, H. Beck, S. Plaisance, Y. Dor, E. Keshet, F. Lupu, B. Nemery, M. Dewerchin, P. Van Veldhoven, K. Plate, L. Moons, D. Collen, and P. Carmeliet, *Nature Med.* **8**, 702 (2002).
[39] G. Semenza, *Biochem. Pharmacol.* **64**, 993 (2002).
[40] L. G. Chicoine, J. W. Avitia, C. Deen, L. D. Nelin, S. Earley, and B. R. Walker, *J. Appl. Physiol.* **93**, 311 (2002).

For the proper assessment of HPN, a comprehensive methodological approach is required.

Chronic Hypoxia Models of PH

Chronic hypoxia models rely on exposure of animals from days to several weeks (usually from 2 to 4 weeks) of hypobaric hypoxia (such as the experiments performed in Denver) or at normobaric levels using N_2 dilution of oxygen so as to create a hypoxic environment. The animals are housed in a Plexiglass chamber open to room air (normoxia group) or maintained at 10% FiO_2 (hypoxia groups) for 14 days. Hypoxia is maintained using a Pro:Ox Model 350 unit (Reming Bioinstruments, Redfield, NY) that controls a fractional concentration of O_2 in inspired gas by a solenoid-controlled infusion of N_2 (Roberts Oxygen, Rockville, MD) balanced against an inward leak of air through holes in the chamber. The chamber is opened once daily for drug administration. Water and rat chow are provided *ad lib.* The animals are maintained at 20–24° in a room with a 12:12-h light–dark cycle. This approach was used by Girgis *et al.*[23] to elucidate the protective role of sinvastatin in HPN.

Either approach creates a simulated altitude of 5000 m (17,000 ft) in a hypobaric chamber, with an inspired partial oxygen pressure of approximately 76 mm Hg. Normoxic controls at Denver altitude are at 1600 m with an inspired partial oxygen pressure of approximately 124 mm Hg, whereas control animals at sea level conditions of oxygen breathe a partial oxygen pressure of approximately 150 mm Hg.

Pulmonary Artery Hemodynamics

Pulmonary artery catheterization provides key data on pulmonary artery pressures and pulmonary vascular resistance *in vivo.* HPN is defined as an increase in pulmonary artery pressure. Although there is no clear definition of critical levels of PH in animals, chronic hypoxia leads to a doubling of mean pulmonary artery pressures from an average of 18 mm Hg to about 34 mm Hg. However, there are substantial species differences.[41] With the additional measurement of the cardiac output and the pulmonary capillary wedge pressure, pulmonary vascular resistance can be calculated (pulmonary vascular = pulmonary artery pressure − capillary wedge pressure / cardiac output) (Fig. 1).

[41] J. T. Reeves, W. W. Wagner, Jr., I. F. McMurtry, and R. F. Grover, *Int. Rev. Physiol.* **20,** 289 (1979).

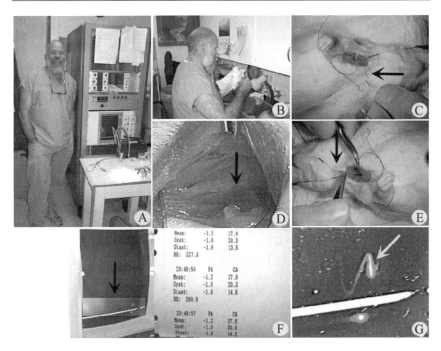

Fig. 1. Cardiopulmonary monitoring section of the Cardiovascular Research Laboratory of the University of Colorado Health Sciences Center. Kenneth Morris has been responsible for this section for more than 25 years, having participated in numerous important publications in the field of HPN (A). The procedure used to determine the pulmonary artery pressure starts with anesthesia (xylocaine and pentabarbital in this example, B). An incision is made along the trachea. The catheter is highlighted by the arrow in C. A close-up of the jugular vein is highlighted by the arrow in D. After a small incision in the jugular vein and ligation of the proximal segment, the catheter is flushed with phosphate-buffered saline containing heparin and then inserted into the jugular vein (E). The other end of the catheter is attached to a pressure transducer. Proper positioning of the catheter into the pulmonary artery is monitored online through a fluorescent screen (F). The highlighted rectangular area in the monitor (arrow) is shown in a close-up in G. Note the wave corresponding to the pulmonary artery pressures (G), which are then registered in a printout as mean, systolic, and diastolic over the course of catheterization.

Isolated Perfused Lung System

The most frequently isolated lung system is the isolated perfused rat lung model originally described by Anton Hauge[42] and modified by Ivan McMurtry.[43] With appropriate adjustment of the blood or artificial

[42] A. Hauge, *Acta Physiol. Scand.* **72,** 33 (1968).
[43] I. F. McMurtry, A. B. Davidson, J. T. Reeves, and R. F. Grover, *Circ. Res.* **38,** 99 (1976).

perfusate flow (ml/min) and ventilation parameters (tidal volume, respiratory rate, and end expiratory pressure), this model can be used for virtually any lung preparation. Studies with isolated rat, guinea pig, hamster, rabbit, and ferret lungs have been reported. The advantage of this model is that the lung can be studied in isolation from the systemic circulation and that the blood or perfusate flow can be manipulated. Thus, under constant flow conditions, any pressure rise constitutes a rise in pulmonary vascular resistance. Both precapillary artery and postcapillary venular constriction can be observed in this preparation, and the left atrial pressure can be controlled manner to mimic left ventricular failure or to study hydrostatic pulmonary edema.[44] Alveolar hypoxia can be implemented by switching the air reservoir from a room airgas mixture to one that contains 10, 5, 3, or 0% oxygen. The magnitude of the hypoxic pressor response depends on a number of experimental conditions, such as temperature of the lung and perfusate, the baseline perfusion pressure, and whether the lung is perfused with blood.[15] It remains remarkable that a normal hematocrit of the perfusate is not required for maximal responses to hypoxia and that the addition of a relatively small number of red blood cells to the artificial salt solution perfusion fluid increases the magnitude of the hypoxia-induced pulmonary vasoconstriction.

Hemolysate, not hemoglobin, potentiates the hypoxic pressor response and alters the vascular reactivity of this preparation.[45] Use of a physiological salt solution without plasma proteins (the oncotic pressure is generated by a chemically inert starch) allows the study of low concentrations of vasoactive substances, as the albumin binding of active compounds is avoided and the molecules have direct access to vascular and extravascular receptors. Lungs perfused with a salt solution have lower baseline perfusion pressures and become edematous more easily than blood-perfused lungs. Applications of this time-proven model are virtually endless and include the study of vasoactive compounds, shear stress-dependent production of mediators, consequences of gene knockouts, chemotaxis of inflammatory cells to the lung microcirculation, or gene expression of immediate-early genes such as cyclooxygenase-2 (COX-2). Figure 2 shows the individual elements of the perfusion and ventilation system.[46] Several investigators have used the isolated lung preparation to explore the mechanism of hypoxic pulmonary vasoconstriction, mostly applying a pharmacological approach using ion channel blockers, inhibitors of nitric oxide,

[44] A. Sakai, S. W. Chang, and N. F. Voelkel, *J. Appl. Physiol.* **66,** 2667 (1989).
[45] N. F. Voelkel, J. D. Allard, S. M. Anderson, and T. J. Burke, *J. Appl. Physiol.* **86,** 1715 (1999).
[46] S. W. Chang and N. F. Voelkel, *Methods Enzymol.* **187,** 599 (1990).

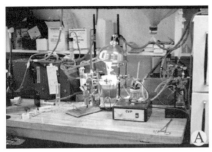

FIG. 2. Isolated perfused lung system in the laboratory of Dr. Norbert Voelkel.

or receptor blockers, but the isolated lung can also be useful to examine nutritional or dietary manipulations or the effects of infections on lung vascular tone control.

Pulmonary Vascular Morphology and Morphometry

As stated previously, pulmonary vascular remodeling is a hallmark of pulmonary hypertensive states. In the author's view, the predominant form of remodeling in mild to moderate PH is vascular medial thickening due to smooth muscle cell hyperplasia/hypertrophy.[5,47] To access medial vascular thickening, most of the studies have resorted to measuring percentage medial thickening or the fraction of peripheral pulmonary arteries that have acquired a muscular media. The quantitative approach has been facilitated by the more recent incorporation of immunohistochemical detection of smooth muscle cell markers, such as smooth muscle α-actin. The protocol used to quantify percentage medial thickening used in our studies is described later.[30]

Immunohistochemistry for smooth muscle marker smooth muscle α-actin: 1/150 dilution; histological sections are heated in $1\times$ citrate buffer with times varying from 15 to 25 min, followed by H_2O_2 block of endogenous peroxidase for 15 min, avidin–biotin block for 10 min each, and block with rabbit or mouse 10% serum for 30 min. The primary antibody is incubated at $37°$ for approximately 30 min, followed by Quickkit (Vector, Burlingame, CA) for 10 min at room temperature. This kit involves a second-step incubation with a universal biotinylated horse antibody for 10 min, followed by streptavidin–peroxidase for 5 min and development with diaminobenzidine with hydrogen peroxide, according to the manufacturer's protocol.

[47] R. M. Tuder and A. L. Zaiman, *in* "Pulmonary Circulation" (A. Peacock and L. Rubin, ed.). In press (2004).

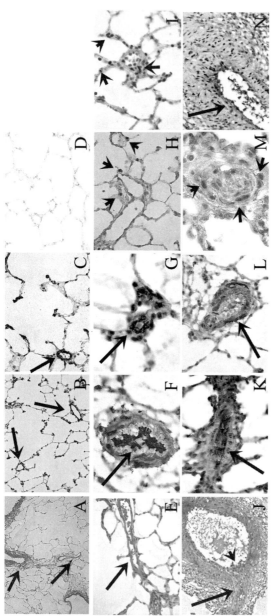

FIG. 3. Morphology and immunohistochemistry of normal (A–D), chronic hypoxic (E–I), and chronically hypoxic rats that have also received the VEGF receptor blocker SU5416 for 3 weeks (J–N). Normal pulmonary arteries progressively taper, losing their medial smooth muscle layer prior to the generation of the last segment of terminal bronchioles in the rat (A, arrows). The normal endothelial cell layer forms a characteristic monolayer, which is highlighted by immunohistochemical detection of the endothelial cell marker factor VIII-related antigen (arrow, B). The lack of muscularization in peripheral branches of the pulmonary artery is best illustrated in sections immune stained for smooth muscle cell actin (arrows, C). The normal rat lung has scarce apoptotic cells as demonstrated by the immunohistochemical detection of active caspase 3 (D). A chronically hypoxic rat lung shows the characteristic muscularization of intraalveolar branches of pulmonary arteries (arrow, E). Endothelial cells in chronically hypoxic rat lungs form a monolayer as in normal lungs. However, due to the component of vasoconstriction and remodeling, the intima appear crenated (arrow), leading to a reduction of vascular lumen and approximation of the intima layer (factor VIII-related antigen immunohistochemistry, F). The peripheral muscularization is best seen in G by smooth muscle α-actin immunohistochemistry (arrow, G). Scanty extracellular cells appear to be in the process of apoptosis, as highlighted by active caspase 3 immunohistochemistry (arrowheads, H). Vascular

Morphometry is performed on lung slides of two randomly chosen animals in each treatment group. The percentage medial thickness is calculated with the formula (external diameter−internal diameter/external diameter) ×100 in slides immunostained with the smooth muscle α-actin antibody. Only vessels sectioned longitudinally or with an approximately circular profile are analyzed for assessment of medial thickness. The diameter of pulmonary arteries is determined by the use of ImagePro software.

Data shown in Fig. 3 represent the findings of rats exposed to normoxia or chronic hypoxia with or without VEGF receptor blocker SU5416 treatment. The average medial arterial thickness in rats under normoxic conditions is 23%. VEGF receptor blockade resulted in a 3% increase in average medial thickness. Under chronic hypoxic conditions, the percentage medial thickness in control rats was raised to 28%, whereas in combination with SU5416 treatment, the medial thickness was 35%.

A potential pitfall is the orientation of the pulmonary vessel in the section plane. We have opted to perform the measurements in almost circular profiles and have excluded oval or longitudinal profiles from our quantitative analysis. Furthermore, ours and most of the studies in the field do not perfuse the pulmonary arteries with a fixative (formaldehyde or paraformaldehyde) at physiological perfusion pressures (approximately 18 mm Hg), which would lead to fixation of pulmonary arteries in their physiological morphology.

An alternative approach is to quantify the number of peripheral muscularized pulmonary arteries, used by Fagan et al.[48] to compare the pulmonary vascular remodeling of eNOS wild-type or eNOS-/-mice in normoxia or chronic hypoxia. Their experimental approach was based on the identification of vessels at the level of alveolar ducts (<25 μm), and circumferential staining with myosin (dark brown) was determined as none,

[48] K. A. Fagan, B. W. Fouty, R. C. Tyler, K. G. Morris, L. K. Hepler, K. Sato, T. D. Le Cras, S. H. Abman, H. D. Weinberger, P. L. Huang, I. F. McMurtry, and D. M. Rodman, J. Clin. Invest. 1–3, 291 (1998).

and septal cells are proliferating since they express the proliferating nuclear cell antigen (arrows, I). Chronically hypoxic rats treated with the VEGF receptor blocker SU5416 develop severe PH with endothelial cell proliferation the following death of endothelial cells early in hypoxic exposure and outgrowth of apoptosis-resistant endothelial cells.[30] The lungs shown marked intima obliteration of pulmonary arteries (arrow, J), most notably in branching points (short arrow, J). The proliferated endothelial cells are highlighted by factor VIII-related antigen immunostaining (arrow, K), which are negative for smooth muscle cell actin (arrow, L). In line with the severity of the PH, there is marked medial thickening of the vascular media, in excess of that observed with hypoxia alone. This process requires ongoing apoptosis of endothelial cells, which express active caspase 3 (arrows, M). Concomitantly, there is active proliferation in the vascular media and endothelium of pulmonary arteries (arrow, proliferating nuclear cell antigen immunohistochemistry).

partial (<75%), and full (>75%). For each animal, at least 10 separate fields were studied and a minimum of 50 vessels per animal characterized. Additionally, at least 10 vessels per animal corresponding to the terminal bronchiole (~40 μm) were identified and images digitized (Nikon Instruments, Melville, NY; and Optitronics Engineering, Goleta, CA), and medial thickness (area between intima and adventitia) and vessel diameter were measured using Image 1.61 software (National Institutes of Health). Neighboring vessels were counted and the number of alveoli per vessel was determined.

Exposure to chronic hypoxia led to a decrease in nonmuscularized vessels from 60 to 40% in control mice, whereas partial or completely muscularized arteries increased from about 40 to 60%. eNOS-/-mice had a larger number of peripheral vessels that were muscularized under normoxia and a larger increase in muscularization under chronic hypoxia than wild-type mice.[48]

A combination of both approaches has been used by Girgis et al.[23] where rats were exposed to normoxia or chronic hypoxia for 15 days and received sinvastatin or vehicle. In their study, standard lung sections through the hilum were stained with a α-smooth muscle actin antibody (1:100, Sigma, St. Louis, MO). Peripheral pulmonary arteries associated with alveolar sacs and ducts were classified as nonmuscular (0–25% of circumference with actin staining), partially muscular (26–75%), and fully muscular (>75% of circumference). Between 50 and 100 vessels were counted for each animal. The percentage medial thickness (% MT) of muscularized arteries measuring from 50 to 200 μm in external diameter (ED) was determined using an Olympus-BHS microscope coupled to an MTI color video camera (DAGE-MTI, Michigan City, IN) and an I Cube video grabber board. Measurements were obtained with ImagePro Plus Software (Media Cybernetics, Silver Spring, MD) after calibration with an Olympus 0.01-mm calibration slide. Only arteries with a circular or quasi-circular outline were examined. An average of three measurements was taken for medial thickness. Percentage medial thickness was calculated as (MT × 2/ED) × 100. A total of 20 arteries in consecutive fields were examined for each animal.

Results showed that rat lungs under normoxia have 82% of nonmuscular arteries, whereas 18% are either partially or completely surrounded by a muscular medial layer. Under chronic hypoxia, nonmuscularized arteries decrease to 32% and muscularized arteries increase to 68%. Sinvastatin treatment of chronically hypoxic rats led to only 49% of partially or completely muscularized pulmonary arteries. Between 469 and 646 vessels were examined in this study. The protection imparted by sinvastatin on vascular remodeling was paralleled by a proportional protection in the rise of pulmonary artery pressures caused by chronic hypoxia.[23]

Right Ventricular Hypertrophy

The measurement of right ventricular hypertrophy correlates very closely with pulmonary artery pressures and pulmonary vascular remodeling in HPN. The heart is removed in block with the lungs and dissected free from its large vessels. The heart is then weighed and the right ventricular free wall (RV) is dissected through the septal plane, leaving the left ventricular plus interventricular septum (LV+S) as a single unit. Both the right ventricular free wall and the septum+left ventricle are weighed. Data are expressed as the ratio of RV/LV+S.

Figure 4 shows the regression analysis of measurements of pulmonary artery pressures and RV/LV+S ratios for rats (six per group) treated with SU5416 or vehicle under normoxia or chronic hypoxia conditions.[30] The correlation coefficient for this experiment is of 0.84 ($p < 0.01$). RV/LV+S S ratios of normoxic control, normoxic+SU5416, hypoxic control, and hypoxic+SU5416 were on average 0.236, 0.24, 0.276, and 0.478, respectively. Figure 5 shows a normoxic control and a chronically hypoxic, SU5416-treated heart side by side. Note the dramatic increase in right ventricular mass and the pronounced right atrial and ventricular dilation. Surprisingly, a close correlation was not found between right ventricular hypertrophy and pulmonary artery pressures in rats exposed to normoxia and treated with SU5416, as these animals showed a mild elevation of pulmonary artery pressures. However, right ventricular hypertrophy and dilation were pronounced in animals exposed to chronic hypoxia and SU5416 treatment. In fact, these rats usually die of their severe PH.

Gene Chip Array

Gene microarray technology[49] now permits the analysis of the gene expression profile of lung tissue obtained from patients with PPH with that found in normal lung tissue. Because the vascular lesions are distributed homogeneously throughout the entire lung, a tissue fragment of the lung is likely representative of the whole lung. RNA extracted from such fragments is likely to provide meaningful information regarding the changes in the gene expression pattern in PPH when compared with structurally normal lung tissue. By examining a sufficient number of lung tissue samples, one can model the range of normality. Methods exist for determining coordination in expression data using cluster expression profiles. Cluster analysis can give clues to the pathogenesis by displaying genes whose expression is altered in a coordinate manner. Finally, an important goal is to discern sets of genes that differentiate between normal and disease

---

[49] D. J. Lockhart and E. A. Winzeler, *Nature* **405,** 827 (2000).

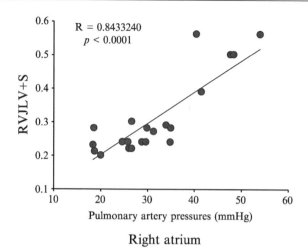

Right atrium

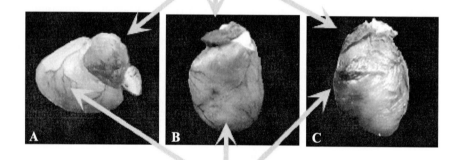

Right ventricle

FIG. 4. (A) Correlation between mean pulmonary arteries pressure and right ventricular mass in multiple experiments reported in.[30] B–E: Heart images of control (B), chronically hypoxic (C) and chronically hypoxic, SU-5416-treated, rats (D).

states—or discrimination analysis. Building discrimination models has a long history in statistical pattern recognition and machine learning and has been applied to cancer paradigms using gene expression data.[50] For our study, we used Affymetrix oligonucleotide microarrays (human FL) to characterize the expression pattern in lung tissue obtained from six patients

[50] T. R. Golub, D. K. Slonim, P. Tamayo, C. Huard, M. Gaasenbeek, J. P. Mesirov, H. Coller, M. L. Loh, J. R. Downing, M. A. Caligiuri, C. D. Bloomfield, and E. S. Lander, Science 286, 531 (1999).

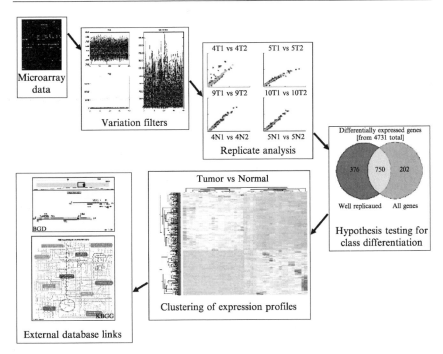

Fig. 5. Stepwise description of the gene microarray approach. Using this schema, we first filter genes by variability testing. Genes which are either 1) never called present or 2) exhibit no variability across all experimental conditions are excluded from further analysis, as they add no meaning to the data set. Replicates are encouraged to decrease variability. Clustering is performed and genes of interest are further examined by links to external databases.

with primary PH (PPH), including two patients with the familial form of PPH (FPPH), and from six patients with histologically normal lungs.[51]

Although the number of patient samples was small, gene dendogram, cluster analysis, and concordant expression differences show that categorical and robust differences exist in the profile of expressed genes among structurally normal lungs, lungs from patients with sporadic PPH, and lungs from patients with FPPH. Whole tissue total RNA expression profiles demonstrate striking differences in the expression signatures between sporadic and familial PPH. Importantly, the differences in expression profiles are complemented by independent gene mutation analysis. In summary, microarray gene expression analysis and profiling is a useful molecular tool that provides a better characterization and understanding of the pathobiology of distinct clinical phenotypes of PH.

[51] M. W. Geraci, M. Moore, T. Gesell, M. E. Yeager, L. Alger, H. Golpon, B. F. Gao, J. E. Loyd, R. M. Tuder, and N. F. Voelkel, *Circ. Res.* **88**, 555 (2001).

Advances in Expression Profiling in Human Diseases

Many of the advances in microarray application are a result of the application of expression profiling in the field of cancer. The study of gene expression in human tumor tissues has advanced our understanding of cancer biology greatly; these advances have in turn led to improvements in the diagnosis and treatment of cancer. This progress has been enabled by the wide application of microarray technology, which allows the efficient and accurate measurement of the expression of thousands of genes simultaneously. When combined with advanced methods for data analysis, microarrays have proven to be extremely useful in the classification of histologically indistinct tumor types[52] and in deducing sets of previously unappreciated tumor subtypes responsible for variations in invasiveness[53] and outcome.[54] Expression profiles have also been used to provide insight into metastatic processes and to reflect metastatic potential[55] and tissue of origin,[54] as well as susceptibility to chemotherapy agents.[56]

A significant challenge to this type of study is the collection of biological material with sufficient homogeneity and of sufficient quantity and quality for microarray study. Biopsy specimens from patients with early stage disease tend to be small, and routine histological preservation (formalin fixation) is incompatible with nucleic acid extraction. To circumvent these issues, our laboratory as well as others have turned to peripheral blood mononuclear cells (PBMCs) as a source of genetic material, which has provided insights in the study of autoimmune diseases such as lupus[57] and arthritis[58]; we have successfully applied PBMC gene expression profiling to the study of PH. Systemic diseases such as PH have been shown to

[52] J. Khan, J. S. Wei, M. Ringner, L. H. Saal, M. Ladanyi, F. Westermann, F. Berthold, M. Schwab, C. R. Antonescu, C. Peterson, and P. S. Meltzer, *Nature Med.* **7,** 673 (2001).
[53] M. Bittner, P. Meltzer, Y. Chen, Y. Jiang, E. Seftor, M. Hendrix, M. Radmacher, R. Simon, Z. Yakhini, A. Ben Dor, N. Sampas, E. Dougherty, E. Wang, F. Marincola, C. Gooden, J. Lueders, A. Glatfelter, P. Pollock, J. Carpten, E. Gillanders, D. Leja, K. Dietrich, C. Beaudry, M. Berens, D. Alberts, and V. Sondak, *Nature* **406,** 536 (2000).
[54] A. Bhattacharjee, W. G. Richards, J. Staunton, C. Li, S. Monti, P. Vasa, C. Ladd, J. Beheshti, R. Bueno, M. Gillette, M. Loda, G. Weber, E. J. Mark, E. S. Lander, W. Wong, B. E. Johnson, T. R. Golub, D. J. Sugarbaker, and M. Meyerson, *Proc. Natl. Acad. Sci. USA* **98,** 13790 (2001).
[55] S. Ramaswamy, K. N. Ross, E. S. Lander, and T. R. Golub, *Nature Genet.* **33,** 49 (2003).
[56] S. Dan, T. Tsunoda, O. Kitahara, R. Yanagawa, H. Zembutsu, T. Katagiri, K. Yamazaki, Y. Nakamura, and T. Yamori, *Cancer Res.* **62,** 1139 (2002).
[57] V. Rus, S. P. Atamas, V. Shustova, I. G. Luzina, F. Selaru, L. S. Magder, and C. S. Via, *Clin. Immunol.* **102,** 283 (2002).
[58] J. Gu, E. Marker-Hermann, D. Baeten, W. C. Tsai, D. Gladman, M. Xiong, H. Deister, J. G. Kuipers, F. Huang, Y. W. Song, W. Maksymowych, J. Kalsi, M. Bannai, N. Seta, M. Rihl, L. J. Crofford, E. Veys, F. De Keyser, and D. T. Yu, *Rheumatology (Oxford)* 759 (2002).

affect the cytokine and chemokine milieu of circulating cells. These changes include alterations in growth factors such as VEGF and potent vasoregulatory agents such as nitric oxide, prostacyclin, and endothelin. Cells that comprise the peripheral blood in patients with PH are exposed continuously to this altered cytokine environment, and we had hypothesized that this exposure would result in informative modulation of gene expression by PBMCs.

Interpretation of Data

In addition to the entire database challenges (e.g., making the controls and units commensurate across comparisons), no traditional distance metric (e.g., Euclidean or Pearson correlation) is appropriate to use on thousands of highly correlated genes drawn on nonnormal distributions. Our collaborators have developed a Bayesian metric that uses the database itself to estimate the underlying joint distribution.[59] We have implemented this metric in a database retrieval tool, which we will deploy on our own datasets and develop further to include searching public databases as they come online.

The classical corrections for multiple testing (e.g., Bonferroni) make very conservative assumptions and therefore lack statistical power. We have adopted a recent innovation in the statistical literature[60] called the "false discovery rate" (FDR), which overcomes this problem. Instead of attempting to fix a family error rate for all of the hypotheses taken together (as Bonferroni does), FDR allows us to set the proportion of null hypotheses rejected incorrectly. Second, it is important to recognize that the expression of many (perhaps even most) genes does not appear to be drawn from a normal distribution. With respect to statistical considerations, we have developed a novel approach to addressing the particular statistical difficulties inherent in analyzing gene chips. In brief, we prefer nonparametric statistics (the Mann–Whitney U test for two classes or its multiclass equivalent, the Kruskal–Wallis test). We then combine this nonparametric test with the false discovery rate correction for multiple testing. We calculate the variances for each gene. The null hypothesis is that these variances represent random and normally distributed noise. We can then compute the statistic $W=(N-1)$ s2/median (s2), where N is the number of observations of the gene, which is approximately χ^2 distributed with N-1 degree of freedom.[61] We calculate a p value for rejecting the null hypothesis that

[59] L. Hunter, R. C. Taylor, S. M. Leach, and R. Simon, *Bioinformatics*. **17**(Suppl. 1), S115 (2001).

[60] Y. Benjamini and J. Lindenstrauss, American Mathematical Society, Providence, 1999.

[61] R. V. Hogg and A. T. Craig, Prentice Hall, Upper Saddle River, 1995.

the gene did not vary and perform the false discovery rate multiple testing correction,[62] setting the false discovery rate to 10%. This results in a list of genes with significantly greater variation than the median variation gene, with at most 10% of that list including genes having true variation less than or equal to median variation. These preprocessing steps screen out genes with low variance and low mRNA level expression measurements. The remainder of the genes are then fed to the Kruskal–Wallis or ANOVA test and FDR correction (10% in this case), producing a final list of genes whose expression levels are significantly different between at least two samples. This strategy for the evaluation of expression data is detailed in elsewhere.[9]

Strategy for Validation of Array Results for Individual Genes

We have adopted a very rigorous validation and follow-up protocol for array data. First, experiments are performed in parallel and in duplicate for the reason stated earlier. Briefly, we know from published data and from our own experience that Affymetrix arrays have an inherent "false-positive" rate of approximately 1%. Therefore, performing arrays in duplicate and only analyzing changes concordant in all two samples decreases the false-positive rate to $(0.01)^2$ or 1:10,000 genes and is the preferred method to analyze data.[49] Second, we validate gene changes by using quantitative polymerase chain reaction on a select population of genes. Third, if reagents are available, we use *in situ* hybridization and immunohistochemistry to detect changes in expression, which assists us in determining whether the transcriptional changes are also seen at the level of protein translation. We have used this strategy to validate findings from microarrays in examining normal human lung and in diseases such as primary PH, emphysema, interstitial lung disease, and lung cancer.[51,63–67]

[62] Y. Benjamini and Y. Hochberg, *J. Res. Stat. Soc.* **B57**, 289 (1995).
[63] G. P. Cosgrove, M. I. Schwarz, M. W. Geraci, K. K. Brown, and G. S. Worthen, *Chest* **121**, 25S (2002).
[64] H. A. Golpon, M. W. Geraci, M. D. Moore, H. L. Miller, G. J. Miller, R. M. Tuder, and N. F. Voelkel, *Am. J. Pathol.* **158**, 955 (2001).
[65] M. W. Geraci, B. Gao, Y. Hoshikawa, M. E. Yeager, R. M. Tuder, and N. F. Voelkel, *Respir. Res.* **2**, 210 (2001).
[66] A. C. Spalding, R. M. Jotte, R. I. Scheinman, M. W. Geraci, P. Clarke, K. L. Tyler, and G. L. Johnson, *Oncogene* **21**, 260 (2002).
[67] M. Sugita, M. Geraci, B. Gao, R. L. Powell, F. R. Hirsch, G. Johnson, R. Lapadat, E. Gabrielson, R. Bremnes, P. A. Bunn, and W. A. Franklin, *Cancer Res.* **62**, 3971 (2002).

[6]   Detection of Oxygen Sensing During Intermittent Hypoxia

By Nanduri R. Prabhakar, Ying-Jie Peng, Jeffrey L. Overholt, and Ganesh K. Kumar

Introduction

Intermittent hypoxia is associated with a number of clinical conditions, including sleep apneas, apneas in premature infants, asthma, and pulmonary fibrosis. Epidemiologic and cross-sectional studies on recurrent apnea patients have identified a strong association between apneas and serious cardiovascular disturbances, including the development of hypertension.[1] Although apneas are associated with hypoxia as well as hypercapnia, studies on experimental animal models suggest that hypoxia is the primary stimulus for developing hypertension and increased sympathetic activity.[2]

Peripheral chemoreceptors, especially carotid bodies, sense the changes in arterial oxygen. Carotid bodies are located at the bifurcation of the common carotid artery and receive sensory innervation from the carotid sinus nerve, a branch of the 12th cranial nerve (glossopharyngeal). Hypoxemia augments the sensory discharge of the carotid bodies, which in turn stimulates breathing and increases sympathetic nerve activity as well as blood pressure via reflex mechanisms involving the central nervous system. Carotid bodies constitute the "frontline" defense system during apneas even more so than other respiratory chemoreceptors (e.g., central CO_2 sensors) because the circulation time from lungs to carotid bodies is relatively shorter ($\sim$6 s) than to the central chemosensitive areas. Thus, during apneic episodes (which usually last anywhere from 10 to 15 s), the carotid bodies have already responded to the hypoxic blood before the central areas are exposed. Consequently, it has been suggested that chronic apnea syndromes lead to altered oxygen sensing in the carotid body, leading to increased sympathetic activity and hypertension (see Prabhakar[3] for references). Therefore, assessing the mechanisms by which recurrent hypoxia affects oxygen sensing by the carotid body is important in understanding the progression of cardiovascular disease associated with recurrent apnea.

---

[1] F. J. Nieto, T. B. Young, B. K. Lind, E. Shahar, J. M. Samet, S. Redline, R. B. D'Agostino, A. B. Newman, M. D. Lebowitz, and T. G. Pickering, *JAMA* **283**, 1829 (2000).

[2] E. C. Fletcher, G. Bao, and C. C. Miller, *J. Appl. Physiol.* **78**, 1516 (1995).

[3] N. R. Prabhakar, *J. Appl. Physiol.* **90**, 1986 (2001).

Rat Model of Intermittent Hypoxia

Studies on human subjects with recurrent apneas provide an indirect assessment of chemoreceptor function. Animal models mimicking the effects of intermittent hypoxia associated with apneas are ideal for more directly assessing the chemoreceptor activity and elucidating the underlying mechanisms. Currently, two animal models are available. In one model,[4] dogs were subjected to repeated episodes of upper airway occlusion during sleep, and in the other,[3] rats were subjected to intermittent hypoxia (~20 s of 5% inspired oxygen, nine episodes per hours, 8 h/day). Of these, the rat model is more ideally suited to study chemoreceptor function during intermittent hypoxia for a number of reasons. First, rats can be subjected with relative ease to intermittent hypoxia lasting hours, days, or weeks. Second, the morphology and physiology of the carotid body have been investigated extensively in this species. Third, cellular mechanisms can be best addressed using this species, as the methods for maintaining dissociated cell cultures of carotid bodies are well established. Finally, rats are more cost effective than dogs. More importantly, rats subjected to intermittent hypoxia (30 days) exhibit increases in blood pressure and sympathetic nerve activity[5] similar to that reported in patients with chronic recurrent apneas.[6]

Carotid bodies are composed primarily of two cell types: type I (also called glomus cells) and type II (also called sustentacular cells). The sensory complex for oxygen is composed of an afferent nerve fiber innervating the glomus cell. Two distinct mechanisms are associated with an increase in sensory discharge by hypoxia, namely transduction and transmission. While the transduction process seems to involve changes in ionic conductance(s) in glomus cells, transmission is associated with the release of neurotransmitters from glomus cells.[7] This article describes the methods for (a) exposing the animals to intermittent hypoxia, (b) monitoring the sensory activity of carotid bodies *in vivo,* (c) recording ionic conductance(s) from glomus cells in the slice preparation, and (d) monitoring transmitter changes in the carotid bodies.

[4] R. J. Kimoff, D. Brooks, R. L. Horner, L. F. Kozar, C. L. Render-Teixevia, V. Champagne, P. Mayer, and E. A. Phillipson, *Am. J. Respir. Crit. Care Med.* **156,** 886 (1997).
[5] E. C. Fletcher, *Respir. Physiol.* **119,** 189 (2000).
[6] V. K. Somers and F. M. Abboud, *Sleep* **16,** S30 (1997).
[7] N. R. Prabhakar and J. L. Overholt, *Respir. Physiol.* **122,** 209 (2000).

Experimental Methods

Exposure of Rats to Chronic Intermittent Hypoxia

The Institutional Animal Care and Use Committee of Case Western Reserve University approved all animal protocols and surgical procedures described in this article.

Materials

Intermittent hypoxia chamber (dimensions $0.62 \times 0.55 \times 0.29$ m³; see Fig. 1A)
Gas tanks (100% N_2, 21% O_2, Praxair)
Animal cages with rat chow and water bottle
O_2 analyzer (Beckman Model OM-11)
CO_2 analyzer (Beckman LB-2)

Procedure

1. Male Sprague–Dawley rats weighing 250–300 g are housed in feeding cages and placed in the chamber. The animals are unrestrained, freely mobile, and fed *ad libitum*.

2. Exposure to intermittent hypoxia is accomplished by flushing the chamber with alternating cycles of nitrogen interspersed with room air so that the inspired O_2 reaches 5% O_2 during hypoxia. The duration of gas flow during each hypoxic and normoxic episode is regulated using timed solenoid valves. The paradigm of intermittent hypoxia consists of 15 s

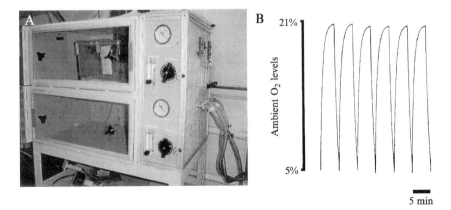

FIG. 1. (A) The intermittent hypoxia chamber used for conditioning animals. (B) Changes in ambient O_2 levels during intermittent hypoxia are shown.

hypoxia followed by 5 min of normoxia, nine episodes per hour, 8 h/day. Ambient O_2 and CO_2 levels in the chamber are monitored continuously by gas analyzers and recorded on a strip chart recorder. An example of the changes in ambient O_2 level during intermittent hypoxia is presented in Fig. 1B.

3. Animals are exposed to intermittent hypoxia between 9:00 AM and 5:00 PM (8 h) for 10 consecutive days.

Note. The lag time to reach 5% O_2 is from 68 to 75 s and from 70 to 85 s to reach 21% O_2 during normoxia. The paradigm of 15 s hypoxia is based on the average duration of apneic episodes encountered in human subjects.

4. Acute experiments are performed in the morning following the 10th day of intermittent hypoxia exposure.

Recording of Carotid Body Sensory Activity *In Vivo*

Surgical Methods

Materials

Urethane (Sigma)
Normal saline
Heparin
Ethyl alcohol
Mineral oil
Pancuronium bromide
Small animal surgical board
Intravenous catheter (PE 50 tubing)
Arterial catheter (PE 60 tubing)
Trachea catheter (PE 240 tubing)
Animal respirator (Harvard Apparatus, Model 683)
High-temperature cautery (Conmed, CO)
2-0 and 4-0 silk threads (Ethicon)
Binocular dissecting microscope (Wild; Germany)
Surgical instruments: Fine surgical scissors, Mayo scissors, fine forceps, dissecting forceps, hemostats, small animal retractors, scissors, forceps, scalpel handle and blade, finer clamps, electric clipper
1-cc, 5-cc syringes
Cotton-tipped applicators
Gauze sponges
Heating pad (Gaymar)
Heating lamp

Surgical Preparation

1. Anesthetization of the rat. A male Sprague–Dawley rat weighing between 250 and 300 g is anesthetized with urethane (1.2 g/kg) in distilled water administrated by an intraperitoneal injection. Supplemental doses (15% of initial dose) of urethane are given as needed as judged by abolition of the hind limb withdrawal reflex in response to noxious pinching of toes.

2. Preparation of the rat. Hair from the surgical field (neck, inguen) is removed using an electric clipper, and the area is disinfected with 70% alcohol. The rat is positioned on its back on an animal surgical board covered by a heating pad. The head of the rat is mounted to a head holder. The limbs are stretched using slings fixed to the operating board. The rectal temperature of the rat is monitored by a thermistor probe and maintained at 37–38° by the heating pad.

3. Intubation of trachea. A midline 2-cm skin incision is made in the subhyoid region with a scalpel. The soft tissue and muscles are blunt dissected to expose the trachea along with the esophagus. Loose ligatures (A 2-0 silk thread) are placed under the trachea and esophagus. A 2-mm incision is made in the trachea using fine surgical scissors. A tracheal catheter (PE 240 tubing) is inserted into the trachea toward the lungs, and the ligature is tightened to secure the catheter in the trachea.

4. Cannulation of artery and vein. A 3-cm longitudinal incision is made in the inguinal region. The femoral artery and vein are exposed. Two loose ligatures (4-0 silk thread) are made around the artery. The posterior ligature is tightened to occlude the artery, and a small incision is made with fine surgical scissors. The arterial catheter, filled with dilute heparin–saline (20 units heparin/ml), is pushed into the artery toward the abdominal aorta, and the catheter is secured by tightening the ligature. A similar procedure is done to cannulate the femoral vein. The incision is covered by warm, moist gauze.

5. Mechanical ventilation. The tracheal cannula is connected to a rodent respirator (Harvard Apparatus). The rat is maintained on artificial respiration with air at a tidal volume of 1.6 ml and a rate of 70 strokes/min. The animal is paralyzed with an intravenous administration of pancuronium bromide (2.5 mg/kg/h).

Note. Ventilating rat with 50% O_2 was found to maintain blood pressure during the course of experiments lasting a few hours.

6. Exposure of carotid sinus nerve. The trachea and esophagus, above the site of the tracheal cannula, are ligated, sectioned, and retracted rostrally. A retractor is placed in the neck region to expose carotid bifurcation. A loose ligature (2-0 silk thread) is placed around the right

external carotid artery and is retracted gently laterally. Using the dissecting microscope, the superior cervical sympathetic ganglion is identified and removed using fine surgical scissors. The carotid sinus nerve is identified, dissected free of connective tissue, and transected where it joins the glossophareyngeal nerve. Nerve branches emanating from bifurcation of the common carotid artery coursing toward the carotid body are cut to eliminate baroreceptor activity. The carotid sinus nerve is then desheathed and immersed in warm mineral oil to prevent drying.

Monitoring Physiological Variables

Equipment. Blood pressure transducer (Grass Instrument, PT300), AC/DC amplifier (Grass Instrument, P122), strip chart recorder (Dash 10, Astro-Med Inc.), and blood gas analyzer (ABL 5, Copenhagen, Denmark).

Procedure

7. Recording arterial blood pressure. The femoral artery catheter is connected to the blood pressure transducer, amplified, and recorded on a strip chart recorder.

8. Measurements of arterial blood gases. Arterial blood samples (200 μl) are collected periodically in heparinized syringes, and arterial pO_2, pCO_2, and pH are analyzed using a blood gas analyzer.

Recording of Carotid Chemoreceptor Activity

Equipment. Unipolar platinum–iridium electrode, ground electrode (silver–silver chloride), AC preamplifier (Grass Instrument, P511K), oscilloscope, audio monitor, and window discriminator.

Procedure

9. The carotid sinus nerve is placed on a unipolar platinum–iridium electrode, and the reference electrode made of silver–silver chloride is placed in the neck muscle. Electrical activity from the sinus nerve is amplified (Grass Instruments P511K), with a bandwidth of 100–3000 Hz, and displayed on an oscilloscope (Tektronix 5B12N). Action potentials above the baseline noise are converted to standardized pulses using a window discriminator (Winston Electronics Co., RAD II-A) and fed into a rate meter (CWE Inc., RIC-830) to display the magnitude of the sensory discharge. Signals from the rate meter, as well as action potentials, along with the blood pressure signals are recorded on a chart recorder. An example of carotid body sensory activity in response to hypoxia (inspired oxygen 12%) in control and intermittent hypoxia (10 days)-exposed rats is shown in Fig. 2.

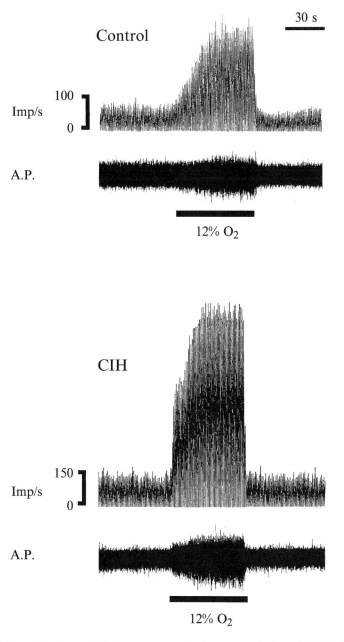

F𝐈G. 2. Examples of carotid body sensory activity in response to hypoxia (12% O_2) in a control and in intermittent hypoxia (CIH; 10 days)-conditioned rats. Imp/s, integrated activity; A.P., raw action potential; solid horizontal bar, duration of hypoxic challenge.

10. The following criteria are used to identify the sensory activity from the carotid body: (a) prompt augmentation of sensory discharge by ventilating the rat with a hypoxic gas mixture (12% O_2 balance nitrogen) and (b) marked decrease in sensory discharge in response to ventilating the rat with 100% O_2. The magnitude of sensory discharge is recorded as impulses per second as analyzed from rate meter output.

11. Normalization of chemoreceptor activity and data analysis. The amplitude of the integrated chemoreceptor activity depends on the contact of the nerve to the electrode and to the extent of desheathing of the nerve. These procedures are expected to vary from experiment to experiment. To compare data within animals and between control and intermittent hypoxia-exposed animals, it is necessary to normalize data. This can be accomplished as follows. At the end of the experiment, the respirator is turned off for about 1 min (asphyxia), which results in an intense increase in carotid body activity. The maximum increase in sensory activity is taken as 100% (asphyxic response). Chemoreceptor activity (impulses/ second) is analyzed during baseline conditions, as well as with different levels of inspired oxygen, and arterial blood samples are collected at the end of each inspired oxygen level for measurements of pO_2. The magnitude of the sensory response is plotted against arterial pO_2 as a percentage of the asphyxic response.

Carotid Body Slice Preparation and Recording of Ionic Currents from Glomus Cells

Preparation of Carotid Body Slices for Electrophysiological Recording

Materials

Urethane (1.2 g/kg in distilled water)
F-12 nutrient mixture (HAM, GIBCO)
Dulbecco's modified Eagle medium (DMEM, GIBCO)
Insulin–transferrin–selenium-X (ITS, GIBCO)
Penicillin–streptomycin (GIBCO-BRL)
Slicing solution (see later)
Agarose (Sigma, type VII-A: low gelling temperature)
90 × 20-mm petri dish
140 × 15-mm petri dish
Sylgard silicone elastomer kit (Dow-Corning)
Falcon six-well plate (Becton-Dickinson)
Two 10-ml beakers (with small stirring bars)
15-ml polypropylene conical tube (Falcon)

Ice bucket
Krazy glue (Elmer's Products)
Leica GZ6 dissecting microscope
NVSL manual vibroslice (WPI Instruments)
Double-edge stainless steel razor blades (Electron Microscopy Sciences)
Isotemp hotplate stirrer (Fisher Scientific)
$37°$ CO_2 incubator
Gas tank 95% O_2 with 5% CO_2

Procedure

1. All solutions are prepared fresh on a weekly basis and stored in a $4°$ refrigerator. Carotid body sections are maintained in a nutrient media consisting of a 50/50 mixture of HAM F-12 and DMEM supplemented with penicillin–streptomycin and ITS prepared under sterile conditions. For isolation and slicing we prepare a solution containing (mM, all from Sigma): NaCl (140), KCl (2), $NaHCO_3$ (26), $MgCl_2$ (5), glucose (10), and $CaCl_2$ (0.1). This solution is then sterile filtered and stored in an airtight glass bottle.

2. Prior to animal preparation, 5 ml of slicing solution and 150 mg of low gelling point agarose (3% agarose solution) are added to each of the two 10-ml beakers containing small stirring bars. This mixture is heated to near boiling with continuous stirring on the Isotemp hotplate until the agarose is dissolved. Beakers are then placed in the 5% CO_2 incubator at $37°$.

3. To minimize contamination, surgical instruments are soaked in 70% ethanol prior to surgical procedures. Rats are anesthetized with an IP injection of 1.2 g/kg urethane. The carotid artery bifurcations are removed rapidly via a ventral approach (see previous section) and placed in ice-cold nutrient media bubbled with 95% O_2/5% CO_2 gas mixture in a 15-ml conical tube. After 1 h, individual bifurcations are then pinned down by the arterial branches in our dissection chamber consisting of a 90-mm petri dish that has been filled partially with polymerized sylguard and covered with the ice-cold slice solution bubbled continuously with the 95% O_2/5% CO_2 gas mixture. We place black tape on top of the sylguard to facilitate viewing of the carotid body tissue, and the smaller dish is placed in a larger 140-mm dish that is filled with crushed ice. The carotid bodies themselves are removed from bifurcation under the dissecting microscope. Each carotid body is removed separately and placed in the middle of the 3% agarose slicing solution in the 10-ml beaker. These beakers are then placed on ice to solidify the agar.

4. The following steps are performed as quickly as possible following solidification of the agarose. Solidified agarose blocks containing the individual carotid bodies are removed from the 10-ml beakers. The blocks are trimmed using a razor blade so that a 5 × 5-mm block remains containing the carotid body tissue. These blocks are mounted side by side using a drop of Krazy glue on the specimen holder of the NVSL vibroslice (specimen holder and bath are stored in a −20° freezer). The specimen holder is placed in the specimen bath filled with the ice-cold slice solution and bubbled with the 95% O_2/5% CO_2 gas mixture. A fresh razor blade is broken in half, rinsed with 95% ethanol to remove residual oil, rinsed with distilled water, and clamped in the vibroslice. The blade is advanced to the top of the block, and sections are made at 200 μm until tissue is reached. Carotid body tissue sections are then made at 120–150 μm until both carotid bodies are sectioned. As each individual section comes off, it is removed using a dropper made of a glass pipette of 6 mm i.d. and transferred to the six-well plate containing nutrient media. Once all tissue is sectioned, sections are allowed to settle in the six-well plate in the 37° incubator for 1 h, and the medium is removed and replaced with fresh nutrient medium. Tissue sections are maintained in this medium at 37° in a 5% CO_2 incubator until use.

Note. We find that sections can be maintained for a week or longer (depending on the amount of contamination), but are best used when left in the incubator for 2 nights.

Membrane Current Recordings from Rat Carotid Body Tissue Slices

Materials

> Patch-clamp amplifier (Axon Instruments 200B)
> Digital interface (Axon Instruments Digidata 1200)
> Pclamp software (Axon Instruments)
> IBM-compatible computer (Dell XPS R400)
> Upright, fixed stage microscope (Olympus BX51W1)
> 40× water immersion objective (Olympus LUMPlanF1/IR)
> 2× mag changer (Olympus)
> Manipulator (SD Instruments MX7630 DL)
> Narishige pipette puller (PP-83)
> Narishige microforge (MF-83)
> Electrode glass (Corning 7052; 1.20 i.d., 1.60 o.d., 75 mm length; Garner Glass Company)
> Extracellular solution (see later)
> Intracellular solution (see later)

0.2-μm syringe filter (Acrodisc, Gelman Laboratory)
Microfil 34-gauge silica needle (World Precision Instruments)
60-ml syringes
Pharmed tubing (1.6 mm i.d., 4.8 mm o.d., VWR Scientific)
Luer connectors

General Electrophysiology

1. Glomus cells can be identified based on electrophysiological characteristics (Fig. 3). Glomus cells, but not type II, express a fast,

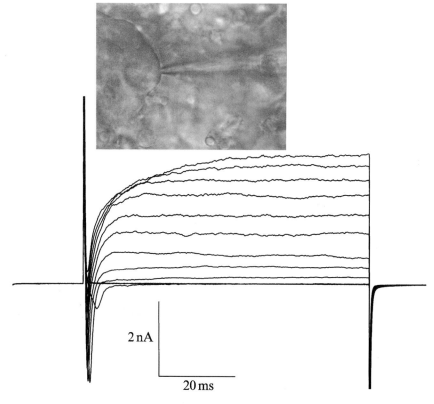

2 nA

20 ms

FIG. 3. Membrane current recording in rat carotid body section. Representative family of outward currents elicited by 75-ms depolarizing voltage steps from −50 to 70 mV in 10-mV increments (HP = −85 mV, solutions specified in text). Note the rapid, inward Na$^+$ current followed by the slower, outward K$^+$ current. *(Inset)* A 120-μm section of rat carotid body showing a patch-clamp electrode (tip approximately 1 μm in diameter) sealed to one glomus cell in a cluster (800 × magnification).

inward Na^+ current followed by a large (nA), outward K^+ current. With a little practice, inadvertent sealing of type II cells is rarely a problem.

2. Electrodes are pulled from borosilicate glass capillary tubing using a two-stage Narishige pipette puller on the day of the experiment. Electrodes have a typical resistance of 4–6 $M\Omega$ when filled with intracellular solution after fire polishing the tips using the Narishige microforge.

3. The recording chamber consists of a linear flow bath that is fed by gravity from a series of 60-ml syringes connected by three-way stopcocks and Pharmed tubing fitted with luer connectors into a single outlet. Each tube is bubbled with an appropriate gas to maintain oxygenation and pH. The solution is removed from the bath by a small suction tube positioned at the outflow end where the bath is grounded by an Ag/AgCl wire electrode. At the beginning of an experiment, a single section is removed from the six-well plate and placed in the bath with a small amount of extracellular solution using the dropper mentioned earlier. The section is positioned near the inflow of the bath and is stabilized by placing two small pieces of silver wire (5 mm length) on the agarose protruding from either side of the carotid body tissue itself. The flow of solution is then started via opening the stopcock containing the control extracellular solution of choice.

4. The intracellular solution is placed in a 10-ml syringe and 4–5 ml is filtered through a syringe filter into a 10-ml beaker. The tip of the electrodes is frontfilled by suction from the solution in the beaker and then backfilled from the syringe using the Microfil needle. Electrodes are placed in the holder in the headstage of the patch-clamp amplifier and lowered into solution. The junction potential and electrode capacitance are nulled using the electronic circuitry of the patch-clamp amplifier, and the electrode is brought into close proximity of a glomus cell cluster in the slice using the micromanipulator. Positive pressure is added at the electrode tip by mouth pressure applied to the tubing connected to the port on the electrode holder when the electrode is in solution. When the electrode is just touching one cell in the cluster, a slight negative pressure is added at the electrode tip to facilitate sealing. Formation of the gigaseal is monitored using a repeating step of 10 mV for 10 ms. When the seal begins to form, a potential of -50 is added at the electrode tip to facilitate sealing.

General Current Recording

5. We have found that ion channel activity can be affected by the pH buffer system used in the extracellular solution.[8] CO_2/bicarbonate-buffered solutions can be somewhat more difficult to use (to maintain the pH around 7.4 it is necessary to bubble the solutions continuously with

5% CO_2); however, this buffering system is expected to be more physiological. The standard extracellular solution for recording K^+ currents contains (mM): NaCl (140), KCl (2), $MgCl_2$ (2), $CaCl_2$ (2.5), $NaHCO_3$ (26), and glucose (10). The standard intracellular solution contains (mM) L-aspartic acid monopotassium salt (120), KCl (20), $CaCl_2$ (1), EGTA (5), HEPES (5), glucose (10), MgATP (5), and TrisGTP (0.1), and the pH is adjusted to 7.2 using KOH. All chemicals are from Sigma.

6. As shown in Fig. 3, using standard conditions, the electrophysiological profile of glomus cells in rat carotid body slices is dominated by a large, outward K^+ current. Contributions to whole cell current from Ca^{2+}-dependent K^+ current and Ca^{2+} current itself under these conditions can be minimized by using ATP-free intracellular solutions that result in rapid rundown of the Ca^{2+} current. Peak currents can be studied easily using 50- to 80-ms steps from a holding potential of -85 mV. Current rundown and drug effects can be monitored using a wash protocol consisting of a 50-ms step to 20 mV repeated every 10 s. Current traces are sampled at 14 kHz and filtered at 5 kHz for offline analysis using the Clampfit module of the Pclamp software package.

Monitoring Transmitter Changes in Carotid Bodies

Measurement of Catecholamines

Materials

High-performance liquid chromatography (Shimadzu)
Electrochemical detector (Bioanalytical System)
Perchloric acid (0.1 N)
Adult male Sprague–Dawley rats (weighing 200–250 g; Charles River Laboratories)
Vibra cell sonicator (Sonics and Materials, Inc., Danbury, CT)
C-18 reversed-phase column (Prodigy; Phenomenox)

Procedure

1. Isolation of the carotid body. Adult Sprague–Dawley rats are anesthetized with urethane as described earlier. Bifurcations at the common carotid arteries are excised rapidly following protocols approved by the Institutional Animal Care and Use Committee. Carotid bodies are isolated from bifurcation under the microscope and placed in oxygenated Ca^{2+}/Mg^{2+}-free modified Tyrode solution at $4°$.

---

[8] B. A. Summers, J. L. Overholt, and N. R. Prabhakar, *J. Neurophys.* **84**, 1636 (2000).

2. Extraction of carotid bodies. Carotid bodies (n = 4 CBs) are homogenized in 300 μl of medium containing 0.1 N HClO$_4$, 0.1 M sodium metabisulfite, and 0.25 mM disodium EDTA using a Vibra Cell sonicator using 80% output power level. Sonication is repeated three times, each for the duration of 30 s at 4° and the resulting homogenate is centrifuged at 12000 g for 15 min at 4°. The clear supernatant is used immediately for the analysis of catecholamines.

3. Separation and detection of catecholamines. Catecholamines are separated using a Prodigy C-18 (Phenomenox) reversed-phase column by isocratic elution with a mobile phase consisting of 7% methanol, 0.1 M sodium acetate, 0.0125 M sodium citrate, 0.1 mM disodium EDTA, and 0.215 mM sodium octyl sulfate at pH 4.5. The elution of catecholamines is monitored using a LC-4C model amperometric detector from the Bioanalytical System. The chromatograms are recorded and analyzed with a Hitachi D-2500 chromatoIntegrator.

Note. Under the experimental conditions, dopamine, dihydroxyphenyl-alanine acetic acid (oxidation product of dopamine), and norepinephrine are eluted at ~9, ~4, and ~2 min, respectively. The detection limit for catecholamines is 250 fmol each.

4. Quantitation of catecholamines. Concentrations of catecholamines are determined from a standard curve correlating their concentration to the corresponding integrated peak size and/or current. Data are expressed as picomoles per milligram protein. The protein concentration of the carotid body extract is determined by the bicinchoninic acid (BCA) method[9] using bovine serum albumin as the standard.

Acknowledgments

This work was supported by grants from the National Institutes of Health, Heart, Lung, and Blood Institute HL-25830 and HL-66448.

[9] C. M. Stoscheck, *Anal. Biochem.* **160**, 301 (1987).

[7] Functional Analysis of the Role of Hypoxia-Inducible Factor 1 in the Pathogenesis of Hypoxic Pulmonary Hypertension

By LARISSA A. SHIMODA and GREGG L. SEMENZA

Introduction

Many lung diseases are associated with reduced alveolar oxygen concentration, resulting in pulmonary hypertension and eventual right heart failure. Indeed, complications of chronic lung disease are the fourth leading cause of mortality in the United States, with current treatment modalities limited primarily to supplemental oxygen, mechanical ventilation, and lung transplant.[1-3] The key to development of new therapeutic approaches to prevent or reverse pulmonary hypertension lies in understanding the processes involved in pulmonary vascular responses to prolonged hypoxia. The factors underlying this response remain poorly understood. The transcription factor hypoxia-inducible factor 1 (HIF-1) plays a critical role in mediating adaptive responses to hypoxia.[4] HIF-1 regulates the expression of dozens of genes important in growth, vascular development, and metabolism.[5] A member of the basic helix–loop–helix family of proteins, HIF-1 is a heterodimer of HIF-1α and HIF-1β subunits. Whereas HIF-1β is expressed constitutively in the lung, HIF-1α protein expression is regulated by the inspired oxygen concentration.[6] This article provides methods to quantify physiological changes that occur in the lung in response to alveolar hypoxia. While we have analyzed the role of HIF-1 in these processes, the methods can be utilized to evaluate any factor of interest, providing the appropriate mouse model is available.

[1] National Center for Health Statistics, Report of Final Morbidity Statistics (1998). Information cited in American Lung Association, Trends in Chronic Bronchitis and Emphysema: Morbidity and Mortality (2000).

[2] D. M. Mannino, *Chest* **121**, 121S (2002).

[3] T. L. Petty, *Chest* **121**, 116S (2002).

[4] A. Y. Yu, L. A. Shimoda, N. V. Iyer, D. L. Huso, X. Sun, R. McWilliams, T. Beaty, J. S. K. Sham, C. M. Wiener, J. T. Sylvester, and G. L. Semenza, *J. Clin. Invest.* **103**, 691 (1999).

[5] G. L. Semenza, *Pediatr. Res.* **49**, 614 (2001).

[6] A. Y. Yu, M. G. Frid, L. A. Shimoda, C. M. Wiener, K. Stenmark, and G. L. Semenza, *Am. J. Physiol. Lung Cell Mol. Physiol.* **275**, L818 (1998).

The Model

Animal Model

The choice of an appropriate animal model is of vital importance to the success of the study. The use of female mice is discouraged as females can be protected from development of hypoxic pulmonary hypertension, most likely due to the involvement of estrogen–dependent mechanisms.[7,8] Another important consideration is the fact that there can be marked strain-specific differences in the development of hypoxic pulmonary hypertension; we have found the C57BL/6J strain (available from Jackson Laboratories, Bar Harbor, ME) to be reliable in producing consistent results. Our studies have utilized adult (8 week old) littermate males bred on a mixed strain 129 and C57BL/6J background that are either heterozygous for a null mutation at the *Hif1a* locus encoding HIF-1α (*Hif1a$^{+/-}$*) or wild type (*Hif1a$^{+/+}$*).[9] Experiments involve a total of four experimental groups: *Hif1a$^{+/+}$* mice maintained under normoxic conditions, *Hif1a$^{+/-}$* mice maintained under normoxic conditions, *Hif1a$^{+/+}$* mice maintained under hypoxic conditions, and *Hif1a$^{+/-}$* mice maintained under hypoxic conditions, with a minimum of three animals per group.

Hypoxic Exposure

A well-established method for inducing hypoxic pulmonary hypertension is to place animals in an environmental chamber and expose them to normobaric hypoxia (10% O_2) for 21 days. Pulmonary vascular responses to chronic hypoxia are fully developed within 6 weeks.[4,10] The normoxic control mice are housed in the same environment (i.e., a chamber gassed with room air). The chamber can be constructed from half-inch-thick Plexiglas (our chamber measures 4 ft long by 2 ft high by 2 ft deep) or from a 55-gal aquarium tank, with a top-loading, hinged Plexiglas lid, which should be lined with weather-stripping foam to provide a tight seal between the lid and the sides of the chamber. The hypoxic chamber is flushed continuously through tubing connected to an inlet port (a plastic or metal tubing connector attached to a hole drilled into the side of the chamber

[7] M. Rabinovitch, W. J. Gamble, O. S. Miettinen, and L. Reid, *Am. J. Physiol.* **240,** H62 (1981).
[8] T. C. Resta, N. L. Kanagy, and B. R. Walker, *Am. J. Physiol. Lung Cell Mol. Physiol.* **280,** L88 (2001).
[9] N. V. Iyer, L. E. Kotch, F. Agani, S. W. Leung, E. Laughner, R. H. Wenger, M. Gassmann, J. D. Gearhart, A. M. Lawler, A. Y. Yu, and G. L. Semenza, *Genes Dev.* **12,** 149 (1998).
[10] A. Hislop and L. Reid, *Br. J. Exp. Pathol.* **57,** 542 (1976).

and glued in place with aquarium epoxy). Tubing connected to an outlet port at the other end of the chamber carries the outflow gas to a fume hood. The desired oxygen concentration (10%) can be achieved either by mixing air and 100% N_2 or by the use of premixed tanks. If air and N_2 are mixed, the chamber is flushed continuously with air and a servo control mechanism (PROOX 110, Reming Bioinstruments, Redfield, NY) is used to regulate O_2 concentration. The sensor of the controller should be placed inside the chamber, and N_2 flow attached to the controller following the manufacturer's instructions. The controller will inject N_2 as needed to maintain the set O_2 concentration. Chamber O_2 and CO_2 concentrations should be monitored continuously (O_2 Analyzer Model 5577, Hudson RCI, Temecula, CA, and LB-2 CO_2 analyzer, Sensormedics, Anaheim, CA), and gas inflow should be adjusted to provide a flow rate adequate to maintain low CO_2 concentrations (<0.5 %) and prevent buildup of ammonia and humidity within the chamber. The flow required will depend on the size of the chamber and the number of animals exposed. Mice can be placed in labeled cages inside the chamber and can be exposed to room air for 10 min twice a week to clean the cages and replenish food and water supplies.

Physiological Measurements

Right Ventricle (RV) Pressure

For measurement of RV pressure, which is considered an estimate of pulmonary artery pressure, mice are anesthetized with sodium pentobarbital (43 mg/kg ip). A small incision is made in the skin above the trachea using a No. 10 scalpel. Blunt dissection is then used to isolate the trachea from the surrounding muscle, and a piece of 4-0 suture is slipped under the trachea. The trachea is cannulated by creating a small incision, being careful not to slice completely through the trachea, and sliding a cannula made from 15-gauge stainless steel tubing (approximately 1 in. long) with a beveled ($45°$) tip into the incision (cannulas can be purchased from Columbus Instruments, Columbus, OH). The cannula is tied in place with the suture and attached via tubing to a rodent ventilator (CIV-101, Columbus Instruments). The lungs are ventilated with 10% O_2, 5% CO_2 balanced N_2 gas at a tidal volume of 0.2 ml and a rate of 90 breaths per minute. Once the mouse is ventilated, a lateral incision is made in the abdomen and the diaphragm is visualized by lifting the xyphoid process. The beating heart should be clearly visible through the diaphragm. A heparinized [0.5 ml heparin (10,000 units/ml) in 10 ml saline] saline-filled 22-gauge needle (1.5 in. long) connected to a pressure transducer (P10EZ, Spectramed Inc., Oxnard, CA) is inserted through the diaphragm into the RV, which lies against the diaphragm, and pressure is recorded on a polygraph (Model 7; Grass Instruments, Quincy, MA) or computer-based recording system (PowerLab 8SP, AD Instruments, Castle Hill, Australia). RV pressure can be distinguished from left ventricular pressure by the smaller pulse pressure (typically 5–12 mm Hg) and lower mean pressure ($\leq$10 mm Hg for normoxic mice and $\geq$15 mm Hg for chronically hypoxic mice). The RV puncture should be verified by postmortem examination. Determination of whether any observed increase in RV pressure is due to reversible vasoconstriction or to vascular remodeling is achieved by allowing animals to recover in room air for 2–6 h prior to, and ventilating with 21% O_2, 5% CO_2, balanced N_2 gas during, RV puncture.

Using RV pressure as an estimate of pulmonary artery pressure, the development of pulmonary hypertension was demonstrated in $Hif1a^{+/+}$ mice exposed to chronic hypoxia, with mean RV pressure increasing from 7.33 $\pm$ 0.49 mm Hg (normoxia) to 18.36 $\pm$ 1.88 mm Hg (3 weeks at 10% O_2).[4] Normoxic $Hif1a^{+/-}$ mice exhibited RV pressures similar to their wild-type counterparts (6.83 $\pm$ 0.48 mm Hg), but the chronic hypoxia-induced increase in RV pressure was blunted significantly in $Hif1a^{+/-}$ mice, reaching only 11.87 $\pm$ 0.95 mm Hg, thus implicating HIF-1 in the pathogenesis of hypoxic pulmonary hypertension.

Hematocrit

After measuring RV pressure, the chest is opened through an incision along the sternum and blood is collected from the heart (right or left ventricle) via a 23-gauge needle attached to a 3-ml syringe containing 0.02 ml heparin. The blood must be drawn slowly; if the pressure applied by suction from the syringe is too great, the venous circulation will collapse, preventing blood from traveling to the heart. Approximately 1 ml of blood can be collected per mouse. The blood is transferred into a 1.5-ml Eppendorf tube, and heparinized capillary tubes (0.5 mm o.d, VWR Scientific) are filled and sealed with Crit-o-Seal clay (VWR Scientific). The capillary tubes are centrifuged in a hematocrit spinner for 5 minutes at 7000 rpm and read on a hematocrit chart (VWR Scientific).

Hematocrit values, which vary with strain, ranged from 40 to 48% in normoxic $Hif1a^{+/+}$ and $Hif1a^{+/-}$ mice.[4,11] Whereas values as high as 70% were observed in chronically hypoxic $Hif1a^{+/+}$ mice (mean = 65.7 $\pm$ 2.1% at 3 weeks), hematocrits were significantly lower in hypoxic $Hif1a^{+/-}$ mice

[11] L. A. Shimoda, D. J. Manalo, J. S. K. Sham, G. L. Semenza, and J. T. Sylvester, *Am. J. Physiol. Lung Cell Mol. Biol.* **281**, L202 (2001).

(61.1 $\pm$ 1.1%), consistent with the known role of HIF-1 in regulating erythropoietin production.

Measurement of RV Hypertrophy

Following RV puncture, the heart and lungs are removed *en bloc* and transferred to a 10-cm dissecting dish filled with ice-cold (4°) HEPES-buffered salt solution (HBSS) containing (in mM) 130 NaCl, 5 KCl, 1.2 MgCl$_2$, 1.5 CaCl$_2$, 10 N-[2-hydroxyethyl]piperazine-N'-[2-ethanesulfonic acid] (HEPES), and 10 glucose, with the pH adjusted to 7.2 with 5 M NaOH. The dissecting dish is prepared prior to beginning experiments by filling a 10-cm petri dish halfway with a silicone elastomer compound (Sylgard 184, Dow Corning) prepared according to manufacturer's instructions and allowed to cure for 72 h before use, creating an elastic bottom to which tissue can be pinned. Using a stereomicroscope under low magnification, the heart is separated from the lungs, and the atria and protruding vessels are dissected free. The left ventricle can be identified by the descending coronary artery, which is clearly visible, running down its length, as well as the very thick wall, whereas the wall of the right ventricle is substantially thinner. At this time, inspection is performed to verify RV puncture, as the hole created by the needle puncture should be clearly visible. The RV wall of the heart is separated carefully from the left ventricle and the septum (LV + S) with dissecting scissors by following where the RV wall is attached to the remainder of the heart, and the two portions are blotted dry with tissue and weighed. This measurement should be made as quickly as possible after sacrifice as the heart will contract over time, making precise dissection difficult.

Absolute values for each portion of the heart will vary with age of the mice; however, we measured values for LV + S in 11-week-old mice in the range of 0.90–0.11 g in all experimental groups, verifying that pulmonary hypoxia does not cause left ventricular hypertrophy.[4,11] In contrast, RV values were greater in hypoxic (0.0326 $\pm$ 0.001 g) than in normoxic (0.0264 $\pm$ 0.001 g) *Hif1a*[+/+] mice, indicating development of RV hypertrophy. The increase in RV weight was blunted significantly in *Hif1a*[+/-] mice (0.0249 $\pm$ 0.0015 g in normoxic *vs* 0.0291 $\pm$ 0.006 g in hypoxic). These results are consistent with the lower RV pressure observed in chronically hypoxic *Hif1a*[+/-] mice and provide further evidence supporting the role of HIF-1 in the pathogenesis of hypoxic pulmonary hypertension.

Pulmonary Vascular Morphometry

For histological studies, lungs are fixed by tracheal instillation of 10% buffered formalin at constant tracheal pressure. The trachea cannula is connected via tubing to a three-way stopcock. A 1-ml Pasteur pipette is modified by gently removing a portion of the tip to allow it to fit into the stopcock, and the pipette is filled with 10% buffered formalin to a level equivalent to 20 cm H_2O (the height should be measured prior to beginning the experiment). Once filled, the stopcock is opened so that formalin is allowed to flow into the lungs. Once the level of formalin stabilizes, indicating full filling of the lung (approximately 10–15 min), the suture is pulled tight while the cannula is removed to close off the trachea, and the heart and lungs are removed *en bloc*. The heart is removed for measurement of RV hypertrophy as described earlier, and the lungs are immersed and stored in 10% buffered formalin for at least 48 h until sectioning. The formalin-fixed lungs are then dehydrated via graded alcohol solutions starting with 70% ethanol at 4° overnight, followed by the following incubations at 4°: 70% ethanol for 30 min, 80% ethanol for 30 min, two incubations in 95% ethanol of 45 min each (changing solution between incubations), and two incubations in 100% ethanol for 45 min each. Following dehydration, the samples should be cleared by two incubations in 100% xylene (reagent grade, Sigma Chemicals) for 45 min at 40°. Full clearing has occurred when the tissue becomes translucent. The sample is then immersed in infiltrating paraffin (Paraplast, Sigma Chemicals) at 58–60° (the paraffin will harden below 58° and the polymer additives will become unstable at temperatures higher than 62°) for a minimum of 2 h, changing the solution three to four times at 30–90 min per incubation. All of the aforementioned steps should be performed on a rotator or agitator with 20 ml of solution. An embedding mold is filled with melted paraffin, and the sample is placed into the mold, orienting the lobe such that cross-sectional cuts can be obtained. After embedding, the samples are sectioned with a microtome into 5-μm slices parallel to the hilum and are stained with Verhoff's Elastic stain kit (Newcomer Supply, Middleton, WI, or IMBE Inc., San Marcus, CA) according to the manufacturer's instructions. For each lung section, intraacinar vessels are identified, and video images are captured via a CCD camera attached to a Nikon microscope and transferred to a computer. Measurements of arterial diameter can be obtained using a computerized image analysis program (Image 1.55; National Institutes of Health, Bethesda, MD). Vessels with an external diameter ≤ 100 μm should be analyzed and classified as nonmuscularized (no visible muscle layer), partially muscularized (muscle present but not completely encompassing vessel), or completely muscularized. Approximately 500 vessels from the lungs of at least three different mice should be scored. For all muscular vessels, the area bounded by the internal elastic lamina should be measured and subtracted from the area bounded by the external elastic lamina to obtain the area of the medial layer, which can then be expressed as a percentage of the total vessel area. Percentage wall thickness can also be calculated as the diameter of the external elastic lamina minus the diameter of the internal elastic lamina divided by the diameter of the external elastic lamina.

In normoxic mice, the greatest proportion of small vessels are nonmuscular. Following exposure to hypoxia, a shift from nonmuscular to partially and completely muscular arterioles was observed in $Hif1a^{+/+}$ mice, which was significantly greater than that observed in hypoxic $Hif1a^{+/-}$ mice.[4] In addition to the increase in the number of muscular vessels, the wall thickness of these vessels was greater in hypoxic $Hif1a^{+/+}$ mice than in $Hif1a^{+/-}$ mice. The blunted vascular remodeling in response to chronic hypoxia in mice partially deficient for HIF-1α is the basis for the lesser degree of hypoxic pulmonary hypertension observed.

Electrophysiological Measurements

While this section provides information regarding the methods used to measure capacitance (cell size) in single smooth muscle cells, it is assumed that the reader is familiar with basic dissection and patch-clamp techniques, as a detailed description of these methods is beyond the scope of this chapter. If information regarding these techniques is required, it is suggested that the "Rat Dissection Manual" (by Bruce D. Wingerd) and "The Axon Guide for Electrophysiology and Biophysics" (available from Axon Instruments, Union City, CA) be consulted.

Pulmonary Arterial Smooth Muscle Cell (PASMC) Isolation

Single PASMCs are obtained by enzymatic digestion from lungs not subjected to histological measurements. Under moderate magnification, intrapulmonary arteries (200–500 μm outer diameter) can be isolated from the upper left and lower right lobes of the lung and cleaned of connective tissue. The arteries are cut longitudinally and, after disrupting the endothelium by very gently rubbing the luminal surface of the vessels with a cotton swab, allowed to recover for 30 min in cold (4°) HBSS. This is followed by incubation for 20 min in reduced Ca^{2+} HBSS (20 μM $CaCl_2$) at room temperature to gradually increase temperature and decrease Ca^{2+} concentration. The tissue is then digested in approximately 2–3 ml of reduced Ca^{2+} PSS containing collagenase (type I; 1750 U/ml), papain (9.5 U/ml), bovine serum albumin (2 mg/ml), and dithiothreitol (1 mM) at 37° for 10 min. The digestion solution should be prepared immediately before use, as the enzymes will lose potency within a short period of time (approximately 1 h). Following digestion, the enzyme solution containing the digested tissue is poured over a fine plastic mesh and washed with a few drops of

Ca^{2+}-free HBSS. The tissue is then transferred to a glass 5-ml beaker containing 1 ml of filtered Ca^{2+}-free HBSS, and single smooth muscle cells are dispersed by gentle trituration (10–20 times) with a P1000 micropipette. The tissue should flow easily through the pipette tip and should not catch on the edges of the tip. After trituration, the remaining tissue is removed with forceps, and the cell suspension is then ready for transfer to the cell chamber for study.

Measurement of Cell Capacitance

During whole cell patch clamp, a step change in membrane potential requires a significant current transient in order to charge the membrane capacitance. The transient can be quantified, with the amplitude corresponding to the size of the cell, and can be reflective of changes in cell size (i.e., hypertrophy) after hypoxia. Once PASMCs are isolated, one or two drops of the cell suspension should be placed in the cell chamber and allowed to settle for 15 min to permit the cells to adhere to the bottom of the chamber. The rest of the suspension can be kept at 4° for use within 2–4 h. For these measurements, the chamber can be as simple as a 35-mm petri dish or as elaborate as a custom-designed chamber with inflow and outflow ports (Warner Instruments, Hamden, CT). Once cells have adhered, the cell chamber is filled with HBSS, and the ground electrode is placed in the solution. Patch pipettes (tip resistance 3–5 MΩ) are pulled from borosilicate capillary tubes with an internal filament (1.5 mm outer diameter; World Precision Instruments, Sarasota, FL) using a Flaming-Brown micropipette puller (Model P-87, Sutter Instruments, Novato, CA). The program and temperature used for pulling pipettes will vary depending on the filament, although in general, we use a four-step pulling program, with each step having a temperature value of 350–410, a pull velocity value of 50, a time value of 150, and an air jet pressure of 700–800. The tips of the pipettes are fire polished under high magnification using a microforge (Model MF-9, Narishige, Tokyo, Japan) to promote gigaohm seals and prevent puncture of the cell during seal formation and are filled using a microfil (World Precision Instruments) attached to a 1-cc syringe, with an internal solution containing (in mM) 35 KCl, 90 potassium gluconate, 10 NaCl, and 10 HEPES, with pH adjusted to 7.2 with 5 M KOH. A silver chloride electrode is placed into the solution in the filled pipette and is attached to the mounted head stage of the amplifier. Electrophysiological measurements are recorded using an Axopatch 200B amplifier (Axon Instruments Inc.) in voltage-clamp mode. The pipette tip is brought close to the cell under high magnification (using a 20X objective), and an electronic micromanipulator is used to bring the tip into contact with the cell surface to form a weak seal. During this time, the cell should be subjected to pulses. Applying slight suction, either by mouth or by syringe, will create a strong seal (gigaohm resistance) between the cell and the pipette tip. Once a tight seal has been achieved, the holding potential is decreased to −60 mV, and pipette potential and capacitance are compensated electronically with the appropriate controls on the amplifier before proceeding from patch to whole cell mode. Once access to the cell is gained (whole cell mode) after the membrane within the seal is ruptured by gentle suction, it is possible to estimate the cell capacitance by eliminating the transient using the whole cell capacitance dial on the amplifier. The absolute value of the capacitance displayed on the dial after the transient is eliminated can be used to estimate the surface area of the cell, assuming that membrane capacitance per unit area is 1 μF/cm^2. A second, perhaps more precise, method for determining cell capacitance is calculating the area under the capacitive current elicited by a 10-mV hyperpolarizing pulse from a holding potential of −70 mV using the Clampfit data analysis program (Axon Instruments Inc.). Voltage-clamp protocols are applied using pClamp software (Axon Instrument Inc.). Data should be filtered at 10 kHz, digitized with a Digidata 1200 A/D converter (Axon Instrument Inc.), and analyzed with pClamp software (Axon Instrument). All measurements can be conducted at room temperature (22–25°) under normoxic conditions.

Average capacitance values measured in freshly isolated PASMCs from normoxic $Hif1a^{+/+}$ and $Hif1a^{+/-}$ mice were 14.4 ± 2.2 and 12.8 ± 3.2 pF, respectively.[11] Exposure to chronic hypoxia increased capacitance in PASMCs from $Hif1a^{+/+}$ mice (23.4 ± 1.2 pF), but not from $Hif1a^{+/-}$ mice (13.2 ± 0.8 pF). These results indicate that PASMC hypertrophy occurs in response to hypoxia and that this response is dependent on HIF-1 activity. The absence of cell hypertrophy in hypoxic $Hif1a^{+/-}$ mice explains, in part, the reduced vascular remodeling observed in these mice.

Conclusions

The development of pulmonary hypertension secondary to chronic lung disease correlates with higher morbidity and mortality. The development of transgenic animals has allowed molecular biology to merge with traditional physiological measurements, providing a powerful combination of techniques by which to explore the complex processes mediating pulmonary vascular responses to prolonged hypoxia. Using these techniques, we have demonstrated that HIF-1 plays a critical role in this disease process, thus providing a potential target for therapeutic strategies aimed at the treatment and prevention of hypoxic pulmonary hypertension.

Section II

Cardiovascular and Blood

[8]   Perceived Hyperoxia: Oxygen-Regulated Signal Transduction Pathways in the Heart

By SASHWATI ROY, SAVITA KHANNA, and CHANDAN K. SEN

Introduction

Cellular O_2 concentrations are maintained within a narrow range (perceived as "normoxia") due to the risk of oxidative damage from excess O_2 (hyperoxia) and of metabolic demise from insufficient O_2 (hypoxia).[1] pO_2 ranges from 90 to below 3 Torr in mammalian organs under normoxic conditions with an arterial pO_2 of about 100 Torr or $\sim$14% O_2.[2] Thus, "normoxia" for cells is a variable that is dependent on the specific localization of the cell in organs and functional status of the specific tissue. O_2 sensing is required to adjust to physiological or pathophysiological variations in pO_2. Whereas acute responses often entail changes in the activity of preexisting proteins, chronic responses invariably involve O_2-sensitive changes in signal transduction and gene expression.[3] Several articles have highlighted the key significance of understanding the fundamentals of O_2 sensing.[4–14] Current work in this field is almost exclusively focused on the study of hypoxia. Reoxygenation, however, has been mostly investigated in the context of oxidative injury, and there is a clear paucity of data describing the O_2-sensitive signal transduction pathways under conditions of oxygenation that mildly or moderately exceed perceived normoxia. While acute insult caused during reperfusion may be lethal to cells localized at the focus of insult, elevation of O_2 tension in ischemic tissue is expected to trigger phenotypic changes in the surviving cells that may be associated with tissue remodeling (Fig. 1).

[1] G. L. Semenza, *Cell* **107,** 1 (2001).
[2] T. Porwol, W. Ehleben, V. Brand, and H. Acker, *Respir. Physiol.* **128,** 331 (2001).
[3] A. Elsasser, M. Schlepper, W. P. Klovekorn, W. J. Cai *et al.*, *Circulation* **96,** 2920 (1997).
[4] R. C. Elphic, M. Hirahara, T. Terasawa, T. Mukai *et al.*, *Science* **291,** 1939 (2001).
[5] K. Kondo, H. Yang, W. Kim, J. Valiando *et al.*, *Science* **292,** 464 (2001).
[6] P. Lundgren, Y. B. Chen, H. Kupper, Z. Kolber *et al.*, *Science* **294,** 1534 (2001).
[7] P. H. O'Farrell and G. L. Semenza, *Cell* **98,** 105 (1999).
[8] G. L. Semenza, *Cell* **98,** 281 (1999).
[9] H. Zhu and H. F. Bunn, *Science* **292,** 449 (2001).
[10] P. Jaakkola, *Science* **292,** 468 (2001).
[11] M. Ivan, K. Kondo, H. Yang, W. Kim *et al.*, *Science* **292,** 464 (2001).
[12] R. K. Bruick and S. L. McKnight, *Science* **294,** 1337 (2001).
[13] A. C. Epstein, J. M. Gleadle, L. A. McNeill, K. S. Hewitson *et al.*, *Cell* **107,** 43 (2001).
[14] D. Lando, *Science* **295,** 858 (2002).

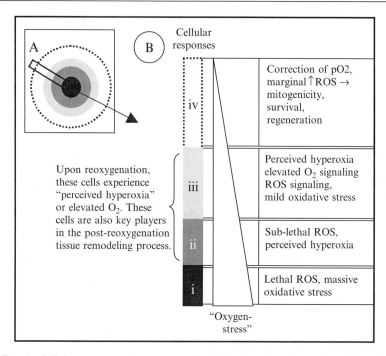

Fig. 1. Cellular responses to reoxygenation following chronic moderate hypoxia: A schematized concept. Chronic hypoxia results in cellular adjustments such that reoxygenation poses "perceived hyperoxic insult" as evident in part in the form of oxidative damage in numerous studies. (A) Diagram of hypoxia-reoxygenated tissue, with the focus of insult represented by the black center. (B) Cross section of A. Focal ischemia is known to be associated with graded oxygenation from near-zero status at focus to levels increasing with distance from focus. Important elements triggered by oxygen, molecular or reactive, during the course of reoxygenation-associated remodeling include (i) cell death or fatal injury at the focal point of insult, making room for regenerating tissues; (ii) nonfatal cellular injury, triggering reparative responses, and (iii) survival of phenotypically altered cells that favor remodeling (physiological or pathological/fibrogenic). Fibrosis denies room to regenerating healthy cells, and (iv) correction of pO_2 of mildly hypoxic cells localized beyond a critical distance from the focus of insult, favoring regeneration and restoration of physiologically functioning of the organ.

Oxygenation of the Heart

Under conditions of systemic normoxia, heart cells receive a limited supply of O_2 representing less than 10%.[15–17] We determined that the pO_2 of murine ventricular myocardium is in the range of 5%.[18] Moderate hypoxia is associated with a 30–60% decrease (~1–3% O_2) in pO_2.[19] In response to mild or moderate compromise in pO_2, adaptive processes in surviving cells allow for physiological functioning of the tissue. These adjustments are evident, for example, in the hibernating myocardium where the organ maintains vital functions in the face of prolonged moderate hypoxia.[20] Although adjustments in metabolism and contractile function have been demonstrated to allow myocardial survival in the face of reduced O_2 supply, the cellular basis of adaptation and the signaling pathways involved in the process have yet to be defined.[21]

Perceived Hyperoxia

In response to chronic moderate hypoxia, cells adjust their normoxia set point such that reoxygenation-dependent relative elevation of pO_2 results in perceived hyperoxia.[18] We hypothesized that O_2, even in marginal relative excess of the pO_2 to which cardiac cells are adjusted, results in the activation of specific signal transduction pathways that alter the phenotype and function of these cells. While acute insult caused during reperfusion may be lethal to cells localized at the focus of insult, the elevation of O_2 tension in the surrounding ischemic tissue triggers phenotypic changes in the surviving cells that may be associated with tissue remodeling.

Oxygenation *In Vitro* Does Not Model *In Vivo* Conditions

Although cells are cultured in the laboratory at an ambient O_2 concentration of 21%, which corresponds to a pO_2 of approximately 150 mm Hg at sea level, cells in the human body are exposed to much lower O_2 concentrations, ranging from ~12% in the pulmonary alveoli to less than 6% (40 mm Hg) in most other organs of the body. Culturing cells at room air is generally considered to be a "normoxic" condition. Studies related to cellular effects of hyperoxia have focused on concentrations of O_2 much higher than 21%.[22] As demonstrated in 1977, human diploid fibroblasts grown at 10% O_2 have a longer life than cells grown at the routine 20.6% O_2.[23] Consistently, the development of preimplantation embryos clearly

---

[15] S. Winegrad, D. Henrion, L. Rappaport, and J. L. Samuel, *Circ. Res.* **85,** 690 (1999).

[16] P. Gonschior, G. M. Gonschior, P. F. Conzen, J. Hobbhahn *et al., Basic Res. Cardiol.* **87,** 27 (1992).

[17] W. J. Whalen, *Physiologist* **14,** 69 (1971).

[18] S. Roy, S. Khanna, A. A. Bickerstaff, S. V. Subramanian *et al., Circ. Res.* **92,** 264 (2003).

[19] E. M. Siaghy, Y. Devaux, N. Sfaksi, J. P. Carteaux *et al., J. Mol. Cell. Cardiol.* **32,** 493 (2000).

[20] G. R. Budinger, J. Duranteau, N. S. Chandel, and P. T. Schumacker, *J. Biol. Chem.* **273,** 3320 (1998).

[21] H. S. Silverman, S. Wei, M. C. Haigney, C. J. Ocampo *et al., Circ. Res.* **80,** 699 (1997).

[22] N. Sitte, M. Huber, T. Grune, A. Ladhoff *et al., FASEB J.* **14,** 1490 (2000).

[23] L. Packer and K. Fuehr, *Nature* **267,** 423 (1977).

favors 7% O_2 over 20% O_2 ambience.[24] Ambient O_2 is dissolved readily in the cell culture medium.[25] If indeed the cells have signaling pathways sensitive to supraphysiological levels of O_2, then it is rational to assume that biology studied at 21% O_2 may not be expected to best represent cellular responses *in vivo*. The study of O_2-sensitive signal transduction pathways will enable us to examine the significance of controlling ambient O_2 conditions and to revisit results published from cells cultured in room air. In addition, we are cautiously optimistic that such studies will allow us to reconcile some of the differences between *in vitro* and *in vivo* biology and allow us to have more meaningful *in vitro* experimental models.

The experimental set-up described here was developed to control oxygen levels during all procedures of cell culture, as well as while performing live cell imaging (Fig. 2).

Cell Culture and Live Cell Imaging Under Controlled O_2 Environment

Some cellular processes (e.g., HIF1 stabilization)[26] have been shown to respond quickly (within minutes) to changes in O_2 tension. To control for this, we have the regulated O_2/CO_2 levels of cells not only in the culture incubators where cells will be grown, but also during all culture procedures (seeding, media change, etc.) and even during the isolation process. Such precise control on O_2 levels is achieved using a specially designed glove box where desired O_2 tension is maintained constantly (Fig. 2L). All cell culture procedures, including isolation, seeding, splitting, and media change, are performed in this sterile glove box. O_2 and CO_2 levels in this glove box are maintained using an electronic O_2 (PRO-OX) and CO_2 feedback regulator (PRO-CO2, BioSpherix, Redfield, NY). These controllers are fitted with sensors that sense O_2/CO_2 levels in the glove box and adjust the level of these gases accordingly within seconds. Cells are maintained in culture incubators where the desired oxygen tension (0.1–99.9%) is maintained using The OxyCycler (BioSpherix), which is a fully automated device that provides oxygen profile control for accurately modeling chronic, acute, or intermittent hypoxia or hyperoxia. The oxygen environment inside the culture incubator is monitored using a gaseous oxygen sensor (BioSpherix). Continuous monitoring of the oxygen tension in cell culture media is done using a dissolved O_2 probe (BioSpherix).

[24] P. A. Batt, D. K. Gardner, and A. W. Cameron, *Reprod. Fertil. Dev.* **3,** 601 (1991).
[25] K. Mamchaoui and G. Saumon, *Am. J. Physiol. Lung Cell. Mol. Physiol.* **278,** L858 (2000).
[26] G. L. Semenza, *Annu. Rev. Cell Dev. Biol.* **15,** 551 (1999).

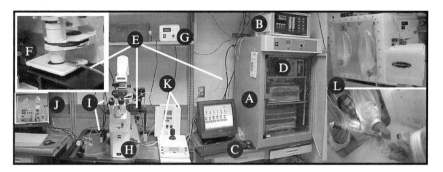

Fig. 2. Cell culture and live cell imaging under a controlled O_2 environment. Cells are grown in an incubator (A) where the gaseous environment can be regulated variably as a function of time using the OxyCycler (Biospherix, Redfield, NY) (B) controlled by PC software (C) capable of cycling (in minutes) and maintaining ambient O_2 and CO_2 concentrations based on feedback information from O_2 sensors installed in the incubator (A). Mixed gas is delivered by pump (D) via tubing (E) to a cell culture enclosure (f, zoomed inset), which is heated by g. Cells in f are imaged by an Axiovert 200 M (Zeiss) fully motorized fluorescence microscope (H) supported by dual (color and B/W) AxioCam digital cameras (I) and rested on an air table. Time-lapse images are collected and analyzed using Axiovision software (Zeiss) installed in a PC (J). The microscope contains necessary hardware/software to image cells grown on standard plastic culture plates (growing on glass coverslips is not necessary). A twin-tip micromanipulator (InjectMan NI2, Eppendorf) and microinjector (Femtojet, Eppendorf) system (K) is attached to the microscope to perform microinjections/manipulations as well as the collection of nuclear materials from single cells. For splitting/seeding of cells under a controlled gas environment, a specialized glove box (L) fitted with an O_2 controller (PRO-OX, Biospherix) is available. Six gas-controlled incubators are available for parallel experimentation. (See color insert.)

The system used to perform live cell imaging (phase contrast as well as fluorescence) under controlled oxygen conditions has been illustrated (Fig. 2). In brief, the system includes an Axiovert 200 M (Zeiss) fully motorized fluorescence microscope supported by an AxiCam digital camera (color; multicolor tissue section imaging; B/W, high-resolution fluorescence imaging) and Axiovision software to perform live cell fluorescence imaging. The microscope is fitted with a chamber where temperature and gas (O_2 and CO_2) conditions can be regulated to observe live cell processes. In this chamber, O_2 concentration between 0 and 95% can be regulated. The microscope contains necessary hardware/software to perform fluorescence resonance energy transfer (FRET) analysis. Apotome (Zeiss), recently developed hardware that provides confocal capability, has been added to the microscope. A twin-tip micromanipulator (InjectMan NI2, Eppendorf) and microinjector (Femtojet, Eppendorf) system is attached to the microscope to perform microinjections/manipulations as well as collections of nuclear materials from single cells. The microscopy system is supported

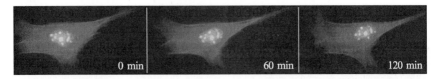

Fig. 3. Live cell imaging using fluorescence microscopy. Cotransfection of actin-EGFP (green fluorescence) and nuclear-targeted DS2-red (golden fluorescence) plasmids are used to visualize cytoskeletal and nuclear changes in live cells. Three images are shown after the indicated time from a 2-h time-lapse series where images were collected in a digital video format. Images were collected using the live cell microscopy system as described (Fig. 2). (See color insert.)

by a high-end Celsius 670 workstation. Time-dependent changes in cell morphology and actin cytoskeleton in specific O_2 environments are imaged on a time-lapse basis and are calculated using Axiovision software.

Images shown in Fig. 3 were collected using the live cell microscopy system as described (Fig. 2). Three images shown in Fig. 3 are from a 2-h time-lapse series. To visualize cytoskeletal and nuclear changes in live cells under controlled oxygen conditions, murine adult cardiac fibroblasts (CF) were cotransfected with actin-EGFP (green fluorescence) and nuclear-targeted DS2-red (golden fluorescence) plasmids (Fig. 3).

Oxygen-Regulated Signal Transduction Pathways in Cardiac Cells

Using the system described (Fig. 2), we investigated oxygen-regulated signal transduction pathways in cardiac fibroblasts (CF). Variations in cell culture O_2 levels proposed later are based on an *in vivo* situation where cardiac cells are exposed to prolonged moderately hypoxic conditions followed by reperfusion/reoxygenation. As stated earlier, under these conditions, cardiac cells will adjust to low O_2 levels and, following reoxygenation, will experience "perceived hyperoxia" (Fig. 1). Based on our own data,[18] we chose 3% O_2 levels that can be considered moderately hypoxic, whereas 10% O_2 would be marginally hyperoxic relative to the pO_2 (5%) to which cells are adjusted *in vivo*. Additionally, the effect in room air, that is, 21% O_2, is also investigated because of the following reasons: (i) to magnify the O_2-sensitive cellular responses that are seen marginally in 10% O_2 and (ii) to compare our results with existing literature where room air (21%) has been considered normoxia.

To perform the experiments, primary cultures of adult murine cardiac fibroblasts are generated and cultured as described.[27] Briefly, ventricles are removed, minced, and incubated in Hank's buffer containing trypsin (0.1 mg/ml) and collagenase (50 units/ml) for 10 consecutive 10-min treatment periods at $37°$. Cells from each digestion period are pooled,

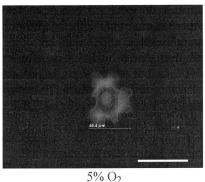
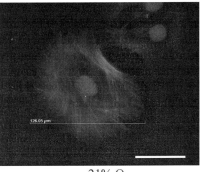

5% O$_2$ 21% O$_2$

FIG. 4. Freshly isolated CF are phenotypically closer to CF cultured at 5% O$_2$ compared to CF cultured at higher O$_2$ tensions. After isolation, CF were cultured at 5 or 21% O$_2$ for 8 days. For morphologic comparisons, CF cultured at various O$_2$ tensions for 8 days were stained with phalloidin (actin, red) and nuclei were stained with DAPI (blue). Imaging was performed using a Zeiss microscope (Fig. 2). Scale bar: 50 μm. (See color insert.)

resuspended in complete media (Dulbecco's modified Eagle medium containing 10% heat-inactivated fetal bovine serum, 100 U/ml of penicillin, and 0.1 mg/ml of streptomycin), and seeded in culture dishes and maintained under standard pH-buffered culture conditions and desired oxygen tensions. After 3 h, nonadherent debris is discarded and attached fibroblasts are maintained in complete media and culture conditions described earlier. Fibroblasts are stained for vimentin, a marker of fibroblast-like cells; the presence of contaminating vascular smooth muscle and endothelial cells is determined using desmin or factor VIII, respectively.

Compared to cells cultured at 3% O$_2$, remarkable reversible growth inhibition and a phenotype indicative of differentiation were observed in cells that were cultured in 10 or 21% O$_2$.[18] A representative figure of such a phenotype is presented in Fig. 4. CF exposed to high oxygen levels (10 or 21% O$_2$) exhibited higher levels of mitochondrial-reactive oxygen species production.[18] The molecular signature response to perceived hyperoxia included the induction of p21, cyclin D1, D2, and G1, Fra-2, transforming growth factor (TGF)β1, lowered telomerase activity, and activation of TGFβ1 and p38MAPK. CF deficient in p21 were resistant to such oxygen sensitivity.[18]

Acknowledgments

Supported by NIH-NIGMS 27345 and NIH-NINDS 42617 to CKS.

[27] S. V. Subramanian, R. J. Kelm, Jr., J. A. Polikandriotis, C. G. Orosz et al., Cardiovasc. Res. **54,** 539 (2002).

[9]   Evaluation of Cytochrome P450-4A ω-Hydroxylase and 20-Hydroxyeicosatetraenoic Acid as an O_2 Sensing Mechanism in the Microcirculation

By Julian H. Lombard, Jefferson C. Frisbee,
Richard J. Roman, and John R. Falck

Vascular Responses to Changes in Oxygen Availability

The vasculature is extremely sensitive to changes in blood and tissue PO_2, and oxygen tension is an important regulator of local blood flow and active tone in resistance vessels. Previous theories of metabolic autoregulation have generally hypothesized that the changes in arteriolar resistance and tissue blood flow that occur in response to reduced PO_2 encountered during arterial hypoxemia, reduced blood flow, or increases in tissue oxygen consumption are mediated through increases in the levels of vasodilator metabolites, for example, adenosine, in the parenchymal tissue.[1] Other investigators have proposed that blood vessels themselves are sensitive to reduced PO_2, independent of parenchymal cell influences. In some vessels, vascular relaxation in response to reduced PO_2 appears to be mediated, in whole or in part, by the release of vasodilator substances from the endothelial cells, for example, cyclooxygenase metabolites[2–4] or NO.[5] Other studies indicate that vascular smooth muscle cells themselves are sensitive to changes in O_2 availability and exhibit increased K^+ channel activity[6,7] in response to reduced PO_2.

In contrast to the vasodilation occurring in response to reduced PO_2, increasing tissue oxygen delivery elicits a vasoconstrictor response in most peripheral vascular beds.[8] This enables the microcirculation to normalize blood flow during increases in tissue perfusion and to shunt flow from areas of tissue that are oxygenated adequately to regions that require additional blood flow. The vasoconstriction that occurs in response to elevated PO_2

[1] F. J. Haddy and J. B. Scott, *Physiol. Rev.* **48,** 688 (1968).
[2] R. Busse, U. Forstermann, H. Matsuda, and U. Pohl, *Pflug. Arch.* **401,** 77 (1984).
[3] K. T. Fredricks, Y. Liu, and J. H. Lombard, *Am. J. Physiol.* **267,** H706 (1994).
[4] J. H. Lombard, Y. Liu, K. T. Fredricks, D. M. Bizub, R. J. Roman, and N. J. Rusch, *Am. J. Physiol.* **276,** H509 (1999).
[5] T. Graser and G. M. Rubanyi, *J. Cardiovasc. Pharmacol.* **20,** S117 (1992).
[6] C. Dart and N. B. Standen, *J. Physiol.* **483,** 29 (1995).
[7] D. Gebremedhin, P. Bonnet, A. S. Greene, S. K. England, N. J. Rusch, J. H. Lombard, and D. R. Harder, *Pflug. Arch.* **428,** 621 (1994).
[8] B. R. Duling, *Circ. Res.* **31,** 481 (1972).

0076-6879/04 $35.00

could be explained by the generation of a vasoconstrictor metabolite by the blood vessels or parenchymal cells (or both). Until recently, few enzyme systems were known that generate vasoconstrictor substances as tissue or intravascular PO_2 increases within the normal physiological range. However, studies[9] have demonstrated that the production of the endogenous vasoconstrictor 20-hydroxyeicosatetraenoic acid (20-HETE) by cytochrome P450-4A ω-hydroxylase depends directly on the concentration of O_2 within the normal physiological range of blood and tissue PO_2, suggesting that increased O_2 availability could cause vasoconstriction as a result of an increase in the production of 20-HETE.

Role of 20-HETE in Regulating Vascular Tone

The possible physiological roles of 20-HETE and cytochrome P450 pathways for arachidonic acid metabolism have been reviewed by Roman.[10] A variety of evidence suggests that 20-HETE and CP450 ω-hydroxylase contribute to the regulation of vascular tone (for review, see Roman[10] and Harder *et al.*[11]). 20-HETE is an endogenous vasoconstrictor compound produced by ω-hydroxylation of arachidonic acid catalyzed by enzymes of the CP450-4A and CP450-4F gene families.[10,11] 20-HETE is formed from arachidonic acid in the renal cortex[12,13] and in a variety of blood vessels, including cat cerebral microvessels,[14] rat renal microvessels,[15] rat cerebral arterioles,[10,16] rat mesenteric resistance arteries,[17] and dog renal arteries.[18] This compound has been demonstrated to constrict a variety of blood vessels, including cerebral microvessels,[14,16] mesenteric

[9] D. R. Harder, J. Narayanan, E. K. Birks, J. F. Liard, J. D. Imig, J. H. Lombard, A. R. Lange, and R. J. Roman, *Circ. Res.* **79,** 54 (1996).

[10] R. J. Roman, *Physiol. Rev.* **82,** 131 (2002).

[11] D. R. Harder, W. B. Campbell, and R. J. Roman, *J. Vasc. Res.* **32,** 79 (1995).

[12] B. W. Escalante, W. C. Sessa, J. R. Falck, P. Yadagiri, and L. Schwartzman, *J. Pharmacol. Exp. Ther.* **248,** 229 (1989).

[13] K. Omata, N. G. Abraham, and M. L. Schwartzman, *Am. J. Physiol.* **262,** F591 (1992).

[14] D. R. Harder, D. Gebremedhin, J. Narayanan, C. Jefcoat, J. R. Falck, W. B. Campbell, and R. J. Roman, *Am. J. Physiol.* **266,** H2098 (1994).

[15] A-P. Zou, J. D. Imig, M. Kaldunski, P. R. Ortiz de Montellano, S. Zhihua, and R. J. Roman, *Am. J. Physiol.* **266,** F275 (1994).

[16] D. Gebremedhin, A. R. Lange, T. F. Lowry, M. R. Taheri, E. K. Birks, A. G. Hudetz, J. Narayanan, J. R. Falck, H. Okamoto, R. J. Roman, K. Nithipatikom, W. B. Campbell, and D. R. Harder, *Circ. Res.* **87,** 60 (2000).

[17] F. Zhang, M. H. Wang, U. M. Krishna, J. R. Falck, M. Laniado-Schwartzman, and A. Nasjletti, *Hypertension* **38,** 1311 (2001).

[18] Y. H. Ma, D. Gebremedhin, M. L. Schwartzman, J. R. Falck, J. E. Clark, B. S. Masters, D. R. Harder, and R. J. Roman, *Circ. Res.* **72,** 126 (1993).

arterioles,[19] renal vessels,[15,18,20,21] and cremasteric arterioles[9,22]; several studies have suggested that 20-HETE may play a role in regulating active tone, blood flow, and vascular resistance in the peripheral circulation.[10,11]

Expression of P450 Protein and Isoforms in the Vasculature

There are substantial differences in the expression of various isoforms of cytochrome P450-4A ω-hydroxylase between tissues, between strains of rats, and with age.[10,23,24] For example, preglomerular microvessels of the rat kidney express mRNA and protein for the cytochrome P450-4A2 isoform[21,23] with smaller amounts of the CYP4A3 isoform, but the CYP4A1 and CYP4A8 isoforms were not detected. In contrast, rat cerebral arteries express P450-4A ω-hydroxylase protein; and mRNA for the CYP4A1, CYP4A2, CYP4A3, and CYP4A8 isoforms have all been reported in these vessels.[16] Studies in our laboratory[25] have demonstrated that cytochrome P450-4A ω-hydroxylase protein is expressed in arterioles and in skeletal muscle cells of the cremaster muscle of Sprague–Dawley rats. While several studies have demonstrated the expression of cytochrome P450 enzymes and the production of 20-HETE in renal and cerebral microvessels,[14,18,21] little is known regarding the specific isoforms present in the vessels and parenchymal cells of the skeletal muscle microcirculation, or in many other vascular beds.

The pattern of CYP450 isoform expression appears to change with development and may be altered in pathophysiological conditions such as hypertension.[10] With regard to development, CYP4A1 and CYP4A3 mRNA and proteins are highly expressed kidneys in neonatal rats,[24] whereas 20-HETE production increases with age[26]; but the levels of these isoforms decline in adult rats.[24] In contrast, CYP4A2 is not expressed in the

[19] M. H. Wang, F. Zhang, J. Marji, B. A. Zand, A. Nasjletti, and M. Laniado-Schwartzman, *Am. J. Physiol.* **280,** R255 (2001).

[20] M. Alonso-Galicia, J. R. Falck, K. M. Reddy, and R. J. Roman, *Am. J. Physiol.* **277,** F790 (1999).

[21] J. D. Imig, A. P. Zou, D. E. Stec, D. R. Harder, J. R. Falck, and R. J. Roman, *Am. J. Physiol.* **270,** R217 (1996).

[22] M. P. Kunert, R. J. Roman, J. R. Falck, and J. H. Lombard, *Microcirculation* **8,** 435 (2001).

[23] O. Ito, M. Alonso-Galicia, K. A. Hopp, and R. J. Roman, *Am. J. Physiol.* **274,** F395 (1998).

[24] D. L. Kroetz, L. M. Huse, A. Thuresson, and M. P. Grillo, *Mol. Pharmacol.* **52,** 362 (1997).

[25] M. P. Kunert, R. J. Roman, M. Alonso-Galicia, J. R. Falck, and J. H. Lombard, *Am. J. Physiol.* **280,** H1840 (2001).

[26] K. Omata, N. G. Abraham, B. Escalante, and M. L. Schwartzman, *Am. J. Physiol.* **262,** F8 (1992).

kidney of newborn rats, but the levels of this isoform increase until it becomes the dominant isoform expressed in the kidney of adult rats.[24] Because various CYP450-4A ω-hydroxylase isoforms appear to have different catalytic activities for the production of 20-HETE,[23,27] it is important to characterize the expression patterns of individual CYP450-4A isoforms in resistance vessels of different vascular beds during normal conditions and during pathophysiological conditions, such as hypertension. Of particular interest is the observation that spontaneously hypertensive rats (SHR), which exhibit an increased level of 20-HETE production, also appear to exhibit a different pattern of expression of P450-4A ω-hydroxylase isoforms than their normotensive controls, with a greater expression of the cytochrome P4504A1 isoform.[17] Because the catalytic activity of the CYP4A1 isoform to form 20-HETE is much greater than that of the CYP4A2 or CYP4A3 isoforms,[27] overexpression of this isoform of the enzyme could lead to an enhanced production of 20-HETE in vessels of the hypertensive rats, with a resulting increase in the sensitivity of the vessel to vasoconstrictor stimuli (possibly including elevated PO_2) that are mediated via 20-HETE formation.

Potential Role of 20-HETE in Mediating Vascular Responses to Increased Oxygen Availability

Inhibition of the formation and action of 20-HETE has a number of functional consequences in individual vascular beds and in whole animals. For example, some studies have indicated that 20-HETE plays a major role in the autoregulation of cerebral blood flow[16] in rats. In that study, elevation of pressure led to an increase in 20-HETE formation in isolated cerebral arterioles, as estimated by gas chromatography/mass spectroscopy; and inhibition of the synthesis or action of 20-HETE attenuated cerebral autoregulation in response to changes in blood pressure.[16]

Several studies have indicated that cytochrome P450-4A ω-hydroxylase and 20-HETE may play a role in oxygen sensing in the microcirculation.[9,25] For example, a concentration of 17-octadecynoic acid (17-ODYA) that is sufficient to block the formation of 20-HETE by renal arterioles[15] inhibits the constriction of skeletal muscle arterioles in response to elevated PO_2 without affecting resting tone or vasoconstrictor responses to norepinephrine.[9,25] Subsequent studies have suggested that 20-HETE plays a role in mediating arteriolar constriction in response to elevated PO_2 in the

[27] X. Nguyen, M. H. Wang, K. M. Reddy, J. R. Falck, and M. L. Schwartzman, *Am. J. Physiol.* **276,** R1691 (1999).

cremaster muscle microcirculation[9,22,25] and in the hamster cheek pouch retractor muscle.[28] Increases or decreases in 20-HETE levels also appear to contribute to O_2-induced constriction and hypoxic dilation, respectively, of isolated skeletal muscle resistance arteries.[29,30] The observation that cytochrome P450 ω-hydroxylase produces a vasoconstrictor metabolite (20-HETE) that depends on O_2 concentration over the physiological range of PO_2 suggests that 20-HETE is a likely candidate for the vasoconstrictor substance that mediates the local control of blood flow during increased O_2 availability.

Experimental Methods

Determination of the Contribution of 20-HETE to the Constriction of Arterioles to Elevated PO_2 in the Microcirculation In Situ

One approach to evaluating the role of CP450-4A ω-hydroxylase in mediating the responses of blood vessels to changes in PO_2 is to measure changes in vessel diameter as PO_2 is changed in the presence and absence of inhibitors that will block either the formation or the action of 20-HETE. Utilization of *in situ* preparations enables the investigator to evaluate the combined influence of CP450-4A ω-hydroxylase in blood vessels and in parenchymal tissue (e.g., skeletal muscle fibers) on microvascular regulation. To evaluate the role of CP450-4A ω-hydroxylase and 20-HETE in mediating the constriction of arterioles in response to elevated PO_2, the constriction of *in situ* arterioles in response to elevation of superfusion solution PO_2 can be compared before and after inhibition of the formation of 20-HETE with 17-octadecynoic acid (17-ODYA), N-methylsulfonyl-12,12-dibromododec-11-enamide (DDMS), or N-hydroxy-N'-(4-butyl-2-methylphenol)-formamidine (HET-0016) or before and after blocking the action of 20-HETE with 20-hydroxyeicosa-6(z),15(z)-dienoic acid [6(Z),15(Z)-20-HEDE].

17-ODYA is a commercially available compound (Sigma Chemical Co.) used to evaluate the role of cP450 metabolites in regulating flow, but it has the disadvantage of also inhibiting epoxygenases and is therefore less specific for selective inhibition of 20-HETE production than DDMS. DDMS is a selective inhibitor that blocks the formation of 20-HETE in

[28] J. H. Lombard, M. P. Kunert, R. J. Roman, J. R. Falck, D. R. Harder, and W. F. Jackson, *Am. J. Physiol.* **276,** H503 (1999).

[29] J. C. Frisbee, R. J. Roman, U. M. Krishna, J. R. Falck, and J. H. Lombard, *J. Vasc. Res.* **38,** 305 (2001).

[30] J. C. Frisbee, U. M. Krishna, J. R. Falck, and J. H. Lombard, *Microvasc. Res.* **62,** 271 (2001).

concentrations of 1–10 μM.[31] HET-0016 is a potent competitive inhibitor of CYP4A and 4F enzymes that is the most selective inhibitor of the formation of 20-HETE currently available.[32] 6(Z),15(Z)-20-HEDE is an inactive analog of 20-HETE used as a competitive antagonist of the actions of 20-HETE.[10,20] If 20-HETE mediates the constriction of arterioles in response to elevated PO_2, blocking the formation or action of 20-HETE with these mechanistically different inhibitors should inhibit arteriolar constriction in response to increased oxygen availability.

Materials

Surgical tools and microdissection instruments (Fine Science Tools, Foster City, CA)

Pentobarbital sodium solution, 50 mg/ml (Abbott Laboratories, North Chicago, IL)

Hypodermic needle 26 G 3/8 and 1-cc syringe (Fisher Scientific, Pittsburgh, PA)

Arterial and venous cannulas made from PE50 polyethylene tubing (Clay-Adams; Parsippany, NJ) and 23-gauge needle (Fisher Scientific); tracheal cannula made from PE 205 polyethylene tubing (Clay-Adams, Parsippany, NJ)

Blood pressure transducer (Maxxim Medical, Athens, TX), blood pressure display unit (Stoelting, Wood Dale, IL), and computer-operated data acquisition system (CODAS; DATAQ Instruments, Toledo, OH) or chart recorder (e.g., Grass polygraph; Grass Instrument Co., Quincy, MA)

Leitz Laborlux microscope

Animal platter with viewing area surrounded with silicone rubber to pin out and observe the superfused *in situ* cremaster muscle preparation

Roller pump or vacuum system to drain superfusate after it flows over the tissue

Video camera (Hitachi Model KP130 AU, Hitachi Denshi Ltd.), television monitor (RCA Model TC1115; RCA Closed Circuit Video Equipment, Lancaster, PA), and video micrometer (For-A Model IV-550 video microscaler, For-A, Tokyo, Japan)

Gas tanks and regulators: 0% O_2, 5% O_2, 10% O_2, 15% O_2, and 21% O_2; with 5% CO_2 and balance N_2

---

[31] M. H. Wang, E. Brand-Schieber, B. A. Zand, X. Nguyen, J. R. Falck, N. Balu, and M. L. Schwartzman, *J. Pharmacol. Exp. Ther.* **284,** 966 (1998).

[32] N. Miyata, K. Taniguchi, T. Seki, T. Ishimoto, M. Sato-Watanabe, Y. Yasuda, M. Doi, S. Kametani, Y. Tomishima, T. Ueki, M. Sato, and K. Kameo, *Br. J. Pharmacol.* **133,** 325 (2001).

Pharmacology organ bath (100 ml) to heat and aerate physiological salt solution (PSS)

Circulating water bath (Haake, Inc.)

Stopcocks, polyethylene tubing, and screw clamps to direct and regulate superfusate flow

Gas-impermeable Teflon tubing to connect gas tanks to pharmacology organ bath

Bicarbonate-buffered PSS: 130 mM NaCl, 4.7 mM KCl, 1.6 mM CaCl$_2$, 1.18 mM NaH$_2$PO$_4$, 1.17 mM MgSO$_4$, 14.9 mM NaHCO$_3$, and 0.026 mM disodium EDTA

Inhibitors: 17-ODYA (Sigma Chemical Co.), DDMS (Dr. J. R. Falck, University of Texas Southwestern Medical Center-Dallas, TX), HET-0016 (Taisho Pharmaceutical Saitama, Japan), 6(Z),15(Z)-20-HEDE (Dr. J.R. Falck, University of Texas Southwestern Medical Center-Dallas, TX), indomethacin (Sigma Chemical Co.), and baicalein (Biomol, Plymouth Meeting, PA).

Protocol

1. Rats are anesthetized with sodium pentobarbital (60 mg/kg; ip). Arterial and venous catheters are installed to measure mean arterial pressure and to administer supplemental anesthetics or experimental drugs as necessary, and the trachea is cannulated to ensure a patent airway.

2. The cremaster muscle is prepared for observation utilizing standard procedures that do not interrupt the deferential feed vessels.[33,34] Briefly, the scrotum is opened and the testicle and surrounding cremaster muscle are carefully freed by cutting the connective tissue between the testicle and the scrotum. Additional connective tissue adhering to the cremaster muscle is removed to improve visibility. The rat is placed on the animal platter and the tip of the muscle is secured with a small pin inserted into the silicone rubber lining the trough surrounding the viewing port. The cremaster muscle is opened via a longitudinal incision and freed from the testicle by cutting through the clear ligament closest to the testicle, which preserves flow through the deferential feed vessels.[34] The testicle is then extirpated or inserted into the abdominal cavity via the inguinal canal. The muscle is spread over the viewing port using four more pins that are spaced evenly around the perimeter of the muscle (Fig. 1). Thin strips of tissue paper are placed on the edges of the muscle to keep them moist and to draw superfusate by capillary action into the trough surrounding the viewing port. The platter is attached to the microscope stage, and the

---

[33] S. Baez, *Microvasc. Res.* **5**, 384 (1973).

[34] M. A. Hill, B. E. Simpson, and G. A. Meininger, *Microvasc. Res.* **39**, 349 (1990).

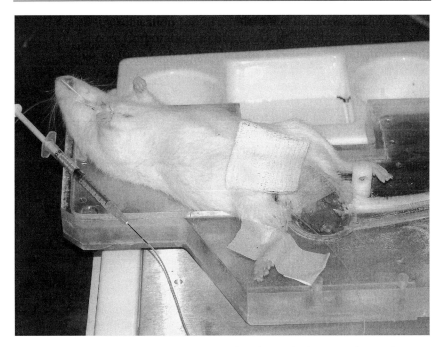

FIG. 1. *In situ* cremaster muscle preparation pinned over viewing port prior to placement on the microscope stage and initiation of superfusion with physiological salt solution.

preparation is superfused continuously with bicarbonate-buffered PSS via a polyethylene spout near the base of the muscle. Superfusate flow and bath temperature are adjusted to maintain a brisk flow rate (3–5 ml/min) and a temperature of 35° on the muscle, which approximates the normal temperature in the scrotum.

3. Arteriolar diameters are measured using television microscopy and a video micrometer.

4. Changes of active tone in response to elevated PO_2 are assessed by measuring arteriolar constriction in response to the elevation of super-fusion solution PO_2. During control conditions, the cremaster muscle is superfused continuously with a physiological salt solution equilibrated with a 0% O_2–5% CO_2–95% N_2 gas mixture to ensure that all oxygen delivery to the tissue is via the microcirculation.[35] Oxygen availability to the tissue is increased by equilibrating the superfusion solution with gas mixtures containing 5% O_2, 10% O_2, and 21% O_2, with 5% CO_2 and the balance N_2.

[35] J. H. Lombard and B. R. Duling, *Circ. Res.* **41,** 546 (1977).

5. To utilize 17-octadecynoic acid to inhibit 20-HETE formation, the compound is dissolved in 100% ethanol to prepare a 10 mM stock solution. The stock is diluted with PSS to prepare 10 ml of a 10 μM solution. The superfusion is stopped for 30 min and 17-ODYA is applied topically with a Pasteur pipette. The preparation is covered with tissue paper to keep it moist and to maintain contact with the inhibitor. After treatment with 17-ODYA or vehicle, superfusion with PSS is restored for 30 min to allow the preparation to recover. To ensure the quality of the preparation, it is important to verify the presence of active tone in the arterioles (as indicated by the occurrence of a large dilation in response to topical application of 1 mM adenosine); the absence of signs of tissue damage, for example, bleeding, leukocyte adhesion, or stasis of flow in microvessels; and the presence of a brisk flow velocity in the arterioles.

6. A more precise indication of the role of 20-HETE in regulating vessel oxygen responses can be obtained by use of DDMS, synthesized in the laboratory of Dr. John R. Falck. For *in vivo* studies, the DDMS is prepared from powder stored at $-85°$. A 10^{-2} M stock solution is initially prepared by diluting the compound in 100% ethanol, and the stock solution is stored at $-85°$. To prepare the working solution, the stock is removed from the freezer just prior to use and placed in a beaker of ice. Fifty microliters of the 10^{-2} M stock solution is removed and diluted to a final concentration of 50 μM with PSS. The tube containing the stock solution is flushed with N_2 gas, recapped, and returned to the $-85°$ freezer. The 50 μM DDMS solution is applied topically for a 30-minute period by dripping it onto the tissue while the preparation is covered with tissue paper to keep it moist and to maintain contact with the inhibitor. After 30 min, continuous superfusion of the tissue is restored with PSS containing 1 μM DDMS or vehicle, and vessel responses to elevated PO_2 are tested after recovery of control diameters and normal flow in the microcirculation.

7. In HET-0016 studies, animals receive daily subcutaneous injections (2 mg/kg) of HET-0016 dissolved in 10% lecithin in saline for 10 days, which produces HET-0016 levels that inhibit 20-HETE formation in rats.[36] Injection of HET-0016 on successive days has the added advantage of allowing any membrane stores of 20-HETE[37] to be depleted prior to the experiment. Alternatively, the HET-0016 can be added to the perfusion

[36] F. Kehl, L. Cambj-Sapunar, K. G. Maier, N. Miyata, S. Kametani, H. Okamoto, A. G. Hudetz, M. L. Schulte, D. Zagorac, D. R. Harder, and R. J. Roman, *Am. J. Physiol.* **282,** H1556 (2002).
[37] M. A. Carroll, M. Balazy, D. D. Huang, S. Rybalova, J. R. Falck, and J. C. McGiff, *Kidney Int.* **51,** 1696 (1997).

and superfusion solutions in a final concentration of 0.1–1.0 μM. This protocol blocks the synthesis of 20-HETE, but will not reduce responses associated with the release of 20-HETE from tissue phospholipid pools.

8. Arteriolar responses to increased superfusate PO_2 can also be tested before and during blockade of the action of 20-HETE by superfusing the preparation with 6(Z), 15(Z)-20-HEDE,[20] synthesized in the laboratory of Dr. John R. Falck. In this approach, the antagonist is added to the superfusion solution in a final concentration of 1–10 μM, and O_2-induced constriction is tested 20 min later. Control studies compare arteriolar responses to elevated PO_2 in animals treated with the vehicle for the antagonist. The efficacy of 6(Z),15(Z)-20-HEDE can be verified by demonstrating the ability of the antagonist to block the response of the arterioles to exogenously applied 20-HETE.[20] In previous studies in our laboratory,[30,38] 1 μM 6(Z), 15(Z)-20-HEDE completely blocked the constriction of isolated resistance arteries to exogenously applied 20-HETE.

9. The addition of exogenous 20-HETE can also be employed to determine whether an increased sensitivity of arterioles to 20-HETE may contribute to the enhanced sensitivity of microvessels to elevated PO_2 and other vasoconstrictor stimuli in hypertension. Arteriolar responses to increasing concentrations of exogenously applied 20-HETE (from 10^{-11} to 10^{-6} M) are determined in the presence of 10 μM 17-ODYA to inhibit endogenous 20-HETE formation, 10 μM baicalein to inhibit lipoxygenase, and 1 μM indomethacin to inhibit cyclooxygenase and prevent the metabolism of exogenous 20-HETE to other metabolites (e.g., a vasoactive cyclooxygenase metabolite). This approach is identical to that employed in previous studies of the response of cremasteric arterioles of normotensive animals to 20-HETE[9] and studies comparing 20-HETE sensitivity in SHR and WKY rats.[22] Vascular reactivity data are fit with the following regression equation: $y = a + b$ (log $[x]$), where y is the constrictor response (decrease in vessel diameter in response to challenge with a specific agonist), a is an intercept term, and $[x]$ is the agonist concentration. For this analysis, the b (slope) coefficient represents the change in microvessel diameter for a logarithmic change in agonist concentration. The sensitivity of arterioles to exogenously added 20-HETE is evaluated by comparing the magnitude of the 20-HETE-induced constriction of arterioles in the two groups and by calculating the β coefficient representing the change in vessel diameter for a logarithmic change in agonist concentration.[39]

[38] J. C. Frisbee, R. J. Roman, U. M. Krishnaj, J. R. Falck, and J. H. Lombard, *Microcirculation* **8**, 115 (2001).

[39] J. C. Frisbee and J. H. Lombard, *Microcirculation* **6**, 215 (1999).

*Determination of the Contribution of 20-HETE to Responses of
Isolated Resistance Arteries to Changes in PO₂*

Studies of isolated resistance arteries allow the investigator to evaluate the role of cytochrome P450-4A ω-hydroxylase within the blood vessel (i.e., independent of parenchymal cell influences) in regulating the response of arteries to changes in PO_2 and other vasoactive stimuli. In this approach, isolated resistance arteries or larger arterioles (e.g., first-order arterioles of the cremaster muscle) are cannulated with micropipettes and maintained at normal perfusion pressure, while the lumen and outside of the vessel are perfused with PSS from different reservoirs, allowing independent investigation of oxygen responses in the vessel lumen versus the outside of the vessel.[3] Studies of cannulated arteries also allow the vascular endothelium to be removed by air perfusion[3,4] in order to determine the relative roles of the endothelium and the vascular smooth muscle cells in mediating the responses of the vessels to vasoconstrictor and vasodilator stimuli.

As in the case of *in situ* microcirculation, the role of CP450-4A ω-hydroxylase in contributing to O_2 sensitivity and other vasoconstrictor responses, for example, myogenic activation in response to transmural pressure elevation, can be assessed in isolated resistance arteries by measuring the diameter of the vessels before and after inhibition of the formation or action of 20-HETE with 17-ODYA, DDMS, HET-0016, or 6(Z), 15(Z)-20-HEDE. If changes in the production of 20-HETE by the vessels mediate changes in active tone in the arteries, blocking the formation or action of 20-HETE with these different inhibitors should inhibit constriction of the vessels in response to increased PO_2 and other vasoconstrictor stimuli mediated by 20-HETE. These inhibitors can also be used to evaluate the role of reduced 20-HETE production in mediating dilator responses to decreased PO_2. In the latter strategy, vessel responses to reduced PO_2 are compared with and without blockade of 20-HETE formation or action, and the contribution of reduced 20-HETE production to the responses is evaluated by comparing the magnitude of the dilation in the two groups, as described previously by Frisbee *et al.*[40]

The responses of isolated resistance arteries to elevated PO_2 have not been studied extensively, whereas vessel responses to reduced PO_2 have been more widely investigated. In isolated middle cerebral arteries and skeletal muscle resistance arteries, vasodilation in response to reduced PO_2 appears to be due primarily to elevations in endothelium-derived vasodilator compounds (especially prostacyclin) rather than changes in

[40] J. C. Frisbee, K. G. Maier, J. R. Falck, R. J. Roman, and J. H. Lombard, *Am. J. Physiol.* **283**, R309 (2002).

the levels of cytochrome P450 products, such as 20-HETE.[3,4,40] However, Frisbee et al.[40] demonstrated that changes in 20-HETE levels make a modest contribution to the dilation of skeletal muscle resistance arteries in response to reduced PO_2 and that changes in 20-HETE production contribute to the altered responses to hypoxia in skeletal muscle resistance arteries of Dahl S hypertensive rats compared to normotensive controls.[38]

Materials

Surgical tools and microdissection instruments (Fine Science Tools)

Pentobarbital sodium solution, 50 mg/ml (Abbott Laboratories)

Hypodermic needle 26 G 3/8 and 1-cc syringe (Fisher Scientific)

Pressure transducers (Maxxim Medical), blood pressure display unit (Stoelting), and computer-operated data acquisition system (CODAS; Dataq Instruments) or other recording system (e.g., Grass polygraph) for initial blood pressure measurements and to monitor inflow and outflow pressures during studies of perfused arteries

Isolated vessel chamber (fabricated in house or purchased from Living Systems, Inc., Burlington, VT).

Glass micropipettes fabricated with a microelectrode puller (Model 700C, David Kopf, Instruments, Tujunga, CA).

Video camera (Hitachi Model KP130 AU), television monitor (RCA Model TC1115; RCA Closed Circuit Video Equipment), and video micrometer (For-A Model IV-550 video microscaler)

Bicarbonate-buffered PSS: 119 mM NaCl, 4.5 mM KCl, 1.17 mM $MgSO_4$, 1.18 mM NaH_2PO_4, 1.6 mM $CaCl_2$, 0.026 mM EDTA, 5.5 mM glucose, and 24 mM $NaHCO_3$

Gas tanks and regulators: 0% O_2, 5% O_2, 10% O_2, 15% O_2, 21% O_2, 50% O_2, and 95% O_2; with 5% CO_2, and the balance N_2

60-ml syringe as PSS reservoir for perfusion solution

Pharmacology organ bath (100 ml) to heat and aerate PSS before flowing onto tissue

Circulating water bath (Haake, Inc.) to maintain perfusate/superfusate temperatures at 37°

Polyethylene tubing for connections between vessel chamber and reservoirs for perfusion and superfusion solutions

Gas-impermeable Teflon tubing to connect gas tanks to perfusion and superfusion solution reservoirs

Air stones for gas equilibration in perfusion reservoir and vessel chamber; glass microscope slides to cover vessel chamber during gas equilibration

Inhibitors: 17-ODYA (Sigma Chemical Co.), DDMS (Dr. J. R. Falck), HET-0016 (Taisho Pharmaceutical), and 6(Z), 15(Z)-20-HEDE (Dr. J. R. Falck)

Protocol

1. Rats are anesthetized with sodium pentobarbital (60 mg/kg, ip), and arterial pressure is measured via a cannula in the carotid artery.

2. Resistance arteries are isolated using microdissection techniques, taking care to avoid stretching. Vessels should be handled only by the connective tissue at the ends. The arteries are cannulated with micropipettes at both ends in a vessel chamber where they can be perfused and superfused with physiological salt solution equilibrated with different concentrations of oxygen (Fig. 2). Inflow and outflow pressures are measured with pressure transducers and a computer-operated data acquisition system and are maintained at values approximating the blood pressure encountered *in vivo,* for example, 80 mm Hg for middle cerebral arteries, 60 mm Hg for first-order cremasteric arterioles, and 100 mm Hg

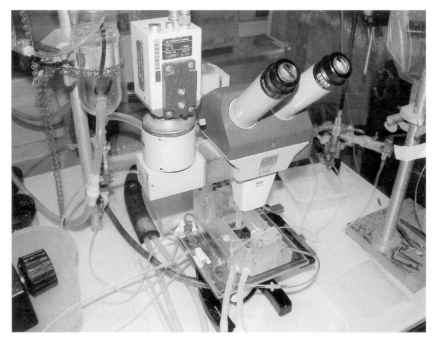

Fig. 2. Microscope and isolated vessel chamber for studies of perfused resistance arteries.

for isolated gracilis arteries. Vessel diameters are measured using television microscopy and a video micrometer. (Fig. 3).

3. Changes of active tone in response to elevated PO_2 are assessed by measuring vessel diameter during simultaneous changes in perfusion and superfusion solution PO_2. During control conditions, the artery is continuously perfused and superfused with a physiological salt solution equilibrated with a 21% O_2–5% CO_2–74% N_2 gas mixture. Oxygen availability to the tissue is increased or decreased by equilibrating the superfusion solution with gas mixtures containing different percentages of O_2, with 5% CO_2 and the balance N_2.

4. To utilize 17-ODYA to inhibit 20-HETE formation, the compound is prepared as described earlier to obtain a 10 mM stock solution, which is then diluted with PSS to a final concentration of 10 μM. The PSS containing 10 μM 17-ODYA (or its vehicle) is added to the vessel chamber while superfusion of the vessel is stopped, and the temperature in the chamber is monitored carefully. The vessels are equilibrated with 17-ODYA for a minimum of 45 min, as the compound is a suicide substrate inhibitor. After treatment with 17-ODYA or vehicle, superfusion with PSS is restored, and the vessel is allowed 30 min to recover.

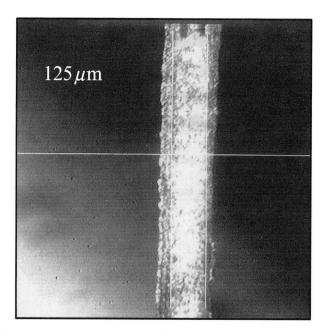

125 μm

FIG. 3. Video micrometer measurement of the diameter of an isolated resistance artery.

Maintenance of active tone in the vessels can be assessed by recording the change in vessel diameter in response to perfusion and superfusion with Ca^{2+}-free relaxing solution to produce maximal dilation of the vessels.

5. As noted earlier, a more precise indication of the role of 20-HETE in regulating vessel O_2 responses can be obtained using DDMS. For studies of isolated arteries, the DDMS solution is prepared as described in the previous section. The DDMS solution (50 μM) is applied to the tissue for 30 min by adding the inhibitor (or its vehicle) to the superfusion solution while flow is stopped, and the temperature is monitored carefully. After 30 min, continuous superfusion of the vessel is restored with PSS containing 1 μM DDMS or vehicle, and the responses of the artery to changes in PO_2 are tested after a 30-min recovery period.

6. For HET-0016 studies, responses of vessels to changes in perfusion and superfusion solution PO_2 are compared in arteries isolated from rats receiving subcutaneous injections of HET-0016 (2 mg/kg) for 10 days, as described previously. As in the case of studies of the *in situ* microcirculation (discussed earlier), HET-0016 can be added to the perfusion and superfusion solutions in a final concentration of 0.1–1.0 μM, although this approach does not eliminate the release of 20-HETE from tissue phospholipid pools.

7. Arteriolar responses to changes in perfusate and superfusate PO_2 can also be tested before and during blockade of the action of 20-HETE by superfusing the preparation with 1–10 μM 6(Z), 15(Z)-20-HEDE.[20] During those studies, the antagonist is added to the superfusion solution, and vessel responses to changes in PO_2 or other vasoconstrictor stimuli are determined after a 30-min equilibration period. Control studies compare vessel responses to changes in PO_2 in animals treated with the vehicle for the antagonist. The efficacy of 6(Z), 15(Z)-20-HEDE can be verified by demonstrating the ability of the antagonist to block the response of the arterioles to exogenous 20-HETE.[20,30,38]

8. MS-PPOH is a selective inhibitor of cytochrome P450-epoxygenase enzymes. It has minimal effects on ω-hydroxylation of arachidonic acid and on the production of 20-HETE.[31] To evaluate the contribution of epoxygenase products to changes in vascular tone during alterations of PO_2, MS-PPOH stock (10 mM in absolute ethanol) can be added to the chamber to achieve a final concentration of 20 μM. The latter approach has been used to demonstrate a contribution of epoxygenase metabolites to hypoxic dilation of isolated skeletal muscle resistance arteries of Dahl S rats on a low salt diet.[39]

9. The role of epoxyeicosatrienoic acids (EETs) acting as endothelium-derived hyperpolarizing factors can also be tested using a 14,15-EET

analog, 14,15-epoxyeicosa-5(Z)-enoic acid (14,15-EEZE).[41] The compound is prepared as a 10 mM/liter stock in 95% ethanol, and vessels are incubated with 14,15-EEZE (10 μM) in the perfusion and superfusion solutions for 20 min. The 14,15-EEZE inhibits relaxations to 14,15-EET, but does not alter the vasoconstrictor response to 20-HETE.[41]

Evaluation of Cytochrome P450-4A ω-Hydroxylase Expression in Arterioles and Parenchymal Tissue by Western Blotting

Because cytochrome P-450 ω-hydroxylase produces 20-HETE in an O_2-dependent manner, localization of the enzyme can provide valuable clues about the potential role of this enzyme in contributing to O_2-dependent changes in blood flow. For example, a long-standing question regarding oxygen sensing in the microcirculation is whether the O_2 sensor is located in the arterioles, in the parenchymal tissue, or both. Kunert et al.[25] used Western blotting techniques to demonstrate that cytochrome P450 protein is found both in isolated arterioles and in parenchymal tissue (isolated skeletal muscle cells) of the cremaster muscle. The latter observation suggests that cytochrome P-450 ω-hydroxylase could contribute to O_2-dependent regulation of blood flow in the microcirculation by sensing changes in O_2 availability in the microvessels and in the parenchymal tissue.

Materials

Anesthesia and Blood Pressure Measurement

Pentobarbital sodium solution, 50 mg/ml (Abbott Laboratories)
Hypodermic needle 26 G 3/8 and 1-cc syringe (Fisher Scientific)
Cannula from PE50 polyethylene tubing (Clay Adams) and 23-gauge needle (Fisher Scientific)
Pressure transducer, blood pressure amplifier, and computer-operated data acquisition system for initial blood pressure measurement

Dissection

Silicone-coated glass petri dish (prepared in our department)
Insect pins (Fine Science Tools)
Extra delicate mini Vannas scissors (Fine Science Tools)
Fine scissors (Fine Science Tools)

[41] K. M. Gauthier, C. Deeter, U. M. Krishna, Y. K. Reddy, M. Bondlela, J. R. Falck, and W. B. Campbell, *Circ. Res.* **90**, 1028 (2002).

Dumont straight-tip microsurgery forceps (Fine Science Tools)
Dumont curved-tip microsurgery forceps (Fine Science Tools)
Dissecting microscope, 25x magnification (Leica, San Dimas, CA)
Methanol (Fisher Scientific)
NaCl (Fisher Scientific)
$MgSO_4$ (Sigma Co.)
KCl (Sigma Co.)
$CaCl_2$ (Sigma Co.)
L-Alanine (Sigma Co.)
Sodium acetate (Sigma Co.)
N-2-Hydroxyethypiperazine-N'-ethanesulfonic acid (HEPES) (Fisher Scientific)

Protein Isolation

Kontes pellet pestle motor (Fisher Scientific)
PowerGen 125 (Fisher Scientific)
Sonic Dismembrator 60 (Fisher Scientific)
Micromax RF Centrifuge (Fisher Scientific)
Optima TL ultracentrifuge (Beckman Coulter, Fullerton, CA)
Microcentrifuge tubes (Fisher Scientific)
KH_2PO_4 (Sigma Co.)
K_2HPO_4 (Sigma Co.)
EDTA (Sigma Co.)
Dithiothreitol (DTT) (Sigma Co.)
Glycerol (Sigma Co.)
Sucrose (Sigma Co.)
EDTA (GIBCO-BRL, Grand Island, NY)
Phenylmethylsulfonyl fluoride (PMSF) (Sigma Co.)
Protease cocktail inhibitor (Sigma Co.)

Protein Determination

Bio-Rad protein assay (Bio-Rad, Hercules, CA)
Fisher brand polystyrene disposable cuvettes (Fisher Scientific)
DU 640 spectrophotometer (Beckman)

Western Blotting

Protean II electrophoresis system (Bio-Rad)
Minitank electroblotting system (Owl Separation Systems, Portsmouth, NH)
Model OSP 500 power supply (Bio-Rad)
10x Tris-glycine running buffer (Bio-Rad)
10.0% SDS–PAGE precast minigel (Bio-Rad)

0.25-μm nitrocellulose membrane (Bio-Rad)

Tris–hydroxymethyl aminomethane (Tris–HCl) (Fisher Scientific)

Tris–hydroxymethyl aminomethane free base (Tris base) (Fisher Scientific)

Glycine (Sigma Co.)

Methanol (Fisher)

Nonfat dry milk (Bio-Rad)

Daiichi P450-4A1 anybody (Gentest, Woburn, MA)

Tween 20 (Bio-Rad)

ProSieve color protein marker (Biowhittaker Molecular Applications, Rockland, ME)

Antigoat secondary antibody (Santa Cruz Biotechnology, Santa Cruz, CA)

Pierce SuperSignal West Pico chemiluminescent substrate (Fisher Scientific)

Kodak Biomax film (Fisher Scientific)

Konica SRX-101 film developer (Wayne, NJ)

Scion image software for Windows (NIH, Bethesda, MD)

Buffer Composition

HEPES PSS, pH 7.4: 140 mM NaCl, 5.4 mM KCl, 1 mM MgCl$_2$, 1.8 mM CaCl$_2$ 10 mM glucose, and 5 mM HEPES

Homogenization buffer, pH 7.7: 250 mM sucrose, 1 mM EDTA, 10 mM potassium phosphate buffer

Microsome resuspension buffer, pH 7.25: 100 mM potassium phosphate buffer, pH 7.25, 1 mM EDTA, 1 mM DTT, and 30% glycerol

Sample buffer: 12 mM Tris–HCl buffer, 5% glycerol, 0.4% SDS, 0.02% bromphenol blue, and 14.4 mM 2-mercaptoethanol (add just before use)

Transfer buffer, pH 8.3: 25 mM Tris–HCl buffer, 190 mM glycine, and 20% methanol

Tris-buffered saline (TBS), pH 8.0: 10 mM Tris buffer and 150 mM NaCl

Procedure

1. Rats are anesthetized with pentobarbital (60 mg/kg, ip), and arterial blood pressure is measured with a cannula in a carotid or femoral artery.

2. The scrotum is opened and the cremaster muscle is pinned out quickly with insect pins in a frozen, silicone-coated glass petri dish. Tissues are flushed immediately with 20 ml of ice-cold PSS and then with cold methanol.

3. Tissues for dissection are maintained in methanol on ice. The vasculature is viewed under low magnification (25x) through a dissecting microscope, and microvessels and parenchymal tissue (e.g., individual skeletal muscle cells) are obtained utilizing microdissection techniques. The isolated vessels and parenchymal tissues are kept in ice-cold methanol and washed twice. After washing, the tissues are quick-frozen in liquid nitrogen and are kept at $-80°$ until ready for the homogenization step. If whole muscle samples are taken, these should be quick-frozen in liquid nitrogen and are kept at $-80°$ until ready for the homogenization while the rest of the samples are dissected.

4. Whole muscle samples are homogenized with a polytron tissue homogenizer while isolated arterioles and skeletal muscle cells are homogenized with a Kontes pellet pestle motor using a small amount of homogenization buffer. Each tube is sonicated for 3 s.

5. Tissue debris and nuclear fragments are removed by low-speed centrifugation (3000 g, $4°$, for 15 min and 21,000 g, $4°$, for 15 min). After the second spin, the protein concentration of the supernatant from tubes containing isolated vessels and skeletal muscle cells is determined using the Bio-Rad protein assay, with albumin as a standard, whereas homogenates from the whole cremaster muscle are centrifuged for a third time. Homogenates from isolated vessels and skeletal muscle cells are aliquoted to about 5 μl so that the total amount of protein in each tube (approximately 20 μg) is sufficient to run at least one Western blot experiment. Samples are stored at $-80°$ until use.

6. Membrane and cytosolic proteins from whole muscle samples are separated by centrifugation at 101,000 g for 90 min at $4°$. The precipitates are redissolved in a resuspension buffer, and the protein concentration is determined using the same method employed for isolated vessels and skeletal muscle cells. Homogenates are aliquoted to about 5 μl so that the amount of protein in each tube (approximately 20 μg) is sufficient to run at least one Western blot experiment. Samples are stored at $-80°$ until use.

7. Electrophoresis is performed on microsomal proteins (5–20 μg) in Laemmli sample buffer. Proteins are separated on a 10% SDS–PAGE precast minigel and transferred to a 0.2 μM nitrocellulose membrane for 2 h at room temperature using a 200-mA current. Equal amounts of protein from each experimental group are loaded on the same gel to eliminate any influence of gel-to-gel variation.

8. Following transfer, the membranes are immersed overnight in a blocking solution consisting of 7.5% nonfat dry milk in Tris-buffered saline with 0.1% Tween 20 (TBST). The membranes are then incubated for 2 h in

1   2   3   4   5

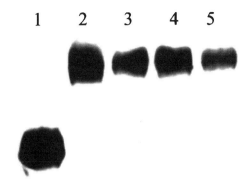

FIG. 4. Western blot showing cytochrome P450-4A ω-hydroxylase protein from whole cremaster muscle of SHR and WKY rats. Lane 1 shows the purified microsomal standard. Lanes 2 and 4 contain 20 μg protein from SHR cremaster muscles, and lanes 3 and 5 contain 20 μg protein from WKY cremaster muscles. The protein from the cremaster muscle runs at a heavier molecular weight than the purified microsomal standard because the enzyme is glycosylated *in vivo*.

a Gentest rat P450-4A antibody (cross-reactive with all 4A isoforms) that is diluted 2000-fold in 2.5% nonfat dry milk in TBST.

9. After incubation with the P450-4A antibody, membranes are washed for 5 min with TBST (4×) and incubated for 2 h in a horseradish peroxidase-coupled secondary antibody (Santa Cruz Biotechnology), which is diluted 1:5000-fold in 2.5% nonfat dry milk in TBS20. After incubation with the secondary antibody, membranes are washed four times with TBST and once with TBS. Bound antibodies are detected by developing with the Pierce SuperSignal Pico chemiluminescent substrate for 5 min. The signals are transferred to imaging film and developed in a Konica SRX-101 film developer (Fig. 4). Relative intensity of the bands migrating in 50–52 and 78–82 kDa (glycosylated isoforms) is quantified by measuring band density using Scion Image–Release Beta 4.0.2 computer software.

Identification of Cytochrome P450-4A Isoforms in Isolated Cremasteric Arterioles and Parenchymal Tissue by Reverse Transcriptase–Polymerase Chain Reaction (RT-PCR)

As noted earlier, several isoforms of cytochrome P-450 ω-hydroxylase are expressed, and these may have different catalytic activities for forming 20-HETE. The relative expression of the various isoforms of cytochrome P-450 ω-hydroxylase differs between tissues and may be altered by

pathophysiological conditions such as hypertension. Highly specific RT—
PCR techniques have been developed by Ito *et al.*[23] that enable specific
cytochrome P450-4A isoforms to be identified in various tissues and/or
during various physiological conditions.

In this section, instructions for RT-PCR procedures are taken from
product literature provided by Qiagen (Hilden, Germany). Primer pairs
for the 4A1, 4A2, and 4A3 isoforms and GAPDH are from Ito *et al.*[23]
and the primer pair for 4A8 isoform is from Gebremedhin *et al.*[16] Thermocycler settings and procedures are taken from Ito *et al.*[23] and should be
followed precisely, as the extremely close (98%) sequence homology between 4A2 and 4A3 isoforms requires higher annealing temperatures to
prevent coamplification of the two isoforms.

Materials

Pressure transducer, blood pressure amplifier, and computer-operated
data acquisition system for initial blood pressure measurement
Dissecting instruments (Fine Science Tools) and dissecting micro-
scope (Leica)
Silicone-coated glass petri dish
Liquid nitrogen or dry ice
DNA thermal cycler 480 (Perkin Elmer)
Hanks balanced salt solution (Biowhittaker, Walkersville, MD)
100% methanol (Fisher Scientific, Fair Lawn, NJ)
Ribonucleoside vanadyl complex (Sigma Co.)
Polypropylene microcentrifuge tubes (ISC BioExpress, Kaysville,
UT)
Eppendorf Biopur (Brinkmann Instruments, Inc., Westbury, NJ)
RNase/DNase-free thin wall tube (Midwest Scientific, Valley Park,
MO)
TRIzol reagent (TRIzol, Invitrogen Life Technologies, Carlsbad, CA)
Choloroform (Fisher Scientific)
Isopropanol (Sigma Co.)
Proteinase K (Sigma Co.)
RNase-free H_2O (Qiagen, Hilden, Germany)
Ethyl alcohol USP-200 (Aaper Alcohol and Chemical Co., Shelby-
ville, KY)
RQ1 RNase-free DNase (including reaction buffer and stop solution)
(Promega, Madison, WI)
Centrifuge (micromaxRF, IEC)
Omniscript RT kit (200) or Sensiscript RT kit (50) (Qiagen)

HotStarTaq master mix kit (1000 U) (Qiagen)

Random decamers (Ambion)

UPERase-In The Ultimate RNase inhibitor (Ambion)

10× TBE buffer (Ambion) and electrophoresis box (Bio-Rad)

Agarose (GIBCO, Life Technologies, Grand Island, NY)

100-bp DNA ladder (Promega)

Ethidium bromide solution (Bio-Rad Laboratories)

Rat glyceraldehyde 3-phosphate dehydrogenase (G3PDH) control amplimer (Clontech Laboratories, Inc., East Meadow Circle, CA)

Primer pairs (Operon Technologies Inc., Alameda, CA): 4A1 forward, 5′-GTA TCC AAG TCA CAC TCT CCA-3′; **4A1 reverse,** 5′-CAG GAC ACT GGA CAC TTT ATT G-3′; **4A2 forward,** 5′-AGA TCC AAA GCC TTA TCA ATC-3′; **4A2 reverse,** 5′-CAG CCT TGG TGT AGG ACC T-3′; **4A3 forward,** 5′-CAA AGG CTT CTG GAA TTT ATC-3′; **4A3 reverse,** 5′-CAG CCT TGG TGT AGG ACC T-3′; **4A8 forward,** 5′-ATC CAG AGG TGT TTG ACC CTT AT-3′; **4A8 reverse,** 5′-AAT GAG ATG TGA GCA GAT GGA GT-3′; **GAPDH forward,** 5′-CAC GGC AAG TTC AAT GGC ACA-3′; and **GAPDH reverse,** 5′-GAA TTG TGA GGG AGA GTG CTC-3′

Procedures

Isolation of Arterioles and Parenchymal Tissue

1. Rats are anesthetized with pentobarbital (60 mg/kg, ip), and the arterial pressure is measured with a cannula in the carotid or femoral artery.

2. After opening the scrotum, the cremaster muscle is isolated quickly with scissors and put into cold (4°) HBSS to wash out blood. After washing, the cremaster muscle is pinned out in a silicone-coated glass petri dish filled with methanol (4°) plus the RNase inhibitor.

3. The petri dish is placed on ice, and microvessels and parenchymal tissue, for example, individual skeletal muscle fibers, are isolated carefully using microdissection techniques. Individual samples are placed in Eppendorf tubes containing the TRIzol reagent, frozen in liquid nitrogen shortly after isolation, and immediately stored at −80° until they are homogenized.

Extraction of RNA from Arterioles and Parenchymal Tissue

1. Add 400 μl of the TRIzol reagent to an RNase/DNase-free tube on ice and put the desired tissue (<100 mg) into the tube for subsequent

homogenization. When homogenization is finished, another 600 μl of the TRIzol reagent is added to the tube to bring the total volume up to 1 ml.

2. Incubate the homogenized sample at room temperature for 15 min. *Note*: When working on skeletal muscle fibers and other tissues that are rich in proteins, 20 μl of proteinase K (10 mg/ml) should be added to the tube and incubated at 55° for 10 min, followed by incubation at room temperature for an additional 15 min.

3. Add 200 μl of chloroform to the sample for each milliliter of Trizol reagent used and shake vigorously for 15 s. Let the sample incubate at room temperature for an additional 2–3 min.

4. Centrifuge the sample (12,000 g at 4° for 15 min) to separate it into three layers: a lower organic layer containing proteins, an interphase layer containing DNA, and an upper aqueous layer, which contains the RNA.

5. Remove the upper aqueous layer using a pipette and transfer it into a new Eppendorf tube. Add 500 μl of isopropanol (adjust volume based on starting amount of the TRIzol reagent) to the tube and invert twice. Let the sample stand at room temperature for 15 min to precipitate the RNA.

6. Pellet the RNA by centrifuging the samples at 12,000 g for 20 min at 4°. The pellet should be white.

7. Remove the supernatant and add 1 ml of 75% ethanol to the pellet to remove any lingering chemicals, including residual phenol. Invert the tube twice and centrifuge it at 12,000 g for 5 min at 4°.

8. To eliminate any genomic DNA contamination, dissolve the RNA pellet in 44 μl of DEPC-treated water and add 5 μl of RQ1 reaction buffer and 1 μl of RQ1 RNase-free DNase to the tube.

9. Incubate samples at 37° for 1 h to allow the DNase to work and then stop the reaction by incubating the samples at 65° for 10 min after adding 1 μl of RQ1 stop solution.

10. Add 500 μl of the TRIzol reagent to the samples and repeat steps 3–6 to reextract the RNA.

11. Suspend the final RNA pellet in 30–50 μl of nuclease-free water and determine the concentration of RNA by measuring the absorbance at 260 nm (A_{260}) in a spectrophotometer as follows: concentration of RNA (μg/μl) = 40 μg/ml $\times$ OD$_{260}$ $\times$ dilution factor/1000 ml. *Note*: This equation is based on a sample of 40 μg RNA/ml producing an OD$_{260}$ of 1.0.

12. After the absorbance measurement, store the RNA at −80°.

First-Strand cDNA Synthesis and No Reverse Transcriptase
(NoRT) Control Reactions

1. Prepare a master mix for each sample of RT reaction using an Omniscript RT kit (RNA between 50 ng and 2 μg) as follows:

Master mix	Volume	Final concentration
10X buffer RT	2 μl	1 X
dNTP mix (5 m*M* of each dNTP)	2 μl	0.5 m*M* each dNTP
Random decamers (10 μ*M*)	2 μl	1 μ*M*
UPERase-In (10 units/μl)	1 μl	10 units (per 20-μl reaction)
Reverse transcriptase	1 μl	4 units (per 20-μl reaction)

2. For the NoRT control master mix (7 μl), utilize the same formula, but omit the reverse transcriptase.

Note: Sufficient master mix should be prepared to complete the required number of samples plus one additional reaction to compensate for losses resulting from pipetting.

Prepare RT and NoRT Reactions

1. Add template RNA (50 ng–2 μg) to individual tubes containing the master mix. The volume of the master mix will be 8 μl for the RT reaction and 7 μl for the NoRT control because the reverse transcriptase is not included in the latter tube. Use RNase-free water to bring the total volume up to 20 μl in each tube.

2. Perform RT and NoRT reactions by incubating the samples at $37°$ for 120 min followed by a 5-min incubation at $95°$. After the $95°$ incubation, cool the samples rapidly on ice. *Note*: The standard reverse-transcription reaction requires a 60-min incubation. However, the manufacturer of the RT kit (Qiagen) indicates that it may be advantageous to increase the incubation time to 2 h when analyzing RNAs with a very high degree of secondary structure. A 2-h incubation period is used for the reverse-transcription reaction for cytochrome P450 isoforms.

PCR Amplification

1. Place an aliquot (4 μl) of RT product (*first-strand* cDNA) or NoRT product, respectively, into a PCR tube suitable for a 50-μl PCR reaction system.

2. Add specific P450-4A primers and HotStarTaq master mix into the PCR tube and adjust the total volume up to 50 μl by distilled water according to the following formula:

Component	Volume	Final concentration
Primer 4A(n)-forward	1.25 μl	0.5 μM
Primer 4A(n)-reverse	1.25 μl	0.5 μM
cDNA	4 μl	< 1 μg/reaction
Distilled water	18.5 μl	
HotStarTaq master mix	25 μl	2.5 units HotStarTaq DNA polymerase 1 × PCR buffer 200 μM of each dNTP
Total volume	50 μl	

Note: n = 1, 2, 3, and 8.

3. To check the quality of the extracted RNA sample and PCR reaction, the housekeeping gene, GAPDH, and the G3PDH control amplimer should be used for each PCR run. To exclude any nonspecific DNA contamination from the experimental reagents, perform negative controls by conducting a PCR reaction with distilled water for each PCR run.

4. Add one drop of mineral oil to each PCR tube and place the samples in the DNA thermal cycler. The PCR program starts with an initial heat activation step at 95° for 15 min. The reaction is then cycled 40 times at 94° (denaturation) for 1 min, 55° (annealing) for 2 min and 72°C (extension) for 1 min. The 4A2 and 4A3 isoforms are cycled using a higher annealing temperature of 70° for 10 cycles, followed by 25 cycles at 60°, in order to prevent coamplification due to the close sequence homology of these two isoforms.[23] All samples are then incubated at 72° for an additional 10 min after the completion of the final cycle.

Agarose Gel Electrophoresis

1. Prepare a 1% agarose gel by adding 0.6 g of agarose to 60 ml of 1X TBE buffer. Heat this mixture in a microwave oven until the agarose dissolves. As the agarose is cooling (50–60°), add 6 μl of ethidium bromide to the solution.

2. Cast the mixture by cooling in a gel tray with a comb and allow it to set for 30 min.

3. Remove the comb, place the gel tray in the electrophoresis box, and fill the box with 1X TBE buffer until the gel is immersed completely.

4. Load the DNA ladder and aliquots of each PCR reaction product (10 μl) plus 2 μl of loading dye into the individual preformed wells.

5. Run the gel at 90 V for 45 min and visualize it under ultraviolet light (Fig. 5). The molecular weight of the PCR product is determined by comparison with the 100-bp DNA ladder. The 4A1, 4A2, 4A3, 4A8, and GAPDH primer pairs amplify fragments of 827, 321, 321, 349, and 970 bp,

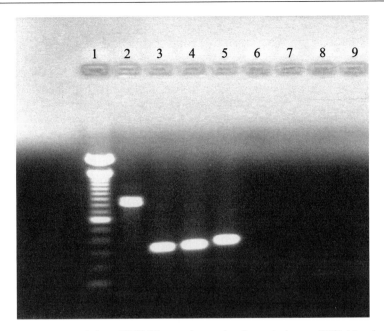

FIG. 5. Agarose gel from RT-PCR experiment showing cytochrome P450-4A ω-hydroroxylase isoforms in rat liver. Lane 1, 100-bp ladder; lane 2, CP450-4A1; lane 3, CP450-4A2; lane 4, CP450-4A3; lane 5, CP450-4A8; and lanes 6–9, negative controls (NoRT reactions) for the various isoforms.

respectively. *Note*: Primer pairs for the 4A1, 4A2, and 4A3 isoforms and for GAPDH are from Ito *et al.*,[23] and the primer pair for the 4A8 isoform is from Gebremedhin *et al.*[16]

Acknowledgments

The authors express their sincere appreciation to Lourdes de la Cruz, Jingli Wang, and Tianjian Huang for their assistance in preparing the manuscript. Supported by NIH HL-29587 (JHL, JCF, RJR), NIH HL65289 (JHL, JCF), NIH HL-72920 (JHL, JCF), NIH GM31278 (JRF), and the Robert A. Welch Foundation (JRF).

[10] Assessment of Roles for Oxidant Mechanisms in Vascular Oxygen Sensing

By Michael S. Wolin, Pawel M. Kaminski, Mansoor Ahmad, and Sachin A. Gupte

The focus of this article is to provide approaches and methods for defining and investigating the role of reactive O_2 species (ROS) and related redox processes in mechanisms of O_2 sensing in vascular tissue. Approaches developed for studies on the effects of interventions on acute changes in force generation are emphasized because they currently provide substantial insight into the integrated function of the O_2 sensor being investigated and on the selectivity of the probes being used.

Definition of the Role of Mediators and Limitations in Energy Metabolism in PO₂-Elicited Responses

Previous studies have provided evidence that non-ROS products of cyclooxygenase (prostaglandins), nitric oxide synthase (NO), and cytochrome P450 (eicosanoids) are often key contributors to PO_2-elicited vascular responses.[1] While these substances are made in multiple cell types in the vessel wall, the endothelium is usually the major source in vascular tissue of these biologically active mediators. Thus, the use of endothelium removal and pharmacological inhibition of these systems with the most selective inhibitors available need to be investigated initially. Since all of these sources of mediators have been shown to have the potential to generate vasoactive levels of endothelium-derived hydrogen peroxide (H_2O_2),[2,3] it is important to examine if endothelium-dependent responses are inhibited by catalase. Since limitations in mitochondrial energy metabolism are also a potential source of PO_2-elicited responses, it is also important to consider if this is a contributing factor, especially at very low levels of O_2. It may be necessary to measure tissue levels of high-energy metabolites such as ATP and creatinine phosphate to evaluate if energy metabolism is limiting because the interactive nature of mitochondrial function with redox systems that regulate oxidant signaling.

[1] M. S. Wolin, T. M. Burke-Wolin, and K. M. Mohazzab-H, *Respir. Physiol.* **115**, 229 (1999).
[2] M. S. Wolin, *Arterioscler. Thromb. Vasc. Biol.* **20**, 1430 (2000).
[3] M. S. Wolin, S. A. Gupte, and R. A. Oeckler, *J. Vasc. Res.* **39**, 191 (2002).

Identification of the Role of ROS in PO_2-Elicited Responses Using
  Mechanistic Probes That Alter Oxidase Activity and Scavenge
  Specific Species

Investigation of the roles of the specific oxidases and ROS that contribute to components of PO_2-elicited responses can be used to document their role in O_2 sensing and signaling mechanisms controlled by the PO_2 sensor.

Probes for Oxidases

Xanthine oxidase (XO), cyclooxygenase, nitric oxide synthase (NOS), mitochondria, cytochrome P450 (P450), and the Nox family of NAD(P)H oxidases have been reported to be significant sources of vasoactive ROS.[2–5] Thus, the use of probes available for investigation of these oxidases will be considered.

Xanthine Oxidase. This enzyme is often found in endothelium, and its oxidase activity is thought to be increased by processes, such as hypoxia–reoxygenation, that cause thiol oxidation and/or limited proteolysis.[6] Since the substrates for this oxidase, hypoxanthine and xanthine, are often limiting, the addition of either of these substrates to a vascular preparation could enhance the role of XO in PO_2-elicited responses. Allopurinol and oxypurinol are rather selective inhibitors of XO at concentrations in the range of 100 μM.

Cyclooxygenase. During the metabolism of arachidonic acid to prostaglandins, the peroxidase reaction of cyclooxygenase generates a reactive intermediate that is able to cooxidize reducing cofactors such as NAD(P)H, and the one-electron oxidized free radical form of this cofactor is able to react with O_2 and generate superoxide.[7] While most of the cyclooxygenase inhibitors do not attenuate the peroxidase reaction directly, inhibition of the cyclooxygenase reaction with probes such as 10 μM indomethacin appears to block the formation of ROS from this enzymatic source.

NOS. The NOS reaction is readily uncoupled by conditions often associated with oxidation of its tetrahydrobiopterin cofactor.[8] Inhibitors such as 0.1 mM nitro-L-arginine (L-NA) or its methylester form

[4] K. K. Griendling, D. Sorescu, and M. Ushio-Fukai, *Circ. Res.* **86**, 494 (2000).
[5] K. K. Griendling, D. Sorescu, B. Lassègue, and M. Ushio-Fukai, *Arterioscler. Thromb. Vasc. Biol.* **20**, 2175 (2000).
[6] D. N. Granger, *Am. J. Physiol.* **255**, H1269 (1988).
[7] R. C. Kukreja, H. A. Kontos, M. L. Hess, and E. F. Ellis, *Circ. Res.* **59**, 612 (1986).
[8] Y. Xia, A. L. Tsai, V. Berka, and J. L. Zweier, *J. Biol. Chem.* **273**, 25804 (1998).

(L-NAME) appear to inhibit superoxide generation by the NADPH oxidase activity of this enzyme. It should be mentioned that it appears that the N^G-methyl-L-arginine inhibitor of NOS does not seem to attenuate superoxide generation.

Mitochondria. Mitochondria appear to generate superoxide from the NADH dehydrogenase reaction and from the semiquinone form of coenzyme Q.[9] While there is general agreement that the proximal region of the electron transport chain generates superoxide when it is in a rather highly reduced form, studies have generated data for diverse interpretations of effects of mitochondrial inhibitors[10,11] because some newer observations in vascular tissue with probes such as rotenone and antimycin A show effects that are often the opposite of what was seen in previous studies with isolated mitochondria. Mitochondria also control cytosolic NAD(H) redox and the availability of NADH for vascular oxidases that use this cofactor.[12] NO also has marked effects on mitochondrial respiratory function that could influence its role in PO_2 sensing.[13] These issues need to be considered when probing the role of mitochondria in PO_2 sensing.

Cytochrome P450. The catalytic reaction of the P450 system generates superoxide, and H_2O_2-derived from this oxidase has been shown to be an endothelium-derived relaxing factor.[14] Probes including sulfaphenazole were used in these studies, and the actions of probes like this agent, together with scavengers of oxidant species on PO_2-elicited responses, could provide evidence for the role of P450-derived ROS.

Nox-type of NAD(P)H Oxidases. Vascular tissue appears to contain NAD(P)H oxidases with distinct Nox-1, Nox-2 (or gp91^{phox}), and Nox-4 subunits.[5] Evidence shows that the activation of Nox oxidases involves the binding of cytosolic subunits including p47^{phox} and p67^{phox} as a result of processes regulated by systems including protein kinase C and the G-protein Rac2.[5] Increasing expression of membrane-bound p22^{phox} and Nox subunits can control the activity in these oxidases in vascular tissue.

[9] J. J. Poderoso, J. G. Peralta, C. L. Lisdero, M. C. Carreras, M. Radisic, F. Schopfer, E. Cadenas, and A. Boveris, *Am. J. Physiol.* **274,** C112 (1998).
[10] E. K. Weir and S. L. Archer, *FASEB J.* **9,** 183 (1995).
[11] G. B. Waypa, J. D. Mack, M. M. Mack, C. Boriboun, P. T. Mungai, and P. T. Schumacker, *Circ. Res.* **88,** 1259 (2001).
[12] M. S. Wolin, T. M. Burke-Wolin, P. M. Kaminski, and K. M. Mohazzab-H, *in* "Nitric Oxide and Radicals in the Pulmonary Vasculature" (E. K. Weir, S. L. Archer, and J. T. Reeves, eds.), p. 245. Futura, Armonk, NY, 1996.
[13] G. C. Brown, *Biochim. Biophys. Acta* **1411,** 351 (1999).
[14] I. Fleming, U. R. Michaelis, D. Bredenkotter, B. Fisslthaler, F. Dehghani, R. P. Brandes, and R. Busse, *Circ. Res.* **88,** 44 (2001).

Oxidases not known to be inhibited by probes discussed previously for other oxidases, which are inhibited by the "DPI" flavoprotein probes diphenyliodonium and diphenylene iodonium in the concentration range of 1–100 μM, appear to be Nox-type oxidases. While there are probes that block superoxide production, which is thought to be due to oxidase assembly including apocynin,[15] these oxidases seem to have basal rates of ROS generation that might not be altered by inhibitors of oxidase assembly during studies on the acute effects of changes in PO$_2$ because regulation may be occurring from changes in substrate availability [O$_2$ and NAD(P)H] in the absence of changes in oxidase assembly.

Using the Modulation of Individual ROS to Define Their Role in PO$_2$-Elicited Responses

Probes and approaches described in the previous section for modulation of the rates of generation of ROS by individual oxidases that might be present could be used as a primary approach to demonstrate roles for oxidases and ROS in PO$_2$ sensing. The role of these oxidases in PO$_2$ sensing could be further supported by examining the effects of modulation of the specific ROS resulting from the oxidases that activate signaling mechanisms and responses that are regulated by changes in PO$_2$.

Superoxide. Superoxide anion is a major product of most of the primary oxidase sources of ROS. Commercially available preparations of SOD can be used to scavenge extracellular superoxide, and cell-permeable forms of SOD (e.g., PEG-SOD) and Tiron (10 mM) are also able to scavenge intracellular superoxide. The cytosolic (SOD-1) and extracellular (SOD-3) forms of superoxide dismutase are Cu,Zn-containing enzymes. Pretreatment of vascular tissue with the Cu-binding agent 10 mM diethyldithiocarbamate for 30 min followed by washout has been an effective method of inhibiting the Cu-containing forms of SOD in vascular tissue, and this treatment will increase superoxide levels.[16] For a prolonged inhibition of this enzyme, it is recommended that Cu be removed from the buffers used because it may be used to regenerate the active form of these enzymes. Scavenging superoxide or inhibiting SOD might not have a major effect on the generation of H$_2$O$_2$ unless there are alternative sources of superoxide removal (e.g., NO) that do not generate peroxide. Nitroblue

[15] R. A. Oeckler, P. M. Kaminski, and M. S. Wolin, *Circ. Res.* **92,** 23 (2003).
[16] P. D. Cherry, H. A. Omar, K. A. Farrell, J. S. Stuart, and M. S. Wolin, *Am. J. Physiol.* **259,** H1056 (1990).

tetrazolium (0.3 mM) has been used to remove superoxide by regenerating O_2 and to provide evidence for its role in PO_2-elicited responses that appear to be mediated through H_2O_2.[17] New probes for the modulation of superoxide are continually being developed. However, these new probes and older agents, including Tiron, diethyldithiocarbamate, and nitroblue tetrazolium, are likely to have additional redox-linked actions, and experiments should be designed to confirm that all probes used have selective actions on PO_2 and superoxide-mediated responses.

Hydrogen Peroxide. Evidence shows that H_2O_2 is often involved in vascular PO_2-elicited responses that involve ROS.[1] In vascular tissue, the enzyme catalase is irreversibly inhibited by pretreatment with the heme peroxidase reaction inhibitor 50 mM 3-amino-1,2,4-triazole followed by washout; however, inhibition of catalase may only have a small effect on increasing peroxide levels due to the presence of glutathione (GSH) peroxidases.[17] GSH peroxidase is inhibited by the presence of 10 mM mercaptosuccinate[18] and by depletion of its substrate glutathione with agents such as diethylmaleate [1 μl/ml (~7 mM), 30-min treatment followed by washout[12]]. Cellular peroxidase activity can be increased with ebselen (e.g., 0.1 mM),[18] which is thought to use NADPH-dependent reductases and GSH to metabolize peroxides in a manner similar to GSH peroxidase. The compound I intermediate of the catalase reaction has been suggested to participate in PO_2-elicited responses, and PO_2 responses can be altered as levels of this intermediate are lowered by agents that react with it, including ethanol and methanol (e.g., 50 mM) and superoxide.[1] In the design of experiments it is important to consider that peroxide-metabolizing pathways may be part of signaling processes that are detecting changes in PO_2 and that changes in peroxide and GSH may have effects on multiple additional vascular signaling systems.

NO-Derived Species. The interaction of oxidants such as superoxide with NO and oxidation of NO to reactive species by molecular O_2 could potentially contribute to PO_2-elicited vascular responses. While it is difficult to describe all of the possible combinations of interactions and approaches to identify these interactions in a concise manner, one can easily detect their role by elimination through inhibition of NO biosynthesis.

Redox. Many PO_2-regulated systems and ROS can have major effects on the redox status of localized pools of NAD(P)H and GSH, and these changes could be part of interactive systems participating

[17] K. M. Mohazzab-H, R. P. Fayngersh, P. M. Kaminski, and M. S. Wolin, *Am. J. Physiol.* **270,** H1044 (1996).

[18] K. M. Mohazzab-H, R. Agarwal, and M. S. Wolin, *Am. J. Physiol.* **276,** H235 (1999).

in PO$_2$ sensing because of their role in controlling ROS levels (biosynthesis and metabolism) and multiple redox-regulated processes. Thus, it is important to consider if these systems are being altered by the interventions used.

Detecting Changes in ROS Originating from Alterations in PO$_2$

There are many methods for the detection for ROS, and each method has its benefits and limitations. Limitations in access to the intracellular environment, perturbation of the physiological function of the vascular preparation, redox interactions or electron transfer to and from probes, and autooxidation of probes generating ROS are some of the issues which need to be considered. Thus, using more than one method of detecting changes in ROS is recommended.

Chemiluminescent Probes

Very sensitive light detection systems can be used for the measurement of changes in ROS caused by alterations in PO$_2$. While a scintillation counter used in the out-of-coincidence mode, which shows counts from one of the two photomultiplier tubes present, is usually sensitive enough,[16] much consideration needs to be given to how PO$_2$ will be altered in scintillation vials because they are usually quite gas permeable. A vascular preparation can be studied simultaneously for changes in force and chemiluminescence if a temperature-controlled vascular preparation can be studied enclosed in a light-tight environment, where the vascular tissue is placed in front of a very sensitive photomultiplier tube. If possible, a cooled photomultiplier tube and photon-counting methods should be used for the analysis of data because these conditions improve signal-to-noise ratios. An apparatus we have constructed for this purpose,[17] and for the measurement of low levels of fluorescence from vascular tissue, is shown in Fig. 1. It is important to measure background chemiluminescence signals in the absence of the vascular tissue accurately because there is usually a large background from the system used for detection, and some probes that detect or modulate ROS levels sometimes generate ROS or background chemiluminescence signals.

Lucigenin is a probe that has been used extensively with vascular preparations, including studies examining changes in PO$_2$, because it is able to detect intracellular superoxide without showing much effect on vascular reactivity.[19] While lucigenin appears to be quite selective for the detection

[19] T. Münzel, I. B. Afanas'ev, A. L. Kleschyov, and D. G. Harrison, *Arterioscler. Thromb. Vasc. Biol.* **22,** 1761 (2002).

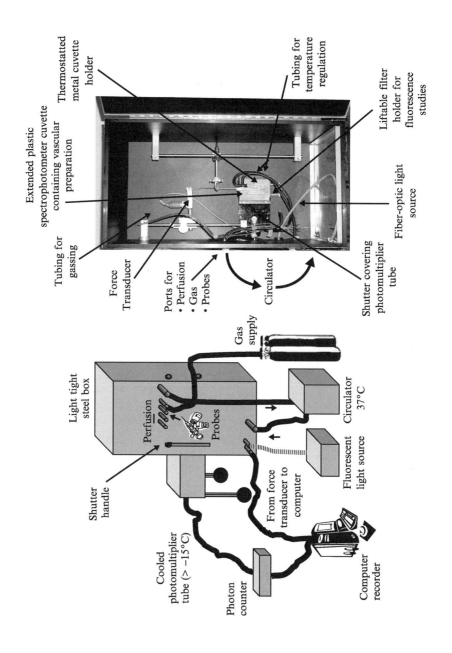

of superoxide, it is important to consider that it can undergo reduction by oxidoreductase enzymes present in tissue, and once it is reduced it can autooxidize generating superoxide.[20] Thus, when lucigenin is used, it is important to minimize this artifactual source of superoxide compared to the physiological sources of superoxide present in the vascular preparation. This is best achieved by using the lowest concentrations of lucigenin (e.g., 1–10 μM) that generates a vascular tissue-derived signal that can be detected.[19] There have been claims that newer superoxide-detecting probes are not susceptible to this autooxidation artifact;[20] however, they are generally very expensive, some show high background signals, which could be a major concern, and their ability to detect intracellular superoxide without complications of redox interactions or alterations in physiological function of vascular preparations needs to be characterized more completely.

Colorimetric Assays

Biochemical assays measuring peroxide metabolism through catalase have been used to detect PO_2-elicited changes in ROS. The effect of PO_2 on peroxide-dependent cooxidation of methanol to formaldehyde by catalase during incubations with vascular segments, followed by a colorimetric assay of formaldehyde present in the incubation buffer, has been used with isolated segments of bovine arteries.[12,16] A somewhat less sensitive method of detecting changes in peroxide metabolism through catalase is to incubate vascular tissue containing a peroxide-dependent irreversible inhibitor of catalase (50 mM 3-amino-1,2,4-triazole) under the conditions to be examined, followed by homogenizing the tissue and measuring its rate of consumption of 10 mM H_2O_2 by rates of decrease in absorbance at

[20] M. M. Tarpey and I. Fridovich, *Circ. Res.* **89,** 224 (2001).

FIG. 1. Apparatus developed for simultaneously measuring force and chemiluminescence from vascular tissue studied *in vitro* under physiological conditions, which can be adapted for the measurement of tissue fluorescence. (Left) A schematic of how a light-tight steel box was adapted for studying the vascular preparation.[17] (Right) The interior of the apparatus, after removal of the black electrical tape, which normally covers the cuvette. For fluorescence (not shown), the bent tip of the fiber-optic light source is mounted in a manner that illuminates the surface of the vascular preparation facing the photomultiplier tube, and filters are selected for the appropriate excitation and emission wavelengths.

240 nm.[21] While these methods give quantitative data on changes in intracellular peroxide, they require adequate amounts of vascular tissue and use incubation conditions that potentially alter vascular function.

Fluorescent Detection

Fluorescent probes, including dihydroethidine (DHE) and 2′-7′-dichlorofluorescein (DCF), have been used to detect changes in ROS as a function of PO_2.[19,20,22] While these probes can be used to provide very impressive results in studies that measure fluorescence resulting from their oxidation to ethidium and fluorescein products, respectively, some important issues need to be considered for use examining changes in PO_2. The primary ROS detected by DHE appears to be superoxide; however, the fluorescence of the ethidium-containing product is markedly dependent of its binding to DNA, thus this probe is potentially useful in examining which cells in a vascular preparation show PO_2-dependent changes in superoxide. Multiple ROS appear to activate fluorescence from DCF, and the combination of peroxide + Fe or peroxynitrite is one of the most active sources of fluorescence from this probe.[23] Multiple PO_2-controlled redox processes can potentially influence the DCF detection reaction, complicating the interpretation of data generated with this probe in evaluating the function of PO_2 sensors.

Radical Detection by ESR

Spin trapping for the measurement of free radical species is a very informative method of detecting species such as superoxide and ROS-derived radical species. However, limitations in access of spin-trapping agents to the intracellular environment, their perturbation of vascular function, and cellular redox interactions with probes used appear to currently be important considerations when using spin-trapping methods with intact cellular preparations.

Studying ROS Signaling Mechanisms Involved in O_2 Sensing

A comprehensive approach to elucidating identification of the oxidase that functions as the actual PO_2 sensor and how the ROS it generates interact with the signaling processes linked to responses that are observed will

[21] T. M. Burke-Wolin and M. S. Wolin, *J. Appl. Physiol.* **66,** 167 (1989).
[22] T. L. Vanden Hoek, L. B. Becker, Z. Shao, C. Li, and P. T. Schumacker, *J. Biol. Chem.* **273,** 18092 (1998).
[23] Y. Tampo, S. Kotamraju, C. R. Chitambar, S. V. Kalivendi, A. Keszler, J. Joseph, and B. Kalyanaraman, *Circ. Res.* **92,** 56 (2003).

most likely be needed for establishing the mechanism of PO_2 sensing. Some of the most sensitive systems linked to changes in basal ROS generation by PO_2 are the metabolizing systems for ROS that have regulatory interactions with cellular signaling mechanisms.[1-3] For example, the metabolism of peroxide by the heme peroxidase activity of cyclooxygenase oxidizes the heme of this enzyme into the redox state needed for producing prostaglandins. Glutathione peroxidase generates oxidized GSH while metabolizing peroxide, and oxidized GSH is used as a substrate for the regulation of proteins through *S*-thiolation. The metabolism of peroxide by catalase can be linked to the stimulation of soluble guanylate cyclase and increases in cGMP. The subcellular localization of oxidases whose activity is controlled by PO_2 can influence which mechanism ROS production is linked to. Thus, the actions of exogenously generated ROS may not exactly match endogenously produced species. Oxidases involved in eliciting PO_2-mediated responses may change with physiological or cell culture conditions, which alter oxidase activity or expression. Thus, cellular and molecular approaches for elucidating PO_2 sensors used for regulation physiological regulation *in vivo* need to be designed carefully to consider the impact of changing oxidase activities and the organization of signaling processes linked to PO_2 sensors.

[11]   Survival Surgery for Coronary Occlusion and Reoxygenation in a Rodent Model

By William A. Wallace, Arturo J. Cardounel,
Savita Khanna, and Chandan K. Sen

Introduction

In order to better understand the pathophysiology and to develop new treatment regimens in myocardial infarction and heart failure, animal models, which simulate human disease, are necessary. Development of animal models based on survival surgery allows the study of tissue remodeling in response to a specific insult. The importance of having a model that is clinically relevant underlies the attractiveness of the widely used coronary artery ligation model in the rat in which the left anterior descending artery (LAD) is temporarily occluded. The progression of heart failure in these rats is similar to the clinical syndrome of heart failure following anterior wall myocardial infarction. LAD occlusion results in acute myocardial ischemia with LV dysfunction, decreased contractility, and elevated filling pressures. Reduction in blood pressure, cardiac output, and stroke volume suggest forward failure with LV dilatation and hypertrophy.[1–4] In addition to the correlations of this model to human disease, pharmacological intervention following LAD occlusion in the rat has been shown to be a useful tool in determining drug efficacy.[5] Therefore, temporary LAD occlusion in the rat serves as a relatively simple and low-cost model for studying the pathophysiology of acute myocardial infarction and secondary heart failure, with clear correlation to the clinical setting. This article describes a simplified approach to study the long-term effects of ischemia–reperfusion in the rat heart. The procedure to conduct such survival surgery with a degree of success and efficiency is described.

Procedure

Male Sprague–Dawley rats (Harlan–Sprague–Dawley) weighing 350 g and greater are used. The procedure itself consists of preoperative anesthesia, endotracheal intubation, left thoracotomy, occlusive ligation of the left anterior descending coronary artery, release of LAD ligature, chest tube placement, thoracotomy closure, extubation, and removal of the chest tube.

Preoperative Anesthesia

The rat is placed in an acrylic induction chamber (Harvard Apparatus, NP 60-5246) and is administered halothane or isofluorane until unconscious. It is then transferred to a heated small animal operating table (Harvard Apparatus, AH 50-1239). The operating table is prewarmed to the 1.5 setting. The rat continues to receive anesthesia delivered via a table-top anesthesia machine and vaporizer (Harvard Apparatus, NP 72-3011; NP-72-3038) via spontaneous respiration through a nose cone. Our laboratory uses a medical air cylinder (21% O_2) to deliver the anesthetic. Some laboratories opt to use pure oxygen instead. If pure oxygen is used, it is important to observe that cardiac biology may be altered by exposure to perceived hyperoxia.[5] The vaporizer is initially opened to its highest setting,[6] but is titrated down to 2.5 over approximately 1 min.

At this time, while the animal continues to breathe anesthesia, it is positioned and shaved. The rat is placed in a supine position, securing the hind

[1] D. Elsner and G. A. Riegger, *Curr. Opin. Cardiol.* **10**(3), 253 (1995).
[2] P. Anversal *et al.*, *J. Clin. Invest.* **89**(2), 618 (1992).
[3] J. M. Hagar, R. Matthews, and R. A. Kloner, *J. Am. Soc. Echocardiogr.* **8**(2), 162 (1995).
[4] J. Kajstura *et al.*, *Circ. Res.* **74**(3), 383 (1994).
[5] S. Roy, S. Khanna, A. A. Bickerstaff, S. V. Subramanian *et al.*, *Circ. Res.* **92,** 264 (2003).
[6] S. Goldman and T. E. Raya, *J. Card. Fail.* **1**(2), 169 (1995).

limbs and tail to the table with tape. The animal's left thorax is shaved with a clipper using a #40 blade (Harvard Apparatus, NP 52-5204), taking care to clean away any excess fur. The time necessary for this preparation should provide sufficient anesthesia for attempting endotracheal intubation of the rodent.

Endotracheal Intubation

This step is technically challenging and should be mastered prior to its use in the coronary occlusion and reperfusion protocol. Due to the dimensions of the rodent's anatomy, surgical loupes are recommended to assist in this step (Designs for Vision, Inc, 2.5X). For our purposes, the rat is approached in a prone, suspended position. Using a claw stand (Harvard Apparatus, AH 50-4589), the rat is suspended by its superior incisors from a loop of 2-0 silk suture from the claw stand, allowing the jaw to open (Fig. 1). Using curved forceps in the right hand, grasp the rodent's tongue and retract to your right. In your left hand, a laryngoscope (Harvard Apparatus, AH 59-6581) with a Miller 0 blade (Harvard Apparatus, AH 59-6771) is inserted into the rat's oral cavity to illuminate the pharynx (Fig. 2). The epiglottis may not be visible at this step.

An arrow radial artery catheterization set (Harvard Apparatus, RA-04020) is used for the endotracheal stylus. The 20-gauge angiocath provided in the set is replaced with a 16-gauge polyurethane angiocath (1 3/4 in.). The angiocath is advanced past the needle tip to prevent injury to the animal's oropharynx, and the spring-wire guide is advanced past the angiocath tip to act as a stylus for passage between the vocal cords (Fig. 3).

Replace the forceps in your right hand with the endotracheal intubation set described in the preceding paragraph. Insert the angiocath with the spring-wire guide extended past the tip into the oral cavity to the pharynx. Press gently on the soft palate of the rat to identify the epiglottis. From the time the rat is suspended to this step, only 15 s should have passed, as the rat is no longer receiving anesthesia. The rat should still be breathing spontaneously, and now with the epiglottis visible, the vocal cords should be visualized opening and closing. Pass the spring-wire guide through the vocal cords, and pass the angiocath over the guide, through the vocal cords, and into the trachea.

Remove the needle and spring-wire guide from the angiocath, taking care not to dislodge it from the trachea. Remove the rat from the silk loop and place back into a supine position. Remove the nose cone breathing tube and connect the angiocath to a small animal ventilator (Harvard Apparatus, NP 55-3438). Initial settings for positive pressure ventilation include a stroke volume of 12 ml/kg and a respiratory rate of 70 cycles/min.

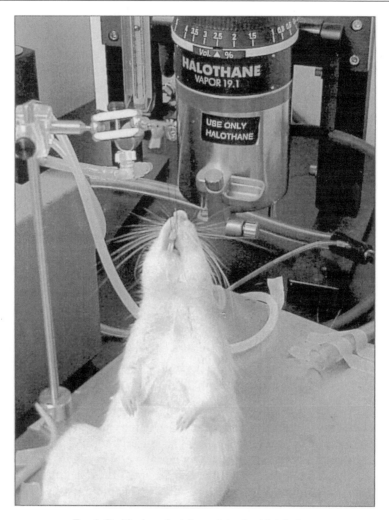

FIG. 1. Positioning of rat for endotracheal intubation.

Connect the anesthesia machine to the ventilator inflow to provide endo-tracheal intubation. Feel the chest of the animal to ensure adequate filling of the lungs, while simultaneously inspecting the animal's left upper abdominal quadrant to exclude inadvertent esophageal intubation. Inad-vertent esophageal intubation would result in gastric dilatation, ineffective ventilation and anesthesia, and great risk for profound hypoxia to the animal. Once proper positioning of the endotracheal tube is verified, secure

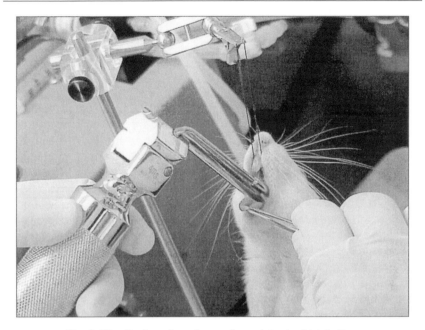

Fig. 2. Visualization of oropharynx for endotracheal intubation.

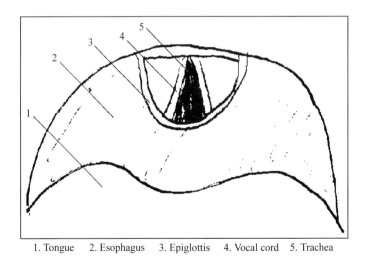

1. Tongue   2. Esophagus   3. Epiglottis   4. Vocal cord   5. Trachea

Fig. 3. Illustration of rat oropharynx. Adapted from *A Color Atlas of Anatomy of Small Laboratory Animals, Volume Two: Rat, Mouse, Hamster*, 1990 Wolfe Publishing.

the endotracheal tube both to the table and to the animal to prevent accidental removal.

Coronary Occlusion and Reperfusion

Use a fiber-optic goose-necked light source (Harvard Apparatus NP 72-0210, NP 72-0267) to illuminate the surgical field. Clean the rodent's chest with povidine/iodine solution. Identify the inferior point of the sternum of the animal (Fig. 4). One centimeter superior to this point make a left anterior transverse thoracotomy, extending toward the animal's left axilla. Medium-weight scissors work well in cutting the skin. Dissect the underlying fascia to identify the pectoralis major. Divide this muscle transversely. Likewise, identify the serratus anterior and pectoralis minor muscles and divide them using dissecting scissors. Identify blood vessels and cauterize them before division to minimize blood loss (Harvard Apparatus, HB 56-1605). At this point you should identify the fourth and fifth ribs and the fourth intercostal space (Fig. 5).

Carefully make a defect in the intercostal muscle superior to the fifth rib, taking care not to injure the underlying lung and heart. Extend the intercostal incision laterally, avoiding the lung. Extend the incision medially, being careful to avoid the internal thoracic artery, located just lateral

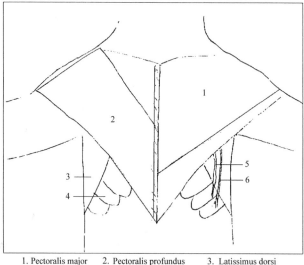

1. Pectoralis major   2. Pectoralis profundus   3. Latissimus dorsi
4. Serratus anterior   5. Anterior epigastric vein   6. Long thoracic nerve

FIG. 4. Illustration of rat chest wall musculature. Adapted from *A Color Atlas of Anatomy of Small Laboratory Animals, Volume Two: Rat, Mouse, Hamster*, 1990 Wolfe Publishing.

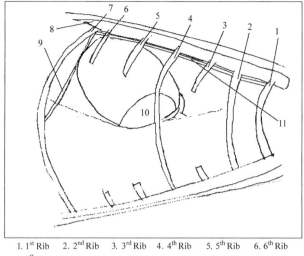

1. 1st Rib   2. 2nd Rib   3. 3rd Rib   4. 4th Rib   5. 5th Rib   6. 6th Rib
7. 7th Rib   8. Sternum   9. Diaphragm   10. Left atrium of heart
11. Internal thoracic artery

FIG. 5. Illustration of rat thoracic anatomy. Adapted from *A Color Atlas of Anatomy of Small Laboratory Animals, Volume Two: Rat, Mouse, Hamster*, 1990 Wolfe Publishing.

to the sternum interior to the ribs. Insert a rib retractor (Harvard Apparatus, AH 834-149-07) and open to expose the heart. The claw stand used for intubation may be used to hold the retractor to the right, out of the way of the operation.

Elevate the pericardium from the heart, a thin membrane, and open it with sharp scissors. The LAD should be identified extending toward the apex of the heart, originating underneath the left atrial appendage. The left atria can be retracted with a microclip (Harvard Apparatus, NP 61-0186) to aid in exposure of the LAD. Using microneedle holders (Harvard Apparatus, NP 60-3986), place a 6-0 Maxon or equivalent monofilament suture on a T-30 taper needle around the LAD. Care must be taken to place the suture through the myocardium at the proper depth, as too deep can result in hemorrhage and too shallow can result in transaction of the coronary artery. Cut the needle from the suture and place both ends of the suture through a polyethelene tube of approximately 1 cm in length (inner diameter 1.2 mm). Snare the LAD securing the tubing in place using a microclip at the distal end of the tubing to prevent the suture from slipping (Fig. 6).

Time of coronary occlusion begins upon securing of the suture in the polyethylene suture. For our purposes, we allow a period of 30 min to pass,

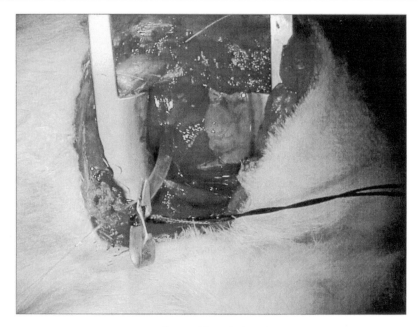

FIG. 6. View of rat heart with occlusion of coronary artery.

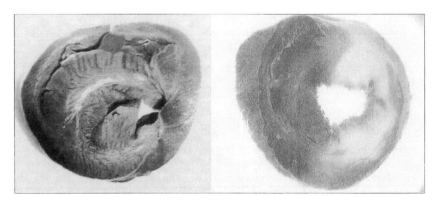

FIG. 7. Cross sections of rat heart without (left) and with (right) ischemia-reperfusion injury.

after which the snare around the LAD is released. This point begins the reperfusion period. The duration of occlusion may be varied depending on specific requirements (Fig. 7).

Thoracotomy Closure

An 18-gauge angiocath (1 3/4 in.) is used as a chest tube with the following modification. Using a 25-gauge needle, the angiocath is pierced three times at the distal tip to allow for additional suction holes. The 18-gauge angiocath is placed through the skin, inferior to the incision, and carefully through the fifth intercostal space, below the incision, into the thoracic cavity. Care must be taken not to injure the heart or lung with the needle. The chest tube is grasped with forceps and brought farther into the thoracic cavity. The angiocath is connected to a 30-ml syringe using a three-way stopcock. Silk suture (2-0) on a T-5 taper needle is used to reapproximate the ribs, placing the suture around the fourth and fifth ribs and tying to secure. Approximately three stitches are necessary for this step.

The muscle layers are then approximated, again using 2-0 silk suture. Reapproximate the deep layers (pectoralis minor, serratus anterior) and then the superficial pectoralis major muscle. Use 9-mm wound clips (NP 34-0554, NP 34-0555) to close the skin. Evacuate the thoracic cavity by aspirating the syringe attached to the chest tube. Use the stopcock to maintain negative pressure in the thoracic cavity when emptying out the syringe. Repeat this procedure as necessary (our laboratory does this twice after closure of the skin and an additional time after the endotracheal tube is removed).

Turn off the anesthesia, but continue ventilating the animal. As the anesthesia wears off, the rodent will begin to breathe spontaneously over the minute ventilation provided by the ventilator. Detach the endotracheal tube from the ventilator, ensuring that the rodent continues to breathe spontaneously. If the animal stops breathing, simply reconnect the endotracheal tube to the ventilator and allow for more recovery time. If the rodent continues to breathe spontaneously, remove the endotracheal tube and evacuate the chest cavity once more, withdrawing the chest tube as you aspirate. Untape the rat's extremities and transfer it back to its cage.

Postoperative Monitoring

The rat will continue to sleep for approximately another 30 min. Check on the animal every 10–15 min until it wakes up fully. Buprenorphine (0.05–2.5 mg/kg sc) injected intramuscularly every 12 h for 24–48 h is usually sufficient as an analgesic. The rat will begin using its left front extremity after approximately 24 h of the surgery. Observe the rat for every hour for approximately 4 h after recovery. Look for signs of pain and/or discomfort, including labored breathing and increased vocalization. The animal may then be safely transferred back to its normal housing facility. Our facility is a climate-controlled environment with 12-h light/dark cycles.

Water and standard rat chow should be provided *ad libitum*. Observe the animals at least twice daily for evidence of labored breathing, decreased food/water intake, or decreased alertness (blunted response to stimuli). Daily monitoring should occur until such time as the surgical clips are removed (7–10 days) or until the heart is harvested. If postoperative complications arise, the animal should be euthanized properly via CO_2 inhalation.

Harvesting of Tissue

After the period of reperfusion has passed, euthanize the rodent and reopen the original incision. Take care not to injure the heart as early adhesions may have already begun forming. Extend the incision across the midline to facilitate removal of the heart. Using scissors, remove the heart from the chest. Using a #10 blade scalpel, remove the injured area of the left ventricle. Typically, there is a small pale area of myocardium where the injury occurred. An area of local intramyocardial hemorrhage where the suture was placed is observed infrequently.

[12] Measurement of *In Vivo* Oxidative Stress Regulated by the Rac1 GTPase

By MICHITAKA OZAKI and KAIKOBAD IRANI

Introduction

The reduction–oxidation (redox) state is emerging as one of the most important determinants of cellular and organ function. The redox status of cells, and ultimately whole tissue, is dependent on a balance between reducing and oxidizing agents within cells. The reducing properties of the cell are affected by the expression of antioxidant enzymes such as glutathione, catalase, thioredoxin, and superoxide dismutase, among others. In contrast, reactive oxygen species (ROS) generated both extracellularly and intracellularly act as oxidizing agents and include superoxide, hydrogen peroxide, and the hydroxyl anion. These ROS by themselves, or secondary species derived from chemical interactions among one another, are primary determinants of the oxidation side of the cellular redox equation. Production of ROS above the antioxidant buffering capacity of the cell results in oxidative stress.

Although the existence of ROS has been known for a long time, their role as signaling intermediaries has only recently come to light. Originally thought of as toxic by-products of aerobic metabolism that indiscriminately attack macromolecules such as proteins, nucleic acids, and lipids, thereby leading to injury and death, it is now appreciated that ROS can also behave in a much more specific and discrete manner. For example, ROS can selectively induce transcription factors such as HSF-1[1] and NF-κB[2] that play a pivotal role in cellular activation, differentiation, death, and survival.

Recognition of the importance of the redox state in cellular function has prompted a furious search for the intracellular sources of ROS and the mechanisms responsible for their production. Different cells in response to different stimuli produce ROS utilizing, to varying extents, different mechanisms. These include, but are not limited to, the mitochondrial electron transport chain, xanthine/xanthine oxidase, and arachidonate metabolism. Our laboratory has focused its energies on exploring the functional significance of ROS production by the plasma membrane NAD(P)H oxidase regulated by Rac1, a member of the Rho family of small GTPases.

Rac proteins are an integral and necessary component of the microbiocidal oxidase of neutrophils and macrophages.[3] Rac1, which is expressed ubiquitously, also regulates ROS production in many nonphagocytic cells. This Rac1-regulated oxidase, which may utilize NADPH or NADH as its substrate, is activated in response to diverse extracellular stimuli such as growth factors,[4] heavy metals,[1] reoxygenation or reperfusion,[5,6] and cytokines.[2,7] ROS dependent on Rac1 affect many cellular processes in a cell-type and stimulus-specific manner, resulting in a variety of phenotypes ranging from survival to apoptosis.[8]

The interest in redox signaling has spurred the development of many techniques to measure ROS generation and oxidative stress, both intracellular and extracellular. In addition, there has been a great deal of recent enthusiasm regarding the fact that almost all cells and tissues possess a NAD(P)H oxidase akin to the Rac-regulated oxidase present in macrophages and neutrophils. The purpose of this article is to describe the methods we have employed to manipulate the activity of the Rac1-regulated NAD(P)H oxidase *in vivo* using adenoviral vectors. In addition, methods detailing the fluorescence and luminescence-based assays that we have used for measuring ROS generation regulated by the Rac1 GTPase are presented.

[1] M. Ozaki *et al.*, *J. Biol. Chem.* **275**(45), 35377 (2000).
[2] D. J. Sulciner *et al.*, *Mol. Cell. Biol.* **16**(12), 7115 (1996).
[3] A. Abo *et al.*, *Nature* **353**(6345), 668 (1991).
[4] A. M. Doanes *et al.*, *Biochem. Mol. Biol. Int.* **45**(2), 279 (1998).
[5] K. S. Kim *et al.*, *J. Clin. Invest.* **101**(9), 1821 (1998).
[6] M. Ozaki *et al.*, *FASEB J.* **14**(2), 418 (2000).
[7] S. S. Deshpande *et al.*, *FASEB J.* **14**(12), 1705 (2000).
[8] K. Irani, *Circ. Res.* **87**(3), 179 (2000).

Recombinant Adenoviruses as Tools for Manipulating Rac1-Regulated
Signaling *In Vitro* and *In Vivo*

Since our work involves the use of primary cell types and whole
animals, we have turned to recombinant replication-deficient adenoviruses
as efficient tools to express exogenous genes. Details about the construction of recombinant replication-deficient adenoviruses are beyond the
scope of this article. The reader is referred to other articles covering
methods of adenovirus construction, propagation, purification, and titration.[9] Using one such method we constructed recombinant adenoviruses
encoding the dominant-negative form of the Rac1 GTPase (AdRac1N17).
We also used an adenovirus encoding the inert *Escherichia coli* LacZ
protein (Adβgal) as a control.

Measurement of Ischemia/Reperfusion-Induced, Rac1-Regulated
ROS Production in Mouse Liver *In Vivo*

This section describes the introduction and detection of Rac1N17 into
mouse liver tissue and the measurement of oxidative stress in hepatic
tissue. In all *in vivo* experiments, ischemia/reperfusion (I/R) is used as a
stimulus that leads to oxidative stress and injury.

Introduction and Detection of Rac1N17 in Liver Tissue

Male C57Black/6 mice, 6 to 8 weeks, are obtained from Charles River,
Maryland. Mice are starved overnight before introduction of adenoviruses.
General anesthesia is induced with inhalation anesthetic, methoxyflurane
(Metofane). Animals are taped by all four feet, and replication-deficient
recombinant adenoviruses are administered intravenously via the tail vein
in a volume of 200 μl (2 $\times$ 10^9 pfu/body) with a 31-gauge needle. All
adenoviruses are dialyzed in four changes of 1 liter dialysis buffer
(10 mM Tris, pH 8.0, 1 mM MgCl$_2$, 140 mM NaCl) using 12- to 14-kDa exclusion limit dialysis membrane (Life Technologies, Rockville, MD) for 1 h
each at 4° just before use to remove glycerol and CsCl completely. No viruses were injected in uninfected control animals. Seventy-two hours after
systemic introduction of viruses, animals are sacrificed by rapid cervical
dislocation. The left and middle lobes of the liver are excised and processed
for immunohistochemistry and Western blotting.

---

[9] D. M. Sullivan and T. Finkel, *Methods Enzymol.* **325,** 303 (2000).

Notes

1. It is important to dialyze out the CsCl before injecting the adenovirus systemically. Not doing so results in significant operative mortality.
2. In our hands, 2×10^9 plaque-forming units (pfu) per animal resulted in expression of the transgene in approximately 50% of cells. Higher pfu led to greater expression but concomitant increase in viral toxicity.
3. It is extremely important to use a null virus or a virus encoding an inert protein such as LacZ as a control in all experiments, particularly those measuring indices of tissue injury. This is because adenovirus infection leads to a small but significant degree of tissue injury.
4. Because of tropism of adenovirus for the liver, the expression of transgenes is limited to hepatic tissue. In our hands, systemic injection of virus resulted in no expression of Rac1N17 or the marker LacZ in heart, kidney, or brain.
5. Although we found good expression at 3 days, others have found that expression of transgenes persists for at least 6–7 days after delivery of virus.[10]
6. The protocol below for isolation of cells is optimized for hepatocytes, although the principle of collagenase-induced dispersion is applicable to other tissues as well.

Adenovirus Results in Efficient Expression of Rac1N17 In Vivo

Using recombinant adenovirus, we were able to express Rac1N17 in hepatic tissue (Fig. 1), offering us a valuable tool for examining the role of Rac1 in regulating I/R-induced ROS generation *in vivo*. It should be noted that 2×10^9 pfu/animal resulted in expression of the Rac1N17 gene product in approximately 50% of parenchymal hepatic cells. The presence of an N-terminal myc epitope tag on Rac1N17 facilitated its detection.

Ischemia/Reperfusion Protocol

Anesthesia is administered to mice as described earlier, and heparin sulfate (100 U/kg body weight) is injected intravenously to inhibit thrombosis. A midline laparotomy is made and after dissection and entering the peritoneum, the liver, the hepatic veins and arteries, and the bile duct are visualized and all vessels (hepatic artery, portal vein, and bile duct) to

---

[10] R. M. Zwacka *et al.*, *Nature Med.* **4**(6), 698 (1998).

A

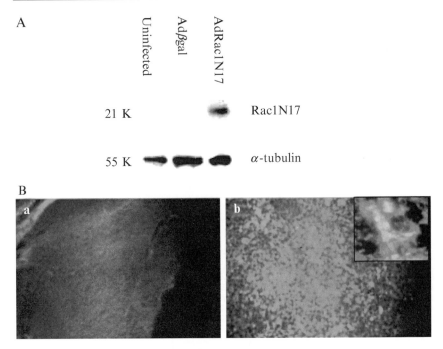

21 K   RaclN17

55 K   α-tubulin

B

FIG. 1. (A) Detection of adenovirally expressed myc-tagged Rac1N17 protein in mouse liver *in vivo*. The myc antibody (9E10, SantaCruz Biotech) was used at 1 μg/ml. (B) Immunohistochemistry of (a) Adβgal-infected and (b) AdRac1N17-infected mice showing expression of myc-tagged Rac1N17 in liver tissue *in vivo*. The 9E10 antibody was used at 20 μg/ml. (Inset) Intracellular expression of Rac1N17.

the left and median liver lobes are clamped. After 60 min of ischemia, these vessels are unclamped and the circulation is restored for 5 min. Sham-operated control mice are subjected to laparotomy without ischemia.

Measurement of Superoxide in Liver Tissue by Lucigenin-Enhanced Chemiluminescence

Lucigenin-enhanced chemiluminescence is used to detect superoxide anion generated in the reperfused liver tissue. Lucigenin (bis-*N*-methylacridinium nitrate) luminesces specifically in the presence of superoxide anion.[11]

Isolation of Hepatocytes. After exposing the liver, the portal vein is cannulated with a 24-gauge catheter and ligated with 4-0 silk. The liver is perfused carefully with 1–2 ml of solution 1 [HBSS (Ca, Mg-free), 1 m*M*

[11] H. Gyllenhammar, *J. Immunol. Methods* **97**(2), 209 (1987).

EGTA, 20 mM HEPES, pH 7.4, preoxygenated and prewarmed up to 37°] to wash out blood and is again perfused with solution 2 (HBSS, 5 mM CaCl$_2$, 0.05% collagenase, 20 mM HEPES, pH 7.4 preoxygenated and prewarmed to 37°) for 5–10 min until it softens. Solutions 1 and 2 are oxygenated by oxygen bubbling (3–5 liter/min) and incubated in a water bath at 37° at least 5 min before use. The left hepatic lobe is excised and cut into 1- to 2-mm pieces. After gentle pipetting in solution 2 for 5 min, the cell suspension is filtered through gauze to remove large fragments of hepatic tissue and centrifuged at 50g for 3 min. The cell pellet is resuspended in HBSS, 20 mM HEPES, pH 7.4 (37°), and used for the assay. This method of hepatocyte isolation yields greater than 90% hepatocytes that are consistently >80% viable, as judged by Trypan blue exclusion.

Detection of Superoxide. Cells (1×10^5) in a volume of 100 μl are added to 1 ml of lucigenin buffer (75 μM lucigenin in HBSS, 20 mM HEPES, pH 7.4). The chemiluminescent signal is measured using a luminometer (Mono-light 2010) and is monitored every 5 min for 30 min. The emitted light units, after subtracting the blank and integrating over 15 min, are used as a measure of superoxide levels. One milliliter of lucigenin buffer added with 100 μl of HBSS with 20 mM HEPES, pH 7.4, is used as blank.

Notes

We found lucigenin chemiluminescence a specific marker of superoxide production. This is supported by the finding that chemiluminescence was completely suppressed with intracellular adenoviral expression of SOD (Fig. 2A). However, other reports have suggested that lucigenin chemiluminescence may be an artifactual measure of superoxide.[12] Refuting this is evidence that lucigenin, when used at low concentrations, is a valid probe for detecting superoxide.[13] Therefore, we recommend its use in the 5 μM range. Using this concentration, we have results similar to those reported.[6]

Measurement of Hydrogen Peroxide Using Amplex Red Reagent

Amplex Red reagent (10-acetyl-3,7-dihydroxyphenoxazine) is a highly sensitive and stable probe for hydrogen peroxide.[14,15] It is available separately or can be purchased as part of a hydrogen peroxide (HRP) assay kit (Molecular Probes, Eugene, OR). In the presence of horseradish peroxidase, the Amplex Red reagent reacts with hydrogen peroxide to produce

[12] M. A. Barbacanne *et al., Free Radic. Biol. Med.* **29**(5) 388 (2000).
[13] Y. Li *et al., J. Biol. Chem.* **273**(4), 2015 (1998).
[14] J. G. Mohanty *et al., J. Immunol. Methods* **202**(2), 133 (1997).
[15] M. Zhou *et al., Anal. Biochem.* **253**(2), 162 (1997).

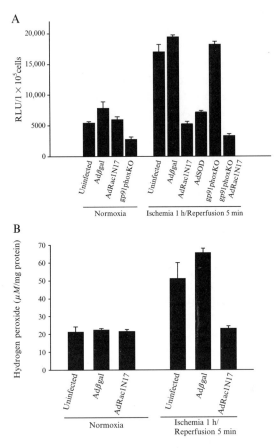

Fig. 2. (A) Inhibition of I/R-induced increase in hepatic superoxide levels in mice infected with AdRac1N17 or AdSOD. *In vivo* superoxide production was measured by lucigenin chemiluminescence. Gp91phoxKO represents knockout animals deficient in the gp91phox component of the neutrophil NADPH oxidase. (B) Inhibition of I/R-induced increase in hepatic hydrogen peroxide levels in mice infected with AdRac1N17. *In vivo* hydrogen peroxide was measured by Amplex Red fluorescence.

highly fluorescent resorufin. This reaction can be used to detect hydrogen peroxide concentrations as small as 50 nM, making this assay much more sensitive than commonly used scopoletin-based assays for hydrogen peroxide. The following protocol is a modification of that provided with the manufacturer's kit and utilizes reagents provided by the manufacturer.

Sample Preparation. Liver is perfused with 1–2 ml phosphate-buffered saline (PBS, pH 7.4)–EDTA (preoxygenated and prewarmed to 37°) via the portal vein. The left lobe is excised and cut into small pieces with a sharp blade. After gentle homogenization in 7 ml of cold PBS (5–7 strokes), the homogenate is diluted 1:20 in PBS and centrifuged (12,000 rpm for 30 s). One hundred microliters of supernatant is used for the assay.

Reagent Preparation. A 400 μM working solution of Amplex Red reagent from its stock solution (20 mM in dimethyl sulfoxide) and a 2 U/ml working solution of HRP from its stock (200 U/ml in 50 mM sodium phosphate, pH 7.4) are made in 50 mM sodium phosphate, pH 7.4. Standard concentrations of hydrogen peroxide ranging from 0 to 10 μM are prepared by serial dilutions of a 9 M stock solution (Sigma). One hundred microliters of the Amplex Red/HRP working solution is aliquoted to each well of a 96-well microtiter plate immediately before the assay and kept protected from light.

Assay. One hundred microliters of the samples, at least in triplicate, is added to the Amplex Red/HRP solution in the microtiter plate and incubated in the dark for 5 min. PBS is used as a blank. The fluorescence intensity of each well is measured using a fluorescence plate reader (Cyto-Fluor, Millipore). The excitation wavelength of resorufin is in the 530- to 560-nm range, and the emission wavelength is 590 nm. Reading at several sensitivities may be obtained. Each reading is subtracted from the blank reading. Using a standard curve plotting fluorescence versus hydrogen peroxide concentration, the concentration of hydrogen peroxide in each sample is calculated. This value is normalized for protein concentration in each sample.

Notes

According to the manufacturer, Amplex Red reagent is not permeable to cell membranes. However, hydrogen peroxide is freely diffusible and can react with the extracellular reagent. Moreover, homogenization results in cell disruption and release of intracellular hydrogen peroxide.

Rac1 Regulates I/R-Induced Oxidative Stress

A marked increase was observed in both hepatic superoxide and hydrogen peroxide levels with I/R in uninfected animals and animals infected with the control Adβgal virus. Notably, infection with Adβgal did result in a small increase in hepatic superoxide levels even under basal normoxic conditions, highlighting the importance of using a virus encoding an inert protein or a null virus in all control experiments. In contrast to uninfected and Adβgal-infected mice, animals infected with AdRac1N17 showed complete suppression of I/R-induced hydrogen peroxide and superoxide accumulation. This demonstrates the pivotal role of Rac1 in regulating oxidative stress *in vivo* induced by I/R.

[13] Proteomic Analysis of the Mammalian Cell Nucleus

By Arun K. Tewari, Sashwati Roy,
Savita Khanna, and Chandan K. Sen

Introduction

Proteomics is the study of the "proteome" or the entire protein complement of a genome, a term first coined in 1997.[1,2] It is now known, however, that the proteome is far more complex than previously suggested by the one gene, one protein adage. The proteome consists of all proteins present in a cell or organism at a given time, including not only those translated directly from genetic material, but also the variety of modified proteins arising from events such as alternative splicing of transcripts and extensive post-translational processing.[1] With the advancement and rapid progress in genome sequencing, many new genes have been discovered; however, several of them have no known function or their exact functional roles are poorly understood. In this respect, proteomics can provide valuable information and correlate genome sequence information and the cellular behavior at the molecular level.

In response to chronic moderate hypoxia, cells adjust their normoxia set point such that reoxygenation-dependent relative elevation of pO_2 results in perceived hyperoxia. We have reported that perceived hyperoxia causes growth arrest and differentiation in cardiac fibroblasts. Such a response to oxygen was associated with an increase in the expression of a number of proteins (e.g., p21, cyclin D, cyclin G) localized in the nucleus.[3] To obtain a global view on the response of nuclear proteins in cardiac fibroblasts to perceived hyperoxia, we utilized a subcellular proteomics approach. The methodological approach is described in this article.

Due to the complex nature of the eukaryotic cells, a single proteomics approach cannot be applied to understand the complete proteome. In order to get the entire subset of proteome, it is essential to characterize the proteins of a particular organelle. A subcellular proteomics approach provides a direct correlation of organelle-specific (e.g., nucleus) gene expression and a possible regulatory mechanism of biological process related to a metabolic state of the cell.

[1] D. K. Arrell, I. Neverova, and J. E. Van Eyk, *Circ. Res.* **88,** 763 (2001).
[2] V. C. Wasinger, S. J. Cordwell, A. Cerpa-Poljak, J. X. Yan *et al.*, *Electrophoresis* **16,** 1090 (1995).
[3] S. Roy, S. Khanna, A. A. Bickerstaff, S. V. Subramanian *et al.*, *Circ. Res.* **92,** 264 (2003).

0076-6879/04 $35.00

The nucleus is the organelle, which hosts most of the genetic information and serves as a hub for gene expression. At present about 964 human nuclear proteins are listed in the SWISS-PROT protein database. Such organelle-specific proteomics-based studies can decipher the different subset of proteins compared to the whole cell lysate. This can be helpful in understanding abnormal protein expression under any disease condition.

This article describes approaches to perform proteomics of the mammalian cell nucleus. Major steps involve separation and identification of nuclear proteins using two-dimensional gel electrophoresis (2DE) and mass spectrometry (MS) followed by data analysis and interpretation using advanced bioinformatic techniques (Fig. 1).

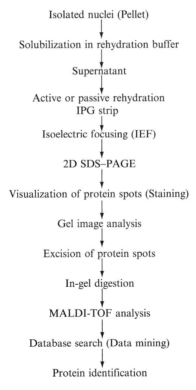

FIG. 1. A schematic flow chart of 2D gel electrophoresis and MALDI-based proteomics assay.

Sample Preparation

The method of sample preparation depends on the specific aim of research. A proper sample preparation method is vital to reduce the complexity of the protein mixture and therefore to the success of the experiment. The protein fraction for 2DE analysis must be devoid of salts, lipids, and nucleic acids and should be prepared in low ionic strength buffer to maintain native charges and the solubility of proteins.

A standard sample preparation method involves protein solubilization in standard 2D solution *(see solution 1)*. This step can be considered a good starting point for sample preparation.[4,5]

Nuclear Protein Solubilization

Cardiac fibroblast nuclei are prepared as described previously and stored at −80° in aliquots.[6]

1. Thaw the nuclei pellet and resuspend in 200 μl of rehydration buffer *(see solution 2)* by vortexing.
2. Add 5 μg/ml DNase I and 5 μg/ml RNase A[4,7] to the sample and incubate at room temperature for 20 min.
3. Sonicate the sample using a sonic dismembrator (Fisher Scientific, Pittsburgh, PA) at maximum power output for 30 s.
4. Remove the insoluble part by a single centrifugation step at 13,000 rpm on a microfuge (Eppendorf).
5. Collect the supernatant in a fresh tube and use for rehydration.

Caution: Nuclear samples contain high amounts of nucleic acids that can cause problems during sample solubilization procedure and subsequently interfere with the separation of proteins by isoelectric focusing (IEF). The nucleic acids can also be removed by using nucleic acid complex-forming agents, for example, spermine or carrier ampholytes.[4] The composition of sample solubilization solution can be determined empirically.

Two-Dimensional Electrophoresis

Two-dimensional electrophoresis is a powerful proteomics method for separating complex mixtures of proteins into individual components depending on the electric charge and size.[8] This results in an array of

[4] R. J. Simpson. Cold Spring Harbor Laboratory Press, New York, 2003.
[5] B. R. Herbert, M. P. Molloy, A. A. Goooley, B. J. Walsh *et al., Electrophoresis* **19,** 845 (1998).
[6] Y. M. Janssen and C. K. Sen, *Methods Enzymol.* **300,** 363 (1999).
[7] M. Chevallet, V. Santoni, A. Poinas, D. Rouquie *et al., Electrophoresis* **19,** 1901 (1998).
[8] P. H. O'Farrell, *J. Biol. Chem.* **250,** 4007 (1975).

protein spots that are assigned in the x and y coordinates, as compared to the protein bands in one-dimensional gel electrophoresis. Each spot represents one protein and thousands of proteins can be separated on a single gel. Also, 2DE is a powerful technique for the detection and analysis of various posttranslational modifications in proteins.[9]

Rehydration

1. Load approximately 180 μl of protein sample (supernatant) in a channel in a rehydration tray.
2. Place a immobilized pH gradient (IPG, pH 3–10, 11 cm) strip gel side down in the same channel of the rehydration tray that contains the protein sample.
3. Overlay the entire IPG strip with 2 ml of mineral oil (Bio-Rad, Hercules, CA).
4. Incubate the strip for 12 to 16 h at room temperature for passive rehydration. (*Note.* The incubation step of the IPG strip can also be carried out under low voltage (50 V) to facilitate the entry of high molecular weight proteins.)

During the entire rehydration process, the strip is overlayed with mineral oil to prevent evaporation of sample and the precipitation of urea.

The First Dimension: Isoelectric Focusing of Proteins

In this method, the separation of proteins is based on their net charge (pI value). Most of the modern protocols utilize dehydrated, precast IPG strips available from a number of suppliers (e.g., Bio-Rad, Amersham Biosciences). After the strips are rehydrated, they are transferred to focusing trays.

1. First wet two paper wicks with 8 μl of nanopure water.
2. Position the wicks over each electrode inside the focusing tray (Bio-Rad).
3. Now place the rehydrated IPG strip over the wicks.
4. Cover the strip with 2 ml of mineral oil.
5. Place the focusing tray inside the protein IEF cell (Bio-Rad).
6. Program the IEF cell as step 1 on 250 V with a linear voltage ramp, step 2 on 8000 V with linear ramp, and step 3 at 8000 V up to a total of 70,000 volt-hours.

---

[9] K. Gevaert and J. Vandekerckhove, *Electrophoresis* **21,** 1145 (2000).

7. After completion of the IEF run, remove the IPG strip with forceps and allow the oil to drain completely.

A typical IEF protocol runs through a series of voltages that increase gradually to a set focusing voltage, which can be held for the required time in hours. The total length of focusing time depends on the size of the IPG strip, pH gradient, sample load, composition of sample, and rehydration solution and should be optimized as per requirements of the experiment.

IPG Equilibration

Immediately after the IEF is over, the IPG strips are equilibrated to solubilize the focused proteins and allow SDS binding as done for one-dimensional SDS–PAGE.

1. Place the IPG strip (gel side up) in channel of rehydration tray.
2. Incubate the strip at room temperature in 4 ml of buffer I *(see solution 3)* for 10 min with gentle agitation.
3. Decant the DTT buffer and place the strip into another channel.
4. Again incubate the strip in buffer II *(see solution 4)* for 10 min.

The equilibration step should also be optimized for each sample.

Second-Dimension SDS–PAGE

1. After equilibration, position the strip on top of the precast 4–20% acrylamide Criterion gels (Bio-Rad) with the plastic backing against the plate.
2. Slide the strip between gel plates and place it directly on top of the second-dimension gel.
3. Overlay the strip with 0.5% molten agarose prepared in SDS–PAGE running buffer containing a small amount of bromphenol blue.
4. Place the gel into a gel running chamber with electrophoresis buffer.
5. Load 8 μl of precision-plus protein molecular weight standards (Bio-Rad) into the single well.
6. Run the gel on 200 V for 60 min at room temperature.

Visualization of Protein Spots by Staining 2D Gels

A large number of protocols are available to detect proteins on 2D gels, and the method of choice depends on the sensitivity of detection, compatibility of mass spectrometry, and objectives of the experiment. This section describes three of the most widely used protocols to stain 2D gels.

Coomassie Brilliant Blue R-250 Staining

a. After the second-dimension SDS–PAGE is over, place the gel in staining solution *(see solution 5)*.
b. Stain with gentle agitation for 3 h to overnight.
c. Destain the gel with an ample volume of Coomassie Blue R-250 (Bio-Rad) destaining solution *(see solution 6)*.
d. Store the destained gels in pure Milli Q water for further analysis.

Sypro Ruby Protein Gel Stain (Bio-Rad)

With the recent development of the ruthenium-based fluorescent dye Sypro Ruby, the sensitivity to detect protein spots on the gels has been enhanced up to the level of silver staining.

a. Fix the gel for 30 min in fixing solution *(see solution 7)*.
b. Remove the fixing solution and incubate the gel in 100 ml of Sypro Ruby stain (Bio-Rad) for 3 h to overnight.
c. Wash the gel for 1 h in an aqueous mixture of 10% methanol and 7% acetic acid.
d. Store the gel in pure Mili Q water for further analysis.

Mass Spectrometry Compatible Silver Staining

A number of kits are available from a variety of the sources such as Bio-Rad, Pierce, and Invitrogen that can be used for this type of staining. Each kit comes with a detailed procedure optimized by the manufacturer.

Image Analysis of Two-Dimensional Gels

Once the 2D gels are ready, the next step of proteomics is gel image acquisition and analysis. Images of the gels, stained with Coomassie, Sypro Ruby, or silver stain, are acquired on the Versadoc 3000 gel imaging system (Bio-Rad) interfaced with computers. Using commercially available 2D analysis software systems, such as PDQuest (Bio-Rad), it is easy to (1) detect and quantify even faint protein spots, (2) make comparisons between 2D gel images, and (3) detect the expression patterns with up- or downregulation of target proteins. A representative gel image of a cardiac fibroblast nuclear protein is shown in Fig. 2.

Once the images are taken, they need to be edited for proper orientation and size by rotation or crop functions. The contrast level and other parameters for the gel images are optimized based on faint spots. The

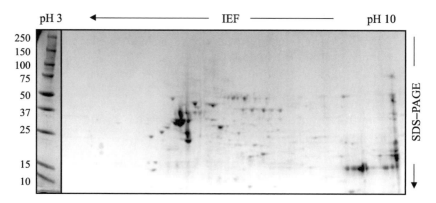

Fig. 2. Separation of cardiac fibroblast nuclear proteins by 2D gel electrophoresis. The first dimension was run on an 11-cm immobilized pH gradient (IPG; pH 3–10 nonlinear) strip. For the second dimension, a 4–20% SDS–PAGE was used. Following electrophoresis, the gel was stained with Coomasie Blue R-250. IEF, isoelectric focusing.

PDQuest program allows all these manipulation for transformations of the gel images.

Spot Detection

The software uses an automated detection system based on unique algorithms. The different detection parameters, such as sensitivity, spot size, background subtraction, and smoothness, are optimized manually to locate the proteins of interests.

Spot Matching

Once the spots are identified on all the gels in the data set, spot matching can be performed. A well-resolved gel is used as a reference gel for this purpose to determine the marker proteins. After the protein spots are selected and matched, data can be normalized, and spot characteristics and statistical data are produced with the help of a software program. Once information about the protein spots on the gel has been generated, it is added to the gel images using an annotation system available with the software.

Spot Cutting

After the gel image analysis, gels are placed on the spot-cutting platform of the Proteome Works spot cutter (Bio-Rad) interfaced with a camera and a computer system. The gels are reimaged, and protein spots are selected for the spot cutter. After assigning each spot, the automated robotically controlled spot cutter cuts the spots and deposits them to an assigned well of a 96-well plate.

In-Gel Digestion

The MassPREP station robotic protein handling system (Perkin-Elmer, Boston, MA) is used for fully automated destaining, reduction, alkylation, and in-gel digestion of proteins. The whole process is controlled and programmed through the MassPREP software system.

a. First destain the excised gel pieces (containing proteins) by incubation in 50 mM ammonium bicarbonate and 15 mM potassium ferricyanide and 50 mM sodium thiosulfate for 15 min at room temperature.

b. Remove the destaining solution and then incubate the gel fragments for 30 min at 56° in 25 μl of 10 mM dithiothratol (DTT) in 100 mM ammonium bicarbonate.

c. Replace the DTT solution by 25 μl of 55 mM iodoacetamide in 100 mM ammonium bicarbonate, and incubate the gel fragments for 20 min at room temperature.

d. Wash the gel pieces for 10 min with 100 μl of 100 mM ammonium bicarbonate and for 10 min with 100 μl of 50 mM ammonium bicarbonate and 30% acetonitrile.

e. Gel pieces obtained from 2DE are only treated with 50 mM ammonium bicarbonate and 30% acetonitrile for 20 min at room temperature because disulfide bond-containing proteins are already reduced and alkylated after IEF.

f. Once the gel pieces are dried, initiate the digestion by the addition of 25 μl of a solution of 50 mM ammonium bicarbonate containing sequencing grade trypsin (Promega, Madison, WI) at 6 ng/μl.

g. Extract all the peptides with 30 μl of 1% formic acid and 2% acetonitrile at room temperature.

Control extractions (blanks) should be performed using pieces of gels devoid of proteins. Digestion of 96 samples takes from 8 to 10 h according to the specificity of the protocol.[10]

Peptide Mass Fingerprinting by Matrix-Assisted Laser Desorption Ionization (MALDI) Mass Spectrometry

MALDI-MS is used to determine the accurate mass of a group of peptides derived from a protein by digestion with a sequence-specific protease, for example, trypsin, and thus generating a peptide mass map or peptide mass fingerprint. Depending on a specific trypsin cleavage site at the amino acids arginine and lysine, the masses of tryptic peptides can be predicted theoretically for all the proteins in the database. The experimentally acquired peptide masses are compared with those obtained theoretically, and the protein can be identified correctly if there are enough peptide matches for a protein in the database. Proteins separated on a 2D gel system provide the required information about protein molecular weights and isoelectric points, which in turn help in the MALDI analysis and final identification.

Before peptide mass fingerprinting, 1 μl of peptide-containing solutions from each sample is mixed with an equal volume of matrix (10 mg/ml α-cyano-4-hydroxycinnamic acid in 50% AcN, 0.1% TFA) and then deposited on a 96-well MALDI target plate and dried at room temperature.

MS measurements are conducted with a matrix-assisted laser desorption ionization/time of flight (TOF) mass spectrometer MS (MicroMass, Altrincham, UK) equipped with a 337-nm nitrogen laser. Analyses are performed in the reflectron mode with an accelerating voltage of 20 kV, a delayed extraction parameter of 100–140 ns, and a low mass gate of 850 Da. The laser power is set slightly above threshold (10–15% higher than the threshold) for molecular ion production. Automated spectral data acquisition is programmed with MssLynx software (Micromass). Spectra are obtained by the summation of several consecutive laser shots. Masses of the peaks are extracted from the spectra and used for protein identification using the SmartIdent peptide mass fingerprint tool. The data search is conducted against SWISS-PROT and TrEMBL databases.[10]

Reagents

Urea, thiourea, DTT, TBP, CHAPS, DTT, leupeptin, pepstatin, aprotinin, DNase I, RNase A, α-cyano-4-hydroxycinnamic acid, acetonitrile, ammonium bicarbonate, potassium ferricyanide, sodium thiosulfate (Sigma, St. Louis, MO)

Trypsin (Promega, Madison, WI)

Tris base, glycerol, methanol, acetic acid (Fisher Scientific, Pittsburgh, PA)

Carrier ampholytes, bromphenol blue, IPG strips, criterion gels, mineral oil, prestained protein standards, iodoacetamide, Coomassie Blue R-250 (Bio-Rad, Hercules, CA)

MS-compatible silver-staining kit (Invitrogen, Carlsbad, CA)

[10] J. R. Chapman, "Mass Spectrometry of Proteins and Peptides: Methods in Molecular Biology." Humana Press, New Jersey, 2000.

Solutions

Solution 1: 8 M urea, 50 mM DTT, or 2 mM TBP (tributylphosphine), 4% CHAPS, 0.2% carrier ampholytes, 0.0002% bromphenol blue

Solution 2: 5 M urea, 2 M thiourea, 4% CHAPS, 40 mM Tris, 2 mM TBP, or 100 mM DTT, 0.2% carrier ampholytes, 0.0002% bromphenol blue, 1 μg/ml leupeptin/pepstatin/aprotinin

Solution 3: 6 M urea, 2% SDS, 0.05 M Tris–HCl, pH 8.80, 20% glycerol, and 2% DTT

Solution 4: 6 M urea, 2% SDS, 0.05 M Tris–HCl, pH 8.80, 20% glycerol, and 2.5% iodoacetamide in place of DTT

Solution 5: aqueous solution of 0.1% Coomassie Blue R-250 (w/v) in 40% methanol (v/v) and 10% acetic acid

Solution 6: aqueous solution of 40% methanol and 10% acetic acid

Solution 7: aqueous solution of 10% methanol and 7% acetic acid.

[14] Oxygen-Dependent Regulation of Erythropoiesis

By JAROSLAV JELINEK and JOSEF T. PRCHAL

Introduction

Hypoxia inducible factor 1, subunit α (HIF-1α) is a central molecule in oxygen sensing and the response to hypoxia. HIF-1α is unstable in normoxic conditions: specific proline hydroxylases convert its proline residues 402 and 564 to hydroxyprolines, the hydroxylated HIF-1α is captured by the von Hippel Lindau protein (pVHL), ubiquitinated, and degraded in proteasomes. Under hypoxic conditions, HIF-1α is not recognized by pVHL; it forms a heterodimer with the HIF-1β subunit and acts as a transcription factor stimulating the expression of genes involved in glucose metabolism, angiogenesis, erythropoiesis, blood coagulation, and other essential processes.[1]

This article describes methods for the quantitative detection of mRNA for hypoxia-regulated genes in blood cells and for assessing the response of red blood cell progenitors to erythropoietin—a hormone playing an essential role in the adaptation to hypoxia. Real data of studies of the congenital defect of the germ-line mutation of *VHL* gene leading to the constitutive

[1] G. L. Semenza, *Curr. Opin. Cell Biol.* **13**, 167 (2001).

upregulation of hypoxia sensing, Chuvash polycythemia,[2] are used as an example.

Separation of Plasma, Platelets, Mononuclear Cells, and Granulocytes from Human Peripheral Blood

Collect blood by phlebotomy in two "yellow top" 8.5-ml Vacutainer tubes (Becton-Dickinson) containing acid citrate dextrose as an anticoagulant. If required, blood samples can be shipped at room temperature and should be processed within 24 h. Use sterile reagents, plastic, and aseptic techniques for all procedures.

Centrifuge the tubes with blood in a swinging bucket rotor at 100g (700 rpm in a Beckman TJ-6 centrifuge) for 10 min at 20°. Collect platelet-rich plasma (a hazy yellow upper fraction) and transfer to a Falcon 15-ml tube (Falcon, Becton-Dickinson, Franklin Lakes, NJ). Leave intact about 2 mm of plasma above the lower red cellular fraction. In patients with polycythemia, the whole blood centrifugation at 100g may not lead to a sufficient separation of plasma. If this occurs, dilute the blood with 2 volumes of phosphate-buffered saline (PBS) with 2 mM EDTA (PBSE) and repeat centrifugation at 100g for 10 min.

Platelets and Plasma

Pellet the platelets from platelet-rich plasma by centrifugation at 400g (1500 rpm in a Beckman TJ-6 centrifuge) for 10 min. Pipette off plasma, aliquot in 2-ml cryotubes, and store it at −80° for further analysis. Suspend the platelet pellet in 1 ml of TRI reagent (Molecular Research Center, Cincinnati, OH) and lyse completely by passing several times through a pipette. Transfer the lysate into a 1.5-ml Eppendorf tube and store at −80° for the preparation of RNA. The cellular fraction is subjected to 1077 g/cm^3 density gradient centrifugation.

Mononuclear Cells and Granulocytes

Prepare six 15-ml Falcon polypropylene tubes with 5 ml of Histopaque-1077 (Sigma, St. Louis, MO). Transfer the cellular fraction remaining after the separation of plasma and platelets in a 50-ml tube (Falcon), dilute to 50 ml with PBSE, mix well by pipetting up and down, and carefully overlay on top of the Histopaque-1077 solution in the 15-ml tubes. Centrifuge the tubes at 400g 20° for 25 min without breaking at the end.

---

[2] S. O. Ang, H. Chen, K. Hirota *et al.*, *Nature Genet.* **32**(4), 614 (2002).

Mononuclear Cells

Mononuclear cells will form a visible gray band between the yellow upper PBSE/plasma phase and the lower clear Histopaque phase. Red blood cells with granulocytes will form a large pellet at the bottom of the tubes. Remove the PBSE/plasma phase by aspiration, collect the mononuclear cells at the interphase by a Pasteur pipette, and transfer them in a 50-ml Falcon tube. Wash the cells by adding PBSE to 50 ml and spinning at 400g for 10 min. Suspend the pellet in 2 ml of RPMI 1640 medium (Vitacell, ATCC, Manassas, VA) containing 20% (v/v) of fetal bovine serum (FBS) (Gibco). Take a 10-μl sample of the cell suspension, mix with 90 μl of 2% acetic acid (colored with methylene blue or gentian violet) to lyse contaminating red blood cells, and count the mononuclear cells in a hemocytometer. Add RPMI 1640 medium with 20% FBS to adjust the mononuclear cell concentration to 3 $\times$ 10^6/ml. Use 1.5 ml of the mononuclear cells for a 15-ml methylcellulose culture of erythroid progenitors (see later). The remaining cells can be used to establish lymphoblastoid cell lines and/or conserved by freezing in a medium containing 10% dimethyl sulfoxide and 90% FBS.

Granulocytes

Remove the Histopaque phase from the tubes after separation of mononuclear cells and transfer the red pellets in two 50-ml Falcon tubes. Add ice-cold RBC lysis solution (155 mM NH$_4$Cl, 1 mM NH$_4$HCO$_3$, 1 mM EDTA) to 50 ml, mix well by inverting, and incubate on ice for 15 min. Centrifuge at 400g for 10 min, and carefully pour off the red cell lysate. Suspend the granulocyte pellets by a vortex mixer and transfer to a 1.5-ml Eppendorf tube. Fill up the tube with RBC lysis solution, spin briefly, and remove the supernatant by aspiration. Repeat this step until the granulocyte pellet is white. Suspend the pellet in 1 ml of TRI reagent (Molecular Research Center, Cincinnati, OH) and lyse completely by passing several times through a pipette. Store at −80° for the preparation of RNA.

In patients with iron deficiency the red blood cells have a lower density and a significant proportion of them may stay at the top of the Histopaque gradient. If this occurs, the contaminating red blood cells need to be lysed with an RBC lysis solution.

Preparation of EBV-Transformed B Lymphoblastoid Cell Lines (B-LCL)

Transfer 10 $\times$ 10^6 blood mononuclear cells suspended in 3.3 ml of RPMI 1640 medium (Vitacell, ATCC, Manassas, VA) with 20% FBS to a T25 tissue culture flask, add 1 ml of medium from the B95-8 marmoset cell line (producing the Epstein–Barr virus), and prefiltered through a filter with a 0.45 μm pore size. Add 10 μl cyclosporin A (Sigma C1832, stock solution 1 mg/ml is made in ethanol); the final concentration is 2 μg/ml. Incubate the flask in an upright position undisturbed for 7 days at 37° in a humidified atmosphere containing 5% CO_2. Start feeding with 0.5–1 ml of medium (RPMI 1640 with 20% FBS) three times a week. Gradually increase the amount of feeding medium as clumps of lymphoblastoid cells begin to grow, and maintain the cell density between 0.5 and 1.0 $\times$ 10^6/ml. Conserve the cells by freezing of several vials containing 20 $\times$ 10^6 cells in 1 ml of a 10% dimethyl sulfoxide/90% FBS solution; store the frozen cells in liquid nitrogen.

Exposure of B-LCL and Other Cells to Hypoxia

Transfer 25-ml aliquots of exponentially growing lymphoblastoid cells in two T75 tissue culture flasks and adjust the cell concentration to 8 $\times$ 10^5/ml. Expose one flask to a hypoxic atmosphere (2% O_2, 93% N_2, 5% CO_2) and the other to normoxia (21% O_2 in air, 5% CO_2). We use a Model 3130 incubator (Forma Scientific, Marietta, OH) for creating hypoxic and Model 3110 for normoxic conditions. Incubate the cultures for 18 h, collect the cells by centrifugation, and lyse in TRI reagent for RNA preparation.

Quantification of Hypoxia-Responsive Gene Expression by Real-Time RT-PCR

RNA Isolation

Isolate total RNA from the cell lysates in TRI reagent following the manufacturer's protocol. Dissolve the RNA in 20–50 μl of nuclease-free water (Promega, Madison, WI) to obtain an approximate concentration of 1 μg of RNA per microliter (RNA sample stock solution). Store the RNA at −80°.

Primers and TaqMan Probes

Primers and probes for mRNA detection (Table I) were designed using Primer Express software (Applied Biosystems, Foster City, CA) and custom synthesized by Applied Biosystems.

Make 20$\times$ concentrated solutions of primers and probes for each gene by mixing 90 μl of 100 μM forward primer, 90 μl of 100 μM reverse primer, 10 μl of 100 μM probe, and 310 μl nuclease-free water.

Store the solutions of primers and probes at −20°.

TABLE I
PRIMERS AND TaqMan MGB PROBES FOR mRNA DETECTION OF
HYPOXIA RESPONSIVE GENES[a]

Gene	Primers and probes (5'→3')
Glucose transporter 1 (SLC2A1)	FP GGC TCC GGT ATC GTC AAC AC
	RP CTG CTC GCT CCA CCA CAA
	Probe CAC TGT CGT GTC GCT G
Vascular endothelial growth factor (VEGF)	FP CCCACTGAGGAGTCCAACATC
	RP GGCCTTGGTGAGGTTTGATC
	Probe CCATGCAGATTATGC
Erythropoietin (EPO)	FP GTACCTCTTGGAGGCCAAGGAG
	RP TATTCTCATTCAAGCT GCAGTGTTCA
	Probe AGAATATCACGACGGGCTGTG
Erythropoietin receptor (EPOR)	FP GACATAGTGGCCATGGATGAAG
	RP CGAGGCCAAAGCAGATGAG
	Probe CTCAGAAGCATCCTCCT
Plasminogen activator inhibitor (PAI)	FP GGACAGACCCTTCCTCTTTGTG
	RP CATCACTTGGCCCATGAAAA
	Probe CCCCACAGGAACAGT
Von Hippel Lindau tumor suppressor (VHL)	FP CGCCGCATCCACAGCTA
	RP CATCGTGTGTCCCTGCATCT
	Probe TCACCTTTGGCTCTTC
α-Globin	FP GCCCTGGAGAGGATGTTCCT
	RP GGTCGAAGTGCGGGAAGTAG
	Probe CTTCCCCACCACCAAG
Glyceraldehyde 3-phosphate dehydrogenase (GAPDH, endogenous control)	FP ATGGAAATCCCATCACCATCTT
	RP CGCCCCACTTGATTTTGG
	Probe CAGGAGCGAGATCC
18S ribosomal RNA (endogenous control)	FP TCGAGGCCCTGTAATTGGAA
	RP CCCTCCAATGGATCCTCGTT
	Probe AGTCCACTTTAAATCCTT

[a] FP, forward primer; RP, reverse primer; 100 μM stock solutions of primers were made in nuclease-free water (Promega, Madison, WI). Probes labeled with the FAM fluorophore and MGB nonfluorescent quencher were supplied as 100 μM solutions by the manufacturer (Applied Biosystems).

Dilution of RNA Samples for Real-Time RT-PCR

Dilute the stock RNA samples 50-fold with nuclease-free water containing 50 ng/μl of yeast tRNA (Invitrogen, Carlsbad, CA) as a carrier. Make a further 1024-fold dilution (two serial 32-fold dilutions) using nuclease-free water with 50 ng/μl of tRNA for the detection of 18S rRNA. One-half microliter of an RNA sample stock solution (500 ng of RNA) is sufficient for the detection of one gene and 18S rRNA as an endogenous control; multiply the amount of RNA and diluent by the number of genes examined.

RT-PCR Reactions

The volume of individual reactions is 20 μl; use triplicate RT-PCR reactions per sample and one PCR reaction without reverse transcriptase (RT) as a control for DNA contamination. Use separate reactions for each target gene and endogenous reference standards (18S ribosomal RNA and/or GAPDH mRNA). Make reaction master mix solutions without and with RT for each gene to be analyzed; the following amounts of reagents are calculated for one 20-μl reaction:

- 10 μl Universal Master Mix 2× concentrated (Applied Biosystems)
- 0.5 μl reverse transcriptase 40× concentrated (omit in PCR without RT)
- 1 μl of 20× concentrated mix of primers and probe for a particular gene (e.g., VEGF). The final concentration of primers is 900 nM; the probe is 100 nM.
- 3.5 μl nuclease-free water (for control, use 4 μl water in PCR without RT). Aliquot 15 μl of reaction master mixes in each well of the 96-well optical reaction plate (Applied Biosystems). Add 5 μl of diluted RNA sample in each well. The RNA amount should be 100-10 ng per reaction for target genes; use 1024-fold dilutions for 18S rRNA. Cover the plate with an optical adhesive cover (Applied Biosystems), and centrifuge at 100g for 1 min.

Perform RT-PCR in the ABI Prism 7000 sequence detection system using the following conditions: reverse transcription at 48° for 30 min, initial denaturation and Taq polymerase activation at 95° for 10 min, 40 cycles of denaturation at 92° for 15 s, and annealing/extension at 60° for 1 min. Use the ABI Prism SDS software to analyze RT-PCR data, and for each well determine the amplification cycle when the fluorescence reaches the threshold (CT). Calculate the mean and standard deviation of the CT value for each triplicate of samples. The CT range of 15–30 amplification cycles gives reliable results; ideally the CT values should be about 20.

To assess the influence of DNA contamination, compare CT values in the PCR without RT (D) with the CT value obtained in RT-PCR (T). If the (D–T) difference is greater than four amplification cycles, the contribution of DNA contamination is negligible (less than 0.1 amplification cycle) and no correction is necessary. If the (D–T) difference is between 0 and 4, the CT value for pure RNA (R) can be calculated from the formula:

$$R = T - \left[\log \left(1 - 2^{(T-D)}\right) / \log (2) \right]$$

To determine the expression of the target gene relative to an endogenous control, calculate the difference between the threshold amplification cycles for the endogenous reference RNA [CT(E), e.g., 18S rRNA] and the mRNA of the examined gene [CT(X), e.g., VEGF], making a correction for the dilution d of the sample in the endogenous reference RNA detection (e.g., $d = 1024$):

$$\Delta CT(E - X) = CT(E) - [\log (d) / \log (2)] - CT(X)$$

The ratio of the mRNA level of the gene (X) relative to the level of endogenous reference (E) can be calculated as

$$2^{\Delta CT(E-X)}$$

Levels of mRNA for the glucose transporter-1 (Glut-1) and vascular endothelial growth factor (VEGF) were increased in B-lymphoblastoid cell lines derived from patients with Chuvash polycythemia (Fig. 1, hatched columns) as opposed to normal controls (Fig. 1, clear columns). The difference was significant in cells cultured in normoxia (21% oxygen), whereas in hypoxia (2% oxygen) Glut-1 and VEGF were upregulated in control and Chuvash cell lines to a similar degree. The ratio of gene/endogenous reference $2^{\Delta CT(E-X)}$ in controls cells cultured at normoxia was set as 100%.

Response of Erythroid Progenitors to Erythropoietin *In Vitro*

Erythropoietin (Epo) is a growth factor essential for erythroid proliferation, differentiation, and cell survival *in vivo* and *in vitro*. The production of Epo is upregulated in hypoxia. The erythropoietin receptor (EpoR) is present in erythroid progenitors (early BFU-E and late CFU-E) and its expression increases during erythroid differentiation. BFU-E progenitors form colonies of differentiated red cells in semisolid media containing Epo. The colonies contain hundreds to several thousands of hemoglobinized cells and have a characteristic shape of several clusters of tightly packed small cells with a pink to red color (Fig. 2). Serum-containing methylcellulose cultures of BFU-E progenitors can be used to demonstrate

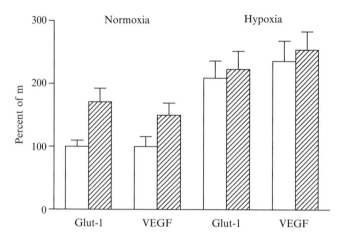

FIG. 1. Upregulation of HIF-1α target genes in patients with a homozygous mutation of the VHL gene.

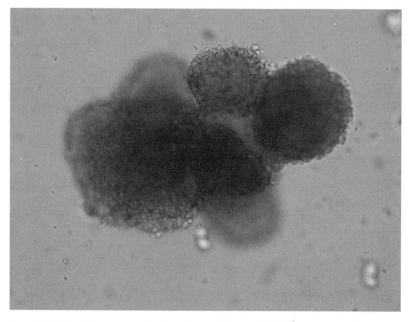

FIG. 2. Multicentric erythroid colony derived from a BFU-E progenitor cell. (See color insert.)

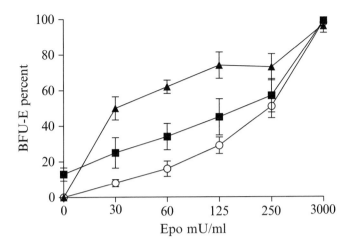

FIG. 3. Dose–response of erythroid progenitors BFU-E to erythropoietin.

the characteristic response of erythroid progenitors to Epo in patients with polycythemia vera (PV) and primary familial and congenital polycythemia (PFCP) (Fig. 3). BFU-E in normal subjects show a minimal response to low Epo concentrations (30–60 mU/ml), whereas patients with PFCP usually form a significant amount of erythroid colonies in the cultures containing 30 to 60 mU Epo per milliliter. The increased response to low concentrations of Epo is also found in Chuvash polycythemia caused by a homozygous mutation of the von Hippel Lindau (VHL) gene[2] and in other polycythemias caused by VHL mutations.[3] Patients with polycythemia vera (PV) form erythroid colonies even without the addition of Epo.[4,5]

Patients with primary familial and congenital polycythemia (PFCP, Fig. 3, closed triangles) have erythroid progenitors characteristically hypersensitive to low concentrations of Epo (30–60 mU/ml); however, their BFU-E do not grow in cultures without Epo. Polycythemia vera patients (Fig. 3, closed squares) show a characteristic formation of erythroid colonies in cultures without added erythropoietin ("endogenous erythroid colonies," EEC). In normal controls (Fig. 3, open circles), no erythroid colonies are formed in cultures without Epo and a small number of BFU-E form colonies at low concentrations of Epo.

[3] Y. D. Pastore, J. J. Jelinek, S. O. Ang, Y. L. Guan, E. Liu, K. Jedlickova, L. Krishnamurti, and J. T. Prchal, *Blood* **101,** 1591 (2003).
[4] J. F. Prchal and A. A. Axelrad, *N. Engl. J. Med.* **290,** 1382 (1974).
[5] C. J. Eaves and A. C. Eaves, *Blood* **52,** 1196 (1998).

Dose Response of Erythroid Progenitor BFU-E to Erythropoietin in Methylcellulose Cultures

Prepare dilutions of erythropoietin (Epogen, Amgen, Thousand Oaks, CA) containing 300, 25, 12.5, 6, and 3 units of Epo in 1 ml of PBSE with 0.5% bovine serum albumin. These dilutions contain 3000, 250, 125, 60, and 30 milliunits (mU) of Epo per 10 μl and can be stored for 1 month at 4°.

Transfer 15 ml of methylcellulose medium "without Epo" (MethoCult H4531, StemCell Technologies) in a 50-ml Falcon tube using a 20-ml syringe with a 16-gauge blunt-end needle (StemCell Technologies, Vancouver, BC, Canada). Add 1.5 ml of medium containing 4.5×10^6 mononuclear cells and suspend by vigorous vortexing.

Pipette 10 μl of each Epo dilution (i.e., 30, 60, 125, 250, 3000 mU of Epo) on duplicate 35×10-mm petri dishes (Falcon). Dispense 1.1 ml of methylcellulose medium with mononuclear cells on the petri dishes with the increasing amount of Epo starting from two dishes without added Epo using a 3-ml syringe with a 16-gauge needle. Tilt and rotate the dishes to spread the methylcellulose medium evenly. Put the six petri dishes with cultures and the one open petri dish with distilled water in a covered 150×15-mm petri dish (Falcon) to maintain high humidity. Incubate for 14 days in a humidified atmosphere of 5% carbon dioxide, 5% oxygen, and 90% nitrogen at 37°. Score colonies of erythroid cells after 14 days of incubation.

Erythroid cells grown in culture can be harvested for the RNA analysis. Add 2 ml of PBSE to petri dishes containing erythroid colonies. Swirl to detach methylcellulose medium and transfer to a 15-ml Falcon tube. Vortex vigorously to dissolve the methylcellulose, fill up the tube with PBSE, and centrifuge at 400g for 15 min. Decant the supernatant, vortex the pellet, add 14 ml of PBSE, and centrifuge at 400g for 5 min. Remove the supernatant by aspiration and lyse the pellet with 1 ml of TRI reagent. Store at $-80°$ until the isolation of RNA.

[15]   Physiologic Responses to Chronic Anemia in Fetal Sheep

By Lowell Davis and Roger Hohimer

Introduction

Fetal sheep made chronically anemic *in utero* adapt by 30% increases in heart weight, 50% increases in stroke volume and cardiac output, and a fivefold increase in resting coronary blood flow.[1] Angiogenesis occurs as myocardial hypoxia-inducible factor-1 and vascular endothelial growth factor levels also increase.[2] These changes are necessary in order to maintain myocardial and systemic oxygen consumption and function when oxygen content is reduced. Importantly, in fetal sheep that are made anemic, maximal coronary conductance in response to adenosine nearly doubles, increasing from 18.2 to 32.8 ml/min/100 g/mm Hg.[1] Even after adjustment for changes in viscosity and hematocrit, coronary blood flow is 40% greater at normal perfusion pressures. Sheep that are made anemic in late gestation but are transfused prior to birth, thus restoring normocythemia, maintain this relative increase in maximal coronary conductance as adults even though resting coronary blood flow and hematocrit are not different.[3] These findings suggest that the coronary vascular tree can be altered by *in utero* events and that these changes may persist into adulthood. Physiologic methods to measure maximal coronary conductance (arterial flow) and associated coronary vascular morphometry are reviewed.

Significance of Maximal Coronary Conductance

Maximal coronary conductance is the slope of the relationship between myocardial blood flow during maximal vasodilation with adenosine infusion and coronary perfusion pressure.[4] It is a physiologic measure of the total resistance vessel area of the coronary circulation and therefore is used to index vascular growth of resistance vessels. In interpreting measurements of coronary conductance across development it is important to note

---

[1] L. Davis, A. Hohimer, and M. Morton, *Am. J. Physiol.* **277,** R306 (1999).

[2] C. Martin, A. Yu, Bjiang, L. Davis, D. Kimberly, A. Hohimer, and G. Semenza, *Am. J. Ob. Gyn.* **178,** 527 (1998).

[3] L. Davis, J. B. Roullet, K. L. Thornburg, M. Shokry, A. Hohimer, and G. Giraud, *J. Physiol. (Lond.)* **15,** 53 (2003).

[4] J. I. Hoffman and J. A. Spaan, *Physiol. Rev.* **70,** 331 (1990).

that maximal coronary blood flow, when expressed per unit weight of heart, has been shown to decrease with age, being higher in the fetus than in the newborn and lower in the adult.[1,2,5] Thus appropriate controls are important.

Fetal Anemia

Fetal sheep are an ideal model to study cardiovascular adaptation to anemia as they have similar cardiac output and size as compared to human fetuses, there is substantive normative data available, and chronic instrumentation is relatively easy. General anesthesia is induced in time-dated pregnant ewes at approximately 117 days (term 145 days gestation) with intravenous diazepam prior to intubation and is maintained with 1.5% halothane and nitrous oxide–oxygen (1:1). Following a midline peritoneal incision, the right side of the fetal neck is exposed through a uterine incision. A polyvinyl catheter, V8 (1.19 mm i.d.) is placed in the carotid artery in each twin and advanced into the ascending aorta. The ear of the first twin selected is notched and the catheter is marked for identification. The uterine incision is closed and the catheters are filled with heparinized saline and tunneled to a pouch on the flank of the ewe. One million units of penicillin is given in the amniotic space of each twin.

Two days following surgery, blood gases, oxygen content, and hematocrit are measured (calibrated at 39°) and one twin is selected for isovolemic hemorrhage, starting on the day following surgery. On a daily basis initially, 75 ml of blood is withdrawn rapidly from the "anemic" twin and replaced with an equal volume of normal saline. The withdrawn blood is stored in sterile citrate phosphate dextrose adenine solution (blood bank) to which penicillin is added. We have found that if the hematocrit is maintained at less than 50% of the control twin for approximately 10 days and the oxygen content is reduced to a nadir of 2.5 ml/dl, coronary conductance will be increased.[3] Arterial blood gases, arterial pH, arterial oxygen content, and hematocrit are measured prior to each hemorrhage and volume replacement in the anemic twin. The subsequent amount of blood withdrawn each day will vary between 25 and 100 ml depending on the hematocrit and oxygen content that is measured. In this manner, fetal coronary blood flow increases gradually as the hematocrit is reduced (Fig. 1). At 138 days, the anemic twin is transfused with approximately 120 ml of stored packed red blood cells through a Y-type blood set (McGaw Inc., Irvine, CA) to remove aggregate material and the catheters are tied off at the skin.

[5] M. F. Flanagan, T. Aoyagi, J. Currier, S. Colan, and A. Fujii, *J. Am. Coll. Cardiol.* **24,** 1786 (1994).

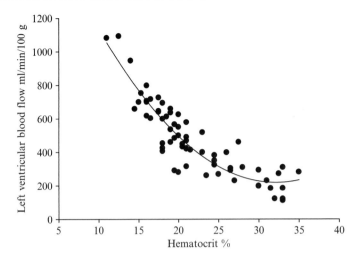

Fɪɢ. 1. Relationship between hematocrit and coronary blood flow.

This is necessary as anemic fetuses often do not survive normal labor. The day following spontaneous delivery, the lambs are weighed, hematocrit and venous blood gas are measured, and 1 ml of iron dextran containing 100 mg of elemental iron is given intramuscularly. A week later the lambs and ewes are returned to the farm to be raised in the usual fashion. An alternative method used to reduce the hematocrit is to withdraw 1 ml/min of fetal blood for 110 min daily until the desired hematocrit is reached.[6] This protocol does not require volume replacement.

Adult Protocol

At 7 months of age, the adult sheep are studied. General anesthesia is as described earlier. Two V8 catheters are placed in the jugular vein and advanced to the right atrium. A V8 catheter is placed in the carotid artery and advanced to the aorta. The descending thoracic aorta is mobilized by blunt dissection through a left fourth intercostal thoracotomy, and a 10-mm inflatable vascular occluder (In Vivo Metric Systems, Heraldsburg, CA) is placed around the descending thoracic aorta distal to the ligamentum arteriosus. A second inflatable vascular occluder (8 mm) is placed around the inferior vena cava above the diaphragm. The pericardial sac is opened and two V5 catheters with 4-mm-long V8 tips are placed in the left

[6] L. E. Shields, J. A. Widness, and R. A. Brace, *Am. J. Obstet. Gynecol.* **174**(1), 55 (1996).

atrial appendage. A Transonic flow probe 3S (Transonic Systems, Ithaca, NY) is placed around the proximal left circumflex coronary artery. The thoracotomy incision is covered but remains open during the remainder of the study. Halothane inhalation is discontinued to limit myocardial depression, and for the remainder of the study, anesthesia is maintained with a fentanyl intravenous infusion at 416 μg min^{-1} with intermittent boluses of ketamine and nitrous oxide (1:1) with oxygen via a ventilator.

Baseline hydrostatic pressures, including aortic arterial and right atrial and left atrial pressures, are measured with Transpac transducers (Abbott Critical Care Systems, Chicago, IL) calibrated prior to use with a manometer and zeroed to atmosphere at the level of the right atrium. The pressures, flow signal, and heart rate are recorded online to permit analysis by each heartbeat.

Left circumflex coronary artery arterial blood flow is first measured in response to coronary arterial pressure changes without adenosine (saline conductance) by slowly inflating either the inferior vena cava or the thoracic aortic occluder over 15 s in a randomized order, releasing the occluder and waiting until flow and pressure return to baseline and then inflating the second occluder. The entire procedure can be accomplished in less than 1 min. Following a recovery period, a dose–response curve to adenosine is generated by increasing the adenosine infusion (5 mg ml^{-1}) infused in the left atrium in steps until maximal circumflex coronary artery flow as measured by the transonic flow probe is reached. The lowest infusion rate that produces maximal steady-state vasodilatation is then selected for use in determining coronary conductance relationships. The pressure–flow relationship during maximal steady-state vasodilatation with adenosine is then recorded by inflating in random order either the aortic or the inferior vena caval occluder followed by the other occluder.

To determine the area of the coronary circulation measured by the flow probe (grams of tissue perfused), heparin is injected into the sheep, following which the sheep is killed with an intravenous injection of sodium pentobarbital. The heart is dissected from the chest cavity, and a needle with a blunt end is secured in the proximal circumflex coronary artery at the site where the flow probe had been placed earlier. The artery is then injected slowly with Evan's blue dye to demarcate the proximal circumflex coronary artery distribution. The myocardium supplied by the circumflex coronary artery is dissected from the other myocardium and weighed. An alternative is to measure coronary blood flow with a cuffed Doppler flow (Crystal Biotech, Hopkinton, MA) placed around the proximal circumflex artery. Doppler flow is converted to blood flow by calibration with microspheres injected at rest and during maximal adenosine infusion.[1] Pressure–flow relationships are described by linear regression in order to

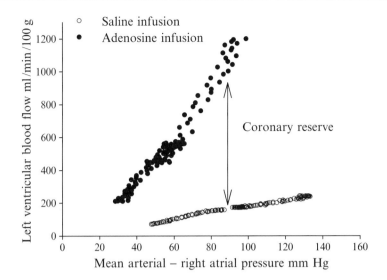

FIG. 2. Representative example of coronary pressure-flow relationships without adenosine (saline conductance) and with maximal adenosine infusion (adenosine conductance). Maximal coronary conductance is the slope of the pressure flow relationship during adenosine infusion and coronary reserve illustrated at 90 mm Hg is the difference between the adenosine and saline infusion.

interpolate resting blood flow at 90 mm Hg. Flow is expressed per 100 g of perfused tissue. Coronary reserve at 90 mm Hg is then determined as the difference between left ventricular blood flow interpolated at 90 mm Hg under resting conditions and during adenosine infusion (Fig. 2).

Coronary conductance can also be performed in chronically instrumented fetuses as well as adult preparations. In these instances, chronic close arterial injection into the left proximal circumflex artery can be used to deliver a smaller dose of adenosine or another drug directly to the coronary circulation. The distal piece of silastic tubing (0.12" i.d. Dow Chemicals) is glued onto a V6 polyvinyl catheter, and the proximal portion of the silastic tubing is swedged on a 27-gauge needle. The needle is drawn through the coronary artery, the silastic tubing is cut and allowed to retract, and hemostasis is achieved with gel foam.

Determination of Capillary Density

Fetal cardiac ventricular capillaries can be characterized using *ex vivo* coronary vascular perfusion and fixation at physiological pressures. Vascular morphometry of capillaries is a standardized method to measure angiogenesis. Pregnant sheep carrying time-bred fetuses are killed with sodium pentobarbital or a commercial euthanasia solution (Euthasol, 12 ml). Quickly, an abdominal and then uterine incisions are made. Each fetus is given 1000 U heparin either through an existing vascular catheter or directly into the umbilical vein while the heart is still beating. The fetal heart is arrested in diastole by injection of a 3-ml bolus of saturated potassium chloride into the umbilical vein. The fetal hearts are carefully excised and rinsed in saline at room temperature. Attention must be paid to a gentle touch to avoid bruising the tissue. It is also important to avoid introducing air into the ventricles or aortic stump. A stainless steel cannula is placed into the aorta just where the brachiocephalic artery originates. The cannula is secured, with care taken not to inadvertently insert it too far and cause damage to the aortic valve. Once again, even small air bubbles must be avoided, as they will prevent uniform perfusion. The superior vena cava, inferior vena cava, pulmonary veins, and pulmonary artery are ligated, and the coronary arteries are perfused from fluid reservoirs placed 25–35 cm above the aortic valve, depending on the age of the fetuses. Young fetuses normally have arterial pressures of 35 mm Hg while at term this averages nearly 50 mm Hg relative to amniotic fluid. Fluids flow out of the heart through catheters placed in the atria and are then cut 5 cm above the aortic valve. Heparinized Ringer's solution containing 1% bovine serum albumin and 1% adenosine is infused for 5 minutes, followed by a freshly made 4% paraformaldehyde solution for 10 min. This process allows the vessels to be maximally dilated when fixed in order to visualize the entire coronary vascular bed. The heart can be dissected into atria, septum, and left and right ventricular free walls.

Several tissue blocks 1 mm in diameter are cut from each ventricle. While as many locations as desired can be sampled, our standard protocol only takes tissue from a site two-thirds of the distance from the apex to the base of the heart. Each block is soaked in fixative overnight, dehydrated in increasing concentrations of alcohol (five steps) followed by progressive (three steps) xylene, and finally the blocks are embedded in paraffin (60°). Randomly chosen blocks are sectioned with a rotary microtome at 6–10 μm thickness. The sections are fixed to slides, stained with hematoxylin and eosin, and examined at 400×, and either photographs recorded or digital images are captured. The final magnification of each image is calibrated directly by the reticule. For each ventricle, at least two photomicrographs are selected of sections that are cut nearly transverse to the capillaries. In some instances, tissue blocks can be reoriented and cut again to obtain transverse sections.

Standard stereologic methods can be employed.[7] Point counting is used to measure capillary surface volume. We have used an overlay grid consisting of lines that create 60 large squares with 209 line intersections and an intersection distance of 20 μm. Capillary volume density relative to the contractile space is defined as the number of intersection points falling on capillary lumens divided by the number of points falling on either lumens or myocardial tissue. Multiple cardiac images can be used to minimize variablility. Additionally, for each image a random number generator is used to select an arbitrary number of transverse grid "squares" to allow an unbiased sampling tissue in a near transverse (relative to fibers and capillaries) plane. We typically use only five squares, although this can be increased if necessary. From this analysis we can calculate numerical capillary density (capillaries/mm^2) for all capillaries whose centers fell within the box. We record as well the minimal and maximal capillary diameters. We assume the minimum diameter to be the true diameter and the larger when they are asymmetric to be due to oblique sectioning. Intercapillary distances can be measured with a digital micrometer and are directly obtained as the shortest distance between two endothelial lumina separated by at least a single myocyte profile. Decreases in average distance may reflect either increased density or increased diameter both, potentially improving oxygen transport. The axial ratio is calculated at the ratio of the maximal to minimal capillary diameter and can used to measure the extent of obliquity. The axial ratio of a perfectly transverse section of a cylinder would be 1. This ratio is used to correct the volume densities and thereby prevent an overestimation of the vascularity when the sections are not perfectly perpendicular to the predominant vascular axis.

An alternative strategy is to use high-resolution scanning techniques that are available and can be calibrated. We have used a commercially available source used for pathology screening to scan 5-μm slides at high resolution (two pixels per micrometer, Aperio Technologies Inc., Vista, CA), and digital images can then be sent by e-mail and viewed using MrSID Geoviewer software (International Land Systems, Inc.). This allows on-screen measurement of dimensions and polygonal areas. A micrometer can be scanned and measured to ensure calibration.

---

[7] J. Smolich, A. Walker, G. Campbell, and T. Adamson, *Am. J. Physiol.* **257** (1989).

[16]   Hypoxia-Inducible Factor 1α-Deficient Chimeric Mice as a Model to Study Abnormal B Lymphocyte Development and Autoimmunity

By Hidefumi Kojima, Brendan T. Jones, Jianzhu Chen, Marilia Cascalho, and Michail V. Sitkovsky

Introduction

Immune cells, in particular T and B lymphocytes, develop during complex differentiation processes in specialized lymphoid organs, the lymph nodes, the spleen, and the gut-associated lymphoid tissue. It is often overlooked that in the majority of areas of lymphoid tissues the oxygen tension is generally much lower (between 0.5 and 4.5%) than in the atmosphere (~20%) and in the blood, suggesting that the metabolism of lymphocytes must adapt to local tissue hypoxia.[1] Furthermore, inflammatory and tumor environments where lymphocytes execute effector functions may have even lower oxygen tensions. These considerations suggest that lymphocytes may have evolved adaptive mechanisms to function under hypoxic conditions. Cellular responses to hypoxia are triggered in large part by hypoxia-inducible factor-1α (HIF-1α). The importance of HIF-1α is manifested by observations that mice with a homozygous HIF-1α deficiency die at midgestation (day 10.5) due to defects of cardiac, vascular, and neural tube formation.[2] However, the inability to develop such mice has also created methodological difficulties in analysis of the possible role of HIF-1α in immune system of adult animals.

In order to investigate the role of HIF-1α in lymphocyte development using a genetic targeting approach, we developed the hypoxia-inducible factor 1α-deficient chimeric mice (using RAG-2-deficient blastocyst complementation system).[3] This allowed the analysis of T and B lymphocytes in the absence of HIF-1α as a model to study abnormal B lymphocyte development and autoimmunity. These studies suggest that HIF-1α deficiency depresses the function of cytotoxic T lymphocytes and blocks B-cell

[1] C. C. Caldwell, H. Kojima, D. Lukashev, J. Armstrong, M. Farber, S. G. Apasov, and M. V. Sitkovsky, *J. Immunol.* **167,** 6140 (2001).

[2] N. V. Iyer, L. E. Kotch, F. Agani, S. W. Leung, E. Laughner, R. H. Wenger, M. Gassmann, J. D. Gearhart, A. M. Lawler, A. Y. Yu, and G. L. Semenza, *Genes Dev.* **12,** 149 (1998).

[3] J. Chen, R. Lansford, V. Stewart, F. Young, and F. W. Alt, *Proc. Natl. Acad. Sci. USA* **90,** 4528 (1993).

development in the bone marrow. B1 lymphocytes of fetal origin, however, accumulate and may produce autoantibodies and autoimmunity.

Strategy to Study the Role of HIF-1α T- and B-Cell Development Using Gene Targeting Technology

Overview of Gene Targeting Techniques In Vivo

Before one can study the effects of a genetic deficiency in HIF-1α, it is necessary to first generate mice with a HIF-1α deficiency. Fortunately, murine genetic technologies have advanced to a level where the specific inactivation of a single gene in live mice is both routine and efficient.

The first engineered "knockout" mice were generated in 1989 when two independent technologies, the culture of embryonic stem cells and the introduction of mutations into mammalian cells by homologous recombination, converged to provide an efficient and relatively convenient method for the specific inactivation of a single gene in a living mouse.[4] Since the first knockouts were generated, hundreds of genes have been specifically deleted by homologous recombination in embryonic stem (ES) cells, often providing interesting and unexpected phenotypes. One drawback of this approach is that the target gene is disrupted in all the cells of the mice from conception. This often will result in severe developmental defects or embryonic lethality of the knockout mice. While such phenotypes can yield interesting information about the function of the gene in question, it tells little about how the gene is utilized in the adult animal. The homozygous knockout of HIF-1α is, in fact, embryonic lethal,[2] necessitating the use of a technique that would result in tissue-specific rather than global inactivation of the gene in question.

One of the earliest techniques that gave rise to tissue-specific inactivation of a gene is the RAG complement system.[3] Using this technique it is possible to generate chimeric mice in which a gene is homozygously deleted in all T and B cells, while at the same time circumventing embryonic lethality. Because of its versatility and efficiency, RAG complementation has been a popular technique used to study gene activities in lymphocytes since the early 1990s.

The techniques used in the RAG complementation system are similar to those used to generate typical complete knockouts, with several notable exceptions. The most prominent of these distinctions is the use of blastocysts from mice that are either RAG-1 or RAG-2 deficient. The RAG

---

genes encode proteins that are specifically expressed in precursor lymphocytes and are required for the assembly of T-cell receptor (TCR) and im-
munoglobulin (Ig) genes. The inactivation of either of these genes results
in a complete inhibition of VDJ recombination, and thus a block of
lymphocyte development.[5,6] While mice lacking in functional RAG genes
are viable and normal in most respects, they completely lack any mature
T or B cells.

Not surprisingly, when RAG-deficient blastocysts are implanted into
pseudopregnant females, the resulting pups are RAG deficient.[3] However,
if normal ES cells are injected into RAG-deficient blastocysts prior to implantation, the resulting pups are chimeric for most tissues, but the mature
T and B cells are completely derived from the normal ES cells. This occurs
because the development of any RAG-deficient lymphocytes is halted in
either the bone marrow or the thymus in an immature state, whereas
cells derived from the injected ES cells are able to develop into mature
lymphocytes.

The RAG-deficient blastocyst complementation takes advantage of this
phenomenon. The targeted gene is first homozygously disrupted in normal
ES cells by homologous recombination. These cells are then injected into
RAG-deficient blastocysts, which are inserted into a pseudopregnant
female. The resulting pups are rescued from embryonic lethality because
of complementation by cells derived from RAG-deficient blastocysts but
all mature lymphocytes are derived from the injected ES cells that contain
functional RAG genes. Lymphocyte function can then be analyzed directly
on the chimeric mice without any further breeding or any concern for
embryonic lethality.

Unlike the generation of a standard knockout, RAG-deficient blastocyst complementation requires homozygous gene disruption in ES cells
injected into the blastocyst. There are two ways to select for cells in which
the recombination event has taken place in both alleles. The first approach
involves a selection in increased G418 concentrations.[7] The targeting
vector is made using a mutant neo gene, which contains a point mutation
that changes asparic acid to glutamic acid at residue 182. Cells that contain
one copy of this mutant neo gene grow normally in 0.4 mg/ml G418, but
poorly in concentrations 8–12 times higher. The few clones that grow at

[5] Y. Shinkai, G. Rathbun, K. P. Lam, E. M. Oltz, V. Stewart, M. Mendelsohn, J. Charron,
M. Datta, F. Young, and A. M. Stall, *Cell* **68**, 855 (1992).
[6] P. Mombaerts, J. Iacomini, R. S. Johnson, K. Herrup, S. Tonegawa, and V. E. Papaioannou,
Cell **68**, 869 (1992).
[7] R. M. Mortensen, D. A. Conner, S. Chao, A. A. Geisterfer-Lowrance, and J. G. Seidman,
Mol. Cell. Biol. **12**, 2391 (1992).

the higher concentration often contain a second neo gene resulting from a homologous recombination event at the second targeted allele. These surviving clones are then subcloned and analyzed by Southern blot or polymerase chain reaction (PCR) to confirm their genetic status.

A second method for generating homozygous mutant ES cells involves the use of a second targeting vector with a different selectable gene.[8] In this technique, heterozygous mutant ES cells are generated using a neo-containing targeting vector. Cells that have undergone homologous recombination are selected for and retransfected using a second targeting vector. This vector will often be the same as the first, but with a different selectable marker, such as a puromycin-resistant gene or hygromycin-resistant gene. As both the normal and the mutant alleles are targeted at the same frequency, only half of the homologous recombination events yield homozygously mutant ES cells. Southern blot or PCR analysis can then confirm the genetic status of the ES cells, and homozygous clones can be injected into blastocysts as described earlier.

The two most significant advantages to RAG-deficient complementation over a total knockout technique are that homozygous lethality is circumvented and that the chimeras can be analyzed directly, saving at least two breeding steps and months of waiting. With these advantages come certain disadvantages. The first is that when using this technique it is necessary to constantly generate new chimeras. The chimeric analysis can initially save time, but it is often more convenient over the long haul to generate knockout mouse colonies, which is impossible if the RAG complementation technique is being used to circumvent embryonic lethality. A second limitation with the technique is that it can only be used to study genes expressed in T and B cells. The third disadvantage to this technique is that the degree of lymphocyte reconstitution can vary from chimera to chimera,[3] which has the potential to obscure interesting phenotypes. Despite these limitations, RAG-deficient complementation has proven to be a powerful technique for studying gene function in lymphocytes, with dozens of genes being inactivated for phenotypic analysis using this technique.[9]

While none of the limitations of the RAG complementation technique would prevent its use in the analysis of HIF-1α deficient mice (as will be obvious later in this article), it is often desirable to inactivate a gene in a temporal manner or in a tissue besides the immune system. In cases such as this, it is often best to generate a conditional knockout mouse using either a Cre/loxP or a Flp/FRT system.[10] Cre and Flp are site-specific

---

[8] A. Cruz, C. M. Coburn, and S. M. Beverley, *Proc. Natl. Acad. Sci. USA* **88,** 7170 (1991).
[9] J. Chen, *Adv. Immunol.* **62,** 31 (1996).
[10] M. Lewandoski, *Nature Rev.* **2,** 743 (2001).

recombinases that recognize the specific DNA target sequences loxP and FRT, respectively. By flanking a gene with these target sequences and controlling the expression of the recombinase in a tissue-specific and/or temporal manner, it is possible to generate mice in which the gene is only inactivated in certain tissues and at certain times.

The major difference between the construction of a total knockout and a conditional knockout lies in the design of the targeting construct. While the specific design can vary, they generally consist of the target gene (or critical part of the target gene) flanked by loxP sites. The selectable marker (neo) is usually flanked by FRT sites. This enables neo to be deleted from the ES cells following selection by the transient expression of Flp recombinase. Once chimeric mice are generated, mice are bred for germline transmission. In order to generate the conditional knockout, germline mutant mice are bred with mice transgenically expressing Cre in a tissue-specific or inducible manner. For example, by placing Cre recombinase under the control of the hCD2 promoter and locus control region, Kioussis and colleagues have generated mice that restrict Cre expression to T and B cells,[11] allowing for the generation of T- and B-cell-specific gene inactivation.

There are several advantages to the generation of conditional knockout mice. Due to the tissue-specific nature of the gene inactivation, embryonic lethality and developmental complications can often be avoided. While the technique is often more technically challenging than the RAG complementation technique, conditional gene inactivation can be obtained in nonlymphoid tissues. Due to the number of crosses required, it takes longer to generate mice for study, but once obtained, more mice can usually be obtained through breeding so no further chimeras are required. Also, it should be noted that because a wide variety of tissue-specific Cre transgenic mice have already been generated, it is often possible to obtain one easily with the desired expression pattern.

A recently developed alternative to these traditional knockout technologies involves the use of short interfering RNA (siRNA), which targets mRNA degradation in a sequence-specific manner.[12] Using either siRNAs or short hairpin RNAs (shRNAs, a single-stranded RNA that is able to loop and base pair with itself), the expression of targeted genes can be reduced by 90%. Unlike traditional knockouts, siRNA-mediated gene suppression is transient.

[11] J. de Boer, A. Williams, G. Skavdis, N. Harker, M. Coles, M. Tolaini, T. Norton, K. Williams, K. Roderick, A. J. Potocnik, and D. Kloussis, *Eur. J. Immunol.* **33,** 314 (2003).
[12] M. T. McManus, B. B. Haines, C. P. Dillon, C. E. Whitehurst, L. van Parijs, J. Chen, and P. A. Sharp, *J. Immunol.* **169,** 5754 (2002).

In order for this technique to be used as an alternative to the more traditional knockout strategies in mice, it is necessary to transgenically express the interfering RNAs. For this purpose, lentivirus, which can be used directly to infect mouse embryos to generate transgenic mice,[13] has been used to express siRNAs.[14] Using this combination of techniques, Rubinson et al.[14] were able to generate transgenic mice in which CD8 expression was reduced by the expression of a CD8-specific shRNA. By using tissue-specific promoters to control the expression of the shRNA, it should be possible to inhibit gene expression in a tissue-specific fashion and avoid embryonic lethality.

While the use of shRNA as a knockout alternative is tempting due to the speed with which mice can be produced, several problems remain with the technology. Because genes are rarely completely inactivated, depending on the gene targeted, a phenotype may not appear. Also, the effectiveness of shRNAs in suppressing gene expression varies from gene to gene, and even from sequence to sequence within the gene. Finally, due to the novelty of this technique, it unknown how effective tissue-specific shRNA expression will prove to be. To study the impact of HIF-1α deficiency in T- and B-cell development and to overcome the embryonic lethality brought by the HIF-1α homozygous deficiency, we performed a RAG-2-deficient blastocyst complementation assay.[3] Chimeric mice were generated from RAG-2-deficient or proficient blastocysts complemented with HIF-1α-deficient RAG-2$^{+/+}$ ES cells. Because the RAG-2 gene is essential to generate both T and B lymphocytes, all T and B lymphocytes in chimeric mice obtained from the RAG-2-deficient blastocyst must be originated from the HIF-1α$^{-/-}$ ES cells that are V(D)J recombination proficient. In chimeras originated from RAG-2-proficient blastocysts, HIF-1α-deficient lymphocytes express the Ly-9.1 antigen, whereas HIF-1α-proficient lymphocytes do not. Expression of the Ly-9.1 antigen thus distinguishes between HIF-1α$^{-/-}$ and HIF-1α$^{+/+}$ lymphocytes.

Chimeric mice generated from RAG-2-deficient blastocysts complemented with HIF-1α-deficient ES cells (RAG-2/HIF-1α chimeric mouse) are normal except for short limbs. The mechanism by which mosaic HIF-1α deficiency affects morphogenesis is now under study.

---

[13] C. Lois, E. J. Hong, S. Pease, E. J. Brown, and D. Baltimore, *Science* **295**, 868 (2002).
[14] D. A. Rubinson, C. P. Dillon, A. V. Kwiatkowski, C. Sievers, L. Yang, J. Kopinja, M. Zhang, M. T. McManus, F. B. Gertler, M. L. Scott, and L. van Parijs, *Nature* **33**, 401 (2003).

Analysis of Autoimmunity in Hif-1α/Rag-2 Chimeric Mice

One of the most striking effects of HIF-1 deficiency in Hif-1α/Rag-2 chimeric mice was the abnormal B-1-cell development in bone marrow and systemic lupus erythematosis (SLE)-like autoimmunity, but quite normal development of T cells in the thymus.[15] B-1 cells are considered to be important mediators of the autoimmunity.[16,17] This section describes methods of evaluation of B-1 cells and autoimmunity in the chimeric animal.

Phenotypical Lymphoid Cell Analysis by Flow Cytometry

Hif-1α-deficient ES cells are of 129 mice origin and Rag-2 deficient blastocysts are of C57BL/6 (B6) mice origin. Because 129 mice are Ly-9.1$^+$, which is expressed on the lymphocyte strain but B6 is not, it was possible to identify the origin of lymphocytes in chimeric mice by Ly-9.1 antigen expression.[15]

Surface antigen expression, or expression of intracellular proteins, can be detected by flow cytometry analysis of single cell suspensions prepared from spleen, lymph nodes, and bone marrow. It is most convenient to study B-1 cells among peritoneal exudate lymphocytes (PEL) in chimeric mice, as B-1 cells reside in the peritoneal cavity mainly. PEL are usually obtained by the injection of 10 ml phosphate-buffered saline (PBS) into the peritoneum of mice, and the PBS, in which PEL are suspended, are harvested slowly from the peritoneum. Cells are then stained with dye-conjugated Abs or with a combination of biotinylated Abs and dye-conjugated streptavidin. B-1 cells are distinct from conventional B cells by the unique surface molecule expression pattern. Usually, B-1 cells express CD5, Mac-1 (CD11b) on their surface. These molecules are not expressed by B-2 cells. In contrast, CD23 is expressed on B-2 cells, but not on B-1 cells. Furthermore, B-1 cells express high levels of IgM but very low levels of IgD as compared to B-2 cells.[16,17]

Analysis of peritoneal lymphocytes in Hif-1α/Rag-2 chimeric mice using combinations of monoclonal antibodies to these molecules revealed dramatic changes in the expression of B-1 cells defining CD5 and B220 surface markers. The CD5^{dull}B220$^{-/dull}$ subset observed in wild-type chimeras (wild-type ES cells were introduced into Rag-2-deficient blastocysts) was

[15] H. Kojima, H. Gu, S. Nomura, C. C. Caldwell, T. Kobata, P. Carmeliet, G. L. Semenza, and M. V. Sitkovsky, *Proc. Natl. Acad. Sci. USA* **99**, 2170 (2002).
[16] K. Hayakawa, R. R. Hardy, M. Honda, L. A. Herzenberg, A. D. Steinberg, and L. A. Herzenberg, *Proc. Natl. Acad. Sci. USA* **81**, 2494 (1984).
[17] K. Hayakawa, R. R. Hardy, and L. A. Herzenberg, *Eur. J. Immunol.* **16**, 450 (1986).

lost, whereas the accumulation of unusual CD5$^+$/B220$^+$ cells was detected in Hif-1α/Rag-2 chimeric mice. These CD5$^+$/B220$^+$ cells in Hif-1α/Rag-2 chimeras express a typical surface antigen pattern of B-1 cells (IgM^{high}/IgD$^{low-inter}$/Mac-1$^+$/CD23$^-$). However, they had much higher levels of B220 expression, leading us to designate these unusual cells as B-1 like. In addition, the increase in cells with the B-1-like phenotype was accompanied by a corresponding decrease in the proportions of other cells (e.g., Mac-1$^-$/CD23$^+$ cells) in the peritoneum of Hif-1α/Rag-2 chimeric mice.

Importantly, as described later, Hif-1α-deficient ES cell origin B cells produce rheumatoid factors (RF). RF binds to some antibodies nonspecifically.[18] Therefore, it is very important to use isotype-matched controls for flow cytometry analysis. Indeed, the nonspecific binding of some rat-IgG to Hif-1α-deficient B cells was observed during analysis (data not shown).

Measurement of Autoantibody Production

Since it has been reported that B-1 cells are involved in autoimmunity and they produce autoantibodies, such as anti-DNA antibody and RF,[19,20] it is important to evaluate autoantibody production in order to assess the autoimmunity in Hif-1α/Rag-2 chimeras.

Titration of Antidouble-Stranded (ds) DNA Antibodies. One of the most typical autoantibodies produced by B-1 cells is the anti-dsDNA antibody. The titer of anti-dsDNA Ab in mouse serum is determined by ELISA as follows. One percent protamine sulfate (Sigma), dsDNA from calf thymus (Sigma) (stock condition; 1 mg/ml, at −80°), horseradish peroxidase (HRP) or alkaline phoshatase (AP)-conjugated antimouse IgG or IgM antibodies, and detection reagents for ELISA are required for the assay. One hundred microliters of 0.001% protamine sulfate is put into a well of a 96-well ELISA plate and kept it for 60–90 min at room temperature. Wash with distilled water three times. After washing, put 50 μl of the dsDNA solution (5 μg/ml solved in 0.15 M NaCl, 0.015 M sodium citrate, pH 8.0) into a well. As control, 50 μl of the 0.15 M NaCl, 0.015 M sodium citrate, pH 8.0, solution without dsDNA should be put into a well. The ELISA plate with the dsDNA solution is left overnight at 37° to dry up. Do not leave it in a humidified environment. After drying up, wash wells with ELISA wash buffer (0.05% Tween 20 in PBS) three times. To block non-specific binding, ELISA wash buffer containing 50% fetal calf serum

---

[18] B. J. Sutton, A. L. Corper, M. K. Sohi, R. Jefferis, D. Beale, and M. J. Taussig, *Adv. Exp. Med. Biol.* **435**, 41 (1998).

[19] M. Murakami and T. Honjo, *Ann. N. Y. Acad. Sci.* **764**, 402 (1995).

[20] P. Casali, S. E. Burastero, M. Nakamura, G. Inghirami, and A. L. Notkins, *Science* **236**, 77 (1987).

(FCS) or bovine serum albumin (BSA) (or whatever is used as the blocking reagent for ELISA) is put into a well and left for 1 h at room temperature. Then wash it with ELISA wash buffer three times. Serially diluted serum from mice are put into a well. Usually, serial dilutions are started from at 1:100 dilutions. Sample serums are diluted with sample dilution buffer (PBS containing 1% BSA or whatever is used as sample dilution buffer for ELISA). Incubate the plate for 1–2 h at room temperature followed by washing with ELISA wash buffer three times. Fifty microliters of detection antibodies, that is, HRP- or AP-conjugated either antimouse IgG or IgM antibodies (for detection of anti-dsDNA IgG and IgM, respectively), is put into a well. Detection antibodies are usually diluted at 1:1000 (or dilution index is indicated by the manufacturers) with sample dilution buffer. Incubate for 1 h at room temperature and wash with ELISA wash buffer three times. Finally, appropriate detection reagents, such as P-phenylendiamine dichloride salt (OPD), are put into wells and the detection reaction is stopped by the addition of a stop reagent such as Na_2SO_4. The timing of termination of reaction is dependent on the system. The titer of anti-dsDNA Ab is determined as the absorbance at the appropriate wavelength. To measure antisingle-strand DNA (ssDNA) Ab, use ssDNA (synthesized DNA or denatured dsDNA) instead of denatured dsDNA.

Titration of Rheumatoid Factors. The titer of RF in serum is assessed by ELISA. RF are antibodies to IgG or IgA.[4] Therefore, we assess the titer of antigoat IgG antibodies, as RF, in serum. One hundred microliters of goat IgG (50 μg/ml) is put into a well of a 96-well ELISA plate and left for 1 hr at room temperature. Wash with ELISA wash buffer three times. As control, blank wells should be prepared. Then, wells including blank wells are filled with blocking buffer three times. Serially diluted serum samples are put into wells. The sample serum is diluted with sample dilution buffer as described earlier. The following steps are exactly the same as the procedure described earlier. The titer of RF is determined as absorbance at the appropriate wavelength.

Measurement of Protein in Urea

Systemic lupus erythematosus (SLE) is usually accompanied by renal disorder. The measurement of protein levels in urine gives us important information on renal disorder.[21] The collection of urine from mice is very easy. When mice are grobbed, they discharge urine. Just collect it. The protein levels in urine samples from mice can be determined using a kit for

[21] E. M. Tan, A. S. Cohen, J. F. Fries, A. T. Masi, D. J. McShane, N. F. Rothfield, J. G. Schaller, N. Talal, and R. J. Winchester, *Arthritis Phenum.* **25,** 1271 (1982).

protein measurement. We use the microprotein kit from Sigma. Using these assays, we showed markedly increased levels of both IgM and IgG anti-dsDNA autoantibodies in the serum of 6- to 8-week-old Hif-1α/Rag-2 chimeras. This was accompanied by extensive deposits of IgM and IgG in the kidneys of Hif-1α/Rag-2 chimeric mice. Furthermore, evidence of an autoimmune disease was also provided by the demonstration of high levels of protein in urine and the accumulation of rheumatoid factor in Hif-1α/Rag-2 chimeras. At this moment, there is no direct evidence that B-1-like cells found in Hif-1α/Rag-2 chimeric mice produce these autoantibodies. It is most likely, however, that they play a key role in the production of auto-antibodies, as a majority of B cells in peripheral lymphoid organs in Hif-1α/Rag-2 chimeric mice are CD5+ (Kojima, unpublished data).

Taken together, these observations suggest that Hif-1α is important for the proper regulation of B-1 cell homeostasis and that Hif-1α deficiency results in autoimmunity. These findings from the analysis of RAG-2/HIF-1α chimeric mice resemble, but are not completely similar to, phenotypes of other SLE autoimmune mice, such as (NZBXNZW)F1 mice[22] and mice with lpr or gld mutations.[23] Thus, the Hif-1α/Rag-2 chimeric mouse could be used as a new model of SLE-like autoimmune mice.

Interpretation of Studies of T- and B-Cell Development in RAG-2/HIF-1α Chimeric Mice

T-Cell Development and Function in HIF-1α-Deficient Chimeric Mice

To determine whether a deficiency of HIF-1α impacted T-cell development, we analyzed the thymuses of RAG-2/HIF-1α chimeric mice by flow cytometry. We found that a lack of HIF-1α expression did not affect significantly the proportions of thymocyte subsets, suggesting that HIF-1α deficiency does not noticeably impair thymocyte development, and the relative numbers of CD4-positive and CD8-positive HIF-1α-deficient T lymphocytes in blood, lymph nodes, and spleen were not significantly different from wild type.[15]

Abnormal B-Cell Development in RAG-2/HIF-1α Chimeric Mice

Conventional B lymphocytes develop in the bone marrow of adult animals. B lymphocyte development is staged according to the heavy and light chain V(D)J recombination events and is accompanied by changes

[22] C. G. Drake, S. J. Rozzo, T. J. Vyse, E. Palmer, and B. L. Kotzin, *Immunol. Rev.* **144,** 51 (1995).
[23] P. L. Cohen and R. A. Eisenberg, *Immunol. Today* **13,** 427 (1992).

in the expression of several surface markers. Prior to heavy chain expression, nascent B cells are called pro-B cells. B cells that succeed in rearranging productively both heavy and light chains and express surface immunoglobulins are called immature B cells.[24]

We also examined development of the bone marrow-independent peritoneal B cell subset (also called B-1 cells). B-1 cells appear to be derived early in ontogeny from fetal liver B-cell progenitors and are thought to be maintained independently in the adult. HIF-1α deficiency did not affect the total number of peritoneal B cells. However, we found a relative increase in the number of HIF-1α-deficient B-1 cells and a relative decrease in the number of conventional B-2 cells. The most dramatic difference was that unlike wild-type B-1 cells, HIF-1α-deficient B-1 cells expressed an unusual high level of B220.[15] The unusual expression of high levels of B220 by peritoneal HIF-1$\alpha^{-/-}$ B-1 cells and the proportional increase in IgMhi/IgDlo HIF-1$\alpha^{-/-}$ B-2 cells may, in fact, reflect "homeostatic" proliferation. Our observations recapitulate to a great extent the phenotype of mice in which B-cell production is interrupted at birth or in the newborn.[25] Because most HIF-1$\alpha^{-/-}$ B cells have a B-1 like phenotype and have been associated with the production of autoantibodies, we predicted an increased level of autoantibodies in the serum of animals with HIF-1$\alpha^{-/-}$ B cells. We found an elevated level of both IgM and IgG antidouble-stranded DNA autoantibodies and rheumatoid factor in the serum of 6- to 8-week-old RAG-2/HIF-1α chimeras. The increased level of autoantibodies was accompanied by extensive deposits of IgM and IgG in the kidneys and by high levels of protein in the urine of RAG-2/HIF-1α, but not RAG-2/WT chimeras.[15]

While there is no direct evidence that B-1-like cells found in RAG-2/HIF-1α chimeric mice produce the autoantibodies, they are a likely source as they comprise the majority of B cells in peripheral lymphoid organs in RAG-2/HIF-1α chimeric mice (Kojima, unpublished data). The phenotype of RAG-2/HIF-1α chimeric mice resembles the phenotypes of SLE autoimmune mice, such as (NZBXNZW)F1 mice[26] or mice with lpr or gld mutations.[27] Thus, HIF-1α-deficient chimeras may be used as an alternative model of SLE-like mice.

[24] R. R. Hardy and K. Hayakawa, *Annu. Rev. Immunol.* **19**, 595 (2001).

[25] Z. Hao and K. Rajewsky, *J. Exp. Med.* **194**, 1151 (2001).

[26] B. C. Seegal, L. Acinni, G. A. Andres, S. M. Beiser, C. L. Christian, B. F. Erlanger, and K. Hsu, *J. Exp. Med.* **130**, 203 (1969).

[27] J. B. Roths, E. D. Murphy, and E. Eicher, *J. Exp. Med.* **159**, 1 (1984).

Conclusion

Understanding the mechanisms of how HIF-1α has an impact on the development of B lymphocytes and on the function of T lymphocytes in response to hypoxic conditions or to other stress stimuli may provide a useful perspective on the regulation of lymphocyte biology and may help in the design of novel immune modulators. The use of animals with targeted disruption of the HIF-1α gene in lymphoid cells may provide a very useful tool to accomplish these goals.

Section III

Ion Channels

[17]   Molecular Strategies for Studying Oxygen-Sensitive K^+ Channels

By KEITH BUCKLER and ERIC HONORÉ

Environmental hypoxia evokes a rapid reflex increase in the respiration rate. This reflex is initiated in the carotid bodies located at the bifurcation of the carotid arteries.[1,2] Upon a decrease in arterial pO_2, chemoreceptor type I carotid body cells release neurotransmitters that activate afferent sensory fibers of the sinus nerve stimulating the brain stem respiratory centers and provoking a reflex increase in ventilation. Similarly, neuroepithelial body cells, which are innervated clusters of amine and peptide containing cells located within the airway mucosa, are transducers of hypoxic stimuli and function as airway chemoreceptors.[3–5] In the perinatal period, before the establishment of sympathetic innervation, hypoxia additionally triggers catecholamine release from adrenomedullary chromaffin cells.[6] This "nonneurogenic" mechanism is essential for the ability of newborns to survive hypoxic stress and modulates respiratory as well as cardiovascular and metabolic responses. In addition to the release of neurotransmitters and the stimulation of respiration, hypoxia also has a profound adaptive effect on the pulmonary circulation.[7,8] Hypoxia-induced vasoconstriction of resistance pulmonary artery smooth muscle (PASM) leads to a redistribution of the nonoxygenated blood toward better ventilated regions of the lung.

These specialized cells share a mechanism in common that transduces hypoxic stimuli into a rapid cellular response: the closing of O_2-sensitive K^+ channels.[1,2,7,8] Hypoxia depolarizes O_2-sensitive cells, increases excitability, provokes the opening of voltage-gated Ca^{2+} channels, increases intracellular Ca^{2+}, and triggers cellular responses, including neurotransmitter release as well as myocyte contraction.

[1] J. Lopez-Barneo, *Trends Neurosci.* **19**, 435 (1996).
[2] C. Peers, *Trends Pharmacol. Sci.* **18**, 405 (1997).
[3] D. Wang, C. Youngson, V. Wong, H. Yeger, M. C. Dinauer, E. Vega-Saenz Miera, B. Rudy, and E. Cutz, *Proc. Natl. Acad. Sci. USA* **93**, 13182 (1996).
[4] X. W. Fu, D. Wang, C. A. Nurse, M. C. Dinauer, and E. Cutz, *Proc. Natl. Acad. Sci. USA* **97**, 4374 (2000).
[5] C. Youngson, C. Nurse, H. Yeger, and E. Cutz, *Nature* **365**, 153 (1993).
[6] R. J. Thompson, A. Jackson, and C. A. Nurse, *J. Physiol.* **498**, 503 (1997).
[7] E. K. Weir and S. L. Archer, *FASEB J.* **9**, 183 (1995).
[8] R. Z. Kozlowski, *Cardiovasc. Res.* **30**, 318 (1995).

This article describes the various methods employed to study O_2-sensitive ion channels in chemosensitive cells. It is divided into five sections, which deal with technical issues including: (i) isolation and culture of chemosensitive cells; (ii) methods for detecting and visualizing ion channel mRNAs and proteins in chemoreceptor cells; (iii) electrophysiological techniques for studying O_2-sensitive K^+ channels; (iv) molecular strategies to interfere with O_2-sensitive K^+ channels; and (v) concludes with a summary of the functional role of one such ion channel, the TASK-like K^+ channel, which plays a key role in O_2 sensing in carotid body type I cells.

Techniques for Isolating and Culturing Chemosensitive Cells

Enzymatic Dissociation and Short-Term Culture of Type I Carotid Body Cells

Various enzymatic procedures have been developed to isolate chemosensitive cells, including type I carotid body cells,[9–11] neuroepithelial body cells,[5] and pulmonary artery myocytes,[12–14] as well as adrenomedullary chromaffin cells.[6] The details of these techniques obviously vary with cell or tissue type and the reader is referred to the relevant publications just cited. By way of example, however, this article describes the technique used to isolate carotid body type I cells.[15–20] Carotid bodies are very small in size and are located close to the bifurcation of the common carotid body. As locating and excising these organs is time-consuming, this operation is performed under anesthesia to avoid postmortem tissue deterioration. In small neonatal animals such as rats the carotid body is often clearly visible and can be directly excised. In older animals, however, the carotid body may be covered in connective tissue and difficult to see so that it is more

[9] J. Urena, J. Lopez, C. Gonzalez, and J. Lopez Barneo, *J. Gen. Physiol.* **93**, 979 (1989).

[10] M. C. Fishman and A. E. Schaffner, *Am. J. Physiol.* **246**, C106 (1984).

[11] M. R. Duchen, K. W. Caddy, G. C. Kirby, D. L. Patterson, J. Ponte, and T. J. Biscoe, *Neuroscience* **26**, 291 (1988).

[12] L. H. Clapp and A. M. Gurney, *Exp. Physiol.* **76**, 677 (1991).

[13] J. M. Post, J. R. Hume, S. L. Archer, and E. K. Weir, *Am. J. Physiol.* **262**, C882 (1992).

[14] E. D. Michelakis, I. Rebeyka, X. Wu, A. Nsair, B. Thebaud, K. Hashimoto, J. R. Dyck, A. Haromy, G. Harry, A. Barr, and S. L. Archer, *Circ. Res.* **91**, 478 (2002).

[15] K. J. Buckler, R. D. Vaughan Jones, C. Peers, and P. C. Nye, *J. Physiol.* **436**, 107 (1991).

[16] K. J. Buckler and R. D. Vaughan-Jones, *Pflug. Arch.* **425**, 22 (1993).

[17] K. J. Buckler, R. D. Vaughan-Jones, C. Peers, D. Lagadic-Gossmann, and P. C. Nye, *Pflüg. Arch.* **425**, 103 (1993).

[18] K. J. Buckler, *J. Physiol.* **498**, 649 (1997).

[19] K. J. Buckler and R. D. Vaughan-Jones, *J. Physiol.* **513**, 819 (1998).

[20] K. J. Buckler, B. A. Williams, and E. Honore, *J. Physiol.* **525** Pt 1, 135 (2000).

convenient to remove the whole carotid bifurcation first and perform any subsequent dissection in a petri dish. The carotid body/carotid bifurcation is dissected rapidly and transferred to ice-cold phosphate-buffered saline (PBS; Sigma). Once in cold saline, further cleaning or dissection of the carotid body can be undertaken. To dissociate single cells the carotid body is incubated in an enzyme mixture typically containing collagenase and trypsin. For neonatal rat carotid body cells, we use a relatively mild enzymatic solution that comprises 0.4 mg/ml collagenase (Worthington type I) and 0.2 mg/ml trypsin (Sigma) in a 9:1 mixture of Ca^{2+}/Mg^{2+}-free PBS: normal PBS. Carotid bodies are incubated in this enzyme solution for 20 min at 35°. Following this initial incubation the carotid bodies are teased apart carefully with forceps under a dissecting microscope and are then incubated for a further 5–10 min. The tissue is then transferred to a centrifuge tube before being triturated with a fire-polished glass pipette. The cell suspension is then centrifuged (500–1000 rpm for 5 minutes), the supernatant is discarded, and the cells are resuspended in culture medium. The cell suspension is then triturated again before plating out onto glass coverslips coated with poly-D-lysine (Sigma). Cells are then maintained in culture medium at 36° in an atmosphere of 5% CO_2 in air before use. Although we normally only use cells on the same day in which they are isolated, longer-term culture is also possible (see later). The culture medium used comprises Ham's F-12 (Sigma) supplemented with 10% heat-inactivated fetal calf serum (GIBCO), insulin (84 mU/ml), penicillin (100 U/ml), and streptomycin (100 μg/ml).

Long-Term Culture of Type I Carotid Body Cells

A long-term culture of carotid body type I cells, which retains the expression of O_2-sensitive K^+ channels, has been reported.[21] Isolated type I cells are obtained by combined enzymatic and mechanical dissociation of carotid bodies dissected from 1- to 14-day-old rat pups, much as described earlier. Following inactivation of the enzyme with serum-containing growth medium, the dispersed cell population is plated into the center of modified culture dishes. In this modification, a central hole is drilled in the bottom of a 35-mm culture dish and sealed by attaching a glass coverslip to the underside. This creates a well that is coated with a thin layer of collagen or matrigel to aid cell attachment. A few drops of the dissociated cell suspension are applied to the well and incubated overnight to allow the cells to attach. The cultures are then flooded with 1.5–2 ml of growth medium and incubated in a humidified atmosphere of 95% air, 5% CO_2

[21] C. A. Nurse and I. M. Fearon, *Microsc. Res. Tech.* **59,** 249 (2002).

at 37°. Typically, a litter of about 12 pups provides dissociated carotid body cells for six cultures, which are fed every 5–6 days. The normal growth medium consists of F-12 nutrient medium (GIBCO) supplemented with the following additives: 10% fetal bovine serum, 0.6% glucose, 80 U/liter insulin (or 20 ng/ml insulin growth factor, IGF-1), 2 mM L-glutamine, and 1% penicillin–streptomycin. Cultured type I cells grow in monolayer clusters enveloped by glial-like, type II, or sustentacular cells. These cells can undergo DNA synthesis, evidenced by the uptake of bromodeoxyuridine (BrdU), and show a limited capacity for cell division. In this long-term culture (approximately 2 weeks), cells retain expression of the TASK-1 and BK K$^+$ channels as revealed by immunocytochemistry or RT-PCR analysis of mRNA extracted from type I clusters after removal from the culture surface.[21]

Thin Tissue Slices of O_2-Sensitive Tissues

An alternative elegant method based on the brain tissue slice/organotypic culture technique has been developed to study arterial chemoreceptors and neuroepithelial body cells (NEB).[22–24] Carotid bodies are isolated from Wistar rats of ages between postnatal days 5 and 20 and placed in 3% (w/v) low melting point agarose (FMC). After quick cooling on ice, the agarose block is glued with cyanoacrylate to the stage of a vibratome chamber and covered by cold, O_2-saturated Tyrode's solution. Slices 100 to 150 μm thick are cut with standard razor blades. The resulting slices (usually four) are washed twice with cold sterile PBS and placed on 35-mm petri dishes with Dulbecco's modified eagle medium (DMEM) (GIBCO/BRL) supplemented with 1% (v/v) penicillin/streptomycin (GIBCO/BRL) and 10% (v/v) fetal bovine serum (FBS) (BioWhittaker). Once isolated, slices are maintained at 37° in a 5% CO_2 incubator for 48–96 h before the experiments. For experiments, a slice is transferred to a recording chamber (200-μl volume) placed on the stage of an upright microscope (Zeiss Axioskop) equipped with long-distance water-immersion objectives. Once in the chamber, the carotid body slice is retained using a lyra made of silver wire with glued fine Nylon threads to hold the slice on the bottom of the chamber.

A lung slice preparation has also been developed to study O_2-sensitive K$^+$ channels in neuroepithelial body cells.[24] Neonatal New Zealand White

---

[22] R. Pardal, U. Ludewig, J. Garcia-Hirschfeld, and J. Lopez-Barneo, *Proc. Natl. Acad. Sci. USA* **97**, 2361 (2000).

[23] R. Pardal and J. Lopez-Barneo, *Respir. Physiol. Neurobiol.* **132**, 69 (2002).

[24] X. W. Fu, C. A. Nurse, Y. T. Wang, and E. Cutz, *J. Physiol.* **514**, 139 (1999).

rabbits of between 1 and 10 days of age are killed by an intraperitoneal eu-thanyl injection (100 mg/kg). The lungs are perfused with Krebs solution (mM: 140 NaCl, 3 KCl, 1·8 CaCl$_2$, 1 MgCl$_2$, 10 HEPES, 5 glucose, pH 7.3, adjusted with HCl) and then embedded in 2% agarose (FMC Bioproducts, Rockland, ME). Transverse lung slices (200–400 μm) are cut with a Vibra-tome (Ted Pella, Inc. Redding, CA). Sectioning is performed with the tissue immersed in ice-cold Krebs solution. For electrophysiological re-cordings, the lung slices are transferred to a recording chamber mounted on the stage of a Nikon microscope (Optiphot-2UD, Nikon, Tokyo, Japan). The "perfusing" Krebs solution has the following composition (mM): 130 NaCl, 3 KCl, 2.5 CaCl$_2$, 1 MgCl$_2$, 10 NaHCO$_3$, 10 HEPES, 10 glucose, pH 7.35–7.4. The slices are incubated with the vital dye Neutral Red (0.02 mg/ml) for 15 min at 37° to identify NEB cells in fresh lung tissue.

Methods for Detecting and Visualizing Ion Channel mRNAs and Proteins in Chemoreceptor Cells

RT-PCR Detection of K^+ Channels in Carotid Bodies

RT-PCR experiments have been used to identify the expression of Kv1.4, Kv3.4, Kv4.1, and Kv4.3 mRNAs, encoding fast-inactivating volt-age-gated K^+ channels, in the rabbit carotid bodies.[25] The homogeneous distribution of Kv4 subunits in chemoreceptor cells, along with their elec-trophysiological properties, suggests that Kv4.1, Kv4.3, or their heteromul-timers are the molecular correlate of the O_2-sensitive voltage-gated K^+ channel originally described in rabbit carotid body type 1 cells.[25] Total RNA is extracted from rabbit CBs using Trizol (GIBCO-BRL). Reverse transcription is carried out using MuLV reverse transcriptase (PE Biosys-tems) at 42° for 60 min. PCR experiments are performed in a thermal cy-cler (GeneAmp 9700 Perkin Elmer) using thin-walled plastic tubes (PE Biosystems). Unique primers for TH and Kv subunit amplification are designed using the Primer 3 website (http://www-genome.wi.mit.edu/cgi-bin/primer/primer3-www.cgi). Control genes used for RT-PCR are TH and the rabbit ribosomal protein L18. Primer sequences and annealing temperatures for RT-PCR are available in Sanchez et al.[25] RT-PCR condi-tions are 30 s at 95° (15 s at 95°, 20 s at annealing temperature, 60 s at 27°) × 35 and finally for 10 min at 72°. A negative control amplification without RT is performed using TH primers to test the genomic contamination of the CB RNA sample. Moreover, primers for several rabbit channel genes

[25] D. Sanchez, J. R. Lopez-Lopez, M. T. Perez-Garcia, G. Sanz-Alfayate, A. Obeso, M. D. Ganfornina, and C. Gonzalez, *J. Physiol.* **542,** 369 (2002).

are designed to encompass at least one intron present in the orthologous mouse and human genes. The specificity of the RT-PCR products is determined by sequencing or by band size comparisons to PCR amplifications from the corresponding plasmid.

In Situ *Hybridization Visualization of* K^+ *Channel mRNAs in Type I Carotid Body Cells*

This section details the *in situ* hybridization method used for visualizing the mRNA encoding the K^+ channel TASK-1 in type I cells.[20] Carotid body cells from 11-day-old to 3-week-old Wistar or Sprague–Dawley rats are prepared using the protocol described earlier. Cells are plated on poly-D-lysine-coated glass culture chamber slides and maintained in Ham's F-12 for 2–3 h in a 37° 5% CO_2 incubator before fixation with ice-cold 4% (w/v) paraformaldehyde in 0.1 M sodium phosphate-buffered solution (pH 7.4) for 30 min. The slides are then washed twice in PBS and stored in 70% ethanol at −4°. Digoxigenin- or fluorescein-labeled RNA sense and antisense TASK probe (nucleotides 521–583; accession number AF006824) are generated with an *in vitro* transcription kit (Boehringer Mannheim), which incorporates one modified digoxigenin-labeled UTP every 20–25th position in the transcript. Prehybridization consists of a 3-min wash in PBS; permeabilization for 5 min in PBS + 0.1% Triton; 2×5-min washes in PBS; 10 min of citric acid–sodium citrate microwave irradiation; and 2×5-min washes in PBS. Hybridization is performed at 75° overnight with a 10-ng probe/ml [in 4× saline–sodium citrate buffer (SCC) with 50% formamide, 10% dextran sulfate, 1% Denhardt's solution, 5% sarcosyl, 500 mg/ml denatured salmon sperm DNA, 250 μg/ml yeast tRNA, 20 mM dithiothreitol, and 20 mM $NaPO_4$]. Slides are washed briefly in 4× SSC, followed by 30 min in SSC+ RNase A (40 μg/ml) at 37°, 30 min at room temperature in 0.1× SSC, and finally for 30 min at room temperature in 0.1× SSC. Immunological detection of labeled mRNA is carried out using antidigoxigenin-alkaline phosphatase. Type I cells are stained using a mouse monoclonal antityrosine hydroxylase (diluted 1/100, Boehringer Mannheim) and a secondary antimouse F(ab)$_2$ fragment labeled with tetramethylrhodamine isothiocyanate (TRITC) (Boehringer Mannheim, diluted 1/10).

Immunolocalization of K^+ *Channel Proteins in Carotid Bodies*

This section describes the immunocytochemistry procedure developed to visualize 2P domain channels in rat carotid bodies.[26] Each rat is anesthetized by pentobarbital (15 mg/kg; intraperitoneal injection) and perfused

---

[26] Y. Yamamoto, W. Kummer, Y. Atoji, and Y. Suzuki, *Brain Res.* **950,** 304 (2002).

transcardially with Ringer's solution (500 ml), followed with Zamboni's fixative (4% paraformaldehyde, 0.5% picric acid in 0.1 M phosphate buffer, pH 7.4, 500 ml). The bifurcation of carotid arteries is dissected out and further fixed with the same fixative for 5 h. The tissues are soaked in 30% sucrose in PBS (pH 7.4) and frozen. They are sectioned serially at a thickness of 10 μm and mounted on glass slides coated with chrome alum-gelatin. The sections are incubated for 60 min with nonimmune donkey serum (1:50) and rinsed with PBS, pH 7.4. The sections are then incubated overnight at 4° with rabbit polyclonal antisera against TASK-1 (1:100; APC-024, Alomone, Jerusalem, Israel), TASK-2 (1:100; APC-037, Alomone), goat polyclonal TASK-3 (1:100; sc-11309, Santa Cruz Biotechnology, Santa Cruz, CA), and TRAAK (1:100; sc-11309, Santa Cruz Biotechnology). After incubation, the sections are washed with PBS again and treated with TRITC-labeled donkey antibody against rabbit IgG (1:100; Jackson Immunoresearch, West Grove, PA) for TASK-1 and TASK-2 and with TRITC-labeled donkey antibody against goat IgG (1:100, Jackson Immunoresearch) for TASK-3 and TRAAK for 2 h at room temperature. After washing with PBS, sections are mounted in carbonated glycerol and examined by a fluorescent microscope (BX-70, Olympus, Tokyo). Negative controls are incubated with PBS and with the antibody preabsorbed overnight with the immunizing peptides (1 μg/μg for each antibody). Because the molecular structures of TASK-1 and TASK-3 are similar, the antibody against TASK-1 is preabsorbed with the TASK-3 antigen and the antibody against TASK-3 is preabsorbed with the TASK-1 antigen and they are used for controls. Immunoreactivities for TASK-1, TASK-2, TASK-3, and TRAAK are observed in the rat carotid body.[26] Sections incubated with PBS instead of primary antiserum or with preabsorbed antibody show no immunoreactivity. Sections incubated with antibodies against TASK-1 and TASK-3 that are preabsorbed with TASK-3 and TASK-1, respectively, show similar results to sections stained for TASK-1 and TASK-3. Although this study demonstrates a positive staining with the anti-TASK-1 antibody, it has to be mentioned that our own control experiments (K. Buckler and E. Honoré, unpublished data), with the same antibody from Alomone on nontransfected COS-7 cells, show an intense staining of the cytoskeleton, suggesting a possible lack of specificity.

Electrophysiological Techniques for Studying O$_2$-Sensitive
    K$^+$ Channels

This section describes the electrophysiological techniques used to study O$_2$-sensitive ion channels as well as more basic considerations regarding the choice of buffer and the methods used to generate hypoxia in isolated cell/tissue experiments.[15–20]

Hypoxic Solutions and O_2 Monitoring

The first consideration is the choice of buffer and temperature. Carotid bodies are strongly temperature sensitive,[27] and it has been reported that a reducing temperature below 36° suppresses the response of isolated type I cells to hypoxia.[28] The carotid body is also highly sensitive to changes in pH and pCO_2. Selecting a buffer that maintains a normal pH both inside and outside of the cell is therefore important. Because bicarbonate ions play a key role in the regulation of intracellular pH, bicarbonate-free buffers can cause very marked perturbations of intracellular pH.[15] In view of these considerations, we have employed bicarbonate-buffered solutions at physiological temperatures wherever feasible. The standard HCO_3-buffered Tyrode solution we have employed is comprised (in mM) of 117 NaCl, 4.5 KCl, 23 $NaHCO_3$, 1.0 $MgCl_2$, 2.5 $CaCl_2$, and 11 glucose and is equilibrated with 5% CO_2 plus O_2 and N_2 (to give the desired O_2 level, see later); the pH at 37° is 7.4–7.45.

In order to have adequate control over O_2 levels it is essential that the perfusion apparatus be constructed of materials that have low gas permeability. Solutions should be contained within gas-impermeant vessels and equilibrated continuously with a gas mix of the desired composition. We typically use glass reagent bottles fitted with a stopper that carries a glass gas distribution tube and has one small hole to allow gas to escape and through which the solution can be withdrawn. This arrangement limits the ingress of atmospheric air. The delivery of gas to the solution reservoir and the delivery of solutions from the reservoir to the experimental chamber should be via gas-impermeable tubes. From the perspective of gas permeability, the best materials include medical-grade stainless steel tubing and glass tubing. Some forms of plastic also have relatively low gas permeability and, provided the tubes are thick walled, give reasonable performance (e.g., Tygon and Pharmed, Norton Performance Plastics, www.tygon.com). Thin-walled plastic tubing and any silicone rubber-based tubing should be avoided. If solutions have to be pumped or passed through taps/valves the construction of these also needs careful consideration if atmospheric O_2 is to be kept out (peristaltic pumps fitted with Pharmed tubing work well but gravity feed is often simpler). For any kind of electrophysiological experiment it is necessary to work with an open

---

[27] R. Gallego, C. Eyzaguirre, and L. Monti Bloch, *J. Neurophysiol.* **42,** 665 (1979).
[28] T. J. Biscoe and M. R. Duchen, *J. Physiol.* **428,** 39 (1990).

perfusion chamber. The ingress of significant amounts of atmospheric O_2 at this stage can be limited by either using a rapid solution flow rate relative to the recording chambers volume (we typically use a flow of 2 ml/min for a chamber volume of about 50–100 μl) and/or a conical collar fitted above the chamber that is filled with argon (see Stern et al.[29] for a useful design).

Even with the aforementioned precautions, however, O_2 invariably gets in somehow; thus if anoxic conditions are required, some form of chemical O_2 scavenger has to be used. Those in most common use are sodium dithionate ($Na_2S_2O_4$) and sodium sulfite (Na_2SO_3). Both should be used sparingly, and it should be noted that when sodium dithionite reacts with O_2 it (a) generates acid and (b) may generate reactive O_2 species.[30] These compounds should therefore only be used in solutions that are well buffered (e.g., bicarbonate or $>= 20$ mM HEPES) and in systems in which the ingress of atmospheric O_2 can be kept to a minimum. Note that it is not possible to "control" for indirect effects of these compounds by combining them with O_2, as they react rapidly with O_2 and one will be left either with no O_2 (if the scavenger is in excess) or no scavenger (if the O_2 is in excess). An obvious point perhaps but it is a mistake that has been made in the past.

The ability to monitor O_2 levels in the recording chamber is important if one wishes to determine any kind of concentration effect relationship or to confirm that solutions do in fact have the desired level of O_2. The commonest method of monitoring O_2 is to use a minaturized Clarke style O_2 electrode (Diamond General; www.diamondgeneral.com) or a separate platinum microelectrode polarized to about -700 mV and an Ag/AgCl reference electrode. These electrodes are relatively easy to use but because they can be a little unstable (drift), they must be calibrated regularly if accurate measurements of low levels of PO_2 are required. Calibration entails using an air-equilibrated solution and a zero O_2 solution (N_2 equilibrated plus 0.5–2 mM $Na_2S_2O_4$ or Na_2SO_3). It is also important to note that these electrodes can be sensitive to temperature and solution flow rates so both calibration and measurements must therefore be performed under identical conditions. A variation on the theme of polarographic O_2 electrodes is to use carbon fiber electrodes voltage clamped to -600 mV, relative to an Ag/AgCl reference.[31] While these have the advantage of being quite

[29] M. D. Stern, H. S. Silverman, S. R. Houser, R. A. Josephson, M. C. Capogrossi, C. G. Nichols, W. J. Lederer, and E. G. Lakatta, Proc. Natl. Acad. Sci. USA 85, 6954 (1988).
[30] S. L. Archer, V. Hampl, D. P. Nelson, E. Sidney, D. A. Peterson, and E. K. Weir, Circ. Res. 77, 174 (1995).
[31] M. H. Mojet, E. Mills, and M. R. Duchen, J. Physiol. 504, 175 (1997).

small, there is a disadvantage in that the usual approach to their calibration is to assume that zero current corresponds to zero O_2. This inevitably means that there must be some uncertainty in the accuracy with which low levels of O_2 can be determined. A more recently developed methodology for measuring O_2 utilizes small fiber-optic probes coated with compounds whose fluorescence is quenched by molecular O_2 (www. presens.de). We have found these probes to be much more stable than O_2 electrodes. They are insensitive to flow, and measurements can be compensated automatically for changes in temperature. They can also be used (and calibrated) interchangeably between both aqueous and gaseous environments.

Finally, in any experiments involving the equilibration of solutions with gases, proper temperature control and monitoring should not be overlooked, as the process of bubbling solutions with gases can lead to evaporative heat loss. Ideally, the temperature of solutions in the experimental chamber should be monitored directly to confirm stability and constancy. Solution temperature can be stabilized/controlled by placing the solution reservoirs in water baths (even working at room temperature) and by using a heating system for the experimental chamber. The worst-case scenario is to use reservoirs that are thermally insulated (e.g., suspended in air) containing small volumes of solution that are bubbled vigorously and then delivered to the experimental chamber rapidly (this can be especially problematic when only hypoxic solutions are bubbled with gas, as these solutions may then have a significantly lower temperature than control solutions).

Electrophysiological Recording

The majority of studies on the electrophysiological properties of O_2-sensitive cells have utilized tight seal whole cell and single channel recording techniques (i.e., patch clamp). This approach is strongly preferred over sharp electrode recording techniques because of the small size and high resting resistance of many of these cells. A full description of these electrophysiological techniques is beyond the scope of this article and the reader is referred to standard texts for basic information.[32,33] We will confine our comments to specific aspects of applying these techniques to the study of O_2-sensitive ion channels.

---

[32] R. H. Ashley, "Ion Channels: A Practical Approach." IRL Press at Oxford Univ. Press, Oxford, 1995.

[33] B. Sakmann and E. Neher, "Single Channel Recording." Plenum Press, New York, 1995.

Whole Cell Recording

There are two main variants of the whole cell recording technique. The conventional method is to rupture the patch of membrane between the cell and the electrode, allowing the free exchange of cellular and electrolyte constituents. This has the advantage of allowing the experimenter some control over the intracellular environment, for example, the ability to control intracellular Ca^{2+} or add pharmacological agents. It also has a distinct disadvantage in small cells wherein the normal intracellular constituents may be lost rapidly and cell signaling pathways disrupted. There are a number of variations on the theme of the patch pipette filling solution for these sorts of experiments, but most are based on the following recipe (in mM): 130–140 KCl, 1–2 $MgCl_2$, 10 HEPES, pH 7.2, with KOH or NaOH, 1–10 EGTA (+$CaCl_2$ to achieve the desired level of intracellular Ca^{2+}), and 0–5 ATP. An alternative method is the perforated patch configuration in which an ionophore is included within the electrode-filling solution.[34] Upon forming a seal with the cell, this ionophore will insert itself in the patch of membrane between the cell and the electrode, rendering it permeable to monovalent cations. This establishes an electrically conductive pathway between the cell and the electrode, allowing electrical recordings, both current and voltage clamp, to be performed without disrupting the intracellular environment (see later).

Using standard or perforated patch whole cell voltage-clamp techniques with depolarizing voltage steps, a number of voltage-gated K^+ channels and Ca^{2+}-activated K^+ channels have been described as being O_2 sensitive in carotid body type I cells,[9,35–39] pulmonary vascular myocytes,[13,40,41] chromaffin cells,[6,42] and PC12 cells.[43] Discrimination between effects of hypoxia upon these two broad classes of K^+ channel is usually achieved by inhibiting Ca^{2+}-activated K^+ channels by one or more of the following approaches, removal of extracellular Ca^{2+}, blocking Ca^{2+} influx

[34] R. Horn and A. Marty, *J. Gen. Physiol.* **92,** 145 (1988).

[35] J. Lopez-Lopez, C. Gonzalez, J. Urena, and J. Lopez-Barneo, *J. Gen. Physiol.* **93,** 1001 (1989).

[36] J. R. Lopez-Lopez, D. A. De Luis, and C. Gonzalez, *J. Physiol.* **460,** 15 (1993).

[37] J. R. Lopez-Lopez, C. Gonzalez, and M. T. Perez-Garcia, *J. Physiol.* **499,** 429 (1997).

[38] A. Stea and C. A. Nurse, *Pflug. Arch.* **418,** 93 (1991).

[39] C. Peers, *Neurosci. Lett.* **119,** 253 (1990).

[40] O. N. Osipenko, A. M. Evans, and A. M. Gurney, *Br. J. Pharmacol.* **120,** 1461 (1997).

[41] A. M. Evans, O. N. Osipenko, and A. M. Gurney, *J. Physiol.* **496,** 407 (1996).

[42] R. J. Thompson and C. A. Nurse, *J. Physiol.* **512,** 421 (1998).

[43] L. Conforti, I. Bodi, J. W. Nisbet, and D. E. Millhorn, *J. Physiol.* **524** Pt 3, 783 (2000).

with Ca^{2+} channel antagonists (e.g., 100–200 μM Cd^{2+}), or the use of selective Ca^{2+}-activated K^+ channel antagonists (e.g., iberiotoxin).

Perforated Patch Whole Cell Recordings of Background Currents in Carotid Body Type I Cells

Background, or leak, currents may be defined loosely as those currents active at, or around, the normal resting membrane potential of a cell. These ionic currents play a key role in setting the resting membrane potential of the cell. They typically tend to be small in amplitude (a few pA in type I cells) and their study requires a carefully considered approach if they are to be studied in isolation from the much larger voltage-activated conductances.

Whole cell recording of TASK-like currents in type I carotid body cells were performed using the perforated patch-recording technique (we have rarely been successful in recording O_2 sensitivity of background K^+ channels using conventional whole cell recording techniques). Electrodes are fabricated from either Corning 7052 (WPI) or Clarke CG150 borosilicate glass capillaries and are fire polished before use (this greatly aids in the formation of high-resistance seals between the electrode and cell membrane, see later). The perforated patch-filling solution is composed of (mM) 55 K_2SO_4, 30 KCl, 2 $MgCl_2$, 1 EGTA, 10 HEPES, pH 7.2–7.3, with NaOH (approximately 7 mM). The solution also contains the pore-forming antibiotic amphotericin B to form the "perforations" in the patch (120–240 μg/ml amphotericin B added from a stock solution of 60 mg/ml in dimethyl sulfoxide). For electrophysiological recordings, a 3 M KCl reference electrode separate from the bath ground (Ag-AgCl pellet) is used or the bath is earthed via a 3 M KCl bridge to the Ag-AgCl pellet. Cells are subjected to repetitive voltage ramps typically from -100 to -40 mV (i.e., outside the voltage range in which large voltage-activated conductances are significantly activated). Membrane potential and current are filtered at 1 kHz (3 dB) and digitized at 2–4 kHz via a CED1401. Successive current records obtained in response to these voltage ramps are then averaged for each experimental protocol to improve signal to noise. Current voltage relationships are then constructed from these averaged records. This approach allows reasonably good resolution of ionic currents active at the resting membrane potential. Electrophysiological recordings are accompanied by simultaneous recordings of intracellular Ca^{2+}, using Indo-1, to monitor the health of the cell. Using this approach, we applied the following criteria for selecting successful experiments: (1) The initial seal between electrode and cell should be higher than 10 GΩ; (2) the intracellular

Ca^{2+} concentration should not be unduly increased by patching, perforating, and clamping the cell to -70 mV (a high proportion of cells failed this test[18]); and (3) in current-clamp recordings, intracellular Ca^{2+} under current clamp conditions ($I = 0$) should not be substantially above the prepatched level (this serves as a test of seal integrity because if seal resistance declines, the cell will become depolarized and Ca^{2+} will enter through voltage-gated Ca^{2+} channels). These problems can be minimized to some extent by using small-tipped electrodes, although care must be taken to ensure that excessive electrode resistance does not lead to series resistance errors. Perforated patch whole cell recordings from carotid body type I cells fulfilling these criteria had membrane potentials in excess of -40 mV, membrane resistances in excess of 1 GΩ, and holding currents at -70 mV of less than 10 pA (typically of 5 pA).

Single Channel Recording

The effects of hypoxia on the properties of O_2-sensitive ion channels can also be studied using single channel recording techniques. As with whole cell recording, different techniques have different advantages and drawbacks. Cell-attached and perforated vesicle techniques maintain the integrity of intracellular signaling pathways, whereas excised patch techniques can disrupt these pathways but allow the experimenter the freedom to study the effects of putative signaling molecules applied to the intracellular (inside out patch) aspects of the membrane on channel activity.

Single channel studies have been performed in carotid body type I cells, PC12 cells, and pulmonary vascular smooth muscle.[20,44–48] In all cases recorded to date the principal effect of hypoxia seems to be a reduction in channel open probability rather than any significant effect on single channel amplitude. Several different types of channel have been described as being O_2 sensitive: large conductance Ca^{2+}-activated K^+ channels,[46] voltage-gated K^+ channels,[44,47] and background TASK-like K^+ channels.[20] The voltage-gated K^+ channels described thus far appear to retain their O_2 sensitivity in excised patches, suggesting some form of membrane-delimited O_2 sensing and signaling.[44,47] The background K^+ channels of

[44] M. D. Ganfornina and J. Lopez-Barneo, *Proc. Natl. Acad. Sci. USA* **88,** 2927 (1991).
[45] A. M. Riesco Fagundo, M. T. Perez-Garcia, C. Gonzalez, and J. R. Lopez-Lopez, *Circ. Res.* **89,** 430 (2001).
[46] C. N. Wyatt and C. Peers, *J. Physiol.* **483,** 559 (1995).
[47] L. Conforti and D. E. Millhorn, *J. Physiol.* **502,** 293 (1997).
[48] S. L. Archer, J. M. Huang, H. L. Reeve, V. Hampl, S. Tolarova, E. Michelakis, and E. K. Weir, *Circ. Res.* **78,** 431 (1996).

type I cells lose O_2 sensitivity in the excised patch configuration, suggesting that cytosolic constituents are necessary for the expression of their O_2 sensitivity,[20] and there have been mixed reports for Ca^{2+}-activated K^+ channels.[45,46] A search for any new O_2-sensitive channels would therefore be best conducted by initially using the cell-attached configuration before proceeding to any studies in excised patches.

For single channel recording of background K^+ channels, we have used the following pipette filling solution containing (mM) 140 KCl, 4 MgCl$_2$, 1 EGTA, and 10 HEPES, pH 7.4 (with KOH, final K^+ concentration of 146 mM). The use of a high K^+ pipette solution displaces the K^+ equilibrium potential across the patch to around 0 mV (assuming intracellular $K^+ = 140$ mM) and thus allows clear resolution of inward current through channels active around the cells resting membrane potential (-60 mV) and at more negative voltages. For voltage-gated K^+ channels it is often more advantageous to work with lower levels of K^+ in the pipette (3–5 mM K^+ with 140 mM NaCl or an impermeant cation, e.g., NMDG) so that channel activity can be recorded at more depolarized potentials. For inside-out patches we have used the following bathing (intracellular) solution (mM): 130 KCl, 5 MgCl$_2$, 10 EGTA, 10 HEPES, and 10 glucose, pH 7.2, with KOH. For cell-attached recordings a normal extracellular medium may be used, but it should be remembered that exposure of the cell to a hypoxic solution could cause membrane depolarization and a rise in internal Ca^{2+}. In order to identify direct effects of hypoxia, as opposed to secondary effects due to changes in $[Ca^{2+}]_i$ or the potential across the patch (membrane potential–pipette voltage), it may therefore be necessary to (a) prevent Ca^{2+} influx by using a Ca^{2+}-free media (+ EGTA) or one containing Ca^{2+} channel antagonists (e.g., 100 μM Ca^{2+} or 2 mM Ni^{2+}) and (b) stabilize the cells resting membrane potential by using a high K^+ bathing solution to depolarize the cell to 0 mV.

As the amplitudes of single channel currents can be quite small for some O_2-sensitive K^+ channels, minimizing electrical noise can be of major importance. Aside from good shielding and earthing of the experimental apparatus, the choice of glass for and design of the patch pipette is also important. While aluminosilicate and quartz glasses are reputed to give the best low noise performance, we have found thick wall borosilicate glass (Clarke CG150) to be perfectly adequate provided that the electrodes are given a liberal coat of Sylgard resin (Dow Corning) to within 50 μm of the tip. Noise reduction can also be aided by minimizing the length and depth of immersion of the electrode, by using short electrode holders, and by obtaining high resistance seals between the electrode and the cell (which is aided by the fire polishing of pipettes immediately prior to use).

Molecular Strategies to Interfere with O$_2$-Sensitive K$^+$ Channels

The Adenovirus Channel Expression Strategy

The viral gene transfer of dominant-negative K$^+$ channel construct has been performed to identify the molecular basis of the transient O$_2$-sensitive K$^+$ current in rabbit type I carotid body cells.[49] The point mutation W326F in Kv4.3 kills channel activity and behaves as a dominant-negative construct when coexpressed with the WT channel in transiently transfected cells.[49] Adenovirus vectors are generated by Cre-lox recombination of purified ϕ5 viral DNA and shuttle vector DNA.[49] The Kv sequences are cloned into the expression vector pGFPIRS. This vector is a modified version of pEGFP-C3, which contains the polio virus internal ribosomal entry site and clones between the EGFP sequence on the 5′ side and the polycloning site on the 3′ site. This vector (pGFPIrKv4.x) produces a single transcript, encoding both the EGFP protein and the Kv4.x protein. The adenovirus shuttle plasmids are cotransfected with pJM17, containing the full human adenovirus serotype 5 genome into HEK293 cells using LipofectAMINE (Life Technologies Inc.). Homologous recombination between the shuttle vector and pJM17 replaces the region of the adenovirus between map units 1.0 and 9.8 with the expression cassette containing the desired cDNA. Successful recombinations are screened either by direct visualization (AdmtGFP and AdhGFP) or by Southern blot analysis of small-scale infections followed by RNase protection assays of RNA made from infected cells. The recombinant products are plaque purified, expanded, and purified on CsCl gradients yielding concentrations on the order of 10^{10} pfu/ml. Isolated rabbit carotid body cells are plated onto poly-L-lysine-coated coverslips placed in six-well dishes with 1 ml of DMEM:F-12 (1 : 1) with 5% FBS and maintained in culture for up to 96 h. After 6–8 h in culture, chemoreceptor cells are infected by replacing their growth media with 1 ml of new media containing 1 μl of the following ecdysone-inducible virus: AdEGI (control), AdKv1.xDN, or AdKv4.xDN for 12 h. The expression is induced by the addition of 10 μM ponasterone A for 24–72 h before the experiments are performed. In voltage-clamp experiments, whereas adenoviral infections of chemoreceptor cells with Kv1.x DN did not modify the O$_2$-sensitive K$^+$ current, infection with the Kv4.x DN suppressed the transient outward current in a time-dependent manner, depolarized the cells significantly, and abolished the depolarization induced

[49] M. T. Perez-Garcia, J. R. Lopez-Lopez, A. M. Riesco, U. C. Hoppe, E. Marban, C. Gonzalez, and D. C. Johns, *J. Neurosci.* **20**, 5689 (2000).

by hypoxia.[49] The Shal K^+ channels therefore underlie the transient outward, O_2-sensitive K^+ current of rabbit CB chemoreceptor cells and this current contributes to cell depolarization in response to low PO_2.[49,50]

The adenovirus strategy has also been used with pulmonary artery (PA) and ductus arteriosus (DA) myocytes.[14,51] Chronic hypoxic pulmonary hypertension is characterized by pulmonary artery vasoconstriction and cell proliferation/hypertrophy. Smooth muscle cell contractility and proliferation are controlled by cytosolic Ca^{2+} levels, which are largely determined by membrane potential and thus by K^+ channels. Myocytes are depolarized in hypertension due to decreased expression and functional inhibition of several voltage-gated K^+ channels, including Kv2.1.[51–53] Adenoviral gene (Ad Kv2.1) transfer increased the expression of Kv2.1 channels and enhanced 4-AP constriction in human PAs.[51] Functional closure of the human ductus arteriosus (DA) is initiated within minutes of birth by O_2 constriction. Tissue culture for 72 h, particularly in normoxia, causes ionic remodeling, characterized by decreased O_2 and 4-AP constriction in DA rings and reduced O_2- and 4-AP–sensitive IK in DA myocytes.[14] Remodeled DA myocytes are depolarized and express less O_2-sensitive channels, including Kv2.1. Kv2.1 adenoviral gene transfer reverses ionic remodeling significantly, partially restoring both electrophysiological and tone responses to 4-AP and O_2. Kv2.1 channel activity thus plays an important role in controlling DA excitability and muscle tone.

The Antibody Strategy

To investigate the contribution of K^+ currents associated with Kv2.1 and Kv1.5 channel subunits, whole cell patch-clamp experiments are carried out in isolated rat pulmonary artery myocytes dialyzed with Kv channel antibodies.[54] Polyclonal antirabbit, Kv antibodies (Kv1.5 and Kv2.1) are obtained from Upstate Biotechnology Inc. (Lake Placid, NY). The immunogen for the Kv1.5 corresponds to a GST fusion protein with the carboxyl-terminal amino acids 542–602 of rat Kv1.5 (distal to the pore-forming S5–S6 region). The immunogen of Kv2.1 is a keyhole limpet fusion protein with amino acids 837–853 of rat Kv2.1. Patch electrodes

[50] M. T. Perez-Garcia, J. R. Lopez-Lopez, and C. Gonzalez, *J. Gen. Physiol.* **113,** 897 (1999).

[51] E. D. Michelakis, J. R. Dyck, M. S. McMurtry, S. Wang, X. C. Wu, R. Moudgil, K. Hashimoto, L. Puttagunta, and S. L. Archer, *Adv. Exp. Med. Biol.* **502,** 401 (2001).

[52] X. J. Yuan, J. Wang, M. Juhaszova, S. P. Gaine, and L. J. Rubin, *Lancet* **351,** 726 (1998).

[53] J. X. Yuan, A. M. Aldinger, M. Juhaszova, J. Wang, J. Conte, Jr., S. P. Gaine, J. B. Orens, and L. J. Rubin, *Circulation* **98,** 1400 (1998).

[54] S. L. Archer, E. Souil, A. T. Dinh-Xuan, B. Schremmer, J. C. Mercier, A. El Yaagoubi, L. Nguyen-Huu, H. L. Reeve, and V. Hampl, *J. Clin. Invest.* **101,** 2319 (1998).

(resistance of 1–3 MΩ) are fire polished and filled with a solution of composition (mM): 140 KCl, 1.0 MgCl$_2$, 10 HEPES, 5 EGTA, and phosphocreatinine, pH 7.2, with KOH. For experiments using antibodies, electrodes are initially dipped in antibody-free intracellular solution before backfilling with a 1:125 dilution of Kv1.5, Kv2.1, or GIRK-1 antibody. The threshold concentration of antibody for an effect (defined as a >10% decrease in IK within 10 min) is 1:250, with a more obvious effect at 1:125. Higher concentrations of antibody cause severe current inhibition but produce an increase in leak. There is no difference in antibody used as provided versus antibody dialyzed to remove azide (vehicle). The antibody (1:125) could be administered rapidly to the cell by diffusion from the patch pipette in the whole cell configuration. Vehicle controls provide time-dependent controls to detect run-down or run-up of K$^+$ currents (as well as to establish the effects of hypoxia and 4-AP on normal PA cells). Aliquoted antibodies are defrosted daily to avoid degradation. Intracellular administration of anti-Kv2.1 inhibits whole cell K$^+$ current and depolarizes the membrane potential of pulmonary artery myocytes.[54] Anti-Kv2.1 also elevates resting tension and diminishes 4-AP-induced vasoconstriction in membrane-permeabilized pulmonary artery rings. Anti-Kv1.5 inhibits hypoxic depolarization and selectively reduces the rise in [Ca^{2+}]$_i$ and constriction caused by hypoxia and 4-AP. However, anti-Kv1.5 by itself neither causes depolarization nor elevates basal pulmonary artery tone. The critical functional role of Kv2.1 in rat PA myocytes has been confirmed using a similar strategy with a AbKv2.1 (Alomone Laboratories Ltd.).[55] In this study, however, AbKv1.5 (1:100 dilution) was found to have no significant effect on the mean IKv at 50 mV when compared to control conditions. A similar strategy has been used with pheochromocytoma PC12 cells to functionally identify the O$_2$-sensitive Kv1.2 subunit.[43,56]

The Antisense Oligodeoxynucleotide Strategy

Small cell lung carcinoma H146 cells (American Tissue Type Cell Collection, Manassas, VA) express a pH- and O$_2$-sensitive K$^+$ channel and have been proposed to be a neuroepithelial body cell model.[3–5,24,57] The antisense oligodeoxynucleotide strategy was used to demonstrate that this

[55] D. S. Hogg, A. R. Davies, G. McMurray, and R. Z. Kozlowski, *Cardiovasc. Res.* **55**, 349 (2002).

[56] P. W. Conrad, L. Conforti, S. Kobayashi, D. Beitner-Johnson, R. T. Rust, Y. Yuan, H. W. Kim, R. H. Kim, K. Seta, and D. E. Millhorn, *Comp. Biochem. Physiol. B Biochem. Mol. Biol.* **128**, 187 (2001).

[57] I. O'Kelly, R. H. Stephens, C. Peers, and P. J. Kemp, *Am. J. Physiol.* **276**, L96 (1999).

current is mediated by the TASK-3 channel subunit.[58] H146 cells are transfected with (1) LipofectAMINE only, (2) 5′-FITC-labeled, phosphothioate-modified antisense directed across the translation start site of human TASK1 and TASK3, or (3) 5′-FITC-labeled, phosphothioate-modified missense, which consists of the same bases as employed in the antisense probe but in "random" order. Cells are seeded in six-well plates at a density of 2×10^6 cells/well in 0.8 ml of serum-free RPMI 1640 medium (Life Technologies, Inc.). The oligodeoxynucleotides are diluted in 0.1 ml of serum-free RPMI 1640 medium and mixed with 0.1 ml of 6% (v/v) LipofectAMINE (Life Technologies, Inc.) in serum-free RPMI 1640 medium. The resulting oligodeoxynucleotide and cationic lipid mixture is incubated at room temperature for 30 min to allow the formation of DNA–liposome complexes, which are then added (0.2 ml) to the cell suspension, mixed gently, and incubated in a humidified atmosphere of $5\%CO_2/95\%$ air at $37°$ for 4 h. Following incubation, 4 ml of complete RPMI 1640 medium is added to each well. Cells are then cultured as normal for up to 5 days. The concentrations and time course of transfection are shown to be optimal at 1 μM and 4–5 days, respectively.

The Transgenic Strategy

This section describes the results of three studies concerning O_2-sensitive K^+ channels using transgenic mice. SWAP Kv1.5 transgenic mice have been used to demonstrate that the voltage-gated Kv1.5 subunit plays a functional role in hypoxic pulmonary artery vasoconstriction.[59] gp91phox knockout mice were used to demonstrate that the NADPH oxidase is the O_2 sensor regulating Kv3.3 channels in pulmonary airway chemoreceptors.[4] Finally, Hif1α(+/−) heterozygote mice were used to demonstrate that electrical remodeling of the O_2-sensitive K^+ channels in pulmonary artery myocytes during chronic hypoxia is under the control of the Hif1 α transcription factor.[60]

SWAP Kv1.5 Transgenic Mice

The construct consists of a 5′ arm of the promoter and 5′UTR of mKv1.5, the rat Kv1.1 K^+ channel (rKv1.1) tagged with the nine amino acid hemagglutinin tag (HA) and cloned into the SmaI site of mKv1.5

[58] M. E. Hartness, A. Lewis, G. J. Searle, I. O'Kelly, C. Peers, and P. J. Kemp, *J. Biol. Chem.* **276**, 26499 (2001).

[59] S. L. Archer, B. London, V. Hampl, X. Wu, A. Nsair, L. Puttagunta, K. Hashimoto, R. E. Waite, and E. D. Michelakis, *FASEB J.* **15**, 1801 (2001).

[60] L. A. Shimoda, D. J. Manalo, J. S. Sham, G. L. Semenza, and J. T. Sylvester, *Am. J. Physiol. Lung Cell. Mol. Physiol.* **281**, L202 (2001).

located at position −6, a neomycin resistance cassette (NeoR), mKv1.5 starting at an *Xba*I site in the 3'UTR, and the thymidine kinase gene for negative selection. Homologous recombination with this construct should yield rKv1.1 driven by the mKv1.5 promoter, although the effect of the NeoR cassette is unknown and any 3' regulatory elements may be lost. These issues may explain why the rKv1.1 was not expressed in the lung (even though the gene was present and rKv1.1 was expressed in the heart). Electroporation of embryonic stem cells, blastocyst injections, and matings to obtain mice heterozygous and homozygous for the targeted allele are performed to create SWAP mice. Male and female SWAP heterozygotes are backcrossed two generations into C57BL/6 and mated to yield the 3- to 8-month-old SWAP homozygotes, heterozygotes, and wild-type litter-mate controls used in these experiments. SWAP mice are indistinguishable from wild-type mice, having a normal appearance, life expectancy, and gender distribution. RT-PCR is used to confirm the absence of Kv1.5 DNA and mRNA in SWAP mice. Immunoblots using antibodies against the HA epitope unique to the rKv1.1 transgene did not detect any rKv1.1 protein in SWAP lungs. Thus, although the SWAP mouse was designed as a gene replacement model, it functions as a targeted deletion or knockout. In isolated lungs and resistance PA rings, HPV was reduced significantly in SWAP versus wild-type mice. Consistent with this finding, PASMCs from SWAP PAs were slightly depolarized and lacked IKv1.5, a 4-AP and hyp-oxia-sensitive component of IK that activated between −50 and −30 mV. It was concluded that a K$^+$ channel containing Kv1.5 α subunits is an important effector of HPV in mice. The remaining component of hypoxic vasoconstriction in these mice may be related to other K$^+$ channels (including Kv2.1) or to a mechanism independent of K$^+$ channels.[61]

gp91phox Knockout Mice

Both the hydrogen peroxide-sensitive voltage-gated K$^+$ channel sub-unit Kv3.3α and membrane components of NADPH oxidase (gp91phox and p22phox) are coexpressed in the NEB cells of fetal rabbit and neonatal human lungs.[3,5,24] Gene targeting is used to generate mice with a null allele of the 91-kDa subunit of the oxidase cytochrome b.[62] The gp91phox sub-unit coding region is disrupted by insertion of a neomycin phosphotransfer-ase (neo) gene. To test whether NADPH oxidase is the O$_2$ sensor in NEB

[61] J. P. Ward and P. I. Aaronson, *Respir. Physiol.* **115,** 261 (1999).
[62] J. D. Pollock, D. A. Williams, M. A. Gifford, L. L. Li, X. Du, J. Fisherman, S. H. Orkin, C. M. Doerschuk, and M. C. Dinauer, *Nature Genet.* **9,** 202 (1995).

cells, patch-clamp experiments are performed on intact NEBs identified by neutral red staining in fresh lung slices from wild-type (WT) and oxidase-deficient mice. In WT cells, hypoxia (pO_2 = 15–20 mm Hg) caused a reversible inhibition (46%) of both Ca^{2+}-independent and Ca^{2+}-dependent K^+ currents. In contrast, hypoxia had no effect on K^+ current in knockout cells, even though both K^+ current components were expressed. These experiments have definitely confirmed that NADPH oxidase acts as the O_2 sensor in pulmonary airway chemoreceptors.[4] However, hypoxic pulmonary artery vasoconstriction is preserved in gp91phox knockout mice.[63]

Hif1α Knockout Mice

During chronic hypoxia, transcription of voltage-gated K^+ channels is downmodulated and myocytes become depolarized.[52,53,60,64] Heterozygote mice [Hif1α (+/−)] were studied to evaluate the role of the Hifα transcription factor in the downmodulation of K^+ channels.[60] Hif1α (+/+) and mice with one null allele at the Hif1α locus [Hif1α (+/−)] mice are mated, and the offspring genotype is determined by PCR.[60] Mice are generated on a C57B6 × 129 genetic background. Male Hif1α (+/+) and Hif1α (+/−) mice (8 weeks old) are placed in a hypoxic chamber and exposed to either normoxia or normobaric hypoxia for 21 days. The chamber is flushed continuously with either room air or a mixture of room air and N_2 (10 ± 0.5% O_2). Chamber O_2 and CO_2 concentrations are monitored continuously (OM-11 O_2 analyzer and LB-2 CO_2 analyzer, Sensormedics, Anaheim, CA). The mice are exposed to room air for 10 min twice a week to clean the cages and replenish food and water supplies. Hypoxia-induced right ventricular hypertrophy and polycythemia are blunted in Hif1α(+/−) mice. Hypoxia increased PASMC capacitance in Hif1α(+/+) mice, but not in Hif1α(+/−) mice. Chronic hypoxia depolarized and reduced K^+ current density in PASMCs from Hif1α(+/+) mice. In PASMCs from hypoxic Hif1α(+/−) mice, no reduction in K^+ current density was observed, and depolarization was significantly blunted. Thus partial deficiency of HIF-1α is sufficient to impair hypoxia-induced depolarization, reduction of K^+ current density, and PASMC hypertrophy.[60]

[63] S. L. Archer, H. L. Reeve, E. Michelakis, L. Puttagunta, R. Waite, D. P. Nelson, M. C. Dinauer, and E. K. Weir, *Proc. Natl. Acad. Sci. USA* **96,** 7944 (1999).
[64] S. V. Smirnov, T. P. Robertson, J. P. Ward, and P. I. Aaronson, *Am. J. Physiol.* **266,** H365 (1994).

Methods for Heterologous Expression of O_2-Sensitive K^+ Channels

The Xenopus Oocyte Expression System

Kv channel subunits identified in PC12 cells are expressed in Xenopus oocytes and hypoxic inhibition is assayed.[43,56] Oocytes are injected with cRNAs obtained as run-off transcripts of Kv1.2 (pcDNA3 plasmid vector) and Kv2.1 cDNAs (pBluescript-SK-plasmid vector). The double-stranded DNA templates are linearized and *in vitro* transcribed to cRNAs with mMessage mMachine kits (for T7 or SP6 promoter from Ambion) according to the manufacturer's protocol. After the transcription reaction is complete, the template DNA is degraded and cRNA is recovered by phenol–chloroform extraction followed by ethanol precipitation. The size of the *in vitro* transcription product, its quantity, and its quality are evaluated by denaturing agarose gel electrophoresis. cRNAs are stored in RNase-free water at $-80°$. Stage IV–V oocytes are isolated as follows. Frogs are anesthetized with 0.2% tricaine methanesulfonate (MS 222). Clumps of oocytes are removed and washed in Ca^{2+}-free ND-96 solution containing (mM) 82.5 NaCl, 2.0 KCl, 1.0 $MgCl_2$, and 5.0 HEPES, pH 7.5, with NaOH. After removal of the oocytes, the frogs are allowed to recover and are returned to their tanks. Single oocytes are dissociated with 3 mg/ml type II collagenase in Ca^{2+}-free ND-96 solution at $20°$. After digestion, the follicular layer is removed mechanically with a fire-polished Pasteur pipette. cRNA (50 nl; 0.2 $\mu g/\mu l$) is injected into the oocyte with a Drummond 510 microdispenser via a sterile glass pipette with a tip of 20–30 μm. After injection, the oocytes are maintained in a solution of the following composition (mM) : 96 NaCl, 2.0 KCl, 1.0 $MgCl_2$, 1.8 $CaCl_2$, 5 HEPES, 2.5 sodium pyruvate, and 0.5 theophylline, with 100 units/ml penicillin and 100 μg/ml streptomycin; pH 7.5 with NaOH. Injected oocytes are stored in an incubator at $19°$ and used for electrophysiological experiments after 24 h. The whole cell current from injected Xenopus oocytes is recorded using the two electrode voltage-clamp technique. The composition of the external solution is (mM) 115 NaCl, 2 KCl, 1.8 $CaCl_2$, and 10 HEPES, pH 7.2, with NaOH. The two electrodes have a resistance of 1–2 MΩ and are filled with 3 mM KCl. Whole cell leak and capacitative currents are subtracted using currents elicited by small hyperpolarizing pulses (P/4). Currents are digitized between 0.5 and 5 kHz after being filtered between 0.2 and 1 kHz. Single channel (cell-attached) voltage-clamp experiments are performed in Xenopus oocytes from which the vitelline membrane has been removed manually after shrinkage in a hyperosmotic medium (mM): 200 potassium aspartate, 20 KCl, 1.0 $MgCl_2$, 5 EGTA, and 10 HEPES, pH 7.3, with KOH. Microelectrodes with resistances of 3–5 MΩ are prepared, fire polished, and coated with Sylgard (Dow Corning). The external solution composition is (mM) 140 KCl, 20 MgCl$_2$, 10 HEPES, and 5 EGTA, pH 7.3, with KOH. The pipette solution composition is (mM) 140 NaCl, 2.8 KCl, 5 HEPES, 1 EGTA, pH 7.3, with NaOH. During electrophysiological experiments, the effect of hypoxia is studied by switching from a perfusion medium bubbled with air (21% O$_2$) to a medium equilibrated with 10% O$_2$ (balanced N$_2$) or 100% N$_2$ with 5 mM sodium dithionite (Na$_2$S$_2$O$_4$; an O$_2$ chelator). The corresponding mean O$_2$ partial pressures (PO$_2$) in the chamber, measured with an O$_2$-sensitive electrode, are 150 mm Hg (21% O$_2$), 80 mm Hg (10% O$_2$), and 0 mm Hg (N$_2$ + Na$_2$S$_2$O$_4$). Kv1.2, unlike Kv2.1, is reversibly inhibited by hypoxia when expressed in Xenopus oocyte.[43]

Expression of Cloned O$_2$-Sensitive K$^+$ Channel Subunits in Cultured Cell Lines

Various cell lines, including HEK293, COS-7, and L cells, have been used to transiently express recombinant K$^+$ channels and assay their O$_2$ sensitivity.[50,65–67] One example of the transfection procedure used with HEK cells is given in detail, so HEK293 cells are maintained in DMEM supplemented with 10% fetal calf serum (GIBCO/BRL), 100 U/ml penicillin, 100 μg/ml streptomycin, and 2 mM L-glutamine. Cells are grown as a monolayer and plated on squared coverslips (24 × 24 mm) placed in the bottom of 35-mm petri dishes at a density of 2–4 × 10^5 cells/dish the day before transfection. Transient transfections are performed using the Ca^{2+}-phosphate method with 1 μg of plasmid DNA encoding the K$^+$ channel subunit. Green fluorescent protein (GFP; 0.2 μg) in a CMV promoter expression plasmid (GFPPRK5) is included to permit transfection efficiency estimates (10–40%) and to identify cells for voltage-clamp analysis. A group of control cells is obtained by analyzing the currents present in cells transfected with GFP alone or in untransfected cells. When expressed in HEK293 cells, Kv4.2 is not sensitive to hypoxia or redox stimulation.[50] However, coexpression of Kv4.2 with Kvβ1.2 forms a K$^+$ channel that is reversibly inhibited by hypoxia.[50] This work suggests that in rabbit type I carotid body chemoreceptor cells, a Shal subunit in association with a Kv β subunit may encode the O$_2$-sensitive transient outward K$^+$ channel.[50] Similarly, the channels Kv1.2, Kv1.2/Kv1.5, Kv2.1, Kv2.1/Kv9.3, and Kv3.1β transiently expressed in various cells, including COS-7 cells, L cells, and HEK 293, have been shown to display O$_2$ sensitivity.[65–67]

[65] A. J. Patel, M. Lazdunski, and E. Honoré, *EMBO J.* **16,** 6615 (1997).
[66] O. N. Osipenko, R. J. Tate, and A. M. Gurney, *Circ. Res.* **86,** 534 (2000).

Functional Role of TASK 2P Domain K^+ Channel Subunits
in Chemoreception

The primary sensory cells of the carotid body respond to hypoxia and acidosis with a depolarization initiating electrical activity, Ca^{2+} entry, and neurosecretion,[18,19,37,49,68,69] A key ionic current involved in mediating these responses in rat type I cells is an O_2 and acid-sensitive background K^+ current.[18–20] This current displays a baseline activity with no voltage and time dependency and shares many of the biophysical and pharmacological properties of the cloned K^+ channel TASK-1, a member of the four transmembrane segments and two P domains family.[20,70–73] Similarly, in H164 cells (see earlier discussion), TASK-3 has been proposed to encode the O_2-sensitive K^+ channels.[58,74–77] TASK-1 and TASK-3 are insensitive to tetraethylammonium and 4-aminopyridine, although they are blocked by barium, quinine, and quinidine. TASK-1 and TASK-3 are opened by inhalational general anesthetics, including halothane, but are blocked by local anesthetics, such as bupivacaine.[71,78] A key feature of TASK channels is their high sensitivity to external pH (pH of 7.3 and 6.7 for TASK-1 and TASK-3, respectively).[70,78–81] *In situ* hybridization and RT-PCR analysis (A. Patel, unpublished data) have shown that TASK-1 mRNA is abundant in type I carotid body cells.[20] Although the specificity of the TASK-1 antibody used (Alomone) is questionable (see earlier discussion), several studies have demonstrated the expression of TASK-1 at the protein level in type I carotid body cells.[21,26] Due to its baseline activity, TASK-1 sets the resting membrane potential and thus controls cell excitability.

[67] J. T. Hulme, E. A. Coppock, A. Felipe, J. R. Martens, and M. M. Tamkun, *Circ. Res.* **85**, 489 (1999).

[68] M. D. Ganfornina and J. Lopez-Barneo, *J. Gen. Physiol.* **100**, 427 (1992).

[69] M. D. Ganfornina and J. Lopez-Barneo, *J. Gen. Physiol.* **100**, 401 (1992).

[70] F. Duprat, F. Lesage, M. Fink, R. Reyes, C. Heurteaux, and M. Lazdunski, *EMBO J.* **16**, 5464 (1997).

[71] A. J. Patel, E. Honoré, F. Lesage, M. Fink, G. Romey, and M. Lazdunski, *Nature Neurosci.* **2**, 422 (1999).

[72] D. Leonoudakis, A. T. Gray, B. D. Winegar, C. H. Kindler, M. Harada, D. M. Taylor, R. A. Chavez, J. R. Forsayeth, and C. S. Yost, *J. Neurosci.* **18**, 868 (1998).

[73] D. Kim, A. Fujita, Y. Horio, and Y. Kurachi, *Circ. Res.* **82**, 513 (1998).

[74] I. O'Kelly, C. Peers, and P. J. Kemp, *Am. J. Physiol.* **275**, L709 (1998).

[75] I. O'Kelly, A. Lewis, C. Peers, and P. J. Kemp, *J. Biol. Chem.* **275**, 7684 (2000).

[76] I. O'Kelly, C. Peers, and P. J. Kemp, *Adv. Exp. Med. Biol.* **475**, 611 (2000).

[77] I. O'Kelly, C. Peers, and P. J. Kemp, *Biochem. Biophys. Res. Commun.* **283**, 1131 (2001).

[78] H. J. Meadows and A. D. Randall, *Neuropharmacology* **40**, 551 (2001).

[79] G. Czirjak and P. Enyedi, *J. Biol. Chem.* **277**, 5426 (2002).

[80] Y. Kim, H. Bang, and D. Kim, *J. Biol. Chem.* **275**, 9340 (2000).

[81] S. Rajan, E. Wischmeyer, G. X. Liu, R. Preisig-Muller, J. Daut, A. Karschin, and C. Derst, *J. Biol. Chem.* **275**, 16650 (2000).

In cell-attached patches of rat type I carotid body cells, hypoxia closes TASK-1-like channels, while it has no effect in the inside-out patch configuration.[20] The loss of O_2 sensitivity upon excision suggests that some cytosolic messenger or cofactor may be required to confer or maintain the O_2 sensitivity of TASK-1.[20] When expressed in transiently transfected COS or HEK293 cells or in mRNA-injected Xenopus oocytes, TASK-1 currents are not altered significantly by hypoxia (5 Torr for 5 min) (K. Buckler and E. Honoré, unpublished data). However, it has been proposed that recombinant hTASK-1 is an O_2-sensitive channel when expressed stably in HEK293 cells (pcDNA3.1-hTASK-1 construct).[82] In this study, a mild hypoxia of 30–40 mm Hg reversibly inhibited TASK-1.[82] This current was additionally inhibited by extracellular acidosis as expected for a TASK-1 channel.[82] Considering that hypoxic downmodulation of the TASK channel in type I carotid body cells is lost upon excision and that TASK-1 is not O_2 sensitive in COS-7, HEK293 cells, and Xenopus oocytes, it is assumed that the O_2 sensor is independent of TASK-1. The fact that the stable HEK293 cell line expressing TASK-1 displays some O_2 sensitivity suggests that the O_2 sensor may only be expressed in a subset of HEK293 cells.[82] Several inhibitors of mitochondrial respiration also mimic the effects of hypoxia and inhibit the TASK-like current in type I carotid body cells (Wyatt and Buckler unpublished data).[19] The activity of background K^+ currents (i.e., the TASK channels) may therefore be related to mitochondrial respiration such that intrinsic differences between cellular metabolism in different cell types could also explain the differential sensitivity to hypoxia.

The various methods described here have allowed the molecular identity of the O_2-sensitive K^+ channels in chemoreceptive cells to be established. Much work remains to be done, however, to understand how these subunits are modulated by O_2 and to evaluate their possible role in several human pathologies, including pulmonary artery hypertension or paraganglioma.[52,53,60,83,84]

Acknowledgment

We are grateful to Dr. A. Patel for critical comments on the manuscript.

[82] A. Lewis, M. E. Hartness, C. G. Chapman, I. M. Fearon, H. J. Meadows, C. Peers, and P. J. Kemp, *Biochem. Biophys. Res. Commun.* **285,** 1290 (2001).

[83] B. E. Baysal, R. E. Ferrell, J. E. Willett-Brozick, E. C. Lawrence, D. Myssiorek, A. Bosch, A. van der Mey, P. E. Taschner, W. S. Rubinstein, E. N. Myers, C. W. Richard, 3rd, C. J. Cornelisse, P. Devilee, and B. Devlin, *Science* **287,** 848 (2000).

[84] E. K. Weir, H. L. Reeve, J. M. Huang, E. Michelakis, D. P. Nelson, V. Hampl, and S. L. Archer, *Circulation* **94,** 2216 (1996).

[18]   Oxygen Sensing by Human Recombinant K⁺ *
Channels: Assessment of the Use of Stable Cell Lines

By PAUL J. KEMP, DAVID ILES, and CHRIS PEERS

Introduction

Under normal circumstances, environmental O_2 levels generally remain constant. As a consequence, terrestrial mammalian life has evolved to perform optimally at an atmospheric oxygen partial pressure (pO_2) close to 150 mm Hg. Gaseous exchange and mixing at the alveoli of the lungs result in a mild O_2 gradient from inspired to alveolar air and as a result there is a reduction in systemic pO_2 to around 100 mm Hg. Most cellular processes proceed most favorably when arterial blood, at this pO_2, is in plentiful supply. However, there are a number of physiological and pathological circumstances where this pO_2 may be altered either acutely or chronically. Physiologically, mean cerebral blood pO_2 is closer to 20 mm Hg[1] and central neuronal activity is adapted to this relatively hypoxic environment. Pathologically, several cardiorespiratory diseases result in chronically reduced systemic and/or pulmonary pO_2. These include apnea of sleep, congestive heart failure, emphysema, and chronic obstructive pulmonary disease, all of which affect a significant proportion of the adult population and may lead to life-threatening conditions such as pulmonary hypertension. At a higher temporal resolution are cellular and physiological responses to acute perturbation of systemic and/or pulmonary pO_2 levels. Thus, acute episodes of hypoxia, which occur during unpressurized ascent to more than 3000 m or during pathophysiological reduction in arterial pO_2, set in motion a number of homeostatic mechanisms that involve the concerted efforts of a number of chemosensory tissues.

Central to the cardiorespiratory responses to reduced O_2 availability of all chemosensory tissue is the rapid inhibition of ion channels by hypoxia (see Lopez-Barneo *et al.*[2] for a review). Thus, acute hypoxic modulation of K⁺ channel activity is central to chemosensing in the carotid body,[3–5]

---

* Nomenclature for the quotation of K⁺ channels in the text uses that which was quoted in the original article. In parentheses are the sequence-defined names followed by the human gene orthologue nomenclature (where available), for example, Kv1.2 (*shaker,* KCNA2).

[1] W. E. Hoffman, F. T. Charbel, G. Gonzalez-Portillo, and J. I. Ausman, *Surg. Neurol.* **51,** 654 (1999).

[2] J. Lopez-Barneo, R. Pardal, and P. Ortega-Saenz, *Annu. Rev. Physiol.* **63,** 259 (2001).

[3] J. Lopez-Barneo, J. R. Lopez-Lopez, J. Urena, and C. Gonzalez, *Science* **241,** 580 (1988).

[4] K. J. Buckler, *J. Physiol.* **498,** 649 (1997).

neuroepithelial body[6,7] and its immortalized cellular counterpart (H146 cells),[8–13] and, although perhaps controversially, the pulmonary circulation.[14–17] In addition, such O_2 sensitivity is believed to play a significant role in the modulation of excitability in several cellular components of the mammalian nervous system.[18–21] O_2-dependent vasodilatation in the systemic vasculature is underpinned by the direct regulation of voltage-activated Ca^{2+} channels[22] and, as such, is outside the scope of this discussion.

Although much progress has been made in understanding O_2 sensing in native tissues, the use of recombinant ion channels in several expression systems has promoted a rapid appreciation of the fundamental interactions of K^+ channels with elements of the O_2 sensory transduction pathway. As members of several K^+ channel gene families have been implicated in O_2 sensing in native systems, recombinant expression of a number of different K^+ channels has been used as proof-of-concept models.[23] Such K^+ channels represent members from three main K^+ channel gene families: tandem P domain K^+ (K_{2P}) channels; large conductance, Ca^{2+}-sensitive K^+ (BK) channels; and voltage-activated K^+ (Kv) channels. This article discusses the impact that recombinant human K^+ channel expression systems have on our understanding of acute O_2 chemotransduction by native and model chemosensory tissues and, in so doing, assesses the use of cells lines stably expressing human K^+ channels in the study of O_2 sensing.

[5] C. Peers, *Neurosci. Lett.* **119,** 253 (1990).

[6] C. Youngson, C. Nurse, H. Yeger, and E. Cutz, *Nature* **365,** 153 (1993).

[7] E. Cutz and A. Jackson, *Respir. Physiol.* **115,** 201 (1999).

[8] P. J. Kemp, A. Lewis, M. E. Hartness, G. J. Searle, P. Miller, I. O'Kelly, and C. Peers, *Am. J. Respir. Crit. Care Med.* **166,** 17 (2002).

[9] M. E. Hartness, A. Lewis, G. J. Searle, I. O'Kelly, C. Peers, and P. J. Kemp, *J. Biol. Chem.* **276,** 26499 (2001).

[10] I. O'Kelly, A. Lewis, C. Peers, and P. J. Kemp, *J. Biol. Chem.* **275,** 7684 (2000).

[11] I. O'Kelly, R. H. Stephens, C. Peers, and P. J. Kemp, *Am. J. Physiol.* **276,** L96 (1999).

[12] I. O'Kelly, C. Peers, and P. J. Kemp, *Am. J. Physiol.* **275,** L709 (1998).

[13] C. Peers and P. J. Kemp, *Respir. Res.* **2,** 145 (2001).

[14] J. M. Post, J. R. Hume, S. L. Archer, and E. K. Weir, *Am. J. Physiol.* **262,** C882 (1992).

[15] E. K. Weir and S. L. Archer, *FASEB J.* **9,** 183 (1995).

[16] J. P. Ward and P. I. Aaronson, *Respir. Physiol.* **115,** 261 (1999).

[17] J. T. Hulme, E. A. Coppock, A. Felipe, J. R. Martens, and M. M. Tamkun, *Circ. Res.* **85,** 489 (1999).

[18] C. Vergara, R. Latorre, N. V. Marrion, and J. P. Adelman, *Curr. Opin. Neurobiol.* **8,** 321 (1998).

[19] E. A. Coppock, J. R. Martens, and M. M. Tamkun, *Am. J. Physiol.* **281,** L1 (2001).

[20] L. D. Plant, P. J. Kemp, C. Peers, Z. Henderson, and H. A. Pearson, *Stroke* **33,** 2324 (2002).

[21] C. Jiang and G. G. Haddad, *J. Physiol.* **481,** 15 (1994).

[22] A. Franco-Obregon, J. Urena, and J. Lopez-Barneo, *Proc. Natl. Acad. Sci. USA* **92,** 4715 (1995).

[23] A. J. Patel and E. Honore, *Eur. Respir. J.* **18,** 221 (2001).

Generating Stable Cell Lines

Choice of Cell Line

We have successfully employed the human embryonic kidney cell line (HEK293) as a system for generating stable clonal lines expressing a variety of human recombinant K⁺ channels. The HEK293 line provides a particularly robust system in which to study ion channel activity for a number of important reasons. First, it is amenable to genetic manipulation using a variety of different transfection protocols, including calcium phosphate shock, electroporation, and lipofection. We find that lipofection using Superfect gives remarkably consistent K⁺ channel expression. Second, HEK293 cells are relatively straightforward to patch clamp, and recording either single channel or whole cell current data for even extended periods is undertaken with relative ease. Third, native untransfected (or sham-transfected) HEK293 cells consistently express a low level of ion channel activity, which results in background currents from these cells being at least one order of magnitude smaller than currents recorded from most stably expressing cells. Finally, but importantly, the low level of channel activity is derived from a relatively small pool of different endogenously expressed ion channels. This fact can be assumed from functional expression studies using electrophysiology but has been made explicit by advances in gene chip array technology. Thus, Fig. 1 shows a profile of cation channel genes expressed in native, untransfected HEK293 cells. Each spot on the gene chip array represents an annealing of HEK293 cDNA to an expressed sequence tag probe for 3′–untranslated regions of 90 human genes encoding cation channel subunits—each gene is spotted in triplicate and *faux*-colored spots represent a signal of at least twice that of background. Positive spots are labeled with the human gene name. Data of these type reinforce the notion that HEK293 cells represent an appropriate vehicle to express many different human K⁺ channel genes.

Cell Culture

Native, untransfected HEK293 cells (American Type Culture Collection, Manassas, VA) are maintained in Earle's minimal essential medium (containing L-glutamine) supplemented with 10% (v/v) fetal calf serum, 1% antibiotic antimycotic, 1% nonessential amino acids, and gentamicin 50 mg ml⁻¹ (Gibco BRL, Paisley, Strathclyde, UK) in a humidified incubator gassed with 5% CO_2/95% air. We *passage* cells every 7 days in a ratio of 1:25 using trypsin-EDTA (Gibco BRL).

A

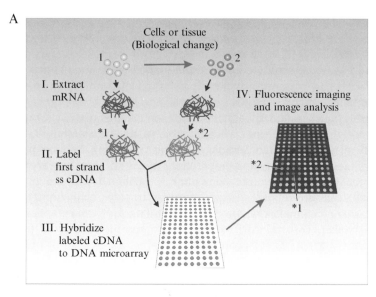

B

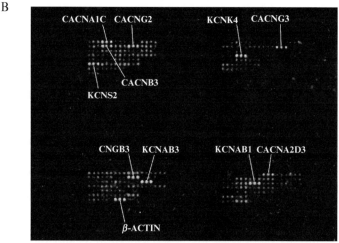

Fig. 1. Using a gene chip microarray to screen for expression cation channel mRNA. (A) Principles of microarraying. Total RNA is extracted from cells or tissues before, during, and after a biological or environmental change, such as adaptation to hypoxia (I). Messenger (m)RNA is used as a template for the synthesis of first-strand cDNA by reverse transcriptase (RT) extension of annealed oligo(dT)primers (II). The cDNA can be labeled either directly through the enzymatic incorporation of fluorochrome-modified nucleotides during the RT reaction or indirectly through the incorporation of aminoallyl-dUTP in the RT reaction, followed by chemical coupling of the fluorochrome label to the aminoallyl-cDNA. Messenger RNAs isolated from cells or tissues under different experimental conditions are labeled with

Choice of Vector

We have obtained consistent transfections using pcDNA3.1-V_5/His_6-TOPO (Invitrogen, Paisley, Strathclyde, UK). This vector has a distinct advantage over others employed in that is also codes for two independent tagging sequences, V_5 and His_6, which are highly immunoreactive and to which there are selective antibodies available commercially (both from Invitrogen). These tagging sequences are particularly useful when working with newly cloned ion channels in that channel-specific antibodies are not required for either immunolocalization of protein or biochemical purification. Thus, together with reverse transcription and polymerase chain reaction (RT-PCR)-based screening, stable expression and cellular localization of the ion channel of interest can be verified quickly using V_5 and/or His_6 immunocytochemistry (see Fig. 4A for an example). Like many mammalian vectors, pcDNA3.1-TOPO contains an antibiotic resistance gene that allows efficient and inexpensive selection of expressing cells using G418.

We have employed pcDNA3.1-TOPO to express stably a number of human K⁺ channel open reading frames in the HEK293 cell line. These include hTASK1 (KCNK3, Genbank accession number AF006823), hTASK2 (KCNK5, Genbank accession number AF084830), and hTASK3 (KCNK9, Genbank accession number AC007869). In addition, we have

---

different fluorochromes, such as Cy3 and Cy5, and pooled together in a single hybridization reaction (III). After hybridization, the fluorescence intensity at each spot is measured at the appropriate emission wavelength for both fluorochromes and the results are processed and analyzed using commercially available image analysis software (IV). The intensity of fluorescence at each spot is, to a large extent, related to the relative abundance of transcripts complementary to the arrayed DNA probes. Consequently, if gene "A" is being expressed under "normal" conditions, but not under "experimental" conditions, then the probe for gene "A" will fluoresce only at the emission wavelength of the fluorochrome used to label "normal" cDNA. Conversely, if gene "B" is activated in response to environmental or biological change, then fluorescence will be detected only at the wavelength of the fluorochrome used to label "experimental" cDNA. Fluorescence will be detected with approximately equal intensity at both wavelengths when the expression levels of the genes represented on the array do not change in response to the environmental or biological changes under study. (B) Basal expression levels of ion channels in native untransfected HEK293 cells. An experimental array of 149 DNA sequences (spotted in triplicate), consisting of expressed sequence tags (ESTs) and PCR-amplified, gene-specific sequences, was designed to measure the expression levels of 90 different human cation channel genes (some of which are amplified from more than one EST of the same gene). Total RNA was extracted from HEK293 cells growing under conditions of normoxia, and equal amounts of RNA were used to generate Cy3- and Cy5-labeled cDNA as described in A. Labeled cDNAs were pooled and hybridized to the array, and fluorescence intensities at each spot were measured. Signal intensities were normalized to the β-actin-positive control and superimposed using ArrayPro4 software. Results suggest that only a minimal array of cation channels are expressed in native untransfected HEK293 cells. (See color insert.)

been kindly gifted (from GlaxoSmithKline, Stevenage, UK) a stable HEK293 cell line expressing hTREK1 (KCNK2, Genbank accession number AF004711[24]), which was generated in a manner analogous to that described later.

Ligation of Cloned Channels into pcDNA3.1-TOPO

Restriction enzymes available for cutting the vector at the multiple cloning site are detailed in the manufacturer's instructions. The enzyme(s) of choice will be determined by the restriction sites available on both the vector and the DNA insert; a combination of *Not*I and *Hind*III is often suitable, but a restriction map of the particular insert must be generated before such experiments can be performed usefully. To prevent recircularization of digested plasmid DNA in the ligation reaction, the 5' end phosphate groups are removed using calf intestinal alkaline phosphatase (New England Biolabs Inc., Hitchin, Herts., UK). Ligation is carried out using the using T4 DNA ligase (Promega, Southampton, Hants., UK) with a 1:3 vector:insert ratio.

Transfection and Selection

Transfection of HEK293 cells is accomplished using the Superfect transfection reagent (Qiagen, Crawley, West Sussex, UK) and selection pressure is maintained with G418 (1 mg·ml^{-1}; Gibco BRL). Twenty-four hours before transfection, wild-type HEK293 cells are plated at *ca.* 80% confluency and maintained in culture as described earlier. The hTASK1-pcDNA3.1-TOPO construct (10 μg) is diluted with medium containing neither serum nor antibiotics. The Superfect transfection reagent (50 μl) is added and the mixture is incubated for 10 min at room temperature to allow lipid/DNA complex formation. The medium is aspirated and cells are washed with Mg^{2+}/Ca^{2+}-free phosphate-buffered saline (PBS). After 10 min, 5 ml of the conditioned medium is added to the transfection solution, which is pipetted onto the cells and incubated at 37° for 3 h. The transfection solution is then removed and the cells are washed once with PBS before the addition of fresh medium. After 48 h, wild-type medium is replaced by selection medium, which contains 1 mg·ml^{-1} G418. Selection medium is replaced every 7 days. After 4 weeks, single colonies are isolated by adhering sterile cloning rings to the bottom of the flask using vacuum grease and cells contained within each ring are isolated by adding 200 μl

[24] C. G. Chapman, H. J. Meadows, R. J. Godden, D. A. Campbell, M. Duckworth, R. E. Kelsell, P. R. Murdock, A. D. Randall, G. I. Rennie, and I. S. Gloger, *Brain Res. Mol. Brain Res.* **82,** 74 (2000).

PBS containing 10% trypsin at 37° for 2 min before trypsin inactivation is accomplished by the addition of 200 μl fetal calf serum. Cells are triturated and transferred into selection medium. Each clone can then be screened for channel expression by patch-clamp analysis of membrane currents, immunocytochemistry, and RT-RCR.

Verification of Expression by RT-PCR

Total RNA is extracted from native and transfected HEK293 cells using the RNeasy mini kit (Qiagen) and treated with RQ-1 RNase-free DNase (1 U·μg^{-1} RNA; Promega), to remove genomic DNA contamination before reextraction using the RNeasy mini kit and RT-PCR using gene-specific primers. Is is noteworthy that treatment with high-fidelity DNase is always an important step in RT-PCR but this is particularly so when attempting to verify stable expression of a human gene transfected into a human host cell; without this step, such data are essentially uninterpretable. Reverse transcription is carried out using AMV reverse transcriptase and oligo (dT$_{15}$) primer (Promega) for 15 min at 42° followed by 5 min at 99°. The hot-start PCR protocol of choice employs the Advantage GC kit (Clontech, Basingstoke, Hants., UK) and the following cycling perameters: denature at 94°/1 min; anneal at X°C (see later)/1 min; extension at 72°/1 min (for 35 cycles) with a final extension period of 10 min at 72°. The following primers have proved very reliable:

hTASK1 sense	5'acgatgaagcggcagaacgtgc3'
hTASK1 antisense	5'caccaaggtgttgatgcgctcgc3', anneal at 65°, 411 bp
hTASK3 sense	5'ctactttgcgatcacggtca 3'
hTASK3 antisense	5'acgtagagcggcttcttctg3', anneal at 58°, 410 bp
hTREK1 sense	5'cctgtgtcaattcgacggagc 3'
hTREK1 antisense	5'ccacactcagtccggtcctg3', anneal at 63°, 728 bp

Products are separated on 2% agarose gels and are visualized with ethidium bromide/ultraviolet transillumination. Sequences of amplicons are always verified by dye terminator PCR with ABI PRISM automated sequencing.

Verification of Expression by Immunocytochemistry

Cells are grown on glass coverslips before being subjected to immunocytochemistry. The cells are washed twice with PBS and fixed for 5 min with 10% formalin and permeabilized by incubation with 0.5% Triton X-100 in PBS for 15 min. Permeabilized cells are refixed in 10% formalin for 5 min, washed three times with PBS, and incubated with blocking solution [5% (v/v) normal goat serum, 0.1% (w/v) NaN$_3$ in PBS] for 3 h.

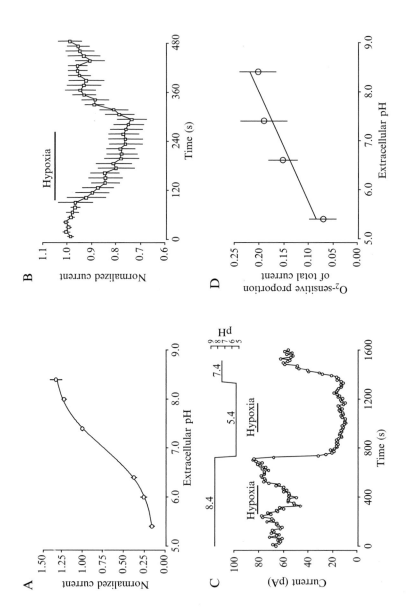

Cells are then incubated overnight in blocking solution using a 1:200 dilution of either anti-V_5 or anti-His_6 (Invitrogen, Groningen, The Netherlands) or channel-specific antibodies (many of which are available from either Alamone, Jerusalem, Israel, or Santa Cruz Technologies, CA). After thorough washing in PBS, the cells are incubated in FITC-labeled second-layer antisera for 3 h. Finally, cells are washed thoroughly in PBS prior to mounting and viewing. We have found that this staining protocol is very robust and that almost all incubations can be decreased in time if necessary. However, the primary antibody incubation must be allowed to proceed for at least 8 h. See Fig. 4A for an example employing anti-V_5 and anti-TREK1 primary antibodies.

Functional Screening Using the Patch-Clamp Technique

Success of the initial electrophysiological screening of clonal cells expressing K⁺ channels is expedited by employing the patch-clamp technique in the whole cell configuration.[25] Although the protocols of choice will vary according to the ion channel under investigation, it is clearly important that only a limited number of definitive electrophysiological characteristics are investigated. We pick between 20 and 30 clones and screen during a 48-h period. For hTASK channel clones, we define positively transfected cell as those that express currents of more than 10 times the amplitude of those in untransfected cells. In addition, we choose cells in which currents are inhibited by 50% or more when extracellular pH is reduced from 7.4 to 6.0 and which activate to more than 120% when extracellular pH is increased to 8.4 (see Fig. 2). hTREK1 are screened by investigating arachidonic acid modulation (see Figs. 3 and 4 and Chapman *et al.*[24]) For patch clamp, cells are grown for 24 h on glass coverslips before being transferred to a

---

[25] O. P. Hamill, A. Marty, B. Sakmann, and F. J. Sigworth, *Pflug. Arch.* **391,** 85 (1981).

FIG. 2. Regulation of K⁺ currents recorded from HEK293 cells stably expressing hTASK1 by pH and hypoxia. (A) Titration curve of extracellular pH versus mean normalized hTASK1 K⁺ currents recorded during a 200-ms step from −70 to 0 mV at extracellular pH values indicated. A normalized current of 1.0 represents the current recorded from each cell when the extracellular pH was 7.4. (B) Mean normalized hTASK1 K⁺ current time series plot recorded at an extracellular pH of 7.4 during normoxia (*ca.* 150 mm Hg) and hypoxia (*ca.* 40 mm Hg) as indicated by the horizontal bar. (C) Time series plot of a typical hTASK1 K⁺ current recorded during perfusion with normoxic (*ca.* 150 mm Hg) and hypoxic (*ca.* 40 mm Hg) solutions at the extracellular pH values plotted on the right axis (and indicated). Periods of hypoxic perfusion are indicated by the horizontal bars. (D) Correlation of hypoxia-sensitive hTASK K⁺ currents with extracellular pH indicating that the pH- and O₂-sensitive currents are one and the same. Taken from Lewis *et al.*[27]

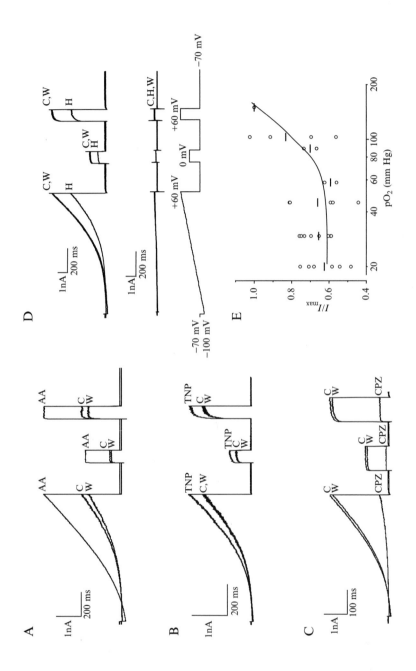

continuously perfused (5 ml·min^{-1}) recording chamber (volume *ca.* 200 μl) mounted on the stage of an inverted microscope (Olympus CK 40, Olympus, UK). Our standard perfusate is composed of (in mM) 135 NaCl, 5 KCl, 1.2 MgCl$_2$, 5 HEPES, 2.5 CaCl$_2$, and 10 D-glucose. Solutions of different extracellular pH values are made by titrating with 1 M NaOH until the required pH is achieved. As a note of caution, using HCl to adjust the pH down from 7.4 is discouraged as it results in artifacts due to chloride-dependent voltage offsets. Whole cell K⁺ currents are generally recorded at room temperature (21 ± 1°) using a pipette solution composed of (in mM) 10 NaCl, 117 KCl, 2 MgSO$_4$, 10 HEPES, 11 EGTA, 1 CaCl$_2$, and 2 Na$_2$ATP (pH 7.2). When filled with this solution, our pipettes are of resistance 3–7 $M\Omega$ (Clarke Electromedical, Cambridge, Cambs., UK). However, lower-resistance pipettes may be required if transfection levels are extremely high, as the resulting large whole cell currents will lead to series resistance voltage errors.

For later studies defining the possible O$_2$ sensitivity of cloned human K⁺ channels, solutions are made hypoxic by bubbling with N$_2$ for at least 30 min while normoxic solutions are allowed either to equilibrate in room air or are bubbled with medical air. Such maneuvers produce no shift in extracellular pH or bath temperature and, in our hands, perfusion with these solutions produces no shift in intracellular pH (when measured in intact cells using the fluorescent pH indicator BCECF). All tubing should ideally be Tygon (BDH, Atherstone, Berks., UK), which is almost completely gas impermeable, yet remains flexible. Patch-clamp experiments that employ altered O$_2$ levels must be paralleled by studies that measure

---

FIG. 3. Regulation of K⁺ currents recorded from HEK293 cells stably expressing hTREK1 by arachidonic acid, membrane distortion, and hypoxia. (A) Typical whole cell hTREK1 K⁺ currents recorded before (C), during (AA), and after (W) addition of 10 μM arachidonic acid. Currents were evoked using the ramp-step protocol shown in D (bottom). (B) Typical whole cell hTREK1 K⁺ currents recorded before (C), during (TNP), and after (W) addition of 400 μM trinitrophenol. Currents were evoked using the ramp-step protocol shown in D (bottom). (C) Typical whole cell hTREK1 K⁺ currents recorded before (C), during (CPZ), and after (W) addition of 10 μM chlorpromazine. Currents were evoked using the ramp-step protocol shown in D (bottom). (D) Typical whole cell hTREK1 currents recorded before (C), during (H), and after (W) perfusion with a hypoxic solution (*ca.* 40 mm Hg). Currents were evoked using the ramp-step protocol shown (bottom). (Middle) A typical current record from an untransfected HEK293 cell, before, during, and after hypoxic perfusion. (E) Plot of normalized current amplitudes (I/I_{max}) versus chamber pO$_2$ levels (on a log$_{10}$ scale). At each pO$_2$ level examined, the fractional inhibition (observed during a hypoxic challenge from 150 mm Hg to the pO$_2$ level shown) is plotted for individual cells (○) and the mean effect is indicated by the horizontal bars. Taken from Kemp *et al.*[29]

pO_2 directly; we employ a polarized carbon fiber electrode.[26] Many experiments that we perform use a single hypoxic stimulus (usually of between 20 and 40 mm Hg). However, graded hypoxia can be simply achieved by splicing gas-permeable silastic tubing of varying lengths into the Tygon perfusion lines. In our experience, measured pO_2 can be increased by *ca.* 20 mm Hg for every additional 5 cm of silastic tubing (see Fig. 3E for an exemplar graded hypoxic response).

Whole cell recordings are made using either Axopatch 200B or 700A amplifiers and a Digidata 1322 A/D interface (Axon Instruments, Forster City, CA). Voltage protocols are generated, and currents are recorded and analyzed off-line using the pClamp 8 suite of software. Our basic voltage protocol to evoke K^+ currents consists of a voltage ramp followed by two voltage steps. Cells are clamped at a holding potential of -70 mV and are then ramped from -100 to either 0 or 60 mV (500 or 1000 ms). Following the ramp, voltage is stepped sequentially from holding to 0 mV and then 60 mV, each for 200 ms. The protocol is repeated at 10-s intervals. Where appropriate, current clamp recording is also carried out using the same solutions as described earlier ($I = 0$ nA). For convenience, we often normalize currents as I/I_{max} (at 0 or 60 mV) and plot successive, mean normalized current amplitudes against time in order to monitor onset and recovery from hypoxia.

Homomultimeric O_2-Sensitive Human K^+ Channels Stably Expressed in HEK293 Cells

Using the protocols described earlier, we have thus far demonstrated that three recombinant K_{2P} channels are sensitive to ambient pO_2; these are hTASK1 (KCNK3[27]), hTASK3 (KCNK9), and hTREK1 (KCNK2[28,29]). Thus, employing the HEK293 human expression system we have demonstrated the acid sensitivity of hypoxic inhibition of hTASK1. Whole cell K^+ currents of cells stably expressing hTASK-1 are, as anticipated, extremely sensitive to extracellular pH, within the physiological range ($IC_{50} \sim 7.0$; Fig. 2A). All cells expressing this signature pH sensitivity are acutely modulated by pO_2; reduction of pO_2 from 150 to <40 mm Hg (at pH 7.4) causes rapid and reversible suppression of

[26] M. H. Mojet, E. Mills, and M. R. Duchen, *J. Physiol.* **504,** 175 (1997).
[27] A. Lewis, M. E. Hartness, C. G. Chapman, I. M. Fearon, H. J. Meadows, C. Peers, and P. J. Kemp, *Biochem. Biophys. Res. Commun.* **285,** 1290 (2001).
[28] P. J. Kemp, A. Lewis, P. Miller, C. G. Chapman, H. Meadows, and C. Peers, *FASEB J.* **16,** A61 (2002).
[29] P. J. Kemp, G. J. Searle, M. E. Hartness, A. Lewis, P. Miller, S. E. Williams, P. Wootten, D. Adriaensen, and C. Peers, *Anat. Rec.* **A270A,** 41 (2003).

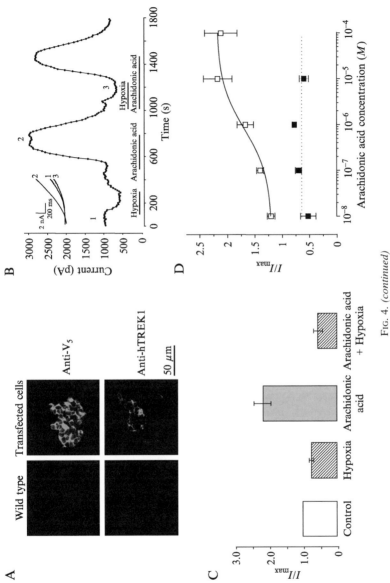

FIG. 4. (continued)

pH-sensitive K^+ currents (Fig. 2B). Furthermore, these two regulatory signals clearly act at the same channel, as the magnitude of the O_2-sensitive current is dependent on the extracellular pH (Figs. 2C and 2D). These data represent the first direct verification that hTASK1 is O_2 sensitive and reinforce the idea that this K^+ channel is key to O_2 sensing in chemoreceptors, as suggested in the carotid body.[29a] They also support the notion that the O_2 dependence of the excitability of certain central neurons, particularly cerebellar granule cells, is underpinned by the hypoxic inhibition of TASK1.[20] We have also shown that hTASK3 is acutely inhibited by hypoxia,[28] which lends weight to the argument that this channel is the O_2-sensitive K^+ channel in the neuroepithelial body model (H146 cells), as suggested previously by our antisense loss-of-function experiments.[9]

Another member of the K_{2P} family, hTREK1 (KCNK2), is highly expressed in human central neuronal tissue.[30] In accordance with this previous study, K^+ currents from HEK293 cells stably expressing hTREK1 (as assessed by immunocytochemical localization with anti-V_5 or anti-TREK1; Fig. 4A) are enhanced by arachidonic acid with maximal enhancement (*ca.* 2.5-fold augmentation) observed at a concentration of 10 μM (Fig. 3A). Membrane distortion using the chemical modulators trinitrophenol (TNP, a chemical crenator; Fig. 3B) and chlorpromazine (CPZ, a chemical cup former; Fig. 3C) evokes reciprocal modulation of this human TREK1 orthologue. Importantly, acute hypoxia ($pO_2 = 40$ mm Hg) evokes a rapid and reversible inhibition of hTREK1 whole cell K^+ currents (Fig. 3D); this effect is graded and maximal at 60 mm Hg (Fig. 3E). Furthermore, acute hypoxia almost completely prevents K^+ current augmentation by either arachidonic acid (Figs. 4B, 4C, and 4D) or membrane

[29a] K. J. Buckler, B. A. Williams, and E. Honore, *J. Physiol.* **525,** 135 (2000).

[30] A. D. Medhurst, G. Rennie, C. G. Chapman, H. Meadows, M. D. Duckworth, R. E. Kelsell, I. I. Gloger, and M. N. Pangalos, *Brain Res. Mol. Brain Res.* **86,** 101 (2001).

FIG. 4. Occlusion of hTREK1 regulation by hypoxia. (A) Immunocytochemical verification of hTREK1 stable expression using anti-V_5 (top) and anti-hTREK1 (bottom) antibodies in transfected (right) and untransfected (left) HEK293 cells. (B) Exemplar time series plot of hTREK1 whole cell K^+ currents during hypoxic challenge (*ca.* 40 mm Hg) and application of 10 μM arachidonic acid or both as indicated by the horizontal bars. Currents were recorded during the final 10 ms of the 200-ms voltage step from -70 to 60 mV. (Inset) Numbered ramp currents corresponding to the points indicated on the time series plot. (C) Mean, normalized whole cell hTREK1 K^+ currents recorded at 60 mV during hypoxic challenge (*ca.* 40 mm Hg) and application of 10 μM arachidonic acid or both as indicated at the bottom of each bar. (D) Mean concentration–response data for arachidonic acid in normoxia (□) and hypoxia (■). Taken from Kemp *et al.*[29] (See color insert.)

stretch (data not shown). The ability of arachidonic acid to augment currents—and of hypoxia to abrogate completely this effect—was also observed in cell-attached patches (data not shown). These important data indicate that hypoxia interacts with hTREK-1, possibly at a site recognized as common to the activation of TREK channels by agents including arachidonic acid and membrane stretch. Because brain pO_2 is as low as 20 mm Hg,[1] these findings also suggest that the recruitment of activated TREK channels to facilitate neuroprotection (cell hyperpolarization) is probably not possible at ambient brain O_2 levels.[29,31,32]

Heteromultimeric O₂-Sensitive Human K⁺ Channels Stably Expressed in HEK293 Cells

Large-conductance, Ca^{2+}-sensitive K^+ (BK) channels have also been stably expressed in HEK293 cells. Ahring *et al.* used a double construct (still based on pcDNA3.1) to coexpress α and β subunits. In this construct, transcription of the two subunits is driven from separate promoters. We were gifted these cells and confirmed, using immunocytochemistry, western blotting, and electrophysiology, that they express high levels of these human proteins, which reinforces the robust nature of this system for expression of a variety of human K^+ channel types. Activity of the BK channel is important for the physiological response to acute hypoxia of several tissues, including the carotid body.[5] To investigate the molecular basis of such O_2 sensitivity, we studied channel activity in excised, inside-out patches from HEK293 cells stably coexpressing the α (*slo1*, KCNMA1) and β (*sloβ*, KCNMB1) subunits of human maxi-K channels.[32a] At activating voltages and in the presence of 300 nM Ca^{2+}_i, single channel activity is acutely and reversibly suppressed upon reducing pO_2 from 150 to >40 mm Hg by over 30% (Fig. 5A). The hypoxia-evoked reduction in current is due predominantly to suppression in NPo (Fig. 5B), although a minor component is attributable to reduced unitary conductance (Fig. 5C). Hypoxia causes an approximate doubling of the time constant for activation but is without effect on deactivation. At lower levels of Ca^{2+}_i (30 and 100 nM), hypoxic inhibition does not reach significance. In contrast, 300 nM and 1 μM Ca^{2+}_i both sustain significant hypoxic suppression of activity over the entire activating voltage range. At these two Ca^{2+}_i levels, hypoxia

[31] P. Miller, P. J. Kemp, A. Lewis, C. G. Chapman, H. J. Meadows, and C. Peers, *J. Physiol.* **548**, 31 (2003).

[32] P. K. Ahring, D. Strobaek, P. Christophersen, S. P. Olesen, and T. E. Johansen, *FEBS Lett.* **415**, 67 (1997).

[32a] A. Lewis, C. Peers, M. L. J. Ashford, and P. J. Kemp, *J. Physiol.* **540**, 771 (2002).

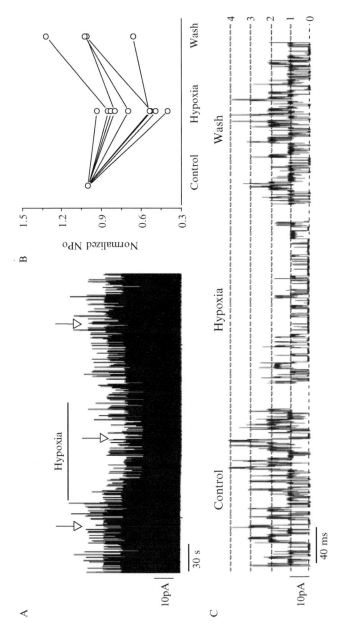

FIG. 5. Hypoxic suppression of single channel activity in HEK293 cells stably coexpressing α and β subunits of human maxi-K channels. (A) Exemplar continuous current recording from an inside-out, excised membrane patch bathed in "physiological" solutions with $[Ca^{2+}]_i$ buffered to 300 nM and held at 50 mV. During the time indicated by the horizontal bar above the trace, perfusate pO$_2$ was reduced from 150 to 40 mm Hg. (B) Normalized NPo values from nine separate patches under the conditions described in A during normoxia (control), hypoxia, and following reestablishment of normoxia (wash). (C) Single channel currents on an extended time base (200 ms per trace) from the recording shown in A at the arrows in normoxia, hypoxia, and following reintroduction of normoxic perfusate, as indicated. Taken from Lewis et al.[32]

evokes a positive shift in the activating voltage (by ~10 mV at 300 nM and ~25 mV at 1 μM). At saturating [Ca^{2+}]$_i$ (100 μM), hypoxic inhibition is absent. Distinguishing between hypoxia-evoked changes in voltage and/or Ca^{2+}$_i$ sensitivity is achieved by evoking comparable channel activity using high depolarizing potentials (up to 200 mV) in the presence of 300 nM or 100 μM Ca^{2+}$_i$ or in its virtual absence (<1 nM). Under these experimental conditions, hypoxia causes significant channel inhibition only in the presence of 300 nM Ca^{2+}$_i$. Thus, because regulation is observed in excised patches, maxi-K channel inhibition by hypoxia does not require soluble intracellular components and, mechanistically, is voltage independent but Ca^{2+}$_i$ sensitive.[32]

Alternative Approaches to the Study of Heteromultimeric O$_2$-Sensitive Human K$^+$ Channels

A huge number of studies have been carried out on recombinant K$^+$ channels over the last decade. However, only a handful of these studies have investigated O$_2$ sensitivity of human K$^+$ channels. Species and expression system notwithstanding, much has been learned from studies on voltage-activated K$^+$ channels (Kv), which have greatly increased our understanding of the molecular basis of hypoxic pulmonary vasoconstriction. Patel and colleagues[33] were able to demonstrate O$_2$ sensitivity of a heteromultimeric, recombinant Kv channel complex. Co transfection of the delayed rectifier Kv2.1 (*shab, KCNB1*) with the electrically silent subunit Kv9.3 (*KCNS3*) produces a K$^+$ current that is acutely inhibited by hypoxia.[33] These experiments were conducted using transient transfection in COS-M6 cells and, interestingly, hypoxic inhibition was only observed in a subset of transfected cells. Whether this functional variation is due to the experimental protocol (transient transfection), the expression system of choice (pCI vector), or the host cell (COS-M6, which may not express endogenously all the components required for the full O$_2$ response) is still a matter of conjecture. However, that the later possibility may hold true is supported by the observations of Hulme *et al.*[17] who found that the mouse L-cell system was able to support consistent hypoxic inhibition of Kv2.1/ 9.3. The mouse L cell endogenously expresses Kvβ2.1, a subunit required for Kv channel trafficking. They further extended these observations by showing that hypoxic inhibits K$^+$ currents of Kv1.2 (*shaker, KCNA2*) and Kv2.1 (*shab, KCNB1*) but was without effect on Kv1.5 (*shaker, KCNA5*) unless it was coexpressed with Kv1.2. All three of these Kv channels are expressed in the pulmonary artery but only coexpression of Kv1.2/Kv1.5 or

[33] A. J. Patel, M. Lazdunski, and E. Honore, *EMBO J.* **16**, 6615 (1997).

Kv2.1/9.3 results in K^+ current inhibition at, or around, resting membrane potential.[17] Hypoxic inhibition of monomeric Kv1.2 and Kv2.1 is only apparent at nonphysiological, positive test potentials.[17] The effect of coassembly of Kv channel α subunits with β subunits has also been demonstrated in transiently transfected HEK293 [for Kvα4.2 (*shal*, KCND2) with Kvβ1.2],[34] which suggests that for full physiological responses to decreased O_2, Kv channels might colocalize with further regulatory proteins. However, it is noteworthy that stably transfected HEK293 cells (in our hands and with K_{2P} or BK channels, at least) always respond to hypoxia, suggesting that the coexpression requirement is more likely a channel-specific phenomenon.

Conclusion

O_2-sensitive K^+ channels have been described and studied in specialized native chemosensory tissues (carotid body, neuroepithelial body, and pulmonary arteriole), immortalized cellular models (e.g., H146 cells), and nonspecialized tissues (central neurons) where they contribute to the cell resting potential. Each tissue and cell type is characterized by a different array of these K^+ channels, and their inhibition by hypoxia contributes to the physiological response to acute episodes of reduced O_2 availability. Although data derived from native tissue are useful, in order for the full picture to emerge they need to be combined with the new information garnered from studies that employ recombinant K^+ channels in strictly defined cellular environments.

Acknowledgments

Much of the work detailed in this article was funded by the British Heart Foundation and The Wellcome Trust and the authors thank these organizations for their support.

---

[34] M. T. Perez-Garcia, J. R. Lopez-Lopez, and C. Gonzalez, *J. Gen. Physiol.* **113**, 897 (1999).

[19] Methods to Study Oxygen Sensing Sodium Channels

By A. K. M. HAMMARSTRÖM and P. W. GAGE

Introduction

Blocked arteries or bleeding in the heart (heart attack) or brain (stroke) deprive cardiac muscle and neurons of O_2 and eventually lead to cell death. Within a short period of altered O_2 tension, cells respond with alterations in membrane potential and detrimental changes in their intracellular environment such as a rise in intracellular Ca^{2+} concentration, lower ATP levels, and changes in pH. There is mounting evidence that an increase in the intracellular Na^+ concentration is an early and fundamental event during hypoxia.[1-3] Na^+ channel blockers such as TTX and lidocaine and low extracellular Na^+ concentrations protect heart muscle and neurons from rises in intracellular Ca^{2+} and ischemic damage.[1] This suggests that voltage-operated Na^+ channel activity is a precursor for cell damage.[1,4] A small, persistent, and TTX-sensitive Na^+ current (I_{NaP}) has been described in cardiac muscle (from most species such as human, dog, rat, rabbit, and guinea pigs) and neurons.[5-14] The increase in I_{NaP} that has been reported previously during hypoxia[15] is a likely trigger for the accumulation of intracellular Na^{+} [16] and abnormal electrical activity during hypoxic episodes. It may also lead to damaging rises in intracellular Ca^{2+} and cell damage.[1]

[1] M. C. Haigney, E. G. Lakatta, M. D. Stern, and H. S. Silverman, *Circulation* **90,** 391 (1994).
[2] E. Carmeliet, *Physiol. Rev.* **79,** 917 (1999).
[3] P. Lipton, *Physiol. Rev.* **79,** 1431 (1999).
[4] D. G. Renlund, G. Gerstenblith, E. G. Lakatta, W. E. Jacobus, C. H. Kallman, and M. L. Weisfeldt, *J. Mol. Cell Cardiol.* **16,** 795 (1984).
[5] G. A. Gintant, N. B. Datyner, and I. S. Cohen, *Biophys. J.* **45,** 509 (1984).
[6] E. Carmeliet, *Biophys. J.* **51,** 109 (1987).
[7] J. B. Patlak and M. Ortiz, *J. Gen. Physiol.* **86,** 89 (1985).
[8] T. Kiyosue and M. Arita, *Circ. Res.* **64,** 389 (1989).
[9] D. A. Saint, Y. K. Ju, and P. W. Gage, *J. Physiol.* **453,** 219 (1992).
[10] V. A. Maltsev, H. N. Sabbah, R. S. Higgins, N. Silverman, M. Lesch, and A. I. Undrovinas, *Circulation* **98,** 2545 (1998).
[11] C. R. French and P. W. Gage, *Neurosci. Lett.* **56,** 289 (1985).
[12] C. R. French, P. Sah, K. J. Buckett, and P. W. Gage, *J. Gen. Physiol.* **95,** 1139 (1990).
[13] C. P. Taylor and B. S. Meldrum, *Trends Pharmacol. Sci.* **16,** 309 (1995).
[14] C. P. Taylor and L. S. Narasimhan, *Adv. Pharmacol.* **39,** 47 (1997).
[15] Y. K. Ju, D. A. Saint, and P. W. Gage, *J. Physiol.* **497,** 337 (1996).
[16] X. H. Xiao and D. G. Allen, *Circ. Res.* **85,** 723 (1999).

There have been many reports in the last decade of various ion channels that respond to O_2 levels.[17,18] For example, various K^+ and Ca^{2+} channels in central neurons and vascular tissue respond to low O_2 levels[19] and several hypotheses of how this O_2 sensing occurs have been put forward. For example, the channel itself, or an associated subunit, may sense changes in the O_2 levels,[17] there may be a NAD(P)H oxidase associated with the channel in the plasma membrane,[20] or there may be other non-haem-containing structures closely associated with the channel.[21] Generally, it is thought that ion channel O_2 sensing involves a redox reaction and/or a haem-containing structure.

We have been exploring where and how this O_2 sensing by I_{NaP} occurs. Results from native channels show that I_{NaP} channels can directly sense O_2 levels in the absence of cytosolic second messengers in inside-out patches from myocytes and hippocampal neurons.[22,23] In addition, we have shown that O_2 sensing involves a redox reaction at, or near, the channel itself.[22]

The minimum ion channel subunit requirement for the O_2 sensing response has been explored in some cases. For example, expressing the rat Kv2.1 alone[24,25] (but not Kv1.5[25]) produces channels that respond to hypoxia. Similarly, the α subunit of L-type Ca^{2+} channels responds to hypoxia when expressed in HEK293 cells.[26] To determine the subunit requirement for the hypoxic response of the Na channel, we expressed recombinant α subunits (human heart; hH1) alone in HEK293 cells and recorded whole cell Na^+ currents from transfected cells.[27]

Both hippocampal neuronal and cardiac voltage-gated Na^+ channels contain a principal α subunit and most likely also a $\beta 1$ subunit and a $\beta 2$ subunit.[28] It has been suggested that the $\beta 1$ subunit may enhance Na^+

[17] J. Lopez-Barneo, *Trends Neurosci.* **19**, 435 (1996).

[18] C. Peers, *Trends Pharmacol. Sci.* **18**, 405 (1997).

[19] J. Lopez-Barneo, R. Pardal, and P. Ortega-Saenz, *Annu. Rev. Physiol.* **63**, 259 (2001).

[20] C. Youngson, C. Nurse, H. Yeger, and E. Cutz, *Nature* **365**, 153 (1993).

[21] C. Jiang and G. G. Haddad, *Proc. Natl. Acad. Sci. USA* **91**, 7198 (1994).

[22] A. K. M. Hammarström and P. W. Gage, *J. Physiol.* **529**, 107 (2000).

[23] R. Khoury, P. W. Gage, and A. K. M. Hammarström, *Proc. Aust. Physiol. Pharmacol. Soc.* **30**, 58P (1999).

[24] A. J. Patel, M. Lazdunski, and E. Honore, *EMBO J.* **16**, 6615 (1997).

[25] J. T. Hulme, E. A. Coppock, A. Felipe, J. R. Martens, and M. M. Tamkun, *Circ. Res.* **85**, 489 (1999).

[26] I. M. Fearon, A. C. Palmer, A. J. Balmforth, S. G. Ball, G. Varadi, and C. Peers, *J. Physiol.* **514**, 629 (1999).

[27] A. K. M. Hammarström, M. Philippa, R. Khoury, and P. W. Gage, *in* "Structure and Function of Ion Channels." Satellite to IUPS 2001.

channel expression in the plasma membrane.[29–31] The biophysical and physiological consequences of coexpressing the $\beta1$ subunit are currently controversial and appear to vary in different expression systems. When the two cardiac subunits are coexpressed in HEK293 or HEK293t cells, there is a depolarizing shift in the channel kinetics compared with the voltage dependence of channel kinetics when the α subunit is expressed alone.[31,32] In contrast, when the two subunits are coexpressed in Chinese hamster oocytes, there is no, or only a modest, hyperpolarizing shift in inactivation.[30,33,34] Interestingly, when the $\beta1$ subunit is coexpressed with the human heart α subunit (hH1), the affinity of resting channels for lidocaine is reduced[35] and the effects of saturated and monounsaturated fatty acids are abolished.[31] Interaction points between the two subunits are thought to include the S5–S6 loops,[36,37] the C terminus of the α subunit,[38] and/or the extracellular domain of the β subunit.[37] In line with this, it is interesting that the C terminus of the α subunit of the L-type voltage-gated Ca^{2+} channel is important for O_2 sensing.[39] Interactions between the $\beta1$ subunits and the α subunit of Na^+ channels may be important for the hypoxic response.

Identification of the link(s) between hypoxia and an increase in I_{NaP} activity may be important for devising ways to prevent or ameliorate two sequelae of ischemia/hypoxia: arrhythmias and irreversible muscle cell damage. The persistent Na^+ channel and/or the O_2 sensor may be prime targets for antiischemic and antiarrhythmic drugs. Such drugs may reduce

[28] L. L. Isom, in "Sodium Channels and Neuronal Hyperexcitability," p. 124. Wiley, West Sussex, 2002.

[29] Y. Qu, L. L. Isom, R. E. Westenbroek, J. C. Rogers, T. N. Tanada, K. A. McCormick, T. Scheuer, and W. A. Catterall, *J. Biol. Chem.* **270**, 25696 (1995).

[30] H. B. Nuss, N. Chiamvimonvat, M. T. Perez-Garcia, G. F. Tomaselli, and E. Marban, *J. Gen. Physiol.* **106**, 1171 (1995).

[31] Y. F. Xiao, S. N. Wright, G. K. Wang, J. P. Morgan, and A. Leaf, *Am. J. Physiol. Heart. Circ. Physiol.* **279**, H35 (2000).

[32] S. N. Wright, S. Y. Wang, Y. F. Xiao, and G. K. Wang, *Biophys. J.* **76**, 233 (1999).

[33] P. B. Bennett, Jr., N. Makita, and A. L. George, Jr., *FEBS Lett.* **326**, 21 (1993).

[34] L. L. Isom, K. S. De Jongh, D. E. Patton, B. F. Reber, J. Offord, H. Charbonneau, K. Walsh, A. L. Goldin, and W. A. Catterall, *Science* **256**, 839 (1992).

[35] J. C. Makielski, J. T. Limberis, S. Y. Chang, Z. Fan, and J. W. Kyle, *Mol. Pharmacol.* **49**, 30 (1996).

[36] N. Makita, P. B. Bennett, and A. L. George, Jr., *J. Neurosci.* **16**, 7117 (1996).

[37] K. A. McCormick, L. L. Isom, D. Ragsdale, D. Smith, T. Scheuer, and W. A. Catterall, *J. Biol. Chem.* **273**, 3954 (1998).

[38] Y. F. Xiao, S. N. Wright, G. K. Wang, J. P. Morgan, and A. Leaf, *Proc. Natl. Acad. Sci. USA* **95**, 2680 (1998).

[39] I. M. Fearon, G. Varadi, S. Koch, I. Isaacsohn, S. G. Ball, and C. Peers, *Circ. Res.* **87**, 537 (2000).

the damage as well as spread of damage that occurs during O_2 deficiency in heart muscle and neurons.

Cell Preparations

One of the secrets for success with patch clamping is to have healthy cells with good cell membranes. As all patch clampers know, it takes time and patience to get any cell preparation to work in one's own laboratory. This article gives a brief outline of the techniques used to prepare cells, but there are other and more detailed descriptions to be found in the literature. The methods described here have been approved by the Animal Experimentation Ethics Committee at the Australian National University, Canberra, Australia.

Dissociation of Ventricular Myocytes

Materials

Ice-cold Tyrode solution [mM: 133 NaCl, 4 KCl, 1.2 NaH$_2$PO$_4$, 1.2 MgCl$_2$, 1 CaCl$_2$, and 10 TES (N-tris(hydroxymethyl) methyl-2-aminoethane-sulfonic acid)]
Two fine forceps (Apollo #5)
Fine scissors
Cotton thread
Cannula (polyethylene tubing i.d. 2.0 mm with fire-polished "lip" to secure the artery)
Langendorf recirculating set-up

Procedure

1. Adult Wistar rats are anesthetized with CO_2 and killed by cervical dislocation. The heart is dissected out and submerged in ice-cold Tyrode solution. The first major aortic branch (appearing about 1 cm downstream) is used for cannulation. The branch is cut off close to the branching point, and the heart is lifted carefully using two forceps and holding on to the artery wall at the branching point, cannulated at the branching point, and secured carefully onto the cannula with cotton thread. Ca^{2+}-free Tyrode solution is perfused through the heart for 5 min at a rate of approximately 9 ml per minute. During this perfusion the blood will be removed from the heart (which will lose its red color) and the solution should therefore not be recirculated.

2. Single ventricular myocytes are obtained by Langendorf perfusion with solutions containing collagenase (100 U·ml^{-1}, CLS II, Worthington Biochemicals Corp.) and protease (0.05 mg ml^{-1}, Type XIV, Sigma) as

described previously.[9] After approximately 0.5 h, the first one-third of the apex is cut off, triturated, and placed in 25 μM Ca^{2+} Tyrode solution. Two subsequent pieces of ventricle are cut off after approximately 5 min (this will vary slightly from day to day but the slice should be somewhat soft, mushy, and easy to dissociate).

3. After isolation (after trituration and a short centrifugal spin the supernatant is kept), cells are stored in 200 μM Ca^{2+} Tyrode solution (at room temperature) for about 1 h before being transferred to the 1 mM Ca^{2+} Tyrode solution. The Tyrode solution is replaced every few hours to prolong survival of the cells.

4. Dissociated cells to be used for electrophysiological experiments are transferred onto poly-L-lysine-coated coverslips in a bath containing bath solution. Cells can survive in Tyrode solution for up to 2 days and for many hours in the bath solution.

Note. It is highly important that air bubbles are not introduced into the heart perfusion line, as this will impede perfusion. It is also important to perform the dissection and set the heart up for perfusion as quickly as possible. The heart should not be forced or pulled too far onto the cannula, as this will impede perfusion. The heart should swell up and lose its blood red color as the Ca^{2+}-free and enzyme solutions are perfusing.

Dissociation of Young Rat Hippocampal Neurons

These methods were adapted by French *et al.*, 1990[12] from the procedure described originally by Kay and Wong.[40]

Materials

Vibratome
Superglue
Razor blades
Fine scissors
50-ml beaker with bubbler containing a fitted "cell cage" [a plastic cuff of slightly smaller diameter than the beaker and about 10–15 mm high with a fine mesh floor (e.g., a small piece of nylon stocking) glued to the bottom]
Several carbogen gas bubblers
Artificial cerebrospinal fluid (ACSF) (124 mM NaCl, 26 mM NaH$_2$CO$_3$, 3 mM KCl, 1.3 mM MgSO$_4$, 2.5 mM NaH$_2$PO$_4$, and 20 mM glucose)
Carbogen; 95% CO_2: 5% O_2 (BOC gases, Australia)

[40] A. R. Kay and R. K. Wong, *J. Physiol.* **392,** 603 (1987).

Procedure

1. Young adult rats (16–21 days old) are decapitated, and their brains are transferred quickly to ice-cold ACSF gassed with carbogen

2. The brain is cut in half using a razor blade, and each cut end is blotted dry and then superglued to the Teflon stage of the Vibratome. Brain slices 500 μm thick are cut in the presence of an ACSF "ice slush" (fresh ASCF is stored in the freezer for 1–2 h before use).

3. The brain slices are stored in the "cell cage" submerged (in ~10 ml) in a 50-ml beaker filled with carbogenated ACSF. After 0.5 h of equilibrating at 35° in a shaking bath, the slices are treated enzymatically for 0.5 h by adding a mixture of 200 units papain (Worthington Biochemicals, New Jersey), 1.1 m*M* cysteine (Sigma), 0.2 m*M* EDTA, and 13.4 μ*M* mercaptoethanol directly to the slices.

4. Following the enzyme incubation, slices are washed several times with fresh ACSF at 35° and allowed a short recovery period (15–30 min at 35°) before the CA1 region is dissected out under a dissecting microscope. Single CA1 neurons are obtained by careful mechanical trituration of three to four pieces of the CA1 region in ~1 ml of bath solution (see later) with a fire-polished glass Pasteur pipette.

5. The cells are allowed 10–15 min to settle to the bottom of the bath before patch clamping.

Note. It is important that the slices are kept under constant conditions (pH and temperature) and not bubbled too vigorously or the slices will tear and the cells will be damaged. Trituration is an important step for achieving healthy cells and it is worthwhile experimenting with various sizes of cut and fire-polished glass Pasteur pipettes and the duration, as well as the force used during the trituration process.

Hippocampal Cell Cultures

The method we use for preparing primary cultures of hippocampal neurons is similar to that described previously.[41]

Materials

Minimum essential medium (MEM; GIBCO) with fetal calf serum (FCS; from Commonwealth Serum Laboratories; at 5–10%), glucose (2%), penicillin/streptomycin (1%), and MITO+ serum extender (0.1%, Collaborative Research Incorporated) added
Petri dishes

[41] L. S. Premkumar, S.-H. Chung, and P. W. Gage, *Proc. R. Soc. Lond. B* **241**, 153 (1990).

Coverslips [treated with poly-L-lysine according to (Sigma) manufacturer's instructions]
Toothpicks
Dissecting board
Fine forceps
Scissors
Scalpel blades
15-ml Falcon tubes

Procedure

1. Newborn Wistar rats are decapitated (CO_2 anesthesia). Using scalpel blades, the brain is cut in half along the midline between the hemispheres. Each brain half is removed carefully from the skull and placed on its side, cut side up.

2. The hippocampal area from each half is identified, gently "lifted" out from the rest of the brain (using the tips of two shaved toothpicks), and submerged in a petri dish containing MEM.

3. As much as possible of the excess tissue is removed from the hippocampus before it is chopped into very small pieces using scalpel blades and triturated mechanically in MEM (in Falcon tubes) to dissociate the cells. Allow the pieces to settle (about a minute or two) after each trituration (two to three times) with fresh MEM.

4. Remove the supernatant (total of ~3 ml) containing suspended cells and place it onto the poly-L-lysine-coated coverslips. The dissociated cells are grown on the glass coverslips for several weeks.

5. Cultures are fed with fresh solution every 2–4 days and used for experiments after 5–12 days.

Expression of Recombinant Channels in L9292/HEK293 Cell Lines

It has been shown previously that Ca^{2+} channels expressed in HEK293 cells retain their O_2 sensitivity.[26] This suggests that the HEK293 cell line is a good model for studying O_2 sensing ion channels, as recombinant channels appear to retain their responsiveness to hypoxia as well as their native biophysical properties in these cells.

Materials

Lipofectin (GIBCO)
Complete Dulbecco's modified Eagle's medium (DMEM) supplemented with 2 mM L-glutamine, 10% fetal bovine serum (FBS), 0.1 mM MEM nonessential amino acids (GIBCO BRL), 1.0 mM sodium pyruvate, and 10 mM HEPES buffer (pH 7.4)

Plasmids
Dynabeads (DynalBiotech, Norway)

Procedure

1. Human embryonic kidney cells (HEK293) or L929 mouse fibroblasts are grown in DMEM (with supplements; see earlier discussion) and
transfected transiently using a liposome-mediated transfection procedure
(Lipofectin, GIBCO) with a plasmid encoding the human heart Na^+
channel (hH1) (gift from Dr. Lederer, USA) and the following procedure.

2. Cells are seeded at subconfluent densities ($\sim 1 \times 10^5$/well) in 6- to 12-
well plates and allowed to settle for 1–2 days. After this time the cells are
transfected with the Na^+ channel expression vector (5–10 μg) using 20 μl
lipofectin (GIBCO) in 1 ml DMEM and incubated for 5 h at 37° (see
GIBCO manufacturer's details for transfection procedure). Cells are then
washed, and fresh medium is added and left in the incubator for another
24–48 h to allow sufficient time for protein expression before patching.

3. The CD4 plasmid (~ 3 μg) is included in the standard transfection
mix, and CD4 antibody–coated dynabeads (M-450, 1 μM/well) are added
to the medium 0.5 h before patching to enable selection of transfected
cells. Cells "rosetted" with beads are identified easily under the
microscope.

Note. Often the HEK293 cells require at least 36 h and the higher
amount of plasmid for good Na^+ channel expression.

Patch Clamping Persistent Na^+ Channels

Bath and Pipette Solutions

Standard whole cell patch-clamp techniques[42] were used to record
Na^+ currents. Solution composition is a compromise between mimicking
normal ion concentrations and osmotic pressures as much as possible using
solutions with good separation between the Na^+ equilibrium potential and
the K^+ and Cl^- equilibrium potentials and providing a good driving force
on Na^+ ions at reasonable membrane potentials. In the solutions we used,
E_{Na} was about 80 mV, whereas E_K and E_{Cl} were about -86 and -76 mV,
respectively. This allowed easy detection of K and Cl single channel currents that can contaminate recordings of Na currents because they were

---

[42] O. P. Hamil, A. Marty, E. Neher, B. Sakmann, and F. J. Sigworth, *Pflug. Arch.* **391,** 85
(1981).

generally in the opposite direction to Na currents. For cell-attached patches, the driving force on sodium across a channel is V_m-Vp-E_{Na}. Bath solutions containing 150 mM K were useful for reducing movement in cardiac myocytes and for making the membrane potential approximately 0 mV. The bath solution for cell-attached and inside-out patch recordings contains 150 mM K-aspartate, 10 mM EGTA, 2 mM $MgCl_2$, 2 mM $CaCl_2$, and 10 mM TES, pH adjusted with NaOH to 7.4. The pipette solution contains (mM) 135 NaCl, 5 KCl, 1 $MgCl_2$, 1 $CaCl_2$, 5 $CoCl_2$, 5 CsCl, and 10 TES adjusted to a pH of 7.4 with NaOH. Under these conditions, the membrane potential would be close to 0 mV when recording from a cell-attached patch.

The bath and pipette solutions for whole cell recordings contain (mM) 135 NaCl, 5 KCl, 1 $MgCl_2$, 1 $CaCl_2$, 5 CoCl, 5 CsCl, and 10 TES (bath solution used for cell storage) and 70 NaF, 50 CsF, 10 K-EGTA, 2 $CaCl_2$, 2 $MgCl_2$, and 10 TES, respectively.

Procedure and Pulse Protocols

1. Whole cell capacitance and series resistance compensation should be made with the controls on the amplifier ($\geq 80\%$), especially for currents greater than 500–1000 pA.

2. Current–voltage relationships are established by applying successive 5- or 10-mV depolarizing steps from a negative holding potential (V_h = -70 to -90 mV). A 300-ms prepulse (to between -120 and -140 mV) preceding each depolarizing pulse reduces any inactivation of the Na channels. Varying the amplitude of the prepulse in 10-mV steps followed by a step to a potential between -30 and -10 mV is used to assess the voltage dependence of inactivation.

3. Currents are analyzed using "in-house" computer techniques and software.[22,43,44] In most instances, the amplitude of I_{NaP} was measured at the end of the 400-ms depolarizing pulse, and transient Na^+ current (I_{NaT}) amplitude was measured at its peak. Lately there have been discussions about what should be classified as an I_{NaP} (Novartis Foundation Symposium 241, "Sodium channels and hyperexcitability," 2002). We define I_{NaP} as the current that persists for 400 ms or more during a prolonged depolarization. It is important to state at what time during a depolarizing pulse measurements were made when comparing results with other laboratories and to reveal on what basis this choice was made.

[43] A. K. Hammarström and P. W. Gage, *J. Physiol.* **510**, 735 (1998).
[44] A. K. Hammarström and P. W. Gage, *J. Physiol.* **520**, 451 (1999).

Isolating I_{NaP}

It is often hard to distinguish a small current such as I_{NaP} in the presence of contaminating Cl^-, K^+, and Ca^{2+} currents, particularly under control conditions when the amplitude of I_{NaP} is very small. We therefore developed the "TTX-subtracted" method to isolate the current.

Procedure. Whole cell current traces are recorded in the same cell before and after exposure to a high concentration of TTX (0.5 μM in neurons and 50–100 μM in cardiac muscle). The traces are then subtracted digitally to isolate whole cell TTX-sensitive Na^+ currents. This subtraction should give a baseline close to zero and a clearer picture of I_{NaP} (see Fig. 1).

Note. It is important to make sure that there are minimal changes in the baseline conditions (e.g., no drift or changes in series-resistance resistance) for this subtraction to work and be valid.

Introducing Hypoxia and Studying O_2 Sensing

Materials

Inflow system (multibarrel pipette, see Fig. 2) attached to a micromanipulator (Narashige) 100% N_2 or 95% N_2: 5% O_2 gas (BOC gases, Australia)

20-ml syringe tubes containing bath solution bubbled with 100% N_2 (bubblers are approximately half the diameter of the 20-ml syringe

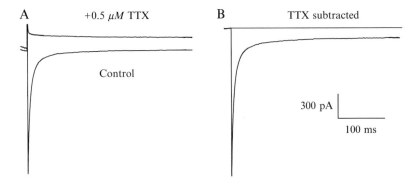

Fig. 1. Currents shown were recorded 15 min after the whole cell seal was obtained with an electrode containing 5 mM sodium cyanide. (A) The two traces show currents recorded before (lower trace) and after exposure of the neuron to 0.5 μM TTX. Currents were generated by a voltage step to Vp = −30 mV from a holding potential of Vp = −100 mV. (B) The TTX-sensitive current isolated by subtracting the two traces in A. The transient current is truncated to show the persistent current in more detail. Scale bars apply for both A and B. The dashed line denotes the zero current level. From Hammarstrom and Gage.[43]

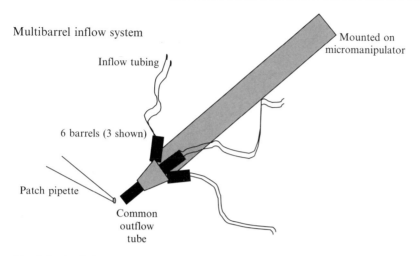

Fig. 2. Bath solution and drugs were applied through a multibarrel drug pipette consisting of six fine metal tubes (200 μm i.d.) feeding into one fine tube (200 μm i.d.). All barrels were gravity fed from 20-ml syringe tubes mounted on the Faraday cage wall. Up to five barrels were filled with varying drug solutions, whereas the sixth barrel was fed with bath solution only. The multibarrel inflow assembly was mounted on a micromanipulator, and the tip of the multibarrel drug pipette was navigated carefully under microscopic control as close as possible to the cell or patch being studied.

tube) and connected to the inflow perfusion system using gas-tight tubing.

O$_2$ electrode (Diamond Electro-Tech 760 Microelectrode, Diamond General Development Corp., Ann Arbor, MI, and Polarographic Amplifier Model 1900, A-M Systems, Inc., Everett, WA)

Procedure

1. Hypoxic solutions (prebubbled with N$_2$ for 10–20 min) or solutions containing drugs are applied through a multibarrel inflow system (Fig. 2) with a fine (200 μm i.d.) tube carefully positioned as close as possible to a cell or patch being studied (using microscope) without touching the patch pipette. This allows a rapid solution exchange close to the patch.

2. The O$_2$ tension in the bath near the outlet of the perfusion tube is reduced to $\leq$45 mm Hg within $\sim$3 min (Fig. 3). The O$_2$ tension at the very outlet of the perfusion tube, close to the patch of membrane, is probably lower than this and would reach equilibrium more quickly than detected by the substantially larger (diameter 0.7 mm) O$_2$ electrode. That is, the perfusion solution would not control the O$_2$ tension around the much

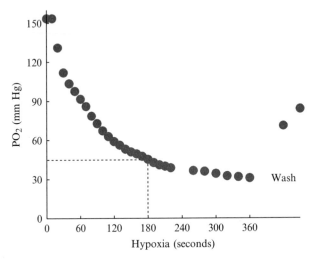

Fɪɢ. 3. Levels of O_2 recorded at various times after commencing the bubbling of the bath solution with 100% N_2. O_2 levels reached levels of less than 30% of normal air within 3 min, which corresponds well with the time of a marked increase in persistent Na^+ activity. From Hammarstrom and Gage.[43]

larger O_2 probe as well as it would at the much smaller tip of the patch pipette.

Native Channels Respond to Hypoxia

In order to determine whether the O_2 sensing by I_{NaP} required cytosolic second messengers, we exposed inside-out patches of membrane to hypoxia: a representative recording is shown in Fig. 4. Before hypoxia there was little channel activity but after 30 s of perfusion with hypoxic solution, channel activity had increased markedly and continued to increase over the next few minutes.

Recombinant Na^+ Channels Respond to Hypoxia

Whole cell Na^+ currents were recorded from a HEK293 cells transfected previously with the hH1 subunit of the cardiac Na^+ channel for 1–2 days. The amplitude of I_{NaP} was usually quite small under control conditions (Fig. 5). In 9 out of 12 HEK293 cells transfected with hH1 there was an obvious slowing of the decay of the Na^+ whole cell current after 3–4 min of hypoxia. During hypoxia, an increase in I_{NaP} was clearly evident near resting membrane potentials (~-60 mV) and the amplitude reached a

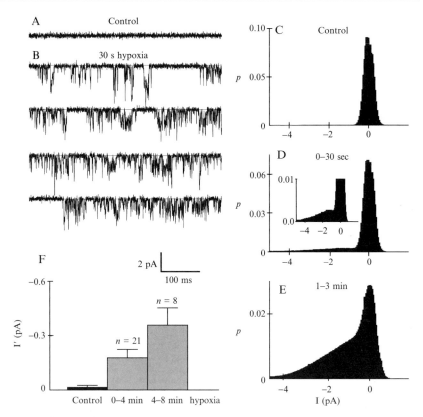

FIG. 4. (A and B) Representative current traces from a hippocampal patch held at Vp of 30 mV in control and after 30 s of hypoxia (noncontinuous traces). Dotted lines show the closed channel current levels. Before hypoxia there was little channel activity (top) but after 30 s of perfusion with hypoxic solution, channel activity had increased markedly and continued to increase over the next few minutes. (C–E) All-points histograms were calculated from full 0.5- to 3-min recordings sampled at 10 kHz obtained from another patch held at Vp = 30 mV, in control, after 0–30 s of hypoxia, and after 1–3 min of hypoxia. The control all-points histogram shows only a baseline peak, reflecting the paucity of I_{NaP} activity during normoxia. The following all-points histograms show progressively more channel activity with exposure to hypoxia (at 30 s and 1–3 min of hypoxia, respectively). (F) The histogram shows the average increase in the mean current during hypoxia measured in control patches ($n = 21$), after 0–4 min ($n = 21$), and 4–8 min ($n = 8$) of hypoxia. Vertical bars show ±1 SEM. From Hammarstrom and Gage.[43]

peak at about −40 mV. The current–voltage curve fitted to the increase in I_{NaP} had a half-maximum, $V_{0.5}$, similar to the control (−55 mV). The E_{Na} of 23 mV is similar to that of the control and is close to that estimated for the Na^+ concentrations used (17 mV). When the I_{NaP} and peak I_{NaT}

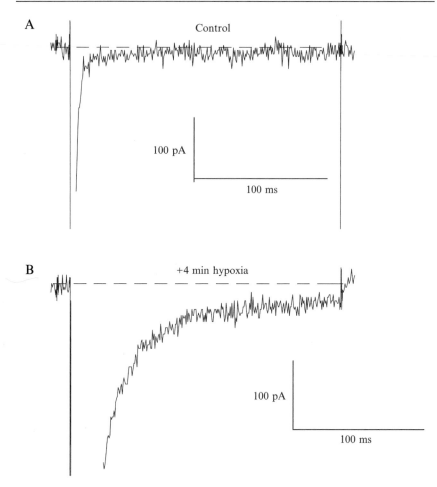

Fig. 5. The current traces shown were recorded from a HEK293 cell transfected with the hH1 subunit. The currents are TTX subtracted and generated by a voltage step to Vp = -40 mV following a 300-ms prepulse to -130 mV from a holding potential of Vp = -80 mV under control conditions (A) and after $\sim$3–4 min of hypoxia (B). Scale bars indicate 100 pA and 100 ms, respectively. Traces have been truncated to show I_{NaP} more clearly.

amplitudes were measured directly, we found that the I_{NaP}/I_{NaT} ratio increased nearly nine-fold from 0.1% in control to 1.04% after 3–5 min of hypoxia (0.0111 $\pm$ 0.0004, $n = 8$ in control compared with 0.0104 $\pm$ 0.0033 during hypoxia, $p = 0.01$). The increase in I_{NaP} during hypoxia was similar, if not larger, in the expression system than in native cells (compare

1.04% of the peak amplitude reported here to 0.62%[15]). I_{NaP} recorded during hypoxia was (like the control current) resistant to inactivation.

In some cells, the amplitude of the transient current tended to decrease during hypoxia but this attenuation was not statistically significant (I_{NaT} hypoxia/I_{NaT} normoxia 0.92 ± 0.28, $n = 8$, $p = 0.77$) similar to previous reports.[15,43]

The regulation of ion channels by O_2 tension was first described in chemoreceptive type I cells of the carotid body, where K^+ channels were inhibited by hypoxia.[45] I_{NaP} appears to be the only channel described so far that increases in activity during hypoxia.

How Do Channels Sense Changes in O_2 Levels?

The fact that only 9 out of the 12 HEK293 cells responded to hypoxia may support the suggestion that an auxiliary mechanism, and not the channel itself, is involved in O_2 sensing. This has also been observed and proposed by others.[19,24,46] Our work on native hippocampal neurons would suggest that the Na^+ channel itself does not sense O_2 levels directly, rather that an auxiliary protein contained within the plasma membrane is involved.[22] Various candidates for the O_2 sensor have been proposed and include NAD(P)H oxidase, Ca^{2+}, ATP, and the mitochondrial electron transfer chain (for a review, see Lopez-Barneo et al.[19]). At present we can only speculate as to the exact identity of the sensor. It is likely to be an endogenous protein contained within or anchored to the plasma membrane and common to most cell types, as it is expressed in HEK293 cells, rat ventricular myocytes, and rat hippocampal neurons.[22,23,27]

Role of I_{NaP} in Hypoxic Cell Damage and Arrhythmias

Because I_{NaP} is resistant to inactivation and is activated during hypoxia, it may well underlie the sustained Na^+ influx that eventually initiates Ca^{2+} influx via reverse mode action of the Na^+/Ca^{2+} exchanger.[1] Furthermore, because I_{NaP} activates near the resting membrane potential, it is probably important in setting the membrane potential and hence the firing rate in myocytes and neurons. The action potential plateau is dependent on a delicate balance between inward and outward currents, and I_{NaP} has a strong influence on action potential plateau duration, at least in ventricular myocytes.[10,47] Computer models suggest that small increases in I_{NaP} cause

[45] J. Lopez-Barneo, J. R. Lopez-Lopez, J. Urena, and C. Gonzalez, Science 241, 580 (1988).
[46] O. N. Osipenko, R. J. Tate, and A. M. Gurney, Circ. Res. 86, 534 (2000).
[47] D. Noble, A. Varghese, P. Kohl, and P. Noble, Can. J. Cardiol. 14, 123 (1998).

significant lengthening of the action potential duration[48] that may trigger lethal arrhythmias. I_{NaP} has been implicated in impaired repolarizations observed in human heart failure patients.[10] Interestingly, the Na^+ channel blocker TTX[10] abolished early after depolarizations in myocytes obtained from heart failure patients. Furthermore, enhanced I_{NaP} underlies arrhythmias in one form of the long QT syndrome.[49,50] The increase in I_{NaP} recorded from native channels[15,22,43] and hH1 recombinant channels[27] during hypoxia may well underlie some of the cell-damaging events (and some arrhythmias) that follow hypoxia. The Na^+ channel itself or the O_2 sensor may be prime anti-ischemic drug targets.

Acknowledgments

This project was supported by the National Heart Foundation and National Health and Medical Research Council of Australia. The authors thank Professor Lederer for the kind gift of the hH1 plasmid and John Curmi, Marina Philippa, and Andrea Everitt for their help and advice. Correspondence should be addressed to AKMH.

[48] B. F. Sakmann, A. J. Spindler, S. M. Bryant, K. W. Linz, and D. Noble, *Circ. Res.* **87**, 910 (2000).
[49] R. Lazzara, *Arch. Mal Coeur Vaiss.* **89**, 51 (1996).
[50] D. Wattanasirichaigoon and A. H. Beggs, *Curr. Opin. Pediatr.* **10**, 628 (1998).

[20] Analysis of Oxygen-Sensitive Human Cardiac L-Type Ca^{2+} Channel α_{1C} Subunit (hHT Isoform)

By Jason L. Scragg, Ian M. Fearon, Stephen G. Ball, Arnold Schwartz, Gyula Varadi, and Chris Peers

Introduction

Rapid and direct reversible hypoxic inhibition of native L-type Ca^{2+} channels has been demonstrated in a variety of cells, including carotid body chemoreceptor cells[1] and vascular (both systemic and pulmonary) smooth muscle cells.[2–4] This observation is of particular physiological importance because it can account for, or at least contribute strongly to, hypoxic

[1] R. J. Montoro, J. Urena, R. Fernandez-Chacon, G. A. De Toledo, and J. López-Barneo, *J. Gen. Physiol.* **107**, 133 (1996).
[2] A. Franco-Obregon, J. Urena, and J. López-Barneo, *Proc. Natl. Acad. Sci. USA* **92**, 4715 (1995).

vasodilation, a systemic vascular response to reduced oxygen (O_2) levels, which allows greater blood flow (and hence increased O_2 delivery) to hypoxic tissues.

Hypoxic inhibition of a recombinant human cardiac L-type Ca^{2+} channel α_{1C} subunit (the hHT splice variant) was first demonstrated by our group in Leeds.[5] Inhibition was observed following stable α_{1C} expression in HEK293 cells, in the absence of auxiliary subunits, and the effects of hypoxia appeared to mimic perfectly the effects of hypoxia on native L-type Ca^{2+} channel activity in smooth muscle cells. Inhibition was reversible, voltage dependent, and appeared to be associated with a slowing of activation kinetics while current inactivation and deactivation were unaffected by hypoxia. O_2 sensing by vascular L-type Ca^{2+} channels is therefore likely to be a property of the pore-forming α_{1C} subunit and does not require modulatory, auxiliary subunits.

In support of observations by Chiamvimonvat et al.,[6] Fearon et al.[7] also confirmed that the recombinant α_{1C} subunit hHT splice variant possessed multiple redox-sensitive sites ("free" sulphydryl groups that may be sensitive to the oxidation state of the cell). Inhibition of Ca^{2+} channel currents was caused by two oxidizing agents (PCMBS and MTSEA) that exert their effects at clearly distinct points on the channel protein. Suppression of channel activity by PCMBS, but not MTSEA, abolished the responses of these channels to hypoxia.[7] These findings support a redox-based mechanism for O_2 sensing by L-type Ca^{2+} channels.

To date, only one study has provided strong evidence for the structural requirements for O_2 sensing by a Ca^{2+} channel. Three C-terminal splice variants of a human cardiac L-type Ca^{2+} channel α_1-subunit (hHT, rHT, and fHT) have been reported.[8] The hHT splice variant, found in the human heart and, importantly, in vascular smooth muscle (where it is more dominant), carries a 71 amino acid insert (amino acids 1787–1857) within its intracellular C-tail region[8] and was the variant used to study hypoxic inhibition as described previously. In contrast, the rHT splice variant, lacking

[3] A. Franco-Obregon and J. López-Barneo, Am. J. Physiol. 271, H2290 (1996).

[4] A. Franco-Obregon and J. López-Barneo, J. Physiol. 491, 511 (1996).

[5] I. M. Fearon, A. C. V. Palmer, A. J. Balmforth, S. G. Ball, G. Mikala, A. Schwartz, and C. Peers, J. Physiol. 500, 551 (1997).

[6] N. Chiamvimonvat, B. O'Rourke, T. J. Kamp, R. G. Kallen, F. Hofmann, V. Flockerzi, and E. Marban, Circ. Res. 76, 325 (1995).

[7] I. M. Fearon, A. C. V. Palmer, A. J. Balmforth, S. G. Ball, G. Varadi, and C. Peers, J. Physiol. 514, 629 (1999).

[8] U. Klöckner, G. Mikala, J. Eisfeld, D. E. Iles, M. Strobeck, J. L. Mershon, A. Schwartz, and G. Varadi, Am. J. Physiol. 272, H1372 (1997).

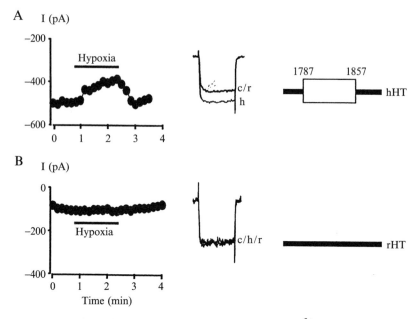

FIG. 1. Hypoxic sensitivity of recombinant human L-type Ca^{2+} channel α_{1C} subunits. (A) Time-series plot of Ba^{2+} currents in a HEK293 cell transiently expressing the hHT isoform of the α_{1C} subunit before, during, and after exposure to hypoxic perfusate. Each point shows the peak current amplitude evoked by successive step depolarizations to 10 mV (100 ms, 0.1 Hz) from a holding potential of −80 mV. Period of exposure to hypoxia (PO_2, 20 mm Hg) is indicated by the horizontal bar. Superimposed examples of currents from the same recording obtained under control conditions (c), during exposure to hypoxia (h), and after recovery (r). (B) Same as in A except that recordings were made in cells expressing the rHT isoform of the α_{1C} subunit. The right-hand side of both A and B shows the portion of the C-terminal region of the α_{1C} subunit present in the isoform examined.

this 71 amino acid insert (Fig. 1), was not O_2 sensitive when expressed transiently in HEK293 cells.[9] Loss- and gain-of-function analysis [swapping analogous regions of hHT and rHT C-tails, thereby generating hHT(−) and rHT(+) clones, respectively] confirmed O_2 sensing to be the property of the 71 amino acid splice insert in the α_{1C} subunit.[9]

Deletion mutational analysis confirmed that the proximal 36 amino acid residues (amino acids 1787–1822) of the 71 amino acid splice insert were required for O_2 sensing (Fig. 2), whereas the distal 23 amino acid residues (amino acids 1833–1855) were not.[9] These data suggested that specific

[9] I. M. Fearon, I. Isaacsohn, S. Koch, G. Varadi, S. G. Ball, and C. Peers, *Circ. Res.* **87**, 537 (2000).

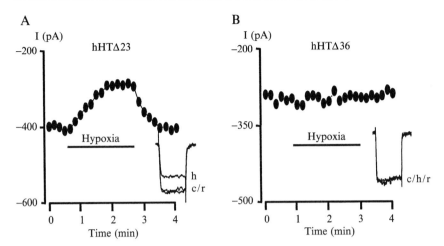

FIG. 2. Hypoxic sensitivity of mutant human L-type Ca^{2+} channel α_{1C} subunits. (A) Time-series plot (as in Fig. 1) of Ba^{2+} currents in a HEK293 cell transiently expressing the hHTΔ23 mutant α_{1C} subunit before, during, and after exposure to hypoxic perfusate. The period of exposure to hypoxia (PO$_2$, 20 mm Hg) is indicated by the horizontal bar. (Inset), Superimposed examples of currents from the same recording obtained before (c), during (h), and after recovery (r) from hypoxia. (B) Same as in A except that recordings were made in cells expressing hHTΔ36 mutant α_{1C} subunits.

stretches or residues, alone or together in the proximal 36 amino acid region, can account for the structural requirements for O$_2$ sensing by L-type Ca^{2+} channels.

Although numerous cellular responses to hypoxia are now well established, the actual mechanism(s) of O$_2$ sensing and the interaction of O$_2$ with ion channels (or closely associated molecules) remain poorly understood. Currently, three possible O$_2$ sensing mechanisms employed by ion channels are suggested to exist; (1) the ion channel itself acts as the O$_2$ sensor; (2) the ion channel is regulated by diffusible products of an O$_2$ sensor, that is, modulation by redox couples or by reactive oxygen species (ROS); and (3) the ion channel associates with an O$_2$ sensing protein that is capable of transducing its redox potential to the channel. All three hypotheses for O$_2$ sensing have received support, and there is theoretical overlap among these hypotheses.

This article (1) discusses the techniques used to generate mutant L-type Ca^{2+} channels and to record their responses to hypoxia following expression in HEK293 cells and (2) aims to serve as a methodological basis for the design of similar experiments aimed at determining the structural requirements for O$_2$ sensing by other ion channels.

Experimental Methodology: Construction of Mutant L-Type
Ca^{2+} Channels

Swapping C-Tail Segments Between hHT and rHT cDNA Clones

Both hHT and rHT cDNAs are truncated at the 5' and 3' ends in order
to remove most of the nontranslated sequences, while their Kozak's consensus sequence is completely retained. The truncated versions are inserted
into pBluescript II KS(−) between the *Hind*III and the *Xba*I restriction
sites. Plasmid DNAs from both cDNAs are transformed in SCS110 cells
(*dam⁻, dcm⁻*; from Stratagene, La Jolla, CA) and cleaved with *Bcl*I and
*Sal*I, and the appropriate 1.8- and 2.0 kb-fragments are isolated from rHT
and hHT, respectively. These C-tail cassettes are exchanged between
the two types of constructs, thus creating the hHT(−) clone in which the
213-bp C-tail insert is removed and the rHT(+) clone into which
the 213-bp segment is inserted. Constructs are transformed in DH5α-
competent cells, and the presence and absence of the insert are verified
in both clones by sequencing. Finally, the full-length cDNAs are liberated
by cleavage with *Hind*III and *Xba*I and are ligated into the corresponding
sites on pHOOK-2 (Invitrogen, San Diego, CA).

*hHTΔ36: Deletion of the Proximal 36 Amino Acids of the 71 Amino
Acid C-Tail Insertion (Amino Acids 1787 to 1822)*

Deletion of this segment is performed by introducing two unique *Nru*I
sites flanking the desired stretch, cutting with *Nru*I, and then religating the
hHT cDNA. Introducing the unique *Nru*I restriction sites is achieved using
the MegaPrimer polymerase chain reaction (PCR) strategy, with the
pBluescript II KS(−)/α$_{1C}$ (hHT isoform) construct as a template.[9] To introduce the upstream *Nru*I site (nucleotide 5353), a specific forward mutagenic primer (nucleotides 5335–5370) is used that carries the mutated
region encoding the *Nru*I site together with a reverse primer (nucleotides
6063–6046). The 728-bp PCR product is then gel isolated and utilized as
a reverse primer (termed "MegaPrimer") in the second PCR reaction together with the forward primer (nucleotides 4270–4287), again with the
pBluescript II KS(−)/α$_{1C}$ (hHT isoform) construct as a template.

The resulting 1.8-kb amplification product (nucleotides 4270–6063) is
subcloned into the *Eco*RV site of pBluescript II KS(−) and transformed
in SCS110 cells, and the presence of the mutation is verified by restriction
analysis and sequencing of the purified plasmid construct. To introduce
the second downstream *Nru*I site (nucleotide 5466), this latter mutant
clone is used as a template in the second set of PCR reactions. To generate
the second MegaPrimer, another specific forward mutagenic primer

(nucleotides 5448–5484) is used that carries the mutated region encoding the *Nru*I site together with a reverse primer (nucleotides 6063–6046). Similarly, this 614-bp PCR product (MegaPrimer) is gel isolated and utilized as a reverse primer in the second PCR reaction together with the forward primer (nucleotides 4270–4287), again with the mutant clone as the template. This 1.8-kb PCR product (nucleotides 4167–6063) is then gel isolated, subcloned into the *Eco*RV site of pBluescript II KS(−), and transformed in SCS110 cells, and the presence of both *Nru*I sites is confirmed by sequencing and restriction analysis following plasmid purification. The desired stretch is deleted following restriction with *Nru*I and the DNA religated.

hHTΔ23: Deletion of the Distal 23 Amino Acids of the 71 Amino Acid C-Tail Insertion (Amino Acids 1833 to 1855)

This deletion is performed by inserting an *Nde*I site at nucleotide 5558 and cleaving with *Msc*I (nucleotide 5496) and *Nde*I, followed by mung bean nuclease treatment (removal of 3′ and 5′ single-stranded extensions to leave blunt ends) and religation. In the first step, the *Nde*I site is introduced by the MegaPrimer PCR strategy as described previously. Again, with the pBluescript II KS(−)/α_{1C} (hHT isoform) construct as a template, a specific forward mutagenic primer (nucleotides 5545–5580) is used that carries the mutated region encoding the *Nde*I site together with a reverse primer (nucleotides 6063–6046). The 518-bp PCR product is then gel isolated and utilized as a reverse primer (termed "MegaPrimer") in the second PCR reaction together with the forward primer (nucleotides 4270–4287), again with the pBluescript II KS(−)/α_{1C} (hHT isoform) construct as a template. The resulting 1.8-kb amplification product (nucleotides 4270–6063) is subcloned into the *Eco*RV site of pBluescript II KS(−) and transformed in SCS110 cells, and the presence of the mutation is verified by restriction analysis and sequencing of the purified plasmid construct. Plasmid DNA (30 μg) is then digested with *Nde*I and treated with 300 units of mung bean nuclease (Stratagene) for 15–30 min in a buffer supplied by the manufacturer. This reaction mixture is then phenol/chloroform extracted, and an additional cleavage is performed with *Msc*I. A 4.8-kb fragment is gel isolated, religated, and transformed in DH5α cells. Plasmid DNAs are sequenced across the junction region to identify positive clones that contain the desired deletion and an in-frame religation.

DNAs from all positive clones (i.e., those that carried the deletion between either the two *Nru*I sites or the *Nde*I and *Msc*I sites) are then transformed in SCS110 cells, purified (from 5-ml overnight cultures), and cleaved with *Bcl*I (nucleotide 4310) and *Avr*II (nucleotide 5971). The liberated

1.7-kb fragments are gel isolated and used to replace the analogous stretch in the wild-type α_{1C} (hHT) cDNA. The entire 1.7-kb DNA fragment and both *Bcl*I and *Avr*II junctions are then sequenced completely for each construct made. The full-length Ca^{2+} channel cDNA is then liberated by cleavage with *Hind*III and *Xba*I and ligated into the corresponding sites of the pHOOK-2 mammalian expression vector (Invitrogen). These final constructs are transformed in DH5α cells, purified on a large scale from a 100-ml overnight culture, and analyzed by restriction and spectrophotometry.

Note. The reader is referred to Sambrook *et al.*[10] for details on the methodology for enzyme restriction and analysis on TAE or TBE agarose gels. All restriction enzymes and DNA molecular weight markers are obtained from New England Biolabs (Herts., England, UK) or Gibco Life Technologies (Paisley, Scotland, UK). Additional materials required include microcentrifuge tubes (1.5 ml), a water bath, DH5α- and SCS110-competent cells, and electrophoresis equipment.

Experimental Techniques

MegaPrimer PCR (Generation of Mutagenic MegaPrimer and Incorporation of the Mutation(s) into Double-Stranded cDNA)

MegaPrimer PCR is ideal as a site-directed mutagenesis strategy because it uses double-stranded DNA as the template and therefore avoids both the need for subcloning into m13-based bacteriophage vectors and single-stranded DNA rescue. When generating the MegaPrimer, good primer design is essential. In our hands, the forward mutagenic primers were typically 36 nucleotides in length (28–36 is optimal) with the mutation near the middle of the primer, whereas the general forward and reverse PCR primers were 18 nucleotides in length. All primers were designed with G or C overhangs at either or both 5' and 3' ends, a G-C content >50% but <65–70% if possible, and little opportunity for primer–dimer formation.

MegaPrimer generation by PCR is a simple three-segment protocol when using *Pfu* DNA polymerase (Table I; a "hot start" at 95° for 5 min is needed to activate the polymerase). The annealing temperature is dependent on both the size and the nucleotide composition of the primer; we found 58° to be the most efficient for MegaPrimer synthesis (Table I). Extension time is dependent on both the polymerase used and the size of the desired product (*Pfu* requires an extension time of 1–2 min/kb of

[10] J. Sambrook, E. F. Fritsch, and T. Maniatis, "Molecular Cloning: A Laboratory Manual." Cold Spring Harbor Laboratory Press, Cold Spring Harbor, NY, 1989.

TABLE I
SUGGESTED CYCLING PARAMETERS FOR MEGAPRIMER PCR USING *Pfu* DNA POLYMERASE

Segment	Number of cycles	Temperature	Duration
1	1	95°	5 min
2	30	58° (annealing)	45 s
		72° (extension)	2 min
		95° (denaturation)	45 s
3	1	58°	2 min
		72°	6 min

template amplified at 72° to achieve maximum synthesis). PCR product amplification (≤1 kb in length) is analyzed routinely on 3% ethidium bromide-stained sieving agarose gels (sieving agarose, Sigma, Poole, Dorset, England, UK).

As a rule, the upstream mutation must be introduced into the cDNA prior to the downstream mutation when utilizing the MegaPrimer PCR strategy. If possible, MegaPrimers should be designed so that they do not overlap previously introduced mutagenic sites. Introduction of the mutation(s) into cDNA with this strategy can also be achieved using *Pfu* DNA polymerase and a similar protocol (extension time may have to be extended when generating a larger product). *Pfu* DNA polymerase (Stratagene), a proofreading DNA polymerase isolated from *Pyrococcus furiosus,* was the polymerase of choice for this site-directed mutagenesis strategy because it exhibits the lowest error rate of any commercially available thermostable DNA polymerase. As an alternative, Vent$_R$ DNA polymerase (extension time of 2 kb/min, New England Biolabs) can be used with large amplification products, but it is less efficient than *Pfu* in its proofreading capability, but is still better than *Taq* DNA polymerase. However, if the product is GC rich, we have found that Proofstart DNA polymerase (a proofreading DNA polymerase from Qiagen, Crawley, England, UK) is very successful (two-segment protocol outlined in the product handbook is sufficient). Constituents of the PCR reaction, and their relative amounts, are added to the reaction mixture in the order suggested by the manufacturer of the polymerase used. However, as with all PCR-based amplifications, optimal conditions are dependent on the quality of the template, primers, dNTPs, DNA polymerase, and thermocycler used and must therefore be determined at the start of the strategy. Additional materials required include dNTPs, 0.5-ml microcentrifuge or 0.2-ml PCR tubes, sterile water, and mineral oil.

Choosing a Plasmid Shuttle Vector for Subcloning PCR Products

pBluescript II KS(−) as a cloning vector was chosen for a number of reasons, including the variable yet unique 21 restriction enzyme recognition sites within its polylinker, which itself interrupts the β-galactosidase coding sequence (*lacZ* gene), enabling blue–white color selection of transformed bacteria. The *LacZ* gene product is responsible for the production of blue colonies when appropriate *Escherichia coli* transformants are grown on media containing X-Gal (a chromogenic substrate for β-galactosidase).

Colonies possessing β-galactosidase activity, produced by pBluescript II KS(−) phagemids (plasmids with a phage origin) having no inserts in their polylinker, appear pale blue at their periphery and dense blue in the center. However, pBluescript II KS(−) phagemids that have inserts in the polylinker produce white colonies (occasionally have a faint blue center but white at the periphery) due to disruption of the coding region of the *lacZ* gene fragment. As an alternative, pGEM plasmids (Invitrogen) are also useful cloning vectors because they too can be used for blue–white screening of cells and have an extensive polylinker.

Transformation of Plasmid DNA in E. coli *(DH5α and SCS110 Cells)*

DH5α-competent cells (Invitrogen) are the cells of choice for straight-forward propagation in bacteria, allowing very high transformation efficiency.

Quick thaw DH5α-competent cells (30 s in a 37° water bath). Put tube on ice and aliquot 50 μl of cells per transformation into a separate 1.5-ml microcentrifuge tube already prechilled on ice. Add the desired volume of plasmid DNA (2 μl ligation mixture or 50–100 ng purified plasmid DNA) to the 50 μl of cells. Mix gently and let stand on ice for 30–45 min. Heat shock cells for 2 min at 42° (water bath). Place cells back on ice for 2–5 min. Add 500 μl of 2YT medium per tube. Incubate at 37° (water bath) for 30 minutes. Spin down cells (13,000 rpm for 20 s). Remove the supernatant gently and resuspend cells in 200 μl fresh 2YT medium. Dilute the sample as desired (1:4, 1:40, and 1:400 in 200 μl 2YT), add to the center of antibiotic-containing 2YT/agar plates, and distribute the solution evenly across the entire surface of the plate using a flame-sterilized spreader. Incubate at 37° for 12–16 h overnight.

Note. The reader is referred to Sambrook *et al.*[10] for details on the methodology for making 2YT/agar plates. Materials required include Bacto-tryptone, Bacto-yeast extract, NaCl, Agar, antibiotic, 37° oven, water bath (37° and 42°), and 100-mm Petri dishes.

Transformation of SCS110 cells (Stratagene) is carried out according to manufacturer's instructions. SCS110 cells are deficient in two methylases *(dam* and *dcm)* found in most strains of *E. coli. Dam⁻* and *Dcm⁻* strains are used to eliminate specific methylation patterns on DNA so it can be cleaved by one or more methylation-sensitive restriction enzymes. *Nru*I and *Bcl*I are both *dam* sensitive, whereas *Msc*I is *dcm* sensitive. Additional materials required to transform SCS110 cells include $MgCl_2$, $MgSO_4$, and glucose; 2-mercaptoethanol is supplied with the cells.

Blue–White Colony Screening (Colorimetric Assay)

Thirty minutes before plating transformed bacteria, add 60 μl X-Gal (20 mg/ml prepared in dimethylformamide, stored at $-20°$) to the center of a plate and, using a flame-sterilized spreader, evenly distribute the solution across the entire surface of the plate. Invert the plate, place in a 37° oven for drying, and then plate the transformed bacteria. The next day, incubate the plates at 4° for 6–8 h or until you are ready to culture white (positive for insert) colonies. Storage at 4° allows the blue color to develop fully, thus enabling better identification of the desired white clones. Materials required include dimethylformamide and X-Gal (5-bromo-4-chloro-3-indolyl β-D-galactopyranoside).

Purification and Concentration of DNA

Purification of DNA (PCR products, plasmids, etc.) from standard or low-melt agarose gels in TAE or TBE buffer can be achieved using either QIAquick or QIAEX II gel extraction kits (Qiagen, all buffers are supplied and a microcentrifuge is required). These procedures (followed according to manufacturer's instructions) are capable of extracting and purifying DNA from 70 to 10 kb in length.

For small-scale preparations, up to 20 μg plasmid DNA from 1.5 ml of a 5-ml overnight 2YT/bacterial culture, use either Wizard Plus miniprep (Promega, Madison, WI, requires a vacuum manifold and microcentrifuge) or QIAprep spin miniprep (Qiagen, microcentrifuge only) DNA purification kits. For large-scale preparations, up to 500 μg from a 100-ml overnight 2YT/bacterial culture, use either Endofree maxi-prep kit (Qiagen) or Wizard Plus midiprep (Promega) DNA purification kits. As a rule, following amplification and purification, always quantify a plasmid DNA concentration using spectrophotometric analysis (A_{260}:A_{280} will also give an indication of its purity) and test its integrity on an ethidium bromide-stained 1% agarose gel following enzyme restriction.

Extraction of DNA with Phenol:Chloroform

Occasionally it is necessary to inactivate or remove enzymes from DNA solutions following one stage of a cloning procedure before proceeding to the next, but without actually purifying the DNA in between. In such a case, extraction first with phenol:chloroform and then with chloroform is the method of choice.

This protocol is designed for a 100-μl nucleic acid (DNA or RNA) reaction volume. First, in a 1.5-ml microcentrifuge tube, add 100 μl of phenol and 100 μl of chloroform to the DNA sample. *Note.* The DNA will tend to partition into the organic phase if the phenol has not been equilibrated adequately to a pH of 7.8–8.0. Vortex the mixture for 5–20 s until an emulsion has formed and then centrifuge at 13,000 rpm for 1 min. Centrifuge again at a higher speed, or for longer, if the organic and aqueous phases are not well separated. Normally, the aqueous phase forms the upper phase. However, if the aqueous phase is dense because of salt (>0.5 M) or sucrose ($>10\%$), it will form the lower phase. The organic phase is easily identifiable because of the yellow color contributed by the hydroxyquinoline that is added to phenol during equilibration. Avoiding the interface, carefully transfer the aqueous phase (DNA solution) to a fresh 1.5-ml microcentrifuge tube and discard both the interface and the organic phase. Although the organic phase and interface can be "back extracted" to achieve the best recovery, we have found this to be of little value. Next, add 100 μl of chloroform to the aqueous phase. Vortex and centrifuge at 13,000 rpm for 1 min. Again, avoiding the interface, transfer the aqueous phase into a clean 1.5-ml microcentrifuge tube with a pipette. Discard the interface and the organic phase. Finally, recover the DNA by precipitation with ethanol.

Precipitation of DNA with Ethanol

Precipitation with ethanol is the most widely used method for concentrating nucleic acids. This protocol is designed for a 100-μl nucleic acid (DNA or RNA) reaction volume. First, adjust the concentration of monovalent cations by adding 10 μl 3 M NaOAc (pH 5.4) to the DNA solution. Mix the solution gently until it turns milky. Add 250 μl of ice-cold ethanol (stored at $-20°$), mix well by vortexing, and store this ethanolic solution on dry ice for 30–45 min to allow the precipitate of DNA to form. DNA can be stored indefinitely in ethanolic solutions at $0°$ or $-20°$. Recover the DNA by centrifugation at 13,000 rpm for 10 min. More extensive centrifugation may be required if low concentrations of DNA (<20 ng/ml) or very small fragments are being processed. Remembering on which side of the microcentrifuge tube the DNA pellet will form, remove the supernatant carefully from the opposite side with a pipette. Taking care not disturb the pellet (which may be invisible), use a pipette tip to remove any drops of fluid that may have adhered to the walls of the tube. It is advisable to set aside the supernatant from valuable DNA samples until recovery of the precipitated DNA has been verified. Next, add 250 μl of ice-cold 70% ethanol (made using −20° absolute ethanol), vortex gently, and centrifuge at 13,000 rpm for 10 min. Remove the supernatant carefully, again avoiding the DNA pellet. Centrifuge for 5 s and remove all traces of ethanol with a pipette tip. Store the open microcentrifuge tube on the bench at room temperature and allow to air dry. Alternatively, use a Speed-Vac for 5 min. Finally, dissolve the DNA pellet (which is often invisible) in the desired volume of H$_2$O and rinse the walls of the tube well with the solution. Quantify the DNA concentration using both spectrophotometric and ethidium bromide fluorescence analysis on an agarose gel. Ethanol precipitation of plasmid DNA is recommended following the purification of plasmid DNA with Wizard Plus miniprep and midiprep purification kits when using ABI PRISM BigDye sequencing technology (Perkin Elmer, Cheshire, England, UK).

Transfection and Culture of HEK293 Cells

In these experiments, wild-type HEK293 cells (ECACC No. 85120602, European Collection of Animal Cell Cultures, Porton Down, England, UK) are transfected transiently with pHOOK-2/α$_{1C}$ mutant constructs using the Superfect transfection agent (Qiagen).[9] The following day, successfully transfected cells are selected by magnetic affinity cell sorting[11] and cultured on glass coverslips (at 37° in a humidified atmosphere of 95% air/5% CO$_2$) for use in electrophysiological studies 48–72 h following transfection. All tissue culture procedures are carried out in a Class II microbiological laminar flow cabinet (Medical Air Technology Ltd, Manchester, England, UK) using sterile techniques.

Unfortunately, since the publication of Fearon *et al.*,[9] the supply of pHOOK vectors and beads required for magnetic cell sorting has been suspended. Possible alternatives include, following subcloning of the L-type Ca^{2+} channel cDNA into a different mammalian expression vector, either transient cotransfection with a GFP reporter plasmid (which will require a fluorescent microscope connected to the electrophysiological equipment) or generation of stable cell lines following selection with the appropriate antibiotic.

---

[11] J. D. Chesnut, A. R. Baytan, M. Russell, M. P. Chang, A. Bernard, I. H. Maxwell, and J. P. Hoeffler, *J. Immunol. Methods* **193**, 17 (1996).

Whole Cell Patch-Clamp Recording

Forty-eight to 72 h following transfection, with coverslip fragments with attached, transfected cells are perfused continually (4 ml/min, bath volume 80 μl) with a solution containing (mM) NaCl, 95; CsCl, 5; MgCl$_2$, 0.6; BaCl$_2$, 20; HEPES, 5; D-glucose, 10; and TEA–Cl 20 (pH 7.4, 21–24°). Whole cell patch-clamp recordings are made using pipettes (resistance 4 to 7 MΩ) filled with (mM) CsCl, 120; TEA–Cl, 20; MgCl$_2$, 2; EGTA, 10; HEPES, 10; and ATP, 2 (pH 7.2 with CsOH). These solutions are used routinely to study Ca^{2+} currents in isolation, using a high concentration of charge carrier (20 mM Ba^{2+}) and substitute ions to minimize contaminating current flow through other channels, particularly K$^+$ channels.

Cells are voltage clamped at -80 mV, and whole cell currents are evoked by 100-ms step depolarizations to various test potentials (0.1 Hz). Series resistance compensation of 70–90% is applied. Currents are filtered at 5 kHz and digitized at 10 kHz. Capacitative transients are minimized by analogue means, and corrections for leak current are made by scaling and subtraction of the average leak current evoked by small hyperpolarizing and depolarizing steps ($\leq$20 mV). Current amplitudes are measured over the last 10–15 ms of each step of depolarization; Ba^{2+} is used as the charge carrier as it displays no inactivation during step depolarizations. Analysis and voltage protocols are performed with the use of an Axopatch 200A amplifier/Digidata 1200 interface (Clampex software, pCLAMP 6.0.3, Axon Instruments Inc., Foster City, CA).

Generating a Hypoxic Environment

Bath hypoxia is achieved by bubbling the reservoir leading to the bath with 100% N$_2$. The reservoir and bath are connected by gas-impermeable Tygon tubing (BDH Laboratory Supplies, Poole, England, UK). The time course of the fall in PO$_2$ in the bath is highly reproducible and always stable within 30 to 60 s of switching from a normoxic (150 mm Hg) to a hypoxic (20 mm Hg) solution. The level of bath hypoxia is measured using a mercury-filled carbon fiber microelectrode polarized to -800 mV and switching from a normoxic environment to a hypoxic one and then back again. The resulting currents are again recorded using an Axopatch 200A amplifier. Acquisition is performed using a Digidata 1200 interface and Fetchex from the pCLAMP 6.0.3 suite (Axon Instruments Inc.).

Section IV

Tumor Biology

[21]   Analysis of von Hippel–Lindau Tumor Suppressor as a Mediator of Cellular Oxygen Sensing

By NORMA MASSON *and* PETER J. RATCLIFFE

Introduction

The hypoxia inducible factor (HIF)[1] controls an extensive transcriptional cascade that plays an important role in a broad array of cellular and systemic responses to hypoxia.[2,3] Recent work has demonstrated that regulation of the HIF α/β heterodimeric DNA-binding complex involves posttranslation hydroxylation of specific amino acid residues that govern the stability and activity of the α subunit.[4–8] These hydroxylations are performed by a series of iron and 2-oxoglutarate dependent oxygenases that have an absolute requirement for dioxygen as cosubstrate, thus providing a direct link between the availability of molecular oxygen and the activity of the HIF transcriptional response. To date, three enzymes (prolyl hydroxylase domain, PHD 1, 2, and 3) have been shown to have HIF-α prolyl hydroxylase activity,[9,10] and one enzyme (factor inhibiting HIF, FIH[11]) has been shown to have HIF-α asparaginyl hydroxylase activity.[12,13]

---

[1] G. L. Wang, B.-H. Jiang, E. A. Rue, and G. L. Semenza, *Proc. Natl. Acad. Sci. USA* **92**, 5510 (1995).

[2] G. L. Semenza, *Genes Dev.* **14**, 1983 (2000).

[3] R. H. Wenger, *FASEB J.* **16**, 1151 (2002).

[4] M. Ivan, K. Kondo, H. Yang, W. Kim, J. Valiando, M. Ohh, A. Salic, J. M. Asara, W. S. Lane, and W. G. J. Kaelin, *Science* **292**, 464 (2001).

[5] P. Jaakkola, D. R. Mole, Y.-M. Tian, M. I. Wilson, J. Gielbert, S. J. Gaskell, A. von Kriegsheim, H. F. Hebestreit, M. Mukherji, C. J. Schofield, P. H. Maxwell, C. W. Pugh, and P. J. Ratcliffe, *Science* **292**, 468 (2001).

[6] N. Masson, C. Willam, P. H. Maxwell, C. W. Pugh, and P. J. Ratcliffe, *EMBO J.* **20**, 5197 (2001).

[7] F. Yu, S. B. White, Q. Zhao, and F. S. Lee, *Proc. Natl. Acad. Sci. USA* **98**, 9630 (2001).

[8] D. Lando, D. J. Peet, D. A. Whelan, J. J. Gorman, and M. L. Whitelaw, *Science* **295**, 858 (2002).

[9] A. C. R. Epstein, J. M. Gleadle, L. A. McNeill, K. S. Hewitson, J. O'Rourke, D. R. Mole, M. Mukherji, E. Metzen, M. I. Wilson, A. Dhanda, Y.-M. Tian, N. Masson, D. L. Hamilton, P. Jaakkola, R. Barstead, J. Hodgkin, P. H. Maxwell, C. W. Pugh, C. J. Schofield, and P. J. Ratcliffe, *Cell* **107**, 43 (2001).

[10] R. K. Bruick and S. L. McKnight, *Science* **294**, 1337 (2001).

[11] P. C. Mahon, K. Hirota, and G. L. Semenza, *Genes Dev.* **15**, 2675 (2001).

[12] D. Lando, D. J. Peet, J. J. Gorman, D. A. Whelan, M. L. Whitelaw, and R. K. Bruick, *Genes Dev.* **16**, 1466 (2002).

[13] K. S. Hewitson, L. A. McNeill, R. M. V., Y.-M. Tian, A. N. Bullock, R. W. Welford, J. M. Elkins, N. J. Oldham, S. Bhattacharya, J. M. Gleadle, P. J. Ratcliffe, C. W. Pugh, and C. J. Schofield, *J. Biol. Chem.* **277**, 26351 (2002).

This article focuses on the regulation of HIF-α stability. Three isoforms of the HIF-α subunit have been identified (HIF-1α, HIF-2α, and HIF-3α). All are subject to VHL-dependent proteolysis following the hydroxylation of specific prolyl residues. For HIF-1α and HIF-2α, prolyl hydroxylation occurs at two sites (P402 and P564 in human HIF-1α) that reside within a central oxygen-dependent degradation domain (ODDD). HIF-3α has been less completely studied but at least one site of prolyl hydroxylation has been defined. Hydroxylation at these sites regulates interaction with the von Hippel–Lindau ubiquitin (VHL) E3 ligase.[4,5,7] The interaction appears similar for all three isoforms, and the assays described here can be used for HIF-1α, HIF-2α, or HIF-3α, henceforth collectively termed HIF-α. VHL E3 is itself a multicomponent ubiquitin ligase that consists of the VHL tumor suppressor protein (pVHL) complexed to elongins B and C, Cul2, and Rbx-1 in a manner analagous to Skp/Cullin/F-box protein interactions in the SCF class of ubiquitin ligases.[14] A single hydroxyproline-binding site on the β domain of pVHL has been defined that makes direct contact with HIF-α.[15,16] Each of the two prolyl hydroxylation sites in HIF-α can independently mediate interaction and VHL-dependent degradation of HIF-α by the ubiquitin/proteasome pathway. We have termed these regions the NODDD and CODDD (N-terminal and C-terminal oxygen-dependent degradation domains, respectively). This article describes methods for assaying these interactions between HIF-α and VHL E3, methods for assaying HIF-α PHD activity by VHL capture, and methods for assaying VHL-dependent ubiquitylation of HIF-α.

Preparation of Cell Extracts with Competent HIF-α Prolyl Hydroxylase/VHL E3 Ligase Activity

Our laboratory uses the following procedure for the preparation of cell extracts, which can be used either as a crude source of HIF-α-PHD activity to modify the substrate *in vitro* or as a source of VHL E3 ligase activity to ubiquitylate the substrate *in vitro*. We have prepared extracts successfully from a wide range of mammalian cell lines, including RCC4, 786-0, Cos-7, HeLa, Hep3B, and U2OS. For both applications, the extract can be prepared in bulk and frozen in aliquots at $-70°$, providing a convenient assay reagent.

[14] W. G. Kaelin, *Nature Rev. Cancer* **2,** 673 (2002).
[15] W. C. Hon, M. I. Wilson, K. Harlos, T. D. Claridge, C. J. Schofield, C. W. Pugh, P. H. Maxwell, P. J. Ratcliffe, D. I. Stuart, and E. Y. Jones, *Nature* **417,** 975 (2002).
[16] J.-H. Min, H. Yang, M. Ivan, F. Gertler, W. G. J. Kaelin, and N. P. Pavletich, *Science* **296,** 1886 (2002).

For bulk preparation, we routinely use 6×15-cm plates of cells (just confluent). The medium is removed and the cells are washed once with phosphate-buffered saline (PBS) followed by two washes (15-ml wash/15-cm plate) in a cold solution ($4°$) of hypotonic extraction buffer (HEB: 20 mM Tris, pH 7.5, 5 mM KCl, 1.5 mM MgCl$_2$, 1 mM dithiothreitol). The dithiothreitol must be added fresh and is particularly critical for extracts to be used in ubiquitylation assays where a reduced state is essential to preserve activity of the endogenous E2 enzymes. The HEB is then removed; the plates are tilted to allow removal of any excess. The cells are scraped off and transferred to a 3-ml Dounce homogeniser at $4°$. A small volume of HEB remains (equivalent to 1–2 packed cell volumes), which is sufficient for homogenization (50 strokes). The homogenate is centrifuged at 13,000 rpm for 5 min in a microfuge at $4°$ to remove cell debris and nuclei. Aliquots of the supernatant are then either used directly or stored at $-70°$.

Other laboratories incorporate an additional ultracentrifugation step at 100,000 g at $4°$ for 4 h to obtain the S100 extract.[4,7,17] The main purpose of this is to remove the proteasome. However, in our experience, the crude cytoplasmic extract described earlier does not cause degradation of the HIF-α substrate synthesized by *in vitro* transcription and translation (IVTT) in subsequent assays and, therefore, we find this step unnecessary. Furthermore, in ubiquitylation assays, the S100 extract is compromised in its ability to utilize the HIF-α NODDD substrate.[6]

VHL Interaction Assays

A variety of coprecipitation techniques are available to study VHL–HIF-α interactions using proteins present in cell extracts or prepared *in vitro*. It is important to remember that interaction between these two proteins is dependent on two separate events: (i) hydroxylation of HIF-α at either of two residues (P402 or P564 in human HIF-1α) and (ii) interaction of VHL with hydroxylated HIF-α sequences. A defect in either process will result in failure of the VHL–HIF-α interaction, which must be considered when interpreting the consequences of mutations assayed in HIF-α, that is, a mutation may result in loss of VHL interaction by preventing the HIF-α–prolyl hydroxylase interaction.

For study of the VHL–HIF-α interaction *in vivo* in tissue culture cells, we have used coimmunoprecipitation from VHL-defective cells (RCC4) that have been engineered to stably express a wild-type VHL gene

[17] M. Ohh, C. W. Park, M. Ivan, M. A. Hoffman, T. Y. Kim, L. E. Huang, N. Pavletich, V. Chau, and W. G. Kaelin, *Nature Cell Biol.* **2,** 423 (2000).

(RCC4/VHL). This permits the effect of VHL mutations to be assayed by comparison with similar transfectants reexpressing specific mutant VHL genes. Complexes are immunoprecipitated using the anti-VHL antibody and are then immunoblotted for the presence of HIFα. Note that RCC4/VHL cells express both HIF-1α and HIF-2α (in contrast to the 786–0/VHL cell line, which only expresses HIF-2α[18]). A related approach utilizing metabolic labeling to visualize both the VHL and associated proteins (including HIF-α) is described. However, while several other VHL-associated proteins are constitutive, recovery of HIF-α is strongly dependent on experimental conditions designed to promote maximal levels of the prolyl-hydroxylated protein. These *in vivo* techniques can be applied as an unbiased screen to study VHL–HIF-α interactions present in other cell types.

For *in vitro* analysis of VHL–HIF-α interaction, we have used assays where either one or both components are synthesised *in vitro* (allowing for separate manipulation of the proteins). In general, we have utilized IVTT VHL and HIF-α proteins synthesized in a rabbit reticulocyte lysate. The reticulocyte lysate contains an endogenous HIF-α prolyl hydroxylase that has been purified and shown to correspond to rabbit PHD2.[19] However, the IVTT reaction is amenable to manipulation, allowing the prolyl hydroxylation status of the synthesized HIF-α protein to be either increased or almost entirely ablated, hence our choice for this system (see Table I). The production of nonhydroxylated HIF-α in IVTT enabled us to develop assays using a separate source of hydroxylase activity, e.g., cell extract or purified recombinant enzyme. In this way the requirements for HIF-α hydroxylation can be assessed with the VHL–HIF-α interaction being used as a readout for the HIF-α hydroxylation status. Other assays are possible utilizing the nonhydroxylated HIF-α protein prepared by bacterial expression or by synthesis in wheat germ extract.[4,7,20]

One disadvantage of using interaction assays based on IVTT VHL is that HIF-α NODDD sequences are unable to interact efficiently.[6] The reason for this is unclear. An alternative assay based on interaction with VHL E3 ligase complex purified from cell extracts is, therefore, described. The VHL E3 assay allows interactions with both NODDD and CODDD to be analyzed and is the method of choice for the study of full-length HIF-α.

Our laboratory has also developed a number of VHL–HIF-α peptide interaction assays. Nonhydroxylated HIF-α peptides can be used to assess

[18] P. H. Maxwell, M. S. Wiesener, G.-W. Chang, S. C. Clifford, E. C. Vaux, M. E. Cockman, C. C. Wykoff, C. W. Pugh, E. R. Maher, and P. J. Ratcliffe, *Nature* **399,** 271 (1999).

[19] M. Ivan, T. Haberberger, D. C. Gervasi, K. S. Michelson, V. Gunzler, K. Kondo, H. Yang, I. Sorokina, R. C. Conaway, J. W. Conaway, and W. G. J. Kaelin, *Proc. Natl. Acad. Sci. USA* **99,** 13459 (2002).

[20] J. Huang, Q. Zhao, S. M. Mooney, and F. S. Lee, *J. Biol. Chem.* **277,** 39792 (2002).

TABLE I
MODULATION OF HIF-α PROLYL HYDROXYLATION CAN BE ACHIEVED BY
ALTERING SYNTHESIS CONDITIONS IN RABBIT RETICULOCYTE LYSATE

IVTT condition	HIF-α CODDD prolyl hydroxylation status
Standard reaction	+
+ 100 μM ferrous chloride	++
+ 100 μM desferrioxamine	−
+ 100 μM cobalt(II)	−
+ 0.1% hypoxia	−

requirements for both hydroxylation and the VHL–HIF-α interaction. However, one major application is that synthetic prehydroxylated HIF-α peptides can be used to directly analyze the determinants of the VHL interaction.[4,5,7]

Coimmunoprecipitation of HIF-α with VHL from [35S]Methionine/ Cysteine-Labeled RCC4/VHL Cells

Reagents

Phosphate-buffered saline, sterile

Dulbecco's modified Eagle's medium (DMEM) lacking methionine and cystine (Sigma)

Redivue PRO-MIX (Amersham [35S]methionine/cysteine, 10 mCi/ml with respect to L-[35S]methionine)

MG132 proteasomal inhibitor, 12.5 mM in DMSO and stored at −20°

Lysis buffer (100 mM NaCl, 0.5% Igepal CA630, 20 mM Tris, pH 7.5, 5 mM MgCl$_2$) stored at room temperature. Immediately before use, the buffer is chilled to 4° and "complete" protease inhibitor (Roche Molecular Biochemicals) is added fresh.

Anti-VHL antibody (IG32, Pharmingen)/Anti-HA antibody (12CA5, Roche Molecular Biochemicals) for use with hemagglutinin-tagged VHL (VHLHA)-stable transfectants

Protein G-Sepharose beads, preblocked with PBS containing 20 mg/ ml bovine serine albumin and stored at 4°

SDS–PAGE reagents

Procedure. Prior to metabolic labeling, cells are washed with PBS and are then incubated at 37° for 1 h in serum-free DMEM lacking methionine and cystine in order to deplete cellular reserves and enhance labeling efficiency. A minimal volume of DMEM lacking methionine and cystine but containing 200 μCi/ml Pro-mix is then added, and the labeling incubation is left at 37° for 4–6 h. For 75-cm^2 dishes, a 4.5-ml volume of medium is sufficient. In order to "see" VHL–HIF-α interactions, proteasomal inhibition must be used to block HIF-α degradation and thus MG132 is also added to the labeling incubation (12.5 μM final concentration). [The retrieval of HIF-α is enhanced further by performing cell labeling in the presence of MG132, under hypoxic conditions (we routinely use 0.1–1% oxygen, 5% CO_2, balance nitrogen in a Napco 7001 incubator, Jouan).] Note that even though hypoxia inhibits prolyl hydroxylase activity, we have consistently observed robust interaction. This is because hypoxic incubation allows accumulation of the HIF-α protein in the cells due to a lack of prolyl hydroxylation and the ability to escape VHL-mediated degradation. However, following harvest under the reoxygenation conditions described later, the accumulated HIF-α protein becomes hydroxylated and available to interact with VHL. Thus the combination of proteasomal inhibition and hypoxic incubation provides the highest output signal. Following labeling, cells are washed once with PBS, transferred to an Eppendorf tube, and resuspended in 1 ml lysis buffer/10^7 cells for 1 h at 4°, followed by centrifugation at 13,000 rpm for 30 min in a microfuge at 4° to remove cell debris. Cell extracts are then precleared overnight at 4° by incubation on a rotator with preblocked protein G-Sepharose beads (50 μl beads/ml lysate). The length of the preclearing step can be reduced, but overnight incubation provides a convenient break in the protocol. For immunoprecipitation, antibody (5 μg antibody/ml lysate) is added and samples are incubated at 4° for 2 h, followed by a 1-h incubation with 20 μl protein G-Sepharose beads on a rotator. The beads (now containing bound VHL and associated proteins) are then washed five times in 1 ml of lysis buffer at 4° (beads are retrieved between washes by centrifugation at 2000 rpm for 15 s in a 4° microfuge). Finally, beads are resuspended in an appropriate volume of Laemmli sample buffer, and samples are resolved by SDS–PAGE. Our laboratory has generally used discontinuous gels (8% acrylamide upper portion, 13% lower portion) in order to easily visualize both VHL and HIF-α proteins from the same sample on the same gel. To enhance the output signal, gels are also processed for fluorography using Amplify (Amersham Pharmacia Biotech). Under standard cell-labeling conditions the method allows visualization of VHL and at least some of its associated proteins (e.g., elongins B and C, Cul-2, fibronectin).[21,22]

[21] M. E. Cockman, N. Masson, D. R. Mole, P. Jaakkola, G. W. Chang, S. C. Clifford, E. R. Maher, C. W. Pugh, P. J. Ratcliffe, and P. H. Maxwell, *J. Biol. Chem.* **275,** 25733 (2000).
[22] M. Ohh, R. L. Yauch, K. M. Lonergan, J. M. Whaley, A. O. Stemmer-Rachamimov, D. N. Louis, B. J. Gavin, N. Kley, W. G. Kaelin, Jr., and O. Iliopoulos, *Mol. Cell* **1,** 959 (1998).

Coimmunoprecipitation of HIF-α and VHL Proteins Synthesized in
Rabbit Reticulocyte Lysate

*Method 1: Utilizing Rabbit Reticulocyte Lysate to Provide a Source of
Prolyl-Hydroxylated HIF-α Protein*

   Reagents

   [^{35}S]Methionine-labeled hemagglutinin-tagged VHL (VHLHA) prepared by coupled IVTT using TnT7 rabbit reticulocyte (Promega).
   Can be stored at $-20°$.
   [^{35}S]Methionine-labeled HIF-α prepared by coupled IVTT using
   TnT7 rabbit reticulocyte (Promega). Maximal levels of prolyl
   hydroxylation can be achieved by supplementing the IVTT reaction
   with 100 μM ferrous chloride. The ferrous chloride solution must be
   prepared fresh and added immediately [we prepare a 1 mM (10$\times$)
   stock in H$_2$O]. The IVTT can be stored at $-20°$.
   NETN buffer: 150 mM NaCl, 0.5 mM EDTA, 20 mM Tris, pH 8.0,
   0.5% Igepal CA630, stored at room temperature
   Anti-HA antibody (12CA5, Roche Molecular Biochemicals), stored
   at 4$°$
   Protein G-Sepharose beads, preblocked with PBS containing 20 mg/
   ml bovine serine albumin and stored at 4$°$. Preblocking is not
   essential, but can reduce nonspecific interactions. Before use the
   beads are preequilibrated by washing once in NETN buffer.
   SDS–PAGE reagents
   Procedure. Four microliters of HIF-α and VHLHA programmed reticulocyte lysates are mixed in 500 μl of NETN buffer for 90 min at 4$°$.
Anti-HA antibody (1 μg) is then added, and incubation is continued at 4$°$
for a further 60 min. Following the addition of protein G-Sepharose
beads (15 μl), incubation is continued on a rotator at 4$°$ for 30 min. Beads
are then washed five times in 600 μl of NETN buffer at 4$°$ (beads are retrieved between washes by centrifugation at 2000 rpm for 15 s in a 4$°$ microfuge). Finally, beads are resuspended in an appropriate volume of
Laemmli sample buffer and samples are resolved by SDS–PAGE. Note
that if the chosen HIF-α sequences are likely to comigrate with VHLHA,
nonradiolabeled VHL can be used.
   This method utilizes HA-tagged VHL protein and anti-HA immunoprecipitation to analyze interactions.[21] The converse experiment is equally
valid; for example, we have also used GAL-tagged HIF-α proteins and
anti-GAL immunoprecipitation (using the aforementioned conditions) to
recover VHL protein.[6] Equivalent results may be obtained using other tags
and appropriately directed immunoprecipitating antibodies.

Method 2: Utilizing Rabbit Reticulocyte Lysate to Synthesize
Nonhydroxylated HIF-α Protein That Can Be Modified In Vitro
with Cell Extract/Recombinant Enzyme

Reagents

[^{35}S]Methionine-labeled VHL prepared by coupled IVTT using TnT7
rabbit reticulocyte (Promega)

[^{35}S]Methionine-labeled GALHIF α prepared by coupled IVTT in the
presence of 100 μM, desferrioxamine (DFO) using TnT7 rabbit
reticulocyte (Promega). The addition of DFO to the IVTT reaction
results in abrogation of HIF-α prolyl hydroxylation.

HEB extraction buffer (added fresh)

Cell extract with competent HIF-α prolyl hydroxylase activity
(preparation using HEB described earlier) and stored at $-70°$.
The extract is thawed just before use and is supplemented with
100 μM ferrous chloride to promote maximal prolyl hydroxylase
activity. The ferrous chloride is added from a 1 mM (10× stock)
solution, prepared fresh in dH$_2$O.

Anti-GAL4 (RK5C1) agarose conjugate (Santa Cruz Biotechnology)
stored at $4°$

NETN buffer, stored at room temperature.

DFO solution, prepared fresh as a 10 mM (100×) stock in H$_2$O

Procedure. The first step is preparation of the immobilized nonhydroxylated HIF-α substrate. Five microliters of GALHIF-α programmed reticulocyte lysate (prepared in DFO) and 20 μl of anti-GAL beads are mixed
with 100 μl of NETN buffer on a rotator for 60 min at $4°$. Beads are then
washed four times in 600 μl of NETN buffer at $4°$ (beads are retrieved
between washes by centrifugation at 2000 rpm for 15 s in a $4°$ microfuge).
A final wash in 600 μl of HEB allows equilibration for the next step.

For enzymatic modification of the immobilized GALHIF-α substrate,
beads are resuspended in 135 μl of cell extract and incubated at $22–30°$
for 60 min on a rotator. We routinely use extract prepared from VHL-
defective cells, for example, RCC4, in order to minimize any potential
ubiquitylation/degradation of the immobilized HIF-α substrate that may
occur during incubation with the cell extract and/or prevent recruitment
of endogenous VHL from the cell extract. However, this is not essential
and other cell extracts can be used successfully. In principle, the length of
incubation (and the temperature) required for maximal hydroxylation can
be determined empirically for each batch of cell extract, but we find that a
60-min incubation period at $22°$ usually provides maximal hydroxylation.
The control for the extract modification is a parallel incubation using only
HEB supplemented with ferrous chloride.

Following the HIF-α modification step, all washes and incubations are performed in the presence of 100 μM DFO to prevent any further HIF-α hydroxylation from occurring. First, the beads are washed three times in 600 μl of NETN buffer containing 100 μM DFO. GALHIF-α beads are then assayed for the ability to bind VHL. [^{35}S]Methionine-labeled VHL (4 μl, prepared in DFO) is mixed with the beads in 100 μl NETN (containing 100 μM DFO) for 120 min at 4°. Beads are then washed four times in 600 μl of NETN buffer (containing 100 μM DFO) at 4° and are resuspended in an appropriate volume of Laemmli sample buffer, and samples are analyzed by SDS–PAGE. To enhance the output signal, the gels are also processed for fluorography using Amplify (Amersham Pharmacia Biotech).

Because this assay allows controlled *in vitro* modification of the HIF-α substrate, it has a great deal of utility and can be used to analyze (i) characteristics/requirements for enzyme modification and (ii) inhibitors of prolyl hydroxylation.

Instead of using cell extract as a source of HIF-α prolyl hydroxylase activity, the modification reaction can also be performed using PHD enzyme produced by IVTT or prepared by bacterial/baculoviral expression.[9]

Reagents

[^{35}S]Methionine-labeled PHD enzyme prepared by coupled IVTT using TnT7 rabbit reticulocyte (Promega)

Ascorbic acid, 100 mM stock (50×), pH 7.4, in dH$_2$O stored at $-20°$

2-Oxoglutarate, 100 mM stock (50×), pH 7.4, in dH$_2$O stored at $-20°$

Ferrous chloride, 1 mM (10×) stock in H$_2$O prepared immediately before use

Procedure. Beads are resuspended in a total volume of 135 μl consisting of 30 μl of IVTT PHD, 2.7 μl ascorbic acid (final concentration 2 mM), 2.7 μl 2-oxoglutarate (final concentration 2 mM), 13.5 μl ferrous chloride (final concentration 100 μM), and 86.1 μl HEB. Incubation conditions and subsequent processing are as described earlier. Alternatively, the 30 μl IVTT PHD can be replaced with an equivalent volume of purified recombinant protein.

Coimmunoprecipitation of HIF-α Protein Synthesized in Rabbit Reticulocyte Lysate with VHL E3 Ligase Complex Retrieved from Cell Extracts

Reagents

[^{35}S]Methionine-labeled GALHIF-α prepared by coupled IVTT in the presence of 100 μM desferrioxamine using TnT7 rabbit reticulocyte

(Promega). The addition of DFO to the IVTT reaction results in abrogation of HIF-α prolyl hydroxylation.

Cell extract with competent HIF-α prolyl hydroxylase activity (preparation using HEB described earlier) and stored at $-70°$. The extract is thawed just before use and is supplemented with 100 μM ferrous chloride to promote maximal prolyl hydroxylase activity. Ferrous chloride is added from a 1 mM (10$\times$ stock) solution prepared fresh in dH$_2$O.

HEB: 20 mM Tris, pH 7.5, 5 mM KCl, 1.5 mM MgCl$_2$, 1 mM dithiothreitol (added fresh)

Anti-HA antibody (12CA5, Roche Molecular Biochemicals) stored at 4°

Extract containing HAVHL E3 is prepared fresh from stably transfected 786-0 HA.VHL cells (or equivalent). The lysate is obtained by resuspending cells in Igepal lysis buffer (10 mM Tris, pH 7.5, 0.25 M NaCl, 0.5% Igepal CA630, 1 ml lysis buffer/$\sim$10^7 cells) at 4° for 5 min followed by centrifugation at 13,000 rpm for 5 min in a microfuge at 4° to remove cell debris.

Protein G-Sepharose beads, preblocked with PBS containing 20 mg/ml bovine serine albumin and stored at 4°

IP wash buffer (125 mM NaCl, 25 mM Tris, pH 7.5, 0.1% Igepal CA630) stored at room temperature

SDS–PAGE reagents

Procedure. The GALHIF-α substrate (4 μl) is first hydroxylated *in vitro* by incubation with cell extract (65 μl) at 30° for 60 min. We routinely use extract prepared from VHL-defective cells, for example, RCC4, in order to minimize any potential ubiquitylation/degradation of the HIF-α substrate that may occur during incubation with the cell extract and/or prevent recruitment of endogenous VHL from the cell extract. However, this is not essential and other cell extracts can be used successfully. The control for the extract modification is a parallel incubation using only HEB supplemented with ferrous chloride. Instead of using the cell extract as a source of HIF-α prolyl hydroxylase activity, the modification reaction can also be performed using PHD enzymes produced by IVTT or prepared by bacterial/baculoviral expression (see procedure given earlier).

During the substrate modification step, the VHL E3-containing extract can be prepared. We use 786-0 HA.VHL cells because the anti-HA antibody allows efficient immunoprecipitation of the HA.VHL E3 ligase. It is essential, therefore, that 786-0 HA.VHL cells (or equivalent) are ready for lysis and processing on the day. Five micrograms of the anti-HA antibody (12CA5) and 500 μl of the 786-0 HA.VHL cell lysate are then added to the HIF-α hydroxylation reaction, and the mixture is incubated at 4° for 1 h.

Importantly, the Igepal lysis buffer does not support HIF-α hydroxylation and thus addition of the 786-0 HA.VHL cell lysate stops any further HIF-α modification. A 12-μl aliquot of protein G-Sepharose beads is then added and incubation is continued at 4° for 30 min with mixing. Beads (now containing the bound VHL E3 and associated HIF-α proteins) are then washed five times in IP wash buffer (600 μl/wash), resuspended in Laemmli sample buffer, and analyzed by SDS–PAGE. To enhance the output signal, gels are also processed for fluorography using Amplify (Amersham Pharmacia Biotech).

The VHL E3 ligase interaction assay is essential for the analysis of HIF-α NODDD (P402) interactions. In contrast to GALHIF-α CODDD, we found that GALHIF-α NODDD proteins do not hydroxylate efficiently in the reticulocyte lysate and, significantly, do not interact efficiently with VHL produced in the reticulocyte lysate.[6] The VHL E3 ligase interaction assay is, therefore, important for the study of NODDD–VHL interactions in isolation or for the study of full-length HIF-α–VHL protein interactions.

VHL–HIF-α CODDD Peptide-Based Assays

Reagents

Biotinylated HIF-α CODDD peptide; we have successfully used a 19-mer (corresponding to residues 556–574 of HIF-1α). This peptide can be modified *in vitro*. Stock solutions of peptide (15 μM in 50 mM Tris, pH 7.5) are stored at −20°.

Cell extract with competent HIF-α prolyl hydroxylase activity (preparation using HEB described earlier) and stored at −70°. The extract is thawed just before use and is supplemented with 100 μM ferrous chloride to promote maximal prolyl hydroxylase activity. The ferrous chloride is added from a 1 mM (10× stock) solution prepared fresh in dH$_2$O.

MG132 proteasomal inhibitor, 12.5 mM in DMSO and stored at −20°

HEB

EBC buffer: 50 mM Tris, pH 7.5, 150 mM NaCl, 0.5% Igepal CA630

Streptavidin Dynabeads M-280 (Dynal ASA).

DFO prepared fresh as a 10 mM (100×) stock in H$_2$O

[^{35}S]Methionine-labeled VHL prepared by coupled IVTT in the presence of 100 μM DFO using TnT7 rabbit reticulocyte (Promega).

SDS–PAGE reagents

Procedure. For *in vitro* modification of HIF-α peptides, we use 15-μl reactions containing MG132 (final concentration 12.5 μM), 1 μl peptide, 10 μl cell extract, and EBC (balance volume). The reaction is incubated at 30° for 60 min. We routinely use extract prepared from VHL-defective cells, for example, RCC4, in order to minimize any potential ubiquityla-tion/degradation of the immobilized HIF-α substrate that may occur during incubation with the cell extract and/or prevent recruitment of endogenous VHL from the cell extract. However, this is not essential and other cell extracts can be used successfully. We do, however, recommend the addition of MG132 to minimize isopeptidase activity in the extract (a problem that varies between batches of cell extract). The control for the extract modification is a parallel incubation in which the extract is replaced by HEB supplemented with ferrous chloride.

Following modification, the peptide is retrieved as follows: 10 μl of streptavidin Dynabeads are added and the mixture is transferred to 4° for 15 min with occasional mixing using a pipette tip. The beads (containing bound HIF-α peptide) are then captured using the magnetic stand pro-vided by the manufacturer (Dynal ASA) and washed twice with 1 ml of EBC.

The HIF-α peptide beads are then assayed for the ability to bind VHL with all incubations and washes now being performed in the presence of 100 μM DFO to prevent any further HIF-α hydroxylation from occurring. [^{35}S]Methionine-labeled VHL (4 μl, prepared in DFO) is mixed with the beads in 100 μl EBC (containing 100 μM DFO) for 30 min at 4°. The beads are then washed four times in 600 μl of EBC buffer (containing 100 μM DFO) at 4°, resuspended in an appropriate volume of Laemmli sample buffer, and samples analyzed by SDS–PAGE. To enhance the output signal, the gels are also processed for fluorography using Amplify (Amersham Pharmacia Biotech).

Similar assays using a hydroxylated peptide (e.g., we have used a 19-mer corresponding to residues 556–574 of HIF-1α bearing a *trans*-4-hydroxy-S-proline substitution at residue P564[5]) and omitting the modi-fication step have been used to assay the hydroxylated HIF-α–VHL interaction directly.

VHL-Dependent Ubiquitylation Assays

Several methods are now available to assay the HIF-α substrate for VHL-dependent ubiquitylation. The first procedures were based on the use of mammalian cell extracts as a source of active E1, E2, and VHL E3 ubiquitylation enzymes.[17,21] More advanced assays reconstitute VHL-dependent ubiquitylation using purified components and allowed specific questions to be asked regarding the requirements for ubiquitylation. The source of purified components, particularly VHL E3, is variable. One assay utilizes a reconstituted recombinant five-subunit VHL E3 complex purified from Sf21 cells after coinfection with multiple baculoviruses.[23] This provided direct biochemical proof that the five subunit VHL E3 complex was sufficient for HIF-α ubiquitylation. Reconstituting the VHL E3 in this way provides an opportunity for manipulation of the individual components. An alternative method utilizing VHL E3 purified from mammalian cells is described in detail later.[6]

In Vitro *Ubiquitylation Using Crude Extract as a Source of VHL E3*

Reagents

Crude cell extract, prepared as described earlier, and stored $-70°$

Ubiquitin, 5 mg/ml in dH$_2$O and stored at $-20°$

Ubiquitin aldehyde (AFFINITI Research Products) 150 μM in dH$_2$O and stored at $-20°$

10× ATP-regenerating system, 20 mM Tris, pH 7.5, 10 mM ATP, 10 mM magnesium acetate, 300 mM creatine phosphate, 0.5 mg/ml creatine phosphokinase, prepared fresh

[35S]Methionine-labeled VHL prepared by coupled IVTT using TnT7 rabbit reticulocyte (Promega)

[35S]Methionine-labeled HIF-α substrate prepared by coupled IVTT using TnT7 rabbit reticulocyte (Promega)

SDS–PAGE reagents

Procedure. Assays are carried out at 30° by mixing 2 μl [35S]methionine-labeled substrate with 40 μl ubiquitylation reaction mix consisting of 27 μl freshly thawed cell extract, 4 μl of 10× ATP-regenerating system, 4 μl of ubiquitin, 0.66 μl ubiquitin aldehyde, and the remaining volume of dH$_2$O. Ubiquitin aldehyde is a potent inhibitor of ubiquitin C-terminal hydrolases. It is not an essential component of the reaction, but its addition will enhance the polyubiquitylation signal obtained. Ten-microliter aliquots can be removed at appropriate times (e.g., 20, 60, and 180 min) and mixed with 5 μl 3× Laemmli SDS sample buffer to stop the reaction. The reaction products are then analyzed by SDS–PAGE and autoradiography allowing direct visualization of the [35S]-labeled substrate and its high molecular weight polyubiquitylated forms. For specific analysis of VHL-dependent ubiquitylation, reactions can be carried out in parallel using an extract from VHL-defective cells, for example, RCC4, and an extract from a stable transfectant reexpressing VHL. An alternative approach is to complement VHL-defective cell extracts with exogenous VHL. In this case, 4 μl of [35S]methionine-labeled wild-type VHL (or mutant VHL) is included in

[23] T. Kamura, S. Sato, K. Iwai, M. Czyzyk-Krzeska, R. C. Conaway, and J. W. Conaway, *Proc. Natl. Acad. Sci. USA* **97**, 10430 (2000).

the ubiquitylation reaction mix at room temperature for 5 min prior to addition of the substrate.

Purified Component Ubiquitylation Assay

Reagents

The purified E1-activating enzyme can be obtained commercially [we have used 0.4 mg/ml E1 (rabbit) in 50 mM HEPES, pH 7.6, AFFINITI Research Products] and stored at $-80°$. Alternatively, we have prepared our own by purification from BL21(DE3) *Escherichia coli* transfected with plasmid-expressing His_6-tagged E1 (mouse). His_6–E1 is purified by Ni^{2+}-agarose affinity chromatography. After dialysis against PBS at $4°$, glycerol is added to 10% (v/v) and aliquots (0.25 mg/ml) are stored at $-80°$.

Purified recombinant E2 ubiquitin-conjugating enzymes can be obtained commercially (we have used 0.46 mg/ml His_6-human CDC34/UbcH3 in 50 mM HEPES, pH 8.0, 50 mM NaCl, 1 mM DTT, 10% glycerol, AFFINITI Research Products) and stored at $-80°$. The alternative option is again to purify recombinant protein expressed in *E. coli*. UbcH5 family members may also be used.[24]

Purified VHL E3 is prepared fresh by anti-HA immunoprecipitation from stably transfected 786-0 HA.VHL cell lysates. The lysate is obtained by resuspending cells in IGEPAL lysis buffer (10 mM Tris, pH 7.5, 0.25 M NaCl, 0.5% Igepal CA630) at $4°$ for 5 min, followed by centrifugation at 13,000 rpm for 5 min in a microfuge at $4°$ to remove cell debris. One milliliter of 786-0 HA.VHL cell lysate ($\sim 10^7$ cells) and 5 μg of anti-HA antibody (12CA5) are incubated at $4°$ for 1 h. A 12-μl aliquot of protein G-Sepharose beads is then added, and incubation is continued at $4°$ for 30 min with mixing. The beads (now containing the bound VHL E3) are washed four times in IP wash buffer (125 mM NaCl, 25 mM Tris, pH 7.5, 0.1% Igepal CA630) with a final wash in HEB.

Ubiquitin, 5 mg/ml in dH_2O and stored at $-20°$

10× ATP-regenerating system, 20 mM Tris, pH 7.5, 10 mM ATP, 10 mM magnesium acetate, 300 mM creatine phosphate, 0.5 mg/ml creatine phosphokinase, prepared fresh

Purified [^{35}S]methionine-labeled GALHIF-α substrate is prepared fresh as follows: First crude [^{35}S]methionine-labeled GALHIF-α substrate is synthesized by coupled IVTT using TnT7 rabbit

---

[24] K. Iwai, K. Yamanaka, T. Kamura, N. Minato, R. C. Conaway, J. W. Conaway, R. D. Klausner, and A. Pause, *Proc. Natl. Acad. Sci. USA* **96,** 12436 (1999).

reticulocyte (Promega). The crude substrate can be prepared in advance and stored at $-20°$. Four microliters of this crude substrate is then immunopurified using 6 μl anti-GAL (RK5C1) agarose conjugate (Santa Cruz Biotechnology) utilizing similar incubation conditions (e.g., volume increased by the addition of IGEPAL lysis buffer) and wash conditions (IP wash buffer and HEB) to the aforementioned for VHL E3 immmunopurification.

SDS–PAGE reagents

Procedure. With most of the purified reagents stored frozen, the main preparation is immunopurification of VHL E3 and GALHIF-α substrate. Therefore, 786-0 HA.VHL cells (or equivalent) must be ready for lysis and processing on the day. The purified component ubiquitylation reaction (40 μl) consists of 4 μl of ubiquitin, 4 μl of 10× ATP-regenerating system, 2 μl of El, 3 μl of E2, 6 μl of VHL E3 immunopurified on protein G-Sepharose beads, 6 μl of GALHIF-α substrate immunopurified on agarose beads, and 15 μl of HEB. The reactions are incubated at $30°$ for 120 min with occasional mixing using a pipette tip. The reactions are stopped by the addition of 20 μl 3× Laemmli SDS sample buffer and analyzed by SDS–PAGE and autoradiography allowing direct visualization of the ^{35}S-labeled substrate and its high molecular weight polyubiquitylated forms. Usually an experiment will consist of multiple substrates and/or multiple conditions and therefore purification of the VHL E3/GALHIF-α substrate must be scaled-up accordingly.

Discussion

The importance of VHL-mediated proteolysis of HIF-α subunits is reflected in the striking stabilization of these proteins and the upregulation of HIF target genes in VHL-defective cells. The assays of VHL/HIF-α interactions and VHL-dependent ubiquitylation of HIF-α described previously have therefore been widely used in understanding the role of VHL in cellular responses to hypoxia and in determining the role of HIF dysregulation in the pathogenesis of VHL-associated cancer.[17,18,21,25,26] Most recently, they have been used to demonstrate the role of hypomorphic VHL mutations in congenital erythrocytosis.[27]

[25] S. C. Clifford, M. E. Cockman, A. C. Smallwood, D. R. Mole, E. R. Woodward, P. H. Maxwell, P. J. Ratcliffe, and E. R. Maher, *Hum. Mol. Genet.* **10,** 1029 (2001).

[26] M. A. Hoffman, M. Ohh, H. Yang, J. M. Klco, M. Ivan, and W. G. J. Kaelin, *Hum. Mol. Genet.* **10,** 1019 (2001).

[27] S. O. Ang, H. Chen, K. Hirota, V. R. Gordeuk, J. Jelinek, Y. Guan, E. Liu, A. I. Sergueeva, G. Y. Miasnikova, D. Mole, P. H. Maxwell, D. W. Stockton, G. L. Semenza, and J. T. Prchal, *Nature Genet.* **32,** 614 (2002).

The methods described also provided assays of HIF prolyl hydroxylase activity and, using purified PHD isoenzymes, can be used to determine the differential characteristics of these isoenzymes with respect to oxygen, Fe(II), and 2-OG concentrations and HIF-α target site selectivity. Nevertheless, in interpreting such results in terms of the physiology of cellular responses to hypoxia, it should be remembered that the concentrations of enzyme, substrate, and cosubstrates may differ substantially from those in the subcellular microenvironments that support physiological HIF hydroxylation. In future studies, an important use of the assays is likely to be in the assessment of compounds that activate the HIF system either by blocking hydroxylation or by blocking the interaction of hydroxylated HIF with VHL.

[22] Analysis of von Hippel–Lindau Hereditary Cancer Syndrome: Implications of Oxygen Sensing

By HAIFENG YANG, MIRCEA IVAN, JUNG-HYUN MIN, WILLIAM Y. KIM, and WILLIAM G. KAELIN, JR.

Introduction

von Hippel–Lindau (VHL) disease is a hereditary cancer syndrome that was first described approximately 100 years ago and is caused by germline mutations that inactivate the VHL tumor suppressor gene, which resides on chromosome 3p25.[1] The cardinal features of this disorder are blood vessel tumors (hemangioblastomas) of the central nervous system and retina, clear cell renal carcinomas, and pheochromocytomas. Tumor development in VHL disease is linked to somatic inactivation of the remaining wild-type VHL allele, thus depriving a susceptible cell of the wild-type VHL gene product (pVHL). In keeping with Knudson's two-hit model, biallelic VHL inactivation is also common in sporadic (nonhereditary) clear cell renal carcinomas, and restoration of VHL function in VHL(−/−) renal carcinomas is sufficient to inhibit tumor growth *in vivo*.

VHL-associated neoplasms accumulate inappropriately high levels of hypoxia-inducible mRNAs, such as vascular endothelial growth factor (VEGF) mRNA and platelet-derived growth factor B (PDGF B) mRNA,

[1] E. Maher and W. G. Kaelin, *Medicine* **76**, 381 (1997).

under well-oxygenated conditions.[2] Many such hypoxia-inducible mRNAs are under the control of the heterodimeric transcription factor hypoxia-inducible factor (HIF), which consists of a labile HIFα subunit (HIF1α, HIF2α, or HIF3α) and a stable HIFβ subunit [HIF1β, also called aryl hydrocarbon receptor nuclear translocator (ARNT)].[3] These observations were linked by Maxwell and colleagues,[4] who found that pVHL-defective tumor cells overproduce HIFα subunits.

It is now appreciated that pVHL contains two subdomains called α and β, which are frequently mutated in VHL disease.[5] The α domain binds to elongin C, which in turn binds to elongin B, Cul2, and Rbx1 (also called ROC1 or Hrt1). This multiprotein complex architecturally resembles so-called SCF (Skp1/Cdc53/F-box protein) ubiquitin ligases. In the presence of oxygen the pVHL β domain binds directly to HIFα subunits and directs their polyubiquitination, which in turn leads to their proteasomal degradation. Hence, cells lacking pVHL fail to degrade HIFα subunits in the presence of oxygen and consequently overexpress HIF target genes.

pVHL binds to a region of HIF1α previously called the oxygen-dependent degradation domain (ODD).[6–8] This interaction depends on hydroxylation of conserved prolyl residues within the ODD by members of the EGLN prolyl hydroxylase family.[5] This enzymatic reaction is inherently oxygen dependent, as the hydroxyl oxygen atom is derived from molecular oxygen and also requires iron, which explains the hypoxia-mimetic effects of iron chelators. Thus, in cells that contain pVHL, changes in oxygen availability lead to changes in HIF proly hydroxylation, which in turn leads to changes in HIF abundance and hypoxia-inducible gene expression.

[2] O. Iliopoulos, C. Jiang, A. P. Levy, W. G. Kaelin, and M. A. Goldberg, *Proc. Natl. Acad. Sci. USA* **93,** 10595 (1996).

[3] G. Semenza, *Genes Dev.* **14,** 1983 (2000).

[4] P. Maxwell, M. Weisner, G.-W. Chang, S. Clifford, E. Vaux, C. Pugh, E. Maher, and P. Ratcliffe, *Nature* **399,** 271 (1999).

[5] W. G. Kaelin, *Nature Rev. Cancer* **2,** 673 (2002).

[6] L. E. Huang, J. Gu, M. Schau, and H. F. Bunn, *Proc. Natl. Acad. Sci. USA* **95,** 7987 (1998).

[7] C. Pugh, J. O'Rourke, M. Nagao, J. Gleadle, and P. Ratcliffe, *J. Biol. Chem.* **272,** 11205 (1997).

[8] P. Kallio, W. Wilson, S. O'Brien, Y. Makino, and L. Poellinger, *J. Biol. Chem.* **274,** 6519 (1999).

Expression and Purification of GST–VBC Complex and
VBC Complex

Materials

Luria–Bertani (LB) medium
Ampicillin, 1000× stock at 100 mg/ml
Kanamycin, 1000× stock at 50 mg/ml
Isopropyl β-D-thiogalactopyranoside (IPTG), 100× stock at 100 mM
Protease inhibitor cocktail tablets minicomplete (Roche)
Cell disruptor EmulSiFlex-05 (Avestin, Canada)
NETN buffer (20 mM Tris, pH 8.0, 100 mM NaCl, 1 mM EDTA, 0.5% NP-40)
Branson Sonifier 250 sonicator with probe type 102 converter (Danbury, CT)
Q-Sepharose resin (Amersham Pharmacia Biotech)
Gravity column (Glass Econo-column; 5.0 cm i.d., Bio-Rad)
Glutathione Sepharose (Amersham Pharmacia Biotech)
Reduced glutathione (Sigma)
Pharmacia FPLC apparatus
20-ml Q-Sepharose fast-flow column (Amersham Pharmacia)
Thrombin (high grade from Sigma)
Heparin sulfate beads (Toyopearl, Montgomeryville, PA)
25-ml SOURCE-Q column (Pharmacia)
Amicon ultrafiltration system with 10-kDa MWCO membrane (Millipore)
Superdex 200 sizing column HR 10/30 (Pharmacia)
30-ml round-bottom polypropylene tubes (Sarstedt)
50-ml conical polystyrene tubes (Falcon)
CaCl$_2$

Production of Pure VBC Complex

Note. This procedure is used to generate the highly pure VBC complex for crystallography studies and was the source of the VBC probe for far Western blot analysis (Fig. 1).[8a]

Procedure

1. Transform BL21(DE3) *Escherichia coli* with a pBB75-derived elongin C(17–112) expression plasmid and a bicistronic, pGEX-4T-

---

[8a] M. Ivan, K. Kondo, H. Yang, W. Kim, J. Valiando, M. Ohh, A. Salic, J. Asara, W. Lane, and W. G. Kaelin, *Science* **292**, 464 (2001).

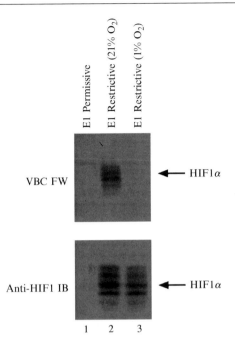

FIG. 1. Far Western blot analysis. VBC far Western and anti-HIF1α immunoblot analysis of ts20 cells grown at a restrictive temperature, which leads to inactivation of E1 ubiquitin-conjugating enzyme under hypoxic or normoxic conditions. From Ivan et al.,[8a] with permission. pVHL only recognizes HIF1α that accumulated in the presence of oxygen.

3-derived plasmid encoding GST-pVHL (54–213) and elongin B.[9] Plate the bacteria onto an LB agar plate containing 100 μg/ml ampicillin and 50 μg/ml kanamycin. Incubate the plate at 37° for 12 to 16 h.

2. Inoculate 50 ml of LB supplemented with the same antibiotics in a 250-ml flask with a single colony and grow the culture for 12 to 16 h at 37° with shaking.

3. Dilute the culture to 6 liters (6 × 1-liter cultures in 2-liter flasks) using 8 ml of the culture per liter and same media as described earlier.

4. Incubate culture at 37° with shaking until OD at 600 nm reaches 0.8 to 1 (typically 4–5 h).

5. Set the temperature to 25° and add fresh IPTG to 1 mM final concentration.

6. Continue the incubation for 12 to 16 h at 25° with vigorous shaking (typically at 250 rpm).

7. Harvest the bacteria by centrifugation at 6000g for 10 min at 4°. Discard the supernatant.

[9] C. E. Stebbins, W. G. Kaelin, and N. P. Pavletich, Science **284,** 455 (1999).

8. Resuspend the cell pellet in 40 ml ice-cold lysis buffer [50 mM Tris–HCl, pH 8.0, 200 mM NaCl, and 10 mM dithiothreitol (DTT)] supplemented with 1 mM phenylmethylsulfonyl Fluoride (PMSF).

9. Pass the cells through a EmulSiFlex-05 cell disruptor at 12,000 psi. Repeat once.

10. Clarify lysates by centrifugation at ~22,000g (e.g., 17,000 rpm in Sorvall SS-34 rotor) for 1 h at 4°.

11. Apply supernatant to 20–30 ml of Q-Sepharose resin in a large-gravity column (Bio-Rad, Glass Econo-column, 5.0 cm i.d.) preequilibrated with lysis buffer. Save flow through (GST–VBC does not bind to column under such conditions).

12. Wash the column with 1 column volume of lysis buffer. Save flow through and combine with flow through from step 11 (steps 11 and 12 mainly removes DNA and some other contaminants).

13. Apply the combined Q-Sepharose flow through pool to 20 ml of Glutathione Sepharose (Pharmacia) in a gravity column (Bio-Rad, Glass Econo-column, 5.0 cm i.d.) preequilibrated with lysis buffer.

14. Wash with 10 column volumes of 50 mM Tris–HCl, pH 8.0, 500 mM NaCl, 10 mM DTT.

15. Elute the GST–VBC with 5 column volumes of 50 mM Tris–HCl, pH 8.0, 150 mM NaCl, 10 mM DTT, 10 mM reduced glutathione (Boehringer Mannheim) by gravity flow (the resin can be regenerated multiple times according to the manufacturer's instruction).

16. Dilute the eluate to 50 mM NaCl in 50 mM Tris, pH 8.0, and 10 mM DTT, and load onto a 20-ml Q-Sepharose fast-flow column on FPLC (this step is designed to concentrate the eluate as well as to remove glutathione).

17. Wash the column with 10 column volumes of 50 mM Tris–HCl, pH 8.0, 50 mM NaCl, and 10 mM DTT.

18. Elute the GST–VBC with 50 mM Tris–HCl, pH 8.0, 400 mM NaCl, 10 mM DTT, collecting 5-ml fractions.

19. Use the Bradford assay to measure the protein concentration in the fractions and calculate the amount of recovered protein. Pool peak fractions. The protein usually comes out in the first 50 ml.

20. Digest GST–VBC overnight at 4° by adding thrombin to 1:50 (w/w) final (the high concentration of thrombin is necessary for complete digestion of the GST–VBC, which has been engineered to have two thrombin sites[9] and CaCl$_2$ to 2.5 mM.

21. In order to remove thrombin, dilute the digested protein solution to 200 mM NaCl, and apply it to a 1-ml heparin-sulfate column packed in a Econo-Pac column (Bio-Rad) preequilibrated in 50 mM Tris, pH 8.0, 200 mM NaCl, 10 mM DTT. Collect the flow through.

22. Apply the flow through to 20 ml of Glutathione Sepharose column preequilibrated in 50 mM Tris, pH 8.0, 200 mM NaCl, 10 mM DTT by gravity (as in step 13). Save flow through.

23. Regenerate the resin according to the manufacturer's instruction and repeat step 22 one more time to ensure complete removal of GST.

24. Monitor the purity and the cleavage of GST–VBC in the flow through by running an aliquot on SDS–PAGE (18% acrylamide). Repeat steps 20 to 23 if necessary.

25. Dilute the protein to 50 mM NaCl with 50 mM Tris, pH 8.0, and 10 mM DTT.

26. Load the diluted sample onto a 25-ml SOURCE-Q column (Pharmacia) on an FPLC.

27. Wash with 2 column volumes of 50 mM Tris, pH 8.0, 10 mM DTT.

28. Elute VBC with a gradient from 0 to 200 mM NaCl in 50 mM Tris, pH 8.0, 10 mM DTT over 300 ml. Measure the protein concentration of the fractions by the Bradford assay and run aliquots on SDS–PAGE.

29. Pool the peak fractions containing VBC, which usually come out at about 150 to 200 mM NaCl. Concentrate the pooled VBC to 20–35 mg/ml using the Amicon ultrafiltration system with a 10-kDa MWCO membrane (Millipore).

30. Run the concentrated VBC on a Superdex 200 sizing column HR 10/30 (Pharmacia) using FPLC with 5 mM bis–Tris–propane, pH 7.0, 200 mM NaCl, and 2 mM DTT.

31. Collect tip of peak in 50- to 100-μl aliquots; flash freeze in liquid nitrogen.

Production of GST–VBC for Pull Down

Procedure

1. Inoculate (e.g., by scraping a frozen glycerol stock with a hot wire loop) 50 ml LB media containing 100 μg/ml ampicillin and 50 μg/ml kanamycin with DH5α E. *coli* transformed with a pBB75-derived elongin C(17–112) plasmid and a bicistronic, pGEX-4T-3-derived plasmid encoding GST-pVHL (54–213) and elongin B.[9]

2. Incubate culture overnight at 37° with shaking.

3. The next morning, dilute the overnight culture into 500 ml of LB media supplemented with the same antibiotics.

4. Incubate culture at 37° with shaking until OD at 600 nm reaches 0.6 (typically 4–5 h).

5. Add IPTG to 1 mM final concentration.

6. Continue incubation at 25° for 16 h with rigorous shaking.

7. Pellet bacteria by centrifugation at 6000g for 10 min at 4°. Discard supernatant.

8. Resuspend pellet in 40 ml ice-cold NETN buffer supplemented with three minicomplete tablets (Roche).

9. Transfer 5-ml aliquots of suspended cells into 30-ml Sarstedt tubes kept on ice.

10. Sonicate bacteria with 3 × 3-s pulses with a Branson Sonifier 250 sonicator to disrupt the cell walls. To prevent overheating, keep 30-ml Sarstedt tub in an ice water bath (e.g., using a 50-ml Falcon tube containing ~25 ml of ice water) while sonicating. Do not oversonicate.

11. Clarify sonicate by centrifugation at 7000g for 30 min at 4°. Pool and save supernatant in a prechilled 50-ml Falcon tube.

12. Wash 1 ml Glutathione Sepharose with 10–15 ml of cold NETN buffer three times (let beads settle by gravity before aspirating NETN).

13. Add Glutathione Sepharose to clarified sonicate and rock for 1 h at 4°.

14. Wash the Glutathione Sepharose with 25 ml of ice-cold NETN buffer three times (spin quickly in a tabletop clinical centrifuge).

15. After the last wash aspirate NETN until the packed bead volume is ~1/2 total volume of remaining liquid [i.e., beads are now a 1:1 slurry (v/v) in NETN]. At this point the Sepharose is ready for the pull-down experiment.

Far Western Blot Analysis with VBC Complex

Materials

Nitrocellulose membrane (Bio-Rad)
Purified VBC complex
Purified anti-VHL IG32 mouse monoclonal antibody (BD Pharmingen)
Goat antimouse alkaline phosphatase-conjugated secondary antibody (Southern Biotech)
Nitroblue tetrazolium (NBT) (Bio-Rad) solution (600 mg dissolved into 8.4 ml N',N'-dimethylformamide and 3.6 ml distilled water)
5-Bromo-4-chloro-3-indolyl phosphate (BCIP) (Bio-Rad) solution (600 mg dissolved into 12 ml N',N'-dimethylformamide)
1× TBS (Tris-buffered saline; 10 mM Tris, pH 8.0; 150 mM NaCl)

Procedure

1. Resolve protein samples by SDS–PAGE.
2. Wet transfer the proteins onto the nitrocellulose membrane.

3. Block the membrane by rocking in 1× TBS containing 4% nonfat powdered milk for 1 h at room temperature.
4. Decant and discard blocking buffer.
5. Gently rock membrane overnight at 4° with 1× TBS containing 4% nonfat powdered milk, 0.01% sodium azide (to prevent bacterial contamination), and purified VBC complex (3–4 µg/ml).
6. Decant and save the VBC cocktail (can reuse 5–10 times; store at 4°).
7. Wash membrane three times with 1× TBS, 5 min each time. Just enough of the liquid should be added so that the whole membrane is covered and is moving in the container during shaking (overzealous washes can lead to loss of signal, probably due to the low affinity of pVHL binding to targets).
8. Rock membrane with 1× TBS containing 4% nonfat powdered milk and monoclonal anti-VHL antibody IG32 (0.5 µg/ml) for 1.5 h at room temperature.
9. Wash the membrane three times again with 1× TBS as in step 7.
10. Rock the membrane with 1× TBS containing 4% nonfat powdered milk and goat antimouse alkaline phosphatase conjugate [1:2000 dilution (v/v)] for 45–60 min at room temperature.
11. Wash the membrane three times with 1× TBS as in step 7. Remove remaining buffer after last wash by aspiration.
12. Mix 200 µl NBT and 100 µl BCIP with 30 ml of alkaline phosphatase buffer (100 mM Tris, pH 9.5, 100 mM NaCl, 5 mM MgCl$_2$) and add to membrane.
13. Rock membrane at room temperature and monitor periodically for the appearance of protein bands. Change the developer if it gets turbid. Typically change buffer once after about 15 min.
14. Stop color development by washing the membrane with distilled water (can also add a few drops of 0.5 M EDTA at this point).

Preparation of S100 Extracts

Materials

Liquid nitrogen bath
Room temperature water bath
Cell scrapers
Beckman SW50.1 rotor
Beckman 5-ml ultraclear centrifuge tubes 13 × 51 mm
1× phosphate-buffered saline (PBS)

Hypotonic solution (20 mM Tris–HCl, pH 7.4, 5 mM MgCl$_2$, 8 mM KCl, 0.5 mM PMSF, 10 μg/ml leupeptin, 1 μg/ml pepstatin, 0.1 mM PABA, 10 μg/ml aprotinin)

Note. The protease inhibitors PMSF, leupeptin, pepstatin, PABA, and aprotinin can be substituted with complete protease inhibitor cocktail tablets (Roche)

Procedure

1. Grow cells in 10-cm tissue culture plates until approximately 90% confluent.
2. Wash cells twice with room temperature PBS.
3. Aspirate residual PBS.
4. Add 500 μl of room temperature PBS per plate.
5. Remove cells from dish by gentle scraping and transfer to a 1.5-ml Eppendorf tube (pool two plates/tube).
6. Pellet cells by centrifugation in a bench-top microcentrifuge at 3000 rpm for 1 min.
7. Aspirate PBS and resuspend cells (by gentle pipeting using a P1000 Pipetman) in 1 ml ice-cold hypotonic solution.
8. Keep tubes on ice for 10 min to induce cell swelling.
9. Pellet cells by centrifugation in a bench-top microcentrifuge at 3000 rpm for 1 min at 4°.
10. Aspirate and discard supernatant.
11. Resuspend cells (by gentle pipeting using a P200 Pipetman) in 100 μl ice-cold hypotonic buffer.
12. Lyse cells using three freeze–thaw cycles (alternating between incubation for approximately 2 min in a liquid nitrogen bath and 2 min in a room temperature water bath).
13. Clarify lysate by centrifugation in a bench-top microcentrifuge at 14,000 rpm for 10 min at 4°.
14. Transfer and pool supernatants into a prechilled Beckman 5-ml ultraclear centrifuge tube.
15. Centrifuge in a Beckman SW50.1 (or equivalent rotor) at 100,000g (32,500 rpm for SW50.1) for 4 h at 4°.
16. Transfer supernatant to a prechilled falcon tube. Determine protein concentration (e.g., by the Bradford method).
17. Transfer 100- to 200-μl aliquots to prechilled Eppendorf tubes.
18. Snap freeze using a dry ice ethanol bath.

In Vitro Ubiquitination Assay

Materials

TNT rabbit reticulocyte lysate (Promega)
10× ubiquitin 80 μg/μl in dH$_2$O (Sigma)
10× ubiquitin aldehyde 1 μg/μl in dH$_2$O (Boston Biochem)
S100 fraction
MG132 100 μM in dH$_2$O (Boston Biochem)
Anti-Gal4 antibody SC-510 (Santa Cruz)
Protein A-Sepharose (Pharmacia)
NETN buffer
Bovine serum albumin (BSA)
Energy regenerating system (ERS) [200 mM Tris–HCl, pH 7.6,
    50 mM MgCl$_2$, 20 mM ATP (Boehringer Mannheim), 200 mM
    creatine phosphate (Boehringer Mannheim), 5 μg/μl creatine
    kinase (Boehringer Mannheim)]

Procedure

1. Make ^{35}S-labeled substrate using a plasmid compatible with *in vitro* translation (e.g., pcDNA-HA-Gal4-ODD)[10] and TNT rabbit reticulocyte lysate (Promega) according to the manufacturer's instructions.

2. In a 1.5-ml Eppendorf tube, add 4 μl 10× ubiquitin (final 8 μg/μl), 4 μl 10× ubiquitin aldehyde (100 ng/μl), 5 μl MG132 (final 12.5 μM), 4 μl of 10× ERS, 4 μl of ^{35}S-labeled substrate, and 50–100 μg of S100 extract (typically 10–20 μl). Bring total volume to 40 μl with ddH$_2$O.

3. Incubate for 90 min at 30°.

4. Add 500 μl ice-cold NETN.

5. Add 1 μg of anti-Gal4 antibody.

6. Rock for 1 h at 4°.

7. Add 30 μl of protein A-Sepharose [prewashed three times with 10 bead volumes of NETN containing 4% BSA and then resupended as 1:1 (v/v) in the same buffer; beads should be allowed to settle by gravity between washes, carefully aspirating the supernatant].

8. Rock for 30 min at 4°.

9. Wash immunoprecipitates five times with ice-cold NETN.

10. After last wash, carefully aspirate residual NETN and add 30 μl of 1× SDS (sodium dodecyl sulfate) sample buffer.

11. Boil sample buffer for 5 min to release bound proteins.

[10] M. Ohh, C. W. Park, M. Ivan, M. A. Hoffman, T.-Y. Kim, L. E. Huang, V. Chau, and W. G. Kaelin, *Nature Cell Biol.* **2**, 423 (2000).

12. Resolve eluted proteins by SDS gel electrophoresis and detect by autoradiography.

Prolyl Hydroxylation Protocols

Materials

Wheat germ extract TNT kit and reticulocyte lysate TNT kit (Promega, Madison, WI)

N-terminally biotinylated, peptide corresponding to HIF1α residues 556–575 DLDLEMLAPYIPMDDDFQLR

Hydroxylation buffer, 10× stock solution (400 mM HEPES, pH 7.4, 800 mM KCl)

IPTG, 100 mM stock solution in water

NETN buffer

Streptavidin-agarose (Pierce Biotechnology, Rockford, IL)

Glutathione-Sepharose (Amersham Pharmacia Biotech)

LB medium with 100 μg/ml ampicillin

Rabbit reticulocyte lysate (RRL, Green Hectares, Oregon, WI, or Promega)

FeCl$_2$, 10 mM stock solution in water, pH 7.4

Ascorbate (Sigma), 20 mM stock in water, pH 7.4

2-Oxoglutarate (Calbiochem), 50 mM stock in water, pH 7.4

Complete protease inhibitor mixture (Roche Molecular Biochemicals, Indianapolis)

EBC buffer: 50 mM Tris–HCl, pH 8.0, 120 mM NaCl, 0.5% IGEPAL

Anti-HIF1α monoclonal antibody (clone H1a-67, Novus Biologicals, Littleton, CO)

Production of Recombinant EGLN1

Production of HA-Tagged EGLN1

Produce HA-tagged EGLN1 in a 50-μl coupled *in vitro* transcription/ translation reaction using pcDNA3-HA-EGLN1[11] and TNT wheat germ extract according to the manufacturer's protocol (TNT, Promega) (Fig. 2).

[11] M. Ivan, T. Haberberger, D. C. Gervasi, K. S. Michelson, V. Gunzler, K. Kondo, H. Yang, I. Sorokina, R. C. Conaway, J. W. Conaway, and W. G. Kaelin, Jr., *Proc. Natl. Acad. Sci. USA* **99,** 13459 (2002).

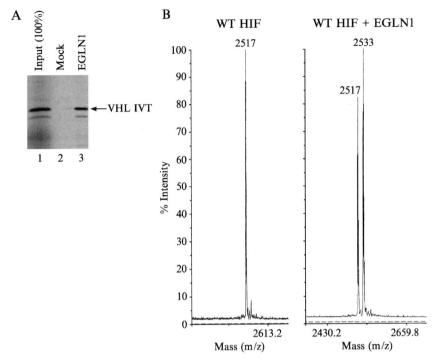

FIG. 2. Hydroxylation of HIF peptide by recombinant EGLN1. The biotinylated HIF1α(556–575) peptide, bound to streptavidin agarose, was incubated with EGLN1 produced in wheat germ extract (A, lane 3), unprogrammed wheat germ extract (A, lane 2), or GST–EGLN1 produced in *E. coli* (B). Hydroxylation was determined based on the ability of the immobilized peptide to bind to ^{35}S-labeled pVHL (A) or by matrix-assisted laser desorption ionization time-of-flight (MALDI-TOF) analysis (B). From Ivan *et al.*,[11] with permission.

Production of GST-Tagged EGLN1

Procedure

1. Inoculate (e.g., by scraping a frozen glycerol stock with a hot wire loop) 5 ml LB media containing 100 μg/ml ampicillin with DH5α *E. coli* transformed with pGEX2TK-EGLN1.[11]
2. Incubate culture overnight at 37° with shaking.
3. The next morning, dilute overnight culture into 100 ml of LB media supplemented with ampicillin.
4. Incubate culture at 37° with shaking for 1 h.
5. Add IPTG to 0.1 m*M* final concentration.
6. Continue incubation at 37° with shaking for 3 h.

7. Aliquot bacterial culture to 4 × 30-ml Sarstedt tubes.
8. Pellet bacteria by centrifugation at 1000g for 10 min at 4°. Discard supernatant.
9. Resuspend each pellet in 10 ml ice-cold NETN buffer supplemented with complete protease inhibitor.
10. Sonicate bacteria with 3× 10-s pulses with a Branson Sonifier 250 sonicator to disrupt the cell walls. To prevent overheating, keep the 30-ml Sarstedt tube in an ice water bath (e.g., using a 50-ml Falcon tube containing ~25 ml of ice water) while sonicating. Do not oversonicate.
11. Clarify sonicate by centrifugation at 10,000g for 20 min at 4°.
12. Save and transfer supernatant to 15-ml Falcon tubes.
13. Wash 1 ml glutathione-Sepharose with 10–15 ml of cold NETN buffer three times (let beads settle by gravity before aspirating NETN).
14. Add 80 μl of 1:1 (v/v) glutathione-Sepharose per Falcon tube.
15. Rock for 1 h at 4°.
16. Wash glutathione-Sepharose three times with 10 ml of cold NETN.
17. After last wash, transfer Sepharose from the four tubes to a single Eppendorf tube.
18. Carefully aspirate residual NETN with a 23-gauge needle.
19. Add ~500 μl of 100 mM Tris, pH 8.0, 120 mM NaCl, and 20 mM reduced glutathione.
20. Rock for 30 min at 4°.
21. Pellet Sepharose by a quick spin in a microcentrifuge.
22. Carefully remove and save supernatant.
23. Eluted GST–EGLN1 is either used immediately or snap frozen in liquid nitrogen with 10% glycerol.

HIF Peptide Hydroxylation Assay

Indirect Method: pVHL Binding

Procedure

1. Incubate 1 μg of the biotinylated HIF peptide with 30 μl of streptavidin-agarose in 1 ml of PBS for 30 min at room temperature with rocking.
2. Wash beads three times with PBS.
3. Wash beads once with 1× hydroxylation buffer. Aspirate residual buffer with a 23-gauge needle.

4. Add 50 μl of rabbit RRL, 10 μl eluted recombinant GST-EGLN1, or 10 μl HA-EGLN1 *in vitro* translate in 200 μl total of 1× hydroxylation buffers supplemented with $FeCl_2$ (100 μM), ascorbate (2 mM), and 2-oxoglutarate (5 mM).
5. Incubate for 2 h at room temperature with tumbling.
6. Wash agarose four times with NETN. (At this point, one can also proceed directly to mass spectrometry analysis.)
7. In parallel, carry out a 50 μl an *in vitro* translation reaction of pRC-CMV plasmid encoding HA-tagged VHL 1–213 in the presence of ^{35}S using the Promega TNT kit according to the manufacturer's instructions.
8. Add 10 μl ^{35}S-labeled pVHL in 500 μl of EBC supplemented with complete protease inhibitor mixture to the washed agarose.
9. Incubate for 1 h at 4° with rocking.
10. Wash agarose four times with NETN.
11. Elute bound pVHL by boiling in SDS-containing sample buffer, resolve by SDS–PAGE, and detect by autoradiography (pVHL migrates with an apparent MW of ∼30 kDa).

Direct Method: Mass Spectrometry

Procedure

1. Wash agarose once with PBS.
2. Elute peptide in 50 μl of 20 mM ammonium acetate, pH 7.0, and 2 mM biotin.
3. Pro564 hydroxylation is confirmed by MS/MS using microcapillary HPLC coupled directly to a Finnigan LCQ DECA quadrupole ion trap mass spectrometer equipped with a nanoelectrospray source. Targeted ion MS/MS of the doubly protonated ion at a mass/charge ratio (m/z) 1267 for the HIF (556–575) peptides is performed with an isolation width of 2.5 Da and a relative collision energy of 30%.

Note. In our hands, HIF hydroxylation can occur in the presence of RRL without the addition of exogenous iron, ascorbate, and 2-oxogluta-rate, which is likely due to the presence of residual amounts of these cofactors in RRL. This is further suggested by the requirement for these cofactors (especially ascorbic acid) when RRL is dialyzed against hydroxylation buffer prior to incubation with the HIF peptide.[11]

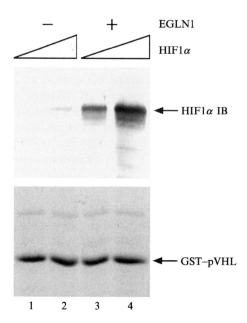

FIG. 3. Hydroxylation of HIF by recombinant EGLN1. Full-length HIF1α (200 ng or 1 μg as indicated by triangles) was incubated with EGLN1 produced in wheat germ extract (lanes 3 and 4) or with unprogrammed wheat germ extract (lanes 1 and 2). Hydroxylation was determined based on binding to immobilized GST–pVHL, elongin B, and elongin C complexes. Bound HIF was detected by anti-HIF1α immunoblot (IB) analysis. From Ivan et al.,[11] with permission.

Full-Length HIF Hydroxylation Assay

1. The production and purification of His-tagged human HIF1α using a baculoviral expression system were described previously.[12]
2. Add 0.2–1 μg of His-HIF1α with 50 μl of HA-EGLN1 *in vitro* translate produced in wheat germ extract in 200 μl total of 1× hydroxylation buffer supplemented with $FeCl_2$ (100 μM), ascorbate (2 mM), and 2-oxoglutarate (5 mM).
3. Incubate for 2 h at room temperature with tumbling.
4. Dilute reaction to a total volume of 500 μl with EBC buffer supplemented with protease inhibitors.
5. Add to 40 μl of Glutathione Sepharose preloaded with the GST–VBC complex (Fig. 3).

[12] T. Kamura, S. Sato, K. Iwain, M. Czyzyk-Krzeska, R. C. Conaway, and J. W. Conaway, *Proc. Natl. Acad. Sci. USA* **97,** 10430 (2000).

6. Incubate for 1 h at $4°$ with rocking.
7. Wash Sepharose four times with NETN.
8. Elute bound HIF1α by boiling in 60 μl SDS-containing sample buffer.
9. Resolve HIF in a 7.5% SDS PAGE gel, transfer onto nitrocellulose for 4 h at $4°$, and detect by immunoblotting with an anti-HIF1α monoclonal antibody (1:500 dilution).

[23] Tumor Hypoxia and Malignant Progression

By PETER VAUPEL, ARNULF MAYER, and MICHAEL HÖCKEL

Introduction

Cells exposed to hypoxic conditions respond by reducing their overall protein synthesis by approximately 50%. Abundant evidence suggests that hypoxia (i.e., the state of oxygen deficiency) can slow down or even completely inhibit (tumor) cell proliferation *in vitro*.[1,2] Furthermore, sustained hypoxia can change the cell cycle distribution and the relative number of quiescent cells, which in turn can lead to alterations in the response to radiation and many chemotherapeutic agents. The degree of inhibition depends on the severity and duration of hypoxia, on the coexistence of other microenvironmental inadequacies (e.g., acidosis, glucose depletion), and on the cell line investigated. The response of cells exposed to hypoxia in terms of the cell cycle is in most cases a G_1/S-phase arrest. Hypoxia levels necessary to induce a disproportionate lengthening of G_1 or an accumulation of cells in this cycle phase are in the range of 0.2–1 mm Hg.[3,4] Above this "hypoxic threshold" the environmental O_2 status appears to have only negligible effects on the proliferation rate. Under anoxia, most cells undergo immediate arrest in whichever phase of the cell cycle they are in.

In addition to hypoxia-mediated changes in tumor cell proliferation, hypoxia can induce programmed cell death (apoptosis) both in normal and in neoplastic cells.[5] p53 accumulates in cells under hypoxic conditions

[1] M. Höckel and P. Vaupel, *J. Natl. Cancer Inst.* **93,** 266 (2001).
[2] A. J. Giaccia, *Semin. Radiat. Oncol.* **6,** 46 (1996).
[3] O. Amellem, M. Loffler, and E. O. Pettersen, *Br. J. Cancer* **70,** 857 (1994).
[4] C. J. Koch, J. Kruuv, and H. E. Frey, *Radiat. Res.* **53,** 43 (1973).
[5] C. Riva, C. Chauvin, C. Pison, and X. Leverve, *Anticancer Res.* **18,** 4729 (1998).

and induces apoptosis involving Apaf-1 and caspase-9 as important downstream effectors.[6] However, hypoxia also initiates p53-independent apoptosis pathways, including those involving genes of the BCL-2[7] family and others.[8] Below a critical energy state, hypoxia/anoxia may result in necrotic cell death, a phenomenon seen in many human tumors and experimental tumor models. Hypoxia-induced proteome changes leading to cell cycle arrest, differentiation, apoptosis, and necrosis may explain delayed recurrences, dormant micrometastases,[9,10] and growth retardation in large tumor masses.[11]

In contrast, hypoxia-induced proteome and/or genome changes in the tumor and/or stromal cells may promote tumor progression via mechanisms enabling cells to overcome nutritive deprivation, to escape from the "hostile" environment, and to favor unrestricted growth.

Sustained hypoxia in a growing tumor may also lead to cellular changes that can result in a more clinically aggressive phenotype.[12-14] During the process of hypoxia-driven malignant progression, tumors may develop an increased potential for local invasive growth,[15,16] perifocal tumor cell spreading,[12,17] and regional and distant tumor cell metastasis.[14,18] Likewise, an intrinsic resistance to radiation and other cancer treatments may be enhanced, resulting in a poor prognosis.[19,20]

[6] M. S. Soengas, R. M. Alarcon, H. Yoshida, A. J. Giaccia, R. Hakem, T. W. Mak, and S. W. Lowe, *Science* **284**, 156 (1999).

[7] S. Shimizu, Y. Eguchi, H. Kosaka, W. Kamiike, H. Matsuda, and Y. Tsujimoto, *Nature* **374**, 811 (1995).

[8] H. M. Sowter, P. J. Radcliffe, P. Watson, A. H. Greenberg, and A. L. Harris, *Cancer Res.* **61**, 6669 (2001).

[9] L. Holmgren, M. S. O'Reilly, and J. Folkman, *Nature Med.* **1**, 149 (1995).

[10] R. Demicheli, M. Terenziani, P. Valgussa, A. Moliterni, M. Zambetti, and G. Bonadonna, *J. Natl. Cancer Inst.* **86**, 45 (1994).

[11] R. T. Prehn, *Cancer Res.* **51**, 2 (1991).

[12] M. Höckel, K. Schlenger, B. Aral, M. Mitze, U. Schäffer, and P. Vaupel, *Cancer Res.* **56**, 4509 (1996).

[13] M. Höckel, K. Schlenger, S. Höckel, B. Aral, U. Schäffer, and P. Vaupel, *Int. J. Cancer* **79**, 365 (1998).

[14] D. M. Brizel, S. P. Scully, J. M. Harrelson, L. J. Layfield, J. M. Bean, L. R. Prosnitz, and M. W. Dewhirst, *Cancer Res.* **56**, 941 (1996).

[15] C. Cuvier, A. Jang, and R. P. Hill, *Clin. Exp. Metastasis* **15**, 19 (1997).

[16] C. H. Graham, J. Forsdike, C. J. Fitzgerald, and S. MacDonald-Goodfellow, *Int. J. Cancer* **80**, 617 (1999).

[17] M. Höckel, K. Schlenger, S. Höckel, and P. Vaupel, *Cancer Res.* **59**, 4525 (1999).

[18] K. Sundfor, H. Lyng, and E. K. Rofstad, *Br. J. Cancer* **78**, 822 (1998).

[19] P. Vaupel, O. Thews, and M. Höckel, *Med. Oncol.* **18**, 243 (2001).

[20] P. Vaupel and M. Höckel, *in* "Recombinant Human Erythropoietin (rhEPO) in Clinical Oncology" (M. R. Nowrousian, ed.), p. 127. Springer, Berlin, New York, 2002.

This article presents current information from experimental and clinical studies, which illustrates the interaction between tissue hypoxia and the phenomenon of malignant progression. As more and more evidence concerning the fundamental biologic and clinical importance of tumor hypoxia emerges, data described here should be considered partially selective and therefore can only represent a "snapshot" of currently available data.

Evidence and Characterization of Tumor Hypoxia

Clinical investigations carried over since the late 1980s have clearly demonstrated that the prevalence of hypoxic tissue areas [i.e., areas with O_2 tensions (pO_2 values) ≤ 2.5 mm Hg] is a characteristic pathophysiological property of locally advanced solid tumors and that such areas have been found in a wide range of human malignancies: cancers of the breast, uterine cervix, head and neck, rectum and pancreas; brain tumors, soft tissue sarcomas, and malignant melanomas.[19–22]

Evidence has accumulated showing that up to 50–60% of locally advanced solid tumors may exhibit hypoxic and/or anoxic tissue areas that are distributed heterogeneously within the tumor mass. The pretherapeutic oxygenation status assessed in cancers of the breast, uterine cervix, and head and neck is poorer than that in the respective normal tissues and is independent of clinical size, stage, histology, grade, nodal status, and a series of other tumor characteristics or patient demographics (Fig. 1). Data do not suggest a topological distribution of the pO_2 values within a tumor. Tumor-to-tumor variability in oxygenation is greater than intratumor variability. Local recurrences have a higher hypoxic fraction than the respective primary tumors, although there is no clear-cut difference between primary and metastatic malignancies.[19–22]

Pathogenesis of Tumor Hypoxia

Hypoxic (or anoxic) areas arise as a result of an imbalance between the supply and the consumption of oxygen (Fig. 2). Whereas in normal tissues or organs the O_2 supply matches the metabolic requirements, in locally advanced solid tumors the O_2 consumption rate of neoplastic as well as stromal cells may outweigh an insufficient oxygen supply and result in the development of tissue areas with very low O_2 levels.

---

[21] P. Vaupel and D. K. Kelleher (eds.), "Tumor Hypoxia." Wissenschaftliche Verlagsgesellschaft, Stuttgart, 1999.
[22] P. Vaupel, S. Briest, and M. Höckel, *Wien. Med. Wschr.* **152,** 334 (2002).

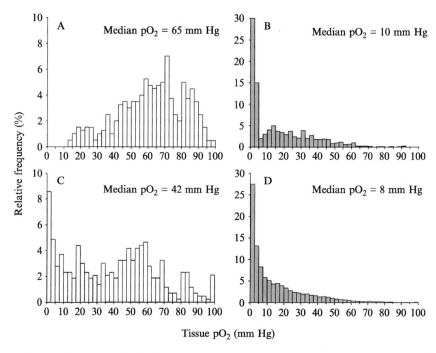

FIG. 1. Frequency distributions (histograms) of oxygen partial pressures (pO_2 values) measured in normal breast tissue (A) and in normal uterine cervix of nullipara (C) compared to locally advanced primary cancers of the breast (B) and of the uterine cervix (D).

Major pathogenetic mechanisms involved in the emergence of hypoxia in solid tumors are (a) severe structural and functional abnormalities of the tumor microvessels (perfusion-limited O_2 delivery), (b) a deterioration of the diffusion geometry (diffusion-limited O_2 delivery), and (c) tumor-associated and/or therapy-induced anemia leading to a reduced O_2 transport capacity of the blood (anemic hypoxia). There is abundant evidence for the existence of substantial heterogeneity in the tissue oxygenation status, predominantly due to the former two mechanisms.

Perfusion-limited O_2 delivery leads to *ischemic hypoxia,* which is often transient. For this reason, this type of hypoxia is also called "acute" hypoxia, a term that does not, however, take into account the mechanisms underlying this condition.

Hypoxia in tumors can also be caused by an increase in diffusion distances so that cells far away (>70 μm) from the nutritive blood vessel receive less oxygen (and nutrients) than needed. This condition is termed *diffusion-limited hypoxia* and is also known as "chronic" hypoxia. In

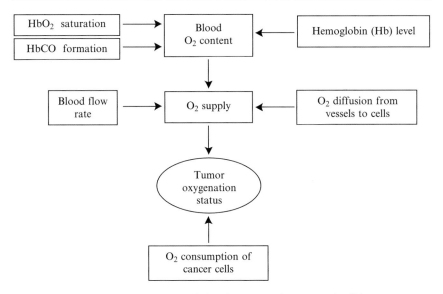

FIG. 2. Key factors determining the oxygenation status of solid tumors.

addition to enlarged diffusion distances, an adverse diffusion geometry (e.g., concurrent vs countercurrent tumor microvessels) can also cause hypoxia.

Tumor-associated or therapy-induced anemia can contribute to the development of hypoxia *(anemic hypoxia)*. Hypoxia is particularly intensified in tumors or tumor areas exhibiting low perfusion rates. A similar condition can be caused by carboxyhemoglobin (HbCO) formation in heavy smokers, which can lead to *toxic hypoxia*, because hemoglobin blocked by carbon monoxide (CO) can no longer transport oxygen. Very often, tumor microvessels are perfused (at least transiently) by plasma only. Where this occurs, hypoxia develops very rapidly around these vessels because only a few tumor cells at the arterial end can be supplied adequately under the given conditions *(hypoxemic hypoxia)*.

There is abundant evidence for the existence of substantial heterogeneity in the development and extent of tumor hypoxia due to pronounced intratumor (and intertumor) variability in vascularity and perfusion rates.[22,23]

While normal tissues can physiologically compensate for these O_2 deficiency states, locally advanced tumors (or at least larger tumor areas) cannot adequately counteract the restriction in O_2 supply and thus the

[23] P. Vaupel, F. Kallinowski, and P. Okunieff, *Cancer Res.* **49,** 6449 (1989).

development of hypoxia. In normal tissues, ischemia can (primarily) be compensated by an increase in the O_2 extraction from the blood and anemia by a rise in the local blood flow rate.

Methods for Detection of Tumor Hypoxia

Assessment of the tumor oxygenation status by invasive and noninvasive procedures has been reviewed previously[1,24-27] (Table I). So far, the most direct method for identifying hypoxia in solid tumors is the *polarographic measurement of O_2 partial pressure* (pO_2) *distributions* using O_2-sensitive microsensors. With this invasive microtechnique, frequency distributions of intratumor pO_2 values can be obtained with a relatively high spatial resolution. However, it has the disadvantage that it can only be used in tumors that are readily accessible to the microsensors, while at the same time it requires considerable operator experience for proper use.

Although some studies in experimental mouse tumors have generated conflicting results in terms of methodological reliability, the Eppendorf pO_2 histography system has nevertheless proven to be a very attractive option for use in the clinical setting. It has become the most widely used method for the polarographic measurement of pO_2 in tumors. After validation of a standard procedure according to the principle of systematic random sampling, pO_2 measurements can be performed in conscious patients before, during, and after treatment along linear tracks with measurement points at a defined distance (usually 0.7 mm) apart from each other. Following pO_2 measurement, core biopsies of approximately 2 mm in diameter can be taken from those tumor areas where pO_2 determination was made and can be processed for routine histology (e.g., for exclusion of measurements in the tumor bed or necrotic tissue areas), immunohistochemical parameters, and vascular morphometry. Ultrasound guidance or computed tomography can be used to assist proper microsensor placement.

The accuracy of polarographic needle electrodes in monitoring very low O_2 levels characteristic of radiation resistance has been evaluated in tumors with a tissue pO_2 of 0 mm Hg. Under ideal experimental conditions, the standard deviation was found to be ±0.3 mm Hg. In the clinical setting, the standard deviation increases to almost 1 mm Hg. By considering the size of the standard deviation, the probability of negative readings

[24] J. D. Chapman, *Radiother. Oncol.* **20**, 13 (1991).

[25] H. B. Stone, J. M. Brown, T. L. Phillips, and R. M. Sutherland, *Radiat. Res.* **136**, 422 (1993).

[26] J. D. Chapman, E. L. Engelhardt, C. C. Stobbe, R. F. Schneider, and G. E. Hanks, *Radiother. Oncol.* **46**, 229 (1998).

[27] P. L. Olive, J. P. Banáth, and C. Aquino-Parsons, *Acta Oncol.* **40**, 917 (2001).

TABLE I
Methods Currently Available or Undergoing Development for Detection of
Tumor Hypoxia[a]

1. Invasive microsensor techniques for direct tissue pO_2 measurements
   Polarographic O_2 sensors
   Luminescence-based optical sensors

2. Electron paramagnetic resonance oximetry

3. Techniques for intravascular O_2 detection
   Cryospectrophotometry (HbO_2 saturations)
   Near-infrared spectroscopy (HbO_2 saturations)
   Phosphorescence imaging

4. Nuclear magnetic resonance spectroscopy and imaging techniques
   ^{1}H-MRI, BOLD effect
   ^{19}F-MR relaxometry

5. Noninvasive detection of sensitizer adducts
   ^{18}F-fluoromisonidazole (PET)
   ^{123}I-iodoazomycin-arabinoside (SPECT)

6. Immunohistochemistry with exogenous hypoxic markers
   Misonidazole (^{3}H-MISO)
   Pimonidazole (PIMO)
   Etanidazole (EF5)
   Nitroimidazole-theophylline (NITP)

7. Immunohistochemistry with endogenous hypoxic markers
   Hypoxia-inducible factor-1α (HIF-1α)
   Carbonic anhydrase-IX (CA-IX)
   Glucose transporter-1 (GLUT-1)
   Vascular endothelial growth factor (VEGF)
   Urokinase-type plasminogen activator system

8. Assessment of lactate levels in cryosections by imaging bioluminescence

[a] Histomorphometric and DNA strand break (after 2–4 Gy) assays as indirect measures of tumor hypoxia are not listed. pO_2, oxygen partial pressure; HbO_2, oxyhemoglobin saturation; MRI, nuclear magnetic resonance imaging, PET, positron emission tomography; SPECT, single photon emission computed tomography; BOLD, blood oxygen level dependent.

being obtained if the true tissue pO_2 value is 0 mm Hg can be calculated. For example, with a standard deviation of 1 mm Hg the probability of a pO_2 measurement being less than or equal to -1 mm Hg is about 16% and the probability of a reading being less than or equal to -2 mm Hg is approximately 2%.[28] Thus negative pO_2 readings (at least down to -1 mm Hg) should be interpreted as being consistent with a tissue $pO_2 = 0$ mm Hg

28 O. Thews and P. Vaupel, *Strahlenther. Onkol.* 172, 239 (1996).

and should be included in the data analysis, especially as an omission of these values would lead to an overestimation of O_2 levels. In order to describe the technical quality of the pO_2 measurements, it is recommended that the fraction of pO_2 readings below 0 mm Hg should be stated as well as the most negative pO_2 value for the respective measurement series.

Since 1996, the assessment of tumor (and normal tissue) oxygenation has also been possible using *fiber-optic O_2 sensors*.[29] The O_2-dependent quenching of a ruthenium luminophor fluorescent light pulse is the underlying measuring principle of this novel method. To our knowledge, this technology has not yet been used systematically in the clinical setting, presumably because spatial monitoring with the currently available devices presents a problem in the clinical routine and because of a limited measurement range.

Electron paramagnetic resonance (EPR) *oximetry* may be a useful tool because it can be used to assess pO_2 in the same tumor region over the course of tumor growth and during response to treatment. This method is minimally invasive (requiring only application of the paramagnetic material) and may in the future allow a direct assessment of the tumor oxygenation status.[30,31] Early clinical studies are in progress (H. Swartz, personal communication).

Cryospectrophotometric measurement of *intravascular oxyhemoglobin saturations* (HbO_2) is an *ex vivo* method that potentially should allow the characterization of the oxygenation status in human tumors.[32] It has a high spatial resolution but sampling is limited to within 10 mm of the tumor surface due to the limitations of *in vivo* tissue freezing. However, as with other techniques [near infrared spectroscopy, phosphorescence imaging, nuclear magnetic resonance imaging (MRI) of fluorinated blood substitutes], only information related to the vascular compartment can be obtained, and the situation in the extravascular space can only be inferred. The technique and equipment necessary for application of near infrared spectroscopy in human tumors are still being developed and data are therefore as yet only anecdotal.

[29] D. R. Collingridge, W. K. Young, B. Vojnovic, P. Wardman, E. M. Lynch, S. A. Hill, and D. J. Chaplin, *Radiat. Res.* **147,** 329 (1997).
[30] H. M. Swartz, G. Bacic, B. Friedman, F. Goda, O. Grinberg, P. J. Hoopes, J. Jiang, K. J. Liu, T. Nakashima, J. O'Hara, and T. Walczak, *Adv. Exp. Med. Biol.* **361,** 119 (1994).
[31] H. J. Halpern, C. Yu, M. Peric, E. Barth, D. J. Grdina, and B. A. Teicher, *Proc. Natl. Acad. Sci. USA* **91,** 13047 (1994).
[32] W. Mueller-Klieser, P. Vaupel, R. Manz, and R. Schmidseder, *Int. J. Radiat. Oncol. Biol. Phys.* **7,** 1397 (1981).

MRI procedures for measurement of the oxygenation status of human tumors have the potential advantage of allowing the assessment of hypoxia in deep-seated tumors. This method is noninvasive and thus does not disturb the tumor vasculature or perfusion. Repeated measurements are possible and sampling errors are thus minimized. The limitations, however, include low sensitivity, poor quantification, and loss of a detailed spatial resolution due to volume averaging (tumor voxels of about 1 ml). Furthermore, the parameter measured is not directly interpretable as tissue pO_2 or O_2 concentration.

In contrast, *immunohistochemistry with exogenous hypoxia markers* such as 2-nitroimidazoles can provide detailed information concerning hypoxia at the cellular level. However, these methods require injection of a compound some hours before tumor resection or biopsy, are invasive (and thus subject to sampling errors), and are only possible in tumors accessible to biopsy. Disadvantages include the need for injection or ingestion of a hypoxia marker compound and the fact that the intensity of binding can be affected by metabolic factors that vary from tumor to tumor.

One further aspect requiring consideration when assessing tumor hypoxia and comparing the degree of hypoxia measured using the different methods is that of necrosis. Investigators need to be aware of whether the method they are using detects hypoxia, regardless of tissue viability, or whether hypoxia is being measured solely in viable cells. Methods detecting hypoxia per se such as polarographic, EPR, or MRI techniques will tend to overestimate the amount of viable hypoxic tissue depending on the degree of necrosis. Methods using hypoxia markers that have to be taken up by viable cells may provide a more accurate assessment, as hypoxic necrotic areas will not be labeled. Care also needs to be taken when comparing the dimensions of the hypoxic fraction from studies using different methods.

Because of problems with the techniques used so far, there are efforts to find alternative methods for estimating tumor hypoxia. One possibility is the use of so-called *endogenous hypoxia markers* (see later). The levels of these markers are increased under hypoxic conditions, and most of them can be detected in tumor sections (even in archived materials) by immunohistochemistry. This approach requires neither the injection of foreign tracer nor any additional procedure beyond that of taking a tumor biopsy.

Immunohistochemistry with endogenous (intrinsic) hypoxic markers encompasses the detection of proteins involved in the hypoxic "stress response" of tumor cells. These include transcription factor hypoxia-inducible factor-1α (HIF-1α), carbonic anhydrase-IX (CA-IX), glucose transporter-1 (GLUT-1), and—albeit less reliably—vascular endothelial growth factor (VEGF).

These proteins are potential surrogate parameters of tumor hypoxia. They appear to be prognostic parameters in many tumor entities, although they are not exclusively hypoxia specific. CA-IX appears to be superior to HIF-1α[33] for clinical use, although it shows a slower response to changes in tumor oxygenation.

Role of Hypoxia in Malignant Progression

When tumors develop, they often become more malignant with time, a process termed tumor progression. Substantial new data suggest that tumor hypoxia or anoxia (i.e., no measurable oxygen) and the HIF system are greatly involved in processes conferring a growth advantage to tumor cells and the development of a more malignant phenotype.[1,34–38] Depending on the level and (perhaps) duration of hypoxia, three mechanisms may be involved in hypoxia-induced tumor propagation: alterations in gene expression with subsequent changes of the proteome and/or changes in the genome and clonal selection.

Hypoxia-Induced Alterations in Gene Expression

HIF-1α-Dependent Mechanisms. Tumor cells respond—at the latest—to oxygen levels below 1% (7 mm Hg) with distinct changes in gene expression.[39] Below this critical threshold this exponentially increasing reaction involves the activation of a genetic program used by mammalian cells to adapt to hypoxic stress. However, while physiologically occurring hypoxia is generally of a transient nature, malignant tumors often exhibit chronically hypoxic areas, partially as a consequence of their markedly elevated proliferation rate and a subsequent increase in oxygen demand. In many malignant tumors, hypoxia-induced changes in the gene expression therefore persist. This is in contrast to the physiological response to hypoxia (e.g., during wound healing). The genetic hypoxia response program is controlled by a number of oxygen-regulated transcription factors, among which hypoxia-inducible factor 1 (HIF-1) is of pivotal importance. HIF-1 is a phylogenetically highly conserved heterodimeric transcription factor, which consists of two subunits. One of these is oxygen regulated (HIF-1α)

[33] J. M. Brown and Q.-T. Le, *Int. J. Radiat. Oncol. Biol. Phys.* **54,** 1299 (2002).
[34] G. L. Semenza, *Crit. Rev. Biochem. Mol. Biol.* **35,** 71 (2000).
[35] G. L. Semenza, *Internal Med.* **41,** 79 (2002).
[36] G. L. Semenza, *Trends Mol. Med.* **8,** S62 (2002).
[37] A. L. Harris, *Nature* **2,** 38 (2002).
[38] T. Acker and K. H. Plate, *J. Mol. Med.* **80,** 562 (2002).
[39] B. H. Jiang, G. L. Semenza, C. Bauer, and H. H. Marti, *Am. J. Physiol.* **271,** C1172 (1996).

and the other, HIF-1β, is expressed constitutively. Because the latter subunit is capable of dimerizing with other proteins,[40] HIF-1α is a key control factor in the transcriptional response to oxygen. Oxygen regulation of HIF-1α protein expression takes place at various posttranslational levels. Among these, the hypoxic inhibition of von Hippel–Lindau protein (pVHL-)mediated proteasomal degradation, which takes place continuously under normoxic conditions, is considered to be the most important step.[41] HIF-1α is involved in the transactivation of more than 40 genes[42] through binding to a DNA sequence termed the hypoxia-responsive element (HRE). These genes can be divided into several groups according to their probable or confirmed effects on tumor pathophysiology (Table II).

One group of genes is important for *cell survival* under hypoxic conditions. Hereby, a major mechanism involves a switch from oxidative phosphorylation to ATP production via glycolysis *(metabolic adaptation)*. HIF-1α transactivates all enzymes of the glycolytic pathway, glucose transporters 1 and 3 (GLUT-1 and 3), and lactate dehydrogenase A.[43] As accumulation of lactic acid implies the development of intracellular acidosis, an overexpression of the carbonic anhydrases (CA) IX and XII would seem inevitable, and indeed, CA-IX has been found to be one of the most strongly transactivated of the HIF-1α targets.[44] This HIF-controlled switch to glycolysis may at the same time, at least partially, meet the increased demand of the tumor for nucleotide metabolites needed for DNA synthesis, which are "by-products" of this metabolic pathway.[45] HIF-1α is also involved in increasing oxygen supply, which, apart from a stimulation of angiogenesis (see later), is achieved by an elevation of hemoglobin levels via the transactivation of erythropoietin, transferrin, transferrin receptor, and the rate-limiting enzymes of heme biosynthesis (5-aminolevulinate synthase) and catabolism (heme oxygenase-1). HREs have also been identified in many *growth factor* genes, including those for VEGF, PDGF, TGF-α, TGF-β, FGF-2, and IGF-2. Vice versa, activated growth factor pathways have been shown to augment HIF-1α-mediated transcription. Pathological activation of these pathways (e.g., by mutations in receptor tyrosine kinases) may be potentiated by tumor hypoxia or hypoxia may be

[40] H. I. Swanson, *Chem. Biol. Interact.* **141,** 63 (2002).
[41] C. W. Pugh and P. J. Ratcliffe, *Semin. Cancer Biol.* **13,** 83 (2003).
[42] G. L. Semenza, *Trends Mol. Med.* **7,** 345 (2001).
[43] N. V. Iyer, L. E. Kotch, F. Agani, S. W. Leung, E. Laughner, R. H. Wenger, M. Gassmann, J. D. Gearhart, A. M. Lawler, A. Y. Yu, and G. L. Semenza, *Genes Dev.* **12,** 149 (1998).
[44] A. Lal, H. Peters, B. St Croix, Z. A. Haroon, M. W. Dewhirst, R. L. Strausberg, J. H. Kaanders, A. J. van der Kogel, and G. J. Riggins, *J. Natl. Cancer Inst.* **93,** 1337 (2001).
[45] S. Mazurek, C. B. Boschek, and E. Eigenbrodt, *J. Bioenerg. Biomembr.* **29,** 315 (1997).

TABLE II
CHANGES IN GENE EXPRESSION AND IN THE PROTEOME FAVORING TUMOR GROWTH
AND PROGRESSION

1. HIF-1α-dependent mechanisms

Metabolic adaptation
Tumor response
Glycolytic enzymes, GLUT-1 and 3 (enhanced glycolysis)
Carbonic anhydrases IX and XII (pH regulation)
Adenylate kinase 3 (energy preservation)

Systemic response
Erythropoietin, transferrin, transferrin receptor, 5-aminolevulinate synthase, heme oxygenase-1 (elevation of hemoglobin levels)

Cell proliferation and survival
PDGF, TGF-α, TGF-β, FGF-2, and IGF-2

Angiogenesis
VEGF, Flt-1, PAI-1 Ang-2, iNOS
Downregulation of thrombospondins 1 and 2

Invasion, metastasis
PAI-1 (HRE identified)
Cathepsin D, matrix metalloproteinase 2, urokinase plasminogen activator receptor (uPAR), fibronectin 1, autocrine motility factor

2. HIF-1α-independent mechanisms

Angiogenesis
EGR-1[a] → tissue factor, NF-κB → COX-2

Suppression of anticancer immune response
e.g., suppressed maturation of dendritic cells

Inhibition of apoptosis
NF-κB → BCL-2
Activation of HSP 70

Invasion, metastasis
NF-κB → COX-2 → uPA, MMP-2
Activation/suppression of adhesion molecules (?)

[a] Early growth response factor 1; →, transactivation.

capable of initiating autocrine growth stimulatory loops, as has been shown for IGF-2.[46] HIF-1α is also involved in the *control of apoptosis* and leads to BCL-2 downregulation in embryonic stem cells[47] and to induction of the

[46] D. Feldser, F. Agani, N. V. Iyer, B. Pak, G. Ferreira, and G. L. Semenza, *Cancer Res.* **59,** 3915 (1999).

[47] P. Carmeliet, Y. Dor, J. M. Herbert, D. Fukumura, K. Brusselmans, M. Dewerchin, M. Neeman, F. Bono, R. Abramovitch, P. Maxwell, C. J. Koch, P. Ratcliffe, L. Moons, R. K. Jain, D. Collen, and E. Keshert, *Nature* **394,** 485 (1998).

proapoptotic factors BCL-2/adenovirus–EIB-19-kDa-interacting protein 3 (BNIP3) and NIP-3-like protein X (NIX) in various human carcinoma cell lines.[8] However, in the latter study, O_2 levels below 0.1% were used, and the resulting conclusion, that HIF-1α is mainly a proapoptotic factor, is contradicted by reports suggesting a strong antiapoptotic potential of the HIF-1α target genes for VEGF[48] and heme oxygenase-1,[49] as well as an increased apoptosis in HIF-1α-deficient cells[50] at 1% O_2.

A further main function of HIF-1α is the stimulation of *angiogenesis,* which is achieved through an increased expression of VEGF, VEGF receptor Flt-1, PAI-1, and Ang-2; some of the growth factors mentioned earlier also have angiogenic properties. Importantly, the resultant tumor microvessels have a pathological architecture and their function is typically not capable of fully restoring the tumor oxygen supply, thus contributing to the aforementioned maintenance of hypoxia and the resulting gene expression changes outlined here. At the same time, this pathological vessel architecture may also facilitate tumor *metastasis.* The importance of HIF-1α for *tumor invasion and metastasis* has not yet been fully elucidated. To our knowledge, the major metastasis-associated gene reported to harbor a HRE is that for PAI-1.[51] Publications have identified the hypoxia-induced expression of cathepsin D, matrix metalloproteinase 2 (MMP-2), urokinase plasminogen activator receptor (uPAR), fibronectin 1, autocrine motility factor, and others,[52,53] together with their abrogation by HIF-1α antisense treatments, although these findings could still indicate an indirect activation until the existence of HREs within these genes is verified.

HIF-1α-Independent Mechanisms. Another factor involved in the regulation of oxygen-dependent transcription is nuclear factor κB (NF-κB). NF-κB activity has been shown to be induced by hypoxia[54] and is also activated by reactive oxygen species (ROS),[55] which are generated during the process of reoxygenation. As outlined earlier, tumor oxygenation is not a

[48] J. H. Baek, J. E. Jang, C. M. Kang, H. Y. Chung, N. D. Kim, and K. W. Kim, *Oncogene* **19,** 4621 (2000).

[49] S. Tanaka, T. Akaike, J. Fang, T. Beppu, M. Ogawa, F. Tamura, Y. Miyamoto, and H. Maeda, *Br. J. Cancer* **88,** 902 (2003).

[50] S. Dai, M. L. Huang, C. Y. Hsu, and K. S. Chao, *Int. J. Radiat. Oncol. Biol. Phys.* **55,** 1027 (2003).

[51] T. Kietzmann, U. Roth, and K. Jungermann, *Blood* **94,** 4177 (1999).

[52] B. Krishnamachary, S. Berg-Dixon, B. Kelly, F. Agani, D. Feldser, G. Ferreira, N. Iyer, J. LaRusch, B. Pak, P. Taghavi, and G. L. Semenza, *Cancer Res.* **63,** 1138 (2003).

[53] H. Niizeki, M. Kobayashi, I. Horiuchi, N. Akakura, J. Chen, J. Wang, J. I. Hamada, P. Seth, H. Katoh, H. Watanabe, A. Raz, and M. Hosokawa, *Br. J. Cancer* **86,** 1914 (2002).

[54] A. Koong, E. Chen, and A. Giaccia, *Cancer Res.* **54,** 1425 (1994).

[55] C. Fan, Q. Li, D. Ross, and J. F. Engelhardt, *J. Biol. Chem.* **278,** 2072 (2003).

predominantly static but rather a dynamic trait, as growing microvessels, although often immature, still continue to reach areas with previously poorly developed vasculature. Net hypoxia is the result of the inability of this incompetent vasculature to meet the ever-expanding demands of the proliferating tumor; even so, focal reoxygenation occurs regularly. NF-κB activation, therefore closely connected with tumor hypoxia, is commonly associated with activation of the proinflammatory cytokines TNFα, IL-1β, IL-8, and monocyte chemoattractant protein-1 (MCP-1).[56] NF-κB has also been shown to play an important role in apoptosis regulation, as it leads to overexpression of the antiapoptotic factor BCL-2.[57] NF-κB additionally upregulates iNOS and COX-2.[58] While iNOS is also a HIF-1α target gene, COX-2 expression seems to be a specific target of NF-κB. COX-2 has angiogenic and growth stimulatory properties, which are mediated via PGE$_2$. COX-2 also activates the genes for uPA and MMP-2, which are associated with tumor invasiveness.[59] The connection of hypoxia and COX-2 overexpression may be of substantial clinical relevance in the future because COX-2 activity can be selectively inhibited by NSAIDs and antineoplastic properties of these drugs have already been demonstrated. Finally, NF-κB has also been shown to activate HIF-1α in response to the depolymerization of microtubules,[60] a commonly occurring event during tissue reorganization processes such as embryonic development and wound healing and also during the spread of malignant tumor cells.

Activator protein-1 (AP-1), like HIF-1 a dimeric transcription factor, has also long been identified as being involved in the hypoxia response,[61] and PDGF-B, endothelin-1 (ET-1), bFGF, and VEGF genes (among many others) contain the corresponding binding sites. Hypoxia has been shown to activate c-jun N-terminal kinase (JNK), leading to an overexpression of c-jun, the most transcriptionally active member of the large group of AP-1 subunits.[62] Prolonged AP-1 activation by hypoxia may depend on HIF-1α,[63] with both of these cooperating in the transactivation of target genes.[64] Such interactions are still not completely understood, the same

[56] C. T. D'Angio and J. N. Finkelstein, *Mol. Genet. Metab.* **71,** 371 (2000).

[57] M. Tamatani, N. Mitsuda, H. Matsuzaki, H. Okado, S. Miyake, M. P. Vitek, A. Yamaguchi, and M. Tohyama, *J. Neurochem.* **75,** 683 (2000).

[58] V. Chiarugi, L. Magnelli, A. Chiarugi, and O. Gallo, *J. Cancer Res. Clin. Oncol.* **125,** 525 (1999).

[59] G. Li, T. Yang, and J. Yan, *Biochem. Biophys. Res. Commun.* **299,** 886 (2002).

[60] Y.-J. Jung, J. S. Isaacs, S. Lee, J. Trepel, and L. Neckers, *J. Biol. Chem.* **278,** 7445 (2003).

[61] R. A. Rupec and P. A. Baeuerle, *Eur. J. Biochem.* **234,** 632 (1995).

[62] E. Shaulian and M. Karin, *Nature Cell Biol.* **4,** E131 (2002).

[63] K. R. Laderoute, J. M. Calaoagan, C. Gustafson-Brown, A. M. Knapp, G.-C. Li, H. L. Mendonca, H. E. Ryan, Z. Wang, and R. S. Johnson, *Mol. Cell. Biol.* **22,** 2515 (2002).

[64] A. Damert, E. Ikeda, and W. Risau, *Biochem. J.* **327**(Pt 2), 419 (1997).

being true for the significance of other transcription factors associated with microenvironmental hypoxia (e.g., HIF-2α, HIF-3α, see Table II for additional details).

Hypoxia-Induced Genomic Instability and Clonal Selection

The changes in gene expression outlined earlier and the resulting tumor cell phenotypic properties depend on the sustained presence of an extrinsic (hypoxic) stimulus. In marked contrast to this, malignancy itself is, in the first instance, an intrinsic trait caused by mutations of key regulatory genes and is largely independent of environmental factors. Tumor progression is often associated with the development and/or aggravation of individual facets of this intrinsic malignancy. As first outlined by Nowell,[65] this process is believed to be the result of an accumulation of sequential genetic changes leading to increasingly aggressive clonal subpopulations. Later, it was speculated that this process might require an elevated mutation rate,[66] whereas others argued that severely selective conditions in the absence of such a "mutator phenotype" are sufficient.[67] Because both phenomena— genetic instability and clonal selection—are not mutually exclusive but rather complementary to each other, it is important to note that hypoxia has been shown to have a substantial influence on both.

Oxygen levels below 0.1% O_2 (0.7 mm Hg) result in diminished nucleotide excision repair (NER), leading to point mutations.[68] The exact mechanism of this observation is unclear, although reduced levels of proteins involved in the NER-system—due to an overall reduction of protein synthesis and an increase in protein degradation at these extremely hypoxic conditions—may play a role. Additionally, low ATP levels that may limit enzyme function and a hypoxia-associated decrease in pH may also have an influence and could theoretically also affect other repair systems. Another repeatedly confirmed consequence of such a severe degree of hypoxia is the activation of a specific endonuclease[69] leading to DNA double-strand breaks.[70] Double-strand breaks can be considered as initiating events in the mutation pattern typically found in malignant tumors

[65] P. C. Nowell, *Science* **194,** 23 (1976).

[66] L. A. Loeb, *Cancer Res.* **51,** 3075 (1991).

[67] I. P. Tomlinson, M. R. Novelli, and W. F. Bodmer, *Proc. Natl. Acad. Sci. USA* **93,** 14800 (1996).

[68] J. Yuan, L. Narayanan, S. Rockwell, and P. M. Glazer, *Cancer Res.* **60,** 4372 (2000).

[69] D. L. Stoler, G. R. Anderson, C. A. Russo, A. M. Spina, and T. A. Beerman, *Cancer Res.* **52,** 4372 (1992).

[70] C. A. Russo, T. K. Weber, C. M. Volpe, D. L. Stoler, N. J. Petrelli, M. Rodriguez-Bigas, W. C. Burhans, and G. R. Anderson, *Cancer Res.* **55,** 1122 (1995).

leading to chromosomal aberrations and—via the initiation of breakage–fusion–bridge cycles—to gene amplification. Initially, these strand breaks were assumed to occur randomly, but have, in the meantime, been shown to occur preferentially at fragile sites,[71] that is, chromosomal locations exhibiting an increased likelihood of breakage. Fragile sites are often located in the vicinity of oncogenes and genes mediating drug resistance.[72] Accordingly, hypoxic cells have been shown to exhibit increased resistance to chemotherapeutic agents[73] and enhanced metastatic potential.[74] Some experiments showed an additional mutational impact of reoxygenation, which may be attributable to the production of ROS ("ischemia–reperfusion syndrome").[75] The relevance of this concept has been demonstrated during the application of repeated "cyclic" episodes of hypoxia and reoxygenation in a mouse fibrosarcoma model where an increase in lung micrometastases was observed.[76] Hypoxia also leads to polyploidy,[77] a genomic state known to be associated with increased resistance and metabolic capacity (Table III).

The role of hypoxia as a selection force has, to date, been demonstrated most convincingly for p53-mediated apoptosis. Graeber et al.[78] mixed transformed mouse embryonic fibroblasts that were either p53 deficient (p53$^{-/-}$) or proficient (p53$^{+/+}$) at a ratio of 1:1000. After seven cycles of hypoxia and subsequent reoxygenation, the p53-deficient cells had outgrown the

TABLE III
HYPOXIA-INDUCED GENOMIC INSTABILITY

Reduced DNA repair mechanisms
  → point mutations[68]

Stimulation of endonuclease activity
  → Double-strand breaks[70]

Initiation of breakage–bridge–fusion cycles
  → Gene amplification[71]

Polyploidy[77]

[71] A. Coquelle, F. Toledo, S. Stern, A. Bieth, and M. Debatisse, Mol. Cell 2, 259 (1998).
[72] A. Coquelle, E. Pipiras, F. Toledo, G. Buttin, and M. Debatisse, Cell 89, 215 (1997).
[73] G. C. Rice, C. Hoy, and R. T. Schimke, Proc. Natl. Acad. Sci. USA 83, 5978 (1986).
[74] S. D. Young, R. S. Marshall, and R. P. Hill, Proc. Natl. Acad. Sci. USA 85, 9533 (1988).
[75] J. M. McCord, N. Engl. J. Med. 312, 159 (1985).
[76] R. A. Cairns, T. Kalliomaki, and R. P. Hill, Cancer Res. 61, 8903 (2001).
[77] E. K. Rofstad, N. M. Johnsen, and H. Lyng, Br. J. Cancer 27 (Suppl.), S136 (1996).
[78] T. G. Graeber, C. Osmanian, T. Jacks, D. E. Housman, C. J. Koch, S. W. Lowe, and A. J. Giaccia, Nature 379, 88 (1996).

proficient cells, as their apoptotic rate was substantially lower. By selecting for p53 deficiency, hypoxia not only rendered the remaining population more resistant to apoptosis, but also more genomically unstable due to the function of p53 in the surveillance of genome integrity. This connection is of appreciable clinical relevance, as hypoxic cervical cancers with a low apoptotic index have been shown to exhibit a very aggressive phenotype.[17] In the meantime, other examples for this paradigm have been described: hypoxia also selects for human papillomavirus (HPV) oncoprotein-expressing cervical epithelial cells with reduced apoptotic potential[79] and for defects in the DNA mismatch repair (MMR) system.[80] The selected MMR-deficient cells showed increased resistance to certain chemotherapeutic drugs, namely 6-thioguanine and cisplatin. Interestingly, resistance to cisplatin has been observed in familial colorectal cancer (HNPCC) cells, which are known to be MMR deficient.[81]

Tumor Hypoxia and Acquired Treatment Resistance

Tumor hypoxia is classically associated with resistance to radiotherapy,[82,83] but has also been shown to diminish the efficacy of certain forms of chemotherapy and of photodynamic therapy (Table IV). Radiosensitivity to sparsely ionizing radiation (X and γ radiation) decreases rapidly when the O_2 partial pressure in a tumor reaches values below 25–30 mm Hg. While this phenomenon is probably multifactorial, one major mechanism is thought to predominate. Molecular oxygen can prevent repair of the radiation-induced DNA damage, thus making it permanent ("fixation"). Hence, the radiation dose required to achieve an equivalent tumoricidal effect is 2.8 to 3 times higher in the absence of oxygen than at normal oxygen levels. Radiation-induced DNA damage does not usually lead to cell death directly, but rather via the triggering of apoptosis. Therefore, the aforementioned selection processes favoring apoptosis-resistant clones may also influence the effectiveness of radiotherapy indirectly.

Under hypoxia, a reduced cytotoxic activity has been demonstrated for a variety of chemotherapeutic substances, among them cyclophosphamide,

[79] C. Y. Kim, M. H. Tsai, C. Osmanian, T. G. Graeber, J. E. Lee, R. G. Giffard, J. A. DiPaolo, D. M. Peehl, and A. J. Giaccia, *Cancer Res.* **57,** 4200 (1997).
[80] S. Kondo, S. Kubota, T. Shimo, T. Nishida, G. Yosimichi, T. Eguchi, T. Sugahara, and M. Takigawa, *Carcinogenesis* **23,** 769 (2002).
[81] N. Claij and H. te Riele, *Exp. Cell Res.* **246,** 1 (1999).
[82] L. H. Gray, A. D. Conger, M. Ebert, S. Hornsey, and O. C. A. Scott, *Br. J. Radiol.* **26,** 638 (1953).
[83] J. C. Mottram, *Br. J. Radiol.* **9,** 606 (1931).

TABLE IV
TUMOR HYPOXIA AND ACQUIRED TREATMENT RESISTANCE[a]

1. Direct effects	
Reduced "fixation" of DNA damage	X and γ rays
Reduced generation of free radicals	Antibiotics (bleomycin, doxorubicin)
2. Indirect effects	
Changes in gene expression ($<1\%$ O_2)	
Proliferation kinetics, cell cycle effects	Vinca alkaloids, methotrexate
DNA repair enzymes $\uparrow$[b]	Alkylating agents, platinum compounds
Glutathione levels $\uparrow$	Alkylating agents, platinum compounds
P-Glycoprotein (MDR)	Adriamycin, paclitaxel
GLUT-1	Vinblastine
Genomic changes ($<0.1\%$ O_2)	
Amplification of resistance-related genes (e.g., DHFR)[c]	Methotrexate
Resistance to apoptosis	Various agents
Mismatch-repair deficiency	Cisplatin

[a] Adapted from Vaupel and Höckel.[20]
[b] Upregulation.
[c] Dihydrofolate reductase.

carboplatin, and other alkylating agents, actinomycin D, adriamycin, bleomycin, vincristine, etoposide, and 5-FU.[84] In contrast, mitomycin C was found to be more effective under hypoxic conditions, an observation that has also been confirmed by others.[85,86] Even so, these investigations were of a qualitative nature, and clear hypoxic thresholds for O_2-dependent anticancer agents, although they presumably exist, are still not available. Thus, additional studies are necessary for the collection of quantitative data on hypoxia-induced chemoresistance, although they may be difficult to perform under *in vivo* conditions. Multiple (direct and indirect) mechanisms are probably involved in the hypoxia-induced resistance to chemotherapeutic agents, including a reduced generation of free radicals (e.g., bleomycin, anthracyclines), an increased production of nucleophilic substances such as glutathione, which can compete with the target DNA for alkylation (e.g., in acquired resistance to alkylating agents), an increased activity of DNA repair enzymes (e.g., alkylating agents, platinum compounds), an inhibition

[84] B. A. Teicher, S. A. Holden, A. al-Achi, and T. S. Herman, *Cancer Res.* **50**, 3339 (1990).
[85] M. Yamagata, T. Kanematsu, T. Matsumata, T. Utsunomiya, Y. Ikeda, and K. Sugimachi, *Eur. J. Surg. Oncol.* **18**, 379 (1992).
[86] K. A. Kennedy, J. M. Siegfried, A. C. Sartorelli, and T. R. Tritton, *Cancer Res.* **43**, 54 (1983).

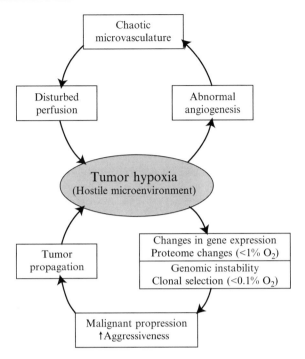

FIG. 3. The "deadly roller coaster" of tumor hypoxia as a consequence of two interwoven vicious circles.

of cell proliferation, and tissue acidosis, which is often observed in hypoxic tumors exhibiting a high glycolytic rate. Furthermore, overexpression of the MDR-1 gene,[87] MMR deficiency (see earlier discussion), and the loss of apoptotic potential can also impart resistance to certain chemotherapeutic drugs.

Tumor Hypoxia as a Prognostic Factor

An adverse prognostic impact of tumor hypoxia in various tumor entities—among them cancers of the uterine cervix, head and neck, and soft tissue sarcomas—has been demonstrated repeatedly.[22] In cervical carcinomas, this impact on prognosis was independent of treatment modality, being found even in cases treated with surgery alone.[12] This finding speaks strongly in favor of a hypoxia-induced enhancement of aggressiveness as a consequence of malignant progression (Fig. 3). Molecular studies investigating the tissue distribution of HIF-1α and of its target proteins

[87] K. M. Comerford, T. J. Wallace, J. Karhausen, N. A. Louis, M. C. Montalto, and S. P. Colgan, *Cancer Res.* **62,** 3387 (2002).

CA-9 and GLUT-1 also almost uniformly showed poorer outcome in cases exhibiting an overexpression of these endogenous markers.[36,88] Substantial evidence therefore exists for the occurrence of hypoxia leading to increased aggressiveness in malignant cells. However, the possibility that this causality is reversed also needs to be considered, in which case the inherently most malignant tumors would necessarily also be the most hypoxic.[89,90] On closer inspection, these two possibilities may not automatically contradict each other, but may rather even be complementary. If hypoxia indeed promotes the malignant phenotype, then the most (inherently) malignant tumor cells would in turn be expected to encourage the development of a hypoxic environment. A combination of both possibilities would mean that hypoxia is the cause of increased aggressiveness, as it promotes tumor progression, while at the same time being the consequence of aggressive malignant growth that leads to defective ("chaotic") vascular morphology and function and other alterations of the nonmalignant part of the tumor, thereby creating an environment adjusted to the pathophysiological demands of the tumor.

Acknowledgment

We thank Dr. Debra Kelleher and Dr. Cornelia Leo for critical reading and editorial help.

[88] J. A. Loncaster, A. L. Harris, S. E. Davidson, J. P. Logue, R. D. Hunter, C. C. Wycoff, J. Pastorek, P. J. Ratcliffe, I. J. Stratford, and C. M. West, *Cancer Res.* **61,** 6394 (2001).
[89] C. N. Coleman, J. B. Mitchell, and K. Camphausen, *J. Clin. Oncol.* **20,** 610 (2002).
[90] P. Okunieff, I. Ding, P. Vaupel, and M. Höckel, *Adv. Exp. Med. Biol.* **510,** 69 (2003).

Section V

Metabolism

[24]   Oxygen-Dependent Regulation of Hepatic Glucose Metabolism

By THOMAS KIETZMANN

Introduction

Aerobic-living animals mainly produce energy, which is present in the form of adenosine triphosphate (ATP), by the aerobic oxidative metabolism of glucose, fatty acids, and amino acids forming CO_2 and urea, respectively. The formation of energy by the aerobic oxidative metabolism of lactate, glycerol, or ethanol, as well as the formation and utilization of ketone bodies, constitutes a minor part. To some extent, energy can also be gained by the anaerobic conversion of glucose to lactate. However, various cells in different organs are specialized in energy metabolism. Because neurons of the central nervous system can only use glucose and ketone bodies as energy substrates, they always require glucose. Erythrocytes and nephrocytes of the renal medulla do not possess mitochondria and thus are incapable of aerobic energy production; as a result, these cells absolutely need glucose. Thus, the permanent provision of glucose is the "leitmotiv" of the energy metabolism of animals.[1]

Aerobic-living organisms generate energy in a redox process where the energy substrates, for example, of food, serve as electron donors and oxygen as the electron acceptor. During combustion the energy substrates are completely dehydrogenated to CO_2 with the formation of reducing equivalents [H], mainly NADH, which are required for the formation of a proton gradient enabling the phosphorylation of adenosine diphosphate (ADP) into adenosine triphosphate (ATP). The generated electrons are then used to hydrogenate O_2 to water within the electron transport chain. This aerobic redox process, called *oxidative phosphorylation,* is the major source of ATP generation, whereas so-called *substrate level phosphorylation* constitutes only a minor part.

Under anaerobic conditions, energy can also be generated in a redox process composed of substrate dehydrogenation and acceptor hydrogenation. Substrates are exclusively carbohydrates, which are only partially dehydrogenated to pyruvate under the formation of NADH. Pyruvate functions as the acceptor for the electrons of NADH under the formation

[1] K. Jungermann and T. Kietzmann, *Annu. Rev. Nutr.* **16**, 179 (1996).

METHODS IN ENZYMOLOGY, VOL. 381

of lactate and regeneration of NAD^+. ATP is formed only by substrate level phosphorylation.[2]

The major nutrient carbohydrate glucose can be utilized by all cells to produce ATP and additionally in hepatocytes, myocytes, and cardiomyocytes to form glycogen, which serves as an energy store. Once glucose has reached the liver via the portal vein, glucose is transported into the cytosol via the glucose transporter GLUT2. Except for GLUT2, all other glucose transporters have a K_M for glucose clearly below the range of the plasma glucose concentration, indicating that glucose uptake into hepatocytes is only a function of glucose supply. Only GLUT4, the major glucose transporter of muscle and adipocyte tissue, is activated by insulin so that glucose uptake into myocytes, cardiomyocytes, and adipocytes is controlled hormonally.[3] Thus the role of the liver in energy metabolism can be simplified in the provision of glucose as a major energy substrate in all metabolic situations, thus functioning as a "glucostat."[1] Therefore, it is necessary that the liver is supplied with substrates and O_2.

Glycolysis: Degradation of Glucose to Acetyl-CoA

Glucose is converted to glucose 6-phosphate by hexokinases, which exist in several isoforms. The glucokinase (GK) (= hexokinase IV) represents the major active form of the hexokinases in the liver, whereas hexokinases I–III are operative in the major other organs.[4] Furthermore, GK expression is restricted to hepatocytes, pancreatic β cells, and some neuroendocrine cells of the gastrointestinal tract.[5] Contrary to other hexokinase family members (hexokinase I–III), the GK (hexokinase IV) has a lower affinity (higher K_M) for glucose with sigmoidal kinetics and is not inhibited by its reaction product glucose 6-phosphate.[6]

After phosphorylation of glucose to glucose 6-phosphate glucose is degraded via the glucose 6-phosphate isomerase reaction, continued by phosphofructokinase and fructose bisphosphate aldolase plus triose phosphate isomerase reactions, followed by glyceraldehydephosphate dehydrogenase, phosphoglycerate kinase, phosphoglycerate mutase, enolase, and pyruvate kinase (PK), and finally by pyruvate dehydrogenase reactions. The net energy outcome from the conversion of one molecule glucose produced by substrate phosphorylation from 1,3-bisphosphoglycerate is 2 ATP.

[2] R. K. Thauer, K. Jungermann, and K. Decker, *Bacteriol. Rev.* **41,** 100 (1977).
[3] H. G. Joost and B. Thorens, *Mol. Membr. Biol.* **18,** 247 (2001).
[4] P. B. Iynedjian, *Biochem. J.* **293,** 1 (1993).
[5] T. L. Jetton, Y. Liang, C. C. Pettepher, E. C. Zimmerman, F. G. Cox, K. Horvath, F. M. Matschinsky, and M. A. Magnuson, *J. Biol. Chem.* **269,** 3641 (1994).
[6] R. L. Printz, M. A. Magnuson, and D. K. Granner, *Annu. Rev. Nutr.* **13,** 463 (1993).

Synthesis and Degradation of Glycogen Stores

The initial reaction carried out by the glucokinase, that is, the phosphorylation of glucose to glucose 6-phosphate, is also the starting point for glycogen synthesis, which is then continued by the glucose phosphate mutase reaction and thereafter by UDP-glucose pyrophosphorylase, which forms UDP-glucose as a starter molecule. Then glycogen synthase elongates the glycogen starter by adding up to 12–14 glucose molecules from UDP-glucose to each 4 position of a terminal glucose of the primer until a 1,6 branch point is reached. The 1,4-1,6-transglucosidase transglucosidase (branching enzyme) then transfers a terminal oligosaccharide of about seven glucose molecules from a 1,4 linkage and forms a new 1,6 branch point at a distance of four monomers from the previous 1,6 branch point.

Glycogen degradation starts with the action of glycogen phosphorylase, which shortens the glucose chains by releasing glucose 1-phosphate with inorganic phosphate. It stops about four monomers from a 1,6 branch point, thus allowing the transglucosidase to transfer a trisaccharide from a 1,6-branched side chain to a 1,4-linked main chain. The left single 1,6-linked glucose molecule is then removed by amyloglucosidase. Glucose phosphate mutase then generates an equilibrium between glucose 1-phosphate and glucose 6-phosphate. Glucose 6-phosphate can only be converted via glucose 6-phosphatase to glucose in the liver. All other glycogen-storing organs channel glucose 6-phosphate into glycolysis.

Gluconeogenesis: De Novo *Synthesis of Glucose*

Glucose, which is needed by the organism and cannot be formed from the glycogen stores of the liver, is synthesized *de novo* mainly in the liver from pyruvate. Lactate and the so-called glucogenic amino acids are the major precursors of pyruvate. Gluconeogenesis begins with the pyruvate carboxylase and phosphoenolpyruvate carboxykinase (PCK) reactions. The latter is dependent on energy in the form of GTP and is considered to be the major rate-generating step in gluconeogenesis. PCK can be located in both the cytosol (PCK1) and the mitochondria (PCK2). The cytosolic to which we will refer as PCK and mitochondrial gene has been cloned.[7,8] The cytosolic PCK is the gluconeogenic key enzyme; it is expressed primarily in the liver and kidney cortex.[9] In the liver, expression is stimulated mainly by glucagon via cAMP under the permissive action

---

[7] S. Modaressi, B. Christ, J. Bratke, S. Zahn, T. Heise, and K. Jungermann, *Biochem. J.* **315,** 807 (1996).

[8] S. Modaressi, K. Brechtel, B. Christ, and K. Jungermann, *Biochem. J.* **333,** 359 (1998).

[9] S. J. Pilkis and D. K. Granner, *Annu. Rev. Physiol.* **54,** 885 (1992).

of glucocorticoids and thyroid hormones; it is inhibited by insulin. In the kidney, expression is enhanced mainly by acidosis via decreases in pH or in bicarbonate levels, which act independently from each other.[10,11]

After phosphoenolpyruvate is formed, it is continued by the enolase, phosphoglycerate mutase, phosphoglycerate kinase, and glyceraldehydephosphate dehydrogenase reactions followed by the fructose bisphosphate aldolase plus triose phosphate isomerase, fructose 1,6-bisphosphatase (FBPase), glucose 6-phosphate isomerase, and glucose 6-phosphatase (G6Pase) reactions. The two antagonistic pathways gluconeogenesis and glycolysis differ between pyruvate and phosphoenolpyruvate and fructose 1,6-bisphosphate, and fructose 6-phosphate, as well as glucose 6-phosphate and glucose, due to energetic and regulatory reasons.

Metabolic Zonation of Liver, Oxygen Gradients, and Modulation of Carbohydrate Metabolism by O_2

In the liver parenchyma, the key enzymes of the glycolytic and gluconeogenic pathway and thus the metabolic capacities are distributed zonally. The glucose-forming enzymes PCK, FBPase, and G6Pase had 2- to 4-fold higher activities periportally, whereas GK and liver type PK had 2.5- to 3-fold higher activities perivenously. Concomitantly, the protein levels were also distributed heterogenously as demonstrated by immunohistochemical techniques.[12] Apparently, periportal cells have a higher capacity for gluconeogenesis, whereas perivenous cells have a higher capacity for glycolysis. These observations constituted the basis for the model of *metabolic zonation*, which was then extended to include not only carbohydrates, but also other major pathways.[13–17]

Due to metabolism in the liver, periportal to perivenous concentration gradients of, for example, substrates and hormones, are formed that may constitute important signals for metabolic zonation. While the zonal concentration gradients of most major carbon substrates, such as glucose or amino acids, are rather shallow, the gradient of oxygen is of special importance; its partial pressure (free concentration) is about 60–65 mm Hg (84–91

[10] A. L. Gurney, E. A. Park, J. Liu, M. Giralt, M. M. McGrane, Y. M. Patel, D. R. Crawford, S. E. Nizielski, S. Savon, and R. W. Hanson, *J. Nutr.* **124,** 1533S (1994).
[11] R. W. Hanson and L. Reshef, *Annu. Rev. Biochem.* **66,** 581 (1997).
[12] K. Jungermann and T. Kietzmann, *Hepatology* **31,** 255 (2000).
[13] D. Haussinger, W. H. Lamers, and A. F. Moorman, *Enzyme* **46,** 72 (1992).
[14] R. Gebhardt and F. Gaunitz, *Cell Biol. Toxicol.* **13,** 263 (1997).
[15] J. J. Gumucio, *Am. J. Physiol.* **244,** G578 (1983).
[16] B. Quistorff, N. Katz, and L. A. Witters, *Enzyme* **46,** 59 (1992).
[17] T. Oinonen and K. O. Lindros, *Biochem. J.* **329,** 17 (1998).

μmol/liter) in the periportal blood and falls to about 30–35 mm Hg (42–49 μmol/liter) in the perivenous blood.[18,19]

The observation that O_2 can significantly modify carbohydrate metabolism has long been known from studies in perfused livers. However, difficulties arise during these studies with the isolated perfused liver, as the pO_2 in the perfusate decreases considerably from the upstream to the downstream zone. Primary hepatocyte cultures have brought about an alternative due to the fact that they can be kept under a defined pO_2 resembeling arterial periportal (experimental gas atmosphere: 13–16% O_2 [v/v]) and perivenous (5–7% O_2 [v/v]) pO_2.

In this system, CO_2 formation increased in proportion to the oxygen concentration up to 6% O_2 (v/v) and then stayed essentially constant. Net glycogen synthesis occurred only above 4% O_2. Net glucose uptake occurred above 2% O_2 and stayed constant above 4% O_2. Net glucose output and net lactate uptake (due to gluconeogenesis) in periportal-like cells increased clearly up to 6% O_2 and then more moderately; net glucose output (due to glycogenolysis) increased below 2% O_2 and net lactate output (due to glucose uptake) below 6% O_2 in perivenous-like cells. Most interestingly, net flow between glucose 6-phosphate and pyruvate in the gluconeogenic direction was enhanced with increasing pO_2 in periportal-like cells and, conversely, was enhanced in the glycolytic direction with decreasing pO_2 in perivenous-like cells. The percentage of futile cycling was decreased in periportal-like cells with increasing pO_2 and in perivenous-like cells with decreasing pO_2. Thus, both periportal and perivenous cells operate at higher efficiencies at their physiological pO_2, that is, with less futile cycling.[20–23]

Moreover, as found in early studies with 48-h primary rat hepatocyte cultures, PCK and GK were induced to higher activities at periportal and perivenous oxygen tensions, respectively. Furthermore, in 24-h hepatocyte cultures, PCK activities were induced by glucagon within 4 h to higher levels at arterial than at perivenous oxygen tensions.[19] These findings were corroborated by the demonstration in 24-h hepatocyte cultures that the glucagon–dependent increase in PCK mRNA was higher under arterial than under venous oxygen tensions and that, conversely, the insulin-dependent increase in GK mRNA was higher under venous than under arterial oxygen tensions (Fig. 1).[24–27]

[18] K. Jungermann and N. Katz, *Physiol. Rev.* **69**, 708 (1989).
[19] M. Nauck, D. Wolfle, N. Katz, and K. Jungermann, *Eur. J. Biochem.* **119**, 657 (1981).
[20] K. Jungermann and T. Kietzmann, *Kidney Int.* **51**, 402 (1997).
[21] K. Jungermann and T. Kietzmann, *Annu. Rev. Nutr.* **16**, 179 (1996).
[22] D. Wolfle and K. Jungermann, *Eur. J. Biochem.* **151**, 299 (1985).
[23] D. Wolfle, H. Schmidt, and K. Jungermann, *Eur. J. Biochem.* **135**, 405 (1983).

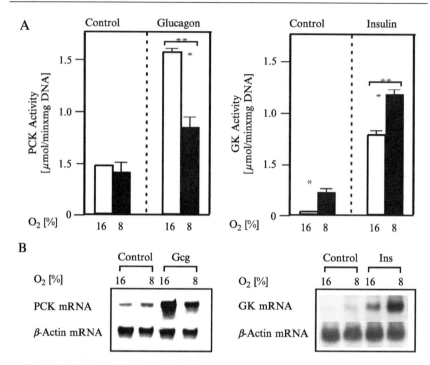

FIG. 1. Modulation of PCK and GK enzyme activity, as well as their mRNA expression by arterial and venous pO$_2$ in rat hepatocyte cultures. (A) Hepatocytes were cultured for 24 h under arterial pO$_2$. Then, medium was changed, when indicated, 1 nM glucagon (Gcg) or insulin (Ins) was added, and the culture was continued for another 4 h for PCK activity and 24 h for GK activity under arterial or venous pO$_2$. PCK and GK enzyme activity values are means ±SEM of culture experiments, each performed in duplicate with hepatocytes from three different preparations. Student's t test for paired values: an asterisk indicates significant differences 16% O$_2$ vs 8% O$_2$, $p \leq 0.05$. Double asterisks indicate significant differences in glucagon or insulin treated at 16% O$_2$ or 8% O$_2$ versus controls at 16% O$_2$ or 8% O$_2$, $p \leq 0.05$. (B) Representative Northern blot. Twenty micrograms of total RNA prepared from cultured hepatocytes was hybridized to digoxigenin-labeled PCK, GK, and β-actin antisense RNA probes. Autoradiographic signals were obtained by chemiluminescence and scanned by video densitometry.

[24] J. Hellkamp, B. Christ, H. Bastian, and K. Jungermann, *Eur. J. Biochem.* **198,** 635 (1991).

[25] T. Kietzmann, H. Schmidt, F. K. Unthan, I. Probst, and K. Jungermann, *Biochem. Biophys. Res. Commun.* **195,** 792 (1993).

[26] T. Kietzmann, U. Roth, S. Freimann, and K. Jungermann, *Biochem. J.* **321,** 17 (1997).

[27] U. Roth, K. Jungermann, and T. Kietzmann, *Biochem. J.* **365,** 223 (2002).

O_2 Responsive Elements and Transcription Factors Controlling Gene Expression

Transient transfection studies with primary hepatocytes showed that modulation by O_2 of the glucagon-dependent induction of the gluconeogenic PCK gene was due to a normoxia response element (NRE) in the PCK promoter at -146/-139[28] (Fig. 2). The factor binding to the NRE is not known yet; however, it was shown that the NRE DNA-binding activity was strictly dependent on the pO_2 and susceptible to redox modifications.[29] The genes for other periportally expressed enzymes, such as tyrosine aminotransferase and serine dehydratase, have promoter elements identical to the NRE of the PCK gene. It remains to be established whether these sequences can function as NREs.

The transcription factor hypoxia-inducible factor 1 (HIF-1) was found to be involved in the hypoxia-induced expression of nearly all glycolytic enzymes. The response to HIF-1 is mediated by the binding of HIF-1 to hypoxia response elements (HRE), which can also be found in a number of other physiologically important genes such as erythropoietin, vascular endothelial growth factor, or plasminogen activator inhibitor 1.[30,31]

Based on these results obtained with primary hepatocytes, this article provides several protocols allowing the measurement of metabolic activities under the control of O_2, as well as studying the modulatory role of O_2 for gene expression in primary hepatocytes.

Experimental Methods

All biochemicals and enzymes should be of analytical grade and can be purchased from different commercial suppliers if not explicitly stated.

Preparation and Culture of Primary Hepatocytes

Animal surgery and isolation of hepatocytes are performed under aseptic conditions via collagenase perfusion of the rat liver according to the method described by Berry and Friend[32] with some modifications. For this, male Wistar rats (200–260 g) are kept on a 12-h day/night rhythm with free access to water and food. Rats are anesthetized with pentobarbital (60 mg/kg body weight) prior to the preparation of hepatocytes.

[28] J. Bratke, T. Kietzmann, and K. Jungermann, *Biochem. J.* **339**, 563 (1999).
[29] T. Kietzmann, J. Bratke, and K. Jungermann, *Kidney Int.* **51**, 542 (1997).
[30] G. L. Semenza, *Crit. Rev. Biochem. Mol. Biol.* **35**, 71 (2000).
[31] T. Kietzmann, U. Roth, and K. Jungermann, *Blood* **94**, 4177 (1999).
[32] M. N. Berry and D. S. Friend, *J. Cell Biol.* **43**, 506 (1969).

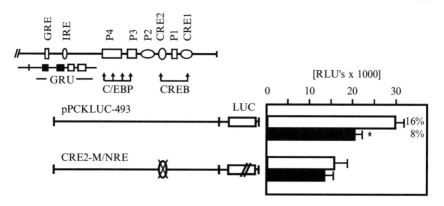

Fig. 2. Mutation of the normoxia response element (CRE2/NRE) abolished modulation by O_2 of the glucagon-dependent induction of the PCK promoter in primary rat hepatocytes. The 5′-flanking region of the rat cytosolic PCK gene with its *cis*-acting regulatory elements.[11] Hepatocytes were transfected by the calcium phosphate precipitation method with luciferase (LUC) gene constructs driven by a 493-bp PCK promoter (pPCKLUC-493) or mutated (M) at the normoxia responsive element (NRE, formerly known as CRE2). Transfected cells were cultured under standard conditions and after 27 h were induced with 10 *nM* glucagon each under arterial (16% O_2 by volume) and venous (8% O_2 by volume) pO_2 for 12 h. Values are means ± SEM. CRE, cAMP regulatory element; CREB, CRE-binding protein; GRE, glucocorticoid response element; GRU, glucocorticoid response unit; IRE, insulin response element; C/EBP, CAAT/enhancer-binding protein; P1–P4, promoter elements; RLU, relative light unit.

Materials

  Perfusion pump
  Oxygenator
  Media and Solutions for Hepatocyte Preparation and Culture. All media and solutions are autoclaved and stored at 4°.

  Krebs–Ringer stock:

	g/liter	Final concentration (mM)
NaCl	7	120.0
KCl	0.36	4.8
MgSO$_4$ × 7H$_2$O	0.296	1.2
KH$_2$PO$_4$	0.163	1.2
NaHCO$_3$	2.016	24.4

Solution needs to be carbogen equilibrated and adjusted to pH 7.35

Preperfusion medium:

	g/liter	Final concentration (mM)
EGTA	0.1	0.25

dissolved in Krebs–Ringer stock solution

Collagenase perfusion medium:

	g/liter	Final concentration (mM)
HEPES	3.356	15
$CaCl_2 \times 2H_2O$	0.588	4
Collagenase	0.500	

The collagenase needs to be dissolved in Krebs–Ringer stock solution, equilibrated with carbogen for 30 min, and then sterile filtered.

Washing buffer:

	g/liter	Final concentration
HEPES/NaOH, pH 7.4	4.77	20 mM
NaCl	7.00	120 mM
KCl	0.36	4.8 mM
$MgSO_4 \times 7H_2O$	0.30	1.2 mM
KH_2PO_4	0.16	1.2 mM
bovine serum albumin	4.00	0.4%

Medium 199 (GIBCO)

Liver Perfusion

1. The liver is *in situ* preperfused in a noncirculating way via the portal vein. To do this, after opening the abdominal cavity and placing the bowels aside, a 21-gauge venule is placed into the portal vein and fixed with 2.0 surgical silk. The catheter is then perfused with 150–200 ml carbogen-gased preperfusion medium with a flow of about 30 ml/min to completely remove blood. Immediately after the start of the preperfusion, the vena cava is incised under the branches of the renal veins to enable outflow of the blood. During the remaining time of the preperfusion, a catheter is inserted and fixed in the vena cava.

2. After preperfsion is finished, the liver is perfused with the collagenase-containing perfusion medium for 7–11 min in a recirculating manner. Furthermore, a ligation of the vena cava inferior closed the perfusion circuit. The pressure of the oxygenated perfusion medium reaching the vena portae should be about 10–15 cm H_2O column.

Generation of Hepatocyte Suspension. After ripping of the liver cap, several pieces of the liver are transferred carefully to a petri dish filled with washing buffer. Parts of the connective tissue and the capsule are removed with the help of sterile forceps, and cell aggregates are filtered through a sterile nylon mesh (pore size 79 μm). Nonparenchymal cells and debris are removed by four to five times selective centrifugation ($20g$, 2 min) steps where the supernatant is discarded. The remaining hepatocyte pellet is then dissolved in M199 (1 g wet weight of cells /50 ml M199) to reach a density of about 10^6/2.5 ml.

Cell Culture. A cell suspension typically 2.5 ml (10^6 cells) is then cultured on 60-mm Falcon plastic dishes and maintained in an atmosphere of arterial pO_2 (16% O_2/ 79% N_2/ 5% CO_2 [by volume]) in medium M199 containing 0.1 nM insulin added as a growth factor for culture maintenance, 100 nM dexamethasone required as a permissive hormone, and, until the first change of medium after 5 h, 4% newborn calf serum. After 5 h the medium is changed and the cells are cultured under periportal pO_2 up to 24 h. The medium is then changed again, and the cells are further cultured in the same medium, except that the serum is removed for another 24 h. The medium is changed again, and the specific experiment under arterial (16% O_2) or venous (8% O_2/87% N_2/5% CO_2 [by volume]) O_2 tensions can be started.

The O_2 values take into account the O_2 diffusion gradient from the media surface to the cells as outlined previously in detail. In a first approximation, one can calculate that hepatocytes cultured at a density of 1×10^6 on a 6-cm dish have a wet weight of about 10 mg and use O_2 at a rate of 0.023 μmol/min. According to the Fick law of diffusion, the O_2 gradient from the medium surface to the cell surface with an average distance of 0.2 cm is then about 3% (v/v) = 30 μM.[19] This problem can be partially circumvented by using gas-permeable culture dishes, which also allow diffusion of O_2 to the bottom of the cells.

Glycolysis and Glycogen Synthesis

Materials

[^{14}C]Glucose
Lactate
Dowex formiate columns
Scintillator
ß counter

Experiment. First, triplicates are taken for each assay point. Before incubation with the respective substrate, the culture plates are washed twice with fresh medium (6 cm diameter with 2.5 ml for 1–1.5 h without hormones). Medium containing 2 mM lactate and [^{14}C]glucose (0.4 μCi/ml) (2 ml for a 6-cm-diameter dish) is then added and after a 20-min pre-incubation the zero time sample can be taken. Incubation can then be continued up to 5 h. A time-linear rate of product formation should be verified for all conditions tested.

Glycolysis. For the measurement of glycolysis, medium samples should be analyzed after glucose and lactate by ion-exchange chromatography. To do this, 100 μl medium is applied to a small glass column (0.5 cm diameter, 2 ml) of Dowex 1 × 8 formiate 200–400 mesh (0.52 g wet weight), and glucose can be eluted with 1.5 ml water. Elution is continued with 0.4 M sodium formate (13.6 g/500 ml H_2O). The first fraction (0.2 ml) can be discarded, and lactate can be recovered quantitatively in the second fraction (1.5 ml). Radioactive fractions can then be mixed with a scintillator and counted directly. The amounts of products formed can be calculated as

$$\text{Calculation}: \frac{\text{cpm}_{\text{empty value}} \times \text{vol(medium/dish)} \times 1000}{\text{cpm}_{\text{glucose 100\%}} \times \text{time[h]} \times \mu\text{g DNA}}$$

$$= \text{\_}\mu M \text{ lactate/h} \times \text{mg DNA}$$

Glycogen Synthesis

Solutions
   0.33 M KOH: 4.625 g / 250 ml H_2O
   10 N KOH: 56.11 g / 100 ml H_2O
   95% ethanol: 95 ml absolute ethanol + 5 ml H_2O

For measuring glycogen synthesis, cells should be washed twice with 0.9% NaCl and either stored at −20° for 12 h or scraped immediately in 0.33 M KOH (6-cm-diameter dish = 600 μl). Sonicate for 10 s and transfer 200 μl of the homogenate to a new tube containing 50 μl glycogen (40 mg/ml H_2O) + 400 μl 10 N KOH, mix vigorously, and incubate at 95° for 30 min. The samples are then cooled on ice and precipitated after the addition of 800 μl 95% ethanol at −20° overnight. The precipitate is then collected by a 10-min centrifugation at 10,000 rpm at 4° in a microcentrifuge. Pellets are then recovered, dissolved in 500 μl H_2O, and precipitated again with 600 μl 95% ethanol for 30 min on ice. The resulting pellet can then be dissolved at room temperature for 30 min in 500 μl H_2O. Usually 400 μl is then counted in scintillation liquid.

Calculation: $\dfrac{\text{cpm}_{\text{empty value}} \times 1.25(400 \text{ of } 500\mu\text{l}) \times 3(\text{from scraped volume of } 600\ \mu\text{l}) \times 1000}{\text{cpm}_{\text{glucose}/\text{m}M\ \text{glucose}(\text{e.g.,}5\,\text{m}M)} \times \text{time[h]} \times \mu\text{g DNA}}$

$= \_\mu\text{mol glucose/h} \times \text{mg DNA}$

Phosphoenolpyruvate Carboxykinase (PCK) Activity

This standard operating procedure describes the measurement of PCK activity according to the method of Seubert and Huth[33] and can be applied to tissue samples and cultured cells.

Materials

   Ultraturrax (Janke and Kunkel, Staufen)
   Photometer
   Lactate dehydrogenase (LDH)
   Pyruvate kinase
   ITP
   Oxalacetate

Solutions

   Lysis buffer (10×, store at $-20°$)

	10 ml	Final concentration (mM)
$MnCl_2$	5 mg	0.25
DTE	15 mg	1
Tris, pH 8.1	606 mg	50

Test solution 1 (store at $4°$)

	50 ml	Final concentration (mM)
Imidazol, pH 7.5	340 mg	100
DTE	7.5 mg	1
NaF	63 mg	30
KCl	112 mg	30
$MnCl_2$	49 mg	5

Immediately before use add 12.5 mM ITP (inosine 5′-triphosphate) (69 mg/10 ml) and 9.1 mM OA (oxalacetate) (12 mg/10 ml) and control pH

---

[33] W. Seubert and W. Huth, *Biochem. Z.* **343**, 176 (1965).

Test solution 2 (10×, store at −20°)

	10 ml	Final concentration (mM)
Triethanol amine, pH 7.5	186 g	100
EDTA	112 mg	3
NADH	38 mg	0.5
ADP	56 mg	1.2
MgSO$_4$	246 mg	10
KCl	276 mg	37

Immediately before use add 1.5 μl LDH per milliliter test solution required.

Preparation of Cytosol. Triplicate or duplicate plates with about 10^6 cells each should be analyzed per point. Cells should be washed twice with 0.9% NaCl and either stored at −20° for 12 h or scraped immediately in lysis buffer (6 cm diameter in 900 μl). The lysate is then transferred to a 15-ml glass centrifuge tube before homogenizing on ice with an Ultraturrax for 25 s. The homogenate is then transferred to an E cup and centrifuged for 20 min at 14,000 rpm in a Sorvall SS34 rotor at 4°. The supernatant constitutes the cytosol in which the PCK activity is measured.

Test. Prewarmed (30°, 2 min) 10-ml glass tubes with 500 μl of test solution 1 are incubated with 500 μl cytosol for exactly 10 min at 30° in a water bath. Please keep in mind that because the processing of each sample takes time, about 15 to 20 s should be allowed between each sample. The reaction is stopped with the addition of 184 μl KBH$_4$ and placing the tube on ice. After the addition of 60 μl 70% perchloric acid, the sample is neutralized with some grains of KHCO$_3$ transferred to a microcentrifuge cup and centrifuged at full speed (14,000 rpm) at 4° in a microcentrifuge. The supernatant can then be used for the optical test at 366 nm (quartz cuvette), in which 300 μl of the sample and 400 μl of test solution 2 (with LDH) are analyzed in the presence of 0.5 μl pyruvate kinase.

The PCK activity can be calculated as

$$\frac{\Delta E \times V_{\text{total}} \times \text{Testvol.} \times V_0 \times 1000}{\varepsilon \times V_{\text{sample}} \times \text{Probevol.} \times \text{min} \times \mu\text{g DNA}} = \mu\text{mol/min/mg}$$

DNA(units/mg DNA)

where ΔE is the extinction difference, V_{total} is the total volume in the cuvette (700 μl), Testvol. is the total volume in test ($0.5 + 0.5 + 0.184 + 0.06 = 1.244$ ml), V_0 is the lysis volume (volume of lysis buffer, e.g., 0.9 ml), ε is the extinction coefficient for NADH = 3.4, V_{sample} is the volume of sample in the cuvette (300 μl), Probevol. is the sample volume in test 1 (0.5 ml), min is minutes test (10 min), and μg DNA is DNA content μg/dish.

Glucokinase (GK) Activity

Glucokinase is assayed in glycyl–glycine using glucose coupled with glucose 6-phosphate dehydrogenase and phosphogluconate dehydrogenase according to an earlier method.[34]

Materials

Ultraturrax (Janke and Kunkel, Staufen)
Photometer
Glucose 6-phosphate dehydrogenase (G6PDH)
6-Phosphogluconate dehydrogenase (6-PGDH)
NADP

Solutions

Lysis buffer:

	50 ml	Final concentration (mM)
HEPES, pH 7.5	600 mg	50
KCl	370 mg	100
MgCl$_2$	51 mg	5
EDTA	19 mg	1
DTE	20 mg	2.5
BSA	500 mg	

Test solution

	10 ml	Final concentration (mM)
Glycylglycine, pH 7.5	198 mg	150
MgSO$_4$ × 2H$_2$O	25 mg	10
ATP	36 mg	6

Test. For GK enzyme activity, triplicate or duplicate plates with about 10^6 cells each should be analyzed per point. Cells should be washed twice with 0.9% NaCl and either stored at $-20°$ for 12 h or scraped immediately in lysis buffer (6 cm diameter in 700 μl). The lysate is then transferred to a

[34] A. Brinkmann, N. Katz, D. Sasse, and K. Jungermann, *Hoppe Seylers. Z. Physiol. Chem.* **359,** 1561 (1978).

15-ml glass centrifuge tube before being homogenized on ice with an Ultra-turrax for 20 s. The homogenate (200 μl) is then transferred to a 1-ml cuvette and mixed with 400 μl of test solution containing 6 μl of freshly made NADP (50 mM, alternatively 10 mM NAD can be used) and 0.5 μl each of G6PDH and 6-PGDH. The reaction is then started by the addition of 25 μl glucose (2 M) solution.

The same reaction can be used to measure hexokinase I–III activity; however, it should be started with a 60 mM glucose solution. The extinction at 366 nm recorded in a conventional photometer is then used to calculate the activity.

Measurement of DNA Content

This method is based on a photometric method[35] and is suitable for the determination of 20–60 μg DNA in a linear range.

Materials

Ultraturrax (Janke and Kunkel, Staufen)
Photometer
Trichloroacetic acid (TCA)
Diphenyl amine

Solutions

DNA standard: 0.1 mg/ml 0.1 N NaOH

0	20	40	60	80 μg DNA
0	200	400	600	800 μl DNA standard
1000	800	600	400	200 μl 0.1 N NaOH
1000	1000	1000	1000	1000 μl TCA mixture

TCA mixture: 10 ml 20% TCA in 1 N HCl, 10 ml H_2O, and 80 ml 10% TCA

Diphenyl reagent: 50 ml   ice acetic acid,
                      1.8 ml   70% perchloric acid,
                      0.26 ml   acetaldehyde (16 mg/ml), and
                                diphenylamine
                      1 g
0.3 M KOH: 1.68 g/100 ml

[35] I. T. Oliver, A. M. Edwards, and H. C. Pitot, *Eur. J. Biochem.* **87,** 221 (1978).

Cells from the 6-cm dishes are scraped with 3 ml 0.3 M KOH into a 10-ml glass tube, homogenized, and incubated for 1 h at 37° with subsequent cooling on ice. One milliliter of 20% TCA dissolved in 1 N HCl is added to the probes and mixed vigorously for 10 to 15 min in ice. After centrifugation for 10 min at 5000 rpm, the remaining pellet, as well as the DNA standards, is dissolved in 2 ml 5% TCA. Then samples are incubated for 20 min at 90° (be careful to avoid loss of fluid due to evaporation) and cooled on ice. One milliliter of the supernatant is then incorporated with 2 ml diphenyl reagent, mixed, and stored in the dark for 16–20 h at room temperature. The extinction of samples at 578 nm is then measured in a conventional photometer, and the amount of DNA can be calculated from the resulting standard curve.

Transfection of Hepatocytes Using the Calciumphosphate Precipitation and Luciferase Assay

A variety of methods for the transfection of DNA into mammalian cells such as virus, liposome, DEAE-dextran, electroporation-mediated, or microinjection protocols are known.[36–41] Among these, one of the oldest methods, if not the oldest, is based on the so-called calcium phosphate precipitation first described by Graham and van der Eb in 1973.[42] This method is used routinely in our laboratory and is very reproducible. A number of modifications in cell culture or length of transfection occured during that time. However, the common basis is that DNA is coprecipitated by small grains of calcium phosphates, which are then applied to the cells and taken up by endocytosis. The precipitate is prepared by mixing a DNA/calcium chloride solution with a phosphate-containing solution.

Gene Constructs. Plasmids are the major source of the material to be transfected. In addition to the necessary components for their replication in bacteria (e.g., antibiotic resistance genes), they should contain a promoter allowing eukaryotic expression, for example, in the studies described here, the PCK gene promoter is followed by the luciferase reporter gene and the SV40 polyadenylation signal.

[36] B. A. Parker and G. R. Stark, *J. Virol.* **31**, 360 (1979).
[37] C. Chen and H. Okayama, *Mol. Cell. Biol.* **7**, 2745 (1987).
[38] C. A. Chen and H. Okayama, *Biotechniques* **6**, 632 (1988).
[39] F. Ginot, J. F. Decaux, M. Cognet, T. Berbar, F. Levrat, A. Kahn, and A. Weber, *Eur. J. Biochem.* **180**, 289 (1989).
[40] D. S. Pasco and J. B. Fagan, *DNA* **8**, 535 (1989).
[41] R. A. Rippe, A. Umezawa, J. P. Kimball, M. Breindl, and D. A. Brenner, *J. Biol. Chem.* **272**, 1753 (1997).
[42] F. L. Graham and A. J. van der Eb, *Virology* **54**, 536 (1973).

Materials.

Luminometer
Dual luciferase assay kit (Promega)
Solutions.

2× HEPES:

	g/100 ml	Final concentration (mM)
HEPES	1.192	50 mM
NaCl	1.636	280 mM
Na$_2$HPO$_4$	0.267	1.5 mM

The pH is adjusted with 5 N NaOH to
exactly 7.05 and stored in aliquots at $-20°$

Calcium chloride 2.5 M:

		Final concentration
CaCl$_2$	36.75 g/100 ml	2.5 M

Autoclave and store aliquots at $-20°$

Transfection mixture per culture
    dish (60 mm):

		Final concentration
Plasmid-DNA	2.5 μg	
H$_2$O	67.5 μl	
CaCl$_2$ 2.5 M	7.5 μl	125 mM
2×HEPES	75 μl	

Rat hepatocyte cultures (about 1×10^6 cells per 6-cm dish in 1.5 ml M199) are transfected transiently with 2.5 μg plasmid DNA containing 500 ng of pRL-SV40 (Promega) to control transfection efficiency and 2 μg of the appropriate promoter *Firefly* luciferase (FL) construct.

The pRL-SV40 contains the *Renilla* luciferase (RL) gene under control of the SV40 promoter, which allows a constitutive expression of the RL gene. In every culture experiment, two culture dishes are transfected per point. DNA can be transfected in 150 μl transfection buffer, composed of 7.5 μl 2.5 M CaCl$_2$, 75 μl 2× HEPES, pH 7.05, and 67.5 μl H$_2$O as a calcium phosphate precipitate. The transfection mixture is set up in a polystyrol tube to avoid sticking of DNA, and after 5–10 min at room temperature a light-gray precipitate should have formed. The 150-μl DNA precipitate is then added dropwise to the medium of fresh plated cells containing 4% newborn calf serum and mixed by slight shaking of the dish. The transfected dishes are then left in the incubator under standard conditions for 5 h.

After removal of media, cells are then cultured again under standard conditions without serum. Twenty-four hours later, cells are treated with fresh media and luciferase activity can be determined in duplicate after culture of the cells under arterial or venous oxygen tensions for up to 72 h.

For Luc activity measurement, cells are washed twice with 0.9% NaCl and incubated for 15 min on a rocking platform with 300 μl passive lysis buffer supplied with the dual luciferase assay kit (Promega). The lysate is then scraped from the plates, vortexed for 20 s, and centrifuged for 2 min. From the supernatant, 20 μl is usually assayed for FL activity in a luminometer (Berthold) by injection of 100 μl luciferase assay reagent II supplied with the dual luciferase assay kit (Promega). After the addition of 100 μl Stop&Glo reagent (Promega), which quenches FL activity, RL activity is recorded again in a luminometer (Berthold).

RNA Preparation and Northern Analysis

All solutions for RNA experiments are prepared using diethyl pyrocarbonate (DEPC)-treated H_2O. The preparation of RNA follows the protocol from Chomzynski and Sacchi[43] with modifications. PCK and GK mRNA levels are then determined in a Northern blot assay by a nonradioactive hybridization technique using digoxigenin (DIG)-labeled antisense RNA as hybridization probes.

Solutions

GTC lysis buffer:

		Final concentration
Guanidinium thiocyanate	47.3 g	4 M
Na-citrate	2.5 ml 1 M	25 mM
N-Lauroyl sarcosine	0.5 g	17 mM
2-Mercaptoethanol	0.7 ml	0.1 M
30% antifoam A	0.33 ml	0.1%
DEPC-treated H_2O	100 ml	

The solution needs to be warmed for 30 min at 65°. The pH is adjusted with 1 N NaOH to 7.0. 2-Mercaptoethanol is added after.

[43] P. Chomczynski and N. Sacchi, *Anal. Biochem.* **162,** 156 (1987).

Phenol (Water-Saturated). Total RNA is prepared from 3×10^6 cells as follows. Cells are washed with 0.9% NaCl, scraped in GTC lysis buffer, and homogenized. RNA is isolated using the combination of phenol/chloroform/isoamylalcohol extraction and precipitation of the resulting upper phase with 1 volume isopropanol.[43] RNA is recovered by centrifugation at 14,000 rpm in a Sorvall SS34 rotor at $4°$. The resulting pellet is dissolved again in GTC lysis buffer, precipitated with isopropanol, and centrifuged again. The RNA pellet is then washed two times in 3 M sodium acetate (pH 5.2) with subsequent centrifugation as described earlier, resuspended in sterile water, and reprecipitated with 2.5 volume absolute ethanol overnight at $-20°$. Final washing is performed with 70% ethanol. The resulting pellet is vacuum dried and dissolved in 0.1% SDS. Twenty micrograms of RNA is usually denatured by formaldehyde, electrophoresed, and blotted according to a standard protocol.[44]

Preparation of DIG-Labeled PCK and GK Antisense RNA Probes

Materials

DIG-11-UTP (Roche)
T3 or T7 RNA polymerase
RNase inhibitor (Roche)
DIG nucleic acid detection kit (Roche)
CSPD (dinatrium 3-(4-methoxyspiro{1, 2-dioxetane-3,2-(5'-chloro)-tricyclo[3.3.1.1^{3,7}]decan}-4-yl)-phenylphosphate)
X-ray film
Densitometer

PCK and GK antisense RNA are synthesized from pBS-PCK and pBS/GK-1 containing a 1200-bp *PstI* cDNA fragment or a 1600-bp *Hind*III/*Bam*HI fragment, respectively, using a T3 RNA polymerase and digoxigenin (DIG) RNA-labeling mixture containing 3.5 mM 11-DIG-UTP, 6.5 mM UTP, 10 mM GTP, 10 mM CTP, and 10 mM ATP. For *in vitro* transcription, the appropriate plasmids pBS-PCK and pBSGK1 are linearized, and the linearized plasmids are purified by phenol/chloroform extraction and precipitated with absolute ethanol. In our hands it did not improve the labeling efficiency when the plasmids were purified from agarose gels. Then the linearized plasmids are transcribed *in vitro* for 1.5 h at $37°$ according to the following scheme: 11 μl linearized plasmid, 2 μl transcription buffer (10×), 2 μl RNA labeling

[44] J. Sambrook, E. F. Fritsch, and T. Maniatis, "Molecular Cloning: A laboratory Manual," 2nd Ed. Cold Spring Harbor Press, Cold Spring Harbor, NY.

mix (10×), 1 μl RNasin (40 U/μl), 2 μl RNA polymerase (20 U/μl), and 2 μl DEPC-treated H_2O.

Thereafter, an additional 0.5 μl RNA polymerase is added, and incubation is prolonged for another hour at 37°. The DIG-labeled RNA is purified and precipitated with 2.5 μl 4 M LiCl and 75 μl absolute ethanol at $-20°$ for at least 2 h. RNA is recovered at 12,000g (10,000 rpm, SS34 rotor) for 10 min at 4°, washed with 80% ethanol, and dried under vacuum. The pellet is then dissolved in 100 μl DEPC-treated H_2O, and the labeling efficiency is checked essentially as outlined in the Roche DIG application manual for filter hybridizations.

Hybridizations are carried out with a 50-ng/ml transcript at 68° according to the DIG nucleic acid detection kit (Roche, Mannheim). The detection of hybrids is performed by an enzyme-linked immunoassay using an antidigoxigenin alkaline phosphatase conjugate. Hybrids are visualized via chemiluminescence with an alkaline phosphatase-labeled antidigoxigenin antibody converting CSPD, thereby emitting light at 477 nm. Luminescent blots are exposed to Hyperfilm-MP (Amersham) and quantified with a video densitometer (Biotech Fischer, Reiskirchen).

Acknowledgments

These studies were supported by the Deutsche Forschungsgemeinschaft SFB 402 Teilprojekt A1 and GRK 335. This article is dedicated to the scientific life of my teacher and long-lasting cooperator Kurt Jungermann (1938–2002).

[25] Assessing Oxygen Sensitivity of the Multidrug Resistance (MDR) Gene

By Katrina M. Comerford and Sean P. Colgan

Introduction

Diminished oxygen supply to tissues (hypoxia) is a common physiologic and pathophysiologic occurrence in nature. Cell and tissue responses to hypoxia are diverse and include changes in metabolic demand, regulation of gene expression, and release of lipid mediators. It is now well accepted that hypoxia contributes significantly to the maintenance and metastases of established tumors.[1] For example, solid tumors have been demonstrated to

---

[1] A. L. Harris, *Nature Rev. Cancer* **2,** 38 (2000).

form hypoxic cores and respond by inducing hypoxia-responsive genes. Due to the combination of tissue mass and the particularly high rate of glycolysis in tumor cells (termed the Warburg effect), hypoxia is considered a property of many tumor types.[2] While most tissues of the body maintain an oxygen gradient spanning a distance of approximately 300–400 μm, studies assessing relative oxygen tensions within tumors have suggested that oxygen concentrations may be as much as 10- to 100-fold decreased at comparable distances from capillary blood supplies.[2]

A major obstacle in the development of effective cancer chemotherapy is tumor development of the multidrug resistance *(MDR1)* phenotype.[3,4] We and others have reported previously that the multidrug resistance gene *(MDR1)* and its gene product P-glycoprotein (P-gp), responsible for the development of the *(MDR1)* phenotype in tumor cells, are hypoxia responsive.[5,6] P-gp is a ~170-kDa member of the ABC type transporter family and functions as an energy-dependent membrane efflux pump, which transports a wide variety of structurally unrelated xenobiotics to maintain cytoplasmic concentrations at subtoxic levels.[3] As such, it is likely that hypoxia-induced *MDR1* and its gene product P-gp contribute to tumor chemotherapy resistance. This article profiles and details methods for investigating the mechanisms of hypoxia-dependent regulation of *MDR1* gene expression in epithelial cells.

Epithelial Hypoxia

For a number of years, we have studied epithelial gene responses to hypoxia. The basic models entail the initial culturing of epithelial cells as polarized monolayers on membrane-permeable supports (Corning-Costar, Cambridge, MA). Because these substrates are expensive and require additional care and manpower, it is suggested that one determines the relative importance of polarity for the individual gene/gene product under investigation. Culturing epithelia on such permeable membrane supports allows for optimal establishment of the polarized epithelial phenotype, and because a number of epithelial functions (barrier function, vectorial ion transport, etc.) occur only in a polarized fashion, it is first necessary to determine

[2] G. L. Semenza, *Crit. Rev. Biochem. Mol. Biol.* **35,** 71 (2000).
[3] M. Kuwano, S. Toh, T. Uchiumi, H. Takano, K. Kohno, and M. Wada, *Anticancer Drug Des.* **14,** 123 (1999).
[4] P. Vaupel, O. Thews, and M. Hoeckel, *Med. Oncol.* **18,** 243 (2002).
[5] K. M. Comerford, T. J. Wallace, J. Karhausen, N. A. Louis, M. C. Montalto, and S. P. Colgan, *Cancer Res.* **62,** 3387 (2002).
[6] M. Wartenberg, F. C. Ling, M. Muschen, F. Klein, H. Acker, M. Gassmann, K. Petrat, V. Putz, J. Hescheler, and H. Sauer, *FASEB J.* **17,** 503 (2003).

the relative importance of growing cells on permeable membrane supports as opposed to solid surfaces (e.g., plastic or glass). In the case of *MDR1*/ P-gp experiments, no differences in hypoxia response were noted between cells grown on solid substrates or on permeable membrane supports.

Once these determinations are made, we then decide what strategy will be used to subject cells to hypoxia. Our basic hypoxia chamber is a commercial airtight glove box with the atmosphere monitored continuously by an oxygen analyzer interfaced with oxygen and nitrogen flow adapters (Coy Laboratories, Ann Arbor, MI). The oxygen analyzer is calibrated regularly within a range of normoxia (equal to pO_2 of 147 torr) to anaerobic hypoxia (pO_2 0 torr accomplished by placing the oxygen sensor in a sealed container flushed with 100% nitrogen). Cultured cells are then brought through the chamber interlock, fed with preequilibrated media, and placed in a humidified modular chamber (Billips-Rothenberg, Del Mar, CA) within the larger glove box. Each modular chamber is flushed constantly with the ambient gas mixture using fish tank pumps (Tetra/ Second Nature, Blacksburg, VA). Gas mixtures within the chamber consist of the desired ambient oxygen (range 1–21%) with a balance made up of nitrogen, carbon dioxide (constant pCO_2 35 mm Hg), and water vapor from the humidified chamber. Samples of equilibrated media are collected and monitored regularly for pCO_2 and pO_2 using a blood gas analyzer (Ciba-Corning, Essex, England). Cultured cells are then subjected to such hypoxia conditions for periods in the range of 1–72 h.

Analysis of mRNA Levels by Genechip Expression Arrays and RT-PCR

Initial insights into changes in expression of *MDR1* gene transcription in response to hypoxia were gained through gene chip expression arrays and confirmed by RT-PCR. The transcriptional profile of epithelial cells exposed to ambient hypoxia can be assessed in quantitative gene chip expression arrays (Affymetrix, Inc., Santa Clara, CA) using total RNA derived from cells exposed to desired periods of hypoxia or normoxia. An important consideration is the inclusion of control normoxic epithelial populations for each time point of hypoxia. For example, for each period of hypoxia, a parallel control normoxia population of cells should be fed and handled in a fashion identical to the hypoxia exposures, as feeding fresh media and/or handling cells has the potential to change mRNA expression profiles. The hypoxia-dependent changes in the expression of *MDR1* demonstrated by the gene chip array are confirmed by RT-PCR. RT-PCR analysis of mRNA levels is performed using DNase-treated total RNA and primers specific for *MDR1* (forward primer 5'-AAC GGA AGC

CAG AAC ATT CC-3' and reverse primer 5'-AGG CTT CCT GTG GCA AAG AG-3', 180-bp fragment). The primer set is amplified using an optimized number of PCR cycles (25 cycles) of 94° for 1 min, 60° for 2 min, min, 72° for 4 min, and a final extension of 72° for 7 min. An aliquot (10 μl) of the PCR reaction is mixed with 5 μl loading buffer (50% glycerol, 100 mM EDTA, 1% SDS, 0.1% bromphenol blue, 0.1% xylene cyanol) and resolved on a 1.5% agarose gel [made in TAE buffer containing 40 mM Tris (pH 8.3), 20 mM sodium acetate, 2 mM EDTA] containing 5 μg/ml of ethidium bromide (Sigma Chemical Co., St. Louis, MO).

Alternatively, mRNA levels can be quantified by real-time PCR (iCy-cler, BioRad, Hercules, CA). This methodology utilizes a PCR-based reaction with quantification of a fluorescent intercalating dye (e.g., with the addition of SYBR Green I, Molecular Probes, Eugene, OR) bound to double-stranded DNA templates derived from specific mRNA. In each case, the relative amount of PCR product is proportional to the fluorescence intensity within each sample, and final calculations (expressed as a fold change in mRNA) are based to the expression of a control gene (e.g., β-actin) and on the efficiency of the primer set under the given experimental conditions.[7]

Western Blotting

After demonstrating that *MDR1* mRNA is increased in response to hypoxia, we then use immunoblotting techniques to determine if P-gp expression is increased in parallel. Following experimental treatment, proteins are isolated from confluent epithelial monolayers grown on 100-mm petri dishes (Falcon). Medium is removed and cells are washed once with ice-cold phosphate-buffered saline (PBS). Attached cells are scraped off the culture dish into a small volume (100 μl) of lysis buffer [150 mM NaCl, 25 mM Tris–HCl, pH 8, 1 mM EDTA, 1% (v/v) NP-40, 500 μM phenyl-methyl sulfonyl fluoride (PMSF) with additional protease inhibitor cocktail (Roche Diagnostics, Indianapolis, IN)], and the cell suspension is transferred to a 1.5-ml microcentrifuge tube and centrifuged at 14,000g at 4° for 5 min. Supernatants are removed to a new microcentrifuge tube, and protein concentration is determined by the DC protein assay (Bio-Rad, Hercules, CA). Samples are then volume normalized according to protein content, and an equal volume of reduced sample buffer [66 mM Tris–HCl, pH 6.8, 2% (w/v) SDS, 10 mM EDTA, 10% (v/v) glycerol, and trace bromphenol blue] is added to the protein samples. Protein samples

[7] M. W. Pfaffl, *Nucleic Acids Res.* **29,** E45 (2001).

are boiled in a heating block for 10–15 min before being resolved by SDS–PAGE (25 μg/lane on an 8% seperating gel and 4% sacking gel). Electrotransfer of proteins onto nitrocellulose membrane (Bio-Rad) is performed using an SDS–methanol buffer (3 g Tris base, 14.4 g glycine, 1 g SDS, and 200 ml methanol/liter) at a constant voltage of 100 V for 1 to 1.5 h [because P-gp is a large protein (170 kDa), electrotransferring for 1.5 h ensures complete transfer of the protein]. A water cooling system is used to reduce heat during transfer. The nitrocellulose membrane is blocked overnight by agitating at 4° in blocking buffer [250 mM NaCl, 0.02% Tween 20, 5% goat serum, and 3% bovine serum albumin (BSA)]. The nitrocellulose is incubated for 3 h with blocking buffer containing the anti-P-gp antibody (5 μg/ml final concentration, rabbit polyclonal antibody; Biogenesis, Poole, UK). The membrane is washed extensively (five times) in wash buffer [500 mM NaCl, 10 mM HEPES, and 0.2% (v/v) Tween 20] and incubated for 45 min min with goat antirabbit peroxidase-conjugated secondary antibody (1 μg/ml, Cappell, West Chester, PA). The nitrocellulose is washed as described earlier before labeled bands are detected by enhanced chemiluminescence (Amersham Biosciences, Piscataway, NJ).

P-glycoprotein Surface Expression

Surface expression of P-gp on intact epithelial is analyzed by cell ELISA. Briefly, cells subjected to desired periods of hypoxia or normoxia are washed with HBSS (Sigma Chemical Co.) and blocked with serum-containing medium for 30 min at 4°. Anti P-gp mAb (clone 3201, which recognizes a cell surface epitope; QED Biosciences, San Diego, CA; used as purified mAb at 20 μg/ml) is added to cells and allowed to incubate for 2 h at 4°. After washing with HBSS, a peroxidase-conjugated sheep antimouse secondary antibody (1 μg/ml diluted in serum-containing medium, Cappel) is added. After washing, plates are developed by the addition of peroxidase substrate [2,2'-azino-bis(3-ethylbenzthizoline-6-sulfonic acid), Sigma Chemical Co., 1 mM final concentration with 1 μl/ml 30% H_2O_2 added just prior to devlopment] and read on a microtiter plate spectrophotometer at 405 nm (Molecular Devices, Framingham, MA). Controls consist of medium only and secondary antibody only.

P-gp Functional Analysis

P-gp function can be assessed a number of ways. One example that we have used is the verapamil-inhibitable efflux of digoxin. The basis of this determination is the quantification of intracellular levels of the P-gp substrate digoxin; thus, the higher the P-gp activity, the lower the intracellular concentration of digoxin. Briefly, epithelial monolayers are subjected to experimental conditions, washed with HBSS, and incubated with digoxin (final concentration, 6 μM; based on pilot experiments, Sigma Chemical Co.) in the absence or presence of verapamil (concentration range: 1–100 μM; Sigma Chemical Co.). Cells are incubated for 60 min (based on pilot experiments) at 37°, washed with HBSS, and cooled to 4°. For digoxin determinations, cells are lysed in ice-cold H_2O and lysates are cleared by Eppendorf centrifugation at 14,000g for 10 min. Digoxin levels in supernatants are assessed by HPLC analysis using a Hewlett-Packard HPLC (Model 1050) with a HP 1100 diode array detector and a reversed-phase HPLC column (Luna 5 μm C18 150 × 4.60 mm; Phenomenex, Torrance, CA). Digoxin is measured with a 20–80% CH_3CN/H_2O elution gradient over 30 min at a pump speed of 1 ml/min. Absorbance is measured at 220 nm. UV absorption spectra are obtained at chromatographic peaks. Digoxin is identified by chromatographic behavior (e.g., retention time, UV absorption spectra, and coelution with internal standards). Controls include monolayers incubated at 4° for the entire period of the assay, and digoxin levels derived from these controls are used to normalize data. All data are then normalized to total protein lysate concentration.

Multicellular Dome Model

Another method to assess P-gp function entails the determination of cytotoxicity to the P-gp substrate doxorubicin. A multicellular dome model was developed using KB cells (ATCC) grown at high density on membrane-permeable supports (see Fig. 1). Briefly, KB cells in suspension are plated at high density (10^7 cells/cm^2) on 0.33-cm^2 collagen-coated permeable supports (pore size of 0.4 μm, Corning-Costar, Cambridge, MA) and allowed to grow as domes on these substrates for 2 weeks. Medium is replaced every other day, and at 2 weeks, multicellular domes are subjected to a 24- or 48-h period of hypoxia or normoxia (as described earlier) in the presence or absence of doxorubicin (Sigma Chemical Co., final concentration, 1 μm). At the termination of the experiment, multicellular monolayers are rinsed extensively and overall viability is measured by the uptake determination of fluorescence following incubation with the esterase-cleavable fluorescent marker BCECF-AM (final concentration, 5 μM; Calbiochem, San Diego, CA). The number of viable, intact cells is determined by measurement of the amount of fluorescent marker BCECF-AM retained over a 30-min period, a sensitive measurement of cytotoxicity. After 30 min, the multicellular spheroids are washed three times in HBSS, and fluorescence

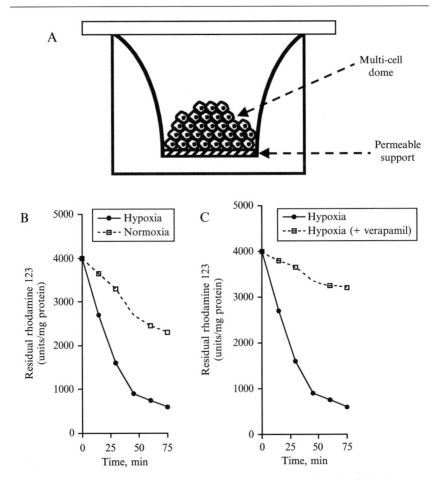

FIG. 1. Analysis of P-gp function in multicellular domes. KB cell multicellular domes are established on membrane-permeable supports (0.4-μm pores). (A) The orientation of cells relative to the membrane support. (B) Multicellular domes are subjected to normoxia (pO$_2$ 147 torr for 48 h) or hypoxia (pO$_2$ 20 torr for 48 h), loaded with the P-gp substrate rhodamine 123, and assessed for the P-gp-dependent efflux of rhodamine 123. (C) Hypoxic multicellular domes are exposed to vehicle control or the Pgp inhibitor verapamil (20 μM) and assessed for rhodamine 123 efflux as described in B.

intensity is measured on a fluorescent plate reader (excitation, 485 nm; emission, 530 nm) as described earlier, with fluorescence being proportional to viability.

Other assays using such multicellular domes can assess P-gp function. For example, efflux of the P-gp substrate rhodamine 123 (Sigma Chemical Co.) has been used widely to study P-gp kinetics. An example of such analysis is shown in Fig. 1B and C. In these assays, multicellular domes are incubated at 37° with the P-gp substrate rhodamine 123 (final concentration, 10 μM, Sigma Chemical Co.) in the presence and absence of verapamil (as described earlier). Multicellular domes are then washed every 15 min with HBSS, and intracellular (residual) rhodamine fluorescence is assessed on a fluorescence plate reader (Cytofluor 2300; Millipore, Inc., Bedford, MA) after each wash. Monolayers not exposed to rhodamine 123 are used to determine background fluorescence. For such analysis, intracellular rhodamine 123 fluorescence is indirectly proportional to P-gp activity.

MDR1 Reporter Assays

The *MDR1* gene promoter is extremely complex, with multiple combinatorial binding sites for a number of transcription factors.[8] In the course of our experiments addressing transcriptional activation of *MDR1,* we determined that the *MDR1* promoter expresses a binding site for the hypoxia-responsive transcription factor hypoxia-inducible factor 1 (HIF-1), a member of the Per-ARNT-Sim (PAS) family of basic helix–loop–helix (bHLH) transcription factors.[9,10] HIF-1 exists as an $\alpha\beta$ heterodimer, the activation of which is dependent on stabilization of an O_2-dependent degradation domain of the α subunit by the ubiquitin-proteasome pathway.[11] As such, the *MDR1* promoter activity of wild-type and mutations is assessed using a transient transfection of luciferase promoter constructs. In this case, Caco2 or BAE cells grown on 30-mm petri dishes are used to assess *MDR1* inducibility by hypoxia. PGL2 basic plasmids (1 μg) expressing a sequence corresponding to the wild-type *MDR1* promoter (-189 to 133), truncations at the 3′ end (-189 to 4), truncations at the 5′ end (-2 to 133), or internal truncations (-119 to 4) of the *MDR1* promoter (a kind gift from Drs. Martin Haas and Bryan Strauss, University of California, San Diego, CA) are cotransfected with a constitutively expressed β-galactosidase reporter gene (0.5 μg/transfection p-Hook-2; Invitrogen) using standard methods of overnight transfection with the Effectene transfection reagent (Qiagen). A subset of cells are transfected with a promoterless vector (pGL2-Basic; Promega Corp., Madison, WI) to control for

[8] S. Labialle, L. Gayet, E. Marthinet, D. Rigal, and L. G. Baggetto, *Biochem. Pharmacol.* **64,** 943 (2002).
[9] G. L. Semenza, F. Agani, D. Feldser, N. Iyer, L. Kotch, E. Laughner, and A. Yu, *Adv. Exp. Med. Biol.* **475,** 123 (2000).
[10] G. L. Semenza, *J. Clin. Invest.* **106,** 809 (2000).
[11] A. L. Harris, *Nature Rev. Cancer* **2,** 38 (2000).

background luciferase activity. After transfection, cells are subjected to hypoxia or normoxia for 24 h, following which luciferase activity is determined (Topcount-NXT; Hewlett-Packard) using a luciferase assay kit (Stratagene). Briefly, cells are washed with ice-cold PBS and harvested by the addition of lysis buffer (200 μl) supplied with the luciferase kit. The cell lysate is transferred to 1.5-ml microcentrifuge tubes and centrifuged at 12,000g for 5 min at 4° to pellet cell debris. The supernatant is removed and assayed for luciferase. All luciferase activity is normalized with respect to constitutively expressed β-galactosidase assessed using a β-galactosidase assay kit (Stratagene).

In Vitro Site-Directed Mutagenesis

Site-directed mutagenesis is a useful tool for studying the regulation of gene expression by transcription factors and for investigating protein structure and function. A number of different mutagenesis methods have been reported previously. We used the GeneEditor *in vitro* site-directed mutagenesis system (Promega) to mutate the HIF-1-binding site (HRE) in the 3' end (−189 to 4) truncation MDR promoter. The GeneEditor uses antibiotic resistance to select for successfully mutated plasmid DNA. Selection oligonucleotides, supplied with the kit, encode mutations that modify the ampicillin resistance gene, resulting in resistance of the target plasmid to the supplied GeneEditor antibiotic selection mix. One limitation to using the GeneEditor *in vitro* site-directed mutagenesis system is that the target vector must use ampicillin as a selectable marker. Annealing a synthetic oligonucleotide that is complementary to the target sequence but includes an area of mismatch sequence generates the mutation. The selection oligonucleotide is annealed to the same strand of the target template as the mutagenic oligonucleotide. Following annealing, the two oligonucleotides are extended with DNA polymerase and ligated with DNA ligase. The subsequent duplex structure is transformed into an *Escherichia coli* host. Resistance to the GeneEditor antibiotic selection mix allows for the selection of successfully generated mutations. Mutations encoding a three nucleotide mutation in the *MDR1* HIF-1 binding site [consensus motif 5'-GC<u>GTG</u>-3' mutated to 5'-GC<u>CAT</u>-3' within the HIF-1 site located at positions −49 to −45 relative to the transcription start site] by PCR introduced a unique NCO1 cleavage site (C<u>CAT</u>GG) and allowed us to screen mutations based on the enzymatic cleavage of plasmid DNA. Oligonucleotides used for the three nucleotide mutations are (mutated sequence in lowercase) 5'-AGG ACA AGC GCC GGG GCc atG GCT GAG CAC AGC CGC TTC-3'. A deletional mutation of the HIF-1 site is also generated using the oligonucleotide 5'-AGG AAC AGC GCC GGG GG CTG

AGC ACA GCC-3′. All mutations are confirmed by sequencing using
PGL-2 basic primers. Hypoxia inducibility in transient transfections using
such mutated constructs is exactly as described earlier.

Antisense Oligonucleotide Treatment of Epithelia

As part of this work to determine the relative role of HIF-1 and Sp1 in
the hypoxia inducibility of HIF-1, we depleted cells of constituitively expressed HIF-1α and Sp1 using antisense oligonucleotide loading. Phos-
phorothioate derivatives of antisense (5′-GCC GGC GCC CTC CAT-3′)
or control sense (5′-ATG GAG GGC GCC GGC-3′) oligonucleotides
are obtained through custom synthesis (Oligo's Etc., Wilsonville, OR).
Sp1 depletion by antisense is accomplished as described earlier using antisense (5′-ATA TTA GGC ATC ACT CCA GG-3′) or control sense (5′-
CCT GGA GTG ATG CCT AAT AT-3′) oligonucleotides. Epithelial cells
are washed in serum-free medium and then in medium containing 20 μg/ml
Effectene transfection reagent (Qiagen, Inc., Valencia, CA) with 2 μg/ml
HIF-1α/SP1 antisense or sense oligonucleotide. Cells are incubated for
4 h at 37° and then replaced with serum-containing growth medium.
Treated cells are subjected to hypoxia or normoxia. To demonstrate
successful depletion of HIF-1α/Sp1 RNA or protein is quantified by RT-
PCR or by Western blot as described previously [Western blotting anti-
HIF-1α polyclonal antibody from Novus, Littleton, CO, and anti-Sp1
monoclonal antibody (clone 1C6) from Santa Cruz Biotechnology, Santa
Cruz, CA]. *Note:* Because antisense oligonucleotides target protein translation, not all antisense oligonucleotide treatments result in noticeable
mRNA changes; therefore, protein is a better determining factor for success. Inhibition of *MDR1* mRNA and/or protein (as described in detail
earlier) can then be determined in HIF-1α- or Sp1-depleted cells.

Conclusion

As summarized briefly within this article, multiple approaches are available to study the function of *MDR1*/P-gp and how such functions relate to
tumor hypoxia. Results of these functional analyses may have broad impli-
cations for studying *MDR1*-expressing tumors and overall chemotherapy
resistance. For example, *MDR1* expression in patients is most prominent
in solid tumors, and *MDR1* correlates positively with the propensity of
tumors for lymph node spread and metastases.[12,13] Consistent with these

[12] L. J. Goldstein, *Eur. J. Cancer* **6,** 1039 (1996).
[13] S. B. Kaye, *Curr. Opin. Oncol.* **10,** S15 (1998).

Tumor oxygenation and *MDR1* expression

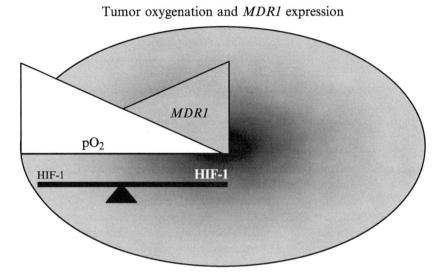

Fig. 2. Model of *MDR1* expression in relationship to HIF-1. In the tumor microenviron-ment, levels of available oxygen (pO_2) decrease with increasing distance to the center of the tumor. Studies indicate that *MDR1*/P-gp expression increases within these areas of hypoxia and that HIF-1-mediated expression may tip the balance toward functional multiple drug resistance.

observations in solid tumors, it is clear that the degree of hypoxia within the tumor microenvironment is sufficient to induce HIF-1. As was shown elsewhere,[14] the majority of solid tumors, particularly those with a propen-sity for P-gp expression, stain positive for nuclear HIF-1α. As such, com-bination therapies aimed at modulating HIF-1 expression (e.g., targeting prolyl- or asparaginyl-hydroxylases HIF-1α)[15,16] in concert with standard chemotherapy may provide a strategy to overcome some tumor resistance. As shown in Fig. 2, a model integrating these features proposes that *MDR1*/P-gp expression is indirectly proportional to levels of available oxygen and that the deeper areas of tumors may have higher propensities for functional expression of multidrug resistance. Given recent understand-ing that HIF-1 may be critical to this gene regulation, this balance may be shifted according to the relative nuclear expression of HIF-1.

[14] K. L. Talks, H. Turley, K. C. Gatter, P. H. Maxwell, C. W. Pugh, P. J. Ratcliffe, and A. L. Harris, *Am. J. Pathol.* **157,** 411 (2000).
[15] C. W. Pugh and P. J. Ratcliffe, *Semin. Cancer Biol.* **13,** 83 (2002).
[16] R. H. Wenger, *FASEB J.* **16,** 1151 (2002).

Acknowledgments

This work was supported by a Health Research Board of Ireland grant (KMC), by Grants DK50189 and HL60569 from the National Institutes of Health, and by a grant from the Crohn's and Colitis Foundation of America (SPC).

[26] Oxygen-Dependent Regulation of Adipogenesis

By LILLIAN M. SWIERSZ, AMATO J. GIACCIA, and ZHONG YUN

Introduction

Molecular oxygen (O_2) is vital to all aerobic organisms through its involvement in energy homeostasis, embryogenesis, and differentiation. When exposed to hypoxia or low O_2 tensions, O_2-sensitive cells increase the expression of a wide variety of genes, including erythropoietin, vascular endothelial growth factor (VEGF), and glycolytic enzymes to stimulate erythropoiesis, angiogenesis, and glycolysis, respectively.[1] The majority of those hypoxia-induced genes are regulated transcriptionally by hypoxia-inducible factor-1 (HIF-1), a member of the basic helix–loop–helix *Per*, *AhR*, and *Sim* (bHLH-PAS) family.[2,3] Evidence is accumulating that O_2 sensing is important during embryonic development and differentiation, as homozygous deletion of either *HIF-1α* or *HIF-1β* is embryonically lethal in mice.[4–6]

Energy homeostasis is maintained in part through storing and releasing energy in the form of fatty acids or their derivatives. Because O_2 sensing plays an important role in energy homeostasis and cell differentiation, O_2 tension can also control adipose tissue function by regulating adipogenesis. Hypoxia inhibits the oxidation of fatty acids in mitochondria, resulting in a relative state of fatty acid abundance in the energy-consuming cells. Consequently, hypoxia may lead to a decreased storage of fat and reduced adipogenesis.

[1] H. F. Bunn and R. O. Poyton, *Physiol. Rev.* **76,** 839 (1996).
[2] G. L. Semenza and G. L. Wang, *Mol. Cell. Biol.* **12,** 5447 (1992).
[3] G. L. Wang, B. H. Jiang, E. A. Rue, and G. L. Semenza, *Proc. Natl. Acad. Sci. USA* **92,** 5510 (1995).
[4] K. R. Kozak, B. Abbott, and O. Hankinson, *Dev. Biol.* **191,** 297 (1997).
[5] N. V. Iyer, L. E. Kotch, F. Agani, S. W. Leung, E. Laughner, R. H. Wenger, M. Gassmann, J. D. Gearhart, A. M. Lawler, A. Y. Yu, and G. L. Semenza, *Genes Dev.* **12,** 149 (1998).
[6] H. E. Ryan, J. Lo, and R. S. Johnson, *EMBO J.* **17,** 3005 (1998).

Adipocyte differentiation results from the sequential induction of transcription factors C/EBPβ, C/EBPδ, C/EBPα, and PPARγ.[7,8] The differentiation of adipocytes is inhibited under hypoxic conditions, which indicates that oxygen is an important regulator of adipogenesis.[9] Hypoxia inhibits the induction of *PPARγ2* and *C/EBPβ* mRNA and dysregulates the expression of *C/EBPδ* mRNA.[9] Overexpression of *PPARγ2* and *C/EBPβ* by use of retroviral infection restores adipogenesis under hypoxia.[9] Use of mouse embryonic fibroblasts deficient in hypoxia-inducible transcription factor 1α indicates that *HIF-1α* is involved in the inhibition of adipogenesis by hypoxia.[9]

In Vitro Cell Culture Systems for Studying Adipocyte Development

The isolation of 3T3-L1 and 3T3-F442A cells from nonclonal Swiss 3T3 cells by Green and colleagues in the 1970s[10–12] led to much of the current understanding of the transcriptional regulation of adipogenesis. Compared with many other cellular lineages, adipogenic differentiation of 3T3-L1 or 3T3-F442A cells *in vitro* recapitulates most of the key features of adipogenesis *in vivo*. The 3T3-L1 and 3T3-F442A cells are already committed to the adipogenic pathway and are considered preadipocytes, although they are morphologically indistinguishable from fibroblasts. Under the appropriate stimuli [insulin, dexamethasone, and the phosphodiesterase inhibitor 3-isobutyl-1-methylxanthine (MIX)] preadipocyte cell lines can be differentiated into cells that resemble mature adipocytes morphologically and ultrastructurally. Other preadipocyte cell lines have also been studied, including TA1[13] and Ob1771.[14] These cells behave in a similar manner to the original 3T3-L1 and 3T3-F442A cell lines, yet there are differences in the differentiation regimen. Mouse embryonic fibroblasts (MEFs) can also be differentiated to adipocytes using PPARγ agonists such as rosiglitazone.[15] However, only a small percentage (about 15%) of the MEF population

[7] S. M. Rangwala and M. A. Lazar, *Annu. Rev. Nutr.* **20,** 535 (2000).

[8] E. D. Rosen and B. M. Spiegelman, *Annu. Rev. Cell Dev. Biol.* **16,** 145 (2000).

[9] Z. Yun, H. L. Maecker, R. S. Johnson, and A. J. Giaccia, *Dev. Cell* **2,** 331 (2002).

[10] H. Green and O. Kehinde, *Cell* **5,** 19 (1975).

[11] H. Green and O. Kehinde, *Cell* **7,** 105 (1976).

[12] H. Green and M. Meuth, *Cell* **3,** 127 (1974).

[13] A. B. Chapman, D. M. Knight, B. S. Dieckmann, and G. M. Ringold, *J. Biol. Chem.* **259,** 15548 (1984).

[14] R. Negrel, P. Grimaldi, and G. Ailhaud, *Proc. Natl. Acad. Sci. USA* **75,** 6054 (1978).

[15] J. M. Lehmann, L. B. Moore, T. A. Smith-Oliver, W. O. Wilkison, T. M. Willson, and S. A. Kliewer, *J. Biol. Chem.* **270,** 12953 (1995).

can be committed to adipogenic differentiation. Most importantly, early passage MEFs should be used, as MEFs exhibit decreased differential capability with increasing passage in culture.

Experimental Procedures

Preparations of Reagents

Cobalt chloride ($CoCl_2$), deferoxamine mesylate (DFO), Oil Red O, insulin (INS), dexamethasone (DEX), and MIX are from Sigma (St. Louis, MO). $CoCl_2$ and DFO are solubilized in sterile water at a concentration of 100 mM and stored at $-20°$ in aliquots. Oil Red O stock solution is prepared by dissolving Oil Red O at 0.3 g per 100 ml isopropanol. A stock solution of rosiglitazone (GlaxoSmithKline Pharmaceuticals) is prepared in dimethyl sulfoxide (DMSO) and stored at $-20°$ in aliquots.

Cell Culture

Mouse embryonic fibroblasts are cultured in Dulbecco's modified Eagle's medium (DMEM) supplemented with 10% fetal bovine serum (FBS), penicillin G (200 units/ml), and streptomycin (150 μg/ml). 3T3-L1 preadipocytes [American Type Culture Collection (ATCC)] are maintained in growth medium (GM) containing 10% bovine calf serum (ATCC) and 1 mM sodium pyruvate in DMEM according to the protocol recommended by ATCC.

Adipocyte Differentiation and Oil Red O Staining

The adipogenesis assay is outlined in Fig. 1A. Based on the previously described procedure,[16] 3T3-L1 cells are maintained in GM for 2 days after confluence. Confluent cells are treated (day 0) with differentiation medium (DM) containing 10% FBS, 10 μg/ml INS, 1 μM DEX, and 0.5 mM MIX in DMEM for 2 days. Cells are then maintained in DMEM containing 10% FBS and 1 μg/ml INS, and the medium is replaced every other day. Differentiated cells can be visualized between day 6 and day 8 by Oil Red O staining as described later. When treated with DM under normoxia (20% O_2), 3T3-L1 preadipocytes change from the fibroblast phenotype (control, Fig. 1B) to mature adipocytes loaded with oil droplets in the cytoplasm (induced, Fig. 1B).

---

[16] Z. Wu, N. L. Bucher, and S. R. Farmer, *Mol. Cell. Biol.* **16,** 4128 (1996).

A

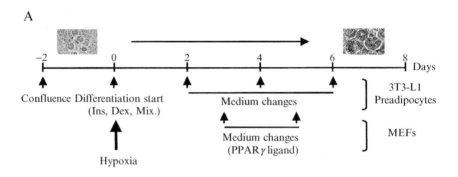

B

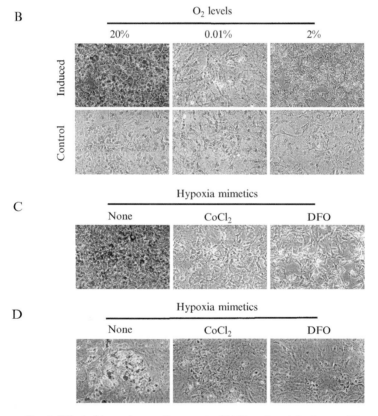

Fig. 1. Effect of hypoxia on adipogenesis. (A) Experimental scheme of the adipogenesis assay. (B) 3T3-L1 cells were induced to differentiate (Induced) or left untreated (Control) at indicated O_2 tensions. Cells were stained on day 6 with Oil Red O and photographed ($\times 20$). (C) 3T3-L1 cells were induced to differentiate in the presence or absence of $CoCl_2$ (50 μM) or DFO (100 μM) for the first 2 days together with IDM and were then maintained in media

MEFs are induced to differentiate 2 days after confluence (day 0) with DM supplemented with 5 μM rosiglitazone.[15,17] The medium is replaced on day 3 with DMEM plus 10% FBS, 1 μg/ml INS, and 5 μM rosiglitazone and is then changed every other day (Fig. 1A). Differentiated cells can be visualized between day 6 and day 8 by Oil Red O staining as described later. Because MEFs are a mixed population of stem cells, about 15% of the MEF population can be induced to undergo adipogenic differentiation (Fig. 1D). Untransformed early passage MEFs should be used to ensure that MEFs possess maximal capabilities to undergo adipogenic differentiation.

To evaluate the effects of hypoxia on adipogenesis, cells are maintained at 20% O_2 in a standard incubator, 2% O_2 in a standard incubator, or 0.01% O_2 in an anaerobic chamber (Bactron Anaerobic/Environmental Chamber, Sheldon Manufacturing, Inc.) immediately following treatment with DM (Fig. 1A). Alternatively, $CoCl_2$ or DFO can be added to DM at a range of 10 to 100 μM either during the initial stage of stimulation or for the entire course of differentiation. The adipogenic conversion of both 3T3-L1 preadipocytes and MEFs is inhibited either under low O_2 micro-environmental conditions (Fig. 1B) or in the presence of hypoxia-mimetic compounds, $CoCl_2$ or DFO (Fig. 1C and 1D).

For visualization of differentiated adipocytes, cells are washed with phosphate-buffered saline (PBS) and stained in freshly prepared Oil Red O staining solution (60% stock solution diluted in water and filtered just before use) for 30 min at 37°. Cells are washed briefly in 60% isopropanol and then rinsed in distilled water for microscopic observation and photography. Alternatively, the cells can be fixed in 2% formaldehyde solution before staining with Oil Red O. For better visualization under the microscope, cells can also be counterstained with Harris hematoxylin (Fisher Healthcare) after staining with Oil Red O.

Northern Blotting Analysis

Total cellular RNA is isolated with TRIzol reagent (Life Technologies). To prepare RNA from hypoxia-treated cells, it is suggested that the cells be placed on ice immediately following removal from hypoxia chambers to minimize the potential effects of reoxygenation. Individual RNA molecules

[17] D. L. Alexander, L. G. Ganem, P. Fernandez-Salguero, F. Gonzalez, and C. R. Jefcoate, *J. Cell Sci.* **111**, 3311 (1998).

without $CoCl_2$ or DFO. Cells were stained on day 6 and photographed ($\times$20). (D) MEFs were treated in DM containing 5 μM rosiglitazone with or without $CoCl_2$ (25 μM) or DFO (25 μM). Cells were stained on day 6 and photographed ($\times$20). Adapted with permission from Yun *et al.*[9] (See color insert.)

are separated by agarose gel electrophoresis, transferred to nitrocellulose membranes (Nytran SuPerCharge, 0.45 μm, Schleicher & Schuell Bioscience), and immobilized by UV cross-linking (UV Stratalinker 1800, Stratagene). Equal amounts of RNA samples transferred to nitrocellulose membrane are verified by methylene blue (40 mg/100 ml in 0.5 M sodium acetate, pH 5.0) staining. Radioactive probes are prepared by random labeling with cDNA as templates (Rediprime II, Amersham Biosciences). Hybridization is carried out in aqueous phase at 65° for 6 to 12 h. The radioactive blot is visualized on a Storm 860 PhosphoImager (Molecular Dynamics, Sunnyvale, CA).

Data presented in Fig. 2 illustrate the effects of hypoxia on the expression of PPARγ2, C/EBPβ, and C/EBPδ, three key transcription factors for adipogenesis.[7,8] At 20% O_2, both *C/EBPβ* and *C/EBPδ* are significantly induced within 2 h of IDM treatment, followed by their gradual decrease (Fig. 2A). *PPARγ2* mRNA is induced by day 2 post-IDM treatment

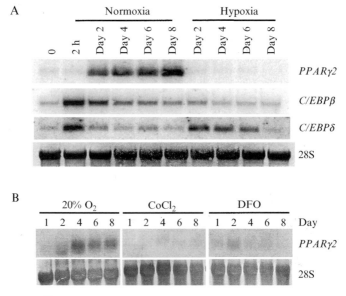

FIG. 2. Effect of hypoxia on the expression of *PPARγ2*, *C/EBPβ*, and *C/EBPδ*. (A) 3T3-L1 cells were induced to differentiate under either normoxia or hypoxia (0.01% O_2). Total cellular RNA was prepared at indicated times after induction. Equal amounts (10 μg/lane) of total RNA were subjected to Northern blotting (A) using ^{32}P-labeled *PPARγ2*, *C/EBPβ*, or *C/EBPδ* cDNA as probes. (B) 3T3-L1 cells were induced to differentiate under the following conditions: 20% O_2, 50 μM $CoCl_2$, or 100 μM DFO. Total RNA (5 μg/lane) was analyzed as in A. Adapted with permission from Yun *et al.*[9]

following the expression of *C/EBPβ* and *C/EBPδ* and remains elevated throughout the rest of the differentiation process (Fig. 2A). Under hypoxia, the induction of *PPARγ2* expression is completely abolished and that of *C/EBPβ* is reduced (Fig. 2A). The *PPARγ2* expression is also repressed in L1 cells treated with $CoCl_2$ or DFO (Fig. 2B). Unexpectedly, *C/EBPδ* expression becomes dysregulated and its mRNA remains elevated under hypoxia (Fig. 2A). These results suggest that negative regulation of *PPARγ2* and/or *C/EBPβ* gene expression is a key mechanism for hypoxia-mediated inhibition of adipogenesis.

Western Blotting Analysis

For Western blotting analysis, cell lysates are prepared on ice immediately following treatment using 25 m*M* HEPES buffer, pH 7.4, containing 1% NP-40, 150 m*M* NaCl, 2 m*M* EDTA, and a protease inhibitor cocktail (Complete, Roche Applied Science). Protein concentrations are measured using the bicinchoninic acid (BCA) assay (Pierce). Equal amounts of total cellular proteins are subjected to SDS–polyacrylamide gel electrophoresis under reducing conditions, electrotransfer to polyvinylidene difluoride (PVDF) membranes (Hybond-P, Amersham Pharmacia Biotech), and then Western blotting analysis. Protein bands can be visualized using enhanced chemofluorescence (ECF) substrates (Amersham Pharmacia Biotech) on a Storm 860 PhosphoImager.

Our previous study[9] indicates that the HIF-1 target gene *DEC1/Stra13*, which contains a bHLH and an Orange domain homologous to those of the HES transcription repressors,[18,19] is directly involved in inhibition of adipogenesis by hypoxia. During L1 cell differentiation under normoxia, *DEC1/Stra13* mRNA is induced approximately twofold within 2 h of IDM treatment (Fig. 3A and 3C), which is consistent with the literature.[20] In contrast, *DEC1/Stra13* mRNA in differentiating L1 cells exhibits approximately a fourfold increase on day 2, followed by a steady decline to the basal level under hypoxia (Fig. 3A and 3C). Interestingly, the DEC1/Stra13 protein level remains elevated (approximately threefold) from day 2 through day 8 under hypoxia as compared to normoxia (Fig. 3B and 3C). The increased stability of the DEC1/Stra13 protein in IDM-stimulated

[18] M. Boudjelal, R. Taneja, S. Matsubara, P. Bouillet, P. Dolle, and P. Chambon, *Genes Dev.* **11**, 2052 (1997).

[19] M. Shen, T. Kawamoto, W. Yan, K. Nakamasu, M. Tamagami, Y. Koyano, M. Noshiro, and Y. Kato, *Biochem. Biophys. Res. Commun.* **236**, 294 (1997).

[20] H. Inuzuka, R. Nanbu-Wakao, Y. Masuho, M. Muramatsu, H. Tojo, and H. Wakao, *Biochem. Biophys. Res. Commun.* **265**, 664 (1999).

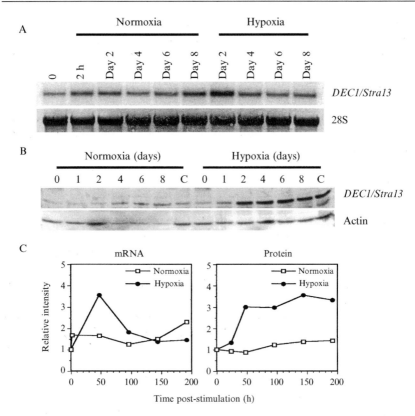

Fig. 3. *DEC1/Stra13* expression during adipogenic differentiation of 3T3-L1 preadipocytes. (A) 3T3-L1 cells were induced to differentiate under either normoxia or hypoxia (0.01% O_2). Equal amounts (10 μg/lane) of total RNA prepared at indicated times after induction were subjected to Northern blotting. The relative levels of *DEC1/Stra13* mRNA were analyzed by densitometry (C). (B) 3T3-L1 cells were induced to differentiate as in A. Cell lysates were prepared at indicated times after treatment and subjected to Western blotting (20 μg protein/lane) with rabbit anti-Stra13 polyclonal antibodies as described. Controls (lane C) were maintained for 8 days under either normoxia or hypoxia without adipogenic stimulation. The relative levels of the DEC1/Stra13 protein were analyzed by densitometry (C). Adapted with permission from Yun *et al.*[9]

L1 cells indicates that both *DEC1/Stra13* mRNA and protein are regulated by hypoxia.

Conclusion

These results demonstrate that hypoxia is a physiological regulator of adipogenesis. Nevertheless, hypoxia can potentially play a significant role in the differentiation of many other cell types, as suggested by the observation that human embryos are located in a low O_2 environment during the first trimester.[21] The regulation of adipogenesis by hypoxia serves as a model for future research in understanding how the cellular microenvironment regulates cell differentiation during both embryogenesis and development of adult stem cells.

Acknowledgments

This work was supported by NIH Grants CA88480 and CA67166 (A.J.G.), Aventis (A.J.G.), the NIH Women's Reproductive Health Research Career Development Program (L.S.), and NIH Cancer Biology Training Grant CA09302 (Z.Y.).

[21] F. Rodesch, P. Simon, C. Donner, and E. Jauniaux, *Obstet. Gynecol.* **80,** 283 (1992).

Section VI

Nervous System

[27] Methods to Detect Hypoxia-Induced Ischemic Tolerance in the Brain

By MYRIAM BERNAUDIN and FRANK R. SHARP

Introduction

Reduced oxygen supply *(hypoxia)* or reduced blood flow *(ischemia)* to the brain is a major cause of morbidity and mortality in both perinatal and adult periods, often resulting in cognitive impairment, seizures, and other neurological disabilities. Although hypoxia–ischemia animal models have increased our understanding of the processes leading to cell death, there are still no pharmacological treatments available to reduce cell death in ischemic brain.

Interestingly, cells can be protected when a noninjurious hypoxic stress is performed several hours or days before a lethal hypoxic–ischemic stress *(preconditioning)*. This phenomenon is called *tolerance*. Ischemic tolerance can be achieved in the brain by several preconditioning sublethal stresses such as hypoxia,[1–5] ischemia itself,[6] hypothermia,[7] hyperthermia,[8] hyperbaric oxygenation,[9] metabolic inhibitors,[10] and spreading depression,[11] as well as cytokines.[12,13] Ischemic tolerance has been also described *in vitro*

[1] J. M. Gidday, J. C. Fitzgibbons, A. R. Shah, and T. S. Park, *Neurosci. Lett.* **168,** 221 (1994).
[2] A. Ota, T. Ikeda, K. Abe, H. Sameshima, X. Y. Xia, Y. X. Xia, and T. Ikenoue, *Am. J. Obstet. Gynecol.* **179,** 1075 (1998).
[3] R. C. Vannucci, J. Towfighi, and S. J. Vannucci, *J. Neurochem.* **71,** 1215 (1998).
[4] B. A. Miller, R. S. Perez, A. R. Shah, E. R. Gonzales, T. S. Park, and J. M. Gidday, *Neuroreport* **12,** 1663 (2001).
[5] M. Bernaudin, A. S. Nedelec, D. Divoux, E. T. MacKenzie, E. Petit, and P. Schumann-Bard, *J. Cereb. Blood Flow Metab.* **22,** 393 (2002).
[6] K. Kitagawa, M. Matsumoto, M. Tagaya, R. Hata, H. Ueda, M. Niinobe, N. Handa, R. Fukunaga, K. Kimura, K. Mikoshiba *et al., Brain Res.* **528,** 21 (1990).
[7] S. Nishio, M. Yunoki, Z. F. Chen, M. J. Anzivino, and K. S. Lee, *J. Neurosurg.* **93,** 845 (2000).
[8] M. Chopp, H. Chen, K. L. Ho, M. O. Dereski, E. Brown, F. W. Hetzel, and K. M. Welch, *Neurology* **39,** 1396 (1989).
[9] K. Prass, F. Wiegand, P. Schumann, M. Ahrens, K. Kapinya, C. Harms, W. Liao, G. Trendelenburg, K. Gertz, M. A. Moskowitz, F. Knapp, I. V. Victorov, D. Megow, and U. Dirnagl, *Brain Res.* **871,** 146 (2000).
[10] F. Wiegand, W. Liao, C. Busch, S. Castell, F. Knapp, U. Lindauer, D. Megow, A. Meisel, A. Redetzky, K. Ruscher, G. Trendelenburg, I. Victorov, M. Riepe, H. C. Diener, and U. Dirnagl, *J. Cereb. Blood Flow Metab.* **19,** 1229 (1999).
[11] S. Kobayashi, V. A. Harris, and F. A. Welsh, *J. Cereb. Blood Flow Metab.* **15,** 721 (1995).

on neuronal cultures by mimicking the ischemic insult with sublethal oxygen and glucose deprivation (OGD). OGD tolerance has been described with preconditioned neuronal cultures with hypoxia,[14] OGD sublethal stress itself,[15–17] or metabolic inhibition.[18]

In several clinical conditions, cerebral ischemia can be anticipated (such as subarachnoid hemorrhage or brain surgery). Therefore, it would be of great interest to develop clinical approaches to induce brain ischemic tolerance. Moreover, ischemic tolerance models might be a useful paradigm to understand the mechanisms that lead to brain protection against ischemia and consequentely to identify new therapeutic targets for stroke.

As hypoxic preconditioning is noninvasive and reproducible, this model has been widely used to study the mechanisms protecting the brain against global hypoxia–ischemia, particularly in newborn rats,[3,19–21] and against focal transient[4] and permanent[5] cerebral ischemia in adult mice. Moreover, hypoxic preconditioning can be mimicked by the iron chelator desferrioxamine (DFX),[20,22] by divalent metal cobalt chloride (CoCl$_2$),[20,21] and by pretreatment with hypoxia-inducible proteins such as erythropoietin (EPO).[13,23] These are preconditioning paradigms that might be applicable to humans.

This article discusses the different models to reproduce and detect ischemic tolerance induced by hypoxic preconditioning. In addition, methods are presented to validate the hypoxia and to assess the hypoxic preconditioning efficiency in the ischemic infarct size reduction.

[12] H. Nawashiro, K. Tasaki, C. A. Ruetzler, and J. M. Hallenbeck, *J. Cereb. Blood Flow Metab.* **17**, 483 (1997).

[13] M. Bernaudin, H. H. Marti, S. Roussel, D. Divoux, A. Nouvelot, E. T. MacKenzie, and E. Petit, *J. Cereb. Blood Flow Metab.* **19**, 643 (1999).

[14] J. Liu, I. Ginis, M. Spatz, and J. M. Hallenbeck, *Am. J. Physiol. Cell Physiol.* **278**, C144 (2000).

[15] U. Bruer, M. K. Weih, N. K. Isaev, A. Meisel, K. Ruscher, A. Bergk, G. Trendelenburg, F. Wiegand, I. V. Victorov, and U. Dirnagl, *FEBS Lett.* **414**, 117 (1997).

[16] L. Khaspekov, M. Shamloo, I. Victorov, and T. Wieloch, *Neuroreport* **9**, 1273 (1998).

[17] M. C. Grabb and D. W. Choi, *J. Neurosci.* **19**, 1657 (1999).

[18] M. Weih, A. Bergk, N. K. Isaev, K. Ruscher, D. Megow, M. Riepe, A. Meisel, I. V. Victorov, U. Dirnagl, and U. Dirnagi, *Neurosci. Lett.* **272**, 207 (1999).

[19] J. M. Gidday, A. R. Shah, R. G. Maceren, Q. Wang, D. A. Pelligrino, D. M. Holtzman, and T. S. Park, *J. Cereb. Blood Flow Metab.* **19**, 331 (1999).

[20] M. Bergeron, J. M. Gidday, A. Y. Yu, G. L. Semenza, D. M. Ferriero, and F. R. Sharp, *Ann. Neurol.* **48**, 285 (2000).

[21] N. M. Jones and M. Bergeron, *J. Cereb. Blood Flow Metab.* **21**, 1105 (2001).

[22] K. Prass, K. Ruscher, M. Karsch, N. Isaev, D. Megow, J. Priller, A. Scharff, U. Dirnagl, and A. Meisel, *J. Cereb. Blood Flow Metab.* **22**, 520 (2002).

[23] M. Sakanaka, T. C. Wen, S. Matsuda, S. Masuda, E. Morishita, M. Nagao, and R. Sasaki, *Proc. Natl. Acad. Sci. USA* **95**, 4635 (1998).

Different Models of Hypoxia-Induced Ischemic Tolerance in the Brain

In Vivo Models of Hypoxia-Induced Ischemic Tolerance

Several *in vivo* models of hypoxia-induced ischemic tolerance have been developed in neonatal and adult rodent brains.

In the *neonatal rat brain,* ischemic tolerance can be achieved by a period of hypoxic preconditioning (3 h at 8% O_2) 24 h prior to an unilateral occlusion of the common carotid artery followed by hypoxia 1.5 h later (3 h at 8% O_2), a well-established tolerance model that was first published by Gidday and colleagues[1] (Figs. 1A and 1B). This model has been widely used[19–21,24] with some changes in the duration of the hypoxic preconditioning (ranging from 2 to 4 h), the time between the ischemic insult and the second hypoxic insult (1.5 to 4 h), and the duration of the second hypoxic insult (45 min to 3 h).[2,3,25]

More recently, two different models of hypoxia-induced ischemic tolerance have been published in the *adult mouse brain* (Figs. 2A and 2B).[4,5] Miller *et al.*[4] showed that hypoxic preconditioning (11% O_2, 2 h) 48 h before a transient focal cerebral ischemia (1.5 h) reduced infarct size significantly (Fig. 2A). The second study[5] showed that hypoxia induces tolerance against permanent focal cerebral ischemia. Normobaric hypoxia (8% O_2 of 1, 3, or 6 h duration) performed 24 h before ischemia reduces infarct volume by approximately 30% when compared to controls (Fig. 2B). When the hypoxia preceded the ischemia by 72 h, the decrease in infarct volume (-12%) was not significant, suggesting that the tolerance tended to disappear. In the study of Miller *et al.*[4] no protection by hypoxic preconditioning was shown against permanent focal cerebral ischemia. The discrepancy between the results of Bernaudin *et al.*[5] and those of Miller *et al.*[4] could be explained by the difference of the degree of hypoxia used (8% instead of 11% O_2), suggesting that the systemic hypoxia stimulus should be stronger to induce tolerance against the permanent occlusion of a major cerebral artery compared to transient occlusion. In addition, the delay between hypoxia and ischemia determines the efficacy of preconditioning: the results of Bernaudin *et al.*[5] suggest that tolerance achieved by exposure to 6 h of hypoxia is short lasting (<72 h). One might suggest that the delay between preconditioning and ischemia should be shorter than 48 h to show a tolerance against the model of permanent ischemia. Therefore, the discrepancy between the results of these two studies

[24] M. Bernaudin, Y. Tang, M. Reilly, E. Petit, and F. R. Sharp, *J. Biol. Chem.* **277,** 39728 (2002).
[25] T. Wada, T. Kondoh, and N. Tamaki, *Brain Res.* **847,** 299 (1999).

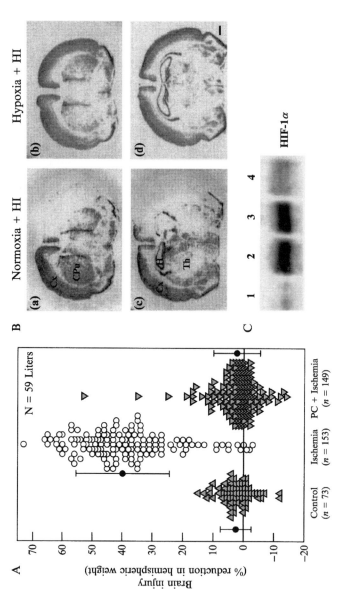

Fig. 1. Hypoxia preconditioning induces ischemic tolerance and HIF-1α expression in the neonatal rat brain. (A) Brain injury in preconditioned animals (hypoxia 8% O_2, 3 h in 6-day-old rats, $n = 149$) is reduced significantly compared to animals that are not preconditioned ($n = 153$) and does not differ significantly from that measured in nonischemic controls ($n = 73$). Each point shown represents one animal. Means and SD for each of the three animal groups are shown as filled symbols to the side of each distribution. Brain injury is quantified as the percentage reduction in hemispheric weight. From Gidday et al.[19] reprinted by permission of Lippincott Williams & Wilkins, Inc. Copyright © 1999 by The International Society of Cerebral Blood Flow and Metabolism. (B) Representative anteroposterior coronal brain sections stained

could also be explained by the difference of the delay between hypoxic preconditioning and ischemia (24 h vs 48 h).

In Vitro *Models of Hypoxia-Induced Ischemic Tolerance*

Ischemic tolerance has been also described *in vitro* on neuronal-enriched cultures by mimicking the ischemic insult with sublethal oxygen and glucose deprivation (OGD)[14–16] or with metabolic inhibitors.[18] Mild hypoxia (8% O_2, 20 min) produced 24 h before a severe hypoxia (2% O_2, 2.5 h) or OGD insult (2.5 h) decreased the neuronal death by 50% approximately.[14] Similar to *in vivo* ischemic-induced ischemic tolerance models, sublethal OGD leads to increased tolerance against subsequent OGD in neuronal-enriched cultures.[15,17] However, this article focuses only on hypoxic preconditioning.

In Vitro *and* In Vivo *Models of Hypoxia-Induced Ischemic Tolerance Can Be Mimicked by HIF-1 Inducers and HIF-1 Target Genes*

In *in vitro* models of preconditioning, as well as during *in vivo* hypoxic or ischemic preconditioning, the reduction of oxygen availability may constitute one of the early cellular signals leading to tolerance. Hypoxia-inducible factor 1 (HIF-1) is an important transcription factor regulating gene expression in response to hypoxia. Therefore, HIF-1 could be a candidate gene for producing hypoxia-induced tolerance to ischemia.[5,20,21,24] Indeed, hypoxic preconditioning induces expression of HIF-1α (Fig. 1C) and its target genes in neonatal[20,21] and adult brain[5] (Fig. 2C). In addition, desferrioxamine (DFX) and cobalt chloride (CoCl₂), two agents that activate HIF-1,[26] also induce tolerance against both hypoxia–ischemia in neonatal rat brain[20,21] (Fig. 3A) and ischemia in adult mice and rat brain.[22] Studies in the neonatal rat brain showed that the level of HIF-1α expression

---

[26] W. Ehleben, T. Porwol, J. Fandrey, W. Kummer, and H. Acker, *Kidney Int.* **51,** 483 (1997).

---

with cresyl violet from animals subjected to ischemia–hypoxia 5 days earlier without (a and c) or with hypoxic preconditioning (b and d). From Jones *et al.*[21] reprinted by permission of Lippincott Williams & Wilkins, Inc. Copyright © 2001 by The International Society of Cerebral Blood Flow and Metabolism. (C) Western blot analysis of HIF-1α protein expression in the cerebral cortex of 6-day-old rats subjected to hypoxic preconditioning or hypoxia–ischemia. Brains are analyzed at the end of each treatment before reoxygenation. Lane 1, untreated control; lane 2, hypoxic preconditioning (hypoxia 8% O_2, 3 h in 6-day-old rats); lanes 3 and 4, contralateral (lane 3) and ipsilateral (lane 4) cortices from rats subjected to ischemia–hypoxia. From Bergeron *et al.*[20] reprinted by permission of John Wiley and Sons, Inc. Copyright © 2000 by the American Neurological Association.

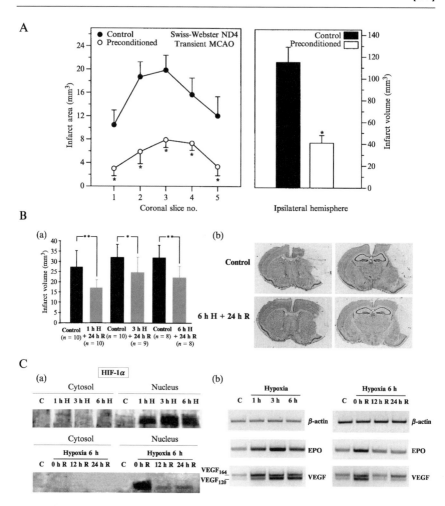

FIG. 2. Hypoxia preconditioning induces ischemic tolerance and HIF-1α expression and its target genes in the adult mouse brain. (A) Effect of hypoxic preconditioning (11% O_2, 2 h) on brain injury in Swiss–Webster ND4 mice ($n = 10$) subjected to *transient* focal cerebral ischemia produced by middle cerebral artery occlusion with an intraluminal filament. (Left) The infarct area for each coronal slice along the anterior–posterior axis and (right) the edema-corrected infarct volume (mean ± SD; $^*p < 0.05$ vs corresponding control group, unpaired t tests). From Miller et al.[4] reprinted by permission of Lippincott Williams & Wilkins, Inc. Copyright © 2001 by Lippincott Williams & Wilkins. (B) Effect of hypoxic preconditioning (8% O_2, 1, 3, or 6 h) on brain injury in Swiss mice subjected to *permanent* focal cerebral ischemia by electrocoagulation of the middle cerebral artery. (a) Infarct volumes were measured 48 h after ischemia (mean ± SD; significantly different from control, ANOVA and PLSD of Fisher, $^*p < 0.05$ and $^{**}p < 0.01$, respectively). (b) Representative coronal brain sections of control mice and mice subjected to hypoxic preconditioning (8% O_2, 6 h) 24 h

correlated with the degree of brain protection afforded after each preconditioning treatment (hypoxia > $CoCl_2$ > DFX) (Fig. 3B). Moreover, pretreatment with HIF-1 target genes such as erythropoietin (EPO) protect the brain against ischemia.[13,23] Similarly, *in vitro* studies showed that DFX[22] and EPO[27] pretreatments protect neurons from OGD damage. Taken together, all these studies suggest that HIF-1 and its target genes could be important mediators of hypoxia-induced tolerance to ischemia.

Detection of Hypoxia-Induced Ischemic Tolerance in the Brain

Detection of Hypoxia

HIF-1 is a heterodimer composed of two protein subunits, HIF-1α and HIF-1β.[28] Whereas HIF-1β is expressed constitutively, HIF-1α expression is tightly regulated by cellular oxygen concentration.[29] Thus HIF-1α determines HIF-1 DNA-binding activity and transcriptional activity during hypoxia. As hypoxic and hypoxic-mimicking preconditioning are known to induce expression of HIF-1α and its target genes in neonatal and adult brains, the expression of HIF-1 and its target genes could be used as an *intrinsic marker of hypoxia.*

Hypoxia tolerance induces a number of known HIF-1 target genes, including EPO, vascular endothelial growth factor (VEGF), glucose-transporter-1 (GLUT-1), and MAP kinase phosphatase-1 (MKP-1),[24] and other known hypoxic inducible genes such as metallothionein-1. In addition, a microarray study showed that hypoxia (8% O_2, 3 h) followed by 0, 6, 18, or 24 h of reoxygenation in the 6- to 7-day-old rat induced a

[27] K. Ruscher, D. Freyer, M. Karsch, N. Isaev, D. Megow, B. Sawitzki, J. Priller, U. Dirnagl, and A. Meisel, *J. Neurosci.* **22,** 10291 (2002).

[28] G. L. Wang, B. H. Jiang, E. A. Rue, and G. L. Semenza, *Proc. Natl. Acad. Sci. USA* **92,** 5510 (1995).

[29] L. E. Huang, Z. Arany, D. M. Livingston, and H. F. Bunn, *J. Biol. Chem.* **271,** 32253 (1996).

before ischemia. Infarct volumes were measured 48 h after ischemia. H, hypoxia; R, reoxygenation. From Bernaudin *et al.*[5] reprinted by permission of Lippincott Williams & Wilkins, Inc. Copyright © 2002 by The International Society of Cerebral Blood Flow and Metabolism. (C) (a) Western blot analysis of HIF-1α in cytosolic and nuclear extracts of the whole brain hemisphere of control mice and mice subjected to normobaric hypoxia (8% O_2). (b) RT-PCR analysis of EPO, VEGF, and β-actin mRNA expression in the whole brain hemisphere of control mice and mice subjected to normobaric hypoxia (8% O_2). C, control; H, hypoxia; R, reoxygenation. From Bernaudin *et al.*[5] reprinted by permission of Lippincott Williams & Wilkins, Inc. Copyright © 2002 by The International Society of Cerebral Blood Flow and Metabolism.

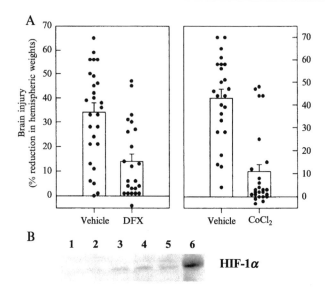

Fig. 3. Pretreatment with DFX or $CoCl_2$ induces ischemic tolerance in the neonatal rat brain. (A) Brain injury in preconditioned animals (a single intraperitoneal injection of 200 mg/kg DFX or 60 mg/kg $CoCl_2$ in 6-day-old rats) is reduced significantly compared to control animals (intraperitoneal injection of vehicle): DFX 14.1% vs vehicle 32.3% ($p < 0.005$) and $CoCl_2$ 10.6% vs vehicle 43.0% ($p < 0.0001$). Each point represents one animal. Mean and SD for each group are represented as histograms overlapping each distribution. Brain injury is quantified as the percentage reduction in hemispheric weight. (B) Western blot analysis of HIF-1α protein in the brain of 6-day-old rats subjected to a single intraperitoneal injection of DFX (200 mg/kg, lanes 3 and 4), $CoCl_2$ (60 mg/kg, lanes 5 and 6), or vehicle (lanes 1 and 2) either 1 h (lanes 1, 3, and 5) or 3 h (lanes 2, 4, and 6) after injection. From Bergeron et al.[20] reprinted by permission of John Wiley & Sons, Inc. Copyright © 2000 by the American Neurological Association.

number of other genes, including adrenomedullin, prolyl 4-hydroxylase, C/EBPδ, 12-lipoxygenase, t-PA, and several ESTs.[24]

Exposure of neonatal rats and adult rats and mice to hypoxia, DFX, or $CoCl_2$ induced an increase in HIF-1α and its target gene expressions in the brain when compared to control animals (Figs. 1, 2, and 3).[5,20–22,30] As shown in Fig. 2C, reoxygenation following exposure to hypoxia largely reduced the increase of HIF-1α activity and its target gene expression.[5,21,24] Therefore, the optimal time point to examine gene expression may be

[30] M. Bernaudin, A. Bellail, H. H. Marti, A. Yvon, D. Vivien, I. Duchatelle, E. T. Mackenzie, and E. Petit, Glia 30, 271 (2000).

immediately after the hypoxic period without allowing any possibility of reoxygenation.

Detection of Ischemic Tolerance in the Brain

In Vivo. Hypoxic–ischemic brain damage in the immature rat tolerance model is largely restricted to the cerebral hemisphere ipsilateral to the common carotid artery occlusion and involves the cerebral cortex, subcortical and periventricular white matter, striatum, thalamus, and hippocampus.[31] Tissue injury takes the form of both selective neuronal death and infarction. The degree of ischemic damage and the neuroprotection afforded by the preconditioning treatment can be assessed by *histological analyses* using cresyl violet (Nissl) staining (Fig. 1B). Moreover, brain damage can be assessed by comparing *hemispheric weights* (Figs. 1A and 3A). Indeed, this method has been correlated with changes in infarct area measured by histological methods, biochemical indexes of ischemic injury, and electrophysiological criteria.[32]

In the murine and rat adult models of ischemic tolerance, focal cerebral ischemia was produced by occluding the middle cerebral artery (MCA) transiently or permanently. Brain damage is restricted to the MCA irrigation territory (cerebral cortex and striatum), but the severity of damage depends on the duration of MCA occlusion and the occlusion site (distal or proximal). In both transient[4,22] and permanent[5] MCA occlusion models, ischemic damage can be assessed by quantifying the *infarct area* using triphenyltetrazolium chloride (TTC) staining[4] or histological analyses using cresyl violet or thionin (Nissl) staining (Figs. 1B and 2B). (Note that TTC is not the preferred method of infarct quantification because the staining changes over time following a stroke and can be perturbed by some metabolic treatments.) Histological staining for infarct volumes is preferred.

In Vitro. Quantification of neuronal injury in neuronal cultures submitted to OGD can be performed by an *ethidium homodimer fluorescent exclusion test,*[14] a commonly used method for cytotoxicity tests. Of note, in neuronal cultures, the lactate dehydrogenase (LDH) activity in the medium correlated robustly with the number of damaged cells.[33] However, hypoxia and iron chelators can increase glycolytic enzyme expression,

[31] R. C. Vannucci, J. R. Connor, D. T. Mauger, C. Palmer, M. B. Smith, J. Towfighi, and S. J. Vannucci, *J. Neurosci. Res.* **55,** 158 (1999).
[32] J. M. Gidday, J. C. Fitzgibbons, A. R. Shah, M. J. Kraujalis, and T. S. Park, *Pediatr. Res.* **38,** 306 (1995).
[33] J. Y. Koh and D. W. Choi, *J. Neurosci. Methods* **20,** 83 (1987).

including LDH-A expression, via HIF-1.[34–36] The Trypan blue exclusion method or the ethidium homodimer method is preferred to quantify neuronal cell death after OGD treatment.

Experimental Methods

Details of the experimental procedures to reproduce models of hypoxia-induced ischemic brain tolerance *in vivo* are given for the immature and adult rat. In addition, protocols for HIF-1 and its target gene expression analyses and for quantification of the hypoxic preconditioning protective effect against ischemia are given. Schematic representations of these two models are shown in Fig. 4.

Animals

All animal experiments should be approved and performed in strict accordance with the local ethics committee and governed by the pertinent national legislation. Animals should be acclimated to the animal quarters at least 3 days prior to study. The animals, maintained on a 12-h light/dark cycle, are given food and water *ad libitum*. In addition, all experimental and control animals are housed in the same room prior to and at the conclusion of the study. All surgical procedures described should be performed by aseptic technique with sterilized instruments and materials.

Immature Rat Model of Hypoxia-Induced Ischemic Tolerance in the Brain

Hypoxic preconditioning is performed on male and female rats at postnatal day 6 (Sprague–Dawley). Rat pups are placed in an 8% O_2 and 92% N_2 humidified atmosphere for a period of 3 h using a hypoxia chamber (BioSpherix, formerly Reming Bioinstruments, NY) consisting of four identical plexiglass chambers, 16 liters in size each, that are arranged adjacent to one another. The inlet O_2, CO_2, and N_2 are fixed using gas tanks of appropriate concentrations and valves to monitor inlet gas concentrations. The exiting gas is monitored continuously for O_2, CO_2, and N_2 as well. Control animals from the same littermates are placed in one of the four chambers and are exposed to ambient oxygen (approximatively 21% O_2)

[34] G. L. Semenza, B. H. Jiang, S. W. Leung, R. Passantino, J. P. Concordet, P. Maire, and A. Giallongo, *J. Biol. Chem.* **271,** 32529 (1996).

[35] J. D. Firth, B. L. Ebert, and P. J. Ratcliffe, *J. Biol. Chem.* **270,** 21021 (1995).

[36] K. Zaman, H. Ryu, D. Hall, K. O'Donovan, K. I. Lin, M. P. Miller, J. C. Marquis, J. M. Baraban, G. L. Semenza, and R. R. Ratan, *J. Neurosci.* **19,** 9821 (1999).

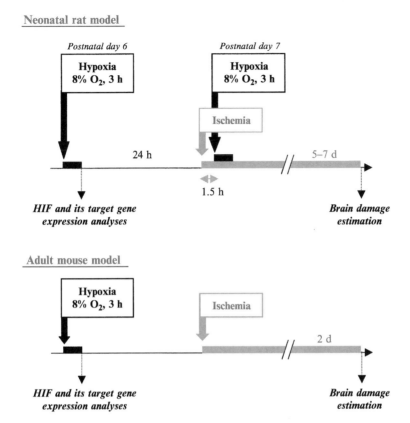

FIG. 4. Schematic representation of hypoxia preconditioning-induced ischemic tolerance in neonatal rat and adult mouse brains.

for the same duration. After 3 h of hypoxia or normoxia, all animals are returned to their dams for 24 h (reoxygenation period).[24] Hypoxic preconditioning can also be produced with 500-ml air-tight chambers partially submerged in a water bath maintained at a constant temperature of 37° and flushed with a humidified gas mixture of 8% O_2 and 92% N_2 delivered into the chamber via inlet and outlet portals at an approximate flow rate of 100 ml/min (Manostat; Barnant, Barrington, IL).[37] Hypoxic preconditioning can be mimicked by a single intraperitoneal injection of DFX (200 mg/kg) or $CoCl_2$ (60 mg/kg) 24 h before the ischemic insult, whereas

[37] M. Bergeron, D. M. Ferriero, H. J. Vreman, D. K. Stevenson, and F. R. Sharp, *J. Cereb. Blood Flow Metab.* **17,** 647 (1997).

control littermates receive a 0.9% sodium chloride vehicle solution (pH 7.4).[20,21] Of note, preconditioning with $CoCl_2$ should be preferentially used because this treatment is more effective than preconditioning with DFX in terms of protective effect (56% of protection with DFX and 75% with $CoCl_2$)[20] (Fig. 3A) and the induction of HIF-1 and its target genes[20,24] (Fig. 3B).

After the reoxygenation periods, ischemia–hypoxia is produced using the Vannuci model, a standard technique for neonatal stroke.[38] This model consists of a permanent, unilateral ligation of the common carotid artery combined with a period of hypoxia. The combination of ischemia and hypoxia induces an infarcted area ipsilateral to the ligation. Ischemia–hypoxia is produced in 7-day-old neonatal rats. Rat pups are anesthetized with isofluorane, and the left common carotid artery is ligated through a ventral midline neck incision. The wound is sutured and animals are returned to their dams for 1.5 h after which time they are exposed to hypoxia (8% O_2, 3 h). Of note, the duration of this second hypoxic period can be decreased to 90 min or less if a moderate degree of injury is preferred. During surgery, animals are kept on a $37°$ heating pad. After variable periods of postocclusion, animals are anesthetized with ketamine (100 mg/kg) and xylazine (20 mg/kg), and their brains are removed for subsequent analyses.

Materials

Animals: dated, pregnant rats are purchased from a commercial breeder (Sprague–Dawley rats, Harlan, IN), and housed in individual cages until the delivery and the time of initial experimentation at postnatal day 6

Hypoxia chamber (BioSpherix, formerly Reming Bioinstruments, NY)

Ketamine (Phoenix Pharmaceutical, Inc., St. Joseph, MO)

Xylazine (Phoenix Pharmaceutical, Inc.,)

Isofluorane (Abbott Laboratories, North Chicago, IL)

Normal saline (Abbott Laboratories)

DFX (Sigma, St. Louis, MO)

$CoCl_2$ (Sigma)

Operating microscope (Zeiss Stemi 2000C, Carl Zeiss, Hallbergmoos, Germany)

Small animal surgical instruments: scalpel, forceps, scissors, surgical suture needles, and thread

Disposable 0.5- and 1-ml syringes (Becton Dickinson and Co., Franklin Lakes, NJ)

Needles (Becton Dickinson and Co.)

[38] J. E. Rice, 3rd, R. C. Vannucci, and J. B. Brierley, *Ann. Neurol.* **9,** 131 (1981).

Adult Mice Model of Hypoxia-Induced Ischemic Tolerance in the Brain

Hypoxic preconditioning (8% O_2 for 1, 3 or 6 h) can be produced exactly as described earlier for the immature rat model.

Focal ischemia is produced 24 h later by permanent occlusion of the left middle cerebral artery under halothane (1–1.5% in N_2O/O_2: 2/1) anesthesia, as reported previously.[5,39] Briefly, a skin incision is made between the orbit and the ear. Under an operating microscope, an incision is made dividing the temporal muscle, and the left lateral aspect of the skull is exposed by reflecting the temporal muscle and surrounding soft tissue. The distal course of the MCA is then visible through the translucent skull. A small craniotomy is performed with a cooled dental drill. The left MCA is then coagulated by bipolar diathermy, the muscle and soft tissue are replaced, and the incision is sutured. Under halothane anesthesia, mice euthanasia is performed, and their brains are removed at various times after ischemia depending on the subsequent analyses.

Materials

> Hypoxia chamber (see earlier description)
> Halothane (Belamont Lab., Neuilly sur Seine, France)
> Normal saline (Abbott Laboratories)
> Operating microscope (Zeiss Stemi 2000C, Carl Zeiss)
> Electrocoagulator (Braun-Aesculap, TB50, France)
> Dental milling machine (Technobox 810, Bien-air, Nederland)
> Small animal surgical instruments: scalpel, forceps, scissors, surgical suture needles, and thread

HIF-1 Expression Analysis

The hypoxic increase of HIF-1 activity occurs mainly through a stabilization of HIF-1α protein that is translocated to the nucleus rapidly. Therefore, an adequate way to study HIF-1 expression changes during hypoxia is to examine HIF-1α protein expression in the nucleus by Western blot analyses on nuclear protein extracts.

Preparation of Nuclear Protein Extracts. Nuclear protein extracts can be prepared from whole brain hemisphere or cerebral cortex tissue lysates depending on the area of interest. Tissue extraction needs to be as quick as possible after hypoxia to reduce the reoxygenation that will decrease HIF-1α expression. Tissue extracts are lysed in the following buffer: 20 mM HEPES, 10 mM KCl, 1 mM EDTA, 1 mM dithiothreitol (DTT),

[39] F. A. Welsh, T. Sakamoto, A. E. McKee, and R. E. Sims, *J. Neurochem.* **49,** 846 (1987).

0.2% NP-40, 10% glycerol, 1 mM phenylmethylsulfonyl fluoride (PMSF), 0.1 mM vanadate, and 1 μg/ml protease inhibitors, pH 7.9, in distilled water (Sigma). After 5 min on ice, samples are centrifuged for 10 min at 13,000g. Supernatants (cytosolic extracts) are kept at $-80°$. Pellets are resuspended in 50 μl of buffer containing 350 mM NaCl, 20% glycerol, 20 mM HEPES, 10 mM KCl, 1 mM EDTA, 1 mM DTT, 1 mM PMSF, 0.1 mM vanadate, and 1 μg/ml protease inhibitors, pH 7.9, in distilled water. The suspension is mixed vigorously and incubated on ice for 30 min. min. Samples are then centrifuged for 10 min at 13,000g and supernatants (nuclear extracts) are kept at $-80°$.

Western Blot. Proteins (50–75 μg for nuclear extracts) are heated at 100° for 3 min in the 2X loading buffer (EC-886, National Diagnostics, Atlanta, GA) and separated on 8% SDS–PAGE gels for 1 h at 150 V. The electrophoresis running buffer contains 25 mM Tris base, 192 mM glycine, and 0.1% SDS. Proteins are transferred onto a nitrocellulose membrane (Optitran BA-S 85, Schleicher and Schell, Keene, NH) in blotting buffer [25 mM Tris base, 192 mM glycine, and 20% methanol (v/v)] for 2 h at 300 mA. Membranes are stained with Ponceau S to verify equal protein loading and transfer. Membranes are blocked with 5% nonfat milk (Bio-Rad, Hercules, CA) in TBST (10 mM Tris–HCl, pH 8.0, 200 mM NaCl, 0.05% Tween 20) for 2 h at room temperature and are then incubated overnight at 4° with one of the following antibodies diluted in the blocking solution: mouse monoclonal antibody for HIF-1α from Stressgen (2 μg/ml, OSA-601; Stressgen Biotechnologies Corp. Inc., Victoria, BC, Canada) or mouse monoclonal antibody for HIF-1α from Novus (6 μg/ml, NB 100–123; Novus Biologicals Inc., Littleton, CO). Antibodies developed by Dr. Semenza and colleagues or Dr. Pousségur and colleagues (HIF-1α rabbit polyclonal antibodies) have shown very good staining.[5,20,21] After washing with TBST (four times, 15 min each), the blots are incubated for 1 h at room temperature with peroxidase-coupled antimouse IgG (1:5,000; Santa Cruz Biotechnologies) or peroxidase-coupled antirabbit IgG (1:80,000; Sigma) diluted in blocking solution and then washed again in TBST. The blots are developed using chemiluminescence reagents. The molecular mass of the HIF-1α protein is around 120 kDa.

Note. Total protein extracts can be used to study HIF-1α expression, but the loading amount of proteins needs to be increased because activated HIF-1α is mainly located in the nucleus, which will dilute the signal.

Materials

   Microcentrifuge tubes (1.5 ml) (USA Scientific, Germany)
   Microcentrifuge (IEC, Needham Hts, MA)

Deionized H_2O

UV wavelength spectrophotomoter (Molecular Devices, SpectraMax Plus, Sunnyvale, CA)

Reagents needed for lysates and Western blot experiments (see earlier description)

Phosphate buffer (Sigma)

Chemiluminescence reagents (RPN 2106, Amersham Pharmacia Biotechnology, Piscataway, NJ)

HIF-1 Target Gene Expression Analysis

Among the well-known HIF-1 target genes are EPO, GLUT-1, and VEGF, which can be used to validate hypoxic preconditioning. Their expression can be studied at the mRNA level using polymerase chain reaction (PCR).

RNA isolation. Brains are removed (whole brain hemisphere or cerebral cortices) as rapidly as possible for RNA isolation. Total RNA is isolated with Trizol reagent (Life Technology, Rockville, MD) according to the manufacturer's protocol. Briefly, the tissue pellets are homogenized in Trizol reagent using a syringe. After extraction with chloroform (an additional step of phenol/chloroform extraction is added to the manufacturer's protocol), RNA is precipitated by isopropyl alcohol and subjected to further purification using an RNeasy minikit (Qiagen, Valencia, CA). The quality and quantity of extracted total RNA samples are examined by loading 5 μg of each sample on a denaturing agarose gel.

Real-Time RT-PCR. TaqMan quantitative reverse transcriptase-RT-PCR can be used to quantitate mRNA levels for selected genes. Two primers and one probe (TaqMan probe) (Applied Biosystems, Foster City, CA) are designed for each gene using Perkin-Elmer PrimerExpress software (Applied Biosystems). Examples for primer and probe sequences already published[24] for EPO, VEGF, and GLUT-1 are listed in Table I with "F" for forward primers, "R" for reverse primers, and "T" for TaqMan probes. TaqMan probes are labeled with VIC on the 5′ nucleotide and TAMRA on the 3′ nucleotide. Assays should be run in triplicates on the Perkin-Elmer ABI 5700 instrument under default conditions (RT: 48° for 30 min; AmpliTaqGold activation: 95° for 10 min and then PCR reaction: 40 cycles of 95°, 15 s and 60°, 1 min). The RT-PCR protocol is done according to the manufacturer's protocol using the TaqMan Gold RT-PCR kit (Applied Biosystems) with 50 ng of RNA for each sample. The abundance of genes of interest is determined relative to the glyceraldehyde-3-phosphate dehydrogenase (GAPDH) transcript using TaqMan GAPDH RNA control reagents kits (Applied Biosystems). The abundance

TABLE I
LIST OF PRIMERS FOR REAL-TIME RT-PCR

Gene name	Primer	Amplicon size
Rat EPO	F:5'-GAATTGATGTCGCCTCCAGA-3' T: 5'-CCAAGCCGCTCCACTCCGAACA-3' R:5'-TGGAGTAGACCCGGAAGAGCT-3'	91 bp
Rat VEGF	F: 5'-GGAATGATGAAGCCCTGGAG-3' T: 5'-CACGTCGGAGAGCAACGTCACTATGC-3' R: 5'-GGTGAGGTTTGATCCGCATG-3' R: 5'-GGTGAGGTTTGATCCGCATG-3'	78 bp
Rat GLUT-1	F: 5'-GGTGTGCAGCAGCCTGTGTA-3' T: 5'-CCATCGGCTCGGGTATCGTCAACAC-3' R: 5'-GACGAACAGCGACACCACAGT-3'	78 bp
Mouse GLUT-1	F: 5'-GGTGTGCAGCAGCCTGTGTA-3' T: 5'-CCATCGGCTCCGGTATCGTCAACAC-3' R: 5'-AATGAGGTGCAGGGTCCGT-3'	113 bp

of each gene can also be determined relative to the 18S ribosomal subunit using TaqMan ribosomal control reagents kits (Applied Biosystems). To verify the presence and the predicted size of amplified fragments, PCR products are separated by electrophoresis and visualized in 3% agarose gels with ethidium bromide.

Materials

Sterile syringes and needles or homogenizer to homogenize the brain for RNA isolation

Diethyl pyrocarbonate (DEPC)-treated water

RNase-free microtubes (0.5 and 1.5 ml), tips, and PCR 96-well microplates with caps (USA Scientific, Germany, and Perkin Elmer)

Reagents needed for RNA isolation and RT-PCR experiments (see earlier discussion)

Real-time PCR cycler (Perkin Elmer, ABI 5700)

Visible wavelength spectrophotomoter (Molecular Devices)

Brain Damage Quantification

Hemisphere Tissue Weight. This method of brain damage quantification can be used for the immature rat model of hypoxia-induced ischemic tolerance. After removal of the brain stem and cerebellum, the forebrain is sectioned at the midline, and left and right hemispheric weights are determined on a microbalance. Hemispheric necrosis is reflected by a reduction

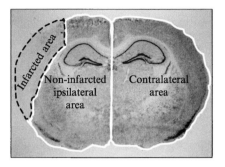

Infarct volume = (Σ contralateral areas-Σ non-infarcted ipsilateral areas) X d
contralateral area = areas of the control hemisphere non-infarcted ipsilateral area = normal tissue of the lesioned hemisphere ipsilateral area = infarcted + non-infarcted ipsilateral d = distance between each section level

FIG. 5. Equation for infarct volume quantification.

in hemispheric weight ipsilateral to the carotid ligation, which is calculated as the ratio of left to right hemispheric weights.

Infarct Volume Measurement. Coronal brain sections (20 μm) are cut on a cryostat, mounted onto gelatin-coated slides, and dried overnight. The slides are immersed in H_2O and then stained for 3–5 min in cresyl violet. Slides are rinsed three times in H_2O and differentiated in 70% (v/v) ethanol with a few drops of acetic acid. The slides are dehydrated in graded ethanols, immersed in xylene, and coverslipped with mounting solution.

Total infarct volumes can be calculated using the method reported by Swanson *et al.*[40] which is a method adaptable to most image analysis systems, highly reproducible, unaffected by edema, and fast. Infarct volumes are determined after integration of the sum of areas of the control hemisphere (contralateral areas) minus the sum of areas of normal tissue of the lesioned hemisphere (noninfarcted ipsilateral areas) multiplied by the distance between each section level analyzed (400 μm) as shown in Fig. 5. The areas of each section can be measured with a Rag 200 image analyzer (Biocom) or the public domain ImageJ processing and analysis software (NIH) in a blind manner.

[40] R. A. Swanson, M. T. Morton, G. Tsao-Wu, R. A. Savalos, C. Davidson, and F. R. Sharp, *J. Cereb. Blood Flow Metab.* **10,** 290 (1990).

Materials

Microbalance (Mettler Toledo AG245)
Cryostat (Microm HM500)
Tissue Tek (Fisher Scientific)
Microscope slides (Fisher Scientific)
Cresyl violet (Fluka, Sigma): three components are made and then mixed: 0.3 g of cresyl violet in 50 ml of H_2O; 3.48 ml of glacial acetic acid in 300 ml of H_2O; and 5.44 g of sodium acetate in 200 ml of H_2O
Cover glass (Corning, Fisher Scientific)
Mounting medium (Permount, Fisher Scientific)
Scanner

Acknowledgments

This study was supported by NS28167, NS42774, AG19561, and a Bugher Award (FRS) from the American Heart Association (National).

[28] Proximal Middle Cerebral Artery Occlusion Surgery for the Study of Ischemia–Reoxygenation Injury in the Brain

By Arturo J. Cardounel,
William A. Wallace, and Chandan K. Sen

Introduction

A variety of experimental models are used in studying ischemic brain injury. The three main classes of *in vivo* rodent models involve global ischemia through multiple vessel occlusion, hypoxia/hypovolemia in order to produce global alterations in cerebral blood flow, or focal ischemia through common carotid artery (CCA) occlusion or craniectomy followed by middle cerebral artery (MCA) occlusion.[1,2] A newer and now widely used approach involves MCA occlusion through the use of an intravascular nylon occluder. This method was developed by Koizumi *et al.*[3] with a more recent modification by Longa *et al.*[2] This has become the most widely used

---

[1] P. Lipton, *Physiol. Rev.* **79**(4), 1431 (1999).
[2] E. Z. Longa *et al.*, *Stroke* **20**(1), 84 (1989).
[3] Y. Y. Koizumi, T. Nakazawa, and G. Ooneda, *Jpn. J. Stroke* **8,** 1 (1986).

model to study the pathophysiology of neuronal injury and to test therapeutic interventions following temporary focal ischemia. The model is relatively easy to perform, minimally invasive, and does not require craniectomy, which may influence blood–brain permeability, intracranial pressure, and brain temperature.[4] However, several inherent complications have been reported using this model. These limitations include (i) insufficient middle cerebral artery occlusion (MCAO) and premature reperfusion,[5] (ii) coagulation and thrombus formation,[3,6] (iii) inadvertent subarachnoid hemorrhage,[2,3,7,8] and (iv) hypothermia during and postsurgery.[9,10] These concerns must be addressed and a standardized model must be developed in order to limit variability and optimize reproducibility. This article describes a MCAO model in which the previous concerns are addressed through laser Doppler flowmetry (LDF) measurements, pre- and postsurgical heparinization, and telemetric intracranial temperature monitoring.

Middle Cerebral Artery Occlusion

Male Wistar rats weighing between 300 and 350 g are anesthetized in an acrylic induction chamber (Harvard Apparatus, NP 60-5246) through administration of halothane or isofluorane until unconscious. The rat continues to receive 1–1.5% halothane delivered via a tabletop anesthesia machine and vaporizer (Harvard Apparatus, NP 72-3011; NP-72-3038) via spontaneous respiration through a nose cone. Halothane is mixed with oxygen-enriched air in order to maintain normal arterial pO_2 and pCO_2 values.

A femoral venous line is established for drug delivery. Heparin (150 IU/kg) is infused at the onset of surgery. A femoral arterial line is inserted for continuous blood pressure measurement. Arterial blood samples are taken prior to and during the surgical and recovery periods for the determination of blood gases, pH, and glucose levels. Body temperature is monitored and maintained at 37° using a homoeothermic blanket with a rectal probe. The brain temperature is monitored throughout the surgical and recovery periods using a minithermister (Minimitter)

[4] W. R. Hudgins and J. H. Garcia, *Stroke* **1**(5), 375 (1970).
[5] L. Belayev *et al.*, *Stroke* **27**(9), 1616 (1996).
[6] C. H. Rabb, *Stroke* **27**(1), 151 (1996).
[7] J. B. Bederson, I. M. Germano, and L. Guarino, *Stroke* **26**(6), 1086 (1995).
[8] Y. Kuge *et al.*, *Stroke* **26**(9), 1655 (1995).
[9] Q. Zhao *et al.*, *Brain Res.* **649**(1–2), 253 (1994).
[10] D. Corbett, M. Hamilton, and F. Colbourne, *Exp. Neurol.* **163**(1), 200 (2000).

implanted in the skull superior to the dura. A probe holder is glued to the skull surface over the cortical area supplied by the middle cerebral artery for LDF. The middle cerebral artery occlusion is performed as described by Longa *et al.*[2] with some minor modifications as described later.

The dorsal skull and ventral neck are shaved with clippers (Harvard Apparatus, NP 52-5204) using a #40 blade (Harvard Apparatus, NP-52-5212), taking care to clean away any excess fur. The rat is placed in the prone position on the operating table. The scalp is cleaned with a providone/iodine solution. Using medium-weight scissors, a midline incision is made on the rat's scalp. The fascia is then cut away and the bregma identified. Just medial to this suture line, the soft tissue and periosteum are removed with a currette, exposing the skull.

A bench-top motor drill (AH 59-7455, AH 59-9860) is then used to thin the bone of the skull at 4 mm lateral and 1.5 mm posterior to the bregma bilaterally to allow for cortical blood flow measurements using a laser Doppler (Moor Instruments). A 1-cm section of PE 50 tubing is secured to the skull with superglue or dental cement in order to hold the laser Doppler in place for the remainder of the procedure. Following placement of the laser Doppler probe, the rat is placed in the supine position.

The midline incision site on the neck is prepared aseptically with a providone/iodine solution. Using medium-weight scissors, a midline incision is made over the neck of the rat. Dissection is performed lateral to the right sternomastoid muscle, and the omohyoid muscle is then identified and divided. Under an operating microscope the right common carotid artery is exposed and dissected free of surrounding connective tissue (Fig. 1).

The external branch of the right common carotid artery is then isolated and the branches of the occipital and superior thyroid arteries are coagulated. The external carotid artery is ligated with a 6-0 braided silk suture at the bifurcation of the lingual and maxillary arteries. A microclip (NP 61-0186) is placed across the external carotid artery at the bifurcation with the common carotid, thereby maintaining blood flow through the internal carotid. Next, a 6-0 silk suture is tied loosely around the mobilized ECA stump. A previously prepared 5-cm length of poly-L-lysine-coated 4-0 monofilament nylon suture, with its tip blunted by heating near a flame, is inserted into the external carotid artery (ECA) lumen through a puncture proximal to the distal ECA ligation.[5] The silk suture around the ECA is then tightened around the intraluminal nylon suture to prevent bleeding and the microvascular clip is removed. The ECA stump is then dissected free and traction is applied in order to advance the suture tip into the internal carotid artery (ICA).

Care must be taken while advancing the sutures as not to advance the suture into the extracranial pterygopalatine artery. This can be avoided

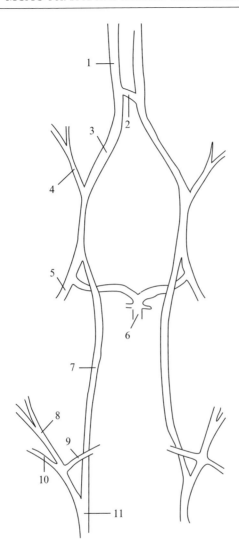

1. Internal ethmoidal artery
2. Anterior communicating artery
3. Anterior cerebral artery
4. Middle cerebral artery
5. Posterior cerebral artery
6. Basilar artery

7. Internal carotid artery
8. External carotid artery
9. Occipital artery
10. Cranial thyroid artery
11. Common carotid artery

FIG. 1. Anatomy of head and neck vasculature in the adult rat.

through visualization of the filament as it is advanced. The pterygopalatine artery runs on a ventral axis, whereas the ICA is located more dorsal; proper illumination of the surgical site will allow visualization of the filament.

The suture is advanced gently through the internal carotid artery a total distance of approximately 18 mm to the origin of the middle cerebral artery. Slight resistance will be felt as the suture reaches the origin of the MCA; proper placement of the filament is verified by a drop in LDF greater than 70%. The occurrence of subarachnoid hemorrhage (SAH) can be determined by measuring LDF on the ischemic and contralateral cortex. A significant drop in LDF on the nonischemic hemisphere suggests SAH, which can be verified later on necropsy.

LDF is monitored prior to and during the ischemic period as well as at reflow. Any animals that may have incomplete ischemia or early reperfusion can thus be disqualified. Following 90 min of ischemia, the rat is re-anesthetized in the induction chamber and transferred back to the operating table in the supine position.

The neck wound is reopened, and the suture is withdrawn back into the external carotid artery, reestablishing blood flow to the internal carotid artery and middle cerebral artery. This should be verified by LDF. The remaining suture is then trimmed to 1–2 mm outside of the vessel.

The wounds are then closed and infiltrated with 0.5% xylocaine. Analgesia is provided as needed with acetaminophen (1 g/kg) by oral gavage. Behavioral assessments are performed at this time and daily thereafter.[11] The animals are then returned to their cages and monitored until fully recovered.

At a time point of choice (usually 24–72 h) postreperfusion, the animals are killed under deep halothane anesthesia for the harvest of brain tissue. The brain temperature is monitored continuously and is maintained at 37° using a heat lamp. This is vital to the reproducibility of the model, as only slight changes in cranial temperature can have a large impact on the severity of ischemic injury.[9,10]

Behavioral Assessment

Ten-Point Neuro Scale

This 10-point scale may be used to evaluate general neurological function. Four points are awarded for a reduction in resistance to contralateral push; three points are awarded if contralateral circling is evident; two

[11] A. C. DeVries *et al.*, *Neurosci. Biobehav. Rev.* **25**(4), 325 (2001).

points are awarded for the appearance of contralateral shoulder adduction, and one point is awarded for contralateral forelimb flexion when suspended vertically by the tail.

Visual Placing

A test of visual acuity and reflexes is performed by lowering the animal slowly toward a table. A positive score is recorded if the animal extends its forepaws before touching the table two out of three times.

Forelimb Strength

The animal is suspended by its forelimbs on a rope suspended between two posts. The time until the rodent falls is measured up to 90 s. A score of 0 is given if the animal falls immediately. A score of 90 is given if the animal does not fall. The exercise is repeated and the best of three trials is recorded.

Bridge Walking

This tests the balance and coordination of the animal. The rodent is placed at the center of a bridge 60 cm long. The bridge is suspended above a foam egg crate. The time required for the rodent to reach a platform on

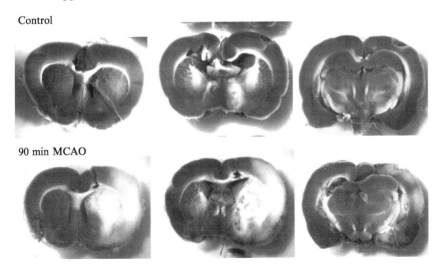

FIG. 2. Infarct size determination by TTC staining. (Top) Normal noninfarcted brain slices. Viable tissue stains red, whereas nonviable infracted tissue is white. (Bottom) A large infarct occupying the cortex and striatum. The infarct is located mainly in the frontal and midcortical regions.

either side within 2 min is recorded. If the animal falls, the fall time is recorded.

Histology

One group of experimental animals is euthanized at 72 h postreperfusion. The brains are removed and frozen rapidly $(-20°)$ in 2-methylbutane and are stored at $-80°$. The brains are then sectioned with a cryostat, 20 μm thick from anterior to posterior, and stained with cresyl violet.

TTC Stain

A second set of experimental animals is euthanized at 72 h postreperfusion. The brains are removed rapidly and 2-mm sections are cut using a rodent brain matrix. The sections are placed in a 2% TTC solution and incubated for 30 min at $37°$. The serial sections are then removed and washed in phosphate-buffered saline. Images are captured by video microscopy using the SNAPPY image capture software, and infarct areas are determined using IT software (Fig. 2).

Acknowledgment

This work was supported in part by NINDS 42617 to CKS.

[29] Carotid Chemodenervation Approach to Study Oxygen Sensing in Brain Stem Catecholaminergic Cells

By OLIVIER PASCUAL, JEAN-CHRISTOPHE ROUX, CHRISTOPHE SOULAGE, MARIE-PIERRE MORIN-SURUN, MONIQUE DENAVIT-SAUBIÉ, and JEAN-MARC PEQUIGNOT

In mammals, an oxygen deficit in the air induces an immediate and sustained increase in ventilation in order to improve the delivery of oxygen from pulmonary airways to the blood circulation and then to the tissues. The neuronal network, which generates and controls the ventilation, is located in the brain stem. This network is vital and still able to generate a respiratory motor output when isolated in a brain stem preparation. Furthermore, lesioning this network suppresses the ventilation of the whole animal. Neuronal respiratory groups include the dorsal group within the nucleus tractus solitarius, which receives inputs from the peripheral arterial chemoreceptors, and the ventral group in the ventrolateral medulla. Although most cells depress their metabolism in order to spare oxygen in response to hypoxia, we found that brain stem neurons are particularly sensitive to hypoxia and are responsive even to a moderate reduction of oxygen availability.[1] Consequently, they cooperate with the peripheral arterial chemoreceptors to maintain the oxygen homeostasis. Catecholaminergic neurons are located in the pons and the medulla oblongata in close vicinity of respiratory premotor neurons.[2] Brain stem catecholaminergic cells are stimulated by hypoxia at different levels, gene expression,[3] protein transduction and activation,[4] and neurotransmitter release. The respiratory premotor neurons possess adrenoceptors,[5] and the norepinephrine released from brain stem catecholaminergic cell groups participates in the modulation of respiratory motor output.[6,7]

Studies in our laboratories have evaluated brain stem oxygen sensing by investigating the ventilatory output in chemodenervated animals and the elementary mechanisms involved in respiratory control. A moderate reduction of oxygen *in vivo* can induce a ventilatory acclimatization partially independent of the peripheral chemosensory afferents.[8,9] We have also used *in vitro* more restricted preparations, such as isolated brain stem, to investigate by electrophysiology the neuronal network involved in the hypoxic modification of respiratory rhythm. Neurochemical and genomic investigations have been associated with measurement of the hypoxic ventilatory responses.

The role of catecholaminergic cell groups in ventilatory responses to hypoxia can be investigated *ex vivo* after the neuropharmacological blockade of catecholamine biosynthesis to determine the catecholamine

[1] O. Pascual, M. Denavit-Saubié, S. Dumas, T. Kietzmann, G. Ghilini, J. Mallet, and J.-M. Pequignot, *Eur. J. Neurosci.* **14,** 1981 (2002).

[2] P. M. Pilowsky, C. Jiang, and J. Lipski, *J. Comp. Neurol.* **301,** 604 (1990).

[3] S. Dumas, J.-M. Pequignot, G. Ghilini, J. Mallet, and M. Denavit-Saubié, *Brain Res. Mol. Brain Res.* **40,** 188 (1996).

[4] P. Schmitt, V. Soulier, J.-M. Pequignot, J. F. Pujol, and M. Denavit-Saubie, *J. Physiol.* **477,** 331 (1994).

[5] J. Champagnat, M. Denavit-Saubié, J. L. Henry, and V. Leviel, *Brain Res.* **160,** 57 (1979).

[6] S. Errchidi, G. Hilaire, and R. Monteau, *J. Physiol.* **443,** 477 (1991).

[7] J. C. Viemari, H. Burnet, M. Bévengut, and G. Hilaire, *Eur. J. Neurosci.* **17,** 1233 (2003).

[8] J. C. Roux, J. Peyronnet, O. Pascual, Y. Dalmaz, and J. M. Pequignot, *J. Physiol.* **522,** 493 (2000).

[9] J. C. Roux, J. M. Pequignot, S. Dumas, O. Pascual, G. Ghilini, J. Pequignot, J. Mallet, and M. Denavit-Saubie, *Eur. J. Neurosci.* **12,** 3181 (2000).

turnover and the activity of tyrosine hydroxylase, the rate-limiting enzyme in catecholamine biosynthesis.

The elementary mechanisms of ventilatory acclimatization can also be found at the genomic level where the induction of mRNA and protein can be studied in *in vivo* and *in vitro* preparations.

This article addresses physiological conditions of hypoxia known as moderate or tolerable hypoxia, which allows long-term exposure lasting for weeks. The exposure to tolerable hypoxia will enable us to define the morphofunctional mechanisms involved in the ventilatory acclimatization of intact or carotid chemodenervated animals. Moreover, exposure to moderate hypoxia induces the activation of brain stem catecholaminergic cells located in cardiorespiratory areas and/or involved in the modulation of cardiorespiratory output. Such hypoxic conditions can be reproduced easily either *in vivo* by subjecting small rodents (rats, mice, guinea pigs) to a gas mixture in which the oxygen level has been lowered below 20.9% or *in vitro* by superfusing isolated brain stem or brain stem slices using an artificial cerebrospinal fluid saturated with a hypoxic gas mixture.

In order to define oxygen sensing mechanisms in carotid chemodenervated animals, this article successively describes the *in vivo* and *in vitro* preparations used to test the effect of a moderate hypoxia on brain stem respiratory areas and then the methods used to evaluate the consequences of hypoxia on ventilation, catecholamine neurochemistry, and gene expression.

Preparations Used to Study Oxygen Sensing in the Brain Stem

The study of oxygen sensing in preparations deprived of chemosensory inputs can be performed *in vivo* or *in vitro*. *In vivo* preparations are based on bilateral carotid sinus transection and enable the measurement of breathing and ventilatory acclimatization to hypoxia in awake animals. This procedure can be followed by *ex vivo* neurochemical and cytochemical measurements of neurotransmitters, protein, and mRNA levels. *In vitro* studies are performed on isolated brain stem or brain stem slices; these preparations are deprived of both peripheral afferents and central inputs coming from rostral brain structures, such as the hypothalamus. The perfused preparation of isolated brain stem in the adult has been developed in the guinea pig,[10] but rat brain stem, although less resistant, can be used as well. The isolated brain stem of the newborn rat is commonly used because it does not need perfusion due to its small volume. However, the immature brain stem provides functional responses, which may differ significantly from the adult brain stem. Alternatively, the slice preparation

[10] M. P. Morin-Surun, E. Boudinot, T. Schafer, and M. Denavit-Saubié, *J. Physiol.* **485,** 203 (1995).

of adult brain stem has been used to study the effect of a reduction of O_2 on the synaptic transmission by measuring the postsynaptic response to afferent stimulation. This preparation enables intracellular recording and provides a useful model for pharmacological studies. Due to its reduced thickness, the diffusion of drug and gas is more homogeneous in slice than in isolated brain stem.

All surgical procedures need to follow guidelines for animal experimentation and approval from institutional animal care committees should be obtained.

In Vivo-*Chemodenervated Preparations in the Rat Model*

In the rat, denervation of the carotid sinus has been found to be followed by a very small increase in arterial blood pressure, which indicates an absence of functional barosensory fibers in the carotid sinus nerve in this species.[11] Chemoreceptor afferents have been found in the aortic nerve of rats, but their contributions to chemoreflex breathing are minor, and thus it is generally accepted that the bulk of peripheral arterial chemosensory inputs to the brain stem is conveyed by the carotid sinus nerve.[12] Because aortic bodies have a low hypoxic sensitivity in rodents[13] and aortic denervation changes the arterial blood pressure, we decided to use the rat model and denervate only the carotid bodies.

The age of the animal also represents a critical parameter; carotid body denervation in newborn rat leads to respiratory failure and strongly increases the mortality of these animals.[14] For these reasons, carotid chemodenervation is best performed in adult animals.

Methods of Chemodenervation.

Materials

Avertin: tribromoethanol (Sigma-Aldrich) 1 g, pentanol 0.5 g, ethanol 6 ml, NaCl 0.9% 67.5 ml for anesthesia

Normal saline

Small animal surgical board

Polyvidone iodine solution (Betadine, Asta Medica, Merignac, France)

Heated blanket

Surgical instruments: fine surgical scissors, fine forceps, surgical clamps

[11] E. M. Krieger, *Circ. Res.* **15,** 511 (1964).

[12] H. N. Sapru and A. J. Krieger, *J. Appl. Physiol.* **42,** 344 (1977).

[13] J. P. Mortola, "A Comparative Perspective," Vol. 1, p. 344. The Johns Hopkins University Press, 2001.

[14] M. A. Hofer, *Am. J. Physiol.* **251**(4 Pt 2), R811 (1986).

Syringe with a 33-gauge needle
2-0 silk suture (Ethicon)

Anesthesia. Ventilatory control depends on complex mechanisms that keep cellular CO_2, O_2, and pH homeostasis under physiological conditions. This regulation is based on chemical control, which acts primarily on peripheral arterial chemoreceptors to modulate breathing. Anesthesia and related conditions may depress respiration and have a sustained effect on ventilatory control. Moreover, carotid afferents provide a tonic excitatory input to medullary respiratory neurons. This input is important for medullary neuronal responsiveness, and sustaining a regular respiratory pattern becomes critical in the absence of tonic excitatory inputs coming from the carotid sinus nerve.[15,16] Therefore, the choice of anesthetic agent during carotid body denervation represents a crucial parameter. In our studies the animals are anesthetized by a single intraperitoneal injection of avertin (1 ml/100 g body weight, 1.4% tribromoethanol) so as to make them areflexic to a nociceptive stimulus (a pinch to the front paw), as checked during the surgical procedure. Avertin has been used successfully as an anesthetic agent in rodents by numerous laboratories worldwide. A reason cited for its popularity is that tribromoethanol provides a rapid and deep anesthesia in rodents. A surgical anesthesia (negative pedal reflex) is achieved within 5 min of intraperitoneal injection, and anesthesia is maintained for 20–30 min, which is sufficient to perform bilateral carotid sinus nerve transection. Finally, the animals are fully mobile within 60–80 min.

Note. Other anesthetics, including pentobarbital, have been tested but mortality was increased during carotid sinus nerve transection.

Surgery. All surgical procedures described should be performed using aseptic techniques with sterilized instruments, materials, and buffers.

1. Male Sprague–Dawley rats (240–260 g, IFFA CREDO, France) are anesthetized with avertin. When a surgical plane of anesthesia is reached, the rat is positioned on its back on an animal surgical board and the body temperature is maintained close to $37°$ using a heated blanket.

2. After the neck is shaved and disinfected with 75% ethanol, a midline 1-cm incision is made in the subhyoid region. The soft tissue and muscles are spread gently in order to expose the carotid artery, taking care to avoid any lesion of the vagus nerve and the carotid arteries. In order to identify

---

[15] A. Serra, D. Brozoski, N. Hedin, R. Franciosi, and H. V. Forster, *J. Appl. Physiol.* **91**, 1298 (2001).

[16] H. V. Forster, L. G. Pan, T. F. Lowry, A. Serra, J. Wenninger, and P. Martino, *Respir. Physiol.* **119**, 199 (2000).

the carotid sinus nerve, a very gentle pinch on the glossopharyngeal nerve allows detection of the carotid sinus nerve branching (Fig. 1). At this time, carotid sinus nerves need to be transected bilaterally at the point of branching off from the glossopharyngeal nerve and at the cranial pole of the carotid body (Fig. 1). In order to determine the effect of denervation independently of the effects of anesthesia and pain, the same procedure is used for sham-operated animals. The operation includes the same midline approach as for the chemodenervated rats, and both carotid bifurcations are then exposed without denervation.

3. After closing the incision with the silk suture, the wounds are painted twice a day with a 10% polyvidone iodine solution (Betadine) in order to prevent any infection.

Note. Different studies have been carried out on carotid sinus nerve sections using various surgical techniques, that is, cutting, crushing, ligaturing, or carotid body removing. Each protocol is able to temporarily eliminate the chemoafferent inputs but does not prevent a possible regeneration of the chemosensory fibers.[17] To study the effects of chronic chemodenervation, transection in two points and removing a large segment of the carotid sinus nerve are necessary in order to preclude any nerve regeneration. Prophylactic antibiotic treatment may be used but is unnecessary if an aseptic technique is adhered to scrupulously during the whole surgical procedure.

In Vitro-*Isolated Brain Stem Preparations*

Isolated brain stem preparations deprived of peripheral influences exhibit respiratory and cellular responses to a reduction of oxygen. Even if *in vivo* and *in vitro* comparison has to be done carefully, *in vitro* brain stem preparations may be useful in studying the hypoxia effect at the elementary levels. They enable stable electrophysiological recordings of spontaneously discharging respiratory neurons as well as their evoked response to synaptic inputs. The respiratory motor output can be recorded when the preparation contains the corresponding neuronal network. Oxygen reduction can be measured with appropriate electrodes simultaneously to the cellular recording. In some preparations, oxygen reduction can be either restricted to the brain stem surface or induced in the deep tissue. Compared to the *in vivo* preparation, a number of various O_2 levels can be produced consecutively inside the brain stem or at its surface.

---

[17] P. Zapata, L. J. Stensaas, and C. Eyzaguirre, *Brain Res.* **113**, 235 (1976).

A   Tail

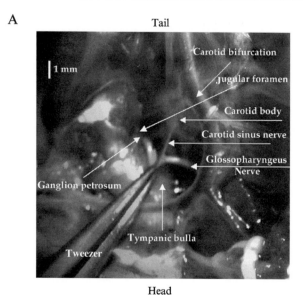

Head

B

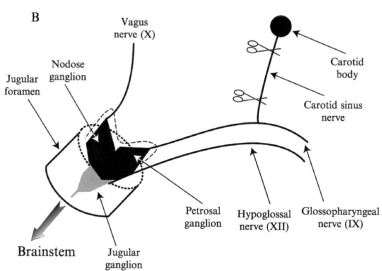

FIG. 1. Carotid sinus nerve transection. (A) Surgical procedure. (B) Schematic representation of nerves coming from the carotid body to the brain stem. The piece of nerve removed during the procedure is materialized by the portion of nerve between the two scissors.

Perfused Isolated Adult Brain Stem

The advantage of using an adult mammalian isolated brain stem preparation is preserving the entire respiratory neuronal circuits and thus allowing a homeostatic response to oxygen reduction. This approach allows stable respiratory recordings and manipulation of parameters such as oxygen level, pH, CO_2, and temperature in the absence of peripheral inputs. This preparation contains the brain stem from the midcollicular level to the first cervical root. It was introduced by Llinas *et al.*,[18] to study cerebellar plasticity *in vitro* and was later extended to study of the brain stem respiratory network.[10,19] It consists of an isolated guinea pig "brain stem–cerebellum en bloc" maintained *in vitro* by perfusion through the vertebral artery with artificial cerebrospinal fluid (ACSF). This preparation provides long-term *in vitro* survival (6–8 h).

The Recording Chamber. The chamber is a glass cup (4 cm in diameter) partly filled with silicone (Rhodorsil RTV 141 A and B, Rhône Poulenc, France) making an inclined plane (30°) in order to compensate for the thickness of the cerebellum and thus get the ventral surface horizontal when the brain stem is placed in the cup. The chamber is placed centrally in a plastic tank containing distilled water, which is heated to 30° via a thermostat resistance to maintain the ACSF temperature in the perfusion catheters. The tank is covered around the chamber. Water is bubbled with carbogen (95% O_2/5% CO_2) in order to maintain a saturated atmosphere around the chamber. Both the brain stem and the chamber are perfused using two catheters (Silastic). They arrive by two holes into the tank, pass through the tank, and then exit into the chamber (Fig. 2). The catheter chamber is fixed with one pin into the silicone chamber. The basilar catheter exits the chamber and is connected to a microloader (Eppendorf), which is passed into an articulated metal guide positioned to arrive at the level of the basilar artery. Perfusions through the catheters are made via a peristaltic pump (Gilson, 2 ml/min for the basilar artery and 2.5 ml/min for the chamber) from saturated ACSF maintained at 29°. A needle connected to a vacuum pump and fixed in the chamber maintains a constant ACSF level (and volume) by aspiration.

Preparation of Artificial Cerebrospinal Fluid. The ACSF solution is prepared daily, before surgical procedure, from a concentrated ACSF solution (10 *N*). The composition of ACSF (in m*M*) is 124 NaCl, 5 KCl, 1.2 KH_2PO_4, 2 $CaCl_2$, and 2.2 $MgSO_4$. One liter of daily ACSF is made with

[18] R. Llinas, Y. Yarom, and M. Sugimori, *Fed. Proc.* **40,** 2240 (1981).
[19] T. Schafer, M. P. Morin-Surun, and M. Denavit-Saubié, *Brain Res.* **618,** 246 (1993).

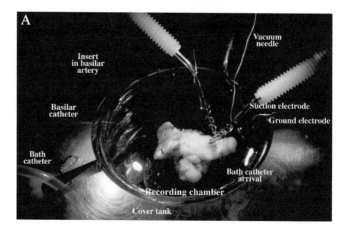

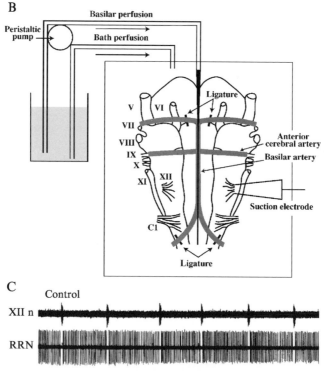

Fig. 2. Adult perfused brain stem preparation. (A) The recording chamber. (B) Schematic representation of the perfusion system of the preparation and position of the recording electrode. (C) Example of recording obtained with such a preparation. The top trace corresponds to the signal recorded from the hypoglossal nerve (XII) and the bottom trace corresponds to an intracellular recording of a respiratory-related neuron (RRN) performed in the respiratory neuronal network of the brain stem.

100 ml of a concentrated solution adjusted to 1 liter with tri-distillate H_2O. Glucose (10 mM) and $NaHCO_3$ (26 mM), which are missing in the concentrated solution to avoid mycelium formation, are then added. After magnetic stirring, ACSF is bubbled immediately with 95% O_2 and 5% CO_2 for 1 h to obtain saturated ACSF. One part is cooled in ice and the other part is maintained at 29° in a thermostat boiler. During the experimental session, saturated ACSF (1 liter) maintained at 29° is always ready.

Surgical Methods. Animals are decapitated rapidly at the second cervical vertebra with a guillotine under light anesthesia induced by sodium pentobarbital (30 mg/kg, ip). This anesthesia increases GABA inhibitory inputs without changing the excitatory inputs that preserved the animal from anoxia induced by decapitation. After decapitation, the head of the animal is immersed into a cup containing saturated cold ACSF maintained at 4°. Cranial bones are broken with cold Gouge forceps at the level of cerebral hemispheres and transcollicular section is performed. Bones over brain stem are broken delicately from the first cervical without touching the brain stem. During these surgical procedures, cold ACSF is dripped onto the exposed brain surface. The dura mater is cut with cold microscissors following the median line of the cerebellum and the brain stem and is turned back laterally. The brain stem is raised from the transcollicular transection with a cold spatula and cranial nerves (9th, 10th, and 12th pairs of cranial nerves) are cut with cold microscissors. Then the brain stem (from colliculus to first cervical) is removed and chilled rapidly in an ACSF cold cup and transferred in the recording chamber. It is placed on the silicone, the ventral surface upward and the cerebellum down on the 30° plane, the transcollicular surface in the front (Fig. 2). The brain stem is secured by pins inserted in the cerebellum. The first cervical spinal cord and the transcollicular section are maintained by a suture bridge (Ethilon) between pins. Under a binocular microscope, the pia mater is dissected with fine forceps, and then an ophthalmic suture (10/0, ethilon) is passed under the basilar artery from the transcollicular level and a coward knot is prepared. A small hole is made in the basilar wall with microscissors. With fine forceps, the catheter, in which a continuous ACSF flux is perfused, is positioned in front of the hole, which is opened with another forceps. The proper catheter position is reached when the ACSF flux cleans the arterial blood. Then a catheter is inserted into the artery and ligatured with a suture. The two vertebral arteries are then also ligatured with a suture. During this surgery, the needle of aspiration is adjusted so that the level of chamber ACSF touches the ventral surface of the brain stem lightly. At the end of the preparation, the chamber ACSF volume is adjusted to cover the brain stem entirely.

Newborn Isolated Brain Stem

This preparation, first described by Suzue,[20] is easier to realize than the adult preparation because it does not require an intracerebral perfusion. However, it is important to keep in mind that the newborn hypoxic response is different from that of the adult.

Material

Composition of ACSF (in mM): NaCl 130; KCl 5.4; CaCl$_2$ 0.8; MgCl$_2$ 1; NaH$_2$PO$_4$ 0.8; NaHCO$_3$ 21; D-glucose 30 equilibrated with 5% CO$_2$ in O$_2$ gas

Recording Chamber. The bottom of a 4-ml dish is covered in silicone (Rhodorsil RTV 141 A and B Rhone Poulenc, France) on which brain stem will be pinned. The chamber is perfused with an ACSF flow of 3.5 to 4 ml/min at 27°.

Surgery. The skull of the newborn rat (1 or 2 days old) is opened under ether anesthesia. After removal of the rostral part of the brain, cold ACSF is dripped onto the exposed brain surface. Then the brain stem with the cerebellum and the rostral part of the spinal cord is transferred in a dish containing cold ACSF. After removal of the cerebellum, the preparation is pinned down with the ventral surface upward for electrophysiological recordings and more precise transections. Using a razor blade driven in the recording chamber, a spinal transection is performed between cervical and thoracic rootlets to preserve the integrity of the C6 (phrenic) level and another section is performed at the rostral part of the pons.

Brain Stem Slices

In brain stem slices, the effect of O$_2$ reduction can be observed both on the brain stem neuronal network and on the synaptic transmission using an electrical stimulation of the tractus solitarius, which contains chemoafferent fibers.

Material

Composition of the ACSF (in mM): 124 NaCl, 5 KCl, 2 CaCl$_2$, 2.2 MgSO$_4$, 1.26 KH$_2$PO$_4$, 26 NaHCO$_3$, 10 glucose equilibrated with 95% O$_2$, 5% CO$_2$

Procedure. Slice studies are performed on Sprague–Dawley male rats (Janvier, LeGenest St. Isle, France) weighing 100–120 g. Rats are anesthetized deeply with ether and decapitated at the second cervical vertebrae.

[20] T. Suzue, *J. Physiol.* **354,** 173 (1984).

After transcollicular section, the brain is chilled rapidly with cold ($4°$) artificial cerebrospinal fluid. The brain is removed and separated from the cerebellum; the dura mater is then removed using fine forceps. During these surgical procedures, cold ACSF is dripped onto exposed brain surfaces. The rostral side of the brain stem is then glued (cyanoacrylate glue) on a plate, and the plate is oriented as the blade first contacts the dorsal side of the brain stem. Coronal slices 450 μm thick are cut with a vibratome (Leica, Wetzlar, Germany) in cold ACSF bubbled with 95% O_2/5% CO_2. Three sections at the level of the obex, containing the nucleus tractus solitarius (NTS), are transferred to a warm ($30°$) recording chamber saturated with 95% O_2/5% CO_2. The upper surface is first exposed to a warm humidified atmosphere 95% O_2/5% CO_2 for 5 min, whereupon the section is immersed completely and superfused continuously at a constant rate (2 ml/min). The recording session begins 1 h after the slice is prepared. Other slices containing the NTS are transferred to a chamber saturated with 95%O_2/5% CO_2 and maintained at $30°$.

Evaluation of O_2 Sensing

Methods used to Test the Physiological Effect of Hypoxia

Exposure to a tolerable hypoxia induces an immediate increase in breathing and this hypoxic ventilatory response is used as an index of O_2 chemosensitivity. Medullary catecholaminergic cell groups are activated by such a hypoxia and play a pivotal role in the neuromodulation of hypoxic ventilatory acclimatization. The hypoxia-induced activation of catecholaminergic neurons is dependent on the gene expression of tyrosine hydroxylase,[3] the rate-limiting enzyme in catecholamine biosynthesis, and is associated with the transcriptional factor hypoxia-inducible factor 1α (HIF-1α). Three methods described here include (i) recording modifications of respiratory motor output in relation to O_2 reduction, (ii) correlating the measurements of respiratory output and catecholaminergic activity in the brain stem, and (iii) evaluating the expression of genes (such as HIF-1α) involved in the catecholaminergic acclimatization to hypoxia.

Acclimatization

Hypoxia tested in each experimental condition is tolerable and is closely related to high-altitude physiology.

Establishment of a Chronic Moderate Hypoxic Environment. Rats are placed in a normobaric Plexiglas chamber and gradually subjected to an atmosphere of 10% O_2/90% N_2 (gas mixture provided by Air liquide, France) with access to food *ad libitum*. The composition of the atmosphere

Schematic representation of a plethysmograph

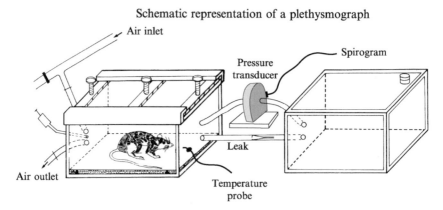

FIG. 3. Schematic representation of a barosensitive plethysmograph.

inside the chamber is monitored continuously and the oxygen content is maintained at $10 \pm 0.5\%$ (this level is progressively reached within about 4 h from the normoxic conditions). During long-term hypoxia, the hyperventilation of rats leads to an increase of the CO_2 content in the chamber. Then, in order to maintain the CO_2 content at $<0.1\%$, exhaled CO_2 is removed by passing the gas mixture from the chamber through soda lime. Expired water is trapped continuously in a chilled glass tank to remove the water produced by respiration. The temperature inside the chamber is kept at thermoneutrality ($26 \pm 1°$). The chamber is cleaned automatically twice a day without opening it in order to maintain hypoxia for at least 2 weeks without any return to normoxia.

Note. The same investigation can be achieved by adding 2–5% CO_2 in the inhaled gas in order to compensate for the decrease in arterial CO_2 pressure elicited by the hyperventilation during sustained hypoxia (normocapnic hypoxia). We chose not to use 5% CO_2 to mimic the physiological response encountered at high altitude, in which case the CO_2 fraction is constant and leads to a respiratory alkalosis.

MEASUREMENT OF VENTILATORY ACCLIMATIZATION. Principle of Ventilation Recording Using a Barometric Plethysmograph[21] (Fig. 3). During each inspiration, the rat inhales a volume of air depending on the ambient temperature, Tc, and the saturation pressure of water vapor, Pc (saturation is obtained by the presence of water in the plethysmograph chamber). In the body, the temperature of the air increases (ambient to body temperature)

[21] D. Jr. Bartlett, *Tenney. Respir. Physiol.* **10,** 384 (1970).

and then, following the physical law $PV = nRT$, the volume of air rejected increases, increasing in turn the internal pressure of the chamber for a short period of time. A differential pressure transducer detects differences of pressure by comparison to a reference chamber.

During each respiratory cycle the amplitude of the variation of pressure (Pt) is analyzed and compared to the amplitude of the variation of pressure during calibration (Pk) with a volume of air injected with a syringe (Vk). $Vt_{(BTPS)}$ represents the tidal volume in milliliters.

$$Vt_{(BTPS)} = Pt/Pk \times Vk \times \frac{Tr(Pb - Pc)}{Tr(Pb - Pc) - Tc(Pb - Pr)}$$

where Pt is the amplitude of the variation of pressure during respiration, Pk is the amplitude of the variation of pressure during calibration, Vk is the volume of calibration (ml), Tr is the body temperature of the rat (311.1 K), Tc is the ambient temperature inside the plethysmograph (K), Pb is the barometric pressure (atmosphere pressure mm Hg), Pc is the saturation pressure of water vapor inside the plethysmograph (mm Hg), Pr is the saturation pressure of water vapor at the body temperature of the rat (50 mm Hg), Fr represents the respiratory frequency in min^{-1}, and Ventilation (V_E) is defined as Vt X Fr.

Ventilation is measured routinely at the same time of the day (10 AM– 4 PM) in awake rats, as anesthesia affects respiration.[22] Plexiglas plethysmographic chambers adapted in volume to the rat (5.4 liter for adult rats) and flushed with humidified air are connected to a reference chamber of the same size. The atmosphere in both chambers is saturated with water vapor. Ambient temperature and O_2 and CO_2 levels inside the experimental chamber are monitored continuously. Prior to placing the animal in the chamber, calibration is performed by injecting 1 ml of air several times. Immediately after calibration, an animal is placed in the chamber and left until it is calm. At this point the flow is interrupted, the inlet and outlet tubes of the chamber are closed, and the pressure fluctuations caused by breathing are recorded with a differential pressure transducer (Celesco, CA). VT (ml), f (min^{-1}), minute ventilation (V_E; ml min^{-1}), TI (s), TE (s), and the total respiratory cycle [Ttot; (s)] are calculated breath by breath by computer analysis of the spirogram. All measurements are done in quadruplicate and are separated by intervals of 10–15 min, depending on the degree of activity of the rat. The mean of these four values is taken as the basal (normoxic) ventilation.

[22] D. Sjogren, S. G. Lindahl, and A. Sollevi, *Anesth. Analg.* **86,** 403 (1998).

Measurement of Ventilatory Response to Acute Hypoxia

Resting ventilation and the ventilatory response to acute hypoxia are measured on postsurgery days D2, D6, D10, and D21 to determine the effects of CSN transection on hypoxic ventilatory responsiveness. The hypoxic ventilatory response is measured after flushing the plethysmograph chamber with a gas mixture containing 7% O_2 and 93% N_2 (Air Liquide, France). Wash out of the plethysmograph requires about 2 min for the 5.4-liter chamber to reach a stable inspired O_2 fraction. This is considered to be the beginning of hypoxia. Once the O_2 level inside the box reaches 7 $\pm$ 0.2%, the rat is allowed to breathe the test gas for a another minute before breathing is recorded. The hypoxic ventilatory response is defined as the difference between minute ventilation levels in acute hypoxia and resting conditions. Subsequent HVR recordings are done in the same way at 4, 7, and 10 min after the beginning of hypoxia. All the recordings are performed at thermoneutrality ($26 \pm 1°$).

Recording of Respiratory Motor Output

In the isolated brain stem, hypoxic exposure could be induced either at the medullary ventral surface, changing the saturated oxygen level of ACSF from the bath perfusion, or into the brain stem using basilar perfusion.

In newborns, *in vitro* respiratory activity is recorded from the phrenic nerve: the fourth cervical rootlet is recorded using a glass suction electrode (diameter 200–220 μm) against a reference in the chamber. The signal is amplified through a Grass amplifier (P511 K).

In adults, *in vitro* respiratory activity is monitored from hypoglossal roots through a glass suction electrode. The glass suction electrode is pulled (Clark Electromedical Instruments, GF, outside diameter 1.2 mm) using a Sutter puller (Model P-97, Sutter Instrument Co), and the tip is broken to obtain a 300-μm diameter. The glass tube is curved under a flame, and a constriction is made at the tip of the glass electrode to block the nerve after its aspiration into the electrode with a syringe. This glass capillary tubing is inserted into the holder of a suction electrode micromanipulator (A-M Systems, Carlsborg). Nerve activity is amplified and filtered (30 Hz–3 kHz) kHz) with a Grass model preamplifier. Half-wave activity of the hypoglossal roots is integrated through a passive RC-integrating circuit (time constant 20 ms). Periodic activity recorded from hypoglossal roots on digital tapes (DTR-1800, Biologic) consists of inspiratory-related bursts of activity followed by expiratory-related low-amplitude tonic activity (Fig. 2). Integrated periodic activity is digitalized by a CED 1401 (Cambridge Electronic Design) to estimate the different respiratory variables: Ti, Te, amplitude of the integrated nerve activity, and instantaneous respiratory frequency are estimated as f = 1/(Ti +Te) for each burst. For each change in O_2 content in saturated ACSF, mean values of the effect are calculated for each respiratory parameter and are expressed as a percentage of control values obtained by averaging 10 consecutive respiratory cycles obtained under control-saturated ACSF.

Recording of Respiratory Neuronal Activity

In vitro, it can be interesting to study the direct effect on the discharge of respiratory neurons of an O_2 reduction on either the brain stem surface or inside the brain. In this case, tubes (Clark Electromedical Instruments, GF, outside diameter 1.2 mm) are pulled using a Sutter puller (Model P-97, Sutter Instrument Co). Microelectrodes are filled with 3 M NaCl (resistance about 10 MΩ) introduced into the brain stem by 2 μM steps. The extracellular signal is amplified 50,000-fold and filtered (100 Hz, 3 kHz) with a Grass preamplifier (P511 preamplifier, Grass Instrument Co).

The absence of pulsations on this *in vitro* preparation allows easy intracellular recordings. Respiratory neuronal activities are recognized by their regular rhythmic discharge in phase with the hypoglossal nerve discharge (Fig. 2). On the integrated trace the peak of activity corresponds to an inspiration.

Intracerebral Measurements of Oxygen Tension

For PO_2 measurements, two types of oxygen-sensitive microelectrodes can be used (either a "Clark" type, tip 20 μm or a combined needle electrodes, cathode 25 μm) connected to an oxygen meter. Electrodes are allowed to stabilize for at least 4 h prior to use. They are calibrated in ACSF equilibrated with 100% N_2 and air at the same temperature as the bath solution before the measurements and are tested after the experiment to check for a possible drift.

Tissue PO_2 in the brain is measured at 100-μm intervals from the ventral to the dorsal medullary surface at the level of the ventral respiratory group (depth 1 mm from the surface and 2.5 mm lateral to the midline). Tests can be performed with potassium cyanide.

Methods Used to Test the Genomic Effect of Hypoxia

Regulations of genes involved in hypoxic-dependent processes are usually studied using culture cells. As we attempted to evaluate the physiological significance of the gene expression during hypoxia, we combined the study of functional modifications and the gene expression induced by a moderate hypoxia on the same animal.

mRNA Detection by a Radioactive Oligonucleotide Probe

Section and Tissue Preparation. Rats are sacrificed by caudal cervical dislocation and brains are removed rapidly. The brains are immediately frozen at $-60°$ in isopentane and then conserved at $-80°$. Sections of 20 μm are performed using a cryostat (Leica, Bron, France) at $-20°$. Only the caudal part of the brain is used. It is glued using Tissue-Tek (Lab-Tek Products, Napervill, Il) and is sectioned from the pyramidal decussation to the locus coeruleus. Sections are thaw mounted onto glass slides (Super Frost plus, Gaffalem, Limeil-Brevannes, France) and stored at $-80°$ until use.

Before use, section are thawed at room temperature and fixed by incubation with 4% paraformaldhehyde (PFA) in 1× phosphate-buffered solution (PBS) for 10 min at $4°$. The sections are rinsed twice with 1× PBS for 15 min, washed in a bath of 95% ethanol (5 min), and air dried.

Oligonucleotide Probe Preparation. The tyrosine hydroxylase (TH) probe used is a 24 oligonucleotide RNA probe complementary to nucleotides 45–69 of the rat TH mRNA.[23] This oligonucleotide is synthesized and purified by Genstet France (Paris).

The HIF-1α probe used is a 24 oligonucleotide RNA probe complementary to nucleotides 240–263 (GenBank accession number Y09507).

These oligonucleotides are 3′ end labeled with [^{35}S]-dATP (Amersham Pharmacia Biotech Orsay, France) using terminal deoxyribonucleotidyl transferase (TdT) (Roche, Meylan, France) to a specific activity of $7×10^8$ cpm/μg.

PREPARATION OF THE PROBE. For 20 μl of probe use 2 pmol of oligonucleotide in H_2O (final volume 13 μl) and then add 4 μl of TdT buffer ×5 (Promega), 2 μl of [^{35}S]-dATP (Amersham), and 1 μl of TdT enzyme (Amersham). Incubate at $37°$ for 45 min. The reaction is stopped by adding 5 μl of 0.5 M EDTA.

PURIFICATION OF THE PROBE. Probes are purified on a P10 column (Bio-rad, Ivry sur Seine, France) prior to use and are stored at $-20°$ in TE (Tris-EDTA) buffer containing 100 mM dithiothreitol (DTT).

In Situ *Hybridization.* The oligonucleotide probe (0.2 pmol) is mixed with 50 μl of *in situ* hybridization buffer (Amersham, RPN 3310). Hybridization buffer containing the oligonucleotide probe is applied on the brain sections and covered with parafilm (FUJI). Sections are incubated in a wet chamber for 15 to 18 h at $42°$. The parafilm is then removed

[23] B. Grima, A. Lamouroux, F. Blanot, N. F. Biguet, and J. Mallet, *Proc. Natl. Acad. Sci. USA* **82,** 617 (1985).

in a solution of standard saline citrate (SSC) 1× (Promega, France), and sections are washed as follows: Twice for 15 min in SSC 1×/DTT 10 mM at 53°, twice for 15 min in SSC 0.5×/DTT 10 mM at 53°, and once for 15 min at room temperature in SSC 0.5× DTT 10 mM. The sections are rinsed briefly in water and dehydrated in ethanol 96%. Slices are dried at air room temperature

Autoradiography. For overall analysis of the signal, autoradiograms of the brain stem are generated by placing ^{35}S-labeled sections against hyperfilm β_{max} X-ray film (Amersham Pharmacia Biotech, Orsay, France) for 1 or 2 weeks. In order to compare directly the optical density of TH and HIF-1α mRNA labeling among chemodenervated, sham, normoxic, and hypoxic animals, sections for each *in situ* hybridization are placed on the same film. Sections are then dipped in nuclear emulsion (Ilford K5 diluted 1:1 with distilled water) for cellular analysis and localization of the maximal labeling. Slices are exposed to emulsion for 6 weeks, revealed, and counterstained with cresyl violet.

mRNA Detection by a Nonradioactive Nucleotide Probe

Section and Tissue Preparation. The rats are anesthetized with 0.1 ml pentobarbital per 100 g of body weight. They are perfused transcardially with chilled saline for 1 min (flow rate, 30 ml/min) followed by a perfusion of chilled PBS containing 4% PFA for 10 min. The brain is removed and incubated in PFA for at least one night at 4°. Then the tissue is rinsed twice for 5 min in PBS–0.1% Tween (PBT). The brains are coated in a solution of gelatin, albumin, and glutaraldehyde solution as indicated below.

Preparation of the Coating Solution. Ninety grams of albumin is dissolved in 200 ml of 0.1 M phosphate buffer, pH 7.3, and kept one night at room temperature. Gelatin (1.5 g) is dissolved in 100 ml of 0.1 M phosphate buffer, pH 7.3. Melt the gelatin, add to the solution of albumin, and adjust to pH 6.5.

Tissue Coating. In a dish, prepare a base by adding 25 μl of 50% glutaraldehyde to 1 ml of gelatin/albumin solution. Wait 10 min until the base is dry. Then place the tissue and coat the tissue with the gelatin, albumin, and glutaraldehyde solution.

Slices of 150 or 250 μm are performed using a vibratome in PBT and progressively dehydrated as follows: 5 min in 25% ethanol/PBT, 5 min in 50% ethanol/PBT, 5 min in 75% ethanol/PBT, and 5 min twice in ethanol 100%. Slices are kept in methanol at −20°.

Hybridization of the Riboprobe. The HIF-1α probe is a 778-base RNA probe complementary to nucleotides 90 to 868 (GenBank accession number Y09507). The probe consists of a RNA sequence containing UTP-dig nucleotides. RNAs are obtained from DNA containing the two bacterian promoters Sp6 and T7 at each extremity. Preparation and purification of the riboprobe are explained in this section.

For *plasmid linearization*, mix the following solutions in an Eppendorf tube: BSA 10%, 20 μl; plasmid 600 μl/λ obtained from Kietzmann, 33 μl; restriction enzyme *Eco*RV, 7 μl; enzyme buffer 10%, 20 μl; and H$_2$O, 120 μl. Incubate this solution for 2 h at 37°.

At the end of the incubation an extraction of the DNA is performed with phenol–chloroform–isoamyl alcohol (25:24:1). An emulsion is obtained by agitation using a vortex for 15 s and then centrifuging for 10 min. The liquid phase containing DNA is collected.

For *purification*, add to the mixture obtained one-half volume of 7.5 M ammonium acetate and 3 volumes of 100% ethanol and then centrifuge for 20 min at 4°. Remove the liquid phase, wash the pellet using 70% ethanol, and remove the ethanol in a Speed-Vac for 5 min.

Linearized DNAs are then suspended in 2 μl of water. DNA (300 μg) can be loaded onto a 0.8% agarose gel containing 1 μg/ml of ethidium bromide to check the linearization step of the DNA. The absorbance ratio (A_{260}/A_{280}) should also be performed to check the purification step; a low ratio indicates contamination by protein (absorbance at 280 nm).

The following solutions are used in *probe preparation*.

Transcription kit (Promega, France)
RNA polymerase T7
UTP-Dig (Boehringer)
Tris–EDTA
DNase I RNase free
4 M LiCl
100% ethanol

The whole preparation of the solution is performed on ice. Mix 7 μl of H$_2$O, 4 μl of transcription buffer (5×), 1 μl of 10 mM GTP, ATP, and CTP, 0.65 μl of 10 mM UTP, 0.35 μl of 10 mM UTP-Dig, 1 μl of RNasin, 1 μl of RNA polymerase, 1 μl of linearized DNA 1 μg/μl, and 1 μl of 0.1 mM DTT.

This solution is incubated for 2 h at 37°. Then the DNA is degraded by adding 2 μl of DNase, digested at 37° for 15 min. The probe is then precipitated by adding 100 μl of TE, 10 μl of 4 M LiCl, and 300 μl of 100% ethanol. Precipitation is performed at −20° for 30 min. Centrifuge for 10 min at 4°. Remove the liquid phase and wash the RNA pellet with 70% ethanol (500 μl). Dry the pellet and suspend it with 100 μl of TE.

Under these conditions the RNA quantity synthesized is around 0.1 μg/μl. The quality of RNA can be checked using an agarose gel at 1%.

The following solution is used for hybridization.

Salt solution 10×: NaCl 114 g, Tris–HCl 14.04 g (pH 7.5), Tris base 1.34 g, $NaH_2PO_4 \cdot 2H_2O$ 7.8 g, $NaHPO_4$ 7.1 g, EDTA (0.5 M) 100 ml, complete to 1 liter with H_2O

The hybridization buffer is composed of salt solution 5×, formamide 50%, dextran sulfate 1%, yeast tRNA (50 μg/ml), and heparin (50 μg/ml)

Prehybridization is performed by incubating the slice in hybridization buffer for 1 h at room temperature. During this time, the probe is denatured for 2 min at 65° and kept on ice. Hybridization is then performed by incubating each slice overnight at 70° in a solution containing 500 μl of hybridization buffer and 1 μg of probe.

The following solutions are used in posthybridization washes.

Solution 1: Formamide 50%/SSC 1×/Tween 20 0.1%

Solution 2: Formamide 50%/SSC 1×

TBST(10×) (for 100 ml):NaCl 16 g, KCl 0.4 g, 50 ml Tris–HCl, pH 7.5, Tween 20 2%

Solution 3: goat serum 10% in TBST 1× and add 2 mM levamisole before use

NTMT: NaCl 100 mM, Tris–HCl, pH 9.5, 100 mM, MgCl 2 50 mM, Tween 20 0.1%

The washing procedure consists of

1 h at 70° with solution 1,

30 min at 70° with solution 1,

three times for 30 min at 65° with solution 2,

three times for 5 min at room temperature with TBST 1×, and

90 min at room temperature in solution 3.

Revelation. Incubate the slice overnight at 4° in Sh anti-dig antibody (Boehringer) 1:2000 in TBST 1×, 1% goat serum. Wash five times for 20 min in TBST 1× at room temperature and three times for 10 min in NTMT containing 2 mM levamisole.

The revelation solution consists of NBT 225 μl, BCIP 175 μl, levamisole 25 mg, and NTMT 50 ml.

Incubate the slices for several hours at room temperature in the dark. Check the revelation under a microscope and stop the reaction by washing twice with PBS, 0.1% Tween 20 and then several washes in PBS during the following day. Slices are mounted on glass in a gelatin glycerol medium (Glycergel, Dako).

Protein Detection by Immunohistochemistry

Material

Pentobarbital
Peristaltic pump
Saline solution 9/1000
Paraformaldehyde 4% in PBS (0.1 M at pH 7.4)
Sucrose 18% in PBS

Tissue Preparation. Animals are anesthetized with pentobarbital 0.1 ml/100 g body weight and perfused transcardially with chilled saline for 1 min followed by PBS (0.1 M at pH 7.4) containing 4% paraformaldehyde for 10 min (flow rate, 30 ml/min) using a peristaltic pump. Because HIF-1α is known to have a rapid turnover and a short half-time decay,[24] particular attention is paid to environmental conditions in order to prevent backward normoxic conditions between the end of the experiment and fixation/freezing procedures. Thereby, anesthesia and fixation of hypoxic animals are carried out with 10% O_2. Brains are removed from the skull, postfixed in the same fixative for 1 h, and placed overnight in PBS containing 18% sucrose. The brains are kept at 4° in PBS until coronal sections (40 μm thick) are cut on a freezing microtome.

Immunohistochemistry. After preincubation in 4% BSA in PBS with 0.2% Triton X-100 and 0.3% H_2O_2 for 30 min at room temperature, sections are then covered with PBS containing primary antisera HIF-1α (1:4000) and TH (1:1000), 1% BSA, and 0.2% Triton X-100 for one night at room temperature. The primary antibodies used are antisera HIF-1α raised against the protein in mice (NB100–105; Novus Biological, Littleton, CO) and tyrosine hydroxylase antisera raised against the protein in rabbits (AHT 1100; Institut Jacques Boy; Reims, France). The following day, sections are rinsed three times for 15 min in PBS. Sections are then incubated in the secondary antibodies. The detection of immunoreactivity is performed using the avidin biotin complex method coupled with alkaline phosphatase for HIF-1α and horseradish peroxidase for tyrosine hydroxylase. Sections are incubated for 2 h with biotinylated sheep antimouse IgG (985035; Silenus, Boronia, Australia; 1:1000) and horseradish peroxidated goat antirabbit IgG (PI-1000; Vector, Burlingame, MA; 1:200) in PBS. Sections are then rinsed twice in PBS for 5 min and the last wash is performed in 0.05 M Tris–HCl, pH 7.5. The tyrosine hydroxylase revelation is performed by incubating the sections in a solution containing 4.5 ml of 0.05 M Tris–HCl, pH 7.6, 4.5 μl 30% H_2O_2, and 0.5 ml of DAB from a

---

[24] G. L. Wang, B. H. Jiang, E. A. Rue, and G. L. Semenza, *Proc. Natl. Acad. Sci. USA* **92,** 5510 (1995).

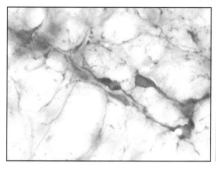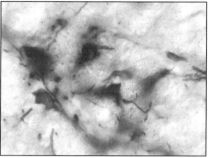

FIG. 4. Protein expression of HIF-1α and tyrosine hydroxylase in the A1C1 area. (Left) Tyrosine hydroxylase expression in A1C1 during normoxia (brown); note that there is no expression of HIF-1α. (Right) Tyrosine hydroxylase (brown) and HIF-1α expression (blue) in the A1C1 area after 6 h of hypoxia.

stock solution at 25%. Revelation is done under a microscope, and the reaction is stopped by washing once for 5 min with 0.05 *M* Tris–HCl, pH 7.6 and twice for 5 min with PBS. HIF-1α revelation is performed by incubating in the alkaline-phosphatase coupled ABC-AP kit (Vector) for 30 min. After a further wash in PBS, alkaline phosphatase is developed in blue alkaline phosphatase substrate kit III (Vector). After washing in PBS, the sections are mounted and coverslipped with mowiol (Fig. 4).

Methods Used to Test the Neurochemical Effect of Hypoxia

Reorganization of the O_2 chemoreflex pathway following bilateral chemodenervation has been studied using neurochemical methods.[8] These neurochemical methods include the *in vivo* measurement of tyrosine hydroxylase activity, the rate-limiting enzyme of catecholamine biosynthesis, and the assessment of norepinephrine turnover after pharmacological blockade of its biosynthesis. These methods provide valuable techniques used to study catecholaminergic neural activities.

Estimation of Norepinephrine Turnover

Norepinephrine turnover in brain stem cell groups is assessed by measuring the exponential rate decline of the norepinephrine content following the pharmacological blockade of tyrosine hydroxylase by α-methylparatyrosine-methyl ester (AMPT).

Materials

AMPT (Sigma, Saint Quentin Fallavier, France)
0.9% saline
1-ml disposable syringes

25-gauge 5/8-in. needles (Becton Dickinson, Meylan, France)
Chronometer
Small animal decapitator (Stoelting, Wood Dale)
Procedure

1. AMPT is weighed and dissolved extemporaneously in saline.

2. Rats are injected intraperitoneally with 25 mg/100 body gram weight 1, 2, or 3 h before being sacrificed. Control rats only receive the vehicle of AMPT (saline 0.9%).[25]

3. Rats are sacrificed by decapitation. *Note.* For a detailed explanation on the way to remove the brain from the skull, see Palkovits and Brownstein.[26] The measurement of catecholamine content can also provide a static evaluation of the neurotransmitter stores. Catecholamine stores are usually performed on saline-treated animals or on untreated animals. The brain is removed quickly, frozen on dry ice, and conserved at $-80°$ until microdissection of catecholaminergic cell groups (see microdissection).

4. Norepinephrine (NE) is assayed by high-pressure liquid chromatography coupled to an electrochemical detection (thereafter referred to as HPLC-ED, see the biochemical analyses section). After injection of AMPT, the decrease of norepinephrine is exponential. The decimal logarithm ($\log_{10}$) of NE concentration is plotted against the time after injection. The slope is obtained using a linear regression model. The fractional rate constant (k) and the half-time ($t_{1/2}$) are calculated from the following expressions: $k = \text{slope}/0.434$ and $t_{1/2} = 0.693/k$.

Estimation of In Vivo *Tyrosine Hydroxylase Activity*

In vivo tyrosine hydroxylase activity[27] is assessed by measuring the accumulation of DOPA, the product of hydroxylation of tyrosine, following the blockade of DOPA decarboxylase by 3-hydroxybenzylhydrazine dichloride (NSD 1015).

Materials

NSD 1015 (Sigma)
0.9% saline

---

[25] An alternative protocol may be used. A single gap between AMPT injection and sacrifice, e.g., 2.5 h, and a control can be sufficient to calculate the turnover rate of NE.

[26] M. Palkovits and M. J. Brownstein, "Maps and Guides to Microdissection of the Rat Brain," 2nd Ed. Elsevier, Amsterdam, 1988.

[27] The blockade of L-amino acid decarboxylase by NSD 1015 also permits the simultaneous measurement of tryptophan hydroxylase activity, the rate-limiting step in serotonin biosynthesis, through the accumulation of 5-hydroxytryptophan.

1-ml disposable syringes
25-gauge 5/8-in. needles (Becton Dickinson)
Chronometer
Small animal decapitator (Stoelting, Wood Dale)

Procedure

1. NSD 1015 is weighed and dissolved extemporaneously in saline.
2. Rats are injected intraperitoneally with 10 mg/100 body gram weight 20 min before sacrifice. Control rats only receive NSD 1015 (0.9% saline).
3. Rats are sacrificed by decapitation. Brain are removed quickly, frozen on dry ice, and preserved at −80° until microdissection of catecholaminergic cell groups (see microdissection).
4. L-Dihydroxyphenylalanine (L-DOPA) is assayed by HPLC-ED (see biochemical analyses). The tyrosine hydroxylation rate is calculated by subtracting, structure by structure, the L-DOPA content in saline-treated rats from those obtained in NSD-treated rats.[28]

Microdissection of Brain Stem Catecholaminergic Cell Groups.[29]

Materials

Dissecting needle 0.96 mm i.d. (ref. 57401, Stoelting)
Crushed or powdered dry ice
Microtubes (Eppendorf) 0.5 ml
Perchloric acid 0.1 M–EDTA–Na$_2$ 2.7 mM
Microtome
Tissue-Tek (Lab-Tek Products, Naperville, IL)
Microscope slides
Dissecting stereomicroscope (optional)

Procedure

1. A sufficient number of 0.5-ml microtubes, used for receiving the punched structures, are prepared and labeled. Then 100 μl of perchloric acid 0.1 M/EDTA–Na$_2$ 2.7 mM is dropped in each microtube.

[28] Because the initial amounts of L-DOPA and 5-HTP in catecholaminergic cell groups before pharmacological blockade with NSD 1015 are negligible (< 10%, unpublished data from our laboratory) relative to the amounts observed after 20 min of blockade, the saline-treated group is not essential.
[29] Some more extensive explanations about the microdissection method can be found in Palkovits and Brownstein.[26]

2. A ball of Tissue-Tek is put on the cooled specimen holder. The frozen brain is then fixed as vertically as possible in the "nose-down" position in the ball of Tissue-Tek and left on a bed of dry ice until fixed.

3. Sectioning is usually performed with a microtome at room temperature. The brain is sliced in serial coronal slices of 480 μm thickness. Sometimes it is necessary to cool the knife or the brain with dry ice. The sections are mounted on numbered glass microscope slides and stored on dry ice. Six to eight consecutive sections can be mounted on the same slide. Four to five slides are needed to collect the entire brain stem. The mounted slices may be punched out immediately or stored at −80° until dissection takes place.

4. The sections must be kept frozen during the dissection process.[29] Therefore, we usually perform our dissection on a 2-cm-thick aluminum plate stored at −80° until use.[30] A cool lamp with adjustable optic fibers is used to illuminate the dissection support from above. To perform dissection of the brain stem catecholaminergic cell groups, we use a 0.96-mm i.d. hollow punching needle equipped with a stylet.[31] The slide is positioned with one hand, whereas the punching needle is held, just like a pen, between the thumb, the middle finger, and the forefinger of the other hand (Fig. 5). The needle is held at a 45° angle and its tip is positioned on the structure to be removed. When the needle tip is in the suitable location, the needle is brought in a vertical position, pressed on the section, and rotated. Then the needle is withdrawn and a clean hole should remain in the section. The structure is removed in the same way on the other sections where it is present.[32] When the structure is fully dissected out, the tip of the needle is brought into a micro tube containing perchloric acid and the tissue disks are pushed out with the stylet.

5. We use the seventh pair of cranial nerves, the pyramidal decussation, and the fourth ventricle as landmarks to confirm the relative position of the different cell groups (Fig. 6). The catecholaminergic cell groups are generally dissected in the following sequence. First the section including

---

[30] A refrigerating table or a glass petri dish filled with (or laid on) powdered dry ice is also suitable to perform the microdissection. Because a black surface will make it easier to distinguish the structures and the anatomical references, dissection support could be painted black or covered with a black tape.

[31] A large variety of valuable homemade dissection needles can be produced using commercial hypodermic needles with tips cut off and beveled.

[32] The dissection procedure should be performed as quickly as possible because frost tends to accumulate on the frozen section, making it difficult to distinguish the position of the nuclei of interest. Then, a brush or the fingertip can be used to sweep the surface of the section. If the sections are colder and too hard to be punched out easily, they can be warmed by touching it a few seconds with a fingertip.

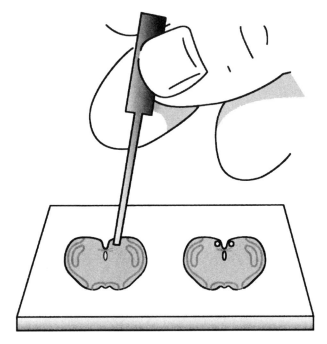

FIG. 5. Schematic representation of the microdissection technique.

the seventh pair of cranial nerves is located. The A6 noradrenergic cell groups (locus coeruleus) are punched out bilaterally from four slices, one before and two after that one containing the seventh pair of cranial nerves. The A5 noradrenergic cell groups are dissected bilaterally in ventrolateral position from the three first slices containing A6. The four sections anterior to the one containing the cranial nerves are counted. The A2C2 cell group is then punched out in the dorsomedian position from six slices. A1C1 cell groups are dissected bilaterally in the ventrolateral medulla on eight slices, one before and one after those containing A2C2.

Biochemical Analyses of Catecholamines and Metabolites

Materials

HPLC-ED
Sonicator (Ultrasons Annemasse, Annemasse, France)
Centrifuge
HPLC-ED Apparatus. The high-performance liquid chromatography system consists of a double piston pump (Shimazu LC A10), a precolumn

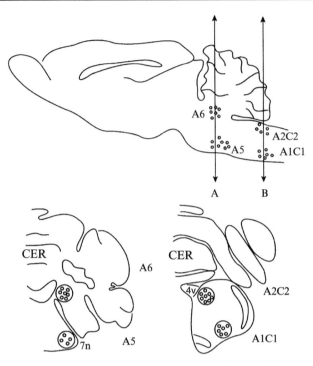

Fig. 6. Schematic localization of the different catecholaminergic cell groups in the brain stem (A1C1, A2C2, A5, and A6). (Top) Parasagittal section of the brain; vertical arrows show the level where coronal sections represented at the bottom have been performed. (Bottom left) Schematic representation of the coronal section performed in A (top). (Bottom right) Schematic representation of the coronal section performed in B (top). CER, cerebellum; 4v, fourth ventricle.

(spherisorb ODS2, 30-μm particles, 20 × 2 mm, Macherey-Nagel, Hoerdt, France), a reversed-phase column (Spherisorb ODS2, 5-μm particles, 125 × 4 mm, Macherey-Nagel, Hoerdt, France), and an electrochemical detector (Eldec 102, Chromatofield, Chateauneuf-les-Martigues, France). The column temperature is maintained at 33°. The mobile phase consists of (in mM) citric acid 27; sodium acetate 50; EDTA-Na$_2$ 1; sodium octyl sulfonate 0.8, and 6% methanol.

Procedure

1. The structures are broken down by ultrasound (1 min, 20 kHz, 40 W) over ice. Excess perchloric acid is removed by 15 μl of 6.5 M KOH, 4.2 M formiate, 15 min over ice. The homogenates are centrifuged (5 min,

8800*g*), and the supernatant is removed. If needed, the centrifuged pellets can be resuspended to allow determination of the protein content.

2. A 10- to 15-μl aliquot of the supernatant is injected through the reversed-phase column at a flow rate of 1 ml·min^{-1}. Catecholamines and their metabolites are measured at 670 mV versus an Ag$^+$/AgCl electrode (detector).

3. The detection limit is calculated by doubling the background noise level, is expressed in term of picomoles of injected amounts, and is usually less than 0.03 pmol for all compounds; the intraassay coefficient is 0.2%.

Note. NE turnover and TH activity are specific markers of the catecholaminergic system. Punching corresponds to a dissection "by excess," that is, all the catecholamine-synthesizing cells of the studied neurochemical structures are removed. Hence, the expression of results per structure is more descriptive of the absolute values of NE turnover and TH activity in catecholaminergic cell groups, which is independent of surrounding non-catecholaminergic tissues punched out together with these cell groups. The total protein content and structure mass are known to fluctuate unspecifically under physiological conditions. Thus, the expression of turnover or TH activity per structure is more relevant than the expression of these results per milligram of total proteins in order to avoid variations following modifications of total protein levels in the structure.

Acknowledgments

This work was supported by ACI BDP 57 and European Community (QLG2-CT-2001 N° 01467), CNRS, University Claude Bernard Lyon I, Région Rhone-Alpes. Olivier Pascual and Christophe Soulage held a fellowship from the Ministère de la Recherche. We are grateful to H. Salin and P. Ravassard for helping us solve technical issues.

[30] Discovery of Oxygen-Responsive Genes in Pheochromocytoma Cells

By KAREN A. SETA, TSUNEO K. FERGUSON, and DAVID E. MILLHORN

Introduction

In recent years, the molecular mechanisms by which cells respond to and adapt to decreased O₂ levels have begun to be elucidated. These responses are complex and involve multiple genes and signal transduction pathways. Our laboratory has pioneered the use of naive PC12 cells as a model system for studying oxygen sensing.[1] In addition, PC12 cells that have been treated with nerve growth factor (NGF) have long been used as a model for sympathetic neurons[2] in experimental models of hypoxia and ischemia. This article describes the use of subtractive suppression hybridization (SSH) to generate custom libraries that are enriched for genes involved in the oxygen sensing response in PC12 cells. When coupled with high-throughput validation procedures, such as microarray analysis and quantitative real-time polymerase chain reaction (PCR), this method provides a powerful set of tools for the elucidation of hypoxia-responsive genes and signal transduction pathways. These methods are also generally applicable to almost any field of study.

Methods

Cell Culture

Propagation of PC12 Cells. PC12 cells from American Type Culture Collection (ATCC) are propagated in 75-mm^2 filter flasks in DMEM/Ham's F-12 (50:50) medium supplemented with 10% fetal bovine serum (FBS), 15 mM HEPES, pH 7.4, 2 mM L-glutamine, 100 U/ml penicillin, and 100 μg/ml streptomycin at 37° in an atmosphere of 21% O_2/5% CO_2/balance room air. Because PC12 cells normally grow as suspension cells, we thaw the initial ATCC stock onto coated plates and passage them several times this way to select for adherent cells (this is necessary for differentiation studies). After that, it is not necessary to continue to use coated plates. Stocks frozen from these selected cells can be thawed directly into uncoated flasks. Medium should be changed every 3–4 days, and cell density should never be allowed to exceed 60–70% confluence.

Hypoxia Induction. Cells are plated onto plates rather than flasks to allow for more rapid gas equilibration and easier cell harvest. Cells are plated in growth medium 16–24 h before the experiment. If medium change is required (e.g., for treatment with pharmacological inhibitors), it is done so that cells are pretreated with the selected agent for 30–60 min prior to hypoxia induction. Cell density ranges from 10^3 to 9×10^4 cells/cm^2 ($\sim$70% confluence) and varies with the particular experiment. Note that both the basal and the hypoxia-induced expression of many genes (e.g., tyrosine hydroxylase, TH) and some biological phenomena (e.g.,

[1] D. Beitner-Johnson, G. E. Shull, J. R. Dedman, and D. E. Millhorn, *Respir. Physiol.* **110**, 87 (1997).
[2] L. A. Greene and A. S. Tischler, *Proc. Natl. Acad. Sci. USA* **73**, 2424 (1976).

cell death) are density dependent in PC12 cells. Also, we get the most reproducible results from hypoxia exposure using cells only between passages 10 and 20 or so (ATCC stock is designated p0).

Hypoxia exposures are carried out in a dual-gas incubator that regulates O_2 levels by displacing room air with nitrogen, such as the Thermo-Forma Model 3130 (Marietta, OH). This model is capable of O_2 levels between 1 and 20%. Other models are available that will reliably go to 0.1% O_2 and some that allow hyperoxia as well as hypoxia. Plexiglass or acrylic chambers that fit into a regular incubator and that are attached to their own gas lines and O_2 monitoring systems can also be used. Methods for measuring O_2 content in tissue culture medium are given elsewhere.[3] We do not routinely measure O_2 levels in the medium itself, and we can detect changes in gene expression and protein activation with very mild atmospheric hypoxia (15% O_2 in the chamber).

NGF Differentiation. PC12 cells are trypsinized and passed through a 28-gauge needle to dissociate clusters. Cells are then seeded in complete medium at 10^3 cells/cm². The next day, cells are washed with 1× phosphate-buffered saline, pH 7.4, and differentiation medium is added. Differentiation medium is the same as complete medium except that 1% FBS + 1% horse serum is substituted for 10% FBS. Nerve growth factor (NGF) is added to a final concentration of 50 ng/ml. Medium and NGF are replaced every other day (no PBS wash). Cells are fully differentiated within 7–10 days.

Custom Subtracted Libraries

Experimental Design. SSH libraries provide a powerful biological approach for obtaining clones that are differentially expressed between two populations. They consist of a subset of genes that are involved in mediating the phenotype of interest. The method can be used with any two populations whose gene expression profiles differ (e.g., treated *vs* untreated, different developmental states, different tissue types). It is important to note that this is an *enrichment* protocol, not a means to isolate a definitive clone set; some genes that are not differentially expressed will be isolated (i.e., false positives), and some genes that are differentially expressed will be lost. The best results will be obtained with samples that have very different expression profiles. For example, SSH is not a cost-effective technique for identifying the expression profile due to the overexpression of a single

---

[3] C. B. Allen, B. K. Schneider, and C. W. White, *Am. J. Physiol. Lung Cell. Mol. Physiol.* **281**, L1021 (2001).

gene in a transfected cell line. Also, an independent validation method is required to verify truly differentially expressed genes. Several validation techniques are described at the end of this article.

A final consideration is sample selection. This is fairly straightforward when the samples being studied are different tissue types, distinct developmental stages, etc. The situation becomes more complicated when the study involves multiple stimuli or a complex continuum of events, such as disease progression. It is generally not cost effective to generate a subtracted library, for example, each time point or dose in a study. Rather, it is better to select a few key time points, doses, or treatments in which the biology is fairly well characterized and/or in which a distinct phenotypic marker is present. The idea is to make the library as general as possible while still being specific to the phenotype of interest. A finely tuned time course and dose–response studies can then be carried out at the level of microarray analysis or another validation procedure. When multiple samples are used for library construction, hybridization and amplification should be carried out separately for each sample. This avoids dilution artifacts that occur when a gene is differentially expressed only temporarily. Samples can be pooled just prior to ligation into vector (see later).

Subtraction Protocol. Our laboratory has produced many SSH libraries using the PCR-Select kit (Clontech, Palo Alto, CA). This protocol combines high subtraction efficiency with an equalized representation of abundant and rare mRNAs.[4-6] The complete protocol can be found at www.clontech.com (protocol No. PT1117-1) and is outlined in Fig. 1. In this protocol, the "tester" is the sample that is being enriched, and the "driver" is the sample that is being subtracted out. Thus, it is possible to enrich for genes specific to each sample simply by switching which one is designated as the tester. In our prototype library, PC12 cells were exposed to hypoxia (1% O_2) or normoxia (21% O_2) for 6 h. The hypoxia sample was the "tester," and the normoxia sample was the "driver." Thus, our prototype library is enriched in genes that are induced by hypoxia.

Total RNA is isolated from each sample using a method of the investigator's choice. We have had good success with TRIREAGENT (Molecular

[4] L. Diatchenko, A. Chenchik, and P. Siebert, *in* "RT-PCR Methods for Gene Cloning and Analysis" (P. Siebert and J. Larrick, eds.), p. 213. BioTechniques Books, MA, 1998.

[5] L. Diatchenko, Y.-F. C. Lau, A. P. Campbell, A. Chenchik, F. Moqadam, B. Huang, S. Lukyanov, K. Lukyanov, N. Gurskaya, E. D. Sverdlov, and P. D. Siebert, *Proc. Natl. Acad. Sci. USA* **93**, 6025 (1996).

[6] N. G. Gurskaya, L. Diatchenko, A. Chenchik, P. D. Siebert, G. L. Khaspekov, K. A. Lukyanov, L. L. Vagner, O. D. Ermolaeva, S. A. Lukyanov, and E. D. Sverdlov, *Anal. Biochem.* **240**, 90 (1996).

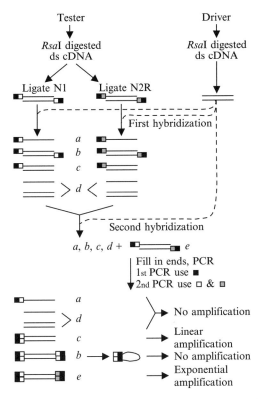

FIG. 1. Schematic of the subtractive suppression hybridization procedure (adapted from the Clontech manual, protocol No. PT1117-1). In suppression PCR, driver/driver hybrids (type *d* molecules) are not amplified because they have no adaptor and therefore no primer-binding sites. Tester/driver hybrids (type *c* molecules) consist of genes that are expressed equally in both populations (i.e., the targets that are being subtracted out). They have an adaptor at only one end and will only be amplified linearly. Intrapool tester/tester hybrids (type *b* molecules) are formed by genes that are expressed abundantly (such as housekeeping genes). They have the same adaptor at both ends. When the cDNA ends are filled in prior to PCR, the 5′ and 3′ ends of each strand of the tester/tester hybrid become complementary to each other. Under the PCR conditions used, these molecules form hairpin loop structures that are not amplified (this is the "suppression" step that equalizes the representation of rare and abundant transcripts). Interpool tester/tester hybrids (type *e* molecules, the differentially expressed genes) contain both types of adaptors and are amplified exponentially.

Research Center, Inc., Cincinnati, OH), following the manufacturer's protocol with two acid phenol/chloroform extractions added prior to ethanol precipitation. Poly(A)$^+$ RNA is then isolated using a procedure such as that found in the MicroPoly(A) pure mRNA isolation kit from Ambion

(Austin, TX). Double-stranded cDNA is synthesized as the first step of the PCR-Select protocol, using 2 μg of the poly(A)$^+$ RNA and substituting Powerscript (Clontech) for the kit-provided reverse transcriptase. If starting material is limited, total (0.05–1 μg) or poly(A)$^+$ (0.025–1 μg) RNA can be amplified using the SMART PCR cDNA synthesis kit (Clontech).

The *Rsa*I digest is performed on tester and driver cDNA to generate short, blunt-ended fragments that will hybridize efficiently. The tester sample is then divided into two pools, and a different adaptor (N1 or N2R, provided in the kit) is ligated to the 5′ end of the digested cDNA in each pool. Each pool is then denatured and hybridized individually to an excess of denatured driver to form the complexes shown in Fig. 1 (first hybridization). The molecules that remain single stranded at this step (type *a* molecules) are the rare and differentially expressed genes. Because the two tester pools are identical (except for the adaptors), the same molecules theoretically remain single stranded in both pools. The two pools are then mixed together with more denatured driver and allowed to hybridize to completion. In the second hybridization, most of the type *a* molecules form hybrids that have both types of adaptors (type *e* molecules), allowing these genes to be selectively amplified by suppression PCR using primers to both N1 and N2R adaptor sites.

To create a cDNA library, the amplified fragments are ligated into a vector and transformed into bacteria. We use pCR II-TOPO (Invitrogen) according to the manufacturer's protocol. Transformation is accomplished by electroporation into DH10B cells, which we have found give higher efficiency than DH5α cells. Use of the TOPO vector eliminates the need to ethanol precipitate samples (and potentially lose genes) prior to electroporation. To determine titer, portions of the electroporation mixture are plated onto LB/agar plates containing 50 μg/ml kanamycin and 36 μl of 50 mg/ml X-Gal (per 150-mm plate) for blue/white color selection. The rest of the electroporation culture is aliquoted and stored as a glycerol stock at $-80°$. Once titer is determined, an aliquot of the frozen stock is plated, and white (insert-containing) colonies are picked into 96-well plates containing 120 μl culture medium (LB + 50 μg/ml kanamycin) so that 10–20% of the titer is picked. Cultures are grown in a humidified chamber at 37° for 16–20 h. Breathe Easy strips (USA Scientific, Inc., Ocala, FL) are placed over the plates to prevent evaporation. Several backup copies of each plate are made, and all plates are stored as glycerol stocks at $-80°$ with a final glycerol concentration of 18–20%.

Quality Control. SSH library construction is simple in concept, but it is a complex multistep procedure that requires strict quality control throughout. The first, and probably most important, step is isolation of high-quality

RNA. We quantify and assess quality of both total and poly(A)$^+$ RNA on a Bioanalyzer 2100 using the RNA 6000 Nano Lab Chip Kit (Agilent, Foster City, CA). For total RNA, the ratio of the area under the peaks for the 28S and 18S bands should be between 1.8 and 2.0. For mRNA, the shape of the curve is compared to the "ideal mRNA" curve in the manual. If a bioanalyzer is not available, total RNA quality can be assessed by standard denaturing formaldehyde agarose gel analysis and UV spectrophotometry. The 28S and 18S ribosomal bands should appear sharp on the gel at ~4.5 and 1.9 kb, respectively, and the 28S:18S ratio should not be less than 1.5:1. The OD$_{260/280}$ should be between 1.8 and 2.1. Due to the small quantities of mRNA generated, UV analysis in a microcuvette is used. For either analysis procedure, the RNA aliquot that is being analyzed should be heated to 37° for 10–15 min prior to analysis. In this way, the presence of contaminants is readily detected before proceeding to more costly, time-consuming steps.

The PCR-Select procedure contains numerous quality control steps throughout the protocol. It is recommended that none of these steps be skipped (including the recommended addition of $[\alpha^{32}P]$dCTP as a tracer in the first-strand cDNA synthesis), even if the researcher is experienced with the protocol. Prior to ligation of the PCR products into a vector, tests for subtraction efficiency and enrichment of known upregulated (or tissue-specific) genes are performed. Both hybridization and PCR-based methods are described in the kit manual.

In a PCR-based subtraction efficiency test, if a gene is efficiently subtracted out, it will appear at an early cycle number in the unsubtracted sample and at a much later cycle, if at all, in the subtracted sample (Fig. 2). If a gene is enriched, it will appear at an earlier cycle in the subtracted sample than in the unsubtracted sample. The kit provides primers for human, rat, and mouse GAPDH, a "housekeeping" gene that is assumed to be unregulated. However, we have found GAPDH is regulated under some conditions in PC12 and other cells. We therefore add 0.5 ng of an exogenous plant gene mRNA (SpotReport 2, Stratagene, LaJolla, CA) per 2 μg of each poly(A)$^+$ sample at the start of the protocol. In this way, the test for a technical aspect of the procedure is independent of biological variability. We have also modified this test from a gel-based procedure, which is very time-consuming, to a real-time PCR-based procedure. The samples are prepared as for the gel-based procedure but are diluted an additional 10-fold. Samples are then run in a SYBR Green-based real-time PCR assay (described later) using a 50° annealing temperature and primers (10 μM each) that form a 146-bp product (forward primer 5'-GACGATGGAACCTGTGTCTACAAC-3'; reverse primer 5'-AAAAAAAAAGGGAACAAAAGAGGAC-3'). If other genes are

18   23   28   33   18   23   28   33   Cycle

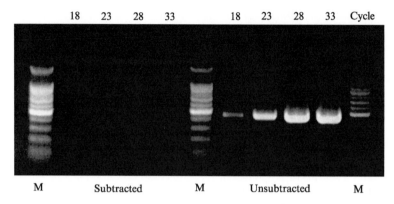

M   Subtracted   M   Unsubtracted   M

FIG. 2. Subtraction efficiency test for SSH showing efficient subtraction of GAPDH. The unsubtracted sample is created by annealing a portion of the N1 and N2R ligation reactions in the absence of driver, as described in the kit protocol. The subtracted sample is a portion of the second hybridization reaction. Both samples are subjected to suppression PCR amplifications. Kit-provided GAPDH primers are then used to analyze the abundance levels of this gene by PCR. Samples are taken at the indicated cycle. M, DNA size marker.

used, the primer pairs should be designed so that they lie within an *Rsa*I fragment. Note that because all *Rsa*I fragments are not always recovered, more than one site may have to be tested.

For PC12 cells undergoing hypoxic treatment, vascular endothelial growth factor (VEGF), TH, and junB are well-known induced genes (see Table I). Figure 3 is a "virtual Northern" blot showing enrichment of VEGF in our prototype library.[7] If an upregulated or tissue-specific gene is not known, exogenous mRNA spikes can be added to mimic such genes. We have not optimized the amount of spike mRNA for this procedure; however, the PCR-Select manual indicates that a twofold greater concentration in the tester sample relative to the driver sample is enough to see enrichment.

Because library sequencing is the most expensive step, it is desirable to eliminate empty wells and wells with multiple colonies. Bacterial PCR using the PCR-Select kit primers (N1 5'-TCGAGCGGCCGCCCGGG-CAGGT-3'; N2R 5'-AGCGTGGTCGCGGCCGAGGT-3') is performed on every clone in the library to verify the presence of a single insert. The PCR is performed in a 96-well format in a final volume of 25 μl per well

[7] D. Beitner-Johnson, K. Seta, Y. Yuan, H.-W. Kim, R. Rust, P. W. Conrad, S. Kobayashi, and D. E. Millhorn, *Parkinson. Relat. Disord.* **7,** 273 (2001).

TABLE I
FREQUENCY OF HYPOXIA-INDUCED GENES ISOLATED IN THE SSH LIBRARY

Gene name	No. copies	No. fragments	Frequency (%)
JunB[a,b]	14	2	5.2
TH[c]	5	7	1.9
VEGF[d]	3	4	1.1
Hexokinase II[e]	2	1	0.7
Bnip3[f]	2	2	0.7
A₂-adenosine receptor[g]	1	1	0.4
Phosphoglycerate kinase[h]	1	1	0.4
Plasminogen activator inhibitor[i]	1	3	0.4
Pyruvate kinase[h]	1	3	0.4
MKP-1[j]	6	3	2.2

[a] M. L. Norris and D. E. Millhorn, *J. Biol. Chem.* **270**, 23774 (1995).

[b] N. R. Prabhakar, B. C. Shenoy, M. S. Simonson, and N. S. Cherniack, *Brain Res.* **697**, 266 (1995).

[c] M. F. Czyzyk-Krzeska, D. A. Bayliss, E. E. Lawson, and D. E. Millhorn, *J. Biol. Chem.* **269**, 760 (1992).

[d] D. Shweiki, A. Itin, D. Soffer, and E. Keshet, *Nature* **359**, 843 (1992).

[e] Y. Niitsu, O. Hori, A. Yamaguchi, Y. Bando, K. Ozawa, M. Tamatani, S. Ogawa, and M. Tohyama, *Brain Res. Mol. Brain Res.* **74**, 26 (1999).

[f] H. M. Sowter, P. J. Ratcliffe, P. Watson, A. H. Greenberg, and A. L. Harris, *Cancer Res.* **61**, 6669 (2001).

[g] S. Kobayashi, D. Beitner-Johnson, L. Conforti, and D. E. Millhorn, *J. Physiol.* **512**(2), 351 (1998).

[h] G. L. Semenza, P. H. Roth, H. M. Fang, and G. L. Wang, *J. Biol. Chem.* **269**, 23757 (1994).

[i] T. E. Fitzpatrick and C. H. Graham, *Exp. Cell Res.* **245**, 155 (1998).

[j] K. A. Seta, R. Kim, H.-W. Kim, D. E. Millhorn, and D. Beitner-Johnson, *J. Biol. Chem.* **276**, 44405 (2001).

in a reaction containing 1× PCR buffer (Invitrogen), 2.5 mM MgCl₂, 0.5 mM dNTPs, 0.625 U Biolase DNA polymerase (Bioline, Randolph, MA), 0.4 μM each primer, and 3 μl of bacterial culture. The samples are analyzed on a 1.5% agarose gel (Fig. 4) using a large-format gel apparatus (Model D3, Owl Scientific, Portsmouth, NH). If greater than 10% of the wells contain multiple clones or no insert, the entire library is rearrayed into new 96-well plates and checked again by PCR. The average insert size of clones is ~500 bp, spanning a range of 50 bp–1.5 kb (both will be smaller if the SMART PCR cDNA synthesis kit is used to amplify the starting material).

Sequence Analysis and Library Curation. Although the TOPO vector provides high efficiency and convenience for library construction, it is not well suited for plasmid-based high-throughput sequencing due to low

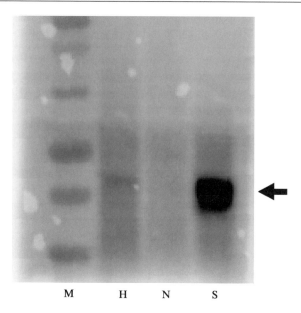

M   H   N   S

Fɪɢ. 3. "Virtual Northern" blot showing enrichment of VEGF following SSH. Un-subtracted samples for both normoxia (N) and hypoxia (H) samples are created by mixing aliquots of the N1 and N2R ligation reactions, as described in the kit protocol. The subtracted sample (S) is a portion of the second hybridization reaction using the hypoxia sample as the tester. All samples are subjected to suppression PCR amplifications. Equal amounts of the PCR products are then subjected to "virtual Northern" blot analysis using a ^{32}P-labeled probe specific for VEGF. M, DNA size marker.

plasmid yield. Therefore, we perform bacterial PCR, in a 96-well format, using modified M13 forward(-40) (5'-GTTTTCCCAGTCACGACGTTG-3') and M13 reverse (5'-CAGGAAACAGCTATGACCATG-3') primers. The entire insert can then be sequenced using the internal T7 or SP6 primer sites. A 50-μl reaction is sufficient for both sequencing and gel analysis to determine quality of the PCR products.

Library sequencing generates reams of sequence and BLAST data that must be managed carefully. Raw sequence data (in text file format) is first trimmed of vector and adaptor sequences with a program written especially for use with our libraries (available upon request). Fortunately, the relatively small size of SSH libraries means that data can be stored using commonly available programs such as Microsoft Excel or Access. We have also created an Excel macro that will search for clones that BLAST to the same gene. These clones can then be mapped to the parent sequence to determine the true representation of that gene in the library (see Fig. 5). This

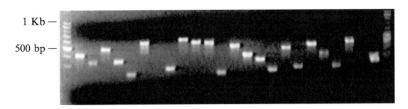

FIG. 4. Purity and size distribution of clones from the SSH library. Bacterial PCR using the N1 and N2R primers is performed on every clone in the library. Samples are analyzed by standard agarose gel electrophoresis with ethidium bromide staining to assess size distribution and presence of a single insert. A portion of a typical gel is shown.

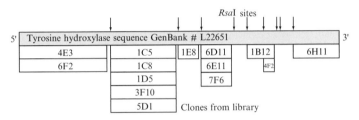

FIG. 5. Analysis of abundance of genes in SSH libraries. Clones that return BLAST hits to the same accession number must be mapped to the parent sequence in order to determine the true representation of the gene in the library. Clones that mapped to TH in our prototype library are shown. The parent sequence (filled bar) is shown, with RsaI sites indicated by arrows. Fourteen clones (open bars) returned BLAST hits to this gene, and their alignment with the parent sequence is shown. The most abundant fragment is represented five times, so there are five copies in this library. To calculate percentage abundance, the number of copies of TH is divided by the sum of the number of copies of all genes.

is necessary because, due to the restriction digest step in the subtraction protocol, a single cDNA molecule in the starting material can be represented by multiple clones that correspond to different restriction fragments (Table I). By using this method, we have found that the redundancy in our prototype library is quite low, with approximately 80% of the genes being represented only once.

Validation

Table I shows several genes known to be regulated by hypoxia and their relative abundance in our prototype library. We have found that abundance in our prototype library corresponds well with induction by hypoxia. This has provided us a valuable starting place for validating our approach.

TABLE II
VALIDATION OF GENES ISOLATED IN THE SSH LIBRARY

Gene name	Microarray[a]	PCR[b]
JunB	+	+
TH	+	+
VEGF	+	+
Hexokinase II	+	+
Bnip3	+	+
A_2-adenosine receptor	+	NT
Phosphoglycerate kinase	1.67-fold	+
Plasminogen activator inhibitor	+	+
Pyruvate kinase	1.99-fold	NT
MKP-1	+	+

[a] A plus indicates induction of >2-fold (average of two experiments).
[b] A plus indicates statistically greater levels in hypoxia samples than in normoxia samples ($p < 0.001$ by t test, n = at least 4); NT, not tested.

For example, MAP kinase phoaphatase-1 (MKP-1) is a gene that had not been previously associated with hypoxia. Six copies of MKP-1 were isolated in our library, which is similar to the abundance level for TH. Microarray, Northern blot, and Western blot analyses confirmed the induction of this gene by hypoxia.[8] Several other genes that were also present at relatively high abundance also showed induction by microarray analysis and are currently under study in our laboratory. We use a combination of microarray analysis, real-time PCR (replacing Northern blots), and Western blots to verify hypoxic induction and to study mechanisms of gene regulation. Table II shows microarray and real-time PCR validation of the genes listed in Table I.

Microarrays. Combining SSH with microarray analysis provides a powerful tool to examine gene expression patterns. By enriching a library specifically for genes involved in the phenotype of interest, many potential false positives are removed and data analysis is simplified. In addition, SSH libraries provide an opportunity for novel gene discovery that does not exist with commercial microarrays. An excellent summary of the methods for producing and analyzing microarrays is available from Hegde and colleagues,[9] so details will not be presented here.

[8] K. A. Seta, R. Kim, H.-W. Kim, D. E. Millhorn, and D. Beitner-Johnson, *J. Biol. Chem.* **276,** 44405 (2001).
[9] P. Hegde, R. Qi, K. Abernathy, C. Gay, S. Dharap, R. Gaspard, J. E. Hughes, E. Snesrud, N. Lee, and J. Quackenbush, *Biotechniques* **29,** 548 (2000).

Microarrays are produced by PCR-amplifying clone inserts using the N1 and N2R primers and then spotting the purified PCR products onto coated slides. Libraries are small enough that multiple libraries (e.g., different time points) can be printed on a single slide, and each clone can be replicated three to four times. Replicates are averaged using an Excel macro written for Axon GenePix scanner output format.

All of the widely used microarray normalization methods[10] assume that the average expression ratio for all the genes on the array is equal to one. This is not the case for microarrays made from SSH libraries because the gene content is biased. To avoid normalization and scanner bias, we include 10 exogenous plant genes (SpotReports 1–10, Stratagene) scattered throughout the body of the array. SpotReports 1, 2, and 3 are also spotted at the top of the array and their mRNAs are added in equal amounts (5, 0.5, and 0.05 ng per 100 μg total RNA, respectively) to each probe-labeling reaction to provide a known set of genes that should have a ratio of 1:1. After the slide is scanned, the SpotReport 1–3 spots at the top of the array are analyzed quickly. If the ratio is not 1:1, the scanner settings are adjusted and the slide is rescanned until the ratio is as close as possible to 1:1 (we generally accept ratios of 0.95–1.05). The SpotReport 1-3 clones in the body of the array are used to demonstrate that there is no positional bias on the arrays. The other seven mRNAs are spiked into the two labeling reactions at different known ratios. When the background-subtracted fluorescence values for each wavelength are plotted, these spots should fall along their predicted lines (see Fig. 6). This system provides a good control for all the technical aspects of microarray analysis (labeling, hybridization, and scanning).

Real-Time PCR. Real-time PCR using SYBR Green I dye is an inexpensive and efficient method for validating gene expression, either individually or for many genes at a time. This method avoids the need for expensive modified oligonucleotides. Total RNA is isolated and DNase treated to remove contaminating genomic DNA. Following cleanup, first-strand cDNA synthesis is performed using oligo(dT) as the primer according to manufacturer's directions and 7.5 μg total RNA in a 40-μl reaction (we use the SuperScript first-strand synthesis system for RT-PCR, Life Technologies, Rockville, MD). SpotReport 2 mRNA is spiked into the cDNA reaction master mix at 0.5 ng per 100 μg total RNA to control for RT and PCR reaction efficiencies. After RNase digestion, the reaction mix is diluted to a final volume of 75 μl (i.e., 0.1 μg/μl RNA equivalent).

[10] J. Quackenbush, *Nature Rev. Genet.* **2,** 418 (2001).

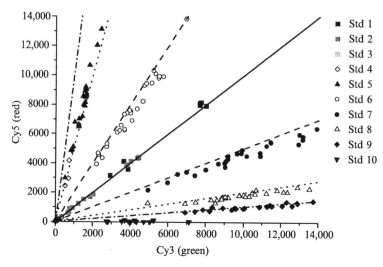

FIG. 6. Quality control standards for SSH microarrays. Plant gene mRNAs (SpotReports 1–10) were spiked into total RNA samples prior to labeling for microarray analysis. SpotReports 1–3 were added in equal amounts to the Cy3 and Cy5 reactions to produce spots with a 1:1 intensity ratio (solid line). SpotReports 6 and 7 were spiked in at 2-fold greater concentrations in the Cy5 and Cy3 reactions, respectively (dashed line). SpotReports 5 and 8 were spiked into the Cy5 and Cy3 reactions with a 5-fold difference (dotted line). SpotReports 4 and 9 were spiked to show a 10-fold difference (dotted and dashed line). SpotReport 10 was added only to the Cy3 reaction. Background-subtracted median pixel intensities are plotted. Each point on the scatter plot represents one spot on the array.

Real-time PCR is performed in a Smart Cycler (Cepheid, Sunnyvale, CA) using the LightCycler DNA Master SYBR Green I dye intercalation assay (Roche, Indianapolis, IN), although there are many other manufacturers for equipment and reagents. Primers should be designed toward the 3' end with a t_m of no more than 2° between them, and the PCR product should be 80–150 bp in length.[11] Between 1 and 5 μl of the diluted cDNA is used in the PCR reaction (this must be optimized for each gene). SYBR Green I has significant fluorescence only when bound to dsDNA, so optical measurements are taken during the 72° extension step in each cycle. Melt curve analysis showing a single sharp peak indicates formation of a single product. PCR products from a representative real-time PCR reaction should be subcloned and sequenced to verify their identity.

A threshold cycle (C_t) is determined for each sample as either the inflection point in the growth curve or as the point where the growth curve

[11] S. A. Bustin, *J. Mol. Endocrinol.* **25,** 169 (2000).

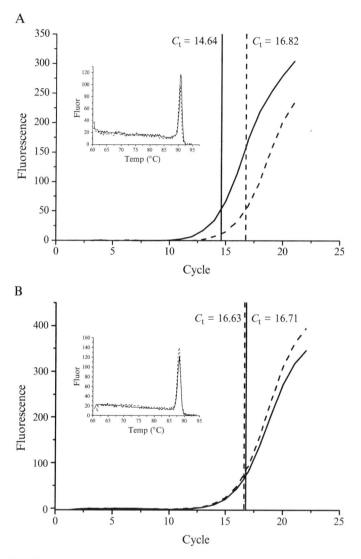

FIG. 7. Validation of gene regulation by real-time PCR. PC12 cells were exposed to normoxia (dashed line) or hypoxia (1% O$_2$, solid line) for 6 h. Total RNA was isolated, and quantitative real-time PCR (SYBR Green I dye intercalation method) was performed as described in the text. Growth curves are plotted for TH (A) and GAPDH (B). Threshold cycles were calculated by the second-derivative method and represent the point at which the growth curves enter log-linear phase. In this experiment, GAPDH is unregulated, whereas TH is strongly induced. Melt curves are shown as insets to indicate the formation of single products.

reaches some predetermined fluorescence value (Fig. 7). The greater the expression level of a particular gene in one sample relative to another, the earlier the C_t will be. Statistical analysis is performed on raw C_t values. Calibration curves can be constructed to convert C_t into fold change or percentage change, if desired.

Acknowledgments

This work was supported by funding from the National Institutes of Health (HL33831, HL59945, HL07571, HL66312, DK58811), the Parker B. Francis Foundation (KAS), and the U.S. Army.

Section VII

General

[31]   Oxygen-Dependent Asparagine Hydroxylation

By Daniel J. Peet, David Lando, Dean A. Whelan,
Murray L. Whitelaw, *and* Jeffrey J. Gorman

Introduction

The capacity of hypoxia-inducible factor (HIF) to activate transcription is regulated in response to the cellular oxygen environment.[1] This regulation is affected by intracellular proline[2,3] and asparagine hydroxylases.[4,5] These enzymes sense intracellular oxygen tension and at normoxia they use dioxygen as a substrate to hydroxylate HIF.

Proline hydroxylation leads to the interaction of HIF with von Hippel–Lindau factor, polyubiquitination, and rapid proteosomal degradation.[2,3] Asparagine hydroxylation occurs adjacent to the C-terminal transcriptional activation domain (CAD) of HIF, repressing transactivation by blocking the interaction of HIF with the transcriptional coactivator p300 and prevents activation of transcription.[5] Under hypoxic stress the hydroxylases are inactivated and HIF is stabilized and activated transcriptionally, resulting in specific gene induction.

Hydroxlation of asparagine and aspartic acid residues has been observed previously within consensus sequences of EGF-like domains of several proteins, but no direct biological function has been attributed to these modifications.[6] The C1r component of the complement system[7–9] and the vitamin K-dependent bovine protein S[10] have been shown to be

---

[1] D. Lando, J. J. Gorman, M. L. Whitelaw, and D. J. Peet, *Eur. J. Biochem.* **270,** 781 (2003).

[2] P. Jaakkola, D. R. Mole, Y. M. Tian, M. I. Wilson, J. Gielbert, S. J. Gaskell, A. Kriegsheim, H. F. Hebestreit, M. Mukherji, C. J. Schofield, P. H. Maxwell, C. W. Pugh, and P. J. Ratcliffe, *Science* **292,** 468 (2001).

[3] M. Ivan, K. Kondo, H. Yang, W. Kim, J. Valiando, M. Ohh, A. Salic, J. M. Asara, W. S. Lane, and W. G. Kaelin, Jr., *Science* **292,** 464 (2001).

[4] D. Lando, D. J. Peet, J. J. Gorman, D. A. Whelan, M. L. Whitelaw, and R. K. Bruick, *Genes Dev.* **16,** 1466 (2002).

[5] D. Lando, D. J. Peet, D. A. Whelan, J. J. Gorman, and M. L. Whitelaw, *Science* **295,** 858 (2002).

[6] J. Stenflo, *Blood* **78,** 1637 (1991).

[7] G. J. Arlaud, A. Van Dorsselaer, A. Bell, M. Mancini, C. Aude, and J. Gagnon, *FEBS Lett.* **222,** 129 (1987).

[8] G. J. Arlaud, A. C. Willis, and J. Gagnon, *Biochem. J.* **241,** 711 (1987).

[9] C. T. Przysiecki, J. E. Staggers, H. G. Ramjit, D. G. Musson, A. M. Stern, C. D. Bennett, and P. A. Friedman, *Proc. Natl. Acad. Sci. USA* **84,** 7856 (1987).

[10] J. Stenflo, A. Lundwall, and B. Dahlback, *Proc. Natl. Acad. Sci. USA* **84,** 368 (1987).

hydroxylated on the β carbon of asparagine to produce the *erythro* isomer. Hydroxylation of the β carbon of aspartic acid, to produce the *erythro* isomer, occurs with several vitamin K-dependent proteins of the blood coagulation system, including protein C[11–13] and factors VII, IX, X, and Z.[12,14–17]

Formation of β-hydroxyasparagine and β-hydroxyaspartic acid by modification of the side chains of protein-bound asparagine and aspartic acid residues is a posttranslational event that is affected by 2-oxoglutarate (OG)-dependent dioxygenase enzymes.[18] The observation of enzymatic incorporation of hydroxylamine into the amide of the side chain of asparagine by *Proteus vulgaris* is an indication of the potential for *N*-hydroxylated asparagine, or the hydroxamic acid aspartohydroxamic acid, to exist in proteins[19]; however, the enzyme responsible for this modification has not been defined.

Characterization of the hydroxylated asparagine and aspartic acid residues within the EGF-like domain-containing proteins has been possible with comparatively insensitive analytical techniques due to the relative ease of obtaining substantial quantities of these proteins from natural sources. By comparison, characterization of the hydroxyasparagine residue of HIF required expression in mammalian cells, very efficient and controlled protein isolation procedures, and the use of very sensitive analytical tools.[5]

This article presents methods for detection, characterization, and assay of oxygen-dependent hydroxylation of asparagine residues. In particular, the role of modern mass spectrometry methods in characterizing this modification in HIF is highlighted.

[11] T. Drakenberg, P. Fernlund, P. Roepstorff, and J. Stenflo, *Proc. Natl. Acad. Sci. USA* **80,** 1802 (1983).

[12] P. Fernlund and J. Stenflo, *J. Biol. Chem.* **258,** 12509 (1983).

[13] D. C. Foster, S. Yoshitake, and E. W. Davie, *Proc. Natl. Acad. Sci. USA* **82,** 4673 (1985).

[14] B. A. McMullen, K. Fujikawa, and W. Kisiel, *Biochem. Biophys. Res. Commun.* **115,** 8 (1983).

[15] B. A. McMullen, K. Fujikawa, W. Kisiel, T. Sasagawa, W. N. Howald, E. Y. Kwa, and B. Weinstein, *Biochemistry* **22,** 2875 (1983).

[16] T. Sugo, P. Fernlund, and J. Stenflo, *FEBS Lett.* **165,** 102 (1984).

[17] L. Thim, S. Bjoern, M. Christensen, E. M. Nicolaisen, T. Lund-Hansen, A. H. Pedersen, and U. Hedner, *Biochemistry* **27,** 7785 (1988).

[18] J. Stenflo, E. Holme, S. Lindstedt, N. Chandramouli, L. H. Huang, J. P. Tam, and R. B. Merrifield, *Proc. Natl. Acad. Sci. USA* **86,** 444 (1989).

[19] N. Grossowicz, E. Wainfan, E. Borek, and H. Waelsch, *J. Biol. Chem.* **187,** 110 (1950).

Detection and Characterization of Protein-Bound
Hydroxyasparagine

Review of the literature on the characterization of β-hydroxylation of asparagine and aspartic acid residues in C1r and the blood coagulation proteins provides strategies for identifying and characterizing these modifications and thereby implying the involvement of an oxygen-dependent dioxygenase. The observation of β-hydroxyaspartic acid in acid hydrolysates is an indication of β-hydroxylation of asparagine or aspartic acid residues of a protein or peptide.[7–10] Acid hydrolysis alone does not differentiate between β-hydroxylation of asparagine and aspartic acid because β-hydroxyasparagine is converted to β-hydroxyaspartic acid upon hydrolysis.[7–10] β-Hydroxyasparagine can be identified using ion-exchange amino acid analysis protocols, but proteins or peptides of interest must be subjected to total enzymatic digestion to avoid conversion of β-hydroxyasparagine to β-hydroxyaspartic acid if direct observation of β-hydroxyasparagine is desired.[9,10] Furthermore, acid hydrolysis alone does not differentiate between potential sites of β-hydroxylation in a protein or peptide sequence, should there be a multiplicity of potential sites. Finally, analysis of acid hydrolysates would not detect the presence of aspartohydroxamic acid because this derivative would be converted to aspartic acid during hydrolysis.

Failure to observe a standard amino acid as a PTH derivative at a particular cycle during Edman degradation sequencing of a mammalian protein is an indication that the amino acid at the position in question is modified posttranslationally, β-hydroxylation may be suspected if an asparagine or aspartic acid is anticipated at the cycle in question according to the nucleotide sequence of the gene that codes for the protein of interest.[13–16] β-Hydroxyaspartic acid can be identified as a PTH derivative[7,14] and differentiated from PTH-β-hydroxyasparagine,[7,9,10] provided the appropriate PTH standards can be obtained.

Mass spectrometry of peptides can provide an indication of β-hydroxylation due to an observed increase of 16 amu in the predicted masses of peptides from the protein sequence of interest.[7,9,11] However, increases of 16 amu are observed more commonly due to the oxidation of protein-bound methionine residues and could also be due to hydroxylation or oxidation of other amino acids, or the presence of aspartohydroxamic acid. A 16 amu mass increase for a particular peptide alone would only support a *prima facie* case for asparagine/aspartic acid hydroxylation. Tandem mass spectrometry can directly confirm the presence of β-hydroxasparagine or β-hydroxaspartic acid in a peptide and determine the location of the modified residue.[7,11,15]

β-Hydroxyasparagine and β-hydroxyaspartic acid contain an asymmetric carbon at the β position of their side chains, and another step in the process of characterizing the specificity of the dioxygenases responsible for β-hydroxylation is to differentiate between the two potential diasterioisomers possible for the modified β carbon. Fortunately, *erythro*-β-hydroxyaspartic acid and *threo*-β-hydroxyaspartic acid can be clearly resolved under ion-exchange separation conditions used for amino acid analysis. Thus the stereoisomers can be identified differentially in acid hydrolysates of proteins or peptides that contain β-hydroxyaspartic acid.[7–12,16] If the presence of β-hydroxyasparagine has been established, it is also possible to differentiate between the stereoisomers by acid hydrolysis and characterization of the resultant stereoisomers of β-hydroxyaspartic acid by ion-exchange separation.[9,10] Postcolumn detection of β-hydroxamino acids is possible with both ninhydrin and *o*-phthaldehyde, with the latter reagent providing better sensitivity.[11,12]

One of the potential limitations with the amino acid analysis approach is the lack of commercial availability of standards. *threo*-β-Hydroxyaspartic acid is the only standard readily available from commercial sources. However, this limitation can be overcome to some degree by the use of acid hydrolysis for the differentiation process. There is really no need to determine the individual elution positions of all of the diasetrioisomers directly, provided that ancillary data have differentiated between β-hydroxylation of asparagine and aspartic acid. Racemization occurs to a minor degree during acid hydrolysis, and a small peak of the alternate diasterioisomer is observed during ion-exchange separation.[9–12] Thus with authentic *threo*-β-hydroxyaspartic acid as a standard, it is possible to determine which isomer predominates in the hydrolysate of the sample of interest.

The stereoisomers of PTH-β-hydroxyaspartic acid can be separated by reversed-phase HPLC,[14] thus Edman degradation can also be used to differentiate between the stereoisomers of β-hydroxyaspartic acid, provided the appropriate standards can be obtained although this has not been demonstrated with β-hydroxyasparagine.

Nuclear magnetic resonance (NMR) is the only single method that will provide comprehensive characterization of β-hydroxyasparagine or β-hydroxyaspartic acid in a protein sequence.[11,15,20] Other methods would have to be used in combination to provide comprehensive characterization of these residues. The various analytical methods, combinations of methods, and the results they yield are presented in Fig. 1. Most of the methods used for the identification of β-hydroxyasparagine and β-hydroxyaspartic

[20] L. A. McNeill, K. S. Hewitson, T. D. Claridge, J. F. Seibel, L. E. Horsfall, and C. J. Schofield, *Biochem. J*, **367**, 571 (2002).

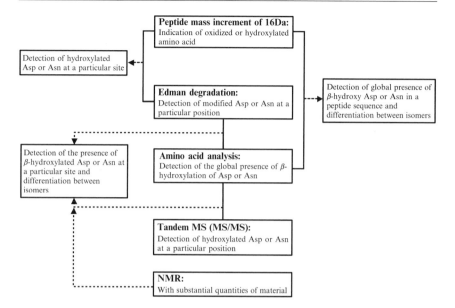

Fig. 1. An outline of analytical strategies for characterization of hydroxylation of asparagine and aspartic acid residues. The central column of text boxes identifies relevant individual analytical techniques, information that can be derived with each technique and potential limitations. Combinations of techniques indicated by solid lines will produce the combined outcomes as indicated by linkages to peripheral text boxes by dashed arrows.

acid in blood-derived proteins required 1–10 nmol of peptide for each analytical step. This applied to the amino acid sequencing and analysis procedures, mass spectrometry, and NMR. Thus, the amount of materials required for methods used previously with blood proteins, including NMR and as combinatorial strategies, would preclude their use in association with cell-based studies.

Fortunately, mass spectrometry techniques have been developed with spectacular improvements in sensitivity. These improvements have been driven by the development of matrix-assisted laser desorption/ionization (MALDI)[21,22] and electrospray ionization (ESI)[23] or, more particularly, nano-ESI.[24] Despite these advances, combination with other

[21] M. Karas and F. Hillenkamp, *Anal. Chem.* **60**, 2299 (1988).

[22] K. Tanaka, H. Waki, Y. Ido, S. Akita, Y. Yoshida, and T. Yoshida, *Rapid Commun. Mass Spectrom.* **2**, 151 (1988).

[23] J. B. Fenn, M. Mann, C. K. Meng, S. F. Wong, and C. M. Whitehouse, *Science* **246**, 64 (1989).

[24] M. Wilm and M. Mann, *Anal. Chem.* **68**, 1 (1996).

correspondingly sensitive analytical techniques is required to complement modern mass spectrometry techniques to achieve comprehensive characterization of hydroxylated asparagine or aspartic acid residues in cell-based studies. Thus, there is a need to establish which would be the most appropriate sensitive complementary technique for use with current mass spectrometry methods in cell-based studies. This article provides evidence that precolumn derivatization-based amino acid analysis is an appropriate complementary method.

Demonstration of a Hydroxylation Site between Residues 846 and 864 of HIF-2α 774–874 Dependent on the Oxygen Environment of Mammalian Cells

Initial attempts to analyze HIF-2α 774–874 isolated from transfected mammalian cells by IMAC involved SDS–PAGE of eluates from IMAC resin followed by in-gel tryptic digestion. However, no ions were evident in the resultant tryptic digests with masses that correlated with tryptic peptides anticipated from HIF-2α 774–874. One reason for the lack of success of this approach may have involved the availability of only very small quantities of materials for analysis (bands were barely detectable by Coomassie staining) and loss of the relatively low molecular weight protein from the gel slices by diffusion during the destaining and subsequent in-gel reduction and alkylation steps.

Successful analysis employed a "one pot" workup in which reduction, alkylation, and tryptic digestion steps were performed successively in solution without any intermediate sample workup steps. This approach was devised to avoid sample losses and relied on the destruction of residual dithiothreitol (DTT) from the reduction step by the iodoacetamide used in the alkylation step and by the resistance of trypsin to iodoacetamide. Residual reagents were removed from the tryptic peptides of HIF-2α 774–874 using batchwise reversed-phase desalting with C18 microcolumns. An important aspect of this approach was that it was possible to identify contaminating proteins by peptide mass mapping and avoid consideration of data emanating from those proteins.

MALDI-TOF-MS was performed on desalted tryptic digests of HIF-2α 774–874 isolated from cells grown (1) in a normoxic atmosphere, (2) in the presence of the hypoxia mimetic 2,2'-dipyridyl (DP), and (3) under hypoxic conditions. Comparison of resultant spectra showed a major difference due to the presence of an intense ion at m/z approximately 2107 in the normoxic sample (Fig. 2A). This ion did not correlate with an expected tryptic peptide of HIF-2α 774–874 but was 16 amu heavier than the peptide spanning residues 846–864 (m/z approximately 2091), which was taken to represent hydroxylation of an amino acid in this sequence that was inhibited by hypoxia (Figs. 2B and 2C). Other peptides were also present in the digests that did not correspond to HIF-2α 774–874 peptides, but these were common to all digests and originated from a protein that was coisolated during IMAC and HPLC of the IMAC eluate.

Experimental

Stable overexpression of the HIF-2α 774–874 is performed in mammalian cells. A stable human embryonic kidney 293T cell line expressing HIF-2α 774–874 fused to an N-terminal 6-histidine tag and a myc epitope is used to prepare protein for analysis.[5] The 293T cells [maintained in 10% fetal calf serum (FCS) in Dulbecco's modified Eagles medium (DMEM) at 37°, 5% CO_2] are grown for 16 h under normoxia (20% O_2), hypoxia [<1% O_2 achieved by placing the culture dishes in an air-tight container with an AnaeroGen sachet (OXOID, Hampshire, UK)], or in 100 μM DP (Sigma, St. Louis, MO). To purify the HIF-2α 774–874 immediately after aspirating off the media, cells are lysed in binding buffer (100 mM Na-phosphate, pH 8.0, 8 M urea, 0.1% NP-40, 0.15 M NaCl, 5 mM imidazole) with protease [0.1 mM phenylmethyl sulfonyl fluoride (PMSF), 0.2 μg/ml aprotinin, 0.4 μg/ml bestatin, 0.5 μg/ml leupeptin, 0.1 μg/ml pepstatin] and phosphatase inhibitors (0.05 mM Na_3VO_4, 0.1 mM NaF, 0.5 mM glycerophosphate) for 15 min at room temperature, filtered through a 0.2-μm filter, and bound to Ni-NDA agarose resin (Scientifix, Australia) for 1 h at room temperature. Because denaturing conditions are significantly more efficient for the purification and prevention of subsequent protein modifications upon cell lysis than native conditions, buffer solutions containing 8 M urea are utilized. After extensive washing of the resin (400 volumes of binding buffer containing 100 mM Na-phosphate, pH 8.0, 8 M urea, 0.5 M NaCl, 20 mM imidazole), the bound protein is eluted with 100 mM Na-phosphate, pH 8.0, 8 M urea, and 200 mM imidazole. As a final purification step, eluted protein is loaded onto a butyl C4 HPLC column (Brownlee, PerkinElmer) in 0.1% (v/v) trifluoroacetic acid (TFA) and is eluted with a gradient to 80% (v/v) acetonitrile/0.1% (v/v) TFA. Fractions are collected and lyophilized under vacuum.

Fusion proteins comprising HIF-2α 774–874 with HexaHis and Cmyc-epitope tags produced under different cellular oxygen conditions are dissolved separately in 0.1 M ammonium bicarbonate (BDH) containing 20 mM dithiothreitol (Progen; 200–007) and are incubated at 56° for 1 h. Iodoacetamide (Fluka; 57670) is subsequently added to a final concentration of 50 mM, and incubation continues for another 30 min at 37° in the absence of light. The reduced and alkylated protein is then digested by

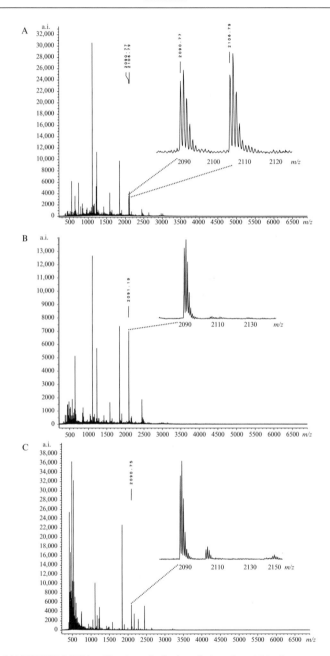

FIG. 2. MADI-TOF-MS identification of a hydroxylation site within the sequence between residues 846 and 864 of HIF-2α. Spectra were obtained with tryptic digests of HIF-2α 774–874

adding an aliquot of sequencing grade trypsin (Boehringer at 1 mg/ml in 1 mM HCl) to a final concentration of 0.01 mg/ml and incubating the digest for 4 h at 37° in the absence of light. Tryptic peptides are desalted using P10 C18 Zip Tips (Millipore; Z10188960) by acidifying the digest through the addition of TFA to 0.5% (v/v) and loading onto a ZipTip conditioned previously with a methanol wetting solution and an activation/equilibration solution of 5% (v/v) aqueous methanol containing 0.1% (v/v) formic acid. Twenty-microliter aliquots of the diluted samples are aspirated through the Zip Tips and out into a waste tube until the entire sample has been transferred to the waste tube. The ZipTips are then washed three times with 20-μl aliquots of 5% (v/v) aqueous methanol containing 0.1% (v/v) formic acid, and the adsorbed peptides are subsequently eluted by repeatedly aspirating 10 μl of 60% (v/v) aqueous methanol containing 0.1% (v/v) formic acid through the ZipTip. All solvents used for ZipTip cleanup are HPLC grade.

Separate aliquots (1–2 μl) of the desalted tryptic digests are mixed with equal volumes of α-cyano-4-hydroxycinnamic acid or 2,6-dihydroxyacetophenone/diammonium hydrogen citrate matrix preparations[25,26] and spotted onto sample targets of a Bruker reflex MALDI time-of-flight (TOF) mass spectrometer.

Peptide mass maps are acquired by accelerating ions, formed by irradiation with a nitrogen laser (337 nm), at 20 kV from the ion source of the mass spectrometer and reflecting the ions at 22.5 kV with a two-stage gridless reflector onto a microchannel plate detector. A 300-ns delay is imposed between laser ionization and acceleration of the ions from the ion source. Data are acquired at a digitization rate of 2 MHz and analyzed using Bruker XMass software. The spectrometer is calibrated externally using a mixture of human angiotensin II (monoisotopic mass 1045.54), human ACTH residues 18–39 (monoisotopic mass 2465.19), and human insulin (average mass 5733.52). The masses of tryptic peptides expected from the reduced and alkylated but otherwise unmodified fusion protein are calculated by simulation of tryptic digestion of the protein sequence using the Prospector

[25] J. J. Gorman, B. L. Ferguson, and T. B. Nguyen, *Rapid Commun. Mass Spectrom.* **10,** 529 (1996).
[26] J. J. Pitt and J. J. Gorman, *Rapid Commun. Mass Spectrom.* **10,** 1786 (1998).

isolated from cells transfected and maintained (A) under normoxic conditions, (B) in the presence of the hypoxic mimetic DP, and (C) under hypoxic conditions. Adapted with permission from Lando *et al.*[5]

program (http://prospector.ucsf.edu/). This program is also used to identify data derived from contaminating proteins.

Characterization of Asparagine 851-Specific Hydroxylation of HIF-2α CAD in Mammalian Cells

The tryptic peptide spanning residues 846–864 of HIF-2α 774–874 contain two proline residues that were potential candidates for hydroxylation in response to oxygen regulation by analogy with proline hydroxylation of the oxygen-dependent degradation domain of HIFs.[2,3] The sequence did not contain any methionine residues that could have produced a 16 amu mass increment due to nonspecific chemical oxidation. Additional analyses were performed on the digests using a Quadrupole-quadrupole (Qq)-TOF-MS with nano-ESI introduction. Several analytical modes are possible using a Qq-TOF-MS. In one analytical mode (TOF-MS), it is possible to obtain spectra showing masses of intact tryptic peptides similar to MAL-DI-TOF-MS spectra. The major difference between spectra from these two instrumental methods is that ESI tends to attach multiple protons to tryptic peptides, thus spectra are more complex with the potential for each peptide to be observed at different m/z values due to the different charge (z) states for each protonated form. Tandem mass spectrometry (MS/MS) sequencing is also possible using the Qq-TOF-MS. In this mode, it is possible to select individual ions, fragment them in a collision cell, and analyze the daughter fragments to determine the amino acid sequence. For very high sensitivity analyses, individual 2-μl aliquots of digests are loaded into separate nano-ESI capillaries and ions are sprayed at approximately 20 nl/min into the Qq-TOF-MS.

Qq-TOF-MS was performed on the digests, and the relevant peptide ions of interest were identified and subsequently selected for MS/MS to determine which, if any, amino acid residue of residues 846–864 was hydroxylated. The ion at approximately m/z 2091 (2090 + 1 proton or M+H$^+$/1) in the MALDI-TOF-MS spectrum of the digest of the normoxic sample was present as a doubly protonated ion at m/z of 1046 in the Qq-TOF-MS spectrum due to adduction of two protons during ionization (2090 + 2 or M+2H^{2+}/2). MS/MS of the peptide at m/z 1046 produced a fragment ion sequence that represented almost complete coverage of residues 846–864 of HIF-2α (Fig. 3A). This was as expected because this ion represented the unmodified peptide. By comparison the putatively hydroxylated peptide was seen as a doubly protonated ion at m/z of 1054 by Qq-TOF-MS. MS/MS of this ion also indicated that it represented residues 846–864 in a hydroxylated form and importantly located the hydroxyl on the single

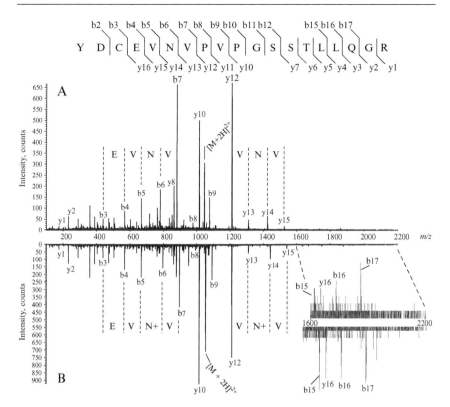

FIG. 3. Identification of asparagine 851 of HIF-2α as a site of oxygen-dependent hydroxylation. Tandem or MS/MS spectra were acquired on ions representing residues 846–864 of HIF-2α in Fig. 2A. Doubly charged versions of the relevant ions formed by nano-ESI on the Qq-TOF-MS were selected and fragmented. Residual intact doubly charged parent ions are indicated with $[M + 2H]^{2+}$ labels at approximate m/z values of 1046 and 1054 for the unmodified (A) and hydroxylated (B) peptides, respectively. Fragments are labeled on spectra according to the correlation of experimental masses with predicted masses of fragmentation products as indicated on the sequence above spectra. Plus symbols on ions in B indicate fragments that were 16 Da heavier than theoretical fragments. Reproduced with permission from Lando et al.[5]

asparagine (Asn851) in the sequence and showed the absence of proline hydroxylation (Fig. 3B).

Experimental

Aliquots (2 μl) of the digests prepared earlier are loaded into nano-ESI capillaries of medium length obtained from Protana, Odense, Denmark (www.proxeon.com), which are subsequently placed into a nano-ESI ion source (Protana) fitted to a Qq-TOF-MS (Applied Biosystems QSTAR-Pulsar). Ions are sprayed from the capillaries with a potential of 850 V applied to the ends of the capillaries. Ions are pulsed into the TOF stage of the spectrometer using default values.

Ions of interest in the TOF-MS spectra are selected by gating the first quadrupole (Q) of the Qq-TOF-MS to transmit a particular ion at low resolution into the second quadrupole (q) or collision cell charged with nitrogen gas and affecting collisions at 40–50 eV. Fragment ion spectra are collected by pulsing the products of collisions of parent ions into the TOF stage of the analyzer. Doubly protonated precursor ions are generally selected for collision, and the mass ranges are set to scan from m/z of 100 to twice the m/z of the precursor ions to detect fragments, and pulse rates into the TOF stage are set at default values for each mass range. Data are collected until the intensities of low mass fragments are satisfactory.

Masses of fragment ions expected for specific tryptic peptides of the HIF fusion proteins are calculated by simulation of the fragmentation process[27,28] using Applied Biosystems BioAnalyst software. Fragment ion spectra of peptides that cannot be reconciled with the HIF fusion proteins are used to search the NCBI database using the Mascot search engine (www.matrixscience.com) with mass error constraints of 0.1 and 0.05 Da for parent and fragment ions, respectively, and allowing for carboxamidomethyl derivatization of cysteinyl residues.

Demonstration of Asparagine-Specific Hydroxylation of HIF CAD by FIH-1

Escherichia coli expressed HIF-2α 774–874 fusion proteins that had been exposed to *E. coli* expressed FIH-1 were purified and subsequently subjected to reduction, alkylation, and tryptic digestion in the same manner as described earlier for fusion proteins isolated from transfected cells. The tryptic digests were also analyzed by MALDI-TOF-MS and Qq-TOF-MS as described for the cell culture-derived fusion proteins. Data obtained in this way confirmed that FIH-1 was able to hydroxylate asparagine 851 of HIF-2α (Figs. 4A and 5A). The finding that this activity was inhibited by the 2-oxoglutarate-dependent dioxygenase enzyme inhibitor dimethyl-oxalylglycine (DMOG), using the same mass spectrometric

[27] R. S. Johnson, S. A. Martin, and K. Biemann, *Int. J. Mass Spectrom. Ion Proc.* **86,** 137 (1988).
[28] P. Roepstorff and J. Fohlman, *Biomed. Mass Spectrom.* **11,** 601 (1984).

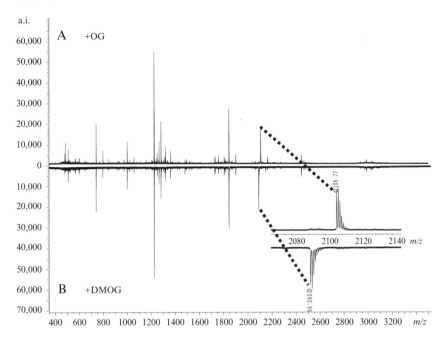

Fig. 4. MALDI-TOF-MS spectra demonstrating a hydroxylation substrate site between residues 846 and 864 of HIF-2α for hydroxylation by recombinant FIH-1. HIF-2α 774–874 was exposed to MBP-FIH-1 in the presence of (A) the cofactor 2-oxoglutarate (OG) or (B) the inhibitor DMOG, and tryptic digests of the products were analyzed by MALDI-TOF-MS. Reproduced with permission from Lando et al.[4]

procedures (Figs. 4B and 5B), was evidence that FIH-1 is an authentic asparagine-specific dioxygenase.

None of the mass spectrometric methods used provided any insight into the position of hydroxylation of the asparagine side chain. Thus, high-energy tandem mass spectrometric techniques, such as postsource decay (PSD) analysis using MALDI-TOF-MS (Fig. 6.) instrumentation and high-energy collision-induced decomposition using a MALDI-tandem-TOF (or TOF/TOF) mass spectrometer (data not shown), were used in an attempt to obtain fragment ion information that would differentiate between aspartohydroxamic acid formation and β-hydroxylation of asparagine 851. Both of these methods produced fragments of the hydroxylated tryptic peptide that confirmed asparagine 851 as the site of modification, but neither method was able to distinguish between the two potential positions of modification of the side chain. This may have been resolved if the fragmentation properties had been determined for a synthetic peptide with the side chain of asparagine 851 in an aspartohydroxamic acid form.

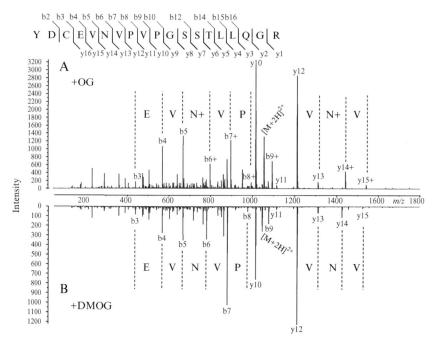

FIG. 5. Demonstration that FIH-1 hydroxylates asparagine 851 of HIF-2α by nano-ESI-Qq-TOF-MS/MS. MS/MS data were acquired on doubly charged peptide ions representing residues 846–864 of HIF-2α from HIF-2α 774–874 samples exposed to MBP-FIH-1 in the presence of (A) OG or (B) DMOG. Details of data acquisition and labeling are as described in the legend to Fig. 3. Reproduced with permission from Lando et al.[4]

Experimental

For *in vitro* hydroxylation experiments, both the HIF-CAD substrate and the FIH-1 asparaginyl hydroxylase enzyme are expressed as fusion proteins in *E. coli* and purified. The mouse HIF-2α 774–874 is generated by polymerase chain reaction (PCR) and cloned into pET-32a (Novagen, WI) for bacterial expression with an NH_2-terminal thioredoxin-6 histidine tag (Trx6His).[4] BL21(DE3) *E. coli* cells transformed with Trx6His HIF-2α 774–874 are induced with 1 mM isopropylthio-β-D-galactoside (IPTG) for 1.5 h at 37°. Cell pellets are lysed by sonication in binding buffer (20 mM Tris–HCl, pH 7.5, 500 mM NaCl, 5 mM imidazole, 40 ml/liter culture) containing 1 mM PMSF and 0.5 mg/ml lysozyme. After centrifugation, lysates are incubated with Ni-IDA agarose (Scientifix, Australia) for 1 h at 4° (1 ml resin/40 ml supernatant), washed with 200 volumes of binding buffer, and the Trx6His HIF-2α 774–874 protein is eluted with binding buffer containing 250 mM imidazole.

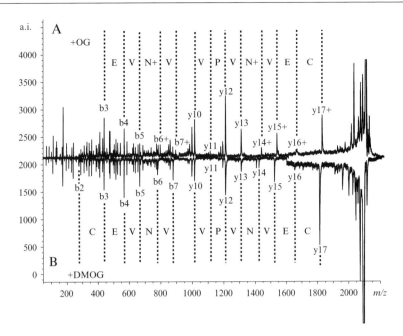

FIG. 6. Demonstration that FIH-1 hydroxylates asparagine 851 of HIF-2α by MALDI-TOF-PSD-MS/MS. Data were produced from singly protonated peptide ions (Fig. 4) representing hydroxylated (A) and unmodified (B) residues 846–864 of HIF-2α produced by exposure of HIF-2α 774–874 to MBP-FIH-1 in the presence of (A) OG or (B) DMOG. The labeling convention used is as for Fig. 3.

The full-length human FIH-1 cDNA is cloned into the pMBP parallel vector,[4] transformed into BL21(DE3) *E. coli* cells, and induced with 0.2 mM IPTG for 4.5 h at 30°. After centrifugation the cell pellets are resuspended in TH buffer (20 mM Tris–HCl, pH 7.9, 150 mM NaCl, 1 mM PMSF, 40 ml/liter of culture) containing 0.5 mg/ml lysozyme, incubated for 30 min on ice, and sonicated. After centrifugation, amylose agarose (Scientifix, Australia) is added to the supernatant (1 ml resin/40 ml supernatant), incubated for 1 h at 4°, washed with 200 volumes of TH buffer, and the MBP-FIH-1 protein is eluted with TH buffer containing 10 mM maltose. Protein samples are buffer exchanged into 20 mM Tris–HCl, pH 8, 150 mM NaCl using a PD-10 column (Amersham Biosciences, Australia). The Trx6His HIF-2α 774–874 is stable at 4° at up to 300 μM, above which the protein precipitates, whereas no problems with solubility or stability are observed with MBP-FIH-1.

In vitro hydroxylation reactions of varying scale are performed with hydroxylation of all the Trx6His-HIF-2α 774–874 substrate essentially complete within 60 min. In these reactions, FIH-1 and all of the other substrates and cofactors are in excess, including OG and Fe^{2+}. A standard reaction uses 10 μg of purified Trx6His-HIF-2α 774–874 and 5 μg of MBP-FIH-1 mixed in 100 μl of hydroxylation buffer (40 mM Tris–HCl, pH 7.5, 10 mM KCl, 1 mM DTT, 1 mM PMSF, 3 mM MgCl$_2$, 4 mM ascorbic acid, 1.5 mM FeSO$_4$) containing 4 μM OG or DMOG, an inhibitor of 2-oxoglutarate-dependent hydroxylases.[2] After a 1-h incubation at 30°, Trx6H HIF-2α 774–874 is affinity purified by binding to Ni-IDA agarose in 20 ml TH buffer. The resin is washed with 200 volumes of TH buffer and then with 200 volumes of 20 mM NH$_4$OAc, pH 8. Bound proteins are eluted with 20 mM NH$_4$OAc, pH 8, containing 250 mM imidazole, and the imidazole is removed by two passes through a PD-10 column equilibrated in 20 mM NH$_4$OAc, pH 8. The purified protein samples are then lyophilized under vacuum.

Sample preparation for MALDI-PSD-MS and MALDI-TOF/TOF-MS are the same as for MALDI-TOF-MS of tryptic digests as described earlier except α-cyano-4-hydroxycinnamic acid is used exclusively as the matrix.

MALDI-PSD analysis is a method of amino acid sequencing by analyzing fragments of peptides that arise due to energy imparted during the MALDI process.[29–32] Metastable fragment ions that form from a particular parent peptide ion, in the field-free region of the TOF analyzer, travel to the reflectron at the same velocity as their parent peptide ion. Thus it is possible to use an electrostatic ion gate in the field-free region before the reflectron to select the parent ions and related fragments with reasonable resolution so that they can be analyzed independently of other parent/daughter combinations. Once the combined parent/daughter repertoires approach the reflectron, only the parents have sufficient energy to be reflected onto the detector. It is necessary to reduce the potential of the reflectron to be able to observe the lighter and less energetic daughters. In the Bruker reflex, it is necessary to use 14 sequentially lowered reflectron potentials to record an entire fragmentation spectrum. The PSD spectrum recorded for hydroxylated residues 846–864 of *E. coli* HIF-2α was acquired with a constant acceleration potential of 28.5 kV and the highest reflectron potential of 30 kV (Fig. 6A). The 14 individual spectra were calibrated using fragment ions of ACTH 18–39 of known energies[33] and

[29] J. J. Gorman, B. L. Ferguson, D. Speelman, and J. Mills, *Protein Sci.* **6**, 1308 (1997).
[30] R. Kaufmann, B. Spengler, and F. Lutzenkirchen, *Rapid Commun. Mass Spectrom.* **7**, 902 (1993).
[31] R. Kaufmann, D. Kirsch, and B. Spengler, *Int. J. Mass Spectrom. Ion Proc.* **131**, 355 (1994).
[32] B. Spengler, D. Kirsch, R. Kaufmann, and E. Jaeger, *Rapid Commun. Mass Spectrom.* **6**, 105 (1992).
[33] J. C. Rouse, W. Yu, and S. A. Martin, *J. Am. Soc. Mass Spectrom.* **6**, 822 (1995).

pasted into a single fragment ion spectrum using Bruker XMass software routines.

MALDI-TOF/TOF-MS uses a collision cell process, similar to product ion analysis with the Qq-TOF-MS described earlier, to generate fragment ions. Parents formed by MALDI in the first TOF stage are selectively bombarded in a collision cell between the two TOF stages, and the residual parent and fragments are accelerated with the same potential into the reflectron of the second TOF stage. Thus, the analytical process is much simpler than for multisegmented MALDI-PSD analysis. MALDI-TOF/ TOF-MS spectra of hydroxylated residues 846–864 of *E. coli* HIF-2α were recorded with an Applied Biosystems 4700 proteomics analyzer.

Demonstration of β-Hydroxylation of HIF CAD by FIH-1 Using Amino Acid Analysis

As discussed earlier, provided that it has been established that an asparagine in a sequence has been hydroxylated, amino acid protocols should be sufficient to characterize the chemical nature of asparagine hydroxylation and perhaps serve as routine tools for detection and assay of the modification. Classical ion-exchange amino acid separation with postcolumn derivatization was found to be too insensitive to detect *in vitro* hydroxylation of recombinant HIF-2α 774–874 by recombinant FIH-1, thus reversed-phase separation of precolumn-derivatized hydrolysates was investigated for this purpose.

Unhydrolyzed DL-*threo*-β-hydroxyaspartic acid eluted at 7.8 min under standard ion-exchange conditions used for amino acid analysis purposes (data not shown). A major peak eluted at the same time when DL-*threo*-β-hydroxyaspartic acid was pretreated by acid hydrolysis. A minor peak (12.6 % by area) was also evident in the hydrolyzed sample at an elution time of 12.6 min that was not seen in the unhydrolyzed sample. Both of these peaks eluted much earlier than any constituent of the standard mixture of common amino acids and neither was present in the reagent or hydrolysis blanks. These results are as expected from previous reports of the relative elution positions of *threo*-β-hydroxyaspartic acid[33] and *erythro*-β-hydroxyaspartic acid and the partial hydrolytic interconversion of isomers of β-hydroxyaspartic acid.[9–12]

The 6-aminoquinoyl carbamyl (AQC) derivative of DL-*threo*-β-hydroxyaspartic acid also eluted earlier than derivatives of the common amino acids during reversed-phase separation. A major peak of absorption at 254 nm and expected fluorescence properties were observed at approximately 8 min in the unhydrolyzed sample (data not shown). This peak eluted well before AQC aspartic acid and was not present in the common amino acid standard mixture or in hydrolysis or reagent blanks. Another peak was evident at approximately 14.6 min by UV absorption at 254 nm, but was essentially nonfluorescent at the emission wavelength employed for detection and was present in the derivatized hydrolysis blank, which indicated that it was the by-product of the derivatization reagent 6-aminoquinoline and not an AQC amino acid.[34] Chromatography of DL-*threo*-β-hydroxyaspartic acid that was derivatized after acid hydrolysis resulted in the appearance of the peak at approximately 8 min and a new peak at approximately 13.6 min (Figs. 7A and 7B), but the relative abundance of the additional peak, which may represent conversion to DL-*erythro*-β-hydroxyaspartic acid, was greater (29% by area) than the suspected interconversion peak observed during ion-exchange separation.

The advantage of the reversed-phase separation of the AQC derivative of β-hydroxyaspartic acid as an indicator of the presence of β-hydroxyaspartic acid or β-hydroxyasparagine in a protein is the relative sensitivity of this approach compared to the ion-exchange method. It was possible to detect approximately 100 fmol of the AQC derivative of DL-*threo*-β-hydroxyaspartic acid in the hydrolyzed sample using fluorescence detection. However, it was not possible to observe the putative interconversion product at this sensitivity with this method. The minimum quantity of DL-*threo*-β-hydroxyaspartic acid detectable with the ion-exchange method was approximately 100 pmol.

Samples of recombinant HIF-2α 774–874 were treated with recombinant MBP-FIH-1 in the presence of OG or DMOG and desalted, and separate portions of the reaction mixtures were subjected to gas-phase acid hydrolysis or tryptic mass mapping. Tryptic mass mapping of OG samples showed stoichiometric hydroxylation of residues 846–864 and no hydroxylation of the DMOG samples (data not shown). Hydrolysates were derivatized with 6-aminoquinoyl-*N*-hydroxysuccinimidyl carbamate and analyzed by reversed-phase HPLC. Chromatograms of OG-derived samples revealed a reagent peak (14.6 min) and AQC derivatives of the common amino acids as detected by absorbance at 254 nm (Fig. 7C). Fluorescence detection revealed relatively small peaks eluting earlier than the reagent peak and the common amino acids. Included in this early eluting category were peaks corresponding to those observed in hydrolyzed DL-*threo*-β-hydroxyaspartic acid (approximately 8 and 13.6 min) and other peaks at approximately 9.8 and 11.2 min (Fig. 7D). The DMOG-derived sample, which was not hydroxylated, only contained the peak at approximately 9.8 min (Fig. 7F). These findings indicate that HIF-2α 774–874

[34] S. A. Cohen and D. P. Michaud, *Anal. Biochem.* **211,** 279 (1993).

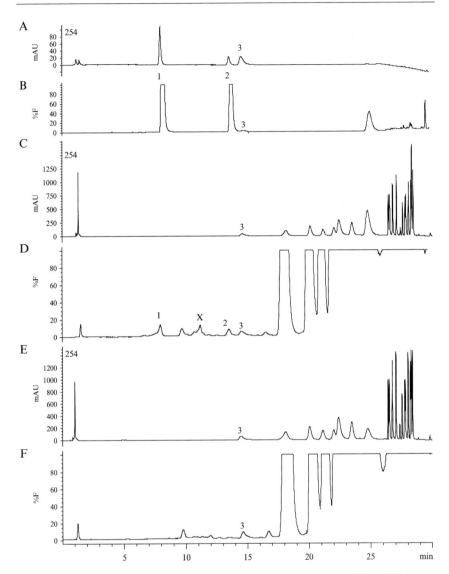

FIG. 7. Demonstration of asparagine β-hydroxylation of HIF-2α 774–874 by FIH-1 using AQC amino acid analysis. Chromatograms of AQC derivatives obtained by the derivatization of hydrolyzed DL-*threo*-β-hydroxyaspartic acid with components detected by (A) UV absorption at 254 nm and (B) fluorescence; hydrolyzed HIF-2α 774–874 after exposure to MBP-FIH-1 in the presence of OG with components detected by (C) UV absorption and (D) fluorescence; and hydrolyzed HIF-2α 774–874 after exposure to MBP-FIH-1 in the presence of DMOG with components detected by (E) UV absorption and (F) fluorescence. Early eluting components labeled 1, 2, 3, and X are peaks referred to in the text with elution times of approximately 8, 13.6, 14.6, and 11.2 min, respectively.

contained a modified amino acid that gave rise to β-hydroxyaspartic acid (peaks at approximately 8 and 13.6 min) upon hydrolysis. The relatively low intensities of these peaks are not surprising as only one residue of β-hydroxyasparagine was expected in the context of 100 amino acid residues of HIF-2α 774–874 plus amino acids of the thioredoxin fusion partner and further amino acids from MBP-FIH-1. The peak at approximately 11.2 min is surprising and may represent hydroxylation of another amino acid in the HIF-2α 774–874 fusion protein and/or MBP-FIH-1. These data may be consistent with NMR data that showed FIH-1 formed *threo*-β-hydroxyasparagine in a synthetic peptide based on the HIF-1α sequence containing a residue equivalent to asparagine 851.[20] If the AQC derivatives observed from HIF-2α 774–874 at approximately 8 and 13.6 min (Fig. 7D) represent *threo* and *erythro* isomers of β-hydroxyaspartic acid, respectively, it could be that the percentage hydrolytic interconversion to the alternate isomer increases with decreasing sample quantity. Alternatively, hydroxylation by FIH-1 may not be stereospecific in the context of a longer protein sequence or the peak at approximately 13.6 min may not represent the *erythro* isomer.

Experimental

A standard solution of 200 mM DL-*threo*-β-hydroxyaspartic acid (Sigma, H2775) is prepared in 1 mM HCl for separation by ion-exchange chromatography without hydrolysis. A standard solution of 2 mM DL-*threo*-β-hydroxyaspartic acid is prepared in deionized water and dried *in vacuuo* for subsequent vapor phase hydrolysis at 110° for 22 h. The hydrolyzed amino acid is dried *in vacuuo* and redried from deionized water prior to reconstitution for ion-exchange separation (as described earlier) or derivatization with 6-aminoquinoyl-N-hydroxysuccinimidyl carbamate.[34]

Unhydroyzed and hydrolyzed DL-*threo*-β-hydroxyaspartic acid samples are reconstituted in 20 μl of 20 mM HCl, diluted with 60 μl of 0.2 M borate buffer (pH 8.8), and mixed by vortexing for 30 s. A 20-μl aliquot of 10 mM 6-aminoquinoyl-N-hydroxysuccinimidyl carbamate in acetonitrile is added to each sample, and the samples are vortexed for 1 min.

Ion-exchange separation of amino acids is performed on a Waters Alliance chromatography system fitted with a Waters cation-exchange amino acid analysis column (Wat 080002) and employing a linear gradient from 0.2 M (pH 2.96) to 1.2 M (pH 6.5) sodium citrate buffer over 80 min min at 65° and ninhydrin detection protocols. Reversed-phase separation of 6-aminoquinoyl carbamate (AQC) derivatives is performed on a Hewlett-Packard 1090M chromatograph fitted with a 1090 diode array detector and a 1046A fluorescence detector. Fluorescence detection employs excitation and emission settings of 250 and 395 nm, respectively. Separation is performed at 1 ml/min and 35° using a 150 × 3.9-mm Waters 4-μm C18 column (AccQ Tag; Wat 052885) and minor variations to recommended protocols. Briefly, the separation involves a series of linear gradients from 100% (v/v) buffer A [140 mM NaOAc/17 mM triethylamine adjusted to pH 4.95 with HOAc] to 4% (v/v) buffer B [60% (v/v) aqueous acetonitrile] over 10 min and retention at 4% (v/v) buffer B for another 5 min; from 4 to 6% (v/v) buffer B over 3 min; to 7% (v/v) buffer B over 5 min; and to 100% (v/v) buffer B at 30 min.

Conclusions

MALDI-TOF-MS mass mapping and various tandem mass spectrometry techniques, especially nano-ESI on a Qq-TOF-MS, have proven to be very effective tools for the detection of asparagine hydroxylation of HIF and locating this modification to a particular sequence position. These technologies are particularly useful when only very small quantities of materials, such as from cell-based experiments, are available for analysis. In order to characterize hydroxylation of asparagine 851 of HIF-2α 774–874 in transfected cells, it was necessary to use an unconventional "one pot" sample workup in which reduction, alkylation, and tryptic digestion of HIF-2α 774–874 were performed without intermediated desalting. Complementary use of high-sensitivity precolumn derivatization amino acid analysis is an excellent adjunct to high-sensitivity mass spectrometry for identification of the presence of β-hydroxylation of asparagine and/or aspartic acid residues and assaying the enzymes that cause this modification. The process of differentiation of the stereospecific nature of β-hydroxylation may also be possible with high-sensitivity amino acid analysis. To confirm this, it will be necessary to determine the elution positions of the *threo* and *erythro* isomers separately in order to see if the apparent later-eluting hydrolytic product of DL-*threo*-β-hydroxyaspartic acid, observed in this study as an AQC derivative, represents conversion to the *erythro* isomer. This method may also provide further insight into the stereospecificity of the hydroxylation of asparagine 851 by FIH-1.

Acknowledgments

Performance of ion-exchange amino acid analysis by Nick Bartone of CSIRO Health Science and Nutrition, Parkville Victoria, Australia, and MALDI-TOF/TOF-MS by Melanie Lin of Applied Biosystems, Framingham, MA, is greatly appreciated.

[32] Deciphering the Oxygen Sensing Pathway by Microscopy

By HELMUT ACKER, CHRISTINE HUCKSTORF, HEINRICH SAUER,
TINO STRELLER, and MARIA WARTENBERG

Introduction

The heterogeneous oxygen partial pressure (PO_2) distribution in mammalian organs ranging from about 0 to 90 Torr under normoxic conditions with an arterial Po_2 of about 100 Torr requires an O_2 sensing signal cascade to adapt cellular functions to PO_2 heterogeneity. Under normoxic conditions, O_2 sensing leads to an optimizing of cell function as exemplified by enhanced phosphoenol pyruvate carboxykinase (PCK) expression in the periportal liver zone with high Po_2 levels and of enhanced glukokinase (GK) expression in the perivenous liver zone with low Po_2 levels.[1] Under arterial hypoxia tissue, PO_2 frequency distribution is left shifted, leading to a hypoxic response of O_2 sensing for adaptation of cell function. The response includes a drastic depression of general protein synthesis[2] but an enhanced expression of an array of proteins involved in the regulation of different functions such as erythropoietin (EPO) in red cell formation,[3] vascular endothelial growth factor (VEGF) in blood vessel formation,[4] lactate dehydrogenase (LDH) in energy metabolism,[5] or plasminogen activator inhibitor-1 (PAI-1) regulating fibrinolysis.[6] Furthermore, potassium channel gating is altered in carotid body (CB) type I cells to release various transmitter exciting synaptically connected nerve fibers for the nervous regulation of ventilation and blood circulation[7,8] in neuroepithelial bodies (NEB) to release serotonin controlling the bronchial muscular tone during hypoxia,[9–11] as well as in vascular smooth muscle cells leading to peripheral blood vessel dilatation[12] or to lung vessel vasoconstriction.[13]

[1] T. Kietzmann, Y. Cornesse, K. Brechtel, S. Modaressi, and K. Jungermann, *Biochem. J.* **354,** 531 (2001).

[2] C. Koumenis, C. Naczki, M. Koritzinsky *et al.*, *Mol. Cell. Biol.* **22**(21), 7405 (2002).

[3] B. L. Ebert and H. F. Bunn, *Blood* **94,** 1864 (1999).

[4] H. F. Bunn and R. O. Poyton, *Physiol. Rev.* **76,** 839 (1996).

[5] B. L. Ebert, J. M. Gleadle, J. F.O'Rourke, S. M. Bartlett, J. Poulton, and P. J. Ratcliffe, *Biochem. J.* **313,** 809 (1996).

[6] A. Görlach, U. Berchner-Pfannschmidt, C. Wotzlaw, R. H. Cool, J. Fandrey, H. Acker, K. Jungermann, T. Kietzmann, *Thrombosis Haemostasis,* **89,** 926–935 (2003).

[7] J. Lopez-Barneo, R. Pardal, and P. Ortega-Saenz, *Annu. Rev. Physiol.* **63,** 259 (2001).

[8] S. Lahiri, *J. Appl. Physiol.* **88,** 1467 (2000).

It is not clear whether these different responses are induced by various O_2 sensing signal cascades. One hypothesis supposes a cascade consisting of mitochondrial or nonmitochondrial heme proteins, respectively, sensing oxygen with a subsequent second messenger formation, for example, reactive oxygen species (ROS), which influence via an iron-mediated Fenton reaction the stability of hypoxia-inducible transcription factor HIF-1α for altering its binding capacity to hypoxic responsive elements of the GK, EPO, VEGF, LDH, and PAI-1 gene or change the thiol status of ion channel proteins.[4,6,14–16] Among the nonmitochondrial heme proteins, isoforms of the neutrophil NADPH oxidase are shown as oxygen sensors for carotid body,[17,18] smooth musculature of the lung,[19] and neuroepithelial bodies.[9,20] Mitochondrial complexes II and III and an unusual cytochrome c oxidase are discussed as oxygen sensors for Hep3B liver tumor cells triggering EPO production,[21] for cardiomyocytes regulating hibernation during hypoxia,[22] and for smooth musculature of lung vessels,[23] as well as for carotid body tissue-enhancing nervous chemoreceptor discharge in dependence on PO_2.[24–27] Another hypothesis supposes an oxygen sensing

[9] X. W. Fu, D. Wang, C. Nurse, M. C. Dinauer, and E. Cutz, *Proc. Natl. Acad. Sci. USA* **97**(8), 4374 (2000).

[10] I. O'Kelly, A. Lewis, C. Peers, and P. J. Kemp, *J. Biol. Chem.* **275**, 7684 (2000).

[11] S. Skogvall, M. Korsgren, and W. Grampp, *J. Appl. Physiol.* **86**, 789 (1999).

[12] A. Franco-Obregon, J. Ureña, and J. Lopez-Barneo, *Proc. Natl. Acad. Sci. USA* **92**, 4715 (1995).

[13] S. L. Archer, H. L. Reeve, E. Michelakis *et al.*, *Proc. Natl. Acad. Sci. USA* **96**, 7944 (1999).

[14] F. Duprat, E. Guillemare, G. Romey *et al.*, *Proc. Natl. Acad. Sci. USA* **92**, 11796 (1995).

[15] M. Taglialatela, P. Castaldo, S. Iossa *et al.*, *Proc. Natl. Acad. Sci. USA* **94**, 11698 (1997).

[16] D. S. Wang, C. Youngson, V. Wong *et al.*, *Proc. Natl. Acad. Sci. USA* **93**, 13182 (1996).

[17] A. R. Cross, L. Henderson, O. T. Jones, M. A. Delpiano, J. Hentschel, and H. Acker, *Biochem. J.* **272**, 743 (1990).

[18] K. A. Sanders, K. M. Sundar, L. He, B. Dinger, S. Fidone, and J. R. Hoidal, *J. Appl. Physiol.* **93**, 1357 (2002).

[19] N. Weissmann, A. Tadic, J. Hanze *et al.*, *Am. J. Physiol. Lung Cell Mol. Physiol.* **279**, L683 (2000).

[20] A. J. Lipton, M. A. Johnson, T. Macdonald, M. W. Liebermann, D. Gozal, and B. Gaston, *Nature* **413**, 171 (2001).

[21] N. S. Chandel, E. Maltepe, E. Goldwasser, C. E. Mathieu, M. C. Simon, and P. T. Schumacker, *Proc. Natl. Acad. Sci. USA* **95**, 11715 (1998).

[22] G. R. S. Budinger, J. Duranteau, N. S. Chandel, and P. T. Schumacker, *J. Biol. Chem.* **273**, 3320 (1998).

[23] E. D. Michelakis, V. Hampl, A. Nsair *et al.*, *Circ. Res.* **90**, 1307 (2002).

[24] B. E. Baysal, R. E. Ferrell, J. E. Willett-Brozick *et al.*, *Science* **287**, 848 (2000).

[25] E. Mills and F. F. Jöbsis, *J. Neurophysiol.* **35**, 405 (1972).

[26] D. F. Wilson, A. Mokashi, D. Chugh, S. Vinogradov, S. Osanai, and S. Lahiri, *FEBS Lett.* **351**, 370 (1994).

[27] T. Streller, C. Huckstorf, C. Pfeiffer, and H. Acker, *FASEB J.* **16**, 1277 (2002).

cascade comprising nonheme iron-binding proteins such as HIF-1 prolyl hy-droxylase (HIF-PH1,2,3,4) and (HIF-1) asparginyl hydroxylase (FIH-1).[28]

This article describes three optical methods used to identify, locate, and image several putative members of the oxygen sensing pathway and their reaction to hypoxia. The three optical methods are based on light absorp-tion photometry,[27] the optical probe technique,[29] and two-photon confocal laser microscopy (2P-CLSM).[30]

Heme Proteins in Living Tissue

The first step in oxygen sensing is mediated by molecules changing their chemical properties in direct dependence of the surrounding oxygen partial pressure (PO_2). These primary oxygen sensors trigger signal cascades and subsequent cellular reactions. As for the carotid body, hypoxia leads to an increase in afferent carotid sinus nerve activity (CSNA). The primary oxygen sensor triggering this increase is yet unknown, but there is large agreement that it is a heme protein, either a mitochondrial component of the respiratory chain[25,26] or a nonmitochondrial such as NADPH oxi-dase.[9,17] All these hemes are characterized by the fact that the absence of O_2 or binding of other ligands such as CO or CN^- may lead to chemical reduction of the heme iron, which results in changed light absorption prop-erties, whereas changes in the absorption spectrum of a heme can be related to its sensitivity to hypoxia or heme ligands.

Absorption Spectrometric Analysis of Carotid Body Tissue

Photometry

Evaluation of O_2-sensitive hemes within the isolated superfused CB is performed by recording CSNA and light absorption difference spectra sim-ultaneously. Tissue from CBs and cervical and nodosal ganglia is obtained from male Wistar rats and cleaned from red cells. Tissue is placed in a superfusion chamber mounted on top of an opaque bench containing small holes of diameters similar to that of the organs. As for superfusion, an iso-tonic salt solution (in mM: NaCl 128, KCl 5.6, glucose 27.5, HEPES 7, $NaHCO_3$ 10, $CaCl_2$ 2.1) is equilibrated with different $O_2/N_2/CO_2$ mixtures to adjust oxygen tensions to various levels at pH about 7.4. Samples of

[28] D. Lando, D. J. Peet, J. J. Gorman, D. A. Whelan, M. L. Whitelaw, and R. K. Bruick, *Genes Dev.* **16,** 1466 (2002).

[29] M. Wartenberg and H. Acker, *Micron* **26,** 395 (1995).

[30] F. Bestvater, E. Spiess, G. Stobrawa *et al., J. Microsc.* **208,** 108 (2002).

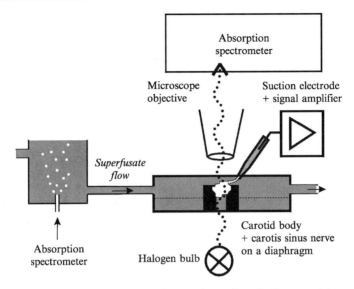

FIG. 1. *In vitro* setup scheme for simultaneous recording of afferent activity and tissue absorption spectra of the rat carotid body.

superfusion medium are taken after changing equilibration for measuring PO_2, PCO_2, and pH by means of a blood gas analyzer (AVL 990, Roche, Mannheim, FRG). Saline flows through the chamber at 60 ml/min. Temperature is maintained at 36°. The system is mounted on the stage of a light microscope (Olympus, Hamburg, FRG) for light absorption measurements.[31] White light from a halogen bulb (12 V, 100 W), passing through an objective (40×), transilluminates the tissue every minute for 10 s. Transillumination-induced absorption changes at different wavelengths are recorded by a photodiode-array spectrophotometer (MCS 210, Zeiss, Köln, FRG) connected to the ocular of a microscope via a light guide (Fig. 1). Absorption difference spectra (ADS) are obtained by the spectral difference between an excitation state and the preceding initial state. The initial state is defined by equilibration of the superfusion medium with 28.8% O_2, 4% CO_2, and 67.2% N_2 (aerobic steady state), whereas the state of excitation is obtained by equilibrating the superfusate with different concentrations of N_2, O_2, CO, or cyanide at constant CO_2 (4%). ADS are evaluated, deconvoluted, and visualized using the software package Tech Plot (Dr. Dittrich, Braunschweig, FRG).[31]

[31] S. Lahiri, W. Ehleben, and H. Acker, *Proc. Natl. Acad. Sci. USA* **96,** 9427 (1999).

Spectral Deconvolution

Hypoxic or ligand stimulation of tissue yields ADS with peaks within the visual light range of 510–630 nm, which usually are superpositions of absorption spectra from a set of basis cytochromes: mitochondrial cytochromes c_{550}, cytochrome b_{563}, cytochrome c oxidase (CCO), cytochrome $a3_{603}$, and the nonmitochondrial cytochrome b_{558} of the NADPH oxidase.[31,32] These components show spectral absorption peaks at 550, 563, 603, and 558 nm, respectively. Their single spectra were obtained from the literature for cytochromes c_{550}, b_{563}, and b_{558}[31] and, for cytochrome $a3_{603}$, from ADS at a high level of CN^- (500 μM).[33] The four cytochromes are sufficient to describe ADS of cervical or nodosal ganglia, but not those of CB. Consequently, the missing component for CB was calculated by the subtraction of ADS of hypoxic CB from ADS of CB superfused with CN^- (500 μM). It was found to have an absorption maximum at 592 nm, thus labeled as cytochrome a_{592}.

Spectral deconvolution describes an ADS as superposition of these five single spectra (Fig. 2). In detail, they were normalized (peak maximum = 1.0) and then varied in size until their sum at each wavelength corresponded as closely as possible to the spectrum. Due to this variation, an absolute weight factor could be assigned to each cytochrome, providing its contribution to the whole spectrum. Given as a percentage of their sum, they become relative weights (RSW), which are independent of the spectrum size, and thus suitable for comparing different recorded spectra.

Characterization of Cytochrome a_{592}

CSNA and ADS may be used for evaluating the role of the five cytochromes during stimulation. CSNA increases during stimulation by hypoxia, CN, or CO, but may be deconvoluted into an early phasic and subsequent tonic component, which may be quantified by phasic and tonic weights, respectively. Analysis of ADS revealed that the RSW remained stable during the tonic phase of CSNA, which allows averaging the RWS of each cytochrome during this period. The results then become exclusively dependent from the stimulus. From here, we could show that at decreasing PO_2 (100, 80, 60, and 0 mm Hg) the RWS of cytochrome a_{592} decreases significantly (from 15% at $PO_2 = 100$ mm Hg to 3% at $PO_2 = 0$ mm Hg), while the RWS of the other cytochromes remain nearly constant. The same

[32] W. Ehleben, B. Bolling, E. Merten, T. Porwol, A. R. Strohmaier, and H. Acker, *Respir. Physiol.* **114**, 25 (1998).
[33] G. L. Liao and G. Palmer, *Biochim. Biophys. Acta Mol. Cell Res.* **1274**, 109 (1996).

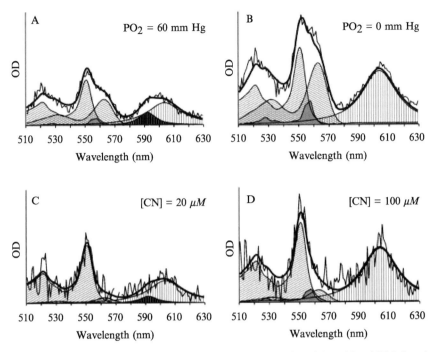

FIG. 2. Examples of ADS deconvolution: Hypoxia (A and B) and CN^- (C and D) induced light ADS of the rat CB are shown as original tracings (noisy curve) and fitted (thick curve) by five different cytochromes: cytochrome c_{550} (filled with diagonal upward lines), cytochrome b_{558} (gray shaded), cytochrome b_{563} (filled with diagonal downward lines), cytochrome a_{592} (dark shaded), and cytochrome a_{603} (filled with vertical lines). Spectra under mild hypoxia and low cyanide require a significant spectral weight of cytochrome a_{592} for fitting.

holds for increasing CN^- (10, 20, 30, 40, 50, and 100 μM). Additionally, the tonic weight of CSNA decreases with increasing stimulation intensity, both at hypoxia and at CN^-. Because both stimuli lead to an increased degree of reduction of the cytochrome iron center, a stable RSW means that the cytochromes are reduced in proportion to the stimulus intensity, whereas a decreasing RSW means that the cytochrome may not change its redox state further, as it is markedly reduced already at the initial state. Consequently, cytochrome a_{592} is characterized by unusual low PO_2 and high CN^- affinity, thus obviously playing an important role in the triggering of CSNA.[27] Cytochrome a of CCO usually peaks at 603 nm instead of at 592 nm as shown here. This shift is described as a result of a single site mutation of arginine 45 hydrogen bonded to the formyl group of heme a to methionine in *Paracoccus denitrificans*.[34] As a yield, a drastically lowered midpoint potential of heme a was measured and, correspondingly,

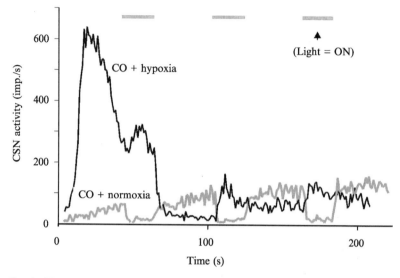

Fig. 3. CSNA signal during CB stimulation with CO + normoxia(PCO = 500 mm Hg, PO_2 = 140 mm Hg, gray line) and with combined stimulation CO + hypoxia (PCO = 500 mm Hg, PO_2 = 60 mm Hg, black line). The projection shows the inverting effect of hypoxia to CSNA due to inhibited CO photodissociation.

a dominance of direct Cu_A–heme a3 electron transfer over the normal electron pathway Cu_A–heme a–heme a3.

During stimulation with CO under normoxia (PCO = 500 mm Hg, PO_2 = 140 mm Hg), we observed the phenomenon of photodissociation,[35] characterized by the inhibition of CSNA during tissue white light transillumination, and by ADS without reduction peaks. Remarkably, this effect is inversed during combined stimulation by CO (PCO = 500 mm Hg) + hypoxia (PO_2 = 60 mm Hg) or CO (PCO = 500 mm Hg) + CN^- (100 μM) (Fig. 3). In this case, ADS revealed a lowered RSW of a_{592} compared to stimulation without CO at the same PO_2. This suggests that this cytochrome should be part of the mitochondrial CCO as an electron transporting element. A second hint for cytochrome a_{592} to be a mitochondrial CCO element was that the CCO blockers CO and CN^- lead to a tonic CSNA increase. In summary, the results characterize in CB tissue a mitochondrial CCO cytochrome a_{592} with low PO_2 affinity, an essential constraint for the triggering of O_2-sensitive CSNA.[36]

[34] A. Kannt, U. Pfitzner, M. Ruitenberg et al., J. Biol. Chem. 274, 37974 (1999).
[35] S. Lahiri and H. Acker, Respir. Physiol. 115, 169 (1999).
[36] P. Nair, D. G. Buerk, and W. Whalen, Am. J. Physiol. 250, H202 (1986).

Characterization of Cytochrome b_{558}

Figure 2 shows a dominant peak of cytochrome b_{558} as part of an NADPH oxidase identified by immunohistochemistry in CB type I cells and macrophages.[37] It was postulated that type I cell NADPH oxidase regulates oxygen sensor ion channel conductivity and gene expression.[17] Different NADPH oxidase isoforms with components p22phox, gp91phox, p47phox, p40phox, p67phox, and Rac1,2 have to be considered. The isoforms (Nox1–4 and the Duox group) concern especially gp91phox as reviewed by Lambeth et al.[38] The neutrophil NADPH oxidase (Nox2) has also been identified in endothelial cells[39] and neuroepithelial bodies.[9] gp91phox knockout mice showed an impaired hypoxic ventilatory control in neonatal animals due to decreased oxygen sensitivity of NEB potassium channel conductivity[9,40] evidencing oxygen sensor function of the NADPH oxidase in this cell type and related strains.[10,41] gp91phox knockout mice showed no impaired oxygen sensing function of pulmonary vasculature smooth muscle cells[13] or carotid body hypoxic drive.[42] However, p47phox knockout mice showed an enhanced carotid body hypoxic drive, assuming a particular Nox isoform for the carotid body.[18]

Oxygen sensing in CB tissue might be hypothesized therefore to be based on an interaction between cytochrome a_{592} as a cytosolic and NADPH oxidase as a plasma membrane oxygen sensor.

Tissue HIF-1α Levels as Visualized by the Optical Probe Technique

Three-dimensional avascular tissues, for example, avascular regions of solid tumors and avascular multicellular tumor spheroids, are characterized by steep oxygen gradients from the tissue periphery toward the center. In contrast, the process of vascularization during embryogenesis is accompanied by the dissipation of oxygen gradients due to development of the cardiovascular system, which is the first organ system to be formed in the early life of the embryo. Estimation of oxygen gradients in solid tumors as well as multicellular tumor spheroids has been performed previously

[37] M. Dvorakova, B. Höhler, R. Vollerthun, T. Fischbach, and W. Kummer, Brain Res. 852, 349 (2000).
[38] J. D. Lambeth, G. Cheng, R. S. Arnold, and W. A. Edens, Trends Biochem. Sci. 25, 459 (2000).
[39] A. Görlach, R. P. Brandes, K. Nguyen, M. Amidi, F. Dehghani, and R. Busse, Circ. Res. 87, 26 (2000).
[40] P. Kazemian, R. Stephenson, H. Yeger, and E. Cutz, Respir. Physiol. 126, 89 (2001).
[41] I. O'Kelly, C. Peers, and P. J. Kemp, Biochem. Biophys. Res. Commun. 283, 1131 (2001).
[42] A. Roy, C. Rozanov, A. Mokashi et al., Brain Res. 872, 188 (2000).

by oxygen-sensitive microelectrodes, which allow determination of the pericellular oxygen pressure.[43,44] To determine oxygen-dependent gene expression in dependence on HIF-1α, the optical probe technique (OPT) has been developed by Wartenberg *et al.*[29,45] This noninvasive method, based on confocal laser-scanning microcopy, was initially designed to assess the diffusion properties of cell membrane–permeant dyes in avascular as well as vascularized tissues and to investigate the transport properties of the multidrug resistance transporter P-glycoprotein. The expression of HIF-1α has been evaluated by means of the OPT during the vascularization of embryoid bodies, which is a three-dimensional tissue cultivated from embryonic stem cells, and closely mimics early postimplantation stages of embryogenesis.[46] It became apparent that the vascularization of embryoid bodies results in a dissipation of oxygen gradients, which is paralleled by the downregulation of HIF-1α (see Fig. 4). This is different from observations in avascular multicellular tumor spheroids of the same size where a sustained upregulation of HIF-1α was observed in the central core region of the spheroidal tissue.[47]

The assessment of protein expression by confocal laser-scanning microscopy in thick specimens is hampered by light attenuation and light scattering in the depth of the tissue, which can be described by a monoexponential decay law. The extinction coefficient is a function of the fluorochrome density and the quantum yield of the particular fluorescence dye.

Correction algorithms for high-magnification objectives with high-numerical apertures (NA) have been developed previously by Visser *et al.*[48] and Roerdink and Bakker.[49] With the use of objectives with high NA, the angle Θ of the converging laser beam with the z axis, which is dependent on the laser light intensity, as well as the length of the light path, has to be taken into account. The width of the light bundle on the specimen is determined by half of the aperture angle ω. The maximum value for Θ corresponds to half of the aperture angle ω. The maximum intensity $I°$ of the light beam hitting the specimen can then be determined by

[43] J. Carlsson and H. Acker, *Int. J. Cancer* **42,** 715 (1988).

[44] W. Mueller-Klieser, *Crit. Rev. Oncol. Hematol.* **36,** 123 (2000).

[45] M. Wartenberg, J. Hescheler, H. Acker, H. Diedershagen, and H. Sauer, *Cytometry* **31,** 137 (1998).

[46] M. Wartenberg, F. Donmez, F. C. Ling, H. Acker, J. Hescheler, and H. Sauer, *FASEB J.* **15,** 995 (2001).

[47] M. Wartenberg, F. C. Ling, M. Muschen *et al.*, *FASEB J.* **7,** 503 (2003).

[48] T. D. Visser, F. C. A. Groen, and G. J. Brakenhoff, *J. Microsc.* **163,** 189 (1991).

[49] J. B. T. M. Roerdink and M. Bakker, *J. Microsc.* **169,** 14 (1992).

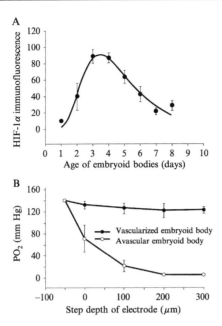

FIG. 4. Transient upregulation of HIF-1α during the growth of ES cell-derived embryoid bodies (A). With increasing growth and age of embryoid bodies, hypoxia occurs in central parts of the tissue. Between days 4 and 5 of cell culture, vasculogenesis and angiogenesis are initiated, which results in complete vascularization of the tissue within 8 days and parallel downregulation of HIF-1α. Vascularization on day 8 of embryoid body culture is followed by dissipation of the steep oxygen gradients measured by oxygen-sensitive microelectrodes in the tissue before vascularization on day 3 of cell culture (B).

$$I^\circ = \int_0^\omega d\Theta \, \sin\Theta \int_0^{2\pi} d\Phi \, I R^2 \cos\Theta \qquad (1)$$

where Φ represents the polar angle of the light beam, I is the intensity of the light reaching the specimen surface orthogonal to the optical axis, and R is the distance between the focus and the spheroidal border of the light cone. The total excitation energy per time within the focus (x, y, z) is then

$$I(x, y, n) = \int_0^\omega d\Theta \int_0^{2\pi} d\Phi \, I^\circ R^2 \cos\Theta \sin\Theta \, \Pi_{l=1}^{n-1} \exp[-C(i, \Theta, \Phi)\Delta z/\cos\Theta]$$
$$(n = 2, \ldots, i_{\max}) \qquad (2)$$

The factor $\Delta z / \cos\Theta$ represents the length of the light path, which forms an angle Θ with the optical axis and penetrates all layers n of the specimen.

$C(i, \Theta, \Phi)$ represents the light attenuation coefficient in this point. Δz represents the thickness of the optical section.

In contrast to Roerdink and Bakker,[49] who used objectives with high NA, we applied low-magnification objectives with low NA. This results in larger working distances and allows one to determine fluorescence information within depths up to 300 μm within the tissue. For objectives with low NA, Eq. (2) can be simplified as follows: $\cos\Theta$ approaches 1, i.e., $\Delta z/\cos\Theta$, approaches Δz. On that condition, the light attenuation coefficient can be considered independent of angles Θ, Φ. The equation then simplifies according to

$$I(x, y, n) = I^{\circ} \exp[-C\Delta z] \tag{3}$$

The light attenuation coefficent C is a function of the fluorochrome concentration within the sample. For the determination of C, calibration experiments with homogeneous stained buffer solutions at a defined fluorochrome concentration (e.g., solutions of fluorochrome-labeled secondary antibody) need to be performed.[29,46] Within stained buffer solutions the logarithm of the measured fluorescence intensity in relation to the penetration depth of the laser beam can be pictured as a linear function. To achieve correction for light absorption and scattering within the tissue, the C value has to be determined experimentally.

Experimental Methods

Materials

Cell Culture of Multicellular Tumor Spheroids and ES Cells

1. Maintenance of mouse ES cells
   Complete growth medium ($1\times$ stored at 4°)
   To 760 ml Iscove's modified Dulbecco's medium (Gibco BRL, Karlsruhe, Germany) add
   10 ml Glutamax I ($100\times$) (Gibco, BRL)
   10 ml penicillin/streptomycin solution (Gibco, BRL)
   10 ml MEM nonessential amino acid solution ($100\times$) (Gibco, BRL)
   100 μM 2-mercaptoethanol (Sigma, Deisenhofen, Germany)
   200 ml fetal calf serum (FCS) (Gibco, BRL)
   Phosphate-buffered saline (PBS) Ca^{2+} and Mg^{2+} free: 137 mM NaCl, 2.7 mM KCl, 10.1 mM Na_2HPO_4, 2 mM KH_2PO_4, pH 7.4, filter sterilized
   Trypsin/EDTA in PBS (Gibco, BRL)

2. Cell culture of multicellular tumor spheroids of the DU-145 cell line. To 860 ml Ham's F10 medium (Biochrom, Berlin, Germany) add

10 ml Glutamax I (100×) (Gibco, BRL)
10 ml penicillin/streptomycin solution (Gibco, BRL)
10 ml MEM nonessential amino acid solution (100×) (Gibco, BRL)
100 μM 2-mercaptoethanol (Sigma, Deisenhofen, Germany)
100 ml FCS (Gibco, BRL)

3. Spinner flasks

Cellspin stirrer system (Integra Biosciences Fernwald, Germany) equipped with 250-ml spinner flasks
Sigmacote solution (Sigma) stored at 4°
5 N NaOH solution

Procedure

Preparation of Spinner Flasks

1. Wash clean spinner flasks with excessive Milli Q-plus water and dry for 1 h at 60°. Siliconize spinner flasks by moistening the interior as well as the mallets with Sigmacote. Excessive Sigmacote is removed from the flasks using a 10-ml glass pipette. Dry the silicon coat in an oven for 1 h at 140°. Rinse spinner flasks three times with 250 ml Milli Q-plus water and autoclave subsequently. Moisten the interior of the flasks with 20 ml of either complete Iscove's medium or Ham's F10 medium prior to the addition of either ES cells or tumor cells; exchange the medium for 125 ml of either complete Iscove's medium or Ham's F10 medium.

2. After the end of the experiment the spinner flasks have to be cleaned prior to inoculation with fresh ES cells and tumor cells. Remove the old medium with residual embryoid bodies and multicellular tumor spheroids. Wash the flasks with 70% ethanol and subsequently with 1 liter of water. Remove the silicon coat by adding 250 ml 5 N NaOH to the spinner flasks for a maximum of 12 h. Remove NaOH and wash the flasks with at least 5 liters of water. The interior of the flasks is cleaned thoroughly with a brush. Subsequently rinse the flasks with 1 liter of Milli Q-plus water.

3. Wash either ES cells grown in 6-cm cell culture petri dishes or DU-145 prostate tumor cells grown in 25-cm^2 cell culture flasks once with 0.2% trypsin and 0.05% EDTA in PBS. Remove the trypsin solution and incubate cells for 5 min with 2 ml 0.2% trypsin and 0.05% EDTA. Triturate the cells with a 2-ml glass pipette until the cell clusters are dissociated and a single cell suspension is achieved (control under a microscope). Prepare spinner flasks (as described earlier) and seed cells at

a density of 1×10^7 cells/ml in either 125 ml complete Iscove's cell culture medium (for ES cells) or complete Ham's F10 medium (for tumor cells). Stir at a speed of 20 rotations per minute. The stirring direction is reversed every four full rotations. Add 125 ml Iscove's complete cell culture medium after 24 h to yield a final volume of 250 ml. Exchange 125 ml of the cell culture medium every day.

Immunohistochemistry

Antibody staining is performed either on whole mount embryoid bodies or multicellular tumor spheroids. The monoclonal anti-HIF-1α (clone mgc3) (Alexis Biochemicals, Grünberg, Germany) is used at a dilution of 1:200. Spheroids and embryoid bodies are washed in PBS and fixed in 4% paraformaldehyde in PBS at 4° for 1 h. Subsequently they are permeabilized in PBS supplemented with 1% (v/v) Triton X-100 (Sigma). Prior to immunostaining, spheroids are incubated for 1 h in 0.01% PBST containing 10% fat-free milk powder to reduce nonspecific binding. Afterward the primary antibody is applied for 2 h at room temperature. After washing three times in 0.01% PBST, tumor spheroids and embryoid bodies are incubated for 60 min in 0.01% PBST supplemented with 10% milk powder and a Cy5 -conjugated F(ab')$_2$ fragment goat antirabbit IgG (H + L) (Boehringer-Mannheim) (concentration 4.6 μg/ml). Excitation is performed using a 633-nm helium-neon laser of the confocal setup. Emission is recorded using a LP655-nm filter set.

Performance of OPT

Based on the features of confocal laser-scanning microscopy to record series of optical sections of discrete depth and defined distance within biological specimen, small regions of interest (ROIs) (600 μm^2, 40×40 pixel) are selected on the surface of the antibody-labeled specimen under investigation and are scanned in the z direction with z scan motor settings of 10 μm (see Fig. 5). Subsequently the mean fluorescence intensity (gray level values) in each optical section along the vertical axis of the tissue, that is, either from the tissue surface toward the center or from the center toward the tissue surface, is determined. Using this methods, a function of fluorescence intensity in relation to the penetration depth of the laser beam within the tissue is achieved. The pinhole settings of the confocal microscope are adjusted to yield a full width half-maximum of the fluorescence intensity (FWHM) of 8 μm. This value is selected because the settings of the confocal pinhole should have a smaller value than the distance of the motor steps in the z direction to avoid overlap

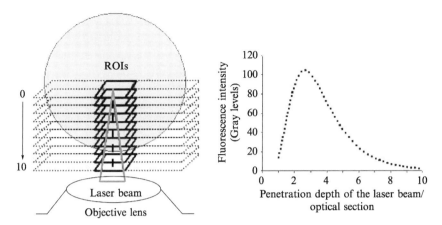

FIG. 5. Principle of the OPT. (Left) A scheme of the objective and confocal laser beam scanning 11 consecutive optical sections (dashed lines) starting from the center of a HIF-1α-stained spherical multicellular tissue and moving toward the periphery. Mean field fluorescence intensity is determined in regions of interest (ROIs) of about 600 μm^2 in the center of each optical section (traced squares). (Right) Tracings of immunofluorescence values plotted versus the penetration depth of the laser beam. The fluorescence decay observed with increased penetration depth of the laser beam in the tissue is due to light absorption and scattering and has to be corrected accordingly.

of the measured fluorescence signals. Each ROI is scanned once in 0.064 s. A whole z series of 16 ROIs is achieved in approximately 7 s, the convolution time of the stepper motor thereby being the rate-limiting step. Subsequently the mean fluorescence intensity I in each ROI is determined and plotted against the penetration depth z of the laser beam within the tissue.

For the correction of light attenuation and scattering, the fluorescence distribution curve derived from the specimen under investigation is contrasted with a calibration curve, which is determined in a fluorochrome solution with the same fluorochrome concentration (e.g., a solution of fluorochrome-labeled secondary antibodies) and the same confocal settings and performance as used for the specimen under investigation. In each measuring point, a quotient between the measured fluorescence intensity in the tissue (I_{meas}) and the fluorescence intensity determined in the fluorochrome solution (I_{cal}) at a distinct penetration depth of the laser beam is determined.

$$D(z) = I_{meas}/I_{cal} \tag{4}$$

This fluorescence distribution coefficient $D(z)$ changes in relation to the penetration depth of the laser beam with the distribution of the fluorochrome in the specimen under investigation. In a homogeneous-stained specimen, all values z equal 1. In specimens where the fluorescence decays toward the center, D approaches values between 0 and 1. Using light attenuation coefficient C and $D(z)$, a correction procedure of the determined fluorescence distribution within the tissue can be performed. Corrected fluorescence intensity values correlate to the measured fluorescence values I as follows:

$$\ln[I_{corr}] = \ln[I] + C \cdot D \cdot z \tag{5}$$

By means of OPT, HIF-1α distribution was assessed within a whole mount multicellular tumor spheroid, which was stained with an antibody directed against HIF-1α and a Cy5-labeled secondary antibody (Fig. 6). The Cy5 fluorochrome has its absorption maximum at red wavelengths allowing fluorescence detection in thick specimens. It is evident that uncorrected fluorescence data suggest an increase in HIF-1α immunofluorescence within the periphery cell layers of the tumor spheroid, which is followed by a fluorescence decay in the center (see Fig. 6A). After correction of fluorescence data for light absorption and scattering, a fluorescence increase of HIF-1α immunofluorescence is achieved that reaches plateau levels within central parts of the tumor spheroid tissue (see Fig. 6B). Fluorescence distribution data obtained with the described correction procedure yield excellent congruence with antibody stainings of cryosections (see Fig. 6C) and correspond to determinations of pericellular oxygen levels obtained with oxygen-sensitive microelectrodes.

Three-Dimensional Imaging of Intracellular Oxygen Sensing

The role of ROS as messengers within the oxygen sensing pathway has been supported by the finding that, similar to a typical response to hypoxia, treatment of healthy human volunteers with the antioxidant N-acetylcysteine (NAC) enhanced the hypoxic ventilatory response (HVR) and blood erythropoietin (EPO) concentration.[50] Thus, NAC or its biochemical derivates, cysteine and glutathione, mimic hypoxia, whereas hydrogen peroxide (H_2O_2) mimicked normoxia in the expression of a number of O_2-dependent genes in cell cultures. An H_2O_2-degrading perinuclear Fenton reaction ($H_2O_2 + Fe^{2+} \rightarrow Fe^{3+} + OH^- + OH\bullet$) has

---

[50] W. Hildebrandt, S. Alexander, P. Bartsch, and W. Droge, *Blood* **99,** 1552 (2002).

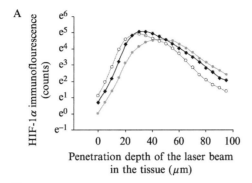

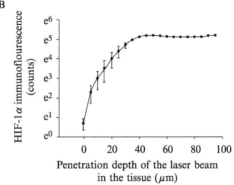

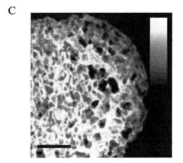

FIG. 6. Correction of HIF-1α immunofluorescence in whole mount multicellular tumor spheroids (A and B), and comparison with HIF-1α distribution in a representative cryosection (C). (A) HIF-1α immunofluorescence in whole mount multicellular tumor spheroids before correction. Immunofluorescence distribution curves obtained from three different tumor spheroids 300–350 μm in size suggest decreasing HIF-1α levels in central parts of the tumor tissue. (B) After mathematical correction, a sustained increase in HIF-1α expression in the center of the tumor spheroids is observed, which represents the true situation as evaluated by the immunostaining of cryosections with HIF-1α. (C) The bar in C represents 50 μm. The bar represents HIF-1α immunofluorescence intensity.

been demonstrated to be involved in O_2 signaling triggering gene expression.[51–53] Also, carotid body nervous discharge has been shown to depend on the Fenton reaction.[54] However, it remained open in which cellular compartment the Fenton reaction takes place and whether transcription factors regulating the O_2-dependent gene expression or ion channels triggering nervous activity are located in this compartment.

Two Photon Confocal Laser-Scanning Microscopy

Intracellular hydroxyl radical (OH•) generation by the Fenton reaction can be detected by irreversible conversion of the nonfluorescent dye dihydrorhodamine (DHR123). Thereby, OH• rearrange the π system of DHR123, yielding fluorescent rhodamine123 (RH123),[52] thus allowing localization of the Fenton reaction.

HepG2 cells transferred to a culture chamber on the microscope stage enabling observation at $37°$ under a variable gas atmosphere (see Fig. 7A) are treated with 30 μM DHR123 (Molecular Probes) for 5 min under normoxic or hypoxic conditions and are then washed without altering the PO_2 as controlled by a polarographic PO_2 catheter electrode (LICOX MCB). Human osteosarcoma cells (U2OS) cells are transfected with the enhanced green fluorescence protein (EGFP) versions of PHD1, PHD2, PHD3, and FIH-1.[55] RH123 and EGFP fluorescence is recorded by 2P-CLSM (experimental setup see Fig. 7B). The laser source consists of a Ti:sapphire laser (Coherent Mira 900-F) pumped by an Ar-ion laser. The output wavelength is tuneable between 720 and 910 nm, the repetition rate is 76 MHz, and the pulse duration is typically 110 fs. Pulses are coupled into a mono-mode optical fiber connected to the scan unit (PCM2000, Nikon, Germany) of a conventional fluorescence microscope (TE300, Nikon, Germany). Dispersion of the optical fiber is precompensated by a dispersion compensator consisting of a pair of gratings (GDC). For image acquisition, a $60\times$ water objective is used (Nikon, Plan Apochromat DIC H; N.A. 1.2), applying an excitation wavelength of 850 nm to excite RH123 fluorescence (emission 530 nm) or EGFP fluorescence (emission 520 nm). Fluorescence is

[51] T. Kietzmann, J. Fandrey, and H. Acker, *NIPS* **15**, 202 (2000).

[52] W. Ehleben, T. Porwol, J. Fandrey, W. Kummer, and H. Acker, *Kidney Int.* **51**, 483 (1997).

[53] T. Kietzmann, T. Porwol, K. Zierold, K. Jungermann, and H. Acker, *Biochem. J.* **335**, 425 (1998).

[54] P. A. Daudu, A. Roy, C. Rozanov, A. Mokashi, and S. Lahiri, *Respir. Physiol. Neurobiol.* **130**, 21 (2002).

[55] E. Metzen, U. Berchner-Pfannschmidt, P. Stengel, J. H. Marxsen, I. Stolze, M. Klinger, W. Q. Huang, C. Wotzlaw, T. Hellwig-Bürgel, W. Jelkmann, H. Acker, and J. Fandrey, *J. Cell Sci.* **116**, 1326 (2003).

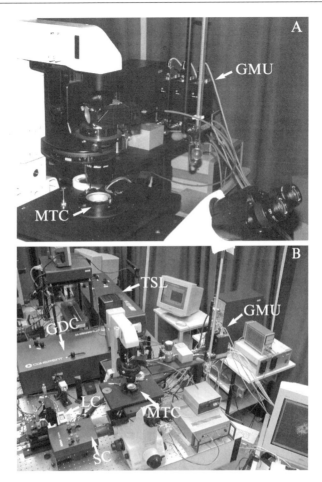

FIG. 7. 2-PCLSM setup for intracellular imaging: (A) microscope tissue chamber (MTC) on a microscope stage supplied by a gas mixing unit (GMU). (B) Ti:sapphire laser (TSL) connected via a grating dispersion compensator (GDC) and mono-mode optical fiber to a laser compressor (LC), which feeds the scan unit (SC) coupled to the microscope with mode locked laser light.

registered by a photo multiplier, digitized, and visualized by EZ 2000 software (Version 2.1.4, Coord Automatisering). The signal-to-noise ratio is determined, and images are deconvolved with Huygens System software (Version 2.2.1, Scientific Volume Imaging) using the maximum likelihood estimation (MLE) method. Data are reconstructed with the Application Visualization System (AVS Waltham). Calculation of isosurfaces is performed on a Unix-based Octane workstation (Silicon Graphics, Mountain View, CA) by use of the application visualization system using a marching cube algorithm (AVS-Express, Waltham) as described.[30]

Specific experimental conditions are required to measure the Fenton reaction in response to hypoxia, such as (1) a nonphototoxic irradiation of the cells to avoid ROS generation during fluorescence excitation; (2) a minimal intracellular DHR123 deposition to minimize secondary RH123 dye distribution via diffusion and channel transport;[56] (3) a start of kinetic measurements under conditions where low levels of ROS are excepted, such as hypoxia, as DHR123 to RH123 conversion is irreversible; (4) a tissue culture during the measurements under physiological conditions to minimize cell stress, and (5) an independent proof of intracellular ROS turnover.

PO_2 Dependence of Intracellular ROS Generation

Figure 8 shows an example of the three-dimensional image of HepG2 cells kept under various PO_2 conditions ranging from anoxia ($PO_2 < 1$ mm Hg), followed by ($PO_2 < 10$ mm Hg), to normoxia (PO_2 about 140 mm Hg), with white hot spots of RH123 generation indicating a localized Fenton reaction surrounding the cell nucleus. Illuminating the cell with blue-green light for 5 s to induce ROS formation by the photoreduction of flavin containing oxidases[57] results in an about 10-fold increase under normoxia, which was attenuated to about 2-fold under hypoxia and not visible under anoxia.

The reliability of these results depends significantly on the application of this highly sophisticated method fulfilling the six criteria as mentioned earlier. 2P-CLSM is not phototoxic due to infrared light.[57] Wavelength tunability enables optimization of the fluorescence excitation, avoiding cross talk of different fluorescence emissions. The 5-min 30 μM DHR123 incubation of cells kept under physiological conditions in our microscope culture chamber gave an optimal dye deposit that was fully convertible to fluorescent RH123 only under normoxia in combination with 5 s of blue light illumination. This drastic ROS increase under normoxia was contrasted by the nearly missing illumination reaction under anoxia, which renders short-term blue light illumination as an ideal proof for intracellular ROS turnover. Under these conditions, hypoxia was accompanied by a

[56] K. I. Hirsch-Ernst, T. Kietzmann, C. Ziemann, K. Jungermann, and G. F. Kahl, *Biochem. J.* **350**, 443 (2000).

[57] P. E. Hockberger, T. A. Skimia, V. E. Centonze et al., *Proc. Natl. Acad. Sci. USA* **96**, 6255 (1999).

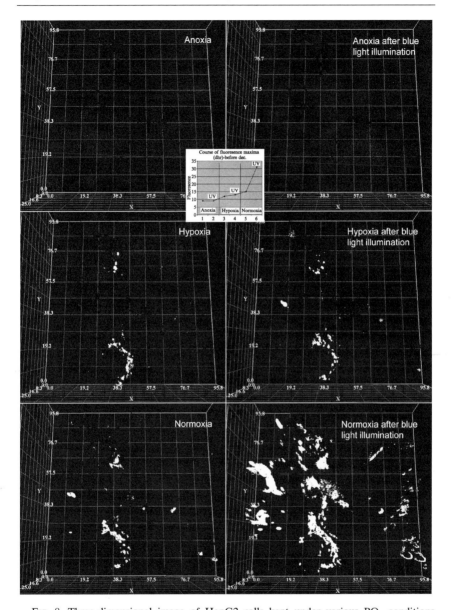

Fig. 8. Three-dimensional image of HepG2 cells kept under various PO₂ conditions starting from anoxia (PO₂ < 1 mm Hg, upper row), followed by (PO₂ < 10 mm Hg, middle row), to normoxia (PO₂ about 140 mmHg, lower row), with white hot spots of RH123 generation indicating a localized Fenton reaction surrounding the cell nucleus. Illuminating the cell with blue-green light for 5 s to induce ROS formation by the photoreduction of flavin containing oxidases (right column) results in an about 10-fold increase under normoxia, which was attenuated to about 2-fold under hypoxia and was not visible under anoxia as shown by the insert.

decrease in ROS levels in agreement with earlier studies.[58] However, other reports described enhanced ROS levels under hypoxia and showed a reversibility of the normally irreversible ROS-induced dye oxidation,[21,59] suggesting that reversibility of the signal was due to fluorescence intensity changes of the oxidized dye but not due to changes in ROS. Thus, data of the present study underline the importance of ROS as possible second messengers in the O_2-signaling system. This is in line with previous studies that showed a perinuclear localization of the Fenton reaction, as well as iron deposits close to the cell nucleus in intimate contact with the ER.[60]

Intracellular Localization of Prolyl Hydroxylases

HIF-1 gained as a transcription factor a central role in the O_2-dependent modulation of gene expression.[61,62] HIF-1 is a dimer composed of the O_2- and redox-sensitive α subunit and the constitutively expressed HIF-1β (ARNT) subunit. Although other HIF-1 isoforms appear to exist,[62] HIF-1α is considered to be the major regulator of physiologically important genes such as GK, EPO, VEGF, LDH, and PAI-1 or carotid body nervous activity[63] maintaining different body functions. Thus, it appears likely that the redox-sensitive HIF-1α may be a target within an O_2 sensing system involving ROS and a Fenton reaction. HIF-1α ubiquitylation and degradation are regulated by the von Hippel–Lindau tumor suppressor (pVHL) E3 ligase complex under normoxia through hydroxylation at prolin residues 402 and 564 located in the HIF-1α oxygen-dependent degradation domain by means of PHD1,2,3, and 4.[64–66] HIF-1α transcriptional activity is suppressed under normoxic conditions by hydroxylation of an asparagine residue within its C-terminal transactivation domain, blocking association with coactivators mediated by the asparaginyl hydroxylase FIH-1.[28] Because the activity of these hydroxylases depends not only on the presence of oxygen, but also on the ferrous state of the bound iron, the redox

[58] J. Fandrey and J. Genius, *Adv. Exp. Med. Biol.* **475**, 153 (2000).

[59] N. S. Chandel, D. S. McClintock, C. E. Feliciano *et al., J. Biol. Chem.* **275**, 25130 (2000).

[60] T. Porwol, W. Ehleben, K. Zierold, J. Fandrey, and H. T Acker, *Eur. J. Biochem.* **256**, 16 (1998).

[61] G. L. Semenza, *Biochem. Pharmacol.* **59**, 47 (2000).

[62] R. H. Wenger, *FASEB J.* **16**, 1151 (2002).

[63] D. D. Kline, Y. J. Peng, D. J. Manalo, G. L. Semenza, and N. R. Prabhakar, *Proc. Natl. Acad. Sci. USA* **99**, 821 (2002).

[64] M. Ivan, K. Kondo, H. Yang *et al., Science* **292**, 464 (2001).

[65] A. C. Epstein, J. M. Gleadle, L. A. McNeill *et al., Cell* **107**, 43 (2001).

[66] P. Jaakkola, D. R. Mole, Y. M. Tian *et al., Science* **292**, 449 (2001).

state of the cell and herewith ROS might have a significant influence on the HIF-1 activity.

For subcellular localization of HIF-1α hydroxylases, chimeric proteins composed of one of the HIF-1α prolyl hydroxylases or FIH-1 fused in frame to the N terminus of EGFP were used.[55] The fusion proteins were then transiently expressed in human U2OS osteosarcoma cells. Intracellular localization was analyzed by fluorescence microscopy as well as by three-dimensional 2PCLSM (see Fig. 9).[55] It was found that hydroxylases showed a distinct pattern of subcellular localization. PHD1 was detectable exclusively in the nucleus, whereas the majority of PHD2 and FIH-1 were found in the cytoplasm. PHD3 was distributed more evenly in the cytoplasm and nucleus. Two out of three HIF-1α prolyl hydroxylases are

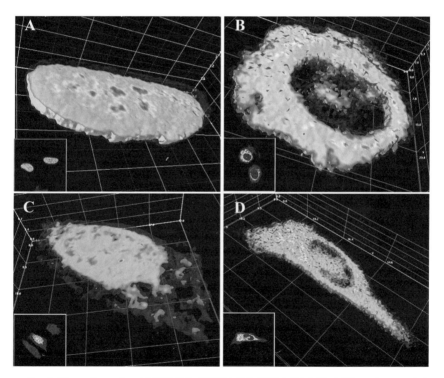

FIG. 9. Three-dimensional 2-PCLSM of PHD1 (A), PHD2 (B), PHD3 (C), and FIH-1 (D). Different EGFP fluorescence intensities of single cells are visualized in gray scales. Therefore, up to 64 optical slices through the transfected cells are recovered by 2P-LSM. After reconstruction of the optical slices, the distribution of EGFP fluorescence within a single cell is visualized three dimensionally. A cut through the cell reveals the inside distribution. An overlay of all optical slices is shown in the inserts.[55]

hypoxia inducible in U2OS cells[55]: PHD2 and PHD3 showed an upregulation of mRNA expression after hypoxia or incubation with the hypoxia mimics desferrioxamine and cobalt. PHD1 was expressed constitutively, confirming results reported previously for HeLa cells.[65] FIH-1 mRNA expression could not be stimulated by hypoxic incubation. These data suggest that PHD2 and PHD3 tightly regulate the HIF level in hypoxia to avoid excessive nuclear HIF-1α accumulation. Prior to characterization of the HIF-1α prolyl hydroxylases it had been demonstrated that the half-life of HIF-1α after reoxygenation depends on the duration of the hypoxic incubation before reoxygenation, that is, a longer hypoxic period shortens the half-life of HIF-1α.[67] This finding may suggest another function of PHD2 and PHD3: during hypoxia, PHD2 and PHD3 are induced while the actual HIF-1α turnover is low because the oxygen supply is limiting. In the case of reoxygenation the half-life of HIF-1α would then be extremely short because of a high PHD capacity within the cells. Because of their distinct intracellular localization, the HIF-1α prolyl hydroxylases PHD1, PHD2, and PHD3 and the asparagine hydroxylase form a cascade of oxygen sensors that tightly control the expression of HIF-1α target genes.[55]

Outlook

The described microscopic techniques are unique tools to reveal nature, regulation, and localization of critical members of the oxygen sensing pathway. Future applications of even more sophisticated microscopic tools, such as the fluorescence resonance energy transfer technique,[68] will enable us to visualize the intracellular compartment and type of interaction between the different signal proteins of the oxygen sensing pathway supplementing and clarifying excellent molecular biology studies as discussed in this article.

Acknowledgments

The project was supported financially by the Rudolf-Bartling Stiftung and BMBF (Federal Government, Germany) Grant 13N7447/5 to H.A.

[67] E. Berra, D. E. Richard, E. Gothie, and J. Pouyssegur, *FEBS Lett.* **491,** 85 (2001).
[68] I. Majoul, M. Straub, R. Duden, S. W. Hell, and H. D. Söling, *Mol. Biotechnol.* **82,** 267 (2002).

[33] Identification of Cyclic AMP Response Element-Binding Protein-Dependent Transcriptional Responses in Hypoxia by Microarray Analysis

By MARTIN O. LEONARD, SUSAN O'REILLY,
KATRINA M. COMERFORD, and CORMAC T. TAYLOR

Introduction

The ubiquitous cellular transcriptional response to hypoxia promotes tissue adaptation to an environment of reduced oxygen tension.[1] A master regulator of this response is hypoxia-inducible factor (HIF)-1.[2] Hypoxia-dependent stabilization, heterodimerization, and transactivation of HIF-1α lead to transcriptional upregulation of a variety of genes, which encode proteins that enhance tissue perfusion and cell survival in low oxygen.[3] More recently, it has been demonstrated that the transcriptional response to hypoxia is not confined to regulation by HIF-1 and that other transcription factors may be involved in a tissue-specific manner. For example, a number of studies have demonstrated a functional role for the cyclic AMP response element-binding protein (CREB) in controlling gene expression in hypoxia.[1] In intestinal epithelial cells, hypoxia-elicited expression of the inflammatory cytokine tumor necrosis factor α (TNF-α) is mediated through phosphorylation-dependent ubiquitination and subsequent degradation of CREB.[4-6] Furthermore, CREB is regulated by moderate hypoxia in PC12 cells.[7] We have used gene microarray analysis to identify a role for CREB in the transcriptional mechanisms that underlie global gene expression in epithelial cells in response to hypoxia.[6,8] This technique also allows the identification of genes previously unidentified as hypoxia responsive. Furthermore, we have utilized a promoter-driven reporter assay to confirm the functional relevance of CREB in hypoxia. Finally, we have

[1] C. T. Taylor and S. P. Colgan, *Pharm. Res.* **16,** 1498 (1999).
[2] G. L. Semenza, *Annu. Rev. Cell Dev. Biol.* **15,** 551 (1999).
[3] G. L. Semenza, *Biochem. Pharmacol.* **64,** 993 (2002).
[4] C. T. Taylor, A. L. Dzus, and S. P. Colgan, *Gastroenterology* **114,** 657 (1998).
[5] C. T. Taylor, N. Fueki, A. Agah, R. M. Hershberg, and S. P. Colgan, *J. Biol. Chem.* **274,** 19447 (1999).
[6] C. T. Taylor, G. T. Furuta, K. Synnestvedt, and S. P. Colgan, *Proc. Natl. Acad. Sci. USA* **97,** 12091 (2000).
[7] D. Beitner-Johnson and D. E. Millhorn, *J. Biol. Chem.* **273,** 19834 (1999).
[8] K. M. Comerford, M. O. Leonard, J. Karhausen, R. Carey, S. P. Colgan, and C. T. Taylor, *Proc. Natl. Acad. Sci. USA* **100,** 986 (2003).

used coimmunoprecipitation/Western blot techniques to identify the posttranslational modification of CREB in hypoxia,[6,8] an event pivotal to functional activity. This article describes in detail the experimental strategies and methodologies utilized in the studies outlined above.

Microarray Technology

Prior to the development of microarray technology, techniques such as serial analysis of gene expression, differential display, and subtractive cloning were used to identify novel gene expression patterns in biological systems.[9] In recent years, the development of microarray technology has fundamentally changed the landscape of gene expression studies. This technique allows for the analysis of global gene expression for thousands of individual genes simultaneously. The two main types of microarray systems currently in use are cDNA microarrays and oligonucleotide microarrays.[9] In our studies we have used oligonucleotide microarrays (Affymetrix Inc., Santa Clara, CA) to investigate global gene expression in epithelial cells in response to hypoxia. These arrays have been designed and manufactured using technology that combines photolithography and combinatorial chemistry with large numbers of distinct oligonucleotides of 21–24 bp on each array. Complementary biotinylated RNA is prepared from sample total RNA, fragmented, and hybridized to an array chip containing oligonucleotides for a large number of known genes and expression-tagged sequences. Staining with streptavidin–phycoerythrin allows for the quantification of gene expression. Bioinformatic analysis of data retrieved, which represents the transcriptomic response to hypoxia, allows for the identification of newly identified hypoxia-responsive genes, as well as the elucidation of the transcriptional regulatory events that underlie induction of the hypoxic phenotype.

Experimental Design

We have analyzed global gene expression in response to hypoxia in renal proximal tubular (HK-2[10]) and colonic crypt (T84[11]) epithelial cells. Confluent monolayers of cells are exposed to a time course of hypoxia (0, 8, 18, and 36 h). Normoxia and hypoxia are defined as 21% and 1% atmospheric O_2, respectively, with the balance being a 95% N_2/5% CO_2 mix. The atmosphere is maintained at 37° and is humidified by circulating

[9] S. Q. Ye, T. Lavoie, D. C. Usher, and L. Q. Zhang, *Cell Res.* **12,** 105 (2002).
[10] M. J. Ryan, G. Johnson, J. Kirk, S. M. Fuerstenberg, R. A. Zager, and B. Torok-Storb, *Kidney Int.* **45,** 48 (1994).
[11] K. Dharmsathaphorn and J. L. Madara, *Methods Enzymol.* **192,** 354 (1990).

atmospheric gas through distilled water in a hypoxia chamber (Coy Laboratories, Grass Lake, MI). One set of samples is also exposed to a period of reoxygenation (21% O_2) for 6 h following 36 h hypoxia. Four independent experimental sets of RNA are generated. Before proceeding with the microarray experiments, confirmation that the cells in each experimental group have perceived hypoxia is carried out. To do this, nuclear accumulation of HIF-1α in parallel experiments is determined by Western blot as a marker of cellular hypoxia.[12] Furthermore, expression of vascular endothelial growth factor (VEGF), a prototypic hypoxia-responsive gene, is determined by routine reverse transcriptase-polymerase chain reaction (RT-PCR) analysis. Having demonstrated that the cells in each individual experiment had perceived hypoxia, the RNA samples are pooled and microarray analysis is carried out.

Total RNA Isolation

Both total cell RNA and purified poly(A)$^+$ mRNA have been used as starting material to make cDNA and subsequent labeled cRNA for microarray analysis. In our studies, we used total RNA isolation with the TRIzol reagent (Life Technologies). Epithelial cells (5×10^5) are seeded on a 35-mm petri dish and cultured until fully confluent. Following exposure to hypoxia, cell culture medium is removed and cells are scraped into the TRIzol reagent (0.1 ml/cm^2). The cell lysate is passed through a pipette several times to ensure complete lysis and transfered to a 1.5-ml microcentrifuge tube. The sample is incubated at room temperature for 5 min. Chloroform (0.2 ml) is added per 1 ml of TRIzol reagent. The sample is vortexed for 15 s, incubated at room temperature for 2–3 min, and centrifuged at 12,000g for 15 min at 4° to separate the aqueous and organic phases. The RNA-containing aqueous phase is removed and transferred to a sterile microcentrifuge tube. RNA is precipitated by the addition of 0.5 ml isopropyl alcohol and incubated at room temperature for 10 min. The sample is then centrifuged at 12,000g at 4° for 10 min. RNA forms a pellet at the bottom of the tube. Following removal of the supernatant, the RNA pellet is washed in 1 ml of 75% ethanol and mixed by vortexing. The sample is centrifuged at 7500g for 5 min at room temperature and the supernatant is removed. The pellet is allowed to dry for 10 min before resuspending in 15 μl of RNase-free water. The sample is then passed up and down a pipette tip several times before incubation at 55–60° for 10 min. Total RNA is quantified by measuring absorbance at 260 nM using a spectrophotometer. High-quality RNA should have a A_{260}/A_{280} ratio between

[12] G. T. Furuta, J. R. Turner, C. T. Taylor, R. Hershberg, K. Comerford, S. Narravula, D. K. Podolsky, and S. P. Colgan, *J. Exp. Med.* **193**, 1027 (2001).

1.8 and 2.1. Separately, 5 μg of RNA is run on a 1% formaldehyde-agarose gel to confirm integrity of the ribosomal RNA bands.

Double-Stranded cDNA Synthesis

Individual RNA samples for each treatment condition are pooled. The first-strand synthesis reaction mixture is prepared using Superscript Choice kit reagents (Invitrogen, Carlsbad, CA). Five micrograms of pooled total RNA is made up to a volume of 11 μl in DEPC-dH$_2$O. One microliter of 1 nM T7-(dT) primer [5'-GGCCAGTGAATTGTAATACGACTC-AC-TATAGGGAGGCGG-(dT)$_{24}$-3'] is added. The sample is incubated at 70° for 10 min, placed on ice, and centrifuged at 14,000g for 30 s. Four microliters of 5× first-strand cDNA buffer, 2 μl of 0.1 M dithiothreitol (DTT), and 1 μl of 10 mM dNTP mix are added. The sample is vortexed and incubated at 42° for 2 min. One microliter of SuperScript II reverse transcriptase is added. The sample is vortexed and incubated at 42° for 1 h. Samples are placed on ice and centrifuged briefly to bring down condensation. To the first-strand reactions, 91 μl of DEPC-dH$_2$O, 30 μl of 5× second strand reaction buffer, 3 μl of 10 mM dNTP mix, 1 μl of DNA ligase (10 U/μl), 4 μl of DNA polymerase I (10 U/μl), and 1 μl of RNase H (2 U/μl) are added. The sample is mixed gently and incubated at 16° for 2 h. Two microliters of T4 DNA polymerase (10 U/μl) is added. Samples are incubated at 16° for 5 min. The reaction is terminated by the addition of 10 μl of 0.5 M EDTA. Double-stranded cDNA is purified using the GeneChip sample cleanup module as per the manufacturer's instructions (Qiagen, Valencia, CA). The cDNA is eluted in a volume of 12 μl. One microliter of cDNA is run on an ethidium bromide gel to confirm integrity. It is important to keep both the columns and the flow through in case cDNA is lost in the procedure and needs to be recovered.

Complementary RNA Synthesis

The synthesis of biotin-labeled cRNA is carried out using the bioarray high-yield RNA transcript labeling kit (ENZO Life Sciences Inc., Farmingdale, NY). To 11 μl of cDNA, 11 μl of dH$_2$O, 4 μl of 10× high-yield reaction buffer, 4 μl of 10× biotin-labeled ribonucleotides, 4 μl of 100 mM DTT, 4 μl of RNase inhibitor mix (20 U/μl), and 2 μl of T7 RNA polymerase (50 U/μl) are added. The sample is mixed gently with a pipette tip and incubated for 4 h at 37°. The reaction contents are mixed gently every 30 min. Biotin-labeled cRNA is washed using the GeneChip sample cleanup module as per the manufacturers instructions (Qiagen). The cRNA yield is determined by measuring the absorbance at 260 nm. The A_{260}/A_{280} ratio should be close to 2.0 for pure RNA. Final quantification of cRNA is calculated as the adjusted cRNA yield to account for the carry-over of unlabeled total RNA and is equal to the total measured RNA minus the fraction of total RNA used as starting material. cRNA must be at a minimum concentration of 0.6 μg/μl for fragmentation. Twenty micrograms of RNA is fragmented by the addition of the appropriate volume of 5× fragmentation buffer [200 mM Tris–acetate, pH 8.2, 500 mM KOAc, 150 mM MgOAc (made up fresh in DEPC-dH$_2$O)] made up to a volume of 40 μl with RNase-free water. Samples are incubated at 94° for 35 min, placed on ice, and stored at $-20°$ until hybridization is carried out. Aliqouts of RNA at each stage of the reaction are saved for ethidium bromide gel analysis to confirm integrity of the samples.

We have found that typical yields of cRNA vary considerably depending on the purity of the T7-(dT)24 primer used in first-strand cDNA synthesis. Primer obtained from Sigma-Genosys (desalted and purified) produces a typical yield of 20 μg, whereas primer obtained from Affymetrix (HPLC purified) usually yields 50–60 μg. Because a minimum of 15 μg of cRNA is required to proceed with hybridization, we recommend using desalted purified primer, as HPLC-purified primer is considerably more expensive.

Target Hybridization and Microarray Scanning

Preparation of target samples for hybridization and scanning is usually carried out as part of a core facility. Therefore, a brief outline of the protocols involved will be discussed here. Fragmented biotin-labeled cRNA prepared from template cDNA is hybridized to test microarrays, which contain a limited number of control oligonucleotide probe sets (Affymetrix). This assesses the integrity of cRNA prior to hybridization to the experimental microarray sets. Reagents are available as part of the GeneChip eukaryotic hybridization control kit (Affymetrix). A number of control cRNAs are spiked into the sample to confirm the efficiency and accuracy of hybridization, washing, and staining procedures. These biotinylated cRNA hybridization controls are typically RNAs of the bacterial proteins bioB, bioC, bioD, and cre. Poly(A) sense strand RNA prepared from template plasmid DNA of the bacterial genes dap, thr, trp, phe, and lys can be included in the original total RNA sample. This can be used to confirm target cRNA sample preparation as well as hybridization, washing, and staining procedure accuracy and efficiency. Inclusion of prelabeled oligo B2 control is also necessary as it stains probe cells of the microarray periphery and allows for grid alignment of the scanned image. Probe sets for housekeeping genes such as GAPDH and β-actin are also included (particularly the 5' probe sets for these genes). The 3'/5' ratio is used as a measure of original RNA sample integrity. Fragmented RNA is hybridized to HG U95A microarrays for 16 h, washed, and then stained for 2 h with the fluorescent detection

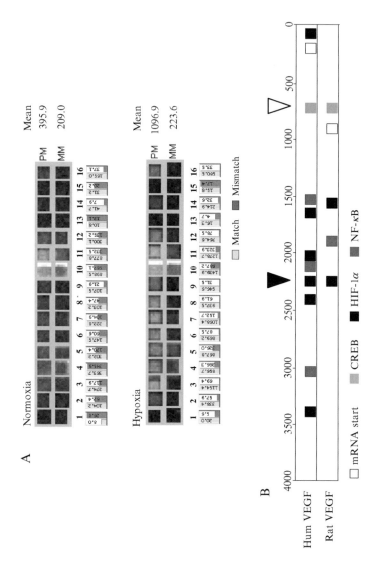

reagent streptavidin–phycoerythrin. The microarray is scanned using an argon–ion laser exciting at 488 nm with detection at 570 nm. Scans are performed in duplicate and a mean fluorescence value is calculated. Photomultiplier tubes collect and convert the fluorescence of each x–y location on the microarray to an electrical current, which is then converted to a numerical value and pixel intensity where it is stored as a dat file.

Data Interpretation and Bioinformatic Analysis

Acquisition and initial analysis of microarray data are typically carried out using Affymetrix Microarray Suite 5.0 software. After the dat file has been acquired, it is inspected visually to check for image artifacts, which may influence the interpretation of results (spots, scratches, intense background). The alignment grid is overlayed onto the dat file in order to define the specific location and boundaries of each probe cell (corresponding to a unique oligonucleotide sequence). Gene expression is calculated from the relative intensity of approximately 16–20 different oligonucleotide intensity values corresponding to one gene. Each oligonucleotide sequence [perfect match (PM)] has a corresponding mismatch (MM) probe cell sequence. This MM sequence has a single base difference compared to the PM. Probe cell intensities for the VEGF gene at normoxia and 8 h of hypoxia are depicted in Fig. 1A and show the specificity of probe cell intensity compared to the mismatch control for each PM oligonucleotide sequence. Absolute and comparison expression analysis is calculated based on these cell intensities (.cel file) using specific algorithms and an output file (.chp file) is created. Absolute expression analysis calculates the presence/absence call for the expression of an individual gene within a population of RNA. An associated signal intensity for each gene on the array is also calculated using a different algorithm, which is estimated by taking the log of the PM intensity minus the MM intensity (stray signal).

FIG. 1. Hypoxia-dependent VEGF expression. (A) HK-2 cells exposed to 8 h of hypoxia demonstrated significantly higher VEGF mRNA expression as determined by microarray analysis. The VEGF identifier 1593 is represented by 16 probe pairs. Scanned images of probe sets with quantification of each individual probe perfect match (PM) and mismatch (MM) are presented. Fluorescence intensity values are given for normoxia and 8 h hypoxia, respectively. (B) Comparison of human and rat VEGF promoter sequences for TF-binding sites. Approximately 3500 bp of rat and human VEGF promoter sequences were aligned using ClustalX 1.8. Sequences were then searched for CREB, HIF-1, and NF-κB binding motifs using Genomatix MatInspector software (www.genomatix.de). Sequences were also screened for genomic repeat sequences using RepeatMask. The conserved HIF-1-binding sequence at approximately 2200 bp is exactly the same HRE demonstrated previously to mediate hypoxic induction for the human gene. Furthermore, a conserved CREB-binding motif (CRE) was observed at approximately 940 bp.

Comparison expression analysis allows the comparison of different cell intensity files within an experiment to calculate alterations in gene expression between samples. An advantage of oligonucleotide microarrays over cDNA microarrays is the ability to compare many experimental chips to a single baseline to create a profile of gene expression over multiple RNA samples. An initial global method of scaling/normalization is used to account for any overall intensity differences between chips. Next, an algorithm is used to calculate difference call, which compares the different intensity values (PM-MM) of each probe pair in a baseline array to its matching probe pair in the experimental array. A second algorithm is used to calculate the signal/log ratio, which is a quantitative measure of the difference call. This strategy cancels out differences in probe-binding coefficients and is therefore more accurate than single array analysis. The output format of the chip file is tabular, which allows for the organization of genes based on difference call together with other parameters.

Gene expression data from epithelial cells exposed to a time course of hypoxia and reoxygenation were compared to normoxia with the creation of individual chip files. These files were then compiled and sorted based on the difference call and signal/log ratio for each time point. This allowed for the selection of genes with a signal/log ratio greater than 1 and less than -1 (twofold difference) in at least one time point. This analysis resulted in the selection of 1571 (12.5%) genes with altered expression at least one time point of hypoxia.

Expression cluster analysis is a mathematical technique used to arrange data into groups based on their pattern of expression profile. There are a number of different types of analysis, including hierarchical clustering and self-organizing maps suitable for different types of data. We chose hierarchical clustering for our data set to organize genes into groups based on their temporal expression profile. Using the software programs Cluster and Treeview (http://rana.lbl.gov),[13] a tab delimited file containing the log signal ratio values of the 1571 filtered genes was normalized across gene expression values, clustered, and visualized (Fig. 2). Within this cohort was a cluster with a closely shared expression profile regulated by oxygen. This cluster was upregulated at all time points of hypoxia and downregulated upon reoxygenation (outlined in Fig. 2). This cluster contained a significant number of genes demonstrated previously as hypoxia responsive, including VEGF and adrenomedullin. Indeed, two separate identifiers corresponding to different oligonucleotide probe sets for VEGF were present within this

[13] M. B. Eisen, P. T. Spellman, P. O. Brown, and D. Botstein, *Proc. Natl. Acad. Sci. USA* **95**, 14863 (1998).

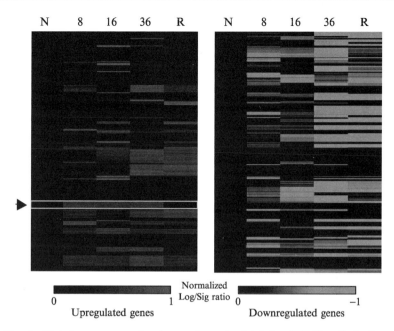

Fig. 2. Global gene expression in renal proximal tubular (HK-2) cells in hypoxia. Global gene expression was analyzed by microarray analysis in HK-2 cells exposed to increasing periods of hypoxia. Complementary RNA was prepared from total cell RNA and analyzed for gene expression by microarray analysis. Genes with a log signal ratio value greater than 1 or less than −1 between any two time points were normalized across gene expression. Using this criteria, 12.9% of expressed genes demonstrated a degree of sensitivity to hypoxia. Cluster and Treeview data were used to separate and cluster these genes into cohorts up- and downregulated in response to hypoxia, respectively. Genes with closely shared expression profiles are clustered together.

cluster. Because VEGF is a prototypic hypoxia-responsive gene, this indicates the accuracy of the experimental data and the clustering analysis, as well as the physiologic relevance of the cluster identified. The cluster also contained genes not demonstrated previously as hypoxia responsive, which demonstrate a tightly hypoxia-dependent expression profile. Thus these data allow for the identification of genes not previously known to be hypoxia responsive.

HIF-1 is the primary regulator of genes such as VEGF in hypoxia. Alternatively, NF-κB and CREB have been identified as hypoxia-responsive transcriptional regulators in a tissue-specific manner.[5,6] A mechanism of identification of potentially important regulatory sequences in genes is the determination of evolutionary conservation. Taking the VEGF promoter as an example, we compared human and rat promoter sequences in order to identify common putative regulatory regions conserved between species. Approximately 3000 bp of the genomic sequence upstream from the NCBI predicted transcription start site from the human and rat VEGF gene was repeat masked (http://searchlauncher.bcm.tmc.edu/sequtil/seq-util.html) before alignment using CLUSTALX1.81 software (http://www.matfys.kvl.dk/bioinformatik/exercise6.html[14]). The sequences were then analyzed using MatInspector software (www.genomatix.de[15]) for CREB, NF-κB, and HIF-1 transcription factor-binding site motifs. This method identified the exact HIF-1-binding site motif demonstrated previously as mediating VEGF expression in hypoxia[16] conserved between human and rat (Fig. 1B). This approach to bioinformatic analysis may be utilized in the identification of putative hypoxia-responsive transcriptional regulators, as well as determining the relative role of already defined hypoxia-responsive factors such as HIF-1.

Reporter Assay

Bioinformatic analysis of microarray data from hypoxic intestinal epithelial cells (T84) identified a cluster of genes not associated previously with induction by hypoxia. An example of such a gene is that of amphiregulin, a growth factor that can bind to the epidermal growth factor receptor and act as a bifunctional growth modulator.[17] Importantly, amphiregulin has an evolutionarily conserved (mouse, rat, and human) cyclic AMP response element close to the transcriptional start site. We identified the importance of CREB in the hypoxia-elicited activation of amphiregulin expression using truncated promoter reporter constructs. These constructs contain varying amounts of the 5' regulatory region,[18] thus allowing delineation of the regions of the promoter that confer sensitivity to hypoxia. Importantly, the promoter regions represented by these constructs do not contain a consensus HIF-1-binding motif (hypoxia response element).

[14] J. D. Thompson, T. J. Gibson, F. Plewniak, F. Jeanmougin, and D. G. Higgins, *Nucleic Acids Res.* **25,** 4876 (1997).
[15] K. Quandt, K. Frech, H. Karas, E. Wingender, and T. Werner, *Nucleic Acids Res.* **23,** 4878 (1995).
[16] J. A. Forsyth, B. H. Jiang, N. V. Iyer, F. Agani, S. W. Leung, R. D. Koos, and G. L. Semenza, *Mol. Cell. Biol.* **16,** 4604 (1996).
[17] G. D. Plowman, J. M. Green, V. L. McDonald, M. G. Neubauer, C. M. Disteche, G. J. Todaro, and M. Shoyab, *Mol. Cell. Biol.* **10,** 1969 (1990).
[18] S. B. Lee, K. Huang, R. Palmer, V. B. Truong, D. Herzlinger, K. A. Kolquist, J. Wong, C. Paulding, S. K. Yoon, W. Gerald, J. D. Oliner, and D. A. Haber, *Cell* **98,** 663 (1999).

Experimental Procedure

HeLa cells are seeded on 60-mm tissue culture dishes and grown at $37°$ in a humidified atmosphere with 5% CO_2 until 80–90% confluent. The cells are then transfected with 2 μg of the amphiregulin promoter–luciferase reporter constructs (pGL2A/B/C) or the promoterless vector pGL2 (generous gifts from Sean Lee, Johns Hopkins, Baltimore, MD[18]). These constructs contain decreasingly sized segments of the amphiregulin promoter ranging from 600 bp upstream from the TATAA box (pGL2A) to approximately 37 bp upstream from the TATAA box (pGL2C). Following overnight transfection, cells are exposed to hypoxia, lysed, and assayed for luciferase activity according to the following protocol.

Transfection

Transfection of HeLa cells is carried out using the Effectene transfection reagent (Quiagen, Valencia, CA). DNA (2 μg) is placed in a sterile 1.5-ml microfuge tube. Reaction buffer (150 μl) is added followed by 16 μl of enhancer, and the samples are vortexed for 1 s before incubation at room temperature for 5 min (optimal DNA/enhancer ratio is 1 μg DNA:8 μl enhancer). Cells are washed once with warm phosphate-buffered saline (PBS) and 4 ml of prewarmed fresh medium is added to the 60-mm dishes. Fifty microliters of the Effectene transfection reagent is then added, and the samples are vortexed for 10 s and incubated at room temperature for 10–15 min (optimal DNA/Effectene ratio is 1 μg DNA:25 μl Effectene). One milliliter of fresh medium is added to each microfuge tube. The medium is then mixed gently with the transfection components by pipetting the solution up and down two to three times with a 1-ml pipette. The mixture is then immediately added dropwise to the cells and the dish is rotated gently. Following transfection, the cells are incubated overnight at $37°$ in a humidified atmosphere of 5% CO_2. Twenty-four hours following transfection the medium is replaced. Forty-eight hours following transfection the cells are treated with (a) preequilibrated normoxic medium (21% O_2) and returned to normoxic growth conditions, (b) preequilibrated hypoxic (1% O_2) medium and placed in a hypoxic environment (1% O_2), and (c) 100 nM PMA for 8 h prior to harvesting (positive control).

Harvesting Whole Cell Extracts

After incubation of the cells in normoxic or hypoxic conditions for 22 h, whole cell extracts are harvested. Medium is aspirated off each 60-mm dish and the cells are washed with 2 ml of ice-cold PBS. Two hundred microliters of lysis buffer (luciferase assay kit, Stratagene, La Jolla, CA) is added to the cells before incubation for 2–3 min at room temperature. The cells are then scraped from the dish, transferred to a microfuge tube, and placed on ice. Each sample is then sonicated for 2–3 s and centrifuged at 14,000 rpm on a bench-top centrifuge for 6 min at 4°. The cell lysate is transferred to a new microfuge tube and allowed to reach room temperature.

Luciferase Activity Assay

The assay is performed at room temperature. One hundred microliters of luciferase substrate (Stratagene) is added to 20 μl of lysate and mixed thoroughly by pipetting in a 5-ml polystyrene test tube. The luciferase activity in each of the whole cell extracts is subsequently assayed for luciferase activity using a Junior LB 9509 luminometer (Berthold Technologies, Germany). All luciferase readings (relative luminescence units) are normalized to the protein content (Bradford method) of the equivalent cell extract. The hypoxic sensitivity of the PGL2A reporter construct, which contains only Wilm's tumor response element and cyclic AMP response element-binding sites, is demonstrated in Fig. 3.

Posttranslational Modification of CREB

Having demonstrated the role of CREB as a hypoxia-dependent transcriptional regulator, we next became interested in the posttranslational modifications of CREB, which determine stability and transcriptional activity in hypoxia. A number of posttranslational modifications can regulate CREB stability and/or activity, including phosphorylation at Ser133,[19] ubiquitination,[6] and sumoylation.[8] In order to investigate the impact of hypoxia on these modifications of CREB, we used a coimmunoprecipitation approach where CREB is immunoprecipitated, separated by SDS–PAGE, and Western blotted with antibodies against the modifying moieties phosphoserine, ubiquitin, or SUMO-1.

Immunoprecipitation

Cells are grown to confluence on 100-mm petri dishes. Confluent monolayers are maintained in normoxia (21% O_2) or exposed to hypoxia (1% O_2) for increasing periods of time (0–48 h) as described earlier. Medium is removed and cells are washed once with ice-cold PBS. Attached cells are scraped off the culture dish in 100 μl of lysis buffer [150 mM NaCl,

[19] B. Mayer and M. Montminy, *Nature Rev. Mol. Cell. Biol.* **2,** 599 (2001).

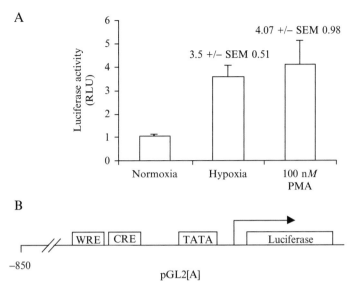

FIG. 3. pGL2[A] truncation of the amphiregulin promoter retains sensitivity to hypoxia. (A) HeLa cells transfected with the pGL2[A]–luciferase reporter construct demonstrated increased luciferase activity following exposure to hypoxia or PMA (100 nM; positive control). (B) A map of the pGL2[A] construct of the amphiregulin promoter demonstrates the presence of Wilm's response element (WRE) and cyclic AMP response element (CRE) regulatory motifs.

25 mM Tris–HCl, pH 8, 1 mM EDTA, 1% (v/v) NP-40, 100 mM phenylmethyl sulfonyl fluoride, and 50 μl of a protease inhibitors cocktail (Sigma, St. Louis, MO)]. Cell lysates are vortexed for 10 s and incubated on ice for 30 min. The cell suspension is transferred to a 1.5-ml microcentrifuge tube and centrifuged at 14,000g at 4° for 5 min. Supernatants are transferred to a new microcentrifuge tube, and the protein concentration is standardized by the DC protein assay (Bio-Rad, Hercules, CA). An equal concentration of protein is transferred to new microcentrifuge tubes and incubated with 20 μl of protein A/G plus agarose beads (Santa Cruz Biotechnology) and 1 μl of anti-CREB antibody (cell signaling) overnight at 4° with rotation. Samples are centrifuged at 14,000g at 4° for 5 min. Supernatants are removed carefully to avoid disturbing the pellet of agarose beads. The beads are then washed with 500 μl of lysis buffer, vortexed, and centrifuged at 14,000g at 4° for 5 min. A volume (60–80 μl) of 2X reducing sample buffer [66 mM Tris–HCl, pH 6.8, 2% (w/v) SDS, 10 mM EDTA, 10% (v/v) glycerol, and 10 μl of 100% bromphenol blue] is added to the beads. The

beads are boiled in a heating block for 10–15 min to elute the target protein (CREB) before being resolved by (8%) SDS–PAGE (25 µg/lane).

Western Blotting

Electrotransfer of proteins onto a nitrocellulose membrane is performed using an SDS–methanol buffer (3 g Tris base, 14.4 g glycine, 1 g SDS, and 200 ml methanol/liter) at a constant voltage of 100V for 1 to 1.5 h. A water cooling system is used to reduce heat during transfer. The nitrocellulose membrane is blocked overnight by agitating at 4° in blocking buffer [1 M NaCl, 10 mM HEPES, pH 8, 0.2% (v/v) Tween 20, and 3% bovine serum albumin]. The nitrocellulose is incubated at room temperature for approximately 1.5 h with blocking buffer containing the antibody to the posttranslational modifying moiety [antiphosphoserine (1:1000 dilution; Zymed), antiubiquitin (1:1000 dilution; StressGen Biotechnologies), or anti-SUMO (0.5 µg/ml; Zymed Laboratories)]. The membrane is washed extensively (five times) in wash buffer [1 M NaCl, 10 mM HEPES, pH 8, and 0.2% (v/v) Tween 20], and incubated at room temperature for 45 min with 1:1000 dilution of species-matched peroxidase-conjugated secondary antibody (cell signaling). The nitrocellulose is washed as described earlier, and labeled bands are detected by standard enhanced chemiluminesence (Amersham) and X-ray radiography techniques.

Summary

While HIF-1 is a master regulator of hypoxia-elicited gene expression in epithelial cells of the kidney and gut, other HIF-1-independent transcriptional pathways likely contribute. This may occur in a tissue-specific manner. Microarray analysis allows us to begin to elucidate such pathways. Reporter assays allow us to determine the functional relevance of transcriptional regulators identified. Finally, coimmunoprecipitation experiments allow us to determine the posttranslational modification of transcription factors in hypoxia, an event pivotal to the induction of the hypoxic genotype and subsequent phenotype.

Section VIII

Wound Healing

[34]   Determination of the Role of Hypoxia-Inducible Factor 1 in Wound Healing

By JONATHAN S. REICHNER and JORGE E. ALBINA

Introduction

Interest in the expression of hypoxia-inducible factor (HIF)-1 during the early inflammatory phase of wound healing originated from the close temporal expression of inducible Nitric Oxide Synthase (iNOS) and HIF-1 in wound cells first noted by this laboratory.[1] Examination of cells isolated from polyvinyl alcohol (PVA) sponge wounds in rats demonstrated a dramatic accumulation of HIF-1α mRNA and protein in cells isolated 1 and 5 days after injury (Figs. 1 and 2). These observations suggested a role for HIF-1 in the regulation of iNOS expression in early wounds and were consistent with reports that iNOS is a hypoxia-responsive gene containing a hypoxia response element in its promoter region.[2] Moreover, the accumulation of vascular endothelial growth factor (VEGF) mRNA in early wounds cells supported the concept that the detected HIF-1α had effective transactivation activity, as HIF-1 is known to regulate the expression of VEGF.[1]

Whereas hypoxia is the canonical regulator and activator of HIF-1α, and wounds are often considered to be relatively hypoxic, the expression of HIF-1α in wound cells preceded the development of hypoxia.[1] Reports indicate that tumor necrosis factor (TNF)-α and interleukin (IL)-1 are capable of inducing HIF-1α protein accumulation in selected cell types.[3,4] Results from this laboratory further showed that TNF-α, but not IL-1β, induces HIF-1α protein accumulation in cultured day 1 wound cells (>90% neutrophils) (Fig. 3).[1] These findings suggest dual pathways of HIF-1 induction within an inflammatory site with early expression stimulated by proinflammatory mediators and with continued HIF-1 expression as the wound becomes hypoxic. Therefore, this article focuses on methodology for the study of HIF-1 expression during inflammation and wound healing. Surgical methods for the implantation of PVA sponges and the

---

[1] J. E. Albina, B. Mastrofrancesco, J. A. Vessella, C. A. Louis, W. L. Henry, Jr., and J. S. Reichner, *Am. J. Physiol.* **281,** C1971 (2001).

[2] G. L. Semenza, *Annu. Rev. Cell Dev. Biol.* **15,** 551 (1999).

[3] R. D. Thornton, P. Lane, R. C. Borghaei, E. A. Pease, J. Caro, and E. Mochan, *Biochem. J.* **350,** 307 (2000).

[4] T. Hellwig-Bürgel, K. Rutkowski, E. Metzen, J. Fandrey, and W. Jelkmann, *Blood* **94,** 1561 (1999).

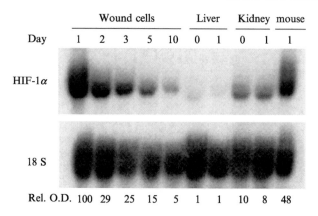

Fig. 1. Evidence that HIF-1α mRNA accumulates in early wound cells. Northern blot analysis of HIF-1α mRNA expression in rat wound cells harvested 1 to 10 days following injury, as well as in liver and kidney sampled from nonwounded animals (0) and on day 1 postinjury (1). Wound cells harvested from mice 1 day after injury also contained abundant HIF-1α mRNA.[1]

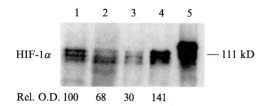

Fig. 2. Western blot analysis of HIF-1α protein expression in rat wound cells harvested 1 (lane 1) and 5 (lane 2) days following injury. Wound cells harvested 1 day after injury were also cultured for 6 h in normoxic culture medium (lane 3) or in culture medium equilibrated with 1% O_2 (lane 4) prior to cell lysis and Western analysis. Extracts from HeLa cells cultured in 1% O_2 for 6 h were analyzed as positive controls (lane 5).[1]

recovery of inflammatory cells and fluids suitable for analysis of HIF-1 expression and content of proinflammatory mediators, respectively, are detailed. Additional methods include the application of pimonidazole HCl[5,6] as a marker of hypoxia within the inflammatory site and the use of immunoblotting and Northern blotting to detect HIF-1 protein and messenger RNA, respectively.

[5] J. A. Raleigh, S.-C. Chou, and M. R. Horsman, *Rad. Res.* **151,** 580 (1999).
[6] C. J. Koch, *Methods Enzymol.* **352,** 3 (2002).

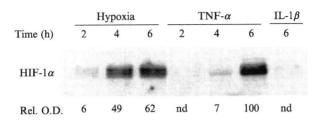

	Hypoxia			TNF-α			IL-1β
Time (h)	2	4	6	2	4	6	6
HIF-1α							
Rel. O.D.	6	49	62	nd	7	100	nd

FIG. 3. TNF-α induces HIF-1α accumulation in wound cells harvested 1 day after wounding. Western blot analysis of HIF-1α protein expression in cultured day 1 wound cells exposed to hypoxic (equilibrated with 1% O_2) culture media or to normoxic medium containing 50 ng/ml rat TNF-α or IL-1β for the times indicated.[1]

Rat Wound Model

Background

The subcutaneous implantation of PVA sponges results in the rapid infiltration of the sponge material by circulating polymorphonuclear leukocytes and monocytes (Fig. 4).[7,8] The magnitude and composition of the infiltrate are well characterized and reflect the changing cellularity of the healing wound with neutrophils predominating during the initial 72 h postwounding (Fig. 5).[7] The PVA model of inflammation affords the ability to mechanically recover inflammatory cells within minutes of sponge harvesting, thereby minimizing possible iatrogenic effects exerted during cell isolation, which may alter phenotype. Furthermore, the fluid contents are recovered easily, permitting coincident quantitation of cytokine/chemokine content. This article presents surgical techniques for the insertion of PVA sponge material for modeling a site of aseptic inflammation in rats.

Surgical Methods

All animal surgery is performed under a protocol approved by the Lifespan Animal Care and Use Committee and conforms to the guidelines established by the National Institutes of Health (Bethesda, MD). All surgical procedures should be performed by aseptic technique using sterilized instruments, materials, and buffers.

---

[7] A. J. Meszaros, J. S. Reichner, and J. E. Albina, *J. Leukocyte Biol.* **65,** 35 (1999).
[8] E. Seifter, G. Rettura, A. Barbul, and S. M. Levenson, *Surgery* **84,** 224 (1978).

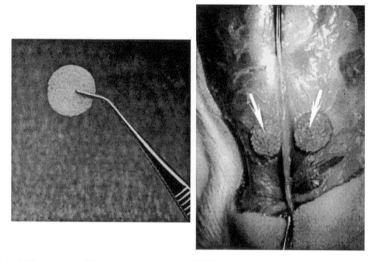

Fig. 4. Placement of the polyvinyl alcohol (PVA) sponges subcutaneously in the dorsum of anesthetized rats. *Left:* View of representative prepared PVA material. *Right:* View of dorsum 10 days after PVA sponge (arrows) implantation.

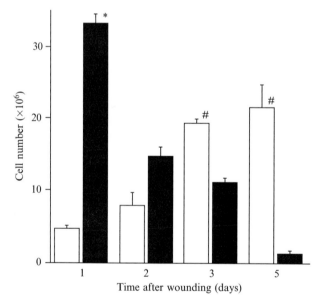

Fig. 5. Composition of wound cellular infiltrate. Absolute number of PMN (filled bars) and macrophages (open bars) determined on days 1, 2, 3, and 5 after wounding from hemacytometer counts (total cells in sponge infiltrate per animal) and differential counts from cytospin. $^*P < 0.05$ vs days 2, 3, and 5 PMN counts; $^\#P < 0.05$ vs day 1 and day 2 macrophage counts. Data are means ± SD from three animals per time point.[7]

Materials

    Polyvinyl alcohol sponges (Ivalon large-pore sponge material from PVA Unlimited, Eudora, IN)

    Sodium pentobarbital (Abbott Laboratories, North Chicago, IL)

    Povidone-iodine (Purdue Frederick, Norwalk, CT)

    Surgical instruments: Fine surgical scissors, Metzenbaum scissors, fine forceps, disposable scalpels, surgical clippers, autoclip wound-closing system (Fisher Scientific), Weitlaner self-retaining retractor (3 × 4 prongs)

    Hank's balanced salt solution (GIBCO, Grand Island, NY)

    Stomacher (Tecmar, Cincinnati, OH)

    0.2% NaCl

    1.6% NaCl

    5-cc syringe

    15-ml conical centrifuge tubes

Procedure

1. PVA sponge preparation. Dry sponge sheets partially, not to hardness, and cut into 1-cm^2 pieces using a paper cutter or in 1-cm-diameter circles using a cork borer. Rinse sponges in a 2-liter beaker overnight with tap water followed by three 10-min rinses with distilled water. Immerse in distilled water, cover with foil, and heat to boiling for 30 min to sterilize. Let cool, aliquot into containers, and cover with sterile saline for storage.

2. A male Fischer rat weighing between 175 and 200 g is anesthetized with 45 mg/kg of sodium pentobarbital administered by ip injection. When a surgical plane of anesthesia is reached, the dorsum is shaved free of fur, prepped with 70% ethanol and 10% povidone-iodine solution, and then rinsed with 70% ethanol.

3. A 3-cm midline full thickness skin incision is made in the dorsum. A Weitlaner retractor is inserted into the incision to keep wound edges open. The apex of the incision is grasped with forceps and is retracted upward. Bloodless pockets are developed using Metzenbaum scissors. Ten sterile PVA sponges are inserted into the pockets, and the incision is closed with skin clips.

Inflammatory Cell Collection

4. At various times after wounding, animals are sacrificed with CO_2 and the incisional wound is reopened using a scalpel. Sponges are retrieved using forceps, placed in Hank's balanced salt solution, and minced with

sterile scissors. Contents are transferred to a sterile plastic sealable pouch, and cells contained within the sponges are isolated by repeated compression using a Stomacher for 30 s.

5. Cells are collected by centrifugation at 700g for 10 min. At this point, any remaining red blood cells can be lysed hypotonically by resuspending the cell pellet in 5 ml 0.2% sterile NaCl in dH$_2$O for 10 s with rescue using an equal volume of 1.6% NaCl solution.

Inflammatory Fluid Collection

HIF-1 induction in early inflammation may be mediated by proinflammatory cytokines rather than hypoxia, such that analysis of the cytokine content of inflammation may be an important parameter in understanding mechanisms of HIF-1 expression. The fluid content of PVA sponges is collected easily by the centrifugation of sample sponges from among the 10 used to initiate the injury. Sponges are placed into the barrel of a 5-cc syringe, which is then inserted into a 15-ml centrifuge tube so that the flange of the syringe barrel rests on the edge of the tube, preventing contact between the tip of the syringe and the bottom of the carrier tube. The tube is centrifuged at 700g to collect inflammatory fluid, which is separated from the cell pellet by careful decanting. Approximately 100 μl fluid can be expected to be recovered from each sponge. It is recommended that sponges be chosen randomly from cephalad to caudal locations as we have noted some slight variation in the cellular and humoral content of the sponges in that regard.

Methods for Analysis: Detection of Pimonidazole HCl Adducts

Significant hypoxia is not detected in small cutaneous wounds in rats 24 h after wounding and the wound is not hypoxic until 4 days after wounding.[9] Careful review of data from Niinikoski[10,11] showed that PO$_2$ levels did not reach their nadir until days 3 to 5 in experimental rat wounds. Our own studies using pimonidazole HCl staining of wound cells, a sensitive measure of hypoxic exposure, found no evidence for hypoxia in wound cells harvested 24 h after wounding and extensive cellular staining in those retrieved 4 days later.[1] Wound cells were harvested from animals injected previously with the hypoxia marker pimonidazole HCl. Pimonidazole (Hydroxyprobe-1) is a substituted 2-nitroimidazole with the chemical

[9] Z. A. Haroon, J. A. Raleigh, C. S. Greenberg, and M. W. Dewhirst, *Ann. Surg.* **231,** 137 (2000).
[10] J. Niinikoski, C. Heughan, and T. K. Hunt, *Surg. Gynecol. Obstet.* **133,** 1003 (1971).
[11] J. Niinikoski and T. K. Hunt, *Surgery* **71,** 22 (1972).

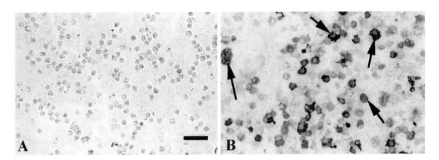

FIG. 6. Wound cells harvested 1 day after injury from pimonidazole HCl-injected rats do not demonstrate pimonidazole adduct staining (A). In contrast, cells harvested 5 days after wounding stain prominently for these compounds (arrows), thus demonstrating local hypoxia (B). Cells found in 1-day-old wounds are greater than 90% polymorphonuclear leukocytes, whereas those retrieved on day 5 are greater than 80% macrophages. Bar: 40 μM.[1]

name of 1-[(2-hydroxy-3-piperdinyl)propyl]-2-nitroimidazole hydrochloride.[5,6] This compound forms intracellular protein adducts under conditions of oxygen tension <10 torr, which can be detected immunologically. This system has been used to establish the time course of hypoxia in wounds.[9] As shown in Fig. 6A, cells harvested from 24-h wounds failed to stain for pimonidazole adducts. In contrast, cells harvested from 5-day-old wounds demonstrated substantial specific staining (Fig. 6B), thus confirming the hypoxic status of the wound at that time.

Materials

Pimonidazole HCl (Hydroxyprobe-1) and Hydroxyprobe-1Mab (Hypoxyprobe kit, Chemicon International, Inc, Temecula, CA)
1% paraformaldehyde in phosphate-buffered saline (PBS)
Permanox, sterile, 60 × 15-mm cell culture dish (Nalgene Nunc International, Rochester, NY)
RPMI 1640 tissue culture media (GIBCO)
Cytocentrifuge (Cytospin, Shandon, Pittsburgh, PA)
Glass microscope slides (Superfrost/Plus, Fischer Scientific, Pittsburgh, PA)
Copeland-style staining jars
PBS
3% H_2O_2
PBS/0.2% Brij 35
Normal goat serum (Sigma)
MOPC 21 (Control IgG1, Sigma)

Universal Elite Avidin-Biotin System (Vector Laboratories, Burlingame, CA)
Vector VIP (Vector Laboratories)
Hydrophobic barrier pen (PAP pen, Z37,782-1, Sigma)
Wright–Giemsa stain (WGHT, Sigma)
Gas-tight modular chamber (Billups-Rothenberg, Del Mar, CA)
Permanox tissue culture plates (NUNC, Naperville, IL)

Procedure

1. Pimonidazole HCl is administered intraperitoneally at 70 mg/kg 3 h before animal sacrifice.

2. Cells harvested from the wounds (as described earlier) are centrifuged onto glass slides and fixed in 1% paraformaldehyde for 20 min for use in immunostaining. Slides are rinsed by five 10-min changes in PBS. Additional slides can be used for morphological identification of the inflammatory cells using Wright–Giemsa staining according to the manufacturer's directions. Prior paraformaldehyde fixation of these slides is not needed.

3. To begin immunohistochemical detection of pimonidazole HCl adducts, endogenous peroxidase activity is quenched by immersion in PBS containing 3% H_2O_2 and cells are permeabilized with PBS/0.2% Brij 35 for 15 min at room temperature. At this point, a hydrophobic barrier is drawn around the cells using a PAP pen to keep staining reagents localized on the cells and to limit the amount of reagents needed. Cells should be completely covered with blocking serum and antibodies (as described in the ensuing steps), which should be accomplished with approximately 125 μl volume.

4. Blocking nonspecific binding of antibody is with 5% normal goat serum in PBS/0.2% Brij 35 for 1 h at room temp.

5. Pimonidazole HCl adducts are detected using the Hydroxyprobe-1Mab1 antibody diluted 1:50 in PBS/0.2% Brij 35 and are incubated overnight in a humidified chamber at 4°. Control slides are similarly treated with MOPC 21 at a concentration of 5 μg/ml (IgG1, Sigma).

6. Following washing, antibody binding is detected using the Universal Elite avidin–biotin system according to the manufacturer's directions.

7. Peroxidase activity is visualized with Vector VIP (or similar diaminobenzidine-containing compound) according to the manufacturer's directions. Color development is monitored continuously under an inverted microscope and is usually complete within 5 min. The reaction is stopped when color begins to appear in isotype control or when the color of test samples seems to stop intensifying. Slides are rinsed in several changes of

PBS to remove excess peroxidase substrate and hence stop the reaction. The recommendation of Vector VIP as a substrate for a horseradish peroxidase-coupled secondary antibody is largely based on its ease of use, as the composite reagents are prediluted and readily combined. In fact, any diaminobenzidine-based reagents work just as well.

8. For culturing of cells under hypoxic conditions, cells are plated in 60×15-mm Permanox culture plates using RPMI 1640 or other CO_2-buffered culture medium. These particular plates are desirable because they permit gas exchange most readily as compared to alternative types of tissue culture plastic. Plates are then placed in a gas-tight modular incubator chamber, which are then gassed for 2 h using certified gases containing 1 or 0% O_2–5% CO_2 balance N_2. They provide a final PO_2 level in the culture media of 8.3 or 0 torr, respectively. The gassing ports are then closed and the chambers are placed in a $37°$ temperature-controlled incubator for the times desired. Normoxic cultures take place at $37°$ in 5% CO_2 in room air. Also, if desired, pimonidazole can be added to culture media and cells reserved during harvesting for evidence of hypoxic exposure.

Methods for Analysis: Detection of HIF-1 by Western Blotting

Materials

> Electrophoresis sample buffer ($2\times$ ESB, 125 mM Tris–HCl, pH 6.8, 4% SDS, 10% glycerol, 0.006% bromphenol blue, 2% β-mercaptoethanol)
> Vertical gel electrophoresis and transblot apparatus (Bio-Rad Mini Protean 3 electrophoresis system with mini trans-blot or similar system)
> Precast polyacrylamide gels
> Transfer buffer (25 mM Tris, 190 mM glycine, 20% methanol)
> Nitrocellulose membranes (0.45 μM)
> PBS
> Nonfat dry milk
> Tween-20 (Sigma)
> Boiling water bath
> Anti-HIF-1α antibody (Novus Biologicals, Littleton, CO)
> Horseradish peroxidase-conjugated rabbit antimouse IgG (Amersham Pharmacia Biotech, Inc., Piscataway, NJ)
> Enhanced ChemiLuminescent reagent (ECL, Amersham Pharmacia Biotech, Inc.)
> X-OMAT AR Kodak autoradiography film (Kodak, Rochester, NY)

Procedure

1. With respect to HIF-1 detection, cultured cells should be collected and lysed in preparation for immunoblotting as rapidly as possible, as the HIF-1 protein is readily degraded upon exposure to ambient oxygen levels. In our laboratory, as soon as the gas-tight chamber is opened, cells are removed by gentle scraping, placed in a microfuge tube, spun in a microcentrifuge for 1 min, and resuspended in 125 μl PBS. At this point an aliquot can be collected for cell count or protein determination in order to normalize gel loading. An equal volume of 2× ESB, preheated to 95°, is added to affect lysis and denaturation, and the samples are placed in a boiling water bath for an additional 7 min to ensure complete protein denaturation by the combined actions of temperature, SDS, and β-mercaptoethanol. Cell pellets of wound cells harvested from PVA sponges are similarly resuspended in 125 μl PBS, sampled for counting or protein determination, and then lysed directly in 95° 2× ESB at approximately 2 × 10^7 cells/250 μl. Lysates are then placed in a boiling water bath for an additional 7 min.

2. Aliquots of 35-μl lysates (approximately 2.8 × 10^6 cell equivalents) are fractionated in a 7.5% SDS–polyacrylamide gel and transferred to a nitrocellulose membrane.

3. Membranes are blocked in PBS, 5% nonfat dry milk, and 0.05% Tween-20 (blocking buffer) for 1 h at room temperature.

4. An anti-HIF-1α antibody (Novus Biologicals) is used for HIF-1α detection at a dilution of 1:1000 in 7 ml blocking buffer. Membranes are incubated for 1 h at room temp on a shaking platform. Blots are washed for 30 min with six changes of PBS.

5. Membranes are then incubated with horseradish peroxidase-conjugated rabbit antimouse IgG (Amersham Pharmacia Biotech, Inc.) diluted 1:2000 in 7 ml blocking buffer. Blots are washed again for 30 min with six changes of PBS.

6. Detection is by chemiluminescence using the ECL reagent according to the manufacturer's directions followed by exposure to X-ray film. The majority of rat HIF-1 protein is detectable at 111 kDa, although three bands are routinely resolved. This is likely due to heterogeneity of the phosphorylation state of HIF-1, as reported by Richard *et al.*[12]

7. Attempts to identify the specific cell types expressing HIF-1α in early wounds by immunostaining have not been successful in

---

[12] D. E. Richard, E. Berra, E. Gothie, D. Roux, and J. Pouyssegur, *J. Biol. Chem.* **274,** 32631 (1999).

our laboratory. It should be kept in mind that an essential criterion for a positive immunostaining reaction is one in which detection is apparent within the nucleus of cells exposed to hypoxia. Staining with anti-HIF-1α antibodies purchased from Novus Biologicals (NB100-105 and NB100-123), BD Transduction Laboratories (San Diego, CA; clone 54), Neomarkers (Fremont, CA; clones OZ12 and OZ15), Santa Cruz Biotechnology, Inc. (Santa Cruz, CA; N18 and C-19), or with those provided generously by other investigators gave inconsistent results, despite application of a number of protocol variations regarding rapid cell fixation and permeabilization.

Methods for Analysis: Detection of HIF-1 mRNA by
    Northern Blotting

Materials

    Total RNA isolation reagent (Ultraspex, Biotecx, Houston, TX, or
        similar product)
    DNase I (Invitrogen Life Technologies)
    Nuclease-free water (Promega, Madison, WI)
    Formaldehyde, 37% stock
    10× MOPS buffer (0.2 M 3-[N-morpholino]propanesulfonic acid),
        0.05 M sodium acetate, 0.01 EDTA)
    RNA loading buffer [7.2 ml deionized formamide, 1.6 ml 10× MOPS
        buffer, 2.6 ml formaldehyde (37% stock), 3.6 ml 0.1% bromphenol
        blue in 50% glycerol]
    Hybridization buffer [50% deionized formamide, 50 mM Tris–HCl,
        pH 7.5, 100 mg/ml dextran sulfate, 58 mg/ml sodium chloride,
        10 mg/ml SDS, 2 mg/ml bovine serum albumin (BSA), 2 mg/ml
        polyvinylpyrrolidone, 2 mg/ml Ficoll (MW = 400,000), 1 mg/ml
        sodium pyrophosphate, 0.25 mg/ml denatured salmon sperm DNA]
    Random primer labeling kit (e.g., Megaprime DNA labeling system,
        Amersham Biosciences)
    Agarose, ultrapure (Life Technologies, Gaithersburg, MD)
    Horizontal gel electrophoresis system (e.g., Horizon 11.14, Life
        Technologies) or similar system
    Turboblotter transfer kit (Schleicher & Schuell, Keene, NH) or
        similar system
    NYTRAN nucleic acid and protein transfer media, 0.2 μM (Schleicher
        & Schuell, Keene, NH)
    20× SSC buffer (GIBCO BRL)
    Stratalinker UV cross-linker (Stratagene, La Jolla, CA)

X-OMAT AR Kodak autoradiography film (Kodak)

Probes

The probe to detect rat HIF-1α mRNA (GenBank accession number Y09507) was generously provided to our laboratory by Dr. Thomas Kietzmann, Institut for Biochimie and Molekulare Zellbiologie, Goettingen, Germany, in a plasmid (pCRII) containing an 800-bp fragment of HIF-1α.[13] The insert was excised by *Eco*RI digestion. For 18S ribosomal RNA, a pUC830 plasmid containing the mouse 18S ribosomal cDNA was obtained from the American Type Culture Collection. *Sph*I and *Bam*HI digestion of the plasmid yielded a 752-bp cDNA insert that detects 18S ribosomal RNA in rat and mouse cells. Probes were radiolabeled with [^{32}P]dCTP by random priming. Measurement of 18S ribosomal RNA was used as a control for variations in the amount of RNA in each lane.

Northern Blot Analysis

Total RNA is isolated from 15–20 × 10^6 cells or 1 g of kidney or liver using Ultraspec (Biotecx), resuspended in RNA loading buffer, and fractionated by 1% agarose/0.66% formaldehyde gel electrophoresis overnight at 30 V with 20 μg mRNA per lane. The gel is removed and soaked for 30 min in 10× SSC on a rotating platform. Gel contents are transferred to a nylon membrane using a Transblotter apparatus according to manufacturer's directions. Transferred mRNA is fixed by UV cross-linking, and the membrane is hybridized in hybridization buffer containing a radiolabeled probe (10 ng/ml) for 18–24 h. The membrane is washed for two 30-min periods with 2× SSC at room temperature followed by two 30-min periods with 2× SSC/1.0% SDS at 65°. The membrane is autoradiographed at −80°. HIF-1 and 18S ribosomal RNA are quantitated by densitometry using NIH Image v1.6.

Acknowledgment

This work was supported by National Institute of General Medical Sciences Grant GM-42859 (to J.E.A.) and allocations to the Department of Surgery by Rhode Island Hospital.

---

[13] T. Kietzmann, Y. Cornesse, K. S. Brechtel, Modaressi, and K. Jungermann, *Biochem. J.* **354**, 531 (2001).

[35]   Measuring Oxygen in Wounds

By Harriet W. Hopf, Thomas K. Hunt, Heinz Scheuenstuhl,
Judith M. West, Lisa M. Humphrey, and Mark D. Rollins

Introduction

Oxygen plays a crucial role in wound healing. It is required for immunity, angiogenesis, collagen deposition, and epithelization. The oxygen supply in wounded tissue is often the limiting factor in healing. Studies of the role of oxygen in wound healing turn on the ability to measure oxygen: in cell and tissue culture, in animal models, and in human wounds. Roy *et al.* described in depth the measurement and control of oxygen levels in cell and tissue culture, while this article focuses on oxygen measurement *in vivo*.

The development of methods for the accurate measurement of wound oxygen tension has led to studies demonstrating that activation of the sympathetic nervous system by such common perioperative stressors as hypothermia, pain, and hypovolemia decreases wound oxygen tension and impairs wound healing. In the past decade, these observations have led to several large, randomized, controlled trials in which improving oxygen delivery to acute surgical wounds (e.g., maintenance of normothermia and administration of high oxygen concentrations to the patient) led to dramatic improvements in wound healing and resistance to infection. Tissue oxygen measurements are relatively simple to make, particularly with current highly stable probes, and their use will undoubtedly lead to further advances in wound care.

The first measurements of tissue oxygen levels were made with microelectrodes. Because tissue perfusion tends to be heterogeneous, multiple measurements over small incremental distances were required to create a map of tissue oxygen gradients. The resulting histogram indicated mean or median tissue oxygen levels. Microelectrodes are not practical for measurements in humans under most circumstances because they require multiple needle punctures. Therefore, in the 1970s, Thomas K. Hunt and others began to develop techniques to measure average tissue oxygen levels from a single site. This article details four of the most important and commonly used methods for oxygen measurement, ranging from microelectrode measurements of wound oxygen gradients to clinically useful measurements in humans.

Materials

The Rabbit Ear Chamber Model

1. Ear chamber: Refer to Figs. 1 and 2 showing cross sections of a chamber as fashioned and used by Silver. The chamber body is transparent Lucite (Perspex). Many designs can be used. They can be simple or sophisticated. There is no commercial source. The upper membrane, held in place by a circlip, may be glass, Teflon, or both. The easiest to use for oxygen measurement is two membranes that have interlocking holes: a permeable Teflon membrane on the inside, topped by an impermeable glass one (see Note 1). The healing tissue grows from the ear cartilage up through the holes into and across the chamber.

2. Oxygen probes: The oxygen probes must have fine tips, a few micrometers for best resolution. They can be handmade by drawing out a fine glass capillary tube over a platinum wire, usually about 25 μm in diameter. It is then electropolished to as small a tip as possible. Experts can make them less than 2 μm. Assembled, membrane-covered, oxygen probes that contain both electrodes and are as small as 0.5 mm can be

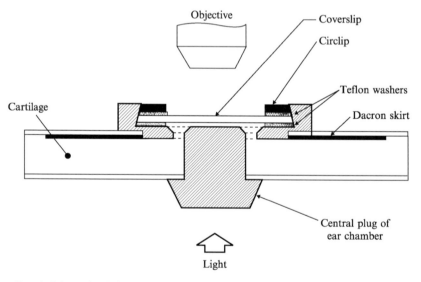

Fig. 1. Schematic of the ear chamber. Note that the healing tissue grows up into the chamber through several round holes. Reprinted with permission from Silver.[1]

[1] I. A. Silver, *in* "Wound Healing and Wound Infection" (T. K. Hunt, ed.), p. 11. Appleton-Century-Crofts, New York, 1980.

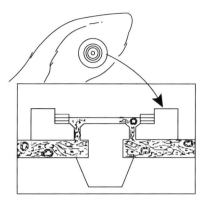

FIG. 2. Partial cross section of the ear chamber as used by Silver. The material can be any material that is transparent to light. Reprinted with permission from Knighton et al.[2]

purchased. Licox (GMS, Germany and Integra Neurosciences, NJ) makes the best of that size (see Notes 2 and 3).

3. Cathode: The cathode is a silver/silver chloride wire of any description.

4. A potentiometer that can be set to 0.6 V.

5. Nanoammeter.

6. Lop-eared rabbit (see Note 4).

7. Sedating drugs may be useful to keep the animal still, but any interference with ventilation will lower the measured oxygen tensions.

8. Punch.

9. Bandaging material.

10. Mineral oil (liquid petrolatum).

11. Holding device to keep the animal still that also must be "homemade."

12. Dissecting microscope with a stage assembly that holds the lower pole of the Lucite plug firmly.

13. Micromanipulator.

14. A Faraday cage is useful if microelectrodes are used (see Note 5). If handled with care, the same animal and chamber can be used many times.

The Wire Mesh Cylinder Model

1. Stainless steel mesh is purchased from Cambridge Wire Cloth Co. (Cambridge, MA). The material is 316SS with 0.010-in.-diameter spacing. The type of metal alloy is important. Perforated plastic tubing is not

[2] D. R. Knighton, I. A. Silver, and T. K. Hunt, *Surgery* **90,** 262 (1981).

satisfactory, although thin meshed plastic of approximately the same properties as described earlier might be useful. We have used silicone mesh for studies requiring magnetic resonance imaging (MRI).

    2. a. Rabbit: 55 × 61 mm of stainless steel mesh rolled around a 3-cm cork borer (steel rod).

      b. Rat or guinea pig: 28 × 33-mm mesh rolled around a 7- to 8-mm rod.

      c. Mouse: 14 × 17-mm mesh rolled around a 2-mm rod.

    3. Polyethylene or polypropylene test tubes with caps that are the same size as the cylinders. Alternately, the caps of Nunc internal threaded cryovials are useful.

    4. The silicone elastomer is manufactured by Dow Corning Medical products MDX4-4210 and is purchased from Factor II Inc. (Lakeside, AZ).

    5. Silastic, as an alternate to the MDX polymer.

    6. Oven.

    7. Glass rod.

    8. Cork borer large enough to contain the cylinders but no larger.

    9. Surgical instruments.

   10. Suture.

   11. Syringes.

   12. 19-gauge spinal needle for introducing the oxygen electrode.

Subcutaneous Wound Tissue Oximetry

Preparation of Tonometer

Silicone tubing (Dow Corning, Midland, MI; 1.2 mm o.d., 0.76 mm i.d.). Available from Fisher Scientific (see Note 6).

20-gauge, 1-in. intravenous catheter with introducer needle (see Note 7).

19-gauge spinal needle, obturator in place with hub cut off or a 12- to 14-g spinal needle.

Placement of Tonometer

Powder free, sterile gloves
Sterile tonometer
Indelible ink marker
Ruler
1% lidocaine (without epinephrine)
Bottle of betadine
Bottle of alcohol
Four sterile towels
#11 blade scalpel

Sterile 4 × 4 gauze
Two small squares of Tegaderm (3M, Minneapolis, MN)

Measurement of PsqO₂

Licox PO_2 computer (Integra Neuroscience/GMS)
C1 Licox oxygen probe (length 15–20 cm, measuring tip 5–8 mm, o.d.
    0.5–0.8 mm) (see Notes 8 and 9)
C8 Licox temperature probe
Small bottle normal saline, sterile
10-ml syringe
16-g auge needle or filter straw
IV tubing with Y connector, including stopcock on one limb (Fig. 3;
    see Note 10)
1-in. Dermapore, silk, or paper tape
Nonsterile gauze 4 × 4's and 2 × 2's

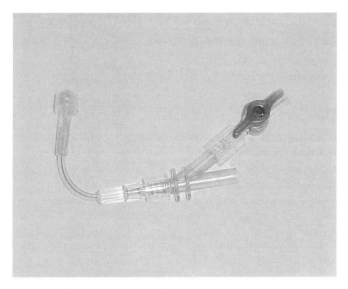

Fɪɢ. 3. Y adaptor. This consists of IV tubing with a Luer connector at the distal end and a
Y connector at the proximal end. One limb has a stopcock and is used for flushing. The
oxygen probe is inserted through the other limb. Reprinted with permission from Hopf *et al.*[3]

[3] H. W. Hopf, T. K. Hunt, H. Scheuenstuhl, J. M. West, L. Humphrey, and M. D. Rollins, *in*
    "Wound Healing: Methods and Protocols" (L. A. DiPietro and A. L. Burns, eds.), p. 389.
    Humana Press, New Jersey, 2002.

Chux or towel
Oxygen supply
Simple face mask
Pulse oximeter

Measurement of Transcutaneous Oxygen in Patients with Ulcers of the Lower Extremity

Electrolyte solution
Transcutaneous oximeter, including oxygen sensors (see Notes 11 and 12).
Attachment device (double stick ring for Novametrix, ring with attachment well for Radiometer)
Alcohol wipes
1-in. Dermapore, paper, or silk tape
Chux or towel
Foam wedge for elevating leg
Oxygen supply
Simple face mask or nonrebreather face mask
Pulse oximeter

Methods

The Rabbit Ear Chamber Model

Healing has six main components: inflammation, epithelization, angiogenesis, contraction, collagen deposition, and resistance to infection. All are influenced by the oxygen supply. None of them occur at very low oxygen tension. Collagen synthesis rises to a peak at about a PO_2 of 250 mm Hg. Resistance to infection peaks at about 600 mm Hg. Each of the cells involved in wound healing, including epithelial, inflammatory, endothelial, and fibroblast, has a range of oxygen tension at which it functions best.

The gradients of oxygen from a point in the vasculature to the center space of a wound are closely packed and steep. The PO_2 falls from near arterial levels at the capillary to near zero at the furthest distance from the capillaries and zero at the avascular center of the wound. These gradients can best be quantified using the rabbit ear chamber. This method is by far the most difficult to perform, but no other technique so clearly defines the heterogeneity of wound oxygen concentrations. It creates a living, healing wound in thin cross section in an apparatus that allows point-by-point PO_2 measurement. Many other substances, particularly those that change upon manipulation of oxygen, may be measured in this system. Measurements using the rabbit ear chamber form the basis for the understanding of oxygen delivery in wounds that has led to improved wound outcomes in surgical patients and those with chronic wounds. Figure 4 shows a profile of oxygen tensions and lactate concentrations in a rabbit ear chamber (see Notes 13–15).

Ear chamber wounds are made by punching a hole in the ear of a lop-eared rabbit into which a transparent "plug" or module is inserted. This "plug" becomes fixed in place and guides healing of the tissues at the edge of the hole through a space about 2 cm in diameter and 60–200 μm thick. Because it is transparent, the module can be mounted on a microscope stage and the growth of vessels though the engineered space can be observed. The top of the "wound space" that the new tissue traverses is a removable transparent membrane made of a material appropriate to the experiment to provide access to the tissue if desired. Magnification can be more than sufficient to see red blood cells flowing through vessels.

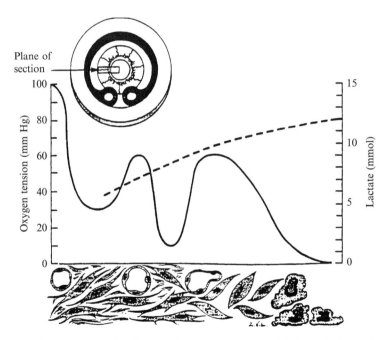

Fig. 4. The wound module. Cross section of the wound module in a rabbit ear chamber (left upper corner). Note that PO_2, depicted graphically above the cross section by the solid line, is highest next to the vessels, with a gradient dropping to zero at the wound edge. Note also that the lactate gradient, denoted by the dotted line, is highest in the dead space and lower (but still above plasma) closer to the vasculature. Reprinted with permission from Silver.[1]

The method was used most frequently by Ian Silver of Bristol University (UK). Most of his papers on this subject were published before 1965 and are difficult to find. A few that describe the method best are listed in Refs. 2, 4–6. Most of the equipment must be handmade. Suzuki and associates in Japan and London have used ear chambers, although not for measuring oxygen[7,8] (see Note 16).

1. Gas-sterilize the entire ear chamber apparatus and implant it using a semisterile technique.

2. Punch a hole in the ear of a rabbit in a place that has the least apparent large vessels (see Fig. 5).

3. Elevate the inner skin from the cartilage. Implant the device by placing the Dacron "skirt" under the inner skin to provide mechanical stability. Rearrange the skin to fit comfortably. No sutures are placed.

4. Bandage the ear securely.

5. At any time after the vessels first appear, the chamber can be used (see Note 17). First, remove the circlip. Next, the glass cover is carefully lifted out (usually leaving the Teflon membrane in place) or the holes in the two glass membranes are aligned.

6. Fill the well of the assembly with mineral oil to prevent access of air.

7. Sedate the rabbit and confine in a holder. Place the ear chamber in the fixation plate on a dissecting microscope.

8. Select the sites for measurement and set the electrodes, mounted on a micromanipulator, in place. The electrodes can be placed either on the surface of the oxygen-permeable Teflon or may pierce through it (see Note 18).

9. When studies are finished for the day, replace the glass cover or seal with a large plug that fills the well precisely, as any exposure to air, even with the Teflon in place, will stop the angiogenesis.

The Wire Mesh Cylinder Model

One of the criticisms of the ear chamber model is the large amount of data generated that are difficult to interpret statistically. One way to avoid the problems of complexity, too much data, and too much diversity

[4] T. K. Hunt, W. S. Andrews, B. Halliday, G. Greenburg, D. Knighton, R. A. Clark, and K. K. Thakral, in "The Surgical Wound" (P. Dineen and G. Hildick-Smith, eds.), p. 1. Lee & Febiger, Philadelphia, 1981.

[5] D. R. Knighton, T. K. Hunt, K. K. Thakral, and W. H. Goodson, III, *Ann. Surg.* **196**, 379 (1982).

[6] T. K. Hunt, M. J. Banda, and I. A. Silver, in "Fibrosis" (D. Evered and J. Whelan, eds.), p. 127. CIBA Foundation Symposium 114, Pitman Publishing Ltd., London, 1985.

[7] Y. Suzuki, K. Suzuki, and K. Ishikawa, *Br. J. Plast. Surg.* **47**(8), 554 (1994).

[8] T. Nakata, K. Takada, M. Komori, A. Taguchi, M. Fujita, and H. Suzuki, *In Vivo* **11**(4), 375 (1998).

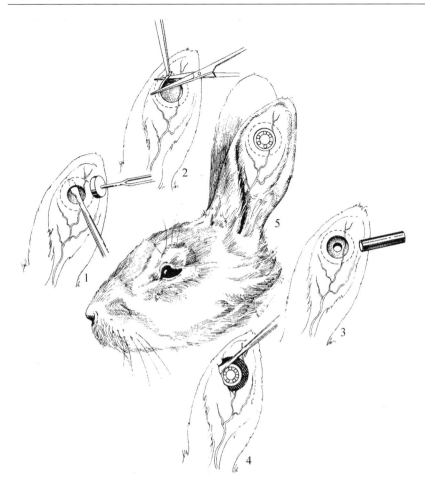

FIG. 5. Implantation of the ear chamber. First, make a small hole in the inner skin of the ear in a place with relatively few visible vessels. Second, undermine the skin to give space for the Dacron skirt that is fixed to the Lucite plug. Mark a spot and punch a hole through the cartilage and dorsal skin. Insert the chamber with the inner skin of the ear over the Dacron skirt, cutting away some skin if necessary. Reprinted with permission from Knighton et al.[2]

is to measure the PO_2 in the dead space, at one definable point, and then calculate the gradients as from arterial or midcapillary level to wound space level. This is done in what is called a "dead-space" wound. This model is perhaps the simplest and least expensive of wound models and can be used in any animal, including humans. Gases, electrolytes, lactate, glucose, cytokines, growth factors, protein, and in-grown collagen (after

14–21 days) have been measured, making this probably the most frequently used model in wound healing.[9–15] Arterial or venous levels drawn at about the same time with serial draws from the wound can be done if wound fluid samples are small in comparison with the volume of the cylinder. This allows studies of the total gradients. The flow of substances both into and out of the wound can be studied. With regard to PO_2 measurements, the PO_2 in the wound is sensitive to arterial PO_2, but the response time for full equilibrium is long, perhaps as long as 45–60 min. An unlimited number of substances can be measured using this model. The cylinders and the implantation method are shown in Fig. 6.

Cylinder Preparation (see Note 19)

STANDARD RAT CYLINDERS

1. Cut the steel mesh into 28 × 33-mm rectangles.

2. Roll on the long axis around a 7- or 8 mm-diameter steel rod and maintain the cylindrical shape by placing in the cap of a polyethylene or polypropylene test tube of the appropriate diameter.

3. Mix the MDX polymer with the curing agent according to the manufacturer's specifications and inject into the cap with the cylinder removed. Replace the cylinder and centrifuge the assembly at about 900g for 10 min. This removes nascent air bubbles in the Silastic, as well as forces the material into the pores of the steel mesh.

4. Incubate the assembly at 75° for 30 min, cool, and repeat the process for the opposite end of the cylinder. Cylinders are autoclaved to sterilize.

SPECIALTY RAT CYLINDERS

1. Specialty cylinders are made from the same dimension mesh rectangles. Tease three wires out of one end along the 28-mm axis, roll the cylinder, and polymerize. Place silastic at the opposite end from the free wires as outlined later. After autoclaving, solid materials may be placed in the cylinder through the open end.

2. Close the open end by pressing a disk of Silastic cut from a 3-mm-thick sheet of polymerized material over the protruding wires. The free

[9] T. K. Hunt, P. Twomey, B. H. Zederfeldt, and J. E. Dunphy, *Am. J. Surg.* **114,** 302 (1967).

[10] B. H. Zederfeldt and T. K. Hunt, *Bull. Soc. Int. Chir.* **1,** 15 (1968).

[11] T. K. Hunt and J. E. Dunphy, *Br. J. Surg.* **56**(9), 705 (1969).

[12] J. Niinikoski, C. Heughan, and T. K. Hunt, *Surg. Gynecol. Obstet.* **133,** 1003 (1971).

[13] T. K. Hunt and M. P. Pai, *Surg. Gynecol. Obstet.* **135,** 561 (1972).

[14] H. P. Ehrlich, G. Grislis, and T. K. Hunt, *Surgery* **72,** 578 (1972).

[15] C. Heughan, G. Grislis, and T. K. Hunt, *Ann. Surg.* **179,** 163 (1974).

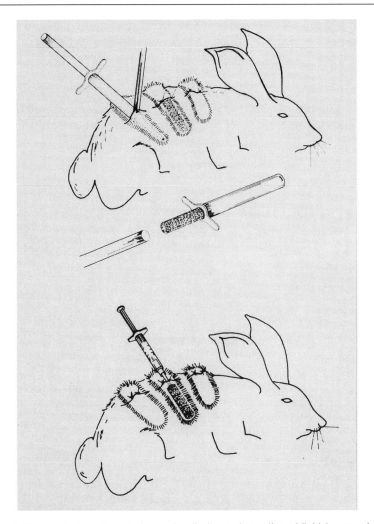

FIG. 6. Method for insertion of wire mesh cylinders and sampling of fluid (see text for full details of placement). Incisions are made in the dorsal skin. A rod is used to insert the cylinder and the incision is closed. After a few days, a syringe is used to sample wound fluid percutaneously. Reprinted with permission, from Hunt et al.[4]

wires grasp the disk strongly enough to use the cylinder like those produced with two polymerized plastic ends (see earlier discussion). Sizes and shapes are not critical for most purposes. The major issue is that they not erode through the skin of the animal.

RABBIT CYLINDERS. These cylinders are manufactured in the same fashion as the rat cylinders from 55 × 61-mm mesh rolled along the short axis. This size cylinder has also been used in sheep and dogs.

Implantation

Four cylinders are implanted in rabbits, guinea pigs, and large rats; two in small rats or mice. The implantation procedure is done under semisterile conditions with the animals under general anesthesia.

1. Make a dorsal, longitudinal incision large enough to insert the cylinder end on.

2. Open the plane between the subcutaneous muscle and the deep fascia for a centimeter or two by spreading a pair of surgical scissors gently.

3. Insert a glass rod or pointed pipette of about the same diameter as the cylinder to enlarge the hole between the fascia and the subcutaneous muscle.

4. Insert a cork borer just large enough to easily hold the cylinder over the glass rod and use a reciprocating motion of the borer and the tube to make a space large enough to hold the cylinder.

5. Remove the glass tube and use it to push the cylinder into position through the cork borer. Enough space from the medial cylinder to the midline should be left so that the skin wound can be closed easily.

6. Repeat the process on the other side through the same incision.

7. Pull the skin down to the spinal ligament using one or two closure sutures during closure of the midline wound so as to separate the cylinders on the two sides. It is important that the cylinders on each side do not touch. If the cylinders do touch, their spaces will communicate and they will lose their identity for statistical purposes.

8. On awakening, the animals seem unaware of the cylinders, provided that the cylinders are not so large that they erode the skin. Dressings are not necessary, and the procedure is well tolerated under general anesthesia.

9. Insert a small needle and syringe though the skin and the Silastic plug to withdraw fluid for analysis (see Notes 20 and 21). Placement of a second, venting needle minimizes bleeding into the cylinder.

10. Measure oxygen tension within the cylinder by introducing an oxygen electrode of the type described in the following section through a 19-gauge spinal needle (see Notes 22–24).

Subcutaneous Wound Tissue Oximetry

In the 1970s Hunt and others began to develop techniques to measure average tissue oxygen from a single site in humans. Hunt developed the silicone tonometer for this purpose (Fig. 7). Silicone is freely permeable to

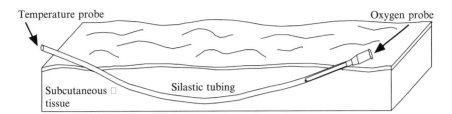

FIG. 7. Schematic drawing of the silicone subcutaneous oxygen tonometer. The oxygen probe is inserted through the catheter hub. The temperature probe is inserted into the protruding end of the tonometer. Adapted from "Musculoskeletal Infection." American Academy of Orthopedic Surgeons, Park Ridge, IL, 1992.[15a]

oxygen. When a length of silicone tubing (1 mm outer diameter) is placed subcutaneously, the PO_2 within the tonometer equilibrates with that of a cylinder of surrounding tissue, thus providing a mean wound tissue oxygen tension equivalent to the mean of the microelectrode-generated histograms. Initially, Hunt infused saline slowly through the tonometer and measured the PO_2 of the effluent (microdialysis). A much more convenient method, however, is to place an oxygen probe within the tonometer to get a continuous measure of tissue oxygen. The tonometer is filled with saline to speed equilibration with the surrounding tissue. The temperature is measured at the same site concurrently because PO_2 is temperature dependent.

Two main types of oxygen probes are currently available: polarographic (Clark) electrodes and optical electrodes (optodes).[16] Polarographic electrodes are based on the observation that when a negatively charged metal and a reference electrode are placed in an electrolyte solution, the current flows in proportion to oxygen concentration, with no current flow in the absence of oxygen. Current technology corrects many of the problems that plagued early polarographic probes, including drift, excessive oxygen consumption at the cathode, and poisoning by substances such as halothane. The electrodes are highly accurate even to oxygen tensions in excess of 800 mm Hg, although their accuracy diminishes somewhat as current decreases toward zero (no oxygen). Optical electrodes use a fluorescent dye (usually ruthenium) in which the fluorescence is quenched by oxygen. Thus the fluorescent output is inversely proportional to oxygen tension. Optodes are highly accurate at low oxygen tensions, but lose accuracy at higher oxygen concentrations (around 300 mm Hg, depending on the design).

[15a] H. W. Hopf and T. K. Hunt, in "Musculoskeletal Infection" (J. L. Esterhai, A. G. Gristina, and R. Poss, eds.), p. 333. Am. Acad. Orthopaedic Surgeons, Park Ridae, IL, 1992.

[16] H. W. Hopf and T. K. Hunt, Adv. Exp. Med. Biol. 345, 841 (1994).

Our laboratory currently uses a polarographic electrode, and the methods described pertain specifically to the Licox probe that we use. However, these methods can easily be adapted to any probe that can be placed securely in the tonometer. Most currently manufactured tissue oxygen electrodes are intended for direct implantation and thus are embedded in a sheath of silicone to increase the averaging area around the electrode. These probes work in a tonometer as well. We continue to use the tonometer because (1) if there is a question of probe drift or malfunction, the probe can be removed to check the calibration and then can be replaced without causing pain to the subject, whereas with a direct implant, a new stick with a sterile probe is required; (2) the probe does not contact the subject directly and the probe need not be sterile, only clean; and (3) multiple measurements can be made over the course of several days without having to leave the probe in place the whole time (the tonometer is easy to secure).

Although the methods described here are used for measuring subcutaneous tissue,[17–21] the same probes can be used with essentially the same technique in a number of tissues, including gut, heart, liver, brain, muscle, and bone.

Preparation of Tonometer

1. Cut silicone tubing into 15-cm lengths.
2. Gently wedge one end of the tubing onto the IV catheter. Wedge the other end onto the 19-gauge spinal needle (if using a 12- to 14-gauge spinal needle, omit this step) (see Notes 25–27).
3. Gas sterilize (do not autoclave).

Placement of Tonometer

METHOD 1. The tonometer should be in place for at least 30 min before actual measurements begin (this can include equilibration time). For about 30 min after the tonometer is placed, readings may reflect arterial PO_2 because of minor bleeding around the tube. The tonometer may be implanted under general or local anesthesia. Placement is identical, except that local anesthetic is not required for placement during general anesthesia.

[17] N. Chang, W. Goodson, F. Gottrup, and T. K. Hunt, *Ann. Surg.* **197,** 470 (1983).
[18] F. Gottrup, R. Firmin, J. Rabkin, B. J. Halliday, and T. K. Hunt, *Crit. Care Med.* **15,** 1030 (1989).
[19] H. W. Hopf, T. K. Hunt, J. M. West, P. Blomquist, W. H. Goodson, III, J. A. Jensen, K. Jonsson, P. B. Paty, J. M. Rabkin, R. A. Upton, K. von Smitten, and J. D. Whitney, *Arch. Surg.* **132,** 997 discussion 1005 (1997).
[20] A. Kurz, D. Sessler, R. Lenhard *et al., N. Engl. J. Med.* **334,** 1209 (1996).
[21] R. Greif, O. Akça, E. P. Horn, A. Kurz, D. I. Sessler *et al., N. Engl. J. Med.* **342,** 161 (2000).

1. Inspect the lateral upper arm for a site without obvious veins, bruises, or broken skin. Mark an entry and exit site a measured 7 cm apart (this is to ensure that the probe tip is at least 2 cm from the exit site and there is no possibility of diffusion of room air to the probe tip).

2. Raise a skin wheal with 1% lidocaine (without epinephrine) at the planned entry and exit sites—the wheal should be at least 1.5 cm diameter. It is not necessary to inject local anesthetic into the subcutaneous tissue between the skin wheals.

3. Prep the arm with a betadine paint solution in a circular fashion.

4. After 1–2 min, use alcohol to wash off the betadine and drape the site with sterile towels.

5. Make a small stab wound with an 11 blade at the planned entry and exit sites. The stab wound should be large enough to allow easy passage of the catheter, but as small and superficial as possible in order to minimize bleeding.

6. Immediately hold point pressure over the sites until the bleeding stops, usually 1–2 min.

7. Using the spinal needle, pull the tonometer through just under the skin so that the catheter is in up to the hub. You want the tonometer to be superficial, but if you see skin puckers, you are too shallow and should pull back and go deeper. The subject should not feel pain—if s/he does, you are probably too shallow.

8. Cut the silicone tubing close to the spinal needle, leaving a fairly long distal segment of silicone tubing. Wipe off any blood with either alcohol or saline. Dry the catheter hub.

9. Cover the tonometer with two small Tegaderm dressings. One should cover the exit site, with about 2 cm covering the silicon tubing. Cut the silicon tubing about 1 cm beyond the Tegaderm edge.

10. The second Tegaderm should cover the entrance site, with about 0.5 cm of the IV catheter hub uncovered. Usually the two Tegaderm will overlap somewhat. If a significant amount of blood accumulates, the dressing should be removed, the site dried, and a fresh dressing applied in the same fashion. Write on the Tegaderm: "Do not remove."

METHOD 2

1. The tonometer in this case consists just of silicone tubing wedged onto the IV catheter. Prepare the entrance and exit sites as outlined earlier.
2. Insert a 12- to 14-gauge spinal needle (see Note 28).
3. Thread the tonometer through the sharp end of the needle (hubless). Remove the spinal needle, leaving the tonometer in place.

Measurement of PsqO$_2$

Each probe has been calibrated at the factory in both nitrogen and room air. Calibration values are listed on the probe package, including the IcO$_2$ (current during room air calibration), IcN$_2$ (current during nitrogen calibration), and TS (temperature coefficient for the probe). These should be input into the proper places on the monitor. Barometric pressure at your site should be entered (760 mm Hg at sea level), along with 209 for oxygen concentration in room air (see Note 29). The calibration temperature and tissue temperature settings should be 00, as you will be measuring actual temperature during both calibration and oxygen measurement. Calibration is performed with both the temperature and the oxygen probes in room air by pushing the blue calibration button (see Notes 30 and 31). For quality assurance, record room air PO$_2$ and temperature immediately after calibration and again after removing the probe from the tonometer (see Notes 32 and 33).

1. Draw up 10 ml of saline into the 10-ml syringe using the 16-gauge needle or the filter straw.
2. Carefully place the calibrated probe through the Y adaptor (see Note 10) and flush with saline to remove all bubbles. Then place the assembly into the hub of the tonometer (see Note 34).
3. Tape the tubing to the patient's arm so that the probe lies flat.
4. Tape the monitor cable or probe/cable connection to the side rail.
5. Flush the assembly with at least 8 ml of normal saline.
6. Place the thermocouple probe through the distal end of the silicon tubing.
7. Advance it as far as it will go. Flush the system gently with about 0.5 ml saline. Figure 8 shows a patient with the entire assembly in place in the arm.
8. Cover the entire arm with a Chux or towel to decrease heat loss.
9. Record the baseline value for PO$_2$ (subcutaneous oxygen) and Tsq (subcutaneous temperature) only after the value has changed ≤ 1 mm Hg in 5 min and it is at least 25 min since probe insertion (see Notes 35–37).
10. Data may be recorded continuously or at specified intervals.
11. For the oxygen challenge, administer 7–10 liters min oxygen via a simple face mask (40–60% oxygen). Record the final value after the oxygen has been on for at least 25 min and PsqO$_2$ has changed ≤ 1 mm Hg in 5 min (see Note 38).
12. After the measurements and any interventions are complete, remove the probes from the tonometer gently and replace in their covers.

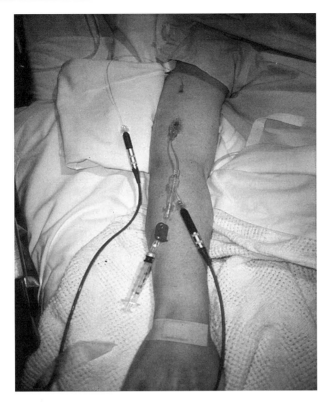

Fɪɢ. 8. A tonometer in a patient's arm. The oxygen probe has been placed through the Y connector, and the temperature probe has been inserted through the distal end of the tonometer until the measurement tip is under the skin. Reprinted with permission from Hopf et al.[3]

13. Record the room air PO_2 after about 5–10 min, when the value is stable (see Note 39).
14. Place a 2 × 2 gauze under the catheter hub and tape a 4 × 4 gauze over the catheter hub and the distal tip of the tonometer if reuse is planned.

Removal of Tonometer

1. Remove the Tegaderm.
2. Pull the catheter hub out slightly until you can grasp *both* the catheter hub and the attached silicone tubing. Normally, you will be able to see the silicone at the catheter hub.

3. While holding the silicone as it attaches to the catheter hub, cut the other free end of the silicone tubing at the exit site and pull the tonometer out by the hub.

4. If you do not see any silicone on the catheter, grasp the tonometer by both ends and pull—if the tubing and catheter are disconnected, the entire tonometer will come out in two pieces. If this protocol is followed, you will never leave silicone under the skin.

5. If necessary, wash the skin with alcohol before dressing with a fresh Tegaderm.

Measurement of Transcutaneous Oxygen in Patients with Ulcers of the Lower Extremity

Subcutaneous oxygen measurement is accurate and simple, and application of the technique in surgical patients has improved outcomes in acute wounds. The oxygen supply is similarly critical and often impaired in chronic wounds. However, use of an invasive technique is often contraindicated or difficult in patients with chronic wounds. Transcutaneous oximetry is a noninvasive measurement that, although less reliable than subcutaneous oximetry, has been used in a parallel fashion in chronic wounds.[22–27] Transcutaneous oximeters use a polarographic electrode embedded on a flat sensing surface to measure PO_2 (Fig. 9). Some also contain a carbon dioxide electrode. Normally, diffusion of oxygen through the skin is so slow that noninvasive measurement is impractical. The transcutaneous probe heats the skin (generally a temperature of 42–44° is selected), which liquefies the stratum corneum and allows increased oxygen diffusion. The probes were originally developed to allow the noninvasive measurement of arterial oxygen tension on the trunk. The device is useful in newborns, where the skin is quite thin, but not in adults, where tissue oxygen consumption becomes a larger factor and the variability of skin thickness makes correction impossible. However, it was discovered that in tissue with impaired perfusion (e.g., on the leg of a patient with peripheral vascular disease), transcutaneous oximetry is a fairly good measure of skin/subcutaneous/wound oxygen tension.

[22] R. Wütschert and H. Bounameaux, *Diabetes Care* **20**, 1315 (1997).

[23] E. M. Burgess, F. Matsen, C. R. Wyss, and C. W. Simmons, *J. Bone Joint Surg.* **64**, 378 (1982).

[24] C. M. Butler, R. O. Ham, K. Lafferty, L. T. Cotton, and V. C. Roberts, *Prosthet. Orthot. Int.* **11**, 10 (1987).

[25] G. S. Dowd, *Ann. R. Coll. Surg. Engl.* **69**, 31 (1987).

[26] K. Ito, S. Ohgi, T. Mori, B. Urbanyi, and V. Schlosser, *Int. Surg.* **69**, 59 (1984).

[27] B. Smith, L. Desvigne, J. Slade, J. Dooley, and D. Warren, *Wound Rep. Reg.* **4**, 224 (1996).

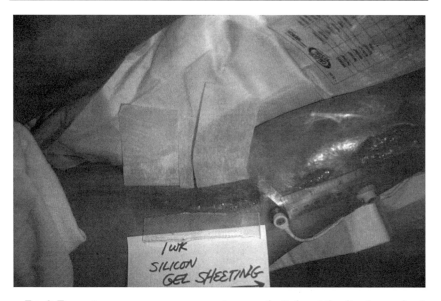

FIG. 9. Transcutaneous oxygen measurement on a patient's leg at the site of a previously nonhealing surgical incision. Probes are placed on clean, flat, intact skin as close to the wound edge as possible.

A number of studies[22–26] have demonstrated that healing is impaired at transcutaneous oxygen levels less than 40 mm Hg and that healing almost certainly will not occur at levels less than 20 mm Hg. A few studies have demonstrated improved healing when tissue oxygen is raised above 20 or 40 mm Hg,[27] but the research in this area lags behind that in acute wounds. However, it seems clear that the results of large, randomized, controlled trials will be similar to those seen in acute wounds.

1. The subject should lie supine on a comfortable surface (see Notes 40–42). Choose six sites: two to three around the ulcer, one on each calf, and one on each foot.

2. Cover the sensors with an electrolyte solution and a gas-permeable membrane. The membrane must be replaced periodically, but can otherwise be cleaned with alcohol between uses. The manufacturer provides clear directions for membrane application. Warm the sensors up for about 10 min prior to application (see Notes 43 and 44). Calibration of the sensors may be done in room air for some probes, whereas others require placement in a calibration chamber with controlled gas concentrations.

3. Wipe the sites clean with alcohol to improve adherence (see Note 45).

4. Place the attachment device on the site or on the probe, which is then applied to the site, depending on the preference of the operator. Place a drop of solution on the sensor (or on the skin) prior to application of the sensor. The sensor cable can be secured with tape (see Note 46). Cover the leg with a Chux, towel, or single bath blanket to minimize heat loss (see Note 47).

5. Equilibration requires 10–20 min. Values should not be recorded until changes are less than 1 mm Hg in 5 min. Choosing a standard equilibration time (such as 10 min) instead of waiting for true equilibration will increase inaccuracy.

6. After the baseline is reached, the subject should undergo an oxygen challenge. The concentration used should be standardized within your laboratory, although various concentrations have been used in published studies. One option is to administer 40–60% oxygen using a simple face mask, with oxygen flow set at 7–10 liters/m. To achieve a higher FiO_2, a nonrebreather mask with a 15-liters/m oxygen flow may be used.

7. Measurements should be made with the subject supine, with legs elevated, and with legs dependent, breathing room air and supplemental oxygen at each position.

Notes

1. Wound healing will not progress if a Teflon membrane is used alone.

2. The same technology can be used to measure other substances. To measure lactate, for example, a larger glass capillary tube is drawn down over a platinum wire and cut so as to leave a small opening at the tip. A small piece of filter paper saturated with a solution of lactate oxidase is placed in this space. The oxidase produces a current proportional to the oxygen derived.

3. Other ion-sensitive electrodes can be made for the measurement of glucose, CO_2, etc. Any microprobe that can sense through or under a Teflon membrane is acceptable. It is also possible to remove the entire membrane and replace it without disturbing the anatomy if necessary.

4. During studies, the rabbit and its ear should be kept warm. If the animal or the ear become cool, tissue oxygen will decrease rapidly. The effect of temperature is a recurring theme in tissue oxygen measurements.

5. The tiny oxygen probes produce so little current that even a person walking into the room during a measurement may produce an error. A Faraday cage is useful to minimize stray currents.

6. The availability of silicone tubing of the recommended size varies. The inside diameter should be no less than 0.7 mm; the outside diameter should be no more than 1.4 mm. The thickness of the silicone should be less than 0.4 mm or equilibration will be delayed.

7. Any brand of IV catheter is fine. Depending on the silicone tubing size, an 18-gauge catheter may be a better fit. The IV catheter can be cut shorter by up to a centimeter using a #20 scalpel. This may be required to accommodate probe length (see Note 8).

8. Note that any oxygen electrode system may be substituted for the Licox device, as long as it is stable and fits the tonometer. The manufacturer of the Licox probes, GMS, was bought by Integra Neuroscience and the design of the probes may change somewhat in the next few months. Thus these directions will have to be adapted to the actual probe dimensions.

9. Make sure you purchase a probe system that allows you to calibrate the probes in room air (rather than requiring that you use a factory-determined calibration value), or you will not be able to adjust for drift that may occur with storage of the probes. The probes must be stored in a saline-filled sheath to prolong the shelf life. However, they must be calibrated in air (ideally 100% humidity) or they will drift. We shake most of the saline out of the sheath and then place the oxygen and temperature probes side by side within the sheath, about 2 cm above the saline.

10. The company that custom built the Y connector/IV extension shown in Fig. 3 (Burron Medical Inc., Bethlehem, PA) either is no longer in business or is no longer willing to custom build the connector for us. However, it should be possible to build these connectors from commercially available products. The IV extension tubing for the Y connector should be of appropriate length so that when the probe is placed through the Y connector, tubing, and IV catheter into the silicone tubing, the probe tip is fully within the silicone tubing (not the IV catheter) but less then 1.5 cm past the tip of the IV catheter (Fig. 8).

11. Radiometer and Novametrix both make excellent transcutaneous oximeters (including oxygen probes) specifically intended for evaluation of the adequacy of lower extremity perfusion and oxygenation.

12. Transcutaneous probes can contain a single oxygen electrode or both oxygen and carbon dioxide electrodes. Most studies have been published using only oxygen values, but carbon dioxide may be worth evaluating.

13. The ear chamber model is remarkably versatile. Gradients across the edges can be demonstrated. The effects of hypovolemia, oxygen breathing, and various drugs or gases can be measured. Implants of bone slices or other materials have been placed to record the architecture and rate of progress of the invading or surrounding vasculature. Additionally, wounds have been allowed to mature and the new tissue, at the chamber's largest diameter, has been reincised. By replacing the coverslip, the evolving vasculature can be studied. Perfusion patterns change; initially the

flow is parallel to the new wound and then later runs across the site of the new wound as the vessels mature. Blood flow often reverses itself from day to day in the new vessels.

14. These data show the problems inherent in blindly placing electrodes into wound tissue, no matter how well the electrodes are engineered. Histograms are needed to analyze such results, and statistical analysis becomes difficult. One weakness of the method is that it is very difficult to apply conventions to the exact site of measurement.

15. It is important to note that in one respect Fig. 4 is misleading. There are areas of zero PO_2 to be found on or in healing tissue, but the mean PO_2 in the space is higher. In other models, the PO_2 in the wound space may quite low. For example, levels are about 5–20 mm Hg in wounds with a significant dead space (e.g., the wire mesh cylinder model; see earlier discussion); however, PO_2 in incised wounds is generally higher than that in the cylinder model.

16. This method is almost an art form. If it is to be undertaken, considerable study and practice are required. Silver and Niinikoski (of Turku, Finland) have the greatest experience. The team of Nakata, Takada, Komori, Taguchi, Fujita, and Suzuki of the Department of Anesthesiology, Tokyo Women's Medical College, Japan, are the most recent users of the ear chamber technique, although they did not use it for oxygen measurements. Their device is similar but not identical to the Silver device.

17. Fluid displaces the air in the chamber after a few days. About a week later, semicircular areas of angiogenesis will appear in the dorsal window as the healing tissue enters the chamber through the holes between the ear cartilage and the chamber. These will grow inward and coalesce into a ring of new vessels that is analogous to a growth ring, and these eventually will fill the entire chamber. Some vessels will enlarge, whereas others will drop out until a mature circulation is developed with only a few vessels of about 0.3 mm size running across it.

18. Using the electrode on the membrane avoids poisoning the probe with protein. Penetrating it usually means that measurements have to be made quickly and that results are likely to be victims of electrode drift.

19. Instructions are given for the preparation of cylinders for use in rats or guinea pigs. The preparation of cylinders for use in rabbits and mice is identical, except for the size of the mesh rectangle and the diameter of the rod. These dimensions were specified earlier.

20. Fluid enters the cylinders slowly, and the skin adheres to them in a few days. In rabbits, adequate fluid for most purposes can be obtained after 3 days. Rats take a little longer. In these respects, guinea pigs are not as good to work with as they may take up to 10 days to fill the chambers. Chambers in guinea pigs are also more often infected.

21. Wound fluids are usually clear tan or brown. Any gross blood or turbidity invalidates most results. In rats and rabbits, infections are rare. Guinea pigs are less immune. It takes an injection of almost a million organisms to infect a rabbit chamber, but almost any bacteria will infect a guinea pig. If the fluid is turbid, measure the PO_2 in a sample several times over a period of 10 min or so. Rapid decline of the PO_2 indicates potential infection. Cells float in the wound space fluid and use a significant amount of oxygen. Oxygen levels should be measured within a few minutes of fluid withdrawal. If consumption appears to be an issue, one can describe or account for it merely by recording the decay rate of oxygen in the cuvette of the blood gas instrument as opposed to the decay due to the electrode itself (measured in a sterile aqueous sample). If rapid decay occurs in the saline control sample, bacteria or white cells are caught in the cuvette, and it should be cleaned.

22. The PO_2 within the wound fluid is usually 7 to 20 mm Hg during room air breathing under general anesthesia, lower on the first days of measurement (mean about 10 mm Hg) than later (15 to 18 mm Hg). The standard deviation is usually about 30% of the mean.

23. For oxygen, it is necessary to use a blood gas syringe that is impermeable to oxygen. Because no hemoglobin, or very little, is present, any contamination with air represents a significant error. For statistical purposes, it is best to sample each chamber only once during the entire experiment, as penetration may alter the wound fluid oxygen as well as many other factors, such as vascular endothelial growth factor. For exploratory purposes, this can be overlooked, as the drift with serial penetrations is relatively small. If insufficient fluid is present, fluid can be taken from the dependent end of the cylinder using a needle in the other end as vent for air to replace the fluid.

24. Alternatively, wound fluid samples can be analyzed for blood oxygen, pH, and carbon dioxide. Such studies must be done with care, however, as the introduction of any air bubbles will falsely raise the PO_2 to around 150 mm Hg. Moreover, blood gas analysis is inaccurate in the absence of red cells due to the loss of buffering capacity. Samples must be analyzed immediately, and may still yield inaccurate results.

25. The cut edges of the spinal needle must be smooth or they may tear the silicone tubing.

26. Leave the introducer needle partly in the catheter for stiffening. Make sure you do not cut the tubing with the needle or kink the catheter. A rolling, pulling motion is most successful. The tubing should reach all the way to the catheter hub or it may come off during placement.

27. The silicone tubing should be at least 3 cm onto the spinal needle or it will slide off during placement. The spinal needle size may need to be adjusted depending on the tubing size.

28. If the large-bore (12- to 14-gauge) spinal needle technique is used, make sure the silicon tubing is sized to fit easily through the needle tip.

29. Barometric pressure is 760 mm Hg at sea level and decreases with increasing altitude. Most hospital blood gas laboratories measure barometric pressure daily. Barometric pressure does not vary much, so once an average barometric pressure is determined for your site, there is no need to measure it each day.

30. Make sure you warm up the machine with the probe connected for at least 15 min before you calibrate. The oxygen and current values should be stable prior to calibration. Make sure you calibrate with the temperature probe and oxygen probe in close proximity. Check that the room air reading you get makes sense. You can calculate the room air oxygen tension for your barometric pressure using the equation: (barometric pressure-47) (0.209), where 47 is water vapor pressure (saturated at $37°$) and 0.209 is the concentration of oxygen in room air. The value should be in the range of 150–160 mm Hg. The calculated value is displayed by the monitor during calibration.

31. The IcO_2 (current) value at the time of calibration should be recorded. The IcO_2 display on the monitor should be updated to reflect the true current rather than the factory calibration value. If there is a need to unplug the machine or disconnect the patient, this can then be done with no loss of calibration. The probes should be left in place in the tonometer and disconnected at the hub from the cable.

32. The oxygen probe should never be left in air for more than a brief period or it will dry out and fail.

33. The probe–cable connection is a straight pull style. Twisting or unscrewing will damage the probe.

34. It is generally easiest to dispel all air bubbles from the Y connector and extension tubing with the assembly in a vertical position, and before connecting it to the tonometer. Make sure that the arm with the tonometer is also filled with saline. This requires loosening the probe–Y connector connection slightly, injecting a small volume from the syringe, and then retightening. After you insert the probe assembly into the tonometer, flush the tonometer at a fairly rapid rate (to dislodge air bubbles), but without excessive force. If you need to refill the saline syringe (sometimes several flushes are required), before you remove the syringe, turn the stopcock off. Fill the hub with saline before reattaching the syringe and opening the stopcock to prevent the introduction of air bubbles. Place gauze over the open end of the tonometer during flushing to keep the patient's arm dry.

35. When you place the temperature probe in the distal end of the tonometer, make sure you watch the wire disappear under the skin. After the probe is inserted, the measured tissue temperature will slowly rise and

then stabilize as the room temperature saline warms to body temperature. If the temperature decreases, check to make sure that the temperature probe has not slipped back—try to advance it again and see if the temperature goes up. The temperature will be displayed only about every minute, so you have to watch for it.

36. It is a good idea to record values every 5 min during equilibration to be sure that a stable baseline has actually been reached.

37. $PsqO_2$ should be 50–70 mm Hg if the subject is well perfused, lower if not. It should not be much higher unless the subject is on oxygen at baseline. $PsqO_2$ over 100 mm Hg (subject breathing room air) suggests an air bubble. If this occurs, the tonometer should be flushed with saline. Routine monitoring of arterial oxygen saturation by a pulse oximeter is helpful for data interpretation. The subcutaneous temperature in well-perfused subjects should be $>34.5°$. It may be as low as $30°$ in poorly perfused subjects.

38. In well-perfused subjects with normal pulmonary function, the administration of 50% oxygen should increase $PsqO_2$ to 90–130 mm Hg. The increase will be smaller, or absent, when perfusion is impaired.

39. At the end of the measurement, if the probe is more than 10% off from the original room air PO_2 reading when it is removed from the patient, the measurement should be rejected or repeated.

40. It is best to have the subject void just before the test, as they will not be able to get up once you start.

41. In the past, a reference sensor was advocated on the chest wall to account for the variability in arterial oxygen, but the introduction of pulse oximetry renders this site unnecessary. It is not feasible to measure multiple sites with only one probe. If only one probe is used, measure the most distal wound edge.

42. The measurement sites should be flat, without obvious vessels or bruising. Thick skin or a callus will give falsely low values. The plantar surface of the foot is rarely an acceptable site. The probe should not be over a bony prominence or a site that will be under pressure. The peri-wound sites should be as close to the open area as possible, but not on macerated or uneven skin. One site should be at the distal edge of the wound, as this is often the lowest value. Sites should be described in detail on the data sheet so that the same sites may be used for follow-up studies. If the wound is hypoxic, the values at the other sites help identify etiology. To assist in identifying the etiology of wound hypoxia, measurements should be made with the subject supine, with legs elevated, and with legs dependent, and breathing room air and supplemental oxygen at each position. However, if venous or arterial insufficiency is already known to be present, or time is a limiting factor, the supine values are most critical.

43. The sensors are generally preset to heat to 44°. If the patient has arterial disease or particularly sensitive skin, the heat can be decreased to 42° as a safety precaution to prevent burns, although this is rarely necessary. The sensors can be left in place for up to 3 h. We have never had a burn and routinely use 44°. When the sensor is removed, the site will be redder than the surrounding skin because of heat-induced vasodilation, but this resolves rapidly.

44. Some monitors allow you to measure the energy required to maintain the sensor surface temperature at 44°. Although no study has evaluated this systematically, it may be a useful measure of skin blood flow, as more energy will be required to maintain the temperature at a higher blood flow.

45. In some cases, the alcohol may not be sufficient to remove products that have been applied to the skin and it will be impossible to affix the sensor securely. If the seal with the skin is not airtight, air will leak in and throw off the values. Isoflurane, a common vapor anesthetic, is an extremely effective solvent and can be used when alcohol fails. Isoflurane can be obtained from the anesthesia work room in your operating room. Because is an extremely potent vapor, you should use very little and make sure to replace the cap immediately.

46. Tape should not be applied directly over the probe, as pressure can cause inaccuracy.

47. Covering the skin will raise skin surface temperature an average of 2°, even in patients with arterial insufficiency.

Acknowledgment

The methods described in this chapter are adapted from Hopf et al.[3]

[36] Regulatory Role of Lactate in Wound Repair

By Q. PERVEEN GHANI, SILVIA WAGNER, HORST D. BECKER,
THOMAS K. HUNT, and M. ZAMIRUL HUSSAIN

Introduction

Healing wounds produce and accumulate large concentrations of lactate.[1–3] Wound healing studies in the last decade revealed that hypoxia and high lactate (10–15 mM) are characteristics of healing.[4,5] It was found that disruption of microcirculation and subsequent increased oxygen consumption led to the production of lactate.[6–8] Under these conditions, some cells, particularly neutrophils, macrophages, and fibroblasts, depend to a large extent on glycolysis for cellular energy and produce large amounts of lactate through aerobic glycolysis.[9] Lactate levels remain elevated even when oxygen tensions in wounds are increased.[1] Studies have shown that high lactate stimulates collagen deposition and angiogenesis.[1,10,11]

We found that lactate downregulates ADP-ribosylation, which is dependent on the NAD$^+$ pool and activities of ADP-ribosyl transferases.[12] During this process, ADP-ribose (ADPR) moiety from NAD$^+$ is transferred enzymatically onto acceptor proteins and modifies their functions.[12] ADP-ribosyl transferases transfer a single ADPR onto the acceptor protein in cytoplasm. In eukaryotes, polyADP-ribose polymerase (PARP) transfers ADPR moieties onto the nuclear proteins and synthesizes polymers of ADPR. The levels of pADPR and ADPR therefore are expected to affect the functions of various nuclear and cytoplasmic proteins and protein expression. Because the depletion of NAD$^+$ downregulates pADPR

[1] T. K. Hunt, W. B. Conolly, S. B. Aronson, and P. Goldstein, *Am. J. Surg.* **135**, 328 (1978).
[2] D. R. Gibson, A. P. Angeles, and T. K. Hunt, *Surg. Forum* **48**, 696 (1997).
[3] T. K. Hunt, *J. Trauma* **30**, S122 (1990).
[4] T. K. Hunt, M. Linsey, M. Sonne, and E. Jawetz, *Surg. Forum* **23**, 47 (1972).
[5] T. K. Hunt and M. P. Pai, *Surg. Gynecol. Obstet.* **135**, 561 (1972).
[6] K. Jonsson, J. A. Jensen, W. H. Goodson, III, J. M. West, and T. K. Hunt, *Br. J. Surg.* **74**, 263 (1987).
[7] N. Chang, W. Goodson, F. Gottrup, and T. K. Hunt, *Ann. Surg.* **197**, 470 (1983).
[8] J. Gosain, J. Rabkin, J. P. Reymond, J. A. Jensen, T. K. Hunt, and R. A. Upton, *Surgery* **109**, 523 (1991).
[9] M. D. Caldwell, J. Shearer, B. Morris, W. Mastrofrancesco, J. Henry, and J. E. Albina, *J. Surg. Res.* **3**, 63 (1984).
[10] J. A. Jensen, T. K. Hunt, H. Scheuenstuhl, and M. J. Banda, *Lab. Invest.* **54**, 574 (1986).
[11] D. R. Knighton, T. K. Hunt, H. Scheuenstuhl, and B. Halliday, *Science* **22**, 1283 (1983).
[12] D. D'amours, S. Desnoyers, I. D'Silva, and G. G. Poirier, *Biochem. J.* **342**, 249 (1999).

and ADPR, we propose that the accumulated tissue lactate may enhance protein expression and activation. Studies have indicated that downregulation of both types of ADP-ribosylation reactions is involved in wound repair.

Effect of Lactate on Collagen Synthesis

That high levels of lactate (15–20 mM) stimulate collagen synthesis by cultured fibroblasts has been known since 1964.[13,14] However, the mechanism was unknown except that the activity of prolyl hydroxylase, the enzyme that converts proline into hydroxyproline in collagen peptide, was enhanced.[14] The regulatory role of lactate became pertinent when Hunt and colleagues[1] observed similar high lactate concentrations in healing wounds. Subsequent studies demonstrated that lactate has several sources in wounds, remains unchanged by changes in oxygen level, and is produced through aerobic glycolysis and as a by-product of the oxidative burst of leukocytes.[1,9,15]

Investigations in our laboratory suggested that lactate stimulates collagen synthesis via at least two separate mechanisms. First, lactate activates collagen promoter activity,[16] which leads to increased procollagen mRNA production[17] and collagen synthesis.[17,18] Based on these results, we hypothesize that collagen gene transcription in fibroblasts is downregulated by polyADP-ribose (pADPR) in the resting state and that high lactate decreases pADPR levels by depleting intracellular NAD^+, the substrate for pADPR. In support of this hypothesis, Gimbel and colleagues[16] reported that inhibitors of PARP, such as nicotinamide and 3-aminobenzamide, also increase collagen promoter action and that the lactate effect is blocked by oxamate, which prevents the reduction of NAD^+ by lactate dehydrogenase (LDH).[18]

Second, independent of its stimulatory action on collagen transcription, lactate also increases the activity of collagen prolyl hydroxylase. Ghani et al.[18] reported that this activation is mediated by the removal of ADPR, which is a potent inhibitor of the enzyme.

These observations suggest the following scenario for the lactate effect on collagen synthesis. A high concentration of lactate lowers the

[13] H. Green and B. Goldberg, *Nature* **204,** 347 (1964).

[14] J. P. Comstock and S. Udenfriend, *Proc. Natl. Acad. Sci. USA* **66,** 552 (1970).

[15] O. Warburg, *Science* **123,** 309 (1956).

[16] M. L. Gimbel, T. K. Hunt, and M. Z. Hussain, *Surg. Forum* **51,** 26 (2000).

[17] Q. P. Ghani, M. Z. Hussain, J. Zhang, and T. K. Hunt, *in* "ADP-Ribosylation" (Poirier and Moreaer, eds.), p. 111. Springer-Verlag, New York, 1992.

[18] M. Z. Hussain, Q. P. Ghani, and T. K. Hunt, *J. Biol. Chem.* **264,** 7850 (1989).

steady-state NAD^+ level, decreasing the levels of nuclear pADPR and cytoplasmic ADPR. Diminution in the pool size of nuclear pADPR results in an enhanced production of procollagen mRNA, which is translated into an increased level of unhydroxylated collagen peptides. Concurrently, lower levels of ADPR in the cytoplasm result in increased prolyl hydroxylase activity, thereby ensuring full hydroxylation of the newly synthesized collagen peptides. Thus by depressing ADP-ribosylation, lactate can act as a dual signal to "turn on" collagen synthesis. This, however, increases the need for higher oxygen and ascorbate. If they are available, collagen deposition proceeds at higher rates.

In summary, lactate, by downregulating ADP-ribosylation, stimulates collagen transcription, procollagen synthesis, hydroxylation, and deposition, provided that enough oxygen is available to allow proline hydroxylation. This appears to account for the paradox that a period of hypoxia and/or lactate enhances collagen synthesis and prolyl hydroxylase activity, whereas hydroxylation and collagen deposition are inhibited by hypoxia.[13,19] The combination of lactate and sufficient oxygen, therefore, appears to promote the optimal healing of wounds.[1,13,18,19]

Effect of Lactate on Angiogenesis

Angiogenesis in the wound is directed mainly by macrophages, which mediate vascular permeability, collagen production,[5,20,21] and elicit angiogenic factors, including vascular endothelial growth factor (VEGF), which is a major stimulus for angiogenesis.[22]

Studies have indicated that the elements of VEGF regulation are similar to those of collagen with respect to ADP-ribosylation. Like collagen, the transcription and synthesis of VEGF are enhanced by the lactate-mediated downregulation of pADPR.[23] Initially, our laboratory reported that hypoxia, lactate, and nicotinamide stimulate macrophage angiogenic activity[10,11,24,25] via a suppression of NAD^+ and the pADPR pool.[24] Subsequently, we found that this angiogenic activity is primarily VEGF mediated.

[19] U. Langness and S. Udenfriend, *Proc. Natl. Acad. Sci. USA* **71,** 50 (1974).

[20] M. Hockel, K. Schlenger, S. Doctrow, T. Kisset, and P. Vaupel, *Arch. Surg.* **128,** 423 (1993).

[21] C. Sunderkotter, K. Steinbrink, M. Goebeler, R. Bhardwaj, and C. Sorg, *J. Leukocyte Biol.* **55,** 410 (1994).

[22] N. Ferrara and T. Davis-Smyth, *Endocr. Rev.* **18,** 4 (1997).

[23] J. S. Constant, J. J. Feng, D. D. Zabel, H. Yuan, D. Y. Suh, H. Scheuenstuhl, T. K. Hunt, and M. Z. Hussain, *Wound Rep. Regen.* **8,** 353 (2000).

[24] D. D. Zabel, J. J. Feng, H. Scheuenstuhl, T. K. Hunt, and M. Z. Hussain, *Lab. Invest.* **74,** 644 (1996).

[25] M. J. Banda, D. R. Knighton, and T. K. Hunt, *Proc. Natl. Acad. Sci. USA* **79,** 7773 (1983).

The finding that both lactate and nicotinamide elicited increased VEGF mRNA and VEGF protein is a strong indication that the downregulation of pADPR may be involved.[23] Lactate and nicotinamide are both known to downregulate pADPR, and upon measurement in cultured macrophages, pADPR was depressed.

Lactate in a normoxic environment reduces NAD^+ via LDH in cultured macrophages,[24] thus creating "pseudo-hypoxia," a biochemical illusion of intracellular hypoxia or energy deficiency. Nicotinamide inhibits the conversion of NAD^+ to pADPR by a feedback inhibition of PARP.[23] The additive angiogenic effects of hypoxia, lactate, and nicotinamide therefore implicate NAD^+ and polyADP-ribosylation in regulating the synthesis and release of VEGF. Prior findings that the lactate effect is inhibited by oxamate, a lactate dehydrogenase inhibitor, and d-lactate, a nonmetabolyzable form, did not elicit angiogenic activity[26] added further support. Additionally, nicotinamide was identified in the angiogenic fraction of a tumor extract that promoted angiogenesis.[27,28]

Effect of Lactate on Angiogenic Activity of VEGF

Studies demonstrated that VEGF from human macrophages is an avid acceptor of ADPR from NAD^+, which renders it angiogenically less potent. The modification of VEGF is reversible, and the removal of ADPR from VEGF returns the angiogenic activity.[29] Further studies revealed that an appreciable amount of macrophage VEGF contains covalently bound ADPR, the level of which is decreased significantly when macrophage cultures are maintained in the presence of 15 mM lactate. The lactate effect is blocked by the simultaneous addition of oxamate, an inhibitor of LDH. In this case, the level of ADPR on VEGF returns to that of the control value.[30]

We postulate that postsynthetic VEGF undergoes reversible mono-ADP-ribosylation, diminishing its angiogenic potential, and that downregulation of ADPR by lactate removes the ADPR-mediated inhibition.[30] This is analogous to the regulation of prolyl hydroxylase activity mentioned earlier. It appears that the conformational alteration of VEGF polypeptide

[26] P. Paty, M. J. Banda, and T. K. Hunt, *Surg. Forum* **39**, 27 (1988).

[27] F. C. Kull, Jr., D. A. Brent, I. Parikh, and P. Cuatrecasas, *Science* **236**, 843 (1987).

[28] Y. R. Smith, B. Klitzman, M. N. Ellis, and F. C. Kull, Jr., *J. Surg. Res.* **47**, 465 (1989).

[29] J. J. Feng, T. K. Hunt, H. Scheuenstuhl, Q. P. Ghani, and M. Z. Hussain, *Surg. Forum* **48**, 700 (1997).

[30] J. J. Feng, Q. P. Ghani, G. Ledger, R. Barkhordar, T. K. Hunt, and M. Z. Hussain, *in* "Molecular and Clinical Aspects of Angiogenesis" (M. Maragoudakis, ed.), p. 129. Plenum Press, New York, 1998.

caused by ADPR linkage is sufficient to depress its angiogenic activity. Results also predict that VEGF molecules normally exist as a mixture of free (angiogenic) and ADP-ribosylated (poorly angiogenic) forms. The ratio of these forms determines the angiogenic potential and is sensitive to metabolic alterations involving a change in ADPR synthesis.

Methods for Lactate Effect

The action of lactate can be assessed by measuring the cellular contents of oxidized NAD, ADPR, and pADPR.

NAD^+ Assay

Analysis of NAD_+ was performed by the enzymatic cyclic assay as described previously following extraction of the nucleotide from cells with cold 0.5 M perchloric acid,[31] as well as by reversed-phase HPLC (RP-HPLC) on a Amersham Pharmacia Biotech (Model Äktabasic 10), equipped with a Resource RPC 3-ml column (6.4 mm Ø interval; 100 mm length) as described by Faraone-Mennella et al.[32]

Immunohistochemical Analysis of pADPR

The content of pADPR in intact cells was measured by immunofluorescence analysis using the 10H antibody according to the method described previously.[33] Following the exposure of cultures to the desired treatments, cells are washed three times with phosphate-buffered saline (PBS) and treated with 1 mM of hydrogen peroxide in culture medium for 10 min at 37° (this step was found essential, which made pADPR detection appropriately sensitive for analysis without affecting cell survival). Cells are then incubated with the 10H primary antibody (Alexis) with a dilution of 1:300 in blocking solution (5% casein in PBS, 0.05% Tween 20) for 30 min min at 37° in a humid chamber. Following a wash with PBS, cells are incubated with 1:100 diluted FITC–coupled secondary antibody (Sigma F4018, goat antimouse, Fab specific) for 30 min at 37°.

The fluorescence of the pADPR–antibody conjugate is evaluated using a DMRBE microscope attached with a fluorescent unit (Leica), QWin software from the image analysis system Quantimet 600 (Leica, Cambridge),

[31] E. Jacobson and M. Jacobson, Arch. Biochem. Biophys. **175,** 627 (1976).
[32] M. R. Faraone-Mennella, A. Gambacorta, B. Nicolaus, and B. Farina, Biochem. J. **335,** 441 (1998).
[33] A. Burkle, G. Chen, J. H. Kupper, K. Grube, and W. J. Zeller, Carcinogenesis **14,** 559 (1993).

and an analogue slow-scan camera (COHU). Thirteen fields at a magnification of 40× are counted for total cell number using a DAPI filter block A4 (Leica) and pADPR-positive cells with a FITC filter block L4 (Leica). The background and the sensitivity of the FITC signal were found to be between 54 and 125 on a scale of 0–255. The percentage of pADPR-positive cells in treated cultures is normalized to pADPR-positive cells of the untreated cultures.

Analysis of pADPR by Western Blotting

Nuclear protein extracts are prepared from 1×10^7 cells according to the method of Andrews and Faller[34] and 65 μg of protein extract is mixed with 2× dilution buffer (12 M urea; 0.5 M Tris base, pH 6.8; 20% glycerol, 4% SDS) and loaded onto a 5% SDS–urea polyacrylamide gel. Electrophoretic separation and blotting are performed according to the procedure described previously.[35] Briefly, a nitrocellulose membrane washed with Dulbecco's phosphate-buffered saline plus 0.05% Tween 20, pH 7.5 (PBST), is incubated in blocking solution [5% casein (v/v) in PBST] for 1 h at room temperature, followed by a second incubation with the 10 H primary antibody (Alexis) diluted 1:250 in blocking solution for 45 min at 37°. After 3× washing with PBST, the membrane is further incubated with the AP-coupled secondary antibody (DAKO, rabbit-anti-mouse) diluted 1:750 in blocking solution for 1 h at room temperature, and washed again 3× with PBST. All incubations and washings are performed with gentle agitation. Colorimetric detection of the conjugate was achieved with NBT and X-phosphate (BCIP) as described by Roche Molecular Biochemicals. The pADPR signal is recorded by a DIANA III camera system. The relative intensity of the signal is assessed by using AIDA image analyzer software (raytest, Straubenhardt, Germany).

Results

We investigated the effect of high concentrations of exogenous lactate in regulating pADPR in cultured fibroblasts and macrophages. During these investigations, we learned that lactate stimulates the production of hydroxyl radicals and hydrogen peroxide *in vitro*.[36] These oxidants counter the lactate-mediated depression of pADPR. In our experiments, therefore, we use catalase and mannitol to quench oxidants that otherwise cause DNA nicking, PARP activation, and increased pADPR synthesis.[12]

[34] N. Andrews and D. Faller, *Nucleic Acids Res.* **19**, 2499 (1991).
[35] P. Adamietz and H. Hilz, *Methods Enzymol.* **106**, 461 (1984).
[36] M. A. Ali, F. Yasui, S. Matsugo, and T. Konishi, *Free Radic. Res.* **32**, 429 (2000).

Nuclear pADPR was measured by Western blot and immunohistochemical analyses using the pADPR-specific antibody. Both methods
provided similar results and demonstrated that the treatment of cultures
with 15 mM lactate significantly downregulates pADPR synthesis. Results
of a typical experiment with fibroblasts are shown. Western blot revealed
three distinct bands located at 170, 118, and 111-kDa (Fig. 1A), indicating
a heterogeneous group of proteins containing pADPR of various chain
lengths.

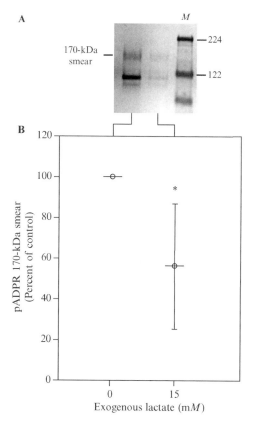

FIG. 1. pADPR synthesis in lactated cultures. Confluent neonatal human dermal
fibroblasts between passage 8 to 12 were treated with 15 mM lactate for 20 h. (A) pADPR
synthesis (170-kDa smear) was evaluated in nuclear extracts by Western blotting as described
in Methods. pADPR value is indicated as percentage of control culture to which no lactate
was added (B). Data represent the mean of five experiments ± standard deviation. Significant
difference between means of control and lactated cultures is indicated as *p < 0.05 (2 tailed,
t-test).

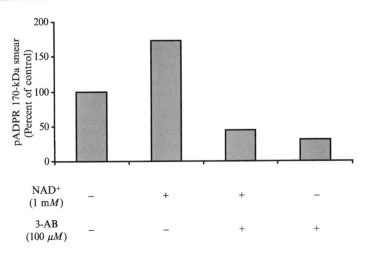

FIG. 2. Synthesis of pADPR in fibroblast cultures exposed to NAD$^+$ and 3-AB. Confluent fibroblasts were treated with 1 mM NAD$^+$ and/or 100 μM 3-AB for 20 h. The synthesis of pADPR was evaluated in nuclear extracts by Western blotting as described in Methods. Values are indicated as percentage of pADPR found in cultures to which no NAD$^+$ and 3-AB were added. Each value is the mean of n = 4 (NAD$^+$) and n = 2 (3-AB).

Quantification showed a 60% reduction in total pADPR (all three bands) and a 44% reduction in pADPR located at the 170-kDa band following lactate exposure (Fig. 1). Further characterization of the assay was achieved by studying the effect of exogenous NAD$^+$ and 3-aminobenzamide (3-AB) on pADPR located at the 170-kDa band. As expected, we found that NAD$^+$ (which translocates intact into fibroblasts) stimulated pADPR synthesis by 73%[37] and 3-AB, which inhibits PARP,[38] reduced pADPR synthesis by 65% compared to untreated cultures, whereas a combination of these compounds also depressed pADPR (Fig. 2). Immunofluorescence analysis displayed a similar decline in pADPR synthesis by lactate. The assay demonstrated the characteristic cobblestone pattern of pADPR in cell nucleus. As shown in Fig. 3, lactate-treated fibroblasts contained only 50% pADPR compared to that exhibited by untreated fibroblasts. Figure 4 shows that the lactate effect on pADPR correlates with depression of the NAD$^+$ pool.[18]

[37] P. Loetscher, R. Alvarez-Gonzalez, and F. R. Althaus, *Proc. Natl. Acad. Sci. USA* **84,** 1286 (1987).
[38] F. R. Althaus and C. Richter, *Mol. Biol. Biochem. Biophys.* **37,** 1 (1987).

− lactate   + lactate

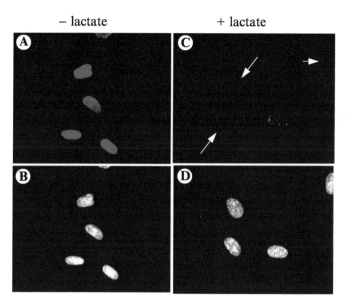

FIG. 3. Immunofluorescence analysis of pADPR production in cultured fibroblasts. 3×10^5 cells were grown on coverslips in the presence of 15 mM lactate and nuclear pADPR was measured by immunofluorescence analysis with anti-pADPR antibody as described in Methods. (A) pADPR in nonlactated cells (B) corresponding DAPI nuclei staining. (C) pADPR in lactated cells (D) corresponding DAPI nuclei staining. No fluorescent signal was observed in the absence of H_2O_2 treatment or antibody.

Conclusions

Based on these observations, the mode of action of lactate can be described as follows. High concentrations of lactate force the LDH-catalyzed conversion of nicotinamide adenine dinucleotide NAD^+ to NADH. The decline of the NAD^+ pool and the subsequent downregulation of NAD-mediated polyadenosine diphosphoribose (pADPR) and adenosine diphosphoribose (ADPR) are crucial mediators that stimulate collagen and VEGF production.[1,10,11]

The implication that downregulation of ADP-ribosylation is involved in promoting VEGF and collagen transcription seems logical. ADP-ribosylation regulates many enzyme activities and functions of nuclear proteins, including those involved in DNA repair, differentiation, and transcription through the modification of transcription factors such as Sp1, p53, Yin Yang 1, and CREB.[12,39] A recent report that the downregulation of

[39] S. L. Oei, M. Griesenbeck, M. Schweiger, and M. Ziegler, *J. Biol. Chem.* **273,** 31644 (1998).

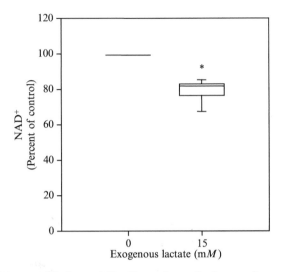

FIG. 4. NAD$^+$ content in lactated fibroblast culture. Confluent cells were treated with 15 mM lactate for 20 h. NAD$^+$ was assayed as described in Methods. Values are indicated as percentage of NAD$^+$ derived from control cultures to which no lactate was added. Each number represents the mean of five experiments. The difference between control and lactated cultures is discernible at *p = 0.008 (2 tailed, Mann-Whitney Rank Sum Test).

polyADP-ribosylation loosens chromatin to facilitate the expression of genes lends a strong support to this mechanism.[40] Earlier, Loetscher et al.[37] suggested that ADP-ribosylation may constitute a link between energy status of the cell to nuclear functions. Results of our studies support this concept. One can speculate that the polyADP-ribosylation of the putative nuclear protein(s) involved in VEGF transcription alters its binding to the VEGF promoter and/or specific transcriptional suppressors or activators, and thereby affects the rate of VEGF transcription.[41–43] The same can be hypothesized for collagen transcription. It is known that Sp1 is essential for collagen promoter function[44] and that the modification of transcription factor Sp1 by pADPR prevents its binding to DNA.[39] However, the precise molecular events remain to be determined.

[40] A. Tulin and A. Spradling, *Science* **299,** 560 (2003).
[41] G. L. Semenza, *Curr. Opin. Cell Biol.* **13,** 167 (2001).
[42] G. Finkenzeller, A. Technau, and D. Marme, *Biochem. Biophys. Res. Commun.* **208,** 432 (1995).
[43] D. Mukhopadhyay, L. Tsiokas, X. M. Zhou, D. Foster, G. Brunge, and V. Sukhatme, *Nature* **376,** 577 (1995).
[44] S. J. Chen, C. M. Artlett, S. A. Jimenez, and J. Varga, *Gene* **215,** 101 (1998).

The metabolic control of collagen deposition and angiogenesis by lactate presents a unifying basis for wound repair. It is remarkable that both forms of ADP-ribosylation reactions that potentially regulate nuclear and posttranslational events of collagen and VEGF production are equally affected by high lactate.

Initiation of collagen synthesis and angiogenesis in wounds can be described as a response to a metabolic demand precipitated in an environment that has little oxygen and/or a high level of lactate. These conditions limit the supply of NAD^+ and, therefore, ADPR and pADPR. In response to this deficit, fibroblasts (and endothelial cells) synthesize and secrete collagen, and macrophages (and endothelial cells) elicit VEGF and keep it in an active form, which stimulates new vessel growth. However, endothelial cells do not respond well to VEGF in hypoxia, despite upregulation of VEGF receptors.[22] Enhanced perfusion caused by new vessel growth subsequently reestablishes normoxia. Endothelial cells now respond actively to VEGF for blood vessel formation. Finally, the endothelial cell response terminates as the macrophage-derived angiogenic signals diminish.

Acknowledgment

Supported by NIH NIGMS Program Project Grant GM27345.

[37] Protocols for Topical and Systemic Oxygen Treatments in Wound Healing

By Gayle M. Gordillo, Richard Schlanger, William A. Wallace, Valerie Bergdall, Robert Bartlett, and Chandan K. Sen

Wound hypoxia, resulting from disruption of the local vasculature, is a key limiting factor in healing.[1-3] The core of the wound is most hypoxic with a progressive increase in the oxygen gradient toward the uninjured tissue at the periphery. The pO_2 of dermal wounds ranges from 0–10 mm Hg centrally to 60 mm Hg at the periphery.[3-5] This is extremely important

[1] G. M. Gordillo and C. K. Sen, *Am. J. Surg.* **186**, 259–263 (2003).

[2] H. Hopf, T. Hunt, J. West, P. Blomquist, W. Goodson, A. Jensen, K. Jonsson, P. Paty, J. Rabkin, R. Upton, K. von Smitten, and J. Whitney, *Arch. of Surg.* **132**, 997 (1997).

[3] T. Hunt, P. Twomey, B. Zederfeldt, and J. Dunphy, *Am. J. Surg.* **114**, 302 (1967).

[4] J. Remensnyder and G. Majno, *Am. J. Pathol.* **52**, 301 (1968).

[5] I. Silver, *Adv. Exp. Med. Biol.* **94**, 769 (1977).

when evaluating wounds that will not heal because oxygen plays a critical role in many aspects of the wound healing response. Recent findings support that molecular oxygen as well as its reactive derivatives may support wound healing.[6,7] Oxidants generated at the wound site serve the role of cellular messengers that orchestrate the healing process.[6–12]

Clinical use of O_2 to promote wound healing began in the 1960s with the administration of systemic hyperbaric O_2 to treat wounds. While the conditions (e.g., pressure, O_2 concentration, frequency, and duration of administration) for systemic hyperbaric O_2 therapy (HBOT) have not been optimized on the basis of randomized clinical trials, HBOT is an FDA-approved therapeutic modality used in wound clinics with an encouraging success rate. Reliance on empiricism and a paucity of data that meets the highest criteria for evidence-based medicine have hindered the general acceptance of O_2 therapy as a standard modality in wound care. Embracing the concept of O_2 therapy depends not only on a favorable clinical outcome, but also on detailed mechanistic insight that explains those outcome results. The use of systemic hyperbaric O_2 therapy presents potential advantages as well as widely recognized risks of oxygen toxicity.[13–16] There is evidence to suspect that the use of pressure and systemic pure O_2 may not be essential in wound care. Elimination of these factors by using sub-pure systemic O_2 under normobaric conditions may significantly minimize the risk of O_2 toxicity. Furthermore, opportunities to treat dermal wounds using topical O_2 therapy warrant further investigation.[17] Given that many growth factors require ROS for their function, it is reasonable to assume that approaches to correct wound pO_2 will serve as an effective adjunct to treat chronic wounds.[8] This article describes some of the currently used protocols for oxygen therapy of wounds.

[6] C. K. Sen, S. Khanna, B. Babior, T. Hunt, E. Ellison, and S. Roy, *J. Biol. Chem.* **277**, 33284 (2002).
[7] C. K. Sen, S. Khanna, G. Gordillo, D. Bagchi, M. Bagchi, and S. Roy, *Ann. N. Y. Acad. Sci.* **957**, 239 (2002).
[8] C. K. Sen, *Wound Repair Regen.* **11**, 431–438 (2003).
[9] C. K. Sen and L. Packer, *FASEB J.* **10**, 709 (1996).
[10] C. K. Sen and L. Packer, *Methods Enzymol.* **352**, 580 (2002).
[11] C. K. Sen and L. Packer, *Methods Enzymol.* **353**, 633 (2002).
[12] C. K. Sen H. Sies, and P. A. Baeuerle, "Antioxidant and Redox Regulation of Genes," p. 556. Academic Press, San Diego, 2000.
[13] R. Arieli and Y. Moskovitz, *J. Appl. Physiol.* **91**, 1327 (2001).
[14] M. Kleen and K. Messmer, *Minerva Anestesiol.* **65**, 393 (1999).
[15] G. Speit, C. Dennog, P. Radermacher, and A. Rothfuss, *Mutat. Res.* **512**, 111 (2002).
[16] L. K. Weaver and S. Churchill, *Chest* **120**, 1407 (2001).
[17] L. Kalliainen, G. Gordillo, R. Schlanger, and C. Sen, *Pathophysiology* **9**, 81 (2003).

Measuring Responses to Oxygen Interventions

It is important to recognize that simply administering supplemental oxygen does not in itself guarantee increased wound oxygenation. The availability of oxygen to wound tissues depends on vascular supply, vasomotor tone, arterial pO_2, and the diffusion distance for molecular O_2. Edema and necrotic debris both increase the diffusion distance for O_2 to reach the wound. Thus, debridement is an important step to diminish obstruction to wound oxygenation. Peripheral vasoconstriction can also significantly limit wound perfusion and oxygenation so that little to no enhancement of wound pO_2 levels are achieved despite breathing supplemental oxygen.[2,18,19] Therefore, for optimal wound perfusion and oxygenation, patients must be warm, have adequate intravascular volume, and have adequate control of pain and anxiety.

Currently, the only clinically feasible method for measuring tissue oxygenation is transcutaneous oxygen measurements (TcOM). This technique entails placement of an electrode on the surface of the skin. The electrode (modified Clark polargraphic electrode) heats the underlying skin to 42–44° to dilate the dermal vascular bed and obtains a measurement of pO_2 in mm Hg. A minimum of three electrodes are used with one placed in the second intercostal space as a reference lead, one electrode placed at the proximal border of the wound, and one electrode placed at the distal border of the wound. The TcOM at the periphery of a wound indicates the relative blood flow/vascularity at the periphery of the wound site and hence the capacity for O_2 delivery to the wound. At present, it is the only quantitative and noninvasive diagnostic procedure available to assess wounds. TcOMs are measured at room air and in response to an oxygen supplement. If the wound is on an extremity, it is tested in a dependent and an elevated position to determine if there is vascular stenosis limiting blood flow to the wound.

The current protocol for TcOM measurements is based on the guidelines established by the Undersea and Hyperbaric Medicine Society (UHMS). The protocol is as follows.

1. Calibrate electrodes as per manufacturer's instructions (Radiometer America, Westlake, OH).
2. Place patient in a semireclining position with room temperature 24° (75.2°F) and breathing room air.
3. Select a reference site.
   a. Usually anterior chest wall second intercostal space.

[18] F. Gottrup, R. Firmin, J. Rabkin, B. Halliday, and T. Hunt, *Crit. Care Med.* **15,** 1030 (1987).
[19] H. Hopf, J. West, and T. Hunt, *Wound Repair Regen.* **4,** A129 (1996).

b. If necessary shave site, debride stratum corneum with adhesive tape, and wipe down with alcohol.

c. Place electrode fixator ring by pressing down firmly to make sure there are no gaps or debris between ring and skin surface.

d. Fill ring with 2–3 drops of couplant solution (Contact Liquid, Radiometer America).

4. Select two sites around wound—one proximal to wound edge and one distal to the wound edge, if possible—and prepare and apply electrode fixator rings as described in steps 3b–3d.

5. Snap electrodes into fixator rings and obtain readings. This usually takes 10–15 min to allow sufficient time for skin to be heated.

6. If the wound is on an extremity, repeat TcOM with extremity elevated for 5 min prior to taking the measurements.

7. Repeat with patient breathing 100% O_2 using a nonrebreathing mask and obtain measurements at 2, 5, and 10 min.

8. Patients receiving hyperbaric oxygen therapy (HBOT) have in-chamber TcOM evaluation performed with their first treatment to document a positive response to supplemental oxygen therapy.

Interpretation of TcOM results and prognosis for healing can be assessed according to guidelines presented in Table I. When performing TcOM evaluation or troubleshooting the interpretation, keep in mind the following conditions that can affect TcOM measurements.

Systemic

a. Comorbid conditions in the patient that affect their ability to oxygenate blood or the ability of their hemoglobin to bind oxygen.

b. Conditions that impair oxygen delivery, for example, limitations in cardiac output or vascular stenosis limiting blood flow.

c. Infection.

Local

a. Increased skin thickness.

b. Obesity.

c. Edema.

d. Cellulite.

Mechanical

a. Pressure on the coupling ring will diminish measurements; do not attempt to fix the coupling rings in place with adhesive tape.

b. Reference lead reads ≤50 mm Hg; the reference lead should always be >50 mm Hg and, if not, the electrodes should be recalibrated.

Caution: TcOM does not measure oxygenation of the wound itself.[1] It indicates the oxygenation status of the intact skin along the periphery of the wounds. Ideally, one would like to get a three-dimensional spatial imaging of the actual wound. Such an approach would identify pockets of hypoxia and provide information regarding compartments of O_2 tension at the actual wound site. We are currently exploring options to noninvasively image O_2 and the redox status of the wound employing electron paramagnetic resonance (EPR) imaging techniques.[20–25]

TABLE I

INTERPRETATION OF TRANSCUTANEOUS OXYGEN MEASUREMENT RESULTS

Oxygenation status	Room air (mm Hg)	100% O_2 at 1 ATA[a] (mm Hg)	100% O_2 at pressure (mm Hg)
Normally perfused skin	50–90	>300	1000–1500
Wound—marginal hypoxia	30–39	65–75 approximately 75% should heal	
Wound—moderate hypoxia	20–29	35–74 approximately 50% heal	
Wound—severe hypoxia	0–19	<25% needs healing	≥200—consider trial HBO <200—dismal prognosis not HBOT candidate

[a] Absolute atmospheres of pressure.

[20] S. J. Ellis, M. Velayutham, S. S. Velan, E. F. Petersen, J. L. Zweier, P. Kuppusamy, and R. G. Spencer, *Magn. Reson. Med.* **46,** 819 (2001).
[21] G. He, Y. Deng, H. Li, P. Kuppusamy, and J. L. Zweier, *Magn. Reson. Med.* **47,** 571 (2002).
[22] G. He, A. Samouilov, P. Kuppusamy, and J. L. Zweier, *Mol. Cell. Biochem.* **234,** 359 (2002).
[23] G. Ilangovan, H. Li, J. L. Zweier, M. C. Krishna, J. B. Mitchell, and P. Kuppusamy, *Magn. Reson. Med.* **48,** 723 (2002).
[24] P. Kuppusamy, H. Li, G. Ilangovan, A. J. Cardounel, J. L. Zweier, K. Yamada, M. C. Krishna, and J. B. Mitchell, *Cancer Res.* **62,** 307 (2002).
[25] S. S. Velan, R. G. Spencer, J. L. Zweier, and P. Kuppusamy, *Magn. Reson. Med.* **43,** 804 (2000).

Oxygen Therapy for Wound Infection Prophylaxis

Normobaric systemic oxygen therapy can be administered in the perioperative period to decrease the incidence of surgical wound infection. This has been demonstrated in a double-blinded randomized controlled trial involving 500 patients undergoing colorectal surgery.[26] In this study, patients were randomized to treatment arms consisting of either 30% oxygen + 70% nitrogen ($n = 250$) or 80% oxygen + 20% nitrogen ($n = 250$) administered with the following protocol.

1. Assigned concentrations were given at the start of anesthesia induction until immediately before extubation.
2. Patients were given 100% O_2 from the time of extubation until the anesthesiologist deemed it safe to resume administration of O_2 at the specified concentrations.
3. For the first 2 h of recovery, O_2 was administered to patients at the specified concentrations *via* a nonrebreathing mask.
4. Oxygen levels were monitored by continuous pulse oximetry and arterial blood gas at 1 and 2 h after recovery from anesthesia.
5. Patients were hydrated aggressively during and after surgery: crystalloid basal infusion rate of 15 ml/kg/h during surgery, blood replaced with crystalloid at a 4:1 ratio or with colloid at a 2:1 ratio, and crystalloids administered at 3.5 ml/kg/h for the first 24 h after surgery and at 2 ml/kg/h for the subsequent 24 h.
6. After 48 h, patients in both groups breathed ambient air or received supplemental oxygen as needed to maintain an oxyhemoglobin saturation >92%.

This protocol is significant because it demonstrates the efficacy of O_2 therapy in preventing a specific wound healing complication, that is, infection. This is also one of the few O_2 treatment protocols that has been validated by a double-blinded, randomized control trial. It also adds a preventive dimension to the concept of O_2 therapy in addition to its therapeutic uses for refractory wounds. The use of aggressive hydration is a key component of this protocol. The ability to prevent infection with oxygen administration is contingent upon optimal perfusion and oxygention, which fail to occur in postoperative patients that are underresuscitated.[2]

Systemic Oxygen Therapy

HBOT has been used since the 1960s to treat refractory wounds and acute conditions related to pressure (e.g., the bends/air or gas embolism)

[26] R. Grief, O. Akca, E.-P. Horn, A. Kurz, and D. Sessler, *N. Engl. J. Med.* **342,** 161 (2000).

and oxygenation (e.g., carbon monoxide poisoning). This modality entails administration of 100% oxygen usually at a pressure of 2–3 atmospheres and is the most commonly used method for the clinical application of oxygen therapy. Patients receive HBOT in either a multiplace or a monoplace chamber. In the multiplace chamber, the patient sits inside a large room and is administered pressurized oxygen through a face mask. In a multiplace chamber, the amount of inhaled oxygen delivered to the wound is contingent upon the vascular supply to the wounded area. This is a critical concept to appreciate in designing studies and interpreting the literature. In this case, the only way that oxygen is delivered to the wound is *via* saturated hemoglobin and dissolved oxygen in the blood. However, the vast majority of patients receiving HBOT are treated in monoplace chambers. Note that under such circumstances, the wound receives oxygen through the systemic as well as topical routes. The inhaled oxygen is carried by the patients' vasculature to the wound and pressurized oxygen comes into direct contact with the surface of the wound. Thus, HBOT protocols using a monoplace chamber incorporate several variables: pressure, systemic oxygen effects, and topical oxygen effects, all of which may influence wound responses.

Selecting patients for HBOT is based on several criteria. The only absolute contraindication to HBOT is untreated pneumothorax. TcOM evaluation must be obtained when initiating HBOT to make sure the patient has the capacity to respond to oxygen therapy. Federal guidelines from the Health Care Finance Administration (HCFA) indicate the approved conditions for HBOT treatment. In addition, the UHMS has a list of approved conditions and developed the standard protocols for HBOT administration. The indications for HBOT and the treatment protocols that are used are those recommended by UHMS and are summarized in Table II. A physician trained in HBOT must supervise these treatments. Obvious complications that can occur during HBOT administration include the following.

1. Middle ear barotrauma. Perforated tympanic membrane more likely with ventilator-dependent patients and young children.

2. Seizures. HBOT lowers the seizure threshold and blood glucose. Diabetic patients should have blood glucose >200 prior to entering the chamber. Patients on seizure medication must have serum drug levels within the therapeutic range.

3. Pulmonary complications. Untreated or occult pneumothorax can be converted to a tension pneumothorax. Apnea/loss of respiratory drive will occur in patients with significant chronic obstructive pulmonary disease defined as room air $pCO2 \geq 55$ on arterial blood gas. Patients with heart failure, defined as an ejection fraction $<30\%$, can develop pulmonary edema.

TABLE II
PROTOCOLS FOR HBOT ADMINISTRATION[a]

Indication	Pressure (ATA)	Duration	# Treatments (tx)	Reevaluate	Comment
Air or gas embolism	2.8	Symptom specific	1–14	After 10–14 tx	Follow US Navy tx tables 6 and 6A
Carbon monoxide (CO)	2.4–3.0	90–120 min	1–10	After 5 tx	qid–bid based on neuro function
CO/cyanide complications	2.5–3.0	90 min	1–10	After 5 tx	Same as CO
Gas gangrene	3.0	90 min	5–10	After 10 tx	tid day 1, bid day 2, then qd
Crush injury	2.0–2.5	90 min	3–12	After 6 tx	tid × 2 days, bid × 2 days, qd × 2 days
Decompression illness	2.8	Symptom specific	1–14	No further response	Follow U.S. Navy tx tables 6 or 7
Select nonhealing wounds	2.0–2.5	90–120 min	10–40	After 30 tx	qd and/or bid
Exceptional blood loss[b]	2.0–3.0	90–120 min	Not specific	Hematocrit = 22.9%	bid initially then qd
Necrotizing soft tissue infection	2.0–2.5	90–120 min	5–30	>30 tx	qd and/or bid
Chronic osteomyelitis	2.0–2.5	90–120 min	20–40	After 40 tx	qd and/or bid
Radiation tissue damage	2.0–2.5	90–120 min	20–60	After 60 tx	bid initially then qd
Compromised graft or flap	2.0–2.5	90–120 min	6–40	After 20 tx	tid first 24 h then bid
Thermal burn[b]	2.0–2.4	90 min adult 45 min peds	5–45	No specified limit	bid initially then qd
Intracranial abscess	2.0–2.5	60–90 min	5–20	No specified limit	qd
Diabetic lower extremity wound	2.0	90 min	30–60	After 30 tx	

[a] tx, treatment; qd, once a day; bid, twice a day; tid, three times a day; qid, four times a day.
[b] Not HCFA/CMS-approved condition.

Additional factors must be taken into consideration before any patient can reap the benefits of HBOT. They must be able to (i) fit into the chamber (obesity is an issue), (ii) come to the HBOT facility to receive daily treatments, and (iii) tolerate the treatments. Claustrophobic patients may become uncomfortable or anxious when spending 2 h at pressure in the monoplace chambers.

Topical Oxygen Therapy

The use of topical oxygen to treat refractory wounds was first described in 1969.[27] Widespread application of this modality has not been feasible until recently when commercial products became available that were designed specifically for this purpose.[17] The concept of topical oxygen therapy is appealing because it can deliver oxygen directly to a wound site without the risks and potentially at a significantly less cost than HBOT. Because a physician does not need to be present and special chambers are not required, topical oxygen therapy can be administered practically anywhere and most patients are treated in their own homes. Given the current geopolitical conditions, it also could be applied under conditions of field combat use.

While the concept of topical oxygen therapy is intriguing, there are little data to support the case for its clinical efficacy. We performed a retrospective analysis of our results for patients treated during the first 9 months with this modality. There were no specific inclusion or exclusion criteria for this study. There were no standardized wound care regimens, but all patients received topical oxygen treatments using the following protocol.

1. Remove dressing from wound and apply topical oxygen device (GWR Medical, Chadds Ford, PA). These are single-use disposable devices that come as either sacral bags (like a large colostomy bag) or boots. They have an adhesive strip for fixation of the device to the patient. Wounds should be debrided/free of necrotic debris. Do not use petroleum-based dressings as any residual will prevent oxygen penetration into the wound.

2. Connect device to oxygen gas cylinder.

3. Initiate oxygen flow at 7–8 liters per minute. The bag should be fully insufflated without any wrinkles. Each device has a release valve to prevent excessive pressure buildup within the bag.

4. Treatments last 90 min and are administered for 4 consecutive days followed by 3 days without treatment.

[27] B. Fischer, *Lancet* **2,** 405 (1969).

Using this protocol in 32 patients with 58 wounds, complete healing was achieved in 42/58 wounds. If the two patients with two wounds who were lost to follow-up are not included, then the overall healing rate for our series was 75%.[17] We were pleased with these results when comparing them to overall results at our wound care center, which reports successful healing in 79% of its patients subjected to specialized wound care. We did not have any complications related to the use of this modality, and all patients with smaller wound dimensions responded positively over time, even if they did not heal completely.

Patient selection criteria are far les rigorous for topical oxygen than for HBO. Contraindications for using the topical oxygen devices include (i) fistulous tract that the end cannot be contacted with a probe and (ii) patient refuses to refrain from smoking while administering the oxygen treatment. There are no known risks for this method of oxygen therapy and the duration of treatment is determined by the prescribing physician. We have not observed any complications related to the use of this modality at our institution. Favorable outcomes in the clinical setting implied that the benefits of oxygen therapy could be achieved with this method of oxygen therapy. There was enough merit in these findings to commit the resources to pursue more mechanistic studies of topical oxygen.

An experimental pig model has been developed for this purpose (Fig. 1). The protocol for implementation is as follows.

1. Premedicate a female Yorkshire pig (80–100 pound) with an intramuscular injection of Telazol (500 mg) and acepromazine (5 mg).

2. Place pig in prone position on operating table and administer general anesthesia via a nose cone using isofluorane and a standard respirator (Harvard Apparatus, NP 72-3001) for the duration of the surgery.

3. Shave the back and sterilely prep and drape the area.

4. Create 2.5 × 2.5 cm full-thickness skin wound defects using a scalpel. Place four wounds on the back in the thoracic area and four wounds in the lumbar area. Arrange the wounds in a 2 × 2 pattern separated by 2.5 cm. Hemostasis is achieved by packing the wounds with gaze until bleeding stops.

5. Treat four wounds in one anatomic location with topical oxygen; the other four wounds serve as untreated controls.

6. Attach the topical oxygen device to the pig and cover the untreated wounds with an occlusive dressing (e.g., Op-site, Tegaderm) and Elasticon tape. Transfer the pig back into its cage.

7. Attach the topical oxygen device to the oxygen cylinder and set the flow at 2 liters/minute. Administer topical oxygen is administered every

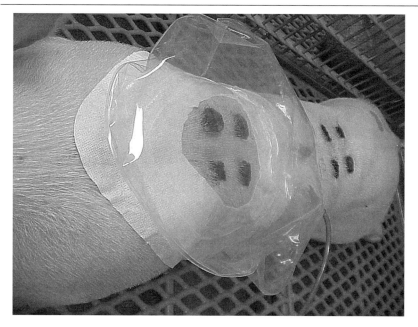

Fig. 1. Pig under general anesthesia receiving topical oxygen treatment. Full-thickness skin wounds are created on the dorsum. Half of the wounds are treated with topical oxygen using the sacral bag and half the wounds are untreated and remain outside the device.

other day for 3 h starting the day of wounding. At the end of treatments the wounds are dressed in the same manner as the control wounds.

This protocol deviates from the human application because we could not place the pig under general anesthesia to receive the topical oxygen treatments as frequently as humans receive treatment. However, it is important to note that none of the protocols used for topical oxygen or HBOT have ever been optimized by a randomized clinical trial to determine optimal concentration, pressure, frequency, or duration of oxygen exposure.

Acknowledgment

This work was supported in part by NIH-GM27345 to CKS.

Section IX

Unicellular Systems

[38] Measurement of Oxidative Stress in Cells Exposed to Hypoxia and Other Changes in Oxygen Concentration

By REINHARD DIRMEIER, KRISTIN O'BRIEN, MARCELLA ENGLE, ATHENA DODD, ERICK SPEARS, and ROBERT O. POYTON

Introduction

It has been reported that the mitochondrial respiratory chain is required for the induction of some hypoxic nuclear genes in both mammalian and yeast cells.[1–3] Because the respiratory chain can produce reactive oxygen species (ROS), which can mediate intracellular cascades, it is reasonable to suspect that ROS act as signaling intermediates in a signal transduction pathway between mitochondrial metabolism and nuclear transcription factors that regulate the expression of oxygen-regulated genes. This would require a change in ROS levels as cells experience a change in oxygen concentration. Previous studies with mammalian cells in culture have reported that when cells are shifted from normoxic to hypoxic conditions, ROS levels increase.[2,4,5] In contrast, other studies with the same mammalian cell lines have reported that ROS levels decrease when cells are exposed to hypoxia.[6] Both sets of studies used fluorescent dyes (either 2′,7′-dichlorofluorescin diacetate or dihydrorhodamine 123) to measure H_2O_2 levels in cells exposed to hypoxia. Because these dyes produce conflicting results in mammalian cells shifted from normoxic to hypoxic conditions, it has been suggested that they lack the necessary precision to monitor ROS levels in cells experiencing changes in oxygen concentration.[7] This would not be surprising because ROS are transient, highly reactive, and probably oxidize macromolecules (proteins, lipids, and nucleic acids) in their immediate vicinity. In this regard, it is important to note that

[1] K. E. Kwast, P. V. Burke, B. T. Staahl, and R. O. Poyton, *Proc. Natl. Acad. Sci. USA* **96,** 5446 (1999).

[2] N. S. Chandel, E. Maltepe, E. Goldwasser, C. E. Mathieu, M. C. Simon, and P. T. Schumacker, *Proc. Natl. Acad. Sci. USA* **95,** 11715 (1998).

[3] F. H. Agani, P. Pichiule, J. C. Chavez, and J. C. LaManna, *J. Biol. Chem.* **275,** 35863 (2000).

[4] T. L. Van deHoeck, L. B. Becker, Z. Shao, C. Li, and P. T. Schumaker, *J. Biol. Chem.* **273,** 18092 (1998).

[5] N. S. Chandel, D. S. McClintock, C. E. Feliciano, T. M. Wood, J. A. Melendez, A. M. Rodriguez, and P. T. Schumacker, *J. Biol. Chem.* **275,** 25130 (2000).

[6] T. Kietzmann, J. Fandrey, and H. Acker, *News Physiol. Sci.* **15,** 202 (2000).

[7] G. L. Semenza, *Cell* **98,** 281 (1999).

it is generally accepted that the most reliable way to assess ROS levels and hence oxidative stress is to measure oxidative damage.[8]

This article presents four different methods for measuring oxidative stress in yeast cells exposed to anoxia. We find that reliable assays involve measuring oxidative DNA and protein damage and measuring the expression of the oxidative stress-induced gene $SOD1$.[9] These assays should be useful for other types of cells (e.g., bacteria and mammalian) exposed to hypoxia or anoxia.[10] We also find that carboxy-H_2-dichlorodihydrofluorescein diacetate, a commonly used fluorescein derivative, is useful for assessing relative ROS levels in yeast cells grown at different steady-state oxygen concentrations and in comparing levels of oxidative stress in respiration-proficient and respiration-deficient cells. However, it is not useful for assessing ROS levels when cells are shifted from one oxygen concentration to another and followed for long periods of time (i.e., hours).

Procedures

Protocol for Exposing Normoxic Yeast Cultures to Hypoxic or Anoxic Conditions

Shifts between different oxygen concentrations are performed conveniently in a fermentor. We have successfully used New Brunswick BioFloIIc and BioFloIII fermentors for this purpose.[11] Inocula for these fermentors are steady-state aerobic "precultures" that have been grown on a reciprocating shaker at 200 rpm. Precultures are grown to steady state on a shaker (200 rpm) at 28° and kept in logarithmic growth phase for at least 10 generations. They are harvested when the cell density reaches 2–4×10^7 cells/ml (80–150 Klett units, see later). It is important to use steady-state precultures to ensure reproducible results.[11]

Growth Media. Precultures and fermentor cultures are grown in SSG-TEA, a semisynthetic galactose growth medium, pH 5.0 (containing per liter, 3 g of Bacto yeast extract, 10 g of galactose, 0.8 g of NH_4SO_4, 1 g of KH_2PO_4, 0.5 g of NaCl, 0.7 g of $MgSO_4 \cdot 7\ H_2O$, 5 μg of $FeCl_2$, and 0.4 g of $CaCl_2$) supplemented with 0.1% (v/v) Tween 80, 20 μg/ml ergosterol, and

[8] R. S. Sohal and R. Weindruch, *Science* **273**, 59 (1996).

[9] R. Dirmeier, K. M. O'Brien, M. Li, A. Dodd, E. Spears, and R. O. Poyton, *J. Biol. Chem.* **277**, 34733 (2002).

[10] V. Grishko, M. Solomon, J. F. Breit, D. W. Killiea, S. P. Ledoux, G. L. Wilson, and M. N. Gillispie, *FASEB J.* **15**, 1267 (2001).

[11] R. O. Poyton, R. Dirmeier, K. O'Brien, P. David, and A. Dodd, *Methods Enzymol.* **381** (42), 2003 (this volume).

350 ppm Dow Corning FG-10 silicon antifoam. To obtain a uniform dispersion, Tween 80, ergosterol, and silicon antifoam are sonicated ($\sim$110 W for 1 min using a Branson Model 250 Sonifier) in solution prior to autoclaving. The galactose is autoclaved separately and added to the media under sterile conditions. Nutritional supplements (e.g., amino acids, nucleotides) are added from sterile stock solutions, as required, to satisfy the auxotrophic requirements of the strain that is used.

Fermentor Cultivation. For shift experiments between normoxic and anoxic or hypoxic conditions, aerobic precultures are inoculated aseptically into the fermentor vessel and grown aerobically by sparging with air at a rate of 4 liters per minute. The temperature, pH, sparge rate, and dissolved oxygen (DO) concentration in the media are maintained and regulated by this fermentor system. Once the culture has reached a cell density of $\sim$2 $\times$ 10^7 cells/ml, the gas entering the fermentor vessel is changed from air to either O_2-free N_2 containing 2.5% CO_2, for anoxic conditions, or a gas mixture containing low oxygen content (e.g., 1–5 μM O_2), for hypoxic conditions. The O_2-free mix of N_2 with 2.5% CO_2 is passed through an Oxyclear O_2 absorber (Lab-Clear, Oakland, CA) to prevent trace O_2 from entering the fermentor. Typically, the dissolved oxygen concentration in the vessel drops precipitously and reaches an anoxic or hypoxic level within 5–10 min after changing the gas. Shift cultures are grown in the dark to prevent photoinhibition of growth.[12] Cell growth is followed by asceptic removal of an aliquot from the fermentor vessel and measuring turbidity with a Klett-Summerson colorimeter fitted with a No. 54 green filter. Cells are harvested at different times after the shift and saved for analysis. During harvest, cells are chilled to 4°, washed twice with ice-cold distilled water, and either processed immediately or frozen in liquid nitrogen.[13] Protocols for growing yeast cells to a steady state at oxygen concentrations that are intermediate between normoxia and anoxia and for shifting cells between these intermediate oxygen concentrations are given elsewhere in this volume.[11]

Dissolved oxygen concentration in the fermentor vessel is measured with a DO meter and oxygen electrode that have been calibrated with air (100%) or O_2-free nitrogen (0%). Actual dissolved oxygen concentrations in air-saturated growth media are calculated according to Henry's law, $[X_i = p_i h / H_i]$, where X_i is the amount (mol) of gas dissolved per mole solution at equilibrium, P_i is the partial pressure of the gas at elevation h, and H_i is the Henry's law constant for the gas at a specific temperature. The

---

[12] E. Sulkowski, B. Guerin, J. Defaye, and P. P. Slonimski, *Nature* **202,** 36 (1964).
[13] P. V. Burke, K. E. Kwast, F. Verts, and R. O. Poyton, *Appl. Environ. Microbiol.* **64,** 1040 (1998).

differential pressure in the fermentor vessel at the 4-liter/min sparge rate used for these experiments is extremely low (less than 0.4% of an atmosphere) and has little effect on the pressure in the vessel (ambient + differential pressure) and X_i. Henry's law gives the dissolved oxygen content in water. To determine its solubility in solution it is necessary to correct for the ionic strength of the solution (e.g., SSG-TEA growth media). This is done by determining the conductivity of the solution. For SSG-TEA media, the conductivity of media at $25°$ is $\sim$12,000–13,000 μSiemens/cm, giving a correction factor[14] of 0.959–0.962. Taking this correction factor into account, we find that the dissolved oxygen concentration of air-saturated SSG-TEA media is 203.4568 ± 0.3151 μM O_2 in our laboratory in Boulder, Colorado.

Assays for Oxidative Stress

Generally speaking, three types of assay have been used for assessing cellular oxidative stress. One makes use of fluorescent dyes (e.g., derivatives of fluorescein or rhodamine) to estimate intracellular ROS levels. The second assesses oxidative damage, caused by ROS, by measuring the accumulation of lipid peroxides (e.g., malonaldehyde and hydroxyalkenals), oxidized nucleosides [e.g., 8-hydroxy-2'-deoxyguanosine (8OH2gG)], or oxidized amino acid side chains on proteins (e.g., o-tyrosine, m-tyrosine, dityrosine, and carbonyl derivatives). The third measures the expression of oxidative stress-induced genes.

Use of Fluorescent Dyes

The fluorescent dyes fluorescein, rhodamine, and their derivatives are frequently used to measure levels of oxidative stress in mammalian cells[15] and yeast.[16–19] Although these fluorescent dyes may be useful for measuring overall levels of oxidative stress, they must be used with care and the proper controls because they are known to react with and become oxidized by a variety of compounds besides the reactive oxygen species (superoxide and hydrogen peroxide) they are assumed to measure. These other

[14] In "Correction Factors for Oxygen Solubility and Salinity." Dissolved Oxygen, U.S. Geological survey TWRI Book 9, p. DO 27-DO 38, 2002.
[15] J. A. Royall and H. Ischiropoulos, Arch. Biochem. Biophys. 302, 348 (1993).
[16] E. Cabiscol, E. Piulats, P. Echave, E. Herrero, and J. Ros, J. Biol. Chem. 275, 27393 (2000).
[17] F. Madeo, E. Fröhlich, M. Ligr, M. Grey, S. J. Sigrist, D. H. Wolf, and K.-U. Fröhlich, J. Cell Biol. 145, 757 (1999).
[18] E. J. Yurkow and M. A. McKenzie, Flow Cytometry 14, 287 (1993).
[19] K. Machida, T. Tanaka, K. Fujita, and M. Taniguchi, J. Bacteriol. 180, 4460 (1998).

compounds include reduced hemoproteins,[20] peroxynitrite,[21] horseradish peroxidase,[22,23] Fe^{2+}, Fe^{2+}-H_2O,[22] and tyrosine.[24] In addition, oxidation of these dyes is sensitive to the levels of the antioxidants catalase,[15,22] superoxide dismutase,[22] and glutathione,[23] and fluorescein oxidation increases with the pH of the growth medium.[23]

We have used the fluorescent dye carboxy-2′,7′-dichlorofluorescein diacetate (carboxy-H_2DCFDA) to measure steady-state ROS levels in both respiratory-proficient and respiratory-deficient yeast strains, in yeast cells cultured both aerobically and anaerobically, and in yeast cells shifted from aerobic to anaerobic conditions.[9] Once carboxy-H_2DCFDA enters the cell, the acetate group becomes hydrolyzed by intracellular esterases. The dye can then become oxidized to the fluorescent product, carboxy-dichloro-fluorescein, which, when excited by light of 500 nm, can be detected at a wavelength of 525 nm with a spectrofluorometer, fluorescent microscope, or flow cytometer. Carboxy-DCF has two negative charges at a physiological pH and thus is thought to be retained within the cell longer than the uncarboxylated form of the dye. However, carboxy-H_2DCFDA is subject to the same difficulties as described earlier and its oxidation may be limited by the activity of intracellular esterases.[24] We have found that carboxy-H_2DCFDA leaks out of the cell, like its uncarboxylated form (H_2DCFDA), making the interpretation of long-term, time-dependent experiments difficult.[9,23,25] Consequently, we believe that this dye is useful for assessing ROS levels in cells grown under steady-state conditions but not during shift experiments between different oxygen concentrations.

To use carboxy-H_2DCFDA for estimating differences in ROS levels between strains grown in different steady-state conditions or in comparing ROS levels in mutant and wild-type cells, the dye is dissolved in dimethyl sulfoxide and added to the culture at a final concentration of 10 μM at least 1 h before measurements are taken. Fluorescence is measured in three 100-μl aliquots, which are diluted 10-fold in 50 mM NaPO4, pH 7.0, and are sonicated briefly to disperse the cells. The fluorescence (wavelength, 515–545 nm) of 5000 cells from each aliquot is then measured using a

[20] T. Ohashi, A. Miztani, A. Murakami, S. Kojo, T. Ishii, and S. Taketani, *FEBS Lett.* **511,** 21 (2002).

[21] N. W. Kooy, J. A. Royall, H. Ischiropoulos, and J. S. Beckman, *Free Radic. Biol. Med.* **16,** 149 (1994).

[22] S. L. B. G. R. Hempel, Y. Q. O'Malley, D. A. Wessel, and D. M. Flaherty, *Free Radic. Biol. Med.* **27,** 146 (1999).

[23] W. Jakubowski and G. Bartosz, *Int. J. Biochem. Cell Biol.* **29,** 1297 (1997).

[24] J. L. Brubacher and N. C. Bols, *J. Immunol. Methods* **251,** 81 (2001).

[25] P. Breeuwer, J.-L. Drocourt, F. M. Rombouts, and Y. Abee, *Appl. Environ. Microbiol.* **60,** 1467 (1994).

BD PharMingen fluorescent-activated cell sorter equipped with a 15-mW argon laser.

Protein Carbonylation

The damaging effects that ROS have on proteins can be caused either by direct reaction of proteins with ROS or by reaction with secondary products produced from ROS. These reactions result in the oxidation of amino acids, the oxidation of the peptide backbone, which eventually may result in protein fractioning or cross linking, and the introduction of carbonyl groups on the side chains of susceptible amino acids. Carbonyl groups (i.e., aldehyde or ketone groups)[26] serve as useful markers for metal-catalyzed protein oxidation that occurs under conditions of oxidative stress. Protein carbonyls are quantitated easily after derivatization with 2,4 dinitrophenyl hydrazine.[27,28] The 2,4-dinitrophenylhydrazine is converted to 2,4-dinitrophenylhydrazone (DNP) by interaction with carbonyl groups. The protein-bound DNP is useful for both quantitating levels of protein carbonylation by HPLC and identifying those proteins that have been carbonylated by immunoblotting with anti-DNP antibodies. Although both types of analysis can be applied to whole cell extracts, we generally apply them to cytosolic and mitochondrial cell fractions.

Preparation of Mitochondrial and Cytosolic Cell Fractions. Yeast cells are harvested and twice washed by centrifugation ($3000g$, 10 min, $4°$) in ice-cold dH_2O. After washing, the cell pellet is resuspended in presphero-plast buffer [0.1 M Tris, 2.5 mM dithlothreitol (DTT) (freshly added), pH 9.3] at a ratio of 0.2 g/ml and incubated in a shaking water bath (100 rpm) at $28°$. After 20 min, cells are centrifuged down in preweighed centrifugation tubes (5000 rpm, 5 min, $4°$) and again washed twice with ice-cold dH_2O ($3000g$, 5 min, $4°$). After washing, the weight of the cell pellet is determined and it is resuspended (0.5 g/ml) in spheroplast buffer [10 mM $NaPO_4$, 1.35 M sorbitol, 1 mM EDTA, 2.5 mM DTT (freshly added), pH 7.5, containing 5 mg zymolyase 20T per ml buffer added shortly before use]. The cell suspension is spheroplasted in a shaking water bath (100 rpm) at $30°$. Spheroplasting is checked by microscopy or by examining cell lysis in the presence of NaSarcosyl. Spheroplasted cells that are diluted with 9 volumes of 2% NaSarcosyl will lyse and produce a clear solution. In contrast, cells diluted with 9 volumes of water will not lyse. Comparison of

---

[26] E. R. Stadtman, *Science* **257,** 1220 (1992).

[27] R. L. Levine, J. A. Williams, E. R. Stadtman, and E. Shacter, *Methods Enzymol.* **233,** 346 (1994).

[28] R. L. Levine, D. Garland, C. N. Oliver, A. Amici, I. Climent, A.-G. Lenz, B.-W. Ahn, S. Shaltiel, and E. R. Stadtman, *Methods Enzymol.* **186,** 464 (1990).

both gives a useful visual marker for lysis. If necessary, incubation with zymolyase can be extended, but not beyond 1 h. The spheroplasts are harvested by centrifugation (5 min at 3000g), washed gently in postspheroplast buffer (1.5 M sorbitol, 1 mM Na$_2$EDTA, pH 7.0), and sedimented at 3000g. The pelleted spheroplasts are resuspended in lysis buffer (0.6 M mannitol, 2 mM Na$_2$EDTA, pH 7.4) and lysed in a Sorvall Omnimixer (3 s at low speed and 25 s at full speed). The cell lysate is then centrifuged for 5 min at 1900g to pellet unbroken cells, nuclei, and debris. The supernatant contains mitochondria as well as cytosolic proteins. It is decanted and centrifuged for 10 min at 12,100g to pellet mitochondria. The resulting supernatant is saved as the cytosolic fraction and frozen in liquid nitrogen. The mitochondrial pellet is washed in lysis buffer, homogenized with a glass-Teflon homogenizer, and centrifuged at 1651g for 5 min to pellet cell debris that was trapped in the pellet. The supernatant, containing the mitochondria, is decanted and centrifuged at 23,000g for 10 min. The mitochondrial pellet is resuspended in Milli-Q H$_2$O and frozen immediately in liquid nitrogen.

Quantitation of Protein–Carbonyl Content in Mitochondria and Cytosolic Fractions. The carbonyl content of mitochondrial and cytosolic protein fractions is quantitated after derivatizing proteins in each fraction with 2,4-dinitrophenyl hydrazine (DNPH). Any DNA contaminating the cytosolic fraction is removed prior to the addition of DNPH by treatment with streptomycin sulfate as follows. A 70-μl aliquot of streptomycin sulfate (10%) is added to 500 μl of cytosol (best results with $\sim$0.5–1 mg of protein) and incubated at room temperature for 15 min. The DNA, precipitated by streptomycin sulfate, is collected by centrifugation (11,000 rpm, 15 min, 4°), and the supernatant is transferred to a 2-ml Eppendorf centrifuge tube containing 1.2 ml ice-cold 100% acetone. This solution is incubated at $-20°$ for 60 min and is then centrifuged at 14,000 rpm, 15 min, 4° to collect the precipitated protein. The protein pellet is washed with 80% (v/v) acetone by centrifugation as described earlier and is then resuspended in 100 μl of HPLC running buffer (6 M guanidinium hydrochloride, 0.5 M KH$_2$PO$_4$, pH 2.5). Mitochondrial samples ($\sim$0.5 mg protein) do not require treatment with streptomycin sulfate. They are sedimented (14,000 rpm, 10 min, 4°) and resuspended in the same volume of HPLC running buffer.

Four aliquots (25 μl each) of each mitochondrial or cytosolic sample are used. Two of these are derivatized with DPNH by mixing with 35 μl of 10 mM DPNH in HPLC running buffer. The other two are not derivatized with DPNH and serve as controls; they are added directly to 35 μl of HPLC running buffer. All four aliquots are incubated for 35 min at room temperature, filtered, and 50 μl of each is fractionated on a Waters

Model 626 HPLC system fitted with a Waters 600S controller and a Waters 996 photodiode array detector using Zorbax GF 450 and Zorbax GF 250 gel filtration columns run in series. The columns are calibrated and eluted (flow rate 1 ml/min) with 6 M guanidinium hydrochloride and 0.5 M KH_2PO_4, pH 2.5. Eluants are monitored for absorbance at 366 nm for DNPH and 276 nm for protein. Applying a molar extinction coefficient for hydrazone of 22,000 M^{-1} and an assumed average molar extinction coefficient of 50,000 M^{-1} for all proteins, carbonylation can be expressed as a ratio of moles of carbonyl per mole of protein. The actual carbonyl content of a sample is then calculated by subtracting the background signal in untreated samples from the signal in DNPH-derivatized samples. In calculations for DNPH-derivatized samples, the absorbance of DPNH at 276 nm (which is 43% of its absorbance at 366 nm) has to be substracted. The calculations of underivatized ($-$) and derivatized ($+$) DNPH samples are given by the following two formulas:

($-$)DNPH :

$$\text{mol carbonyl/mol protein} = 50,000 \ M^{-1} \times A_{366}/22,000 \ M^{-1} \times A_{276}$$

($+$)DNPH :

$$\text{mol carbonyl/mol protein} = 50,000 \ M^{-1} \times A_{366}/22,000 \ M^{-1} \times [A_{276} - (0.43 \times A_{366})]$$

SDS–PAGE of DNP-Derivatized Proteins. Whole cell, mitochondrial, and cytosolic protein samples for one-dimensional SDS–poyacrylamide gel electrophoresis are prepared as follows. One volume of sample containing 5 μg of mitochondrial protein, or 10 μg of whole cell or cytosolic protein, is added to an equal volume of 12% SDS. One additional volume of 20 mM DNPH in 10% trifluoroacetic acid is added, and the mixture is incubated for 35 min at room temperature. The derivatization reaction is stopped by adding 1.05 volumes of 2 M Tris base, 30% glycerol (final concentration of 0.52 M). Derivatized protein samples are separated on 10% polyacrylamide gels [resolving gel: 10% (w/v) 32:1 acrylamide:bis-acrylamide, 0.1% (w/v) SDS, 0.4 M Tris, pH 8.8; stacking gel: 3.5% (w/v) 32:1 acrylamide:bis-acrylamide 0.1% (w/v) SDS, 0.125 M Tris, pH 6.8] Gels are run at 110 V until the yellow front of underivatized DNPH reaches the bottom of the gel. The effect of exposing yeast cells to anoxia on mitochondrial protein carbonylation is shown in Fig. 1.

Two-Dimensional Gel Electrophoresis of Carbonylated Proteins. Aliquots of mitochondria (125 μg protein) or cytosol (70–90 μg protein) are diluted five-fold in lysis buffer containing 9 M urea, 2.5 M thiourea, 5%

Time after shift (min)

0   30   90   120   150   180   210   240   270   300

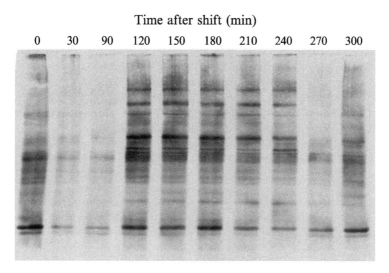

FIG. 1. Pattern of carbonylated yeast mitochondrial proteins after a shift from normoxic to anoxic conditions. JM43 cells were maintained in steady-state aerobic growth in the fermentor for at least six generations, and then the process gas was shifted from air to oxygen-free nitrogen. Cells were harvested at the times indicated and mitochondria were prepared. They were derivatized with 2,4-dinitrophenylhydrazine (DNP). The derivatized proteins were separated by SDS–PAGE, blotted to nitrocellulose filters, and detected with anti-DNP antibodies. Five milligrams of mitochondria were loaded onto each lane. (Lane 1, time of the shift. Lanes 2–9 are 30, 90, 120, 150, 180, 210, 240, and 270 min, respectively, after the shift.)

(w/v) CHAPS, 12.5 mM DTT, and 1% (v/v) carrier ampholytes, pH 3–10, and are incubated for 1 h at room temperature. Rehydration buffer [7 M urea, 2 M thiourea, 4% (w/v) CHAPS, 10 mM DTT, 0.5% (v/v) Triton X, 1% (v/v) carrier ampholytes, pH 3–10, and a trace of bromphenol blue] is added to bring the total volume up to 250 μl. Samples are incubated at room temperature for 20 min and are then used to rehydrate 13-cm Immobiline Drystrips (Amersham Pharmacia Biotech) with a pH 3–10 linear gradient. The strips are rehydrated overnight in an Immobiline Drystrip reswelling tray (Amersham Pharmacia Biotech). First-dimension isoelectric focusing is carried out at 20° using a Multiphor II flatbed system (Amersham Pharmacia Biotech) at the following voltages: during the first minute, voltage is increased linearly to 300 V, is held at 300 V for the next 4 h, is increased linearly from 300 to 3000 V during the next 5 h, and is held at 3000 V for the next 10 h. After isoelectric focusing, the Immobiline strips are derivatized with DNPH using a modification of the method of

Reinheckel *et al.*[29] They are incubated in a solution of 10 mM DNPH in 10% trifluoroacetic acid for 20 min with agitation and then incubated with agitation for 10 min at room temperature in each of the following solutions: (1) 150 mM Tris–HCl (pH 6.8), 8 M urea, 20% (v/v) glycerol, and 2% (w/v) SDS; (2) 150 mM Tris–HCl (pH 6.8), 8 M urea, 20% (v/v) glycerol, 2% (w/v) SDS, and 1% (w/v) DTT; and (3) 150 mM Tris–HCl (pH 6.8), 8 M urea, 20% (v/v) glycerol, 2% (w/v) SDS, 4% (w/v) iodoacetamide, and a trace amount of bromphenol blue. The Immobiline strips are loaded onto a 20 × 20-cm polyacrylamide gel composed of 8–18% (w/v) 37.5:1 acrylamide:bis-acrylamide, 0.37 M Tris (pH 8.8). The 20 × 20-cm gradient gels are poured using a gradient mixer GM-1 (Amersham Pharmacia Biotech) and a Protean II xi multigel casting chamber (Bio-Rad). The gels are run at 10° on a Protean II xl m multicell unit (Bio-Rad) at a constant 15 mA per gel for approximately 17 h in a running buffer containing 40 mM glycine, 200 mM Tris, and 0.1% SDS.

Western Immunoblotting of Carbonylated Proteins Separated by Either One- or Two-Dimensional Electrophoresis. After electrophoresis, one or two-dimensional gels are subjected to western immunoblotting with an anti-DNPH antibody. Proteins are transferred onto Hybond ECL (Amersham Pharmacia Biotech) nitrocellulose membranes using a Semi-Phor blotting apparatus (Hoeffer Scientific Instruments). Oxidatively modified proteins are detected with anti-DNP antibodies (DAKO) (at a dilution of 1:4000) and secondary horseradish peroxidase-linked antibodies (at a dilution of 1:25000), followed by a chemiluminescence reaction using a chemiluminescence detection kit (Perkin Elmer Life Sciences). Depending on the amount of protein loaded and the carbonyl content of the sample, exposure times will vary so taking a series of different exposure times is advisable. The identification of carbonylated mitochondrial proteins by two-dimensional electrophoresis is shown in Fig. 2.

Detection of 8-OH-dG in Mitochondrial and Nuclear DNA

ROS cause several different oxidative chemical modifications to the nucleotide bases of mitochondrial and nuclear DNA. One of these, 8-hydroxy-2′-deoxyguanosine (8-OH-dG), which is derived from 2-deoxyguanosine, is assayed conveniently using HPLC in conjunction with an electrochemical detector. The level of this oxidatively modified deoxynucleoside in unstressed aerobic yeast cells is low.[9] However, because most of the ROS that are present in these cells are generated by mitochondrial respiration, the

[29] T. Reinheckel, S. Korn, S. Mohung, W. Augustin, W. Halangk, and L. Schild, *Arch. Biochem. Biophys.* **375**, 59 (2000).

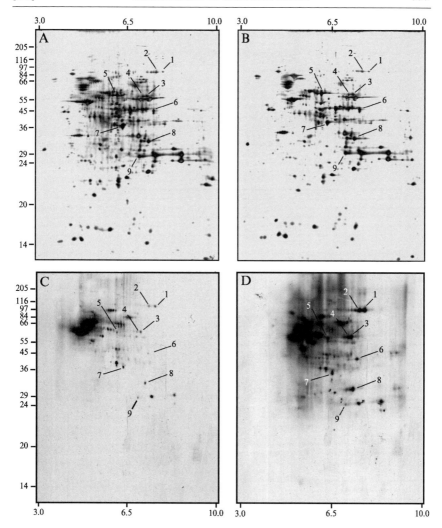

FIG. 2. Two-dimensional electrophoresis of yeast mitochondrial proteins. Samples were taken before (A and C) and 210 min after (B and D) a shift. (A and B) Silver stains of the entire gel and (B and D) immunoblots with anti-DNP serum of A and C, respectively. Those proteins exhibiting an average increase in carbonylation levels of more than twofold are indicated on both silver stains and immunoblots.

level of 8-OH-dG in mitochondrial DNA(mtDNA) is higher (6- to 7-fold) than its level in nuclear DNA (nDNA). Upon exposing cells to anoxia, the levels of 8-OH-dG increase transiently (Fig. 3), with average increases for mtDNA and nDNA of 3.5 and 2.5-fold, respectively.

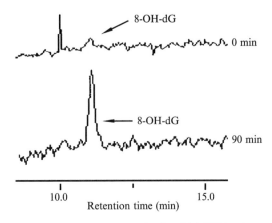

Fig. 3. Detection of 8-OH-dG in yeast mitochondrial DNA after a shift to anoxia. Mitochondrial DNA was isolated, hydrolyzed, and analyzed for 8-OHdG from cells before a shift (top trace) and 90 min after a shift from normoxia to anoxia (bottom trace). 8-OH-dG was detected with an electrochemical detector.

Isolation of Nuclear DNA. Nuclear DNA is isolated by using any of the commercially available kits sold for the isolation of yeast DNA (e.g., Y-DER Yeast DNA extraction kit from Pierce) and following the instructions given therein.

Isolation of Mitochondrial DNA. Mitochondrial DNA is isolated directly from purified mitochondria using a modification of the procedure of Querol and Barrio (1990).[29a] Mitochondria, isolated as described earlier, are pelleted for 20 min at 15,000g in a corex tube, resuspended in 2 ml of 10 mM EDTA, 50 mM NaCl, 0.5% SDS, pH 7.5, with 50 μg/ml proteinase K, and incubated at 37° for 3 h. After incubation, 0.5 ml of 5 M potassium acetate is added and samples are incubated at $-20°$ for 15 min. Samples are then centrifuged (17,000g, 10 min, 4°), the resultant supernatant is transferred to a fresh Corex tube, and mitochondrial DNA is precipitated by adding an equal volume of isopropanol. This suspension is kept at room temperature for 10 min before DNA is harvested by centrifugation (17,000g for 10 min) at room temperature. Subsequently, the DNA is washed with 70% (v/v) EtOH and resuspended in 200–300 μl of dH$_2$O.

DNA Hydrolysis. Prior to analysis, nuclear or mitochondrial DNA (100 μg in 50 μl of dH$_2$O) is boiled for 6 min at 95° and then immediately put on ice. When cool, 30 μl of nuclease P1 (1 μg/μl) and 50 μl of 0.5 M

[29a] A. Querol and E. Barrio, *Nucleic Acid Res.* **18,** 1657 (1990).

sodium acetate, pH 5.2, are added and the mixture is incubated at 37° for 2 h. Then, 2 units of alkaline phosphatase (2 μl of a 1-U/μl solution) is added and incubation is continued for another 2 h at 37°. The samples are then filtered through a 0.2-μm HPLC filter.

Determination of 8-Hydroxy-2'-Deoxyguanosine and 2-Deoxyguanosine Levels in DNA Samples. Levels of 8-hydroxy-2'-deoxyguanosine and 2-deoxyguanosine are quantified using the isocratic high-pressure liquid chromatography method described by McCabe *et al.*[30] Approximately 100 μg of DNA from each sample, hydrolyzed with nuclease P1 and alkaline phosphatase, is injected onto a YMC basic S3, 4.6 × 150-mm column. The mobile phase is 100 mM sodium acetate in 4% methanol, pH 5.2. All reagents are HPLC grade, and Milli-Q water is purified further using a Waters Sep-Pak solid-phase extraction column. Isocratic analysis takes place with a Shimadzu LC-600 pump connected to a Spectrasystem AS3500 autosampler. The oxidized guanine adduct, 8-OH-dG, is detected using a CoulArray electrochemical detection system (ESA, Inc., Chelmsford, MA) at a potential of approximately 300 mV. The eighth cell in the array is adapted to a connection with a Shimadzu SPD-6A UV detector set at 254 nm. Nonoxidized 2'-deoxyguanosine (dG) is quantified by UV detection. Sample adduct concentrations are calculated from standard curves of 8-OH-dG, 0.1–1.5 pmol, and dG, 0.5–15.0 nmol. All data are analyzed using ESA (Chelmsford, MA) CoulArray for Windows software and are expressed as the ratio 8-OH-dG/10^5 dG. All reagents are HPLC grade. Milli-Q water is purified further using a Waters Sep-Pak solid-phase extraction column.

Induction of SOD1

Yeast genes that encode stress proteins are aerobic genes whose expression is maximal in cells grown to steady state in air and reduced in cells grown to steady state under hypoxic or anoxic conditions.[31,32] They are also induced by oxidative stress brought about by exposure to oxidants such as H_2O_2 and paraquat.[32–34]At least one of these genes, *SOD1*, the gene for Cu,Zn-superoxide dismutase, is induced during a shift to anoxia.[9] Because of this and because *SOD1* transcript levels are relatively high, *SOD1*

[30] D. R. McCabe, I. N. Acworth, M. L. Maidt, and R. A. Floyd, *Proc. Am. Assoc. Cancer Res.* **38**, 336 (1997).

[31] K. E. Kwast, P. V. Burke, and R. O. Poyton, *J. Exp. Biol.* **201**, 1177 (1998).

[32] J. Lee, A. Romeo, and D. J. Kosman, *J. Biol. Chem.* **271**, 24885 (1996) 24885.

[33] A. P. Gasch, P. T. Spellman, C. M. Kao, O. Carmel-Harel, M. B. Eisen, G. Storz, D. Botstein, and P. O. Brown, *Mol. Biol. Cell* **11**, 4241 (2000).

[34] D. L. Croteau and V. A. Bohr, *J. Biol. Chem.* **272**, 25409 (1997).

expression is very useful as an indicator of oxidative stress.[33,35,36] *SOD1*
transcript levels are quantitated most easily by Northern blot analysis with
normalization to the transcript from *ACT1*, which is not induced by either
oxidative stress or anoxia.

RNA Isolation. Total yeast cell RNA is isolated by a modification of the
procedure of Elder *et al.*[37] Cells, solutions, tubes, and rotors are kept at 4°
throughout all steps. Frozen pellets, ranging from 0.8 to 1.9 g (wet weight),
are resuspended in 5 ml STES buffer [0.5 *M* NaCl, 0.2 *M* Tris–Cl (pH 7.5),
0.01 *M* EDTA (pH 8.0), 1% (w/v) SDS, 0.1% (v/v) diethyl pyrocarbonate
(DEPC)]. After the addition of 10-g sterile glass beads (0.50 mm) and 5 ml
phenol:chloroform:isoamyl alcohol (PCI, 25:24:1, pH 4.5), cells are disrupted by vortexing for 2 min. Phase layers are separated by centrifugation
(5 min, $4000g_{max}$, 4°), and the aqueous phase is collected. Phenol extraction
(the addition of PCI with vortexing and centrifugation) is repeated at least
two more times or until the phase interface is clear. To precipitate the nucleic acid, 2.5 volumes ethanol and 0.1 volumes 3 *M* sodium acetate, 0.10%
(v/v) DEPC are mixed gently into the isolated aqueous phase and incubated at −20° for at least 12 h. The nucleic acid precipitate is recovered
by centrifugation (30 min, $16,000g$, 4°). The pellet is then washed with
5 ml 70% (v/v) ethanol and is centrifuged again (15 min, $16,000g$, 4°). To
remove excess ethanol, the pellet is allowed to air dry before being dissolved in 100 μl 20 m*M* EDTA, 0.1% (v/v) DEPC (pH 8.0). At this point
the RNA is quantitated spectrophotometrically[38] and divided into aliquots.
Each aliquot is reprecipitated as described earlier and then dissolved in
20 m*M* EDTA, 0.1% (v/v) DEPC (pH 8.0), 69% (v/v) formamide to obtain
a concentration of 2 μg RNA/μl. Yield is variable, but averages about
1400 μg RNA per gram (wet weight) of cells.

Electrophoresis and Northern Blotting. 5X MOPS formaldehyde buffer
[0.1 *M* MOPS, 40 m*M* sodium acetate, 5 m*M* EDTA, 1.1 *M* formaldehyde,
0.1% (v/v) DEPC, pH 7.0] is added to 30 μg RNA to a 1X concentration in
a total volume of 20 μl and denatured at 65° (15 min). Following denaturation, 2 μl 10X loading dye [50% (v/v) glycerol, 1 m*M* EDTA (pH 8.0),
0.25% (w/v) bromphenol blue, and 0.25% (w/v) xylene cyanol] is added
to the sample before being separated on 1.8% (w/v) agarose gels. 1X

[35] C. Godon, G. Lagniel, J. Lee, J.-M. Buhler, S. Kieffer, M. Perrot, H. Boucherie,
M. B. Toledano, and J. Labarre, *J. Biol. Chem.* **273,** 22480 (1998).
[36] A. Dodd, unpublished results, 2002.
[37] R. T. Elder, E. Y. Loh, and R. W. Davis, *Proc. Natl. Acad. Sci. USA* **80,** 2432 (1983).
[38] J. Sambrook, F. Fritsch, and T. Maniatis, *in* "Molecular Cloning: A Laboratory Manual."
Cold Spring Harbor Laboratory, Cold Spring Harbor, NY, 1989.

MOPS formaldehyde buffer is used in the gels and in the running buffer.[39] Gels are run for 3–4 h at 110 V. Blotting is performed as recommended by the manufacturer of the transfer membrane (GeneScreen, NEN, Boston, MA). To remove formaldehyde from the finished gel, it is soaked four times in 5 volumes (500 ml per 100-ml gel) dH_2O [0.1% (v/v) DEPC] for 5 min with shaking. The gel is then equilibrated with 5 volumes 10X SSC [1.5 M NaCl, 0.15 M sodium citrate dihydride, 0.1% (v/v) DEPC] for 30 min with shaking. A capillary blot is assembled and left to transfer over-night.[38] The following day, the RNA is fixed to the membrane by UV cross-linking with a UV Stratalinker 1800 (Stratagene, La Jolla, CA), followed by baking for 20 min at 80°.

Hybridization. Polymerase chain reaction (PCR)-based probes are made corresponding to *SOD1,* encoding Cu/ZnSOD; and *ACT1,* encoding actin as described by Dirmeier *et al.*[9] Radiolabeled probes are made by random primer labeling of PCR fragments (100 ng) using $[\alpha$-$^{32}P]dCTP$ with a random primer extension labeling kit (Sigma, St. Louis, MO) according to the manufacturer's directions. Excess dCTP is removed using ProbeQuant G-50 microcolumns (Amersham Pharmacia, Piscataway, NJ). For the hybridization step, ULTRAHyb buffer (Ambion, Austin, TX) is used according to the manufacturer's directions. Blots are prehybridized for 30 min at 42° before adding radiolabeled probe to a concentration of 6.5×10^6 cpm/ml. Blots are allowed to hybridize for 24 h. Following hy-bridization, blots are washed twice with 2X SSC, 0.1% (w/v) SDS (42°, 5 min), followed by two washes with 0.1X SSC, 0.1% (w/v) SDS (55°, 5 min). Signal intensity is measured with a Storm 860 PhosphorImager (Amersham Pharmacia, Piscataway, NJ). The relative signal strength of *SOD1* is normalized to the level of *ACT1* mRNA, as oxygen levels do not affect *ACT1* transcription.[40]

Acknowledgment

Some of the work reported in this paper was supported by Research Grants GM30228 and HL63324 from the National Institutes of Health.

[39] S. S. Tsang, X. Yin, C. Guzzo-Arkuran, V. S. Jones, and A. J. Davison, *Biotechniques* **14,** 380 (1993).
[40] C. Dagsgaard, L. E. Taylor, K. M. O'Brien, and R. O. Poyton, *J. Biol. Chem.* **276,** 7593 (2001).

[39] Evaluation of Oxygen Response Involving Differential Gene Expression in *Chlamydomonas reinhardtii*

By José A. Del Campo, Jeanette M. Quinn, and Sabeeha Merchant

Chlamydomonas as a Model Organism

The unicellular green alga *Chlamydomonas* has, for decades, offered experimental advantages for the study of chloroplast function and biogenesis, photosynthesis, flagellar motility and assembly, photoreceptor biochemistry, and sexual mating.[1-6] Among these are (1) the ability to manipulate the nuclear and also both organellar genomes,[1,7,8] (2) facultative photosynthetic growth because of the ability of *Chlamydomonas* to use acetate for heterotrophic growth, (3) heterothallic mating types, which permit classical genetic approaches for the dissection of important biological problems,[9] and (4) considerable genomic information through EST[10] and shotgun genome sequencing projects (http://genome.jgi-psf.org/chlre1/chlre1.home.html).

Chlamydomonas has long been used as a model (see Fig. 1) for the study of nutrient-responsive signal transduction, especially in the context of the function of photosynthetic apparatus. The classical areas of interest have included inorganic (sulfur, nitrogen, phosphorus) nutrient utilization,[11,12]

[1] J. D. Rochaix, M. Goldschmidt-Clermont, and S. Merchant (eds.), *in* "The Molecular Biology of Chloroplasts and Mitochondria in *Chlamydomonas*." Kluwer Academic Publishers, 1998.

[2] A. R. Grossman, *Curr. Opin. Plant Biol.* **3**, 132 (2000).

[3] E. H. Harris, *Annu. Rev. Plant Physiol. Plant Mol. Biol.* **52**, 363 (2001).

[4] R. M. Dent, M. Han, and K. K. Niyogi, *Trends Plant Sci.* **6**, 364 (2001).

[5] C. D. Silflow and P. A. Lefebvre, *Plant Physiol.* **127**, 1500 (2001).

[6] O. A. Sineshchekov and E. V. Govorunova, *Trends Plant Sci.* **4**, 201 (1999).

[7] B. L. Randolph-Anderson, J. E. Boynton, N. W. Gillham, E. H. Harris, A. M. Johnson, M. P. Dorthu, and R. F. Matagne, *Mol. Gen. Genet.* **126**, 357 (1993).

[8] J. E. Boynton and N. W. Gillham, *Methods Enzymol.* **264**, 279 (1996).

[9] H. Harris, *in* "The Chlamydomonas Sourcebook: A Comprehensive Guide to Biology and Laboratory Use." Academic Press, San Diego, 1989.

[10] J. Shrager, C. Hauser, C. W. Chang, E. H. Harris, J. Davies, J. McDermott, R. Tamse, Z. Zhang, and A. R. Grossman, *Plant Physiol.* **131**, 401 (2003).

[11] J. P. Davies and A. R. Grossman, *in* "The Molecular Biology of Chloroplasts and Mitochondria in *Chlamydomonas*" (J. D. Rochaix, M. Goldschmidt-Clermont, and S. Merchant, eds.), p. 613. Kluwer Academic Publishers, 1998.

[12] A. Grossman and H. Takahashi, *Annu. Rev. Plant Physiol. Plant Mol. Biol.* **52**, 163 (2001).

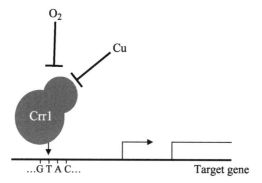

FIG. 1. A model showing known elements involved in copper deficiency and hypoxia responses. The putative DNA-binding protein Crr1 is required for the activation of target genes when cells are facing copper deficiency and/or hypoxic conditions. The core of the CuRE elements (see text) is the sequence GTAC.

as well as trace metal requirements and mechanisms of homeostasis.[13–15] Metabolism and gene expression involving CO$_2$ and O$_2$ have also been long-standing topics of research activity,[16–18] with recently renewed emphasis from the perspective of energy metabolism[19] and the identification of regulatory mutants.[20–23]

    Chlamydomonas reinhardtii, like other organisms, responds to changes in oxygen supply through the alteration of metabolism. Some species of this genus are found in naturally oxygen-deficient habitats such as peat bogs and sewage lagoons,[9] where a hypoxic response is probably critical for survival. Even laboratory cultures become oxygen depleted quite

[13] S. Merchant, *in* "The Molecular Biology of Chloroplasts and Mitochondria in *Chlamydomonas*" (J. D. Rochaix, M. Goldschmidt-Clermont, and S. Merchant, eds.), p. 597. Kluwer Academic Publishers, 1998.

[14] S. La Fontaine, J. M. Quinn, S. S. Nakamoto, M. D. Page, V. Gohre, J. L. Moseley, J. Kropat, and S. Merchant, *Eukaryot. Cell* **1,** 736 (2002).

[15] P. Rubinelli, S. Siripornadulsil, F. Gao-Rubinelli, and R. T. Sayre, *Planta* **215,** 1 (2002).

[16] P. E. Bryant, *Int. J. Radiat. Biol. Relat. Stud. Phys. Chem. Med.* **17,** 533 (1970).

[17] M. R. Badger, A. Kaplan, and J. A. Berry, *Plant Physiol.* **66,** 407 (1980).

[18] A. M. Geraghty and M. H. Spalding, *Plant Physiol.* **111,** 1339 (1996).

[19] M. L. Ghirardi, L. Zhang, J. W. Lee, T. Flynn, M. Seibert, E. Greenbaum, and A. Melis, *Trends Biotechnol.* **18,** 506 (2000).

[20] J. M. Quinn, M. Eriksson, J. L. Moseley, and S. Merchant, *Plant Physiol.* **128,** 463 (2002).

[21] H. Fukuzawa, K. Miura, K. Ishizaki, K. Kucho, T. Saito, T. Kohinata, and K. Ohyama, *Proc. Natl. Acad. Sci. USA* **98,** 5347 (2001).

[22] K. Van, Y. Wang, Y. Nakamura, and M. H. Spalding, *Plant Physiol.* **127,** 607 (2001).

[23] Y. Xiang, J. Zhang, and D. P. Weeks, *Proc. Natl. Acad. Sci. USA* **98,** 5341 (2001).

rapidly as a consequence of respiration if they are not sufficiently agitated.[20,24] Oxygen-responsive gene expression is, therefore, probably an integral aspect of everyday metabolism for *Chlamydomonas,* as well as for other organisms. In addition, there are also special metabolic pathways occurring only in anaerobic *Chlamydomonas* cells; the most publicized of which is photosynthetic hydrogen production.[25]

Hydrogen Production in Low Oxygen

In the early 1930s, Gaffron and co-workers discovered the ability of unicellular green algae to produce hydrogen gas upon illumination (reviewed by Melis and Happe[25]). Historically, hydrogen evolution activity in green algae was induced upon prior anaerobic incubation of the cells in the dark. Under such conditions, an iron-containing hydrogenase enzyme (encoded by the *HydA* gene in *C. reinhardtii*) was induced, catalyzing light-mediated H_2 evolution. Oxygen is a powerful inhibitor of hydrogen evolution at the enzyme level.[19,26] It has also been shown that regulation of the hydrogenase gene takes place at the transcriptional level.[27] The idea of using H_2 gas from green algae as an alternative fuel source has been promoted recently.[28] The pathway involves two stages, photosynthesis and H_2 production, dependent on sulfur availability, with sulfur deprivation serving as a metabolic switch. When sulfur is available for cells, green algae perform normal photosynthesis (water oxidation, oxygen evolution, and biomass accumulation). In the absence of sulfur, photosynthesis in *C. reinhardtii* slips into a hydrogen production mode if oxygen is simultaneously removed. Electrons for H_2 production may originate either at photosystem II (PSII) upon photooxidation of water or at the plastoquinone pool upon oxidation of the cellular endogenous substrate (e.g., starch degradation). Electrons are transported via photosystem I (PSI) to ferredoxin, which serves as the physiological electron donor to the hydrogenase. The signal transduction components involved in metabolic switching in this pathway are completely unknown. The ability to undertake combined molecular and classical genetic studies in this model system should facilitate the discovery of the relevant regulatory factors.

[24] P. M. Wood, *Eur. J. Biochem.* **87,** 8 (1978).
[25] A. Melis and T. Happe, *Plant Physiol.* **127,** 740 (2001).
[26] T. Happe, A. Hemschemeier, M. Winkler, and A. Kaminski, *Trends Plant Sci.* **7,** 246 (2002).
[27] T. Happe and A. Hamiski, *Eur. J. Biochem.* **269,** 1022 (2002).
[28] A. Melis, L. Zhang, M. Forestier, M. L. Ghirardi, and M. Seibert. *Plant Physiol.* **122,** 127 (2000).

Connections between Oxygen and Copper Nutrient Homeostasis

Copper is an essential micronutrient for all organisms because of its function in enzymes that serve as catalysts of oxygen chemistry and redox reactions. *Chlamydomonas* has a well-characterized response to nutritional copper deficiency.[13] The best characterized of these is the replacement of plastocyanin (the most abundant copper protein in photosynthetic cells) with a heme-containing c-type cytochrome.[24,29] This occurs through transcriptional activation of the *Cyc6* gene encoding cytochrome c_6, a heme-containing functional substitute for plastocyanin,[30] and induced degradation of apoplastocyanin in the thylakoid lumen,[31] which occurs probably to ensure the redistribution of copper to cytochrome oxidase.[32] This transcriptional response requires copper response elements (CuREs)[33] containing the sequence GTAC, which forms the core of these CuREs, and Crr1, a putative DNA-binding protein (Fig. 1).[34] Crr1 is also required for the transcriptional activation in copper-deficient cells of *Cpx1,* encoding coproporphyrinogen III oxidase,[35] and also *Crd1,*[36] encoding the aerobic oxidative cyclase in chlorophyll biosynthesis.[37] In copper-replete cells, Crd1 is replaced by Cth1. The inhibition of Cth1 accumulation under copper-deficient conditions also involves Crr1.[36]

Wood[24] noted first that hypoxic *Chlamydomonas* cells accumulate cytochrome c_6 even in copper-replete medium. The response of the *Cyc6* gene to hypoxia is mediated at the transcriptional level via its CuREs and requires Crr1 function.[20,33] Subsequently, it was discovered that each of the Cu deficiency targets mentioned previously also responds to hypoxia (e.g., Fig. 2) in a pathway involving the CuRE and Crr1 function.[38] The hypoxic and nutritional copper signaling pathways in *Chlamydomonas* therefore seem to share some common components.[20] A trivial explanation for the similar output, that is, that hypoxic cells are internally copper deficient, was ruled out because oxygen-deprived cells are able to synthesize

[29] S. Merchant and L. Bogorad, *EMBO J.* **6,** 2531 (1987).

[30] J. M. Quinn and S. Merchant, *Plant Cell* **7,** 623 (1995).

[31] H. H. Li and S. Merchant, *J. Biol. Chem.* **270,** 23504 (1995).

[32] S. S. Nakamoto, *in* "Compartmentalized Copper and Iron Enzymes in *C. reinhardtii:* Venus Versus Mars." Ph.D. Dissertation, University of California, Los Angeles. Department of Chemistry and Biochemistry, 2001.

[33] J. M. Quinn, P. Barraco, M. Eriksson, and S. Merchant, *J. Biol. Chem.* **275,** 6080 (2000).

[34] J. Kropat and S. Merchant, unpublished results (2003).

[35] M. Eriksson *et al.,* unpublished results (2003).

[36] J. L. Moseley, M. D. Page, N. P. Alder, M. Eriksson, J. Quinn, F. Soto, S. M. Theg, M. Hippler, and S. Merchant, *Plant Cell* **14,** 673 (2002).

[37] S. Tottey, M. A. Block, M. Allen, T. Westergreen, C. Albrieux, H. U. Scheller, S. Merchant, and P. E. Jensen, *Proc. Natl. Acad. Sci. USA* **100,** 16119 (2003).

[38] J. M. Quinn, S. S. Nakamoto, and S. Merchant, *J. Biol. Chem.* **274,** 14444 (1999).

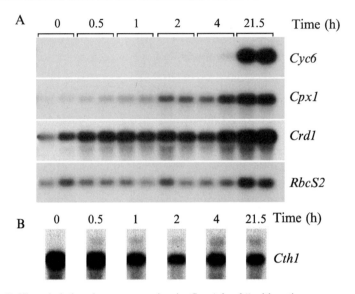

FIG. 2. Hypoxia-induced gene expression in *C. reinhardtii* with a time course response. Wild-type strain CC125 was grown in copper-supplemented (6 μM) Tris–acetate–phosphate medium under normal aeration to a concentration of 1×10^6 cells/ml in a shaker (300 rpm) at room temperature using normal room lighting (approximately 1–15 μmol m^{-2} s^{-1}). Twenty milliliters was removed for RNA preparation ($t = 0$). The remaining culture was bubbled with 98% N$_2$ + 2% CO$_2$ (0% O$_2$). Samples for RNA preparation were removed at the indicated times. RNA samples were probed with *Cyc6, Cpx1, Crd1,* and *RbcS2* (as a control) (A). Except for *Cth1* (B), all samples were analyzed in duplicate cultures.

holoplastocyanin *de novo*, indicating that copper is available inside the cells. Quinn *et al.*[20] tested the idea that the connection between Cu and the hypoxic response results from modification of the Cu(II)/Cu(I) ratio by the oxygen status of the environment, assuming that the copper sensor is specific for Cu(II) vs Cu(I). In this case, the hypoxic response of *Cyc6, Cpx1, Crd1,* and *Cth1* would simply mimic the response to Cu deficiency. However, this is not the case.[20] The *Cyc6* gene responds strongly to Cu deficiency and more weakly to hypoxia, whereas the *Crd1* and *Cpx1* genes respond strongly to hypoxia and less so to Cu deficiency. Furthermore, a specific hypoxia response element was identified in the *Cpx1* gene.[33] This HyRE is related to the CuRE by virtue of the GTAC core sequence, but it does not function as a CuRE.[20] Quinn *et al.*[20] concluded that the hypoxic response shares signal transduction components with the Cu deficiency response. The rationale for the intersection of the regulatory pathways is that hypoxic conditions create copper deficiency in nature by driving copper ions to insoluble Cu(I) species, which precipitate and perhaps have limited bioavailability. The response of Cu deficiency targets to hypoxia might be a mechanism for anticipating and preparing for Cu deficiency in a microaerobic growth environment.

The pattern of cytochrome c_6 and coprogen oxidase accumulation parallels the transcriptional activation of the genes, indicating that the hypoxic response can also be detected and analyzed at the protein level. However, in order to monitor changes in gene expression, it is more appropriate to analyze the effects at the level of transcript abundance, as the half-life for the various mRNAs is shorter than for the corresponding proteins.

The physiological relevance of the Crr1-dependent hypoxic response of one target gene, *Crd1*, was confirmed by the demonstration of a conditional chlorotic phenotype in hypoxic but not aerobic *crd1* mutant cells.[20] The *crd1* mutant strain is useful for the study of the Crr1-dependent hypoxic gene expression pathway because its remarkable chlorotic phenotype in +Cu TAP medium in low air can be used as a physiological marker, and expression of the gene can be used as a molecular marker (see Fig. 2).

At this point, a number of questions remain about the hypoxic response in *Chlamydomonas*. First, for the *Cpx1* gene, the copper response element appears to be necessary and sufficient for the nutritional copper response and is also necessary for the hypoxic response. A second element, with a GTAC core, found in the *Cpx1* promoter is required for the oxygen deficiency response. The second element, called a HyRE, is not necessary for the copper deficiency response. What is the relationship between the CuRE and the HyRE at the mechanistic level? Second, a well-known anaerobically induced gene, *Hyd1*, is normally regulated in the *crr1* mutant, indicating a Crr1-independent hypoxia/anoxia sensing mechanism in *C. reinhardtii*. What are the regulatory components required for the regulation of *HydA*? Given the repertoire of molecular techniques on hand today that can be applied to this experimental model, identification of the hypoxic signaling components should be achievable. With this intention in mind, this article describes the setup of experiments involving the growth of *Chlamydomonas* cells in liquid cultures at different oxygen concentrations.

Growth Media

Chlamydomonas was first introduced into the biochemistry of photosynthesis by Levine and co-workers.[39] In a previous volume of this series,[40] information concerning the culturing of *Chlamydomonas* cells, as well as detailed descriptions about mutagenesis procedures and photosynthesis

---

[39] R. P. Levine, *Science* **162**, 768 (1968).
[40] A. San Pietro (ed.), *Methods Enzymol.* 23 (1971).

studies, can be found. For details concerning the preparation of growth media, both liquid and solid, and particularly copper-free media, see Quinn and Merchant.[41] For general information concerning the cultivation of *Chlamydomonas* cells, including various types of growth media, see Harris.[9]

Culturing

Several different strains have been tested for the response to oxygen concentration in liquid cultures, including CC125, CC425, and 2137 as wild-type strains, and the mutant *crr1-1,* which is affected in the regulation of copper metabolism (see earlier discussion). As diagrammed in Fig. 3, to generate different oxygen concentrations for the culturing of *Chlamydomonas* cells, we used a mixture of three different gases: CO_2, compressed air, and nitrogen. The use of CO_2 in the gas mixture is important because CO_2 is a nutrient and of course also buffers the medium. Variation in air content relative to nitrogen changes both CO_2 and O_2 levels. By providing CO_2 at a fixed concentration (0.2 to 2% depending on the experiment), the key variable in the experiment is O_2.

CO_2 concentration in air (0.036% or 350 ppm) is considered to be low CO_2[11] conditions from a physiological point of view,[42] leading to low growth rates and low biomass accumulation in algal cultures because CO_2 is a substrate for photosynthesis. The ratio of CO_2/O_2 is also metabolically relevant[43,44] because the active site of Rubisco discriminates poorly against O_2. Therefore, in air, a significant fraction of ribulose-1,5-bisphosphate is oxygenated instead of carboxylated,[43,44] leading to the photorespiratory pathway for salvage of phosphoglycolate. To avoid photorespiration, algae induce a carbon concentration mechanism (CCM), involving carbonic anhydrases (CA)[42,45–48] and other components as well. The CCM is repressed by high CO_2 (e.g., 5%).[21–23] The effect of O_2 on regulation of the CCM is not known, but is probably a question worth addressing. Nevertheless,

[41] J. M. Quinn and S. Merchant, *Methods Enzymol.* **297,** 263 (1998).

[42] A. Kaplan and L. Reinhold, *Annu. Rev. Plant Physiol. Plant Mol. Biol.* **50,** 539 (1999).

[43] J. Berry, J. Boynton, A. Kaplan, and M. Badger, *Carnegie Inst. Wash. Yearbook* **75,** 423 (1976).

[44] W. R. King and Andersen, *Arch. Microbiol.* **128,** 84 (1980).

[45] M. Rawat and J. V. Moroney, *Plant Physiol.* **109,** 937 (1995).

[46] J. Karlsson, A. K. Clarke, Z. Y. Chen, S. Y. Hugghins, Y. I. Park, H. D. Husic, J. V. Moroney, and G. Samuelsson, *EMBO J.* **17,** 1208 (1998).

[47] R. P. Funke, J. L. Kovar, and D. P. Weeks, *Plant Physiol.* **114,** 237 (1997).

[48] M. Eriksson, J. Karlsson, Z. Ramazanov, P. Gardestrom, and G. Samuelsson, *Proc. Natl. Acad. Sci. USA* **93,** 12031 (1996).

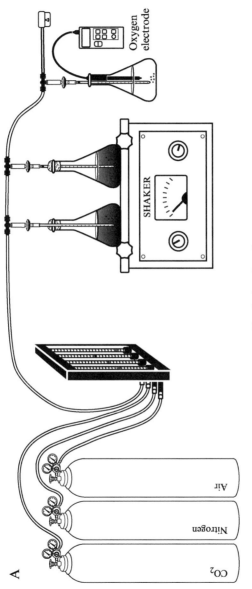

FIG. 3. (*continued*)

B

1. Flow-meters (front view)

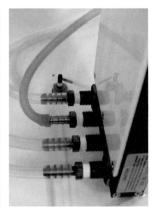

2. Rear view

3. Culture flask connection

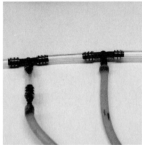

4. Silicone tube connections

5. Filter connection

6. Culture flask

FIG. 3. (A) Scheme for the experimental setup used to grow *Chlamydomonas* cells under different oxygen concentrations. The amount of each gas is monitored on a multichannel gas flowmeter. The mixture of different gases is provided to the cultures using silicone tubing. The oxygen concentration in the cultures is measured using a Clark-type oxygen electrode in a flask at the end of the series containing TAP medium without cells. Note that it is preferable to measure the O_2 content in each flask individually, but there may be technical difficulties in

because of the critical metabolic function of CO_2 and the known responses to CO_2 (e.g., modification of phototactic behavior[49]), we have fixed the CO_2 concentration in experiments in our laboratory. We have not considered the effect of the CO_2/O_2 ratio, although the ratio is known to be critical for photosynthetic carbon metabolism. Ideally, the CO_2 concentration should be fixed to correspond to air levels. However, this low level is technically difficult to achieve consistently and reproducibly when the CO_2 is supplied from a pure CO_2 gas tank. Therefore, for low air experiments (hypoxic conditions), we used two different mixtures: 97.8% nitrogen, 2% air, and 0.2% CO_2 or 96% nitrogen, 2% air, and 2% CO_2. It may be possible to provide HCO_3^{1-} in the medium in place of CO_2 and vary only O_2 and nitrogen in the gas mixture. We have not tried this option.

Other variables to consider include light intensity and acetate for heterotrophic growth because these affect photosynthesis and respiration and hence O_2 content during the experiment. It is also possible to induce hypoxic conditions by growing *Chlamydomonas* cells without vigorous agitation, for example, in Erlenmeyer flasks fitted with cotton plugs (200 ml medium in 250-ml flasks) where the cells are kept suspended by low basal stirring using a magnetic stir bar.[20,24] Note that growth under photoautotrophic conditions (illumination + CO_2) will generate O_2 in the cells and in the growth medium. Therefore, continuous bubbling or flushing is necessary to maintain control of hypoxic conditions. Happe and Kaminski[50] flushed the cell suspension with argon in the dark in order to promote anaerobic conditions for inducing hydrogenase activity in *Chlamydomonas* cultures.

Bubbling and Flask Setup

In order to deliver the desired mixture of gases for bubbling the cultures, we use three different gas flowmeters (Aalborg Instruments & Controls, Inc.) in two scales (one from 0 to 25 ml of gas per minute and another from 0 to 5 liters per minute of gas) because the amount needed for one gas is quantitatively different from the other (e.g., 99.8% air vs 0.2% CO_2). The output mixture has a total flow of 2.5 liters/min^{-1} liter of culture^{-1}. All gases are driven using flexible plastic tubes (Tygon) from the tanks to the flowmeters and from these to the cultures (see Fig. 3). The diameter for

[49] W. Nultsch, *Arch. Microbiol.* **112,** 179 (1977).
[50] T. Happe and A. Kaminski, *Eur. J. Biochem.* **269,** 1022 (2002).

maintaining sterile conditions for extended periods. It is useful to place the electrode at various points of exit from the gas supply in order to evaluate the effect of position of each flask. It is also preferable to mount the flasks in parallel as diagrammed rather than in series. (B) Details for connections of the different components.

the tubes may vary according to the regulator output, the flowmeter input and output, the type of filters, and so on. Once the gas mixture exits the multichannel flowmeter, the outlet pipe is split to the same number of branches to correspond to the number of flasks that are needed (see Fig. 3). Ideally, an individual flowmeter for each culture flask should be used, which allows bubbling of all flasks using exactly the same flow rate. A trouble commonly encountered with bubbled cultures is bacterial contamination. To avoid this, gas to the cultures is filtered through 0.45-μm filters (Pall Gelman Laboratory) and supplied through a sterile disposable pipette. An additional 125-ml flask, containing 50 ml of rich medium (e.g., LB), is aerated in the same way as the *Chlamydomonas* cultures. This is included for monitoring contamination during the course of the experiments. Normally, we grow *Chlamydomonas* in 250-ml Erlenmeyer flasks containing 100 ml of liquid TAP medium. The culture flasks are placed on a shaker at 100 rpm for additional agitation, as bubbling alone is not enough to keep the cells from settling. In our experience, cultures can be kept in axenic conditions for at least 5 days of growth. They multiply from 4×10^5 cells/ml at inoculation to 2×10^7 cell/ml after day 5 for wild-type CC125 in 99.8% air + 0.2% CO_2 corresponding to a doubling time in the log phase of 17 h. The doubling time for *Chlamydomonas* grown in optimal conditions in TAP medium, using a shaker at 220 rpm, 24°, and a light intensity of about 60 μmole m^{-2} s^{-1}, ranges from 6 to 8 h.

Other Culture Conditions

Cultures are maintained at room temperature (22–24°) with ambient illumination (10–15 μmole m^{-2} s^{-1}) provided by overhead standard ceiling fluorescent lights. Due to the ability of *Chlamydomonas* cells to grow heterotrophically, wild-type cultures in TAP medium[9] reach a cell density of 1×10^7 cells/ml in 2% air + 0.2% CO_2 in 4 days (stationary phase) under these conditions. For standard culture conditions in our laboratory (without bubbling, at 24° in a shaker incubator, 220 rpm, 60 μmole m^{-2} s^{-1}), the cell density reached for the same wild-type *Chlamydomonas* strain in the same time is 2.1×10^7 cell/ml. A higher light intensity can be used by placing lamps (fluorescent type, to avoid unwanted heating) close to the cultures, which is important for species that are incapable of growing heterotrophically (like some diatoms).[51]

[51] L. A. Zaslavskaia, J. C. Lippmeier, C. Shih, D. Ehrhardt, A. R. Grossman, and K. E. Apt, *Science* **292**, 2073 (2001).

Monitoring Oxygen Concentration

The oxygen content of the cultures is measured with a standardized oxygen electrode (Orion Research, Inc.). To avoid contamination of the experimental cultures, the oxygen concentration is measured in a parallel flask containing TAP medium to verify that the correct amount of oxygen is delivered to the cultures. The electrode can be calibrated by bubbling air in an aqueous solution (21% oxygen or 100% air). The second point for calibration (0% oxygen) can be achieved by bubbling nitrogen gas or by adding dithionite, which consumes oxygen.

Different Ways for Bubbling

It is possible to set up different systems for bubbling the cultures. As shown in Fig. 4, we use a sterile pipette and a cotton plug placed in the flask. This system needs to be assembled in a laminar flow hood to keep the culture sterile once the medium has been sterilized and cells inoculated. Alternatively, a rubber stopper with two holes can also be used. Through one hole, a glass tube (or a pipette) is added for bubbling the culture. In the other one, a short curved glass tube serves as exhaust for the air. We recommend using a sterile cotton plug on the exhaust tube, which minimizes the chance of contamination when cultures are removed for analysis.

Sample Preparation

Total RNA is prepared and analyzed by hybridization as described.[41] It is important to remember for the preparation of RNA from hypoxic cells that the $t_{1/2}$ of the mRNA encoding *Cyc6* (about 45 min) can be comparable to the RNA preparation time and for *Cpx1* appears to be shorter. Five micrograms of total RNA is loaded per lane. Probes for *Cpx1*, *Cyc6*, and *RbcS2* (encoding the small subunit of Rubisco) are prepared as described.[38] For the detection of *Crd1* transcripts, the 55×10^2-bp *Xho*I/*Pst*I fragment from pCrd1-5[52] is used. A gene-specific probe for *Cth1* is made from a 345-bp *Xho*I fragment corresponding to 4305 to 4650 of the *Cth1* genome sequence.[36] Specific activities of probes ranged from 3 to 6×10^8 cpm μg^{-1} DNA for the experiments shown here. The blots are exposed at $-80°$ to film (XRP-1; Eastman-Kodak, Rochester, NY)

[52] J. Moseley, J. Quinn, M. Eriksson, and S. Merchant, *EMBO J.* **19**, 2139 (2000).

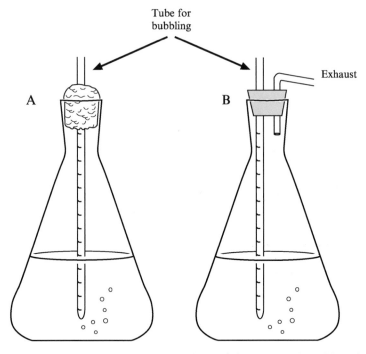

FIG. 4. Alternative options for bubbling cultures. (A) A cotton plug with an inserted pipette is used. (B) A rubber stopper where two holes have been made: one for the tube that reaches the cell suspension and the second one for a exhaust tube in order to avoid high pressure inside the flask.

with two intensifying screens and are typically developed after overnight exposure.

Concluding Remarks: Linking Hypoxia and Copper-Deficient Responses through the *crr1* Mutant

The copper- and oxygen-responsive expression of *Cyc6*, *Cpx1*, and *Crd1* is mediated via a common pathway. Evidence for this common pathway is (1) regulation of all three genes by copper or oxygen requires the *trans*-acting locus *CRRI*,[20,36] (2) the induction of *Cyc6* and *Cpx1* expression by copper or oxygen deficiency can be blocked by the same inhibitor, that is, mercuric ions,[20,53,54] and (3) the *cis* elements mediating copper- and oxygen-responsive expression have the same core sequence. The core of

[53] K. L. Hill, H. H. Li, J. Singer, and S. Merchant, *J. Biol. Chem.* **266**, 15060 (1991).
[54] K. L. Hill and S. Merchant, *EMBO J.* **14**, 857 (1995).

the two CuREs of *Cyc6* was identified initially by deletion analysis of the *Cyc6* promoter[30] and extensive site-directed mutagenesis to be the tetranucleotide sequence GTAC.[33] Each *Cyc6* CuRE is independently capable of conferring copper-responsive expression.[30,33] The same GTAC core was found to define a single CuRE in the *Cpx1* promoter required for copper-responsive expression of this gene.[33,38] Mutation of a second downstream GTAC sequence did not affect copper-responsive expression. When the oxygen-responsive expression of *Cpx1* was analyzed, it was found that mutation of either the GTAC core of the *Cpx1* CuRE or the downstream GTAC sequence would eliminate oxygen-responsive expression of *Cpx1*.[20] This second GTAC thus defines the core of a *cis* element specific for oxygen-responsive expression and is therefore designated a hypoxia-responsive element. This difference in the *cis* element requirement for copper- vs oxygen-responsive expression of *Cpx1* correlates with the increased accumulation of *Cpx1* transcripts relative to *Cyc6* transcripts in oxygen-deficient cells compared to copper-deficient cells.[20] Therefore, it is clear that while copper- and oxygen-responsive expression of these genes is mediated by a common pathway, mechanistic details do differ, and these details potentially relate to the function of these gene products under the different growth conditions. This difference between the responses to copper vs oxygen deficiency highlights the need for the study of oxygen-responsive expression in *C. reinhardtii*.

Acknowledgments

We thank the members of the group, especially Janette Kropat and Stephen Tottey, for their comments on the manuscript. This work was supported by the National Institutes of Health (GM42143) (SM) and a postdoctoral fellowship from the Spanish Ministry for Education (JC).

[40]   Resonance Raman and Ligand-Binding
Analysis of the Oxygen-Sensing Signal Transducer
Protein HemAT from *Bacillus subtilis*

By Shigetoshi Aono, Hiroshi Nakajima,
Takehiro Ohta, and Teizo Kitagawa

Introduction

HemAT is a soluble signal transducer protein responsible for aerotaxis control in Bacteria and Archaea.[1,2] The chemotaxis including aerotaxis is mediated by a signal transduction system consisting of methyl-accepting chemotaxis proteins (MCPs), a histidine kinase CheA, a response regulator CheY, a coupling protein CheW, and the two enzymes that mediate sensory adaptation by covalently modifying the MCPs, CheR, and CheB.[3–6] HemAT is a soluble signal transducer protein, although a typical MCP is an integral membrane protein with an amino-terminal periplasmic substrate-binding sensor domain and a carboxy-terminal cytoplasmic-signaling domain.[7,8] The difference of the location in the cell between HemAT and other typical MCPs reflects the different nature of the corresponding effector molecules sensed by MCPs, that is, O_2, which is sensed by HemAT, can be diffused across the cell membrane, whereas other typical effector molecules such as a sugar, amino acid, and metal ions cannot.

HemAT consists of two domains, the sensor and signaling domains, as other MCPs do. The amino acid sequences of the sensor and signaling domains of HemAT show the homology of those of myoglobin and of the cytoplasmic-signaling domain of Tsr that is an MCP from *Escherichia coli*, respectively.[1] The sensor domain in HemAT contains a b-type heme that acts as the active site for sensing O_2.[1,2,9] HemAT in oxy, deoxy, and

[1] S. Hou, R. W. Larsen, D. Boudko, C. W. Riley, E. Karatan, M. Zimmer, G. W. Ordal, and M. Alam, *Nature* **403**, 540 (2000).
[2] S. Hou, T. Freitas, R. W. Larsen, M. Piatibratov, V. Sivozhelezov, A. Yamamoto, E. A. Meleshkevitch, M. Zimmer, G. W. Ordal, and M. Alam, *Proc. Natl. Acad. Sci. USA* **98**, 9353 (2001).
[3] J. B. Stock and G. S. Lukat, *Annu. Rev. Biophys. Chem.* **20**, 109 (1991).
[4] J. S. Parkinson and E. C. Kofoid, *Annu. Rev. Genet.* **26**, 71 (1992).
[5] L. F. Garrity and G. W. Ordal, *Pharmacol. Ther.* **68**, 87 (1995).
[6] D. F. Blair, *Annu. Rev. Microbiol.* **49**, 489 (1995).
[7] G. L. Hazelbauer, *Curr. Opin. Struct. Biol.* **2**, 505 (1992).
[8] J. J. Falke and G. L. Hazelbauer, *Trends Biochem. Sci.* **26**, 257 (2001).
[9] S. Aono, T. Kato, M. Matsuki, H. Nakajima, T. Ohta, T. Uchida, and T. Kitagawa, *J. Biol. Chem.* **277**, 13528 (2002).

CO-bound forms shows similar electronic absorption spectra to those of myoglobin,[1,9] which is consistent with the amino acid sequence homology between them.

As the binding of oxygen to the heme is the first and a crucial step for sensing O_2 by HemAT, it is important for understanding the function of HemAT to study the oxygen-binding kinetics for HemAT. The heme in HemAT is the active site for sensing O_2. Therefore, elucidating the environmental and coordination structures of the heme is also important in understanding the structure–function relationships of HemAT. Resonance Raman spectroscopy is the most suitable method for elucidation of the heme environmental structure because only the vibrational modes of the heme can be selectively observed.

It is established that resonance Raman spectra of hemoproteins in the high-frequency region contain a few marker bands sensitive to the oxidation state (ν_4) and the spin and coordination states (ν_2 and ν_3) of the heme iron.[10] Analyzing these bands of HemAT compared to those of other hemoproteins gives some useful information on the nature of the active site for sensing O_2 in HemAT. The low-frequency region of resonance Raman spectra of hemoproteins contains information on the several in-plane and out-of-plane modes of heme, including heme propionate modes, vinyl modes, and ligand modes. In particular, the assignment of a ligand–Fe stretching mode is useful because it directly reflects the strength of the Fe–ligand bonds and thus the nature of interactions of the ligand with the surrounding amino acid residues.

For the ligand modes of HemAT from *Bacillus subtilis,* HemAT-Bs, ν(Fe-His), ν(Fe-O_2), ν(Fe-CO), and ν(C-O) modes can be observed at 225, 560, 494, and 1964 cm^{-1}, respectively. The ν(Fe-His) mode is observed only for five coordinate ferrous hemes. The important determinants of ν(Fe-His) frequency are the hydrogen-bonding status and steric distortion of the axially coordinated histidine.[11,12] The anionic character of the axial histidine increases by a hydrogen bonding to it, by which the ν(Fe-His) mode is shifted up to a higher frequency. The ν(Fe-His) frequency is lowered by a geometrical distortion of the axial histidine due to a strain from the polypeptide chain.

The ν(Fe-CO) and ν(C-O) modes are useful to identify not only the nature of the distal heme pocket interacting the bound CO, but also the

[10] T. G. Spiro and X.-Y. Li, *in* "Biological Application of Raman Spectroscopy" (T. G. Spiro, ed.), Vol. III, p. 1. Wiley, New York, 1988.
[11] T. Kitagawa, *in* "Biological Application of Raman Spectroscopy" (T. G. Spiro, ed.), Vol. III, p. 97. Wiley, New York, 1988.
[12] K. Nagai, T. Kitagawa, and H. Morimoto, *J. Mol. Biol.* **136,** 271 (1980).

proximal ligand of the CO-bound heme. It is established that the Fe–CO and C–O stretching modes display an inverse correlation.[13,14] The correlation between ν(Fe-CO) and ν(C-O) frequencies depends on the nature of the proximal ligand, that is, CO-bound hemes with an imidazole/histidine as the *trans* ligand exhibit a correlation line different from those with a thiolate or imidazolate. In the case of HemAT-Bs, the correlation between ν(Fe-CO) and ν(C-O) is on the correlation line characteristic of CO-bound hemes with an imidazole/histidine as the *trans* ligand.

Construction of an Expression System for HemAT-Bs

Materials

    Bacillus subtilis
    TE buffer (pH 8.0)
    Sense primer (5′-gaaggagatccat**atg**ttatttaaaaaagacagaaaacaagaaacagc-3′)
    Antisense primer (5′-ggatcccatatgttattattcttctgtcaggatgacaagcgaat-caacgg-3′)
    dNTP mixture (2.5 mM each, TAKARA)
    Ex *Taq* DNA polymerase (TAKARA)
    Ex *Taq* buffer (TAKARA)
    QIAEX II gel extraction kit (Qiagen)
    TOPO TA cloning kit (Invitrogen)
    0.2-ml polymerase chain reaction (PCR) tube
    Thermal cycler (Applied Biosystems, Gene Amp 2400)

Procedure

    1. The *hemAT-Bs* gene is prepared by PCR. Chromosomal DNA of *Bacillus subtilis,* which is used as a PCR template, is isolated by the method of Saito and Miura.[15] The two synthetic deoxyoligonucleotides whose sequences were shown earlier are used as sense and antisense PCR primers. These primers are dissolved in TE buffer (pH 8.0) with the concentration of 50 μM. Although the original translational initiation codon for HemAT-Bs is reported to be TTG in *B. subtilis,* the translational initiation codon is changed to ATG in the sense primer, which is underlined in the sequence shown earlier.

---

[13] M. Tsubaki and Y. Ichikawa, *Biochim. Biophys. Acta* **827,** 268 (1985).
[14] N.-T. Yu and E. A. Kerr, *in* "Biological Application of Raman Spectroscopy" (T. G. Spiro, ed.), Vol. III, p. 39. Wiley, New York, 1988.
[15] H. Saito and K. Miura, *Proc. Natl. Acad. Sci. USA* **72,** 619 (1963).

2. The reagents for PCR are mixed in a 0.2-ml PCR tube. The prescription of the PCR reaction mixture is shown. PCR is performed by 30 cycles of the following temperature program: 1.3 min at 93°, 2 min at 55°, and 2.5 min at 72°.

0.3 μg chromosomal DNA

1 μM each sense and antisense primers

0.2 mM each dNTP

0.25 units Ex *Taq* DNA polymerase

5 μl Ex *Taq* buffer

Add H_2O to a final volume of 50 μl

3. The PCR product is purified by agar gel electrophoresis using a 0.7% agar gel prior to its subcloning. The QIAEX II gel extraction kit is used for purification of the PCR product from an agar gel. The purified PCR product is cloned into a pCR4-TOPO vector with the TOPO TA cloning kit (Invitrogen) according to the supplier's protocol. The plasmid containing the *hemAT-Bs* gene with the correct direction relative to the *lac* promoter in the vector is selected as an expression vector for HemAT-Bs and is named pCR-HemAT.

Expression of HemAT-Bs in *E. coli*

Materials

> *Escherichia coli* JM109
> Tryptone
> Yeast extract
> NaCl
> Ampicillin
> Isopropyl-β-D-thiogalactopyranoside (IPTG)
> 500-ml culture flask
> 10-liter glass jar fermentor
> Diaphragm pump
> Rotary shaker (SANYO GALLENKAMP, MIR-220R)
> Centrifuge (Beckeman Coulter, Avanti HP-25)

Procedure

Escherichia coli JM109 is used as a host for the expression of HemAT-Bs. pCR-HemAT/*E. coli* JM109 is grown in LB medium (10 g tryptone, 5 g yeast extract, and 10 g NaCl in 1 liter H_2O) containing 50 μg/ml of ampicillin throughout this work. The preculture is carried out in a 500-ml culture flask containing 100 ml of LB medium for 6 h at 37° and 180 rpm using a rotary shaker. Precultured medium is inoculated in 6 liters of fresh medium prepared in a 10-liter glass jar fermentor. Cultivation in the jar fermentor is carried out at 37° with stirring and aerating at 500 rpm and 5 liter/min, respectively. Aeration is performed by means of a diaphragm pump. After 6 h of cultivation at 37°, 1 mM of IPTG is added to induce the expression of HemAT-Bs. Cultivation is continued at 25° for 14 h after adding IPTG. Cells are harvested by centrifugation (10,000g, 15 min) and stored at −80° until use.

Purification of HemAT-Bs

Materials

   Tris–HCl buffer (pH 8.5)
   Phenylmethylsulfonyl fluoride (PMSF)
   NaCl
   K_2HPO_4
   Q-Sepharose (Amersham Biosciences K. K.)
   Superdex 200pg (Amersham Biosciences K. K.)
   Hydroxyapatite (SEIKAGAKU Co.)
   Centricon 50 (Amicon)
   ÄKTA$_{FPLC}$ (Amersham Biosciences K. K.)
   Centrifuge (Beckman Coulter, Avanti HP-25)

Procedure

   1. Harvested cells expressing HemAT-Bs are resuspended in 50 mM Tris–HCl buffer (pH 8.5) containing 1 mM PMSF and are disrupted by sonication. After unbroken cells and cell debris are removed by centrifugation, the supernatant is applied to a column (2.6 × 30 cm) of Q-Sepharose equilibrated with 50 mM Tris–HCl buffer, pH 8.5.

   2. Adsorbed proteins on the Q-Sepharose column are eluted by a linear gradient of NaCl from 0 to 0.6 M at a flow rate of 1 ml/min after washing the column with a 4 bed volume of 50 mM Tris–HCl buffer, pH 8.5. Fractions containing HemAT-Bs are combined and applied to a column (1.6 × 30 cm) of hydroxyapatite equilibrated with 50 mM Tris–HCl containing 1 M KCl, pH 8.5.

   3. Adsorbed proteins on the hydroxyapatite column are eluted at a flow rate of 1 ml/min by increasing linearly the concentration of K_2HPO_4 in the elution buffer from 0 to 0.3 M. Fractions containing HemAT-Bs are combined and concentrated up to ca. 1 ml by a Centricon 50.

4. The concentrated sample is applied to a column (1.6 × 80 cm) of Superdex 200pg equilibrated with 50 mM Tris–HCl buffer containing 0.1 M NaCl, pH 8.5. The column is run at 0.2 ml/min using 50 mM Tris–HCl buffer containing 0.1 M NaCl, pH 8.5, as an elution buffer. One to 2 mg of purified HemAT-Bs is obtained from ca. 10 g of wet cells of pCR-HemAT/*E. coli* JM109.

Autoxidation of HemAT-Bs

Materials

   Tris–HCl buffer (pH 8.5)
   Optical quartz cell (1 cm of the optical path length)
   Incubator (37°)
   UV/vis spectrophotometer (Hitachi U-3300)

Procedure

The rate of autoxidation of the oxy form of HemAT-Bs is measured in air-saturated 50 mM Tris–HCl buffer (pH 8.5) at 37°. An optical quartz cell with a 1-cm optical path length containing the oxy form of HemAT-Bs is maintained in an incubator at 37°, and electronic absorption spectra are recorded at regular intervals of time with a Hitachi U-3300 UV/vis spectrophotometer. The difference of the absorbance at 425 nm is plotted against the incubation time, where the reaction curve is analyzed to be a first-order reaction.

Oxygen Binding and Dissociation Kinetics

Materials

   Tris–HCl buffer (pH 8.0)
   Sodium dithionite
   Stopped-flow spectrophotometer (Unisoku RSP-1000-02)
   Laser flash photolysis system (Unisoku TSP-1000)

Procedure

1. The oxygen dissociation reaction is analyzed by stopped-flow spectroscopy as described later. Sodium dithionite (5 mM) is prepared in 50 mM Tris–HCl buffer (pH 8.0) under argon atmosphere. One hundred microliters each of the sodium dithionite solution and ca. 3 μM HemAT-Bs in the oxy form are mixed by means of a stopped-flow spectrophotometer.

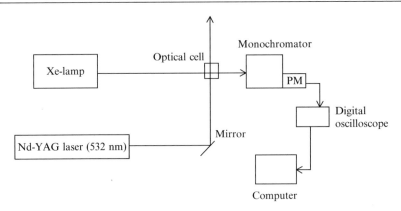

Fig. 1. Schematic diagram of the laser flash photolysis system.

The deoxy form of HemAT-Bs is formed by mixing the oxy form of HemAT-Bs with sodium dithionite. The reaction is monitored at 413 nm. Acquired data by a computer system are analyzed as reported previously.[16]

2. The oxygen-binding reaction is analyzed by laser flash photolysis, as described later. A block diagram of a laser flash photolysis system used is shown in Fig. 1. A Nd-YAG laser pulse at 532 nm with a 20-ns pulse width is used as an exciting light source. A xenon lamp is used as a monitor light source. The exciting laser and monitor light beams cross at right angles. The monitor light passing through the sample cell enters a monochromator, to which a photomultiplier tube is attached. The output signal of the photomultiplier tube is monitored on a digital oscilloscope.

3. Photodissociation of oxygen bound to the heme takes place to form deoxy HemAT-Bs by laser flash photolysis of the oxy form of HemAT-Bs. Photodissociated oxygen escapes from the protein matrix to solvent and then bimolecular recombination takes place between deoxy HemAT-Bs and O_2. The oxy form of HemAT-Bs is recovered by the recombination of O_2, which is monitored at 433 nm. Ten to 15 decay curves are averaged and then analyzed to obtain the rate constant of the oxygen binding. The observed decay is a pseudo first-order reaction. The rate constant of oxygen binding is obtained by dividing the pseudo first-order reaction rate constant by the concentration of oxygen in the sample solution.

[16] L. Kiger, A. K. Rashid, N. Griffon, M. Haque, L. Moens, Q. H. Gibson, C. Poyart, and M. C. Marden, *Biophys. J.* **75,** 990 (1998).

Experimental Procedures for Resonance Raman Measurements
of HemAT

Instrument of Resonance Raman Spectroscopy

1. Resonance Raman spectra are measured with a system illustrated in
Fig. 2A. The system consists of a light source (laser), sample illumination
part (Fig. 2C), and a monochromator/detector part (Fig. 2A). As a light
source, a continuous wave Kr^+ ion laser (Spectra Physics, Model 2016) is
used. The 413. 1-nm emission of the laser is focused to 0.1 mmϕ at the
sample, where the strength of light is 1 mW for the measurements of oxy
and deoxy forms, but 0.1 mW for the measurement of the CO-bound form
to avoid photodissociation of CO.

2. The sample is contained in a spinning cell, which is illustrated in Fig.
2C in the form attached to the spinning motor. The air-tight spinning cell is
made of quartz with a diameter of 5 mm and a height of 5 mm. When the
30-μl sample solution is put into the cell and spun at 500 rpm, the shape of
the solution becomes as illustrated in Fig. 2C. This cell is screwed to the
shaft of the motor (VEXTA, Model BL215A-AF), while the authors'
ingenuity is taxed to its coupling part.[17] Because the cell is held by two O
rings to the cell holder, the spin axis of the cell is retained stable without
swinging. This is quite important in this kind of experiment. Because the
spinning motor is held on a XYZ microstage, the exciting laser light can be
led to the limit of the inside surface of the cell.

3. The scattered light is focused onto the entrance slit of a home-
designed monochromator by a camera lens (50 mm, f = 1.2),[18] where the
notch filter and polarization scrambler are placed in the optical path.
The notch filter is used to remove Rayleigh scattered light, and the
polarization scrambler is used to destroy linear polarization of the scattered
light in order to avoid possible intensity errors due to different spectrometer
throughputs of the light-polarized parallel and perpendicular to grooves of
the grating. The focal length of the monochromator (Ritsu Oyo Kogaku,
MC-100DG) is 100 cm. This monochromator has two interchangeable
blazed holographic gratings (500 nm, 1200 grooves/mm and 900 nm, 1200
grooves/mm). These gratings, which are at one and the other sides of the
grating plate shown in Fig. 2B, are used at first- and second-order
geometry for normal ($\sim$1 cm^{-1}/channel) and higher ($\sim$0.4 cm^{-1}/channel)

[17] M. Aki, T. Ogura, K. Shinzawa-Itoh, S. Yoshikawa, and T. Kitagawa, *J. Phys. Chem. B* **104,**
10765 (2000).
[18] T. Ogura, S. Takahashi, S. Hirota, K. Shinzawa-Itoh, S. Yoshikawa, E. H. Appelman, and
T. Kitagawa, *J. Am. Chem. Soc.* **115,** 8527 (1993).

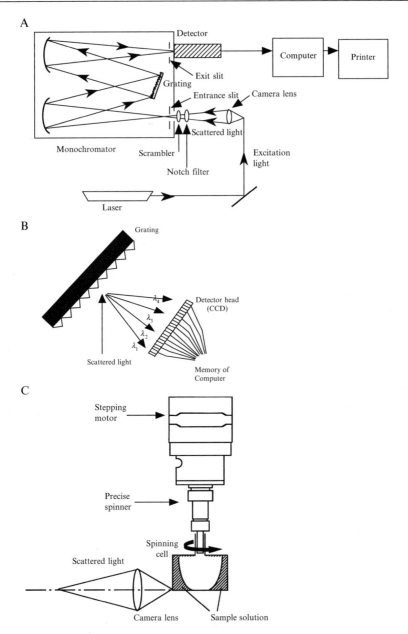

FIG. 2. Schematic diagram of the Raman measurement system. (A) Whole system, (B) spectral dispersion in a polychromator, and (C) spinning cell and sample illumination part. The rate of the spinning motor is variable, and its position (cell position relative to excitation light) can be controlled in a micrometer order with a XYZ microstage.

resolution, respectively. The former is used in this experiment. The dispersed light is detected with a liquid nitrogen-cooled charge-coupled device (CCD) detector (Princeton Instruments, Model LN/CCD-1100-PB). The mechanical width and height of the exit slit are set at 150 and 20 mm, respectively, meaning that the spectral slit width was $6.0\ cm^{-1}$. The frequency width per one channel is $0.90\ cm^{-1}$ in this measurement.

4. Sensitivity variations of individual channels of the detector are corrected by dividing the raw Raman spectrum by a white light spectrum. No other corrections are made to the observed spectra in this study. Because dispersed spectra are linear to the wavelength, the peak positions of Raman bands are determined in terms of wavelength first and their shifts from the excitation wavelength are calculated in terms of wave numbers second. After that, the precise values of wave number shifts of each channel are calibrated by observing Raman bands of indene and an aqueous solution of potassium ferrocyanide as frequency standards, and a curve for small corrections of the frequency shifts on each channel is generated with a polynomial function. Thus, the final value of frequency shift of a given band is determined. This procedure provides the frequency of intense isolated Raman bands with an accuracy of $\pm 1\ cm^{-1}$. These spectral treatments are performed with a program "IGOR Pro" (Wavemetrics Inc.).

Sample Preparation

Materials

$^{13}C^{16}O$ (ICON Inc., 99 atom% for ^{13}C)
$^{18}O_2$ (ISOTEC Inc., 98.2 atom % for ^{18}O)

Procedure

The sample solution is put into the Raman spinning cell with a syringe and then it is connected to a vacuum line through an air-tight syringe. The reduced form is obtained by adding 3 μl of the dithionite solution (4 mg sodium dithionite in 0.5 ml of distilled water) to the solution of purified HemAT in the Raman cell, the inside of which is replaced with N_2 gas in advance. The CO form is obtained through a repeated evacuation of gas inside the cell (to ca. 0.1 mm Hg) followed by the incorporation of CO to it (to ca. 1 atm) for three times. The CO–isotope adduct is obtained from the ordinary CO form ($^{12}C^{16}O$) by evacuating $^{12}C^{16}O$ and incorporating $^{13}C^{16}O$, which is purchased from ICON Inc. (99 atom% for ^{13}C). Ordinary and $^{18}O_2$-substituted oxy forms are obtained by incorporating $^{16}O_2$ or $^{18}O_2$

instead of CO, as is the case for the preparation of the CO form. $^{18}O_2$ is purchased from ISOTEC Inc. (98.2 atom % for ^{18}O), whereas $^{16}O_2$ is normal oxygen gas.

Acknowledgments

This work was supported by Grants-in-Aid for Scientific Research (12147203, 14380303, and 14655304 to S.A. and 14001004 to T.K.) from the Ministry of Education, Science, Sports, and Culture in Japan.

[41] Analysis of Fumarate Nitrate Reductase Regulator as an Oxygen Sensor in Escherichia coli

By RUTH A. SCHMITZ, STEPHANIE ACHEBACH, and GOTTFRIED UNDEN

Introduction

In facultatively anaerobic bacteria, the availability of molecular oxygen (O_2) has profound effects on metabolism and cell composition.[1-4] The shift from aerobic to anaerobic metabolism and the resulting regulatory processes in most bacteria occur mainly at the transcriptional level. Due to the significance of the aerobic/anaerobic shift for bacterial growth and the large number of regulated systems, many different regulatory systems are available in bacteria for O_2 sensing; the regulation in one organism often results from the action of different cooperating regulatory systems. Some sensors react directly with O_2, whereas others react indirectly via redox enzymes, which oxidize or reduce the sensor. Examples for the former type are fumarate nitrate reductase regulator (FNR), FNR-like protein (FLP), and FixL, which contain redox-sensitive groups for direct reaction with O_2, such as a [4Fe-4S] cluster in FNR, vicinal cysteine thiols in FlpA, and heme B in FixL for O_2 binding.[1,4-7] The indirect O_2 sensors

[1] J. Green, C. Scott, and J. R. Guest, *Adv. Microb. Physiol.* **44,** 1 (2001).
[2] R. P. Gunsalus, *J. Bacteriol.* **174,** 7069 (1992).
[3] G. Unden, S. Becker, J. Bongaerts, G. Holighaus, J. Schirawski, and S. Six, *Arch. Microbiol.* **164,** 81 (1995).
[4] G. Unden and J. Schirawski, *Mol. Microbiol.* **25,** 205 (1997).
[5] P. J. Kiley and H. Beinert, *FEMS Microbiol. Rev.* **22,** 341 (1998).
[6] C. Scott, J. R. Guest, and J. Green, *Mol. Microbiol.* **31,** 1383 (2000).
[7] W. Gong, B. Hao, S. S. Mansy, G. Gonzalez, M. A. Gilles-Gonzalez, and M. K. Chan, *Proc. Natl. Acad. Sci. USA* **95,** 15177 (1998).

often interact with electron transport components or oxydases, resulting in the oxydation or reduction of the sensor. This type of sensing is verified in ArcB and PrrB representing sensory histidine kinases, Aer, and NifL from *Klebsiella*.[8–11]

One of the prototypes for O_2 sensing is the FNR protein of *Escherichia coli*. FNR is widespread in many classes of bacteria and was one of the first sensors with a known mechanism for O_2 sensing.[1,3–5] Active FNR, present under anoxic conditions, contains a [4Fe-4S] cluster ligated by four essential cysteine residues, which is converted to a [2Fe-2S] cluster upon interaction with O_2.[12–14] In the active dimeric state, FNR binds to DNA at target promoters containing FNR consensus sites and activates the transcription of target genes. Some genes are also repressed by FNR. Upon prolonged interaction with O_2, the [2Fe-2S] cluster disintegrates completely and apoFNR is formed. FNR is located in the cytoplasm, and thus O_2 has to diffuse into the cytoplasm for direct interaction with the sensor.[3,4,15]

To study various aspects of FNR function, *in vivo* and *in vitro* methods have been developed. The methods may be relevant also for other O_2 sensors or for FNR-like sensors in other bacteria.

Analysis of FNR as an O_2 Sensor *In Vivo*: Determining Regulatory O_2 Tensions ($pO_{0.5}$)

The functional state of FNR is tested *in vivo* using reporter gene fusions. Optimal reporter gene systems (e.g., FFp*melR'*-'*lacZ*) have been developed.[16] The synthetic FFp*melR* promoter contains a consensus FNR site and is subject to regulation by FNR but no other regulators. The promoter-'*lacZ* fusions should be used preferentially in single copy after insertion into the chromosome by phage λ. Reporter gene fusions consisting of FNR-regulated promoters from bacteria (e.g., *frdA'*-'*lacZ*, *nirB-lacZ* in *Escherichia coli* MC4100λJ100 and JCB331) are used to verify the effect

[8] D. Georgellis, O. Kwon, and E. C. Lin, *Science* **292,** 2314 (2001).

[9] J. I. Oh and S. Kaplan, *EMBO J.* **19,** 4237 (2000).

[10] B. L. Taylor, I. B. Zhulin, and M. S. Johnson, *Annu. Rev. Microbiol.* **53,** 103 (1999).

[11] R. Grabbe and R. A. Schmitz, *Eur. J. Biochem.* **270,** 1555 (2003).

[12] N. Khoroshilova, H. Beinert, and P. J. Kiley, *Proc. Natl. Acad. Sci. USA* **92,** 2499 (1995).

[13] N. Khoroshilova, C. Popescu, E. Münck, H. Beinert, and P. J. Kiley, *Proc. Natl. Acad. Sci. USA* **94,** 6087 (1997).

[14] J. Green, B. Bennett, P. Jordan, E. T. Ralph, A. J. Thompson, and J. R. Guest, *Biochem. J.* **316,** 887 (1996).

[15] S. Becker, G. Holighaus, T. Gabrielczyk, and G. Unden, *J. Bacteriol.* **178,** 4515 (1996).

[16] A. I. Bell, K. L. Gaston, J. A. Cole, and S. J. W. Busby, *Nucleic Acids Res.* **17,** 3865 (1989).

of O_2 and FNR on gene expression.[2,15,17,18] Reporter gene fusions, which are subject to control by additional factors, have to be tested under standardized conditions to avoid interference.

Aerobic growth is performed in Erlenmeyer flasks with baffles and vigorous shaking (>200 rpm). The medium in the flasks amounts to maximally 7% of the flask, and the bacteria are used for the measurement of expression at an optical density $OD_{578} \leq 0.6$ to avoid O_2 limitation.[18,19] Anaerobic growth is performed in rubber-stoppered flasks, which are degassed for three cycles by a vacuum pump and gassed with N_2. The activities are determined in triplicate each from three independent growth experiments. Generally, the bacteria have to be grown for at least eight generations under identical conditions before measurement of β-galactosidase to enable adaptation of the bacteria.

Growth in the Oxystat

For studies at defined O_2 tensions (pO_2) in the range from 0 mbar O_2 (anoxic condition) to 210 mbar O_2 (oxic condition), growth was performed in an oxystat (Biostat MD fermenter, Braun Biotech, Germany) in batch cultures (1.5 liter) with constant stirring (400 rpm) (Fig. 1).[15,17,18] The pO_2 of the medium is measured continously with an O_2 electrode (Ingold, type 19/320) and is maintained at a constant level by supplying air and N_2 to the vessel via two valves (valves I and II, respectively), which are controlled by the O_2 electrode and the regulatory device of the fermenter. When the pO_2 falls below 95% of the set value, valve I opens and sterile air is supplied until the set value is reached. The flow of air is adjusted manually or automatically from 0.16 to 1.6 liter min^{-1} during growth to compensate for increased O_2 consumption. The flow of N_2 (0.1 liter min^{-1} is decreased or switched off as required. For growth of *E. coli*, preferably M9 medium,[20] supplemented with acid-hydrolyzed caseine (1 g $liter^{-1}$) and L-tryptophane (0.5 g $liter^{-1}$),[19] or Luria-Bertani (LB) broth[21] if required, is used. Carbon sources such as glucose, succinate, acetate, or others are added at 20 mM. The medium is inoculated ($OD_{578} = 0.02$) from subcultures grown overnight under the same conditions. Bacteria are then

[17] S. Becker, D. Vlad, S. Schuster, P. Pfeiffer, and G. Unden, *Arch. Microbiol.* **168,** 290 (1997).

[18] Q. H. Tran, T. Arras, S. Becker, G. Holighaus, G. Ohlberger, and G. Unden, *Eur. J. Biochem.* **267,** 4817 (2000).

[19] Q. H. Tran, J. Bongaerts, D. Vlad, and G. Unden, *Eur. J. Biochem.* **244,** 155 (1997).

[20] J. H. Miller, "Experiments in Molecular Genetics." Cold Spring Harbor Laboratory Press, Cold Spring Harbor, NY, 1992.

[21] J. Sambrook, E. F. Fritsch, and T. Maniatis, "Molecular Cloning: A Laboratory Manual," 2nd Ed. Cold Spring Harbor Laboratory Press, Cold Spring Harbor, NY, 1989.

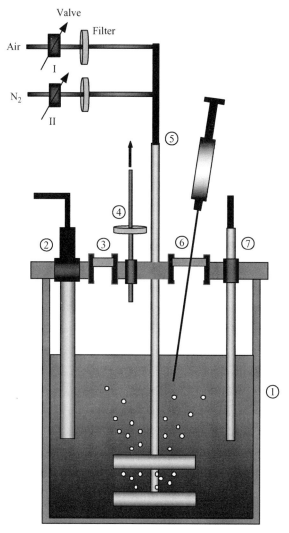

FIG. 1. Oxystat for growth of bacteria at constant pO_2.[15,17,22] The glass vessel (1) of the fermenter (Biostat MD, 1.5-liter fermenter) with the O_2 electrode (2), an inlet (3) for media and inoculum, a gas outlet (4) with a sterile filter, the stirrer (5) with a gas supply for sterile air and N_2, an outlet (6) with a stopper and syringe, and a pH electrode (7) are shown. In addition, the fermenter contains a temperature sensor and devices for supplying media or substrates.

grown to $OD_{578} = 0.3$ to 0.6 (OD_{578} of 1 corresponding to 1.5×10^9 cells ml^{-1}). Higher cell densities are not used to avoid O_2 and other limitations. The oxystat can be used to grow the bacteria at constant pO_2 values from 0.2 to 212 mbar with fluctuations within 5% of the set value over the entire experiment. An O_2 tension of 210 mbar O_2 (or 2.1×10^4 Pa) of air-saturated water (or dilute buffers) corresponds to an approximate concentration of 210 μM O_2.

The Switch Point ($pO_{0.5}$) of FNR-Dependent Regulation

Strains of *E. coli* carrying FNR-dependent reporter gene fusions are grown for eight or more generations ($OD_{578} \sim 0.4$) in the oxystat at defined pO_2. Samples are drawn with a syringe without disturbing the adjusted pO_2. Syringes with long cannula are used, and the samples are withdrawn via a rubber-stoppered exit (Fig. 1). The sample is transferred directly into glass tubes for measurement of the β-galactosidase according to a standard procedure in triplicate.[20] After 20 and 40 min, additional samples are taken for repeating β-galactosidase measurement. In new experiments, bacteria are grown and tested for β-galactosidase at different pO_2 values, thus covering the complete range from anoxic conditions to oxic conditions.

Plotting β-galactosidase activities against the pO_2 in the medium shows an exponential decrease in activity with increasing pO_2 (Fig. 2A). The pO_2 corresponding to the half-maximal decrease in activity is termed $pO_{0.5}$, which is a characteristic value for distinct FNR-dependent genes. Most of the FNR-regulated genes of *E. coli* have a $pO_{0.5}$ in the range of 1 to 5 μM (e.g., *frdA'-'lacZ* or *narG'-'lacZ*) (Table I),[4,15] but some have distinctly higher $pO_{0.5}$ values (21 mbar O_2 for *nirB'-'lacZ*).[18] Using this method, $pO_{0.5}$ values can be determined reproducibly, and factors affecting the function of FNR as a O_2 sensor can be identified by their effect on the $pO_{0.5}$ value.

The method is also useful for determining the regulatory switch point for other O_2-dependent processes.[4,15] In this way the pO_2 for the switch from aerobic respiration to fermentation of *E. coli* can be measured by the appearance of typical fermentation products (formate, ethanol, succinate) or the expression of a gene for a fermentative enzyme (ethanol dehydrogenase, *adhE* gene). The fermentation products are determined by HPLC of culture supernatants, which are withdrawn from the oxystat with a syringe as described for *lacZ* reporter strains. $pO_{0.5}$ values for the onset of fermentation are even lower ($pO_{0.5} = 0.2$ to 1 mbar O_2) than those for FNR-dependent regulation (Table I).

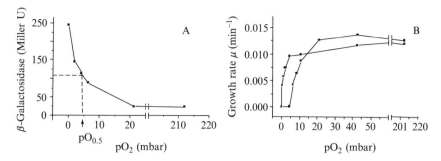

FIG. 2. Effect of pO_2 in the medium (A) on the expression of *frdA'-'lacZ* (fumarate reductase) of *E. coli* and (B) on the growth of *P. putida* with benzoate (O_2 consumption in cytoplasm) or succinate (O_2 consumption at cytoplasmic membrane by respiration). (A) The expression of *frdA'-'lacZ* of *E. coli* MC4100λJ100 was measured in growing bacteria in the oxystat, and the pO_2 value for half-maximal expression ($pO_{0.5}$) is indicated. (B) Rate constants (μ) for growth on benzoate (●) and succinate (■) were determined from growth experiments performed at different pO_2 values in the medium. Data from Becker *et al.*[15] and Arras *et al.*[22]

TABLE I
COMPARISON OF $pO_{0.5}$ VALUES FOR VARIOUS REGULATORY AND METABOLIC PROCESSES
IN THE OXYSTAT[a]

O_2-dependent function	$pO_{0.5}$ (mbar O_2)[b]
Expression of FNR-dependent gene fusions (*E. coli*)	
frdA'-'lacZ (fumarate reductase gene, anaerobic electron transport)	4.7
dmsA'-'lacZ (DMSO reductase gene, anaerobic electron transport)	1.0
FFp*melR-lacZ* (synthetic promoter with FNR consensus)	2.9
Fermentative functions (*E. coli*)	
adhE'-lacZ (expression of ethanol dehydrogenase gene)	0.8
Ethanol formation	0.4
Formate formation	1.0
Catabolic O_2-consuming reactions in the cytoplasm (*Pseudomonas putida*)	
Growth on benzoate	8.2

[a] $pO_{0.5}$ values are determined as described in Fig. 2.
[b] Values from Becker *et al.*[15,17] and Arras *et al.*[22]

$pO_{0.5}$ and Intracellular O_2-Consuming Reactions

In a similar way, the $pO_{0.5}$ for reactions consuming O_2 in the cytoplasm can be determined, such as for oxygenases in aromate degradation.[22] During growth on aromatic compounds, O_2 consumption takes place in

[22] T. Arras, J. Schirawski, and G. Unden, *J. Bacteriol.* **180**, 2133 (1998).

634

the cytoplasm at catabolic rates, and growth depends on O_2 diffusion into the cytoplasm. The limiting pO_2 of aromate metabolism can be determined with an oxystat as the $pO_{0.5}$ value for growth on aromatic compounds. The experiment is performed as described for measuring the expression of FNR-dependent genes, but the growth rates (μ) are determined by the increase in OD_{578}. The bacteria (e.g., *Pseudomonas putida*) are grown in mineral medium in the oxystat on aromatic or other substrates.[22] Growth is followed as the OD_{578} or another growth-related parameter. For inoculation ($OD_{578} = 0.01$), subcultures are used that are grown on the same substrate and under the same growth conditions.

Growth rates are calculated from the exponential growth phase according to $\mu = \ln(OD_{578,t_2}/OD_{578,t_1}) \cdot (t_2 - t_1)^{-1}$, where t_2 and t_1 are the times for measurement, and OD_{578,t_2} and OD_{578,t_1} are the optical densities measured at 578 nm at t_2 and t_1, respectively. Growth and growth rates are determined in the oxystat for pO_2 values from 0 to 210 mbar. For benzoate, maximal growth rates are obtained at pO_2 values > 40 mbar O_2. Below 20 mbar O_2, the growth rates decrease. The values for μ are plotted against the pO_2, and the $pO_{0.5}$ is defined as the pO_2 value, resulting in a half-maximal decrease of growth rate μ. For *P. putida,* $pO_{0.5}$ values for growth on aromatic compounds are in the range of 8 mbar O_2, compared to 1 to 2 mbar for nonaromatic compounds such as succinate, which consumes O_2 at the cytoplasmic membrane by aerobic respiration (Fig. 2B and Table I).[22]

Relation of $pO_{0.5}$ Values to Intracellular O_2 Concentrations

The cytoplasmic O_2 tension of bacterial cells has not been determined in detail so far. The $pO_{0.5}$ values give the O_2 tensions where sufficient O_2 diffuses into the cytoplasm to affect O_2-dependent regulators (FNR) or to supply O_2 as the substrate for oxygenases. $pO_{0.5}$ values obtained in this way can be compared to calculations for the supply of the cytoplasm with O_2 by diffusion under the same conditions (Table II).[3] The O_2 supply to the cell by diffusion can be estimated from the diffusion coefficients of O_2 in water (or medium), phospholipid membranes, and the cytoplasm (protein solution) and from the cell dimensions. Due to the high diffusion coefficients and the small cell dimensions, the supply of O_2 to the cell occurs at high rates. The rates exceed those for O_2 consumption by the cell (respiration and other reactions, including nonspecific reduction in the cytoplasm) largely under aerobic and microaerobic conditions. Only at a very low O_2 tension ($pO_2 \leq 0.11$ mbar) does the O_2 consumption equal the supply according to this calculation, and the cytoplasm might be anoxic, whereas at higher pO_2 values it should be oxic. This calculation also suggests that

TABLE II

COMPARISON OF REGULATORY O_2 TENSIONS (pO_2) FOR VARIOUS REGULATORY AND
CATABOLIC REACTIONS IN *E. coli* OR OTHER BACTERIA TO THE RATES OF O_2 DIFFUSION TO THE
BACTERIAL CYTOPLASM [a]

O_2 in medium (mbar)	Relevant condition	Rates [mmol O_2 min^{-1} (g protein^{-1})]	
		O_2 diffusion	O_2 consumption
210	Aerobic growth	720	0.39
8.2	Growth on benzoate ($pO_{0.5}$, *P. putida*)	14.1	1.65
5	Switch anaerobic ET ($pO_{0.5}$, FNR)	17.1	0.36
1	Switch fermentation ($pO_{0.5}$)	3.4	0.36
0.11	Equivalence O_2 diffusion/consumption	0.36	0.36

[a] Regulatory $pO_{0.5}$ values are from Table I. The O_2 supply of the cytoplasm by diffusion ($J_x = - D \cdot A \cdot \delta_c \cdot \delta_x^{-1}$) was calculated with the following parameters: A (surface of the bacteria) = 0.5×10^6 cm^2 phospholipid $\cdot$ g protein$^{-1,23}$; δ_c (external pO_2); δ_x (radius of the bacterial cell) = 1 μm. For the diffusion coefficient of O_2 in the protein-rich cytoplasm, D = 1.15×10^{-5} cm^2 s^{-1} was taken,[24] i.e., half the value of D in water. The cytoplasmic membrane is assumed to represent no diffusional barrier to O_2.[25] O_2 consumption (V_{max}) was determined with an O_2 electrode and glucose as the substrate in cell suspensions of *E. coli* at the respective O_2 concentrations. The value includes O_2 consumption by respiration, as well as all other cellular processes consuming O_2. For *P. putida*, relevant data are found in Arras *et al.*[22]

at $pO_{0.5}$ values of FNR-dependent regulation and of benzoate metabolism ($\geq$ 1 and 8.2 mbar O_2, respectively) sufficient O_2 is present in the bacteria to inactivate FNR and to supply oxygenases with O_2. Determination of $pO_{0.5}$ values therefore provides physiologically relevant data for the characterization of regulatory (and catabolic) switch points of O_2-dependent cellular reactions.[3,4,26]

Identification and Characterization of fnr *Mutants*

Mutants deficient in *fnr* (encoding FNR) can be identified and checked by testing *in vivo* for β-galactosidase activity. Alternatively, the significantly reduced molybdate reductase activity in *fnr* mutant strains allows a rapid and simple screening for identifying *fnr* mutant strains *in vivo* on agar

[23] F. C. Neidhardt, J. Ingraham, and M. Schaechter, "Physiology of the Bacterial Cell." Sinauer, Sunderland, 1990.
[24] M. J. Johnson, *J. Bacteriol.* **94,** 101 (1967).
[25] W. K. Subczynski, J. S. Hyde, and A. Kusumi, *Proc. Natl. Acad. Sci. USA* **86,** 4474 (1989).
[26] G. Unden and J. Bongaerts, *Biochim. Biophys. Acta* **1320,** 217 (1997).

plates.[27] *E. coli* wild-type and mutants are grown under a nitrogen atmosphere in an anaerobic container on solid LB medium supplemented with 10 mM Na$_2$MoO$_4$, 0.5% glucose, and 0.6 mM sodium sulfide. After incubation at 37° for 2 days, single colonies of wild-type *E. coli* turn deep blue, whereas *fnr* mutant strains remain white. This phenotype of *fnr* mutants was originally observed for *E. coli;* however, similar phenotypes may exist for other bacteria with FNR homologous proteins.

Analysis of FNR as an O$_2$ Sensor *In Vitro*

The FNR protein can be isolated from overexpression strains. Due to the lability of FeS clusters, most of the FeS clusters are lost and apoFNR is obtained during isolation of the protein. For functional analysis, apoFNR has to be reconstituted *in vitro* to obtain active [4Fe-4S]·FNR.[12–14,18] The [4Fe-4S] cluster is electric paramagnetic resonance (EPR) silent and can be identified by Mössbauer spectroscopy. Formation of the [4Fe-4S] cluster is accompanied by changes in the UVvis spectrum, which is used to indicate the formation of [4Fe-4S]·FNR. The formation of [4Fe-4S]·FNR has to be verified by functional assays, such as binding to FNR consensus DNA sites, which can be detected by restriction site protection or gel retardation. Active [4Fe-4S]·FNR is very sensitive toward the presence of O$_2$ and has to be handled under anoxic conditions in an anaerobic cabinet.

Application of Anoxic Conditions for Handling Active FNR or for Growth of Bacteria

When mentioned, anoxic conditions are applied by various methods. The anaerobic cabinet (Coy Laboratories) contains an atmosphere devoid of O$_2$ consisting of 95% N$_2$ (5.0) and 5% H$_2$. Residual O$_2$ introduced into the cabinet is consumed by H$_2$ using catalyst RO-20 (BASF, Ludwigshafen), and the actual portion of H$_2$ amounts to 2 to 3% in the gas phase. Buffers used for reconstitution or handling reconstituted FNR are kept in glass bottles sealed with gas-tight rubber stoppers and are made anoxic by three cycles of degassing on a vacuum pump (15 min each) followed by gassing with N$_2$ (5.0). Buffers are then kept for 12 h in open bottles in the anaerobic cabinet with stirring before use.

For the growth of bacteria under anoxic conditions, gas-tight glass bottles are used, which are sealed with rubber stoppers. The medium is degassed via a vacuum pump for three cycles (20 min each) and then kept

---

[27] B. Frey, G. Jänel, U. Michelsen, and H. Kersten, *J. Bact.* **171,** 1524 (1989).

under an atmosphere of N_2 (5.0). All additions and removals of samples or media are done with syringes via the rubber stopper.

Isolation of Anaerobic and Aerobic FNR

For overproduction of FNR, *E. coli* CAG627pMW68 is used.[18] Plasmid pMW68 codes for a glutathione *S*-transferase (GST)–FNR fusion protein. The *N*-terminal GST tag is used for the isolation of FNR by glutathione affinity chromatography. GST and FNR are linked by a recognition site for cleavage by thrombin. Expression of the fusion is controlled by a isopropyl-β-D-thiogalactoside (IPTG)-inducible tac promoter. The isolated plasmid is stored at $-20°$ and is used for fresh transformation of the expression strain when the yield of FNR decreases. Bacteria are grown aerobically in 4 × 300 ml 2 × YTG medium (16 g liter^{-1} peptone, 10 g liter^{-1} yeast extract, 5 g liter^{-1} NaCl) in 2-liter Erlenmeyer flasks with 2% glucose and 100 mg liter^{-1} ampicillin at 37°. At $OD_{578} = 0.7$, IPTG (1 m*M*) is added and bacteria are grown for another 2 h and are sedimented, and the wet cell mass (about 11 g) is stored at $-75°$.

For the isolation of anaerobic FNR,[18] all buffers are made anoxic by degassing under vacuum for three cycles, 15 min each, and maintained in an anaerobic cabinet. If not stated otherwise, all steps are performed in an anaerobic cabinet at 20°. Bacteria (about 11 g wet weight) are resuspended in 30 ml anoxic buffer A (50 m*M* Tris–HCl, pH 7.6, 10% glycerol) and filled into a French pressure cell. Outside the cabinet, the suspension is passed three times through the cell at 8270 kPa and is collected in a rubber-stoppered bottle, which is flushed with N_2. Refilling of the French pressure cell is performed in an anaerobic cabinet. After clearing of the homogenate by centrifugation (12,000*g* for 20 min), 30 ml of anoxic buffer A is added to the supernatant. The mixture is passed through a glutathione–Sepharose 4B column (1.5 ml bed volume) equilibrated with buffer A, resulting in binding of the FNR–GST fusion protein to the column. The column is rinsed with 20 ml buffer B (buffer A with 200 m*M* NaCl) until the eluate is free of protein. Then 1 ml buffer C (50 m*M* Tris–HCl, pH 7.6) containing 20 U thrombin (Amersham) is applied to the column for 2 h at 20°. FNR is eluted with 3 ml buffer C. FNR obtained in this way contains two additional residues (Gly-Ser) at the *N*-terminal end derived from the thrombin recognition sequence. The eluate (3 ml) contains about 1.5 mg pure FNR, which is not contaminated by GST–FNR and other proteins. "Aerobic" FNR is prepared by the same procedure as for anaerobic FNR, but all buffers are air saturated and all steps are performed under air. FNR obtained by aerobic and anaerobic preparations has $M_r \sim 30,000$ in SDS

TABLE III
MOLECULAR PROPERTIES OF ANAEROBICALLY PREPARED APOFNR AND OF RECONSTITUTED [4FE-4S]·FNR[a]

	ApoFNR	[4Fe-4S]·FNR
M_r in SDS–PAGE	30,000	30,000
Fe ions (mol/mol FNR)	≤0.5	4
Labile sulfide (mol/mol FNR)	<0.02	4
Cysteine thiolates (mol/mol FNR)	2.8 to 4.0	5.0

[a] Fe and sulfide in apoFNR were determined chemically (S. Achebach and G. Unden, unpublished results)[28] in [4Fe-4S]·FNR chemically or derived from the presence of a [4Fe-4S] cluster by Mössbauer spectroscopy.[12,14] Cysteine thiolates were determined with dithiobisnitrobenzoate or by alkylation with iodoacetate (unpublished results).

gels and is essentially devoid of Fe and acid-labile sulfur ("apoFNR") (Table III).

Isolation of Cysteine Desulfurase from Azotobacter vinelandii (NifS$_{Av}$)

In vitro reconstitution of [4Fe-4S]·FNR from apoFNR requires the formation of sulfide by cysteine desulfurases (NifS$_{Av}$ from *A. vinelandii* or IscS from *E. coli*). For overproduction of NifS$_{Av}$, *E. coli* BL21DE3(pNifS) containing the NifS$_{Av}$ expression plasmid pNifS is grown aerobically in 300 ml of LB broth with 100 mg liter^{-1} ampicillin at 30°.[29] Cells are induced at OD$_{578}$ = 0.9 with 1 mM IPTG for 2 h at 30°. After induction (OD$_{578}$ ~ 1.6), cells are sedimented and stored in the same way as CAG627pMW68. NifS$_{Av}$ is isolated from cell homogenates under aerobic conditions by the precipitation of contaminating proteins with 45% ammonium sulfate.[29] NifS$_{Av}$ amounts to about 50% of the protein in the preparation (about 11 mg protein). Before use for reconstitution, the NifS$_{Av}$ protein solution is made anoxic by degassing for 30 min.

Reconstitution of [4Fe4S]•FNR from apoFNR

[4Fe-4S]·FNR is reconstituted under anoxic conditions from anaerobically prepared apoFNR by a modified procedure of Khoroshilova *et al.*[12] as described.[18] Aerobically prepared apoFNR has to be made anoxic before reconstitution by careful degassing. Reconstitution of apoFNR is most efficient within the first 9 h after preparation. Reconstitution of

[28] Green, M. Trageser, S. Six, G. Unden, and J. R. Guest, *Proc. R. Soc. Lond. B Biol. Sci.* **244,** 137 (1991).

[29] L. Zheng, R. H. White, V. L. Cash, R. F. Jack, and D. R. Dean, *Proc. Natl. Acad. Sci. USA* **90,** 2754 (1993).

apoFNR by assembly and insertion of the [4Fe-4S] cluster require anoxic incubation of the sample with cysteine and cysteine desulfurase (sulfide production), a reducing agent (dithiothreitol), and Fe(II). Reconstitution is performed in an anaerobic cabinet.

Buffer C and water are degassed for 45 min and are then incubated with stirring for 12 h in an anaerobic cabinet to obtain anoxic solutions. L-Cysteinium chloride (88 mg) and $(NH_4)_2Fe(SO_4)_2$ (40 mg) are kept in separate reaction tubes in the anaerobic cabinet. Before use, the samples are dissolved in 1 ml of water each, resulting in 0.5 and 0.1 M solutions. From dithiothreitol, a 1 M stock solution is prepared in anoxic water, and aliquots are stored at $-20°$. The reconstitution mixture ($20°$) contains 300 μg freshly prepared apoFNR, 0.3 mM $(NH_4)_2Fe(SO_4)_2$, 2 mM L-cysteine, 2.5 mM dithiothreitol, and 15 μg NifS$_{AV}$ (or 15 μg isolated IscS from *E. coli*) filled to 600 μl with buffer C. The mixture is transferred carefully into a 1-ml UV cuvette (d = 1 cm) and is sealed with a gas-tight rubber stopper. After the start of the reaction, a spectrum (250 to 550 nm) is recorded every 2 min in a diode array spectrometer (Zeiss S10 with WI-NASPECT program). Reconstitution of the [4Fe-4S] cluster in FNR is followed as the absorption of 420 nm after normalization for the protein content at 280 nm (Fig. 3A). From the repeated measurement, reconstitution kinetics are obtained (Fig. 3C). The formation of [4Fe-4S]·FNR is characterized by the formation of an intense yellow-brownish color typical

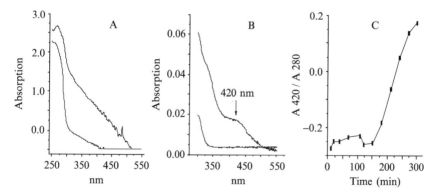

FIG. 3. Reconstitution of [4Fe-4S]·FNR from apoFNR. (A) UV/vis spectra of apoFNR and apoFNR after reconstitution with NifS$_{Av}$, Fe(II), cysteine, and dithiothreitol. (B) Difference spectrum of reconstituted apoFNR ([4Fe-4S]·FNR) *versus* nonreconstituted apoFNR, showing the band at 420 nm, and (C) kinetics of reconstitution followed at 420 nm. Reconstitution (A) of aerobic apoFNR (0.5 g/liter) was performed for 300 min as described in the text. The increase of FNR reconstitution was followed by recording the absorption at 420 nm and is given as the A_{420}/A_{280} ratio. Data from Tran *et al.*[18] and S. Achebach and G. Unden, unpublished results.

for the [4Fe-4S] cluster in FNR. Formation of the yellow-brownish color in the reconstitution depends completely on the presence of apoFNR, NifS$_{Av}$ or IscS, Fe(II), and cysteine, each. At high concentrations (40 μM) FNR starts to precipitate, and samples with strong precipitation are discarded. Formation of reconstituted [4Fe-4S]·FNR is controlled by testing specific binding of the protein to DNA containing the FNR consensus site in a ClaI restriction site protection assay, gel retardation, or in vitro transcription.

DNA-Binding Assays for FNR In Vitro

The functional state of FNR in vitro is tested most commonly by the capacity of FNR for specific binding to DNA at FNR consensus sites (restriction site protection assay or gel retardation) and transcriptional control of FNR-dependent promoters (in vitro transcription assay).

FNR-Binding Boxes

From the analysis of FNR-regulated promoters, a consensus binding sequence was derived, the FNR box, which is composed of two half-sites with a strictly conserved spacing of four nucleotides: **TTGATNNNNATCAA**.[16,30] Each half-site is assumed to bind one FNR monomer of the functional dimer. In many positively FNR-regulated promoters, the two half-sites are centered at around −41.5 bp upstream of the transcriptional start site.[31] The FNR-binding site of FNR-repressed promoters, however, is located at different positions relative to the start site, for example, at −50.5 bp upstream of the transcriptional start in the ndh promoter or −117.5 bp relative to the main transcriptional start of the cydAB operon.[32,33] First predictions of FNR-binding sites in front of FNR-regulated genes are obtained by comparative promoter analysis, when the transcriptional start site of the gene is known. In order to confirm the presence of the proposed FNR-binding sites, binding of reconstituted FNR to the binding site is studied in vitro (see later). The increasing number of bacterial genome sequences allows computer analysis for potential FNR-binding sites in complete bacterial genomes. In a comparative computer analysis of

---

[30] K. Eiglmeier, N. Honore, S. Iuchi, E. C. C. Lin, and S. T. Cole, Mol. Microbiol. **3**, 869 (1989).

[31] J. R. Guest, J. Green, A. S. Irvine, and S. Spiro, in "Regulation of Gene Expression in Escherichia coli," p. 317. R. G. Landes & Co., Austin, TX, 1996.

[32] J. Green and J. R. Guest, Mol. Microbiol. **12**, 433 (1994).

[33] P. A. Cotter, S. B. Melville, J. A. Albrecht, and R. P. Gunsalus, Mol. Microbiol. **25**, 605 (1997).

potential FNR-binding sites, 17 new operons were identified as potential members of the *E. coli* FNR regulon.[34]

Binding to FNR Consensus DNA Sites

The ability of reconstituted FNR protein to bind to a synthetic consensus FNR-binding site (TTGAT-N_4-ATCAA) can be demonstrated by restriction site protection assays performed inside an anaerobic cabinet using anoxic solutions.[18,35] For this assay, a *Cla*I or *Rsa*I restriction site is incorporated between two synthetic FNR half-sites. Binding of FNR protects the synthetic FNR consensus site from cleavage by the restriction endonucleases. When FNR is inactivated by O_2, the DNA fragment is no longer protected from cleavage.

In this assay, 1–10 μM of reconstituted FNR protein or anaerobic cell extracts of a strain expressing FNR is incubated with 0.2 μg of the respective template DNA in a total volume of 25 μl in restriction buffer (33 mM Tris/acetate, pH 7.9, 10 mM magnesium acetate, 66 mM potassium acetate, 0.1 mg bovine serum albumin ml^{-1}, and 2 mM dithiothreitol) at room temperature for 15 min to allow FNR to bind to the synthetic binding site. Restriction enzyme (10–20 U) is added, and the reaction mixture is incubated at 37° for 30 to 60 min, allowing restriction of the DNA. After the addition of 1.5 μl stop solution (0.25 M EDTA, 1% SDS, 5% glycerol, and 0.25% xylene cyanol), the resulting DNA fragments of the restriction reaction are analyzed by gel electrophoresis outside the anaerobic cabinet under air atmosphere using a 0.8% agarose gel. After staining with ethidium bromide, an apparent binding constant of FNR to the binding site can be calculated by densitometry analysis of the respective protected and nonprotected DNA bands. Binding of FNR to FNR-specific target–DNA can also be verified by gel mobility shift assays as described for FNR and the aerobically active mutant FNR–D154A.[28,36,37]

In Vitro *Transcription Assays*

The function of transcriptional regulators can be tested *in vitro* by their ability to activate or repress transcription by RNA polymerase using isolated components.[36,38–41] In *in vitro* transcription assays, supercoiled

[34] A. V. Gerasimova, D. A. Rodionov, A. A. Mironov, and M. S. Gel'fand, *Mol. Biol. (Mosk)* **35**, 1001 (2001).
[35] S. B. Melville and R. P. Gunsalus, *Proc. Natl. Acad. Sci. USA* **93**, 1226 (1996).
[36] E. V. Ziegelhoffer and P. J. Kiley, *J. Mol. Biol.* **245**, 351 (1995).
[37] W. Meng, J. Green, and J. R. Guest, *Microbiology* **143**, 1521 (1997).

circular plasmids or linearized derivatives containing the regulated promoter and a transcriptional terminator are used as DNA templates. In order to analyze transcriptional activation by FNR, the transcription assays are performed under anoxic conditions inside an anaerobic cabinet in a total volume of 20 μl containing 20 mM Tris–HCl, pH 8.0, 10 mM magnesium chloride, 120 mM potassium chloride, 5% (v/v) glycerol, and 2 mM dithiothreitol. Reconstituted FNR protein (30 nM), 1 U σ^{70}-RNA polymerase (Epicenter Technologies, Madison, WI), and 20 nM DNA template are incubated for 5 min at room temperature to allow FNR binding to the target site and initiation of open complex formation. The synthesis of transcripts is initiated by the addition of the nucleotide mixture to a final concentration of 0.3 mM ATP, 0.3 mM CTP, 0.3 mM GTP, 0.05 mM UTP, and 5 μM [α-^{32}P] UTP (3000 Ci/mol) and shifting the assay to 37°. After 10 to 30 min, allowing the transcription to proceed, transcription is terminated by the addition of EDTA to a final concentration of 40 mM followed by phenol extraction and ethanol precipitation outside the anaerobic cabinet. The products of each reaction are dissolved in 10 μl loading buffer containing 80% (v/v) deionized formamide, 0.1% (w/v) SDS, 8% (v/v) glycerol, 8 mM EDTA, and 0.2% (w/v) bromphenol blue and xylene cyanol. The samples are denatured for 3 min at 80°, and 1- to 2-μl aliquots from each reaction are analyzed by electrophoresis in denaturing 6% (w/v) polyacrylamide gels. After placing the gels on Whatmann 3MM paper and drying under vacuum at 80°, the gels are visualized by autoradiography or using a phosphoimager (e.g., the Molecular Dynamics phosphoimager or Fuji BAS 1500 image analyser). [α-^{32}P]UTP-labeled transcripts are analyzed by quantitative densitometry using the respective phosphimager software (e.g., ImageQuant 1.2 software, Molecular Dynamics).

Using a plasmid as the DNA template, which contains a constitutive σ^{70}-dependent promoter, offers the possibility to quantify changes in transcriptional levels of the FNR-dependent transcript relative to the constitutive transcript from the plasmid vector.[40,42]

In addition to the analysis of transcriptional activation by *E. coli* FNR *in vitro,* transcriptional analysis offers the possibility to study the coordinated regulation of transcription by multiple regulators, as shown for the *napF* promoter (Fig. 4), which is regulated by FNR, NarP, and NarL.[40]

---

[38] R. A. Schmitz, *FEMS Microbiol. Lett.* **157,** 313 (1997).

[39] K. Klopprogge and R. A. Schmitz, *Biochim. Biophys. Acta* **1431,** 462 (1999).

[40] A. E. Darwin, E. Ziegelhöffer, P. J. Kiley, and V. Stewart, *J. Bacteriol.* **180,** 4192 (1998).

[41] H. J. Wing, J. Green, J. R. Guest, and S. J. Busby, *J. Biol. Chem.* **275,** 29061 (2000).

[42] M. Tan and J. E. Engel, *J. Bacteriol.* **178,** 6975 (1996).

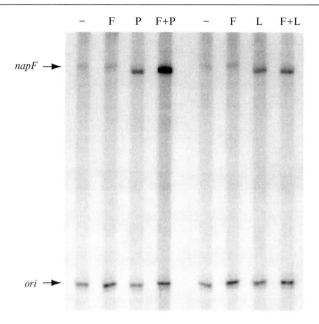

Fig. 4. *In vitro* transcription from the *napF* promoter. Each multiple-round transcription assay mixture contained 20 n*M* supercoiled pVJS2111, 50 n*M* RNA polymerase holoenzyme ($E\sigma^{70}$), and no further additions (−), 0.5 μM FNR (D154A) monomers (F), 6 μM phosphorylated MBP–NarP monomers (P), or 12 μM phosphorylated MBP–NarL monomers (L). The *napF* and constitutive plasmid *ori* transcripts are labeled. From Darwin *et al.*[40] Printed with permission from the American Society for Microbiology.

FNR-Like Proteins in Other Bacteria

FNR-like proteins have been identified in several gram-negative and gram-positive bacteria, some of which differ with respect to their cysteine residues and the presumed coordination and presence of iron–sulfur clusters.[43–47] Only little is known about the regulatory mechanisms of the FNR-like proteins compared to *E. coli* FNR. However, in principle, most of the methods described here for *in vivo* and *in vitro* functional analyses of *E. coli* FNR may be used for FNR-like proteins.

[43] S. Spiro, *Antonie Van Leeuwenhoek* **66,** 23 (1994).
[44] H. Cruz Ramos, L. Boursier, I. Moszer, F. Kunst, A. Danchin, and P. Glaser, *EMBO J.* **14,** 5984 (1995).
[45] A. Klinger, J. Schirawski, P. Glaser, and G. Unden, *J. Bacteriol.* **180,** 3483 (1998).
[46] D. O. Gostick, J. Green, A. S. Irvine, M. J. Gasson, and J. R. Guest, *Microbiology* **144,** 705 (1998).
[47] K. U. Vollack, E. Härtig, H. Korner, and W. G. Zumft, *Mol. Microbiol.* **31,** 1681 (1999).

In order to detect new and yet unknown FNR-like proteins in facultative anaerobic bacteria, three alternative strategies can be used: (i) A
random mutagenesis approach and subsequent screening for a growth
defect under anaerobic fermentative conditions followed by identification
of the respective mutated gene; (ii) identification of FNR homologous proteins by their ability to restore anaerobic growth of an *E. coli fnr* mutant
strain on glycerol with nitrate as the terminal electron acceptor (heterol-
ogous complementation); and (iii) in case the genome sequence of the respective organism is available, comparative sequence analysis can be
performed using the FNR amino acid sequence and the Genetics Computer Group (GCG) program package.[48] Functional analysis of identified
potential FNR-like proteins can be performed *in vivo* and *in vitro* as
described earlier.

Acknowledgments

Work performed in the laboratories of the authors was supported by grants of the DFG
and by the Fonds der Chemischen Industrie to R. A. Schmitz and G. Unden. We thank
V. Stewart for providing Fig. 4.

[48] J. Devereux, P. Haeberli, and O. Smithies, *Nucleic Acids Res.* **12,** 387 (1984).

[42] Experimental Strategies for Analyzing Oxygen Sensing in Yeast

By Robert O. Poyton, Reinhard Dirmeier, Kristin O'Brien,
Pamela David, and Athena Dodd

Introduction

The intracellular levels and often activities of a large number of proteins in *Saccharomyces cerevisiae* are affected by oxygen.[1-3] Among them
are respiratory chain proteins, enzymes involved in the synthesis of heme,
sterols, and unsaturated fatty acids, and enzymes involved in the oxidative

[1] R. O. Poyton, *Respir. Physiol.* **115,** 119 (1999).
[2] R. O. Poyton, R. P. Dirmeier, and K. O'Brein, *in* "Oxygen Sensing: Responses and
Adaptation to Hypoxia" (S. Lahiri, H. Prabhakar, and G. Semenza, eds.), p. 23. Dekker,
New York, 2002.
[3] K. E. Kwast, P. V. Burke, and R. O. Poyton, *J. Exp. Biol.* **201,** 1177 (1998).

stress response. The effect of oxygen on the intracellular concentrations of many of these proteins has been shown to occur at the level of transcription of their genes. Most of these oxygen-regulated genes can be placed into one of two groups: *aerobic* genes, which are transcribed optimally in the presence of air, and *hypoxic* genes, which are transcribed optimally under anoxic or microaerophilic conditions. As discussed later, attempts to estimate the number of oxygen-regulated genes in *S. cerevisiae* by microarray analysis[4,5] are still in their infancy. Nonetheless, it is clear that a substantial percentage (up to 15%) of those genes on the nuclear genome are oxygen-regulated. Oxygen may also affect the expression of proteins via translational control and may affect the folding or assembly of hemoproteins by affecting the availability or redox state of their prosthetic groups.

A striking example of the influence of oxygen on the level and activity of an enzyme comes from considering the effects of oxygen on the *expression* of yeast cytochrome c oxidase genes and the *function* of its oxygen-regulated isoforms.[6,7] This enzyme contains polypeptide subunits encoded by both nuclear and mitochondrial genomes. The three largest subunits (I, II, and III) are encoded by mitochondrial genes (*COX1, COX2,* and *COX3*); they form the catalytic core of the enzyme. Oxygen affects the expression of two of these genes (*COX1* and *COX2*) posttranscriptionally. The other polypeptide subunits are encoded by nuclear genes; some of them modulate catalysis, whereas others function in the assembly or stability of the holoenzyme. Active preparations of yeast cytochrome c oxidase contain at least six subunits (IV, Va or Vb, VI, VII, VIIa, and VIII) encoded by nuclear *COX* genes (*COX4, COX5a* or *COX5b, COX6, COX7, COX9,* and *COX8,* respectively). *COX5a* and *COX5b* encode interchangeable isoforms, Va and Vb, of subunit V.[8] The other nuclear-encoded subunits are encoded by single copy genes. Oxygen affects the expression of all of these nuclear genes at the level of transcription. All of these genes, except *COX5b,* are *aerobic* genes expressed optimally in the presence of air; they are turned off at very low oxygen concentrations (0.25 to 0.5 μM).[9,10] Conversely, *COX5b* is a *hypoxic* gene that is repressed by air

[4] J. J. M. Ter Linde, H. Liang, R. W. Davis, H. Y. Steensma, J. P. Van Dijken, and J. T. Pronk, *J. Bacteriol.* **181,** 7409 (1999).

[5] K. E. Kwast, L.-C. Lai, N. Menda, D. T. James, S. Aref, and P. V. Burke, *J. Bacteriol.* **184,** 250 (2002).

[6] R. O. Poyton and J. E. McEwen, *Annu. Rev. Biochem.* **65,** 563 (1996).

[7] P. V. Burke and R. O. Poyton, *J. Exp. Biol.* **201,** 1163 (1998).

[8] C. E. Trueblood and R. O. Poyton, *Mol. Cell. Biol.* **7,** 3520 (1987).

[9] P. V. Burke, D. C. Raitt, L. A. Allen, E. A. Kellogg, and R. O. Poyton, *J. Biol. Chem.* **272,** 14705 (1997).

[10] P. V. Burke, K. E. Kwast, F. Everts, and R. O. Poyton, *Appl. Environ. Microbiol.* **64,** 1040 (1998).

and turned on at very low oxygen concentrations (0.2 μM).[7,10] This inverse regulation of *COX5a* and *COX5b* by oxygen is especially interesting because these isoforms have differential effects on holocytochrome c oxidase activity. By altering an internal step in electron transport between heme a and the binuclear reaction center (composed of heme a_3 and Cu_B), the "hypoxic" isoform Vb enhances the catalytic constant (i.e., turnover number) fourfold relative to the "aerobic" isoform, Va.[11,12] Hence, these subunit isoforms allow cells to assemble functionally different types of holocytochrome c oxidase in response to different oxygen concentrations.

Interestingly, full respiratory function is achieved in yeast cells grown at oxygen concentrations ranging from 220 μM O_2 (= 159 torr or 21% O_2) (normoxia) to 5 μM O_2 (hypoxia). It is only when cells are grown at very low levels of oxygen (between 0 and 2.5 μM) that the level of respiratory proteins is affected by oxygen concentration.[13] This somewhat surprising finding is most likely attributable to the effects of oxygen tension *per se* on the expression of respiratory protein genes. Studies have demonstrated that the expression of many of these genes is determined by the actual concentration of oxygen and not merely by its presence or absence.[1,9] This has been called *oxygen concentration-dependent* gene expression. By examining the *oxygen concentration-dependent* expression of a number of genes, these studies have also demonstrated that there are at least four classes of oxygen-dependent genes in yeast: (1) *aerobic* genes with hypoxic isoform counterparts, (2) *aerobic* genes without hypoxic isoform counterparts, (3) *hypoxic* genes with aerobic isoform counterparts, and (4) *hypoxic* genes without aerobic isoform counterparts. In general, those genes that have counterparts that are oppositely regulated by oxygen are regulated much more stringently by oxygen concentration than those genes that lack them. The finding that these four classes of genes are regulated differently in response to oxygen concentration lends support to the idea that eukaryotic cells, even those as simple as *S. cerevisiae,* possess multiple mechanisms for sensing oxygen.

Although much progress has been made in understanding the transcriptional machinery involved in oxygen-regulated gene expression in yeast and other eukaryotes, the underlying mechanism(s) of oxygen sensing and the signaling pathways that connect the oxygen sensor(s) to the transcription machinery in eukaryotes are still poorly understood. Because *S. cerevisiae* lends itself well to genetic and biochemical studies and because its genome has been sequenced completely, it is well suited for broadly

[11] R. A. Waterland, A. Basu, B. Chance, and R. O. Poyton, *J. Biol. Chem.* **266,** 4180 (1991).
[12] L. A. Allen, X. J. Zhao, W. Caughey, and R. O. Poyton, *J. Biol. Chem.* **270,** 110 (1995).
[13] P. J. Rogers and P. R. Stewart, *J. Bacteriol.* **115,** 88 (1973).

based studies on oxygen sensing at the cellular level. The following sections review experimental strategies that are used with this yeast to assess the effects of oxygen concentration on gene expression and metabolism, identify oxygen-responsive transcription factors, address the nature of the oxygen sensor, and examine molecular components of the signaling pathway.

Procedures

Steady-State Balanced Growth at Different Oxygen Tensions

Two different types of study have been useful for addressing how *S. cerevisiae* cells sense and respond to oxygen. The first involves growing cells to steady state in different oxygen environments. This type of study led to the discovery of *oxygen concentration-dependent* gene expression.[1] The second type of study involves shifting cells between different oxygen concentrations. This type of study has been useful in looking at the kinetics of induction or repression of oxygen-responsive genes, as well as the stability of mRNA transcripts from oxygen-regulated genes. It is also useful for identifying the cellular and molecular components of pathways involved in regulating oxygen-responsive genes. For example, it was used to identify a role for the respiratory chain in the induction of some hypoxic genes.[14]

Early studies on the effects of oxygen on yeast physiology and gene expression focused on two conditions: aerobic growth in the presence of oxygen and anaerobic growth in the presence of nitrogen or some other inert gas (e.g., argon). For aerobic growth, cells can be grown on solid growth media in petri plates and in liquid media either in batch cultures or in a fermentor, as described later. For anaerobic growth, cells are grown either on solid media (inside of anaerobic jars filled with N_2) or in fermentor vessels that are either saturated with N_2 and closed off (i.e., nonsparged) or left open and sparged with a mixture of O_2-free N_2 and 2.5% CO_2. A feature that distinguishes nonsparged from sparged cultures is that CO_2 is allowed to build up in nonsparged cultures but is blown off in sparged cultures. This prevents heterotrophic CO_2 fixation into the citric acid cycle, which slows growth considerably.[9]

To examine the effects of oxygen tension *per se* on cells it is necessary to use a different experimental strategy. Here, culture conditions are set up so that oxygen is the limiting nutrient and the effects observed are due to oxygen concentration and not some other variable, such as carbon source or growth rate. To ensure that oxygen is the limiting nutrient, cells are

---

[14] O. Maaloe and N. O. Kjeldgaard, *in* "Control of Macromolecular Synthesis," p. 56. W. A. Benjamin, Inc., 1966.

grown in fermentors, as described later, and the oxygen concentration in the fermentor vessel is monitored using a dissolved oxygen electrode. Oxygen limitation is signaled by a decrease in dissolved oxygen concentration in the culture. This type of protocol employs oxygen gas mixtures that sparge cultures at a constant rate. The sparging of cultures with gas mixtures containing different oxygen concentrations can be used for all oxygen concentrations between air ($\sim$220 μM) at sea level and $\sim$0.5–1 μM O_2. It is impractical for oxygen concentrations below 0.5 μM because oxygen at such low concentrations is consumed nearly as fast as it is provided, and oxygen becomes limiting when the cell density is extremely low. For very low oxygen tensions, an alternative paradigm must be used. To achieve good cell yields for cultures grown at very low oxygen concentrations, a computer-controlled feedback system for gas flow is used to regulate the oxygen concentration in the gas entering the fermentor.[10] It allows the oxygen concentration in the gas entering the fermentor to increase with cellular demand. This type of regulated control allows for good yields before the oxygen concentration in the fermentor vessel becomes limiting.[10]

In addition to the aforementioned considerations, cells should be in steady-state balanced growth[15] at the time of harvest in order to obtain reproducible results. To achieve steady-state balanced growth, the cell density of the culture must be kept low enough so that the process of cell growth does not change the chemical or physical properties of the growth medium and cells should be maintained for several generations in the oxygen concentration of interest.

Growth Media and Conditions. Inocula for aerobic growth, anaerobic growth, or growth at intermediate oxygen concentrations are "precultures" that are grown to steady-state balanced growth at 28° on a reciprocating shaker at 200 rpm in air. To achieve balanced growth, these precultures are kept in midlogarithmic phase, by dilution into fresh growth media as necessary, for at least six generations. Precultures and fermentor cultures are grown in SSG-TEA, a semisynthetic galactose growth medium, pH 5.0 (containing per liter, 3 g of Bacto yeast extract, 10 g of galactose, 0.8 g of NH$_4$SO$_4$, 1 g of KH$_2$PO$_4$, 0.5 g of NaCl, 0.7 g of MgSO$_4$ · 7 H$_2$O, 5 μg of FeCl$_2$, and 0.4 g of CaCl$_2$), supplemented with 0.1% (v/v) Tween 80, 20 μg/ml ergosterol, and 350 ppm Dow Corning FG-10 silicon antifoam. To obtain a uniform dispersion, the Tween 80, ergosterol, and silicon antifoam are sonicated ($\sim$110 W for 1 min using a Branson Model 250 Sonifier) in media prior to autoclaving. The galactose is autoclaved separately and added to the media under sterile conditions. Nutritional

[15] K. E. Kwast, P. V. Burke, B. T. Staahl, and R. O. Poyton, *Proc. Natl. Acad. Sci. USA* **96**, 5446 (1999).

supplements (e.g., amino acids, nucleotides) are added from sterile stock solutions, as required, to satisfy the auxotrophic requirements of the strain that
is used. For aerobic growth in batch culture, cells are grown in Erlenmeyer
flasks fitted with a side arm that is accommodated by a Klett-Summerson
colorimeter. Growth is conveniently followed by inserting the side arm into
the Klett-Summerson colorimeter, fitted with a No. 54 green filter. An alternative method for following cell growth is to withdraw a sample and measure
absorbance at 550 nm. Cell density is proportional to the colorimeter
readings or absorbance up to $\sim 1 \times 10^8$ cells per milliliter.

Fermentor Cultivation. For unregulated growth in different oxygen concentrations down to ~ 1 μM, oxygen cells are grown in different gas mixtures
in either a New Brunswick BioFloIIc or a BioFloIII fermentor set up to
maintain a pH of 5.0, a temperature of 28°, and an agitation rate of 200 rpm.
rpm. Generally, we use vessels with working volumes of 3.5–5.0 liters. The
dissolved oxygen concentration is monitored with a polarographic oxygen
probe attached to an appropriate DO meter (e.g., Ingold 4300 transmitter).

The fermentor vessel is sparged with air for aerobic cultures, 2.5% CO_2
in O_2-free N_2 for anaerobic cultures, and gas mixtures that contain different percentages of oxygen in O_2-free nitrogen for intermediate oxygen concentrations. Gas flow rates are maintained at 4 liters/min. Initial oxygen
concentrations in these sparge gases are calculated assuming that air is
composed of 21% O_2, which is equivalent to 220 μM at sea level. The dissolved oxygen concentration in the fermentor vessel is measured with a
DO meter and an oxygen electrode that has been calibrated with air
(100% saturation) or O_2-free nitrogen (0% saturation). Actual dissolved
oxygen concentrations in gas-saturated growth media are calculated
according to Henry's law; $[X_i = p_i h / H_i]$, where X_i is the amount (mol) of
gas dissolved per mole solution at equilibrium, P_i is the partial pressure
of the gas at elevation h, and H_i is the Henry's law constant for the gas at
a specific temperature. The differential pressure in the fermentor vessel at
the 4-liters/min sparge rate used for these experiments is extremely low
(less than 0.4% of an atmosphere) and has little effect on the pressure in
the vessel (ambient + differential pressure) and X_i. Henry's law gives the
dissolved oxygen content in water. To determine its solubility in solution,
it is necessary to correct for the ionic strength of the solution (e.g., SSG-
TEA growth media). This is done by determining the conductivity of the
solution. For SSG-TEA media, the conductivity of the media at 25° is
$\sim 12,000–13,000$ μSiemens/cm, giving a correction factor[16] of 0.959–0.962.
Taking this correction factor into account, we find that the dissolved

---

[16] *In* "Correction Factors for Oxygen Solubility and Salinity." Dissolved Oxygen, U.S.
Geological Survey TWRI Book 9, p. DO 27-DO 38, 2002.

oxygen concentration of air-saturated SSG-TEA media is $203.5 \pm 0.3 \ \mu M$ O_2 in our laboratory in Boulder, Colorado.

Aerobically grown precultures are used as inocula, and the starting cell density in the fermentor vessel is adjusted to allow cells to experience at least six mass doublings before harvest. Cells are harvested when the dissolved oxygen concentration in the fermentor drops to 80% of its initial value. Prior to harvest, cells are chilled quickly to 4° by passage through a cooling coil/salted ice bath into chilled centrifuge bottles. They are harvested by centrifugation ($3000g$, 10 min, 4°), washed twice with ice-cold distilled water, and either processed immediately or flash frozen in liquid N_2. For cells that are to be used for RNA isolation and analysis, diethylpyrocarbonate is added to the distilled water that is used in the washing steps (see later). During anaerobic growth, the fermentor vessel is covered with aluminum foil to prevent photoinhibition of anaerobic cells.[17]

Protocols for Shifting Cells between Oxygen Concentrations

Shifts between oxygen concentrations are also performed in a fermentor using the same conditions for growth described earlier. Although it is possible to perform programmed shifts between any two oxygen concentrations, two types of shifts that have been especially useful for addressing how *aerobic* or *hypoxic* genes are induced have involved shifting cells from anaerobic to aerobic conditions and *vice versa*. To follow the induction of *aerobic* genes, cells are grown under anaerobic conditions (2.5% CO_2 in O_2-free N_2) for at least six generations (cell density $\sim 2 \times 10^7$ cells/ml) and then the sparge gas is changed to air. Dissolved oxygen concentrations in the fermentor are followed during the shift and should increase rapidly to its aerobic value within ~ 5 min. To follow the induction of *hypoxic* genes, cells are grown in the fermentor vessel in air until the dissolved oxygen concentration in the fermentor reaches 80% of its aerobic value or until the cell density reaches 2×10^7 cells/ml, and then the sparge gas is changed from air to 2.5% CO_2 in O_2-free N_2. The dissolved oxygen concentration is monitored and should fall rapidly, reaching its anaerobic value within ~ 10 min. Samples are taken for analysis at different times after each type of shift.

Assays for Assessing Oxygen Effects on Respiratory Metabolism

Useful indicators of the effects of oxygen on yeast cell metabolism are the levels of mitochondrial respiration and mitochondrial cytochromes. The overall level of respiration is measured easily with an oxygen

---

[17] E. Sulkowski, B. Guerin, J. Defaye, and P. P. Slonimski, *Nature* **202,** 36 (1964).

electrode, and the level of respiratory cytochromes is measured spectro-photometrically. These are measured most conveniently in whole cells, as described later. They can also be applied to isolated mitochondria.[11,12]

Measurement of Whole Cell Respiration. Rates of oxygen uptake by whole cells are measured at 30° with a Strathkelvin oxygen electrode system. The assay solution consists of 3 ml of 40 mM NaPO$_4$ buffer, pH 7.4, 0.7% (w/v) glucose, and 5–15 mg (wet weight) yeast cells. Cyanide-sensitive respiration is assayed in an identical solution, adjusted to 1 mM KCN.

For measuring whole cell respiration, the assay buffer is maintained in an aerated state by sparging with filtered house air. Alternatively, the buffer can be aerated in the oxygen electrode chamber by stirring for 3 min min before the chamber is sealed. The solution in the chamber is stirred at a moderate speed to keep the solution homogeneous during the assay. The chamber is sealed by the addition of the oxygen electrode, and data collection is initiated once the electrode output has stabilized. Cells are then introduced into the chamber with a syringe and rates of respiration are measured. The amount of cell suspension added is determined by the activity of the sample. A maximum oxygen consumption of 30% over the 3-min data collection period yields reproducible data. To determine respiratory chain-specific respiration, also known as KCN-sensitive respiration, the rate of oxygen consumption in the presence of 1 mM KCN is subtracted from the rate of oxygen consumption in the absence of KCN.

Spectrophotometric Measurement of Respiratory Cytochrome Concentrations in Intact Cells. Cytochrome concentrations are determined from optical spectra recorded with an Aminco DW-2000 double-beam dual-wavelength spectrophotometer. The dual-beam mode allows both the scanning beam and a reference beam to pass through the sample cuvette, thereby compensating for light scattering from the turbid sample and drift due to settling of the sample during data collection. A typical spectrum of yeast cells is scanned from 400 to 700 nm using a reference wavelength of 577 nm. The reference wavelength (monochromator 2) is set at an isosbestic point wavelength for the desired cytochrome absorptions. For room temperature spectra, we use a slit width of 3 nm and a scan speed of 2 nm/s. Room temperature difference spectra (reduced minus oxidized) are used to quantitate cytochrome levels. Low temperature ($-196°$) difference spectra, taken in liquid N$_2$, are used to obtain a better resolution of cytochrome c and c$_1$. Although low temperature spectra offer a higher resolution of cytochromes and hence are qualitatively better than room temperature spectra, they cannot be used reliably for the quantitation of cytochromes because the path length amplification obtained upon freezing and devitrification is variable.

For room temperature spectra, cells are suspended at ~0.3 g cells/ml in 40 mM KPO$_4$, pH 7.4, and are incubated with 1% (w/v) glucose for 5 min to reduce cytochromes. They are placed in 1-cm pathlength cuvettes and are scanned from 400 to 700 nm in the dual-wavelength mode with 577 nm set as the reference wavelength (monochromator 2). This is the glucose-reduced spectrum, which is used to generate a baseline. Cells are then oxidized by adding 0.2% H$_2$O$_2$ to the cuvette from a 30% (v/v) stock solution. The cells and H$_2$O$_2$ are mixed by inverting the cuvette, and the oxidized spectrum is taken immediately. Finally, cells are reduced with 5 mg/ml of fresh sodium dithionite and a reduced spectrum is taken. Difference spectra (glucose reduced minus oxidized or dithionite reduced minus oxidized) are calculated using Aminco software. In most cases, these two difference spectra are nearly identical. However, dithionite, a chemical reductant, usually reduces cytochromes more completely than glucose, a metabolic reductant. Concentrations of cytochromes can be determined using extinction coefficients of 16.5 mM^{-1} cm^{-1} (Δ605–630 nm, reduced minus oxidized) for cytochrome aa_3; 24.6 mM^{-1} cm^{-1} (Δ550–540 nm, reduced minus oxidized) for cytochrome c, and 25.6 mM^{-1}cm^{-1} (Δ562–577 nm, reduced minus oxidized) for cytochrome b.

For low temperature spectra, cells are prepared as described earlier for room temperature spectra except that the cell suspension is adjusted to contain 30% (v/v) ethylene glycol, placed in a 2-mm pathlength cuvette, and frozen in liquid N$_2$. Samples are devitrified by removing them from liquid N$_2$ and placing them at room temperature for approximately 2 min. The devitrification process will be evident as the sample will become completely opaque. When the sample has devitirfed, it is placed in a cuvette holder and then in an unsilvered Dewar flask, which is placed in the low temperature accessory of the Aminco DW2000 spectrophotometer. Liquid N$_2$ is added to the flask to bring the temperature to $-196°$, and the first, glucose reduced, spectrum is recorded as described earlier, but with a 0.8-nm slit width. The sample is thawed, removed from the cuvette, oxidized with 0.2% H$_2$O$_2$ for 2 min at room temperature, refrozen in liquid N$_2$ as described earlier, and scanned from 400 to 700 nm. To obtain the dithionite-reduced sample, thaw the sample, reduce it with 5 mg/ml sodium dithionite for 2 min at room temperature, freeze in liquid N$_2$, and scan the spectrum from 400 to 700 nm.

Assays for Assessing Oxygen Effects on Gene Expression

Currently, the most common way to address how yeast cells sense oxygen is to analyze how a change in oxygen concentrations affects the expression of oxygen-regulated genes. By studying several of these genes it has been possible to identify common oxygen-regulated promoter elements and the transcription factors that bind them. Focus is now on (1) how these transcription factors work, in combination with other transcription factors, to fine tune expression to specific oxygen concentrations, (2) the nature of the upstream oxygen sensor(s), and (3) what pathways connect upstream oxygen sensors to transcription. Several different methods have been used to follow changes in gene expression in response to growth in different oxygen environments. Some of these follow changes in steady-state transcript levels (e.g., by microarray analysis or Northern blot hybridization), promoter activity (e.g., lacZ-promoter fusions), and protein levels (by two-dimensional electrophoresis).

Microarray Analysis: A Tool for Identifying Potential Oxygen-Regulated Genes

Although microarray analysis has a great deal of potential for identifying oxygen-regulated genes and studying the pathways that connect them to oxygen sensors, few studies have so far used this approach and there is no general consensus from these studies concerning the number of genes that are oxygen regulated. Two studies have looked at the entire yeast nuclear genome ($\sim$6100 genes) and compared transcript levels from cells grown aerobically with those grown anaerobically.[4,5] Whereas the first study[4] identified 219 *aerobic* genes (genes expressed at higher levels in air than in its absence) and 140 *hypoxic* genes (genes expressed at higher levels in anaerobic than in aerobic cells), the second study[5] identified 736 *aerobic* genes and 340 *hypoxic* genes. It is not clear why these two studies have produced such strikingly different results. Part of the discrepancy may be in differences between strains, differences in the protocols that were used, and differences in the level of change that was considered significant. Based on our own studies, we propose that a twofold difference in transcript levels between aerobically and anaerobically grown cells be used as a minimum threshold. Applying this criterion to the data of Kwast *et al.*[5] yields 182 *aerobic* genes and 183 *hypoxic* genes. Interestingly, only 44 genes from each of these lists were found by Ter Linde *et al.*[4] to change by a factor of twofold or more. Interestingly, the level of change exhibited by these genes also varies between the two studies. Regardless of the reason for this difference in level of change, it is clear that microarray analysis by itself cannot be relied upon to establish the identity of a gene with respect to oxygen regulation without supporting information from the other methods of analysis mentioned previously. Moreover, because changes in transcript levels deduced from microarray studies do not always parallel changes in protein levels obtained from two-dimensional analysis of the yeast proteome, it is clear that microarray analysis by itself cannot and should not used to construct complex regulatory schemes.[5]

Northern Blot Analysis

We have found that the most reliable and most convenient method for identifying an oxygen-regulated gene and establishing how much its transcription changes in cells exposed to different oxygen environments is Northern blot hybridization using gels on which the load for *ACT1*, a gene whose expression is unaffected by oxygen concentration, is held constant. This analysis uses whole cell RNA; it does not require the isolation of poly(A)[+]RNA. Currently, the list of oxygen-regulated yeast genes analyzed by this method is limited. However, by comparing this list with those oxygen-regulated genes that have been identified by microarray analysis, it is possible to generate a reliable short list of oxygen-regulated yeast genes that can be used as markers for studies aimed at identifying the oxygen sensors and signal transduction pathways (Table I). Total yeast cell RNA is isolated, electrophoresed, and blotted as described elsewhere.[18] Abbreviated versions of each protocol are presented next.

RNA Isolation. Cells, solutions, tubes, and rotors are kept at 4° throughout all steps. Fresh or frozen cell pellets from fermenter grown cultures are resuspended at 1–2 g wet weight in 5 ml STES buffer [0.5 M NaCl, 0.2 M Tris–Cl (pH 7.5), 0.01 M EDTA (pH 8.0), 1% (w/v) SDS, and 0.1% (v/v) diethyl pyrocarbonate (DEPC)]. Ten grams of sterile glass beads (0.50 mm) and 5 ml phenol:chloroform:isoamyl alcohol (PCI, 25:24:1, pH 4.5) are added and cells are disrupted by vortexing 2 min. Phase layers are separated by centrifugation (5 min, 4000 g_{max}, 4°), and the aqueous phase is collected. It is extracted at least two more times with PCI until the phase interface is clear. Nucleic acids are precipitated by the addition of 2.5 volumes of ethanol and 0.1 volumes of 3 M sodium acetate, 0.10% (v/v) DEPC. The nucleic acid precipitate is recovered by centrifugation (30 min, 16,000g_{max}, 4°), washed with 5 ml 70% (v/v) ethanol, and centrifuged again (15 min, 16,000g_{max} 4°). To remove excess ethanol, the pellet is allowed to air dry before being dissolved in 100 μl 20 mM EDTA, 0.1% (v/v) DEPC (pH 8.0). RNA is quantitated spectrophotometrically[19] and aliquoted. Each aliquot is reprecipitated as described earlier and is then dissolved in 20 mM EDTA, 0.1% (v/v) DEPC (pH 8.0), 69% (v/v) formamide to obtain a concentration of 2 μg RNA/μl.

[18] R. Dirmeier, K. O'Brien, M. Engle, A. Dodd, E. Spears, and R. O. Poyton, *Methods Enzymol.* **381,** 38 (2003).

[19] J. Sambrook, F. Fritsch, and T. Maniatis, *in* "Molecular Cloning: A Laboratory Manual." Cold Spring Harbor Laboratory, Cold Spring Harbor, NY, 1989.

TABLE I
OXYGEN-REGULATED YEAST GENES IDENTIFIED BY BOTH MICROARRAY AND
NORTHERN BLOT ANALYSES[a]

Aerobic gene		Hypoxic gene	
ORF (Open Reading Frame)	Gene	ORF	Gene
YGL187C	COX4	YIL111W	COX5B
YNL052W	COX5A	YEL039C	CYC7
YHR051W	COX6	YBR085W	AAC3
YMR256C	COX7	YHR007C	ERG11
YLR395C	COX8	YDR044W	HEM13
YDL067C	COX9	YJR047C	ANB1
YJR048W	CYC1	YGL055W	OLE1
YPR065W	ROX1	YHR042W	CPR1
YEL034W	TIF5-1α		
YBL030C	AAC2		
YMR056C	AAC1		
YDR256C	CTA1		
YGR088W	CTT1		
YJR104C	SOD1		
YHR008C	SOD2		
YGR234W	YHB1		

[a] Each of the genes listed has been identified as an *aerobic* or *hypoxic* gene by Northern analysis[3] and by either one or both of the microarray studies[4,5] mentioned in the text.

Electrophoresis and Northern Blotting. 5X MOPS formaldehyde buffer [0.1 M MOPS, 40 mM sodium acetate, 5 mM EDTA, 1.1 M formaldehyde, 0.1% (v/v) DEPC, pH 7.0] is added to 30 μg RNA to a final concentration of 1X in a total volume of 20 μl. It is denatured at 65° (15 min). Following denaturation, 2 μl 10X loading dye [50% (v/v) glycerol, 1 mM EDTA (pH 8.0), 0.25% (w/v) bromphenol blue, 0.25% (w/v) xylene cyanol] is added to the sample before being separated on 1.5% (w/v) agarose gels. 1X MOPS formaldehyde buffer is used in the gels and in the running buffer. Gels are run for 2–3 h at 110 V. Blotting is performed as recommended by the manufacturer of the transfer membrane (GeneScreen, NEN, Boston, MA). To remove formaldehyde from the finished gel, it is soaked four times in 5 volumes (500 ml per 100 ml gel) dH$_2$O [0.1% (v/v) DEPC] for 5 min with shaking. The gel is equilibrated with 5 volumes 10X SSC (1.5 M NaCl, 0.15 M sodium citrate dihydride, 0.1% (v/v) DEPC) for 30 min with shaking. A capillary blot is assembled and left to transfer overnight. The following day, RNA is fixed to the membrane by UV crosslinking with a UV Stratalinker 1800 (Stratagene, La Jolla, CA), followed by baking for 20 min at 80°.

Hybridization. Polymerase chain reaction (PCR)-based probes are made corresponding to the oxygen-regulated gene of interest and *ACT1*. Probes are radiolabeled by random primer labeling of PCR fragments (100 ng) using $[\alpha\text{-}^{32}\text{P}]$-dCTP with a random primer extension labeling kit (Sigma, St. Louis, MO) according to the manufacturer's directions. Excess dCTP is removed using ProbeQuant G-50 microcolumns (Amersham Pharmacia, Piscataway, NJ). For the hybridization step, ULTRAHyb buffer (Ambion, Austin, TX) is used according to the manufacturer's directions. Blots are prehybridized for 30 min at 42° before adding radiolabeled probe to a concentration of 6.5×10^6 cpm. Blots are allowed to hybridize for 24 h and then are washed twice with 2X SSC, 0.1% (w/v) SDS (42°, 5 min) and twice with 0.1X SSC, 0.1% (w/v) SDS (55°, 15 min). Signal intensity is measured with a Storm 860 PhosphorImager (Amersham Pharmacia, Piscataway, NJ). The relative signal strength of transcripts from each oxygen-regulated gene is normalized to the level of *ACT1* mRNA.

Proteome Analysis

Although results from microarray studies have suggested that there are a large number of oxygen-regulated yeast genes, it has not yet been demonstrated that the levels of the protein products of these genes are affected by oxygen. Clearly, it is important to analyze the effects of oxygen on the yeast proteome, as well as the yeast transcriptosome, before drawing any conclusions about the effects of oxygen on biosynthetic pathways.[5] The importance of this type of parallel study can be illustrated by looking at the mitochondrial and promitochondrial proteomes in aerobic and anaerobic yeast cells, respectively. Although it has been reported that oxygen has a profound effect on a large number of nuclear genes that encode mitochondrial proteins,[5] the mitochondrial proteome in JM43 cells is remarkably similar to the promitochondrial proteome in JM43 cells.[2] This suggests that the functional changes that distinguish mitochondria and promitochondria in this and other strains are affected by the effects of oxygen on either the levels or the activities of relatively few proteins.[2]

Proteome analysis has been simplified greatly by the development of two-dimensional (2-D) gel electrophoresis protocols, matrix-assisted laser desorption ionization time-of-flight mass spectrometry (MALD-TOF), and databases that allow for tentative or definitive protein identification based on protein size, isoelectric point, and peptide masses derived from MALDI – TOF.[20] In overview, the analysis involves excising proteins from

[20] K. A. Resing, *Ann. N. Y. Acad. Sci.* **971,** 608 (2002).

2-D gels, digestion with a site-specific protease (e.g., trypsin), cocrystallization of the peptides with an organic matrix, and ionization with a MALDI laser. The time it takes to traverse the flight tube is proportional to the mass-to-charge ratio of each peptide. The measured peptide masses are then used to search a database of predicted peptide masses from known proteins. The most frequently used databases include PeptideSearch at the European Molecular Biology Laboratory (narrador.embl-heidelberg.-de/GroupPages/PageLink/peptidesearchpage.html), Protein Prospector from the University of California at San Francisco (prospector.ucsf.edu), ProFound at Rockefeller University (prowl.rockefeller.edu/cgi-bin/Pro-Found), and PeptIdent of the Swiss Institute of Bioinformatics (us.expasy.org/tools/peptident.html). Protein identity is established by the number of peptide matches. MALDI – TOF is particularly powerful for identifying proteins from organisms such as *S. cerevisiae,* for which the entire genome has been sequenced.[21] Protein-specific antibodies can also be used to identify protein spots on two-dimensional gels by Western immunoblotting. Specific protocols for the preparation of cell lysates, two-dimensional electrophoresis, MALDI-TOF, and Western immunoblotting are described as follows.

Cell Lysis: Preparation of Whole Cell, Cytosolic, and Mitochondrial Lysates. In preparing whole cell lysates, it is important to address problems related to proteolytic degradation, efficient solubilization of proteins, and removal of materials such as nucleic acids that interfere with isoelectric focusing, the first dimension in 2-D electrophoresis. These factors, together with cell breakage protocols, are considered elsewhere.[22] The preparation of mitochondria and cytosolic fractions are given elsewhere.[18]

Two-Dimensional Gel Electrophoresis of Cell Proteins. Aliquots (50–100 μg protein) of whole cell lysates or cell fractions are diluted fivefold in lysis buffer containing 9 M urea, 2.5 M thiourea, 5% (w/v) CHAPS, 12.5 mM dithiothreitol (DTT), and 1% (v/v) carrier ampholytes, pH 3–10, and incubated for 1 h at room temperature. Rehydration buffer [7 M urea, 2 M thiourea, 4% (w/v) CHAPS, 10 mM DTT, 0.5% (v/v) Triton X-100, 1% (v/v) carrier ampholytes, pH 3–10, and a trace of bromphenol blue] is added to bring the total volume up to 250 μl. The samples are incubated at room temperature for 20 min and are then used to rehydrate 13-cm Immobiline Drystrips (Amersham Pharmacia Biotech) with a pH 3–10 linear gradient. The strips are rehydrated overnight in an Immobiline

[21] A. Shevchenko, O. N. Jensen, A. V. Podtelejnikov, F. Sagliocco, M. Wilm, O. Vorm, P. Mortensen, A. Shevchenko, H. Boucherie, and M. Mann, *Proc. Natl. Acad. Sci. USA* **93,** 14440 (1996).
[22] A. Blomberg, *Methods Enzymol.* **350,** 559 (2002).

Drystrip reswelling tray (Amersham Pharmacia Biotech). First-dimension isoelectric focusing is carried out at 20° using a Multiphor II flatbed system (Amersham Pharmacia Biotech) at the following voltages: during the first minute, voltage is increased linearly to 300 V, it is held at 300 V for the next 4 h, it is increased linearly from 300 to 3000 V during the next 5 hours, and it is held at 3000 V for the next 10 h. The strips are then incubated with agitation for 10 min at room temperature in each of the following solutions: (1) 50 mM Tris–HCl (pH 6.8), 6 M urea, 30% (v/v) glycerol, and 2% (w/v) SDS; (2) 50 mM Tris–HCl (pH 6.8), 6 M urea, 30% (v/v) glycerol, 2% (w/v) SDS, and 1% (w/v) DTT; and (3) 50 mM Tris–HCl (pH 6.8), 6 M urea, 30% (v/v) glycerol, 2% (w/v) SDS, 4% (w/v) iodoacetamide, and a trace amount of bromphenol blue. The Immobiline strips are loaded onto a 20 X 20-cm polyacrylamide gel composed of 8–18% (w/v) 37.5:1 acrylamide:bisacrylamide, 0.37 M Tris (pH 8.8). The 20 × 20-cm gradient gels are poured using a Gradient Mixer GM-1 (Amersham Pharmacia Biotech) and a Protean II xi multigel casting chamber (Bio-Rad). The gels are run at 10° on a Protean II xl m multicell unit (Bio-Rad) at a constant 8 mA per gel for approximately 17 h in a running buffer containing 0.19 M glycine, 25 mM Tris, and 0.1% SDS.

After electrophoresis, proteins are visualized by silver staining.[23] The gels are fixed in 40% (v/v) ethanol, 12% (v/v) acetic acid, and 0.0185% (v/v) formaldehyde for 20 min to overnight and washed twice in 50% (v/v) ethanol for 10 min each and once in 30% (v/v) ethanol for 10 min. They are pretreated with 0.02% (w/v) sodium thiosulfate for 1 minute and rinsed with water three times for 30 s each. The gels are impregnated with a 0.2% (w/v) silver nitrate solution with 0.0185% (v/v) formaldehyde for 20 min to 2 h, rinsed with water twice for 10 s each, and developed for 2 to 2.5 min with 6% (w/v) sodium carbonate, 0.0185% (v/v) formaldehyde, and 0.0004% (w/v) sodium thiosulfate. The developing reaction is stopped with 12% (v/v) acetic acid for 10 min. The gels are then rinsed twice with water for 20 min each.

Protein Identification by MALDI-TOF Mass Spectrometry. Silver-stained spots are cut out of the gels for in-gel digestion and are destained with 1 ml of 50 mM sodium thiosulfate and 15 mM potassium ferricyanide, followed by fourth washes in 1 ml Milli-Q H$_2$O. The spots are then equilibrated for 20 min in 500 μl 100 mM ammonium bicarbonate and incubated for 20 min in 500 μl 50% (v/v) acetonitrile, 50 mM ammonium bicarbonate. The spots are dried with a Speed-Vac, rehydrated for digestion with 5 μg/ml porcine trypsin (Promega, Madison, WI) in 25 mM ammonium bicarbonate, and incubated at 37° overnight. The peptides are extracted from

---

[23] H. Blum, H. Beier, and H. J. Gross, *Electrophoresis* **8**, 93 (1987).

the gel matrix by sonication for 20 min and are then concentrated using Zip Tips (Millipore Corporation, Bedford, MA). Cysteines are treated with iodoacetamide to form carboxyamidomethyl cysteine and methionine is considered to be in an oxidized form.

Peptide mass fingerprinting is performed using a Voyager-DE STR (Perkin Elmer) operating in delayed reflector mode at an accelerating voltage of 20 kV. The peptide samples are cocrystallized with matrix on a gold-coated sample plate using 0.6 μl matrix (α-cyano-4-hydroxy-transcinnamic acid) and a 0.6-μl sample. Spectra are calibrated internally using trypsin autolysis products. Monoisotope peptide masses are assigned and then used in database searches. A convenient database for identifying monoisotopic peptide masses is PeptIdent (http://ca.expansy.org/tools/peptident.html). One missed cleavage is allowed and a minimum of four matching peptides is required for a match to be considered successful. A pI range of ± 2.00 and a molecular weight range of $\pm 20\%$ are used for the database search. Several criteria are used to identify the proteins of interest; these include the extent of sequence coverage, the number of peptides matched, a match with the theoretical pI and apparent molecular weight, M_r, as determined from the gel.

Not all proteins can be identified with MALDI-TOF, particularly if a given spot contains a mixture of proteins. In these cases, it is advisable to obtain peptide sequences, usually using electrospray ionization tandem mass spectrometry (ESI). New instruments have been developed that combine MALDI-TOF with tandem mass spectrometry, eliminating the need for an additional sample preparation that is required with ESI.[24] A MALDI-TOF spectrum is obtained, and then one or more peptides are selected for fragmentation with a collision gas. The peptide sequences are then searched in one of the databases listed earlier or with Mascot from Matrix Science (www.matrixscience.com).

Western Immunoblotting of Proteins Separated by One- or Two-Dimensional Electrophoresis. After electrophoresis, two-dimensional gels can also be subjected to Western immunoblotting with antibodies to the protein(s) of interest. Proteins are transferred onto Hybond ECL (Amersham Pharmacia Biotech) nitrocellulose membranes using a Semi-Phor blotting apparatus (Hoeffer Scientific Instruments). The protein antigens are detected with the appropriate primary antibodies and secondary horseradish peroxidase-linked antibodies followed by a chemiluminescence reaction using a chemiluminescence detection kit (Perkin Elmer Life Sciences).

[24] W. V. Bienvenut, C. Deon, C. Pasquarello, J. M. Campbell, J.-C. Sanchez, M. L. Vestal, and D. F. Hochstrasser, *Proteomics* **2,** 868 (2002).

Genetic Methods

As a facultative anaerobe with well-developed nuclear and mitochondrial genetic systems, *S. cerevisiae* lends itself well to genetic studies aimed at identifying oxygen sensors and the signal transduction pathways that connect them to oxygen-responsive transcription factors. Indeed, the techniques of classical genetics and recombinant DNA methodologies offer a variety of genetic options for identifying molecular components of oxygen sensing and signal transduction. These have been discussed extensively in previous volumes in this series. They include mutant isolation and mapping,[25] construction of synthetic dosage lethals[26] or two hybrid screens[27] to identify interacting proteins, and placement of genes in a pathway by analysis of their epistatic relationships.[28] A number of different screens and selections have already been used to identify transcription factors that function in *trans* to regulated oxygen-responsive genes. These have been complemented by promoter resection studies aimed at identifying *cis*-binding sites. This article provides the experimental rationale that has been used for genetic selections for genes that are involved in the repression of *hypoxic* genes, genetic selections and screens for genes that are involved in the activation of *aerobic* genes, and promoter resections of oxygen-responsive genes.

Selections for Hypoxic Repressors. Genes that repress the expression of *hypoxic* genes in aerobic cells can be identified by mutating cells with standard mutagens and then selecting for growth on selective media. The strategy used to identify the Rox1p/Reo1p repressor[29,30] provides a good example for how this can be done. Rox1p/Reo1p is a transcriptional repressor that acts globally to repress the expression of several nuclear *hypoxic* genes in aerobic yeast cells.[5,31,32] Some of its target *hypoxic* target genes encode proteins involved in the mitochondrial respiratory chain. Two of these are *CYC7* and *COX5b*. Both of these genes encode an isoform that has an aerobic counterpart, encoded by *CYC1* and *COX5a*, respectively. The aerobic and hypoxic isoforms are partially interchangeable. Rox1p/

[25] F. Sherman, *Methods Enzymol.* **350,** 3 (2002).
[26] V. Measday and P. Heiter, *Methods Enzymol.* **350,** 316 (2002).
[27] M. Fromont-Racine, J.-C. Rain, and P. Legrain, *Methods Enzymol.* **350,** 513 (2002).
[28] H. Endoh, S. Vincent, Y. Jacob, E. Real, A. J. M. Walhout, and M. Vidal, *Methods Enzymol.* **350,** 525.
[29] C. V. Lowry, *Proc. Natl. Acad. Sci. USA* **81,** 6129 (1984).
[30] C. E. Trueblood and R. O. Poyton, *Genetics* **120,** 671 (1988).
[31] J. J. M. TerLinde and H. Y. Steensma, *Yeast* **19,** 825 (2002).
[32] M. Becerra, L. J. Lombardia-Ferreira, N. C. Hauser, J. D. Hoheisel, B. Tizon, and M. E. Cerdan, *Mol. Microbiol.* **43,** 545 (2002).

Reo1p represses *CYC7* and *COX5b* in aerobic cells. Consequently, the level of respiration in strains that carry a wild-type copy of the *ROX1/REO1* gene is insufficient to support growth on a nonfermentable carbon source. Given this background, one can use the following strategy to isolate mutants in *ROX1/REO1* and other as yet unidentified genes involved in the aerobic repression of *hypoxic* genes. First, use a strain carrying a *null* mutation in either *CYC1* or *COX5a*. Second, mutate the strain. Third, select for cells that can grow on a nonfermentable carbon source, such as lactate or glycerol plus ethanol. This strategy requires that the genes for either *CYC7* or *COX5b* be expressed at high enough levels to support respiratory growth

Selections Used for Oxygen-Regulated Trancriptional Activators. The transcription factor Hap1p acts together with heme to activate the transcription of a number of *aerobic* yeast genes and may also function to activate the expression of a number of *hypoxic* genes.[3,32] The strategies used to identify *HAP1* provide a useful approach for selections of other genes required for the activation of either *aerobic* or *hypoxic* genes. Because Hap1p regulates both *CYC1* and *CYC7,* one can again use the ability of cells to grow on nonfermentable carbon sources, either lactate or glycerol, in a genetic selection. Cells carrying a deletion in *CYC1* can still express *CYC7,* but at levels that are insufficient to support growth on lactate. One can select for mutations that upregulate the expression of *CYC7*[33] and hence support growth on lactate containing media, using the same strategy described in the previous section. Alternatively, one can select for mutants that activate *CYC1* by using strains carrying a deletion in *CYC7,* screening for mutants that have a substantially reduced expression of *CYC1,* and selecting for those that act in *trans* and fail to grow on nonfermentable carbon sources.[34]

Promoter Resections of Oxygen-Responsive Genes. A powerful approach for identifying new *trans*-acting factors involved in oxygen-regulated gene expression is the use of promoter–reporter gene fusions to identify additional *cis* factors that mediate the effects of oxygen concentration. This approach is illustrated nicely by studies aimed at identifying the *cis* regulatory elements involved in the expression of *COX6,* a gene that is regulated by both oxygen and carbon sources.[35–37] This study made use of

[33] L. Clavilier, G. Pere-Aubert, and P. P. Slonimski, *Mol. Gen. Genet.* **104,** 195 (1969).
[34] L. Guarente, B. Lalonde, P. Gifford, and E. Alani, *Cell* **36,** 503 (1984).
[35] J. D. Trawick, C. Rogness, and R. O. Poyton, *Mol. Cell. Biol.* **9,** 5350 (1989).
[36] J. D. Trawick, N. Kraut, F. R. Simon, and R. O. Poyton, *Mol. Cell. Biol.* **12,** 2302 (1992).
[37] S. Silve, P. R. Rhode, B. Coll, J. Campbell, and R. O. Poyton, *Mol. Cell. Biol.* **12,** 4197 (1992).

COX6–lacZ promoter fusions. The authors used deletions, linker-scanning mutations, and synthetic constructs to modify the *COX6* promoter in order to localize important *cis* elements. A similar strategy has been used to identify the low oxygen response element (LORE) in *OLE1*.[38] It is important to stress the importance of using linker-scanning and other point mutations in addition to deletions for these studies. Deletions by themselves may give misleading information in these kinds of studies because of DNA context effects.

Acknowledgment

Some of the work reported in this paper was supported by Research Grants GM30228 and HL63324 from the National Institutes of Health.

[38] M. J. Vasconcelles, Y. Jiang, K. McDaid, L. Gilooly, S. Wrentzel, D. L. Porter, C. E. Martin, and M. A. Goldberg, *J. Biol. Chem.* **276,** 14374 (2001).

Section X

Physical Detection of Oxygen

[43]   Measuring Tissue PO$_2$ with Microelectrodes

By DONALD G. BUERK

Large O$_2$ cathodes were used for invasive physiological measurements in tissue over 60 years ago, based on the principle of electrochemical reduction. However, it soon became apparent that there could be serious errors,[1] including depletion of tissue O$_2$, polarographic current sensitivity to blood flow or convective mixing (stirring artifact), spatial averaging of the signal due to the tip size, and damage to the microcirculation or to individual cells during penetration so that the true tissue PO$_2$ may not be determined accurately. O$_2$ electrodes were miniaturized in several different laboratories in an attempt to resolve some of these problems. Measurements with excellent spatial and temporal resolution can be obtained with microelectrodes. However, as a consequence of miniaturization, polarographic currents are smaller, requiring careful electromagnetic shielding and the use of more sensitive electrometers. Smaller is not necessarily better, as it is still possible to have significant PO$_2$ measurement errors with some microelectrode designs. In addition, microelectrode fabrication techniques can be tedious and their fragility under experimental conditions can be frustrating. Nonetheless, PO$_2$ microelectrodes have proven to be extremely valuable tools for understanding O$_2$ metabolism and transport to different organs and tissues, and a variety of robust experimental methods have been developed to provide accurate, quantitative information, as described next.

O$_2$ Electrode Theory

The O$_2$ electrode became a widely used research and clinical sensor after U.S. Patent #2,913,386 was awarded to Clark, Jr., on Nov. 17, 1959 for his membrane-covered electrode design. Electrochemical theory, practical design considerations and construction materials, temperature sensitivity, instrumentation, and many other technical aspects of the Clark-type membrane-covered O$_2$ electrode are described in several books.[2-5] The basic

---

[1] I. A. Silver, *in* "Chemical Engineering in Medicine" (R. F. Gould, ed.), p. 343. Am. Chem. Soc., Washington, DC, 1967.

[2] I. Fatt, "Polarographic Oxygen Sensors." C. R. C. Press, Cleveland, OH, 1976.

[3] M. L. Hitchman, "Measurement of Dissolved Oxygen." Wiley, New York, 1978.

[4] J. P. Hoare, "The Electrochemistry of Oxygen." Wiley Interscience, New York, 1968.

[5] V. Linek, V. Vacek, J. Sinkule, and P. Benes, "Measurement of Oxygen by Membrane-Covered Probes." Ellis Horwood Ltd., Chichester, UK, 1988.

principles are also applicable to PO_2 microelectrodes. A theoretical analysis by Schneiderman and Goldstick[6] identified major criteria that are critical for optimizing the design of recessed cathodes, such as the recessed Au cathode PO_2 microelectrode developed by Whalen et al.[7] Previous mathematical models in the literature were often based on bare or membrane-covered etched needle-type microelectrodes since they were in common use. A major problem with bare electrodes is caused by the depletion of O_2 from the surrounding medium due to the electrochemical reduction reaction at the cathode. This can significantly lower O_2 tension in a relatively large region of tissue even if the tip is small and cause a large spatial averaging effect. Sometimes this region is referred to as the "catchment volume" for an electrode. Based on a bare spherical electrode model, it can be calculated that 95% of the O_2 diffusion field would be contained within ~6 times the tip radius. A membrane with low O_2 permeability and sufficient thickness can reduce the "catchment volume" and increase the accuracy of the measurement by limiting the diffusion field in the tissue surrounding the tip. In contrast, the mathematical model for recessed cathodes[6] predicts that O_2 diffusion fields are almost completely confined within the recess (length, L), especially when it is relatively deep compared to the cathode diameter (d). For example, when $L/d = 10$, 95% of the diffusion field is contained within the recess. The "catchment volume" for a recessed PO_2 microelectrode is much smaller than for a bare cathode with the same tip size. Even a shallow recess ($L/d = 1$) has a substantial effect, with 67% of the diffusion field contained within the recess. A membrane inside the recess can further reduce the diffusion field. Consequently, it is easier to control the diffusion field with a recessed cathode than a membrane-covered cathode.

Computer simulations by Schneiderman and Goldstick[6] examined how recessing affects current sensitivity and minimizes artifacts that can be encountered with O_2 microelectrodes. These include stirring artifacts due to convective mixing and calibration errors due to differences between diffusion (D) and solubility (S) coefficients for O_2 in tissue and the calibration fluid. The product of O_2 diffusion and solubility coefficients DS is the O_2 permeability of the tissue, which is also known as the Krogh diffusion constant. Generally, DS in tissue is less than DS in water so a lower current would be measured in tissue compared to the calibration current for the same PO_2 in an aqueous medium. Increasing L/d reduces calibration

[6] G. Schneiderman and T. K. Goldstick, J. Appl. Physiol. 45, 145 (1978).
[7] W. J. Whalen, J. Riley, and P. Nair, J. Appl. Physiol. 23, 798 (1967).

errors, but it might also be prudent to use a calibration fluid with D and S similar to tissue. Grote et al.[8] used a ethylene glycol, borax, and phosphate buffer solution to calibrate recessed microelectrodes in their rat brain studies. Crawford and Cole[9] filled their recesses with a hydrated polymer and reported improved performance from tests with agar gels and measurements in canine femoral arteries. There is a practical trade-off for increasing L/d or applying membranes with low O$_2$ permeability, as the current sensitivity will be lower and the time response will be slower. The model by Schneiderman and Goldstick[6] predicts that good current sensitivity, rapid time response, lack of stirring artifact, and minimal calibration errors can be achieved with tip diameters <5 μm and $L/d = 10$.

PO$_2$ Microelectrode Designs and Fabrication Methods

Microelectrodes for PO$_2$ measurements in blood and tissues have been fabricated from a variety of metals, including Pt, Au, stainless steel, and stainless steel coated with Au, Pb, and a 90% Pt 10% Ir alloy.[10] Au and Pt are the most common noble metals used for O$_2$ cathodes. Pt is used more commonly because it can be sealed more easily in glass than in Au. However, Au has a wider plateau on the polarographic current–voltage curve than Pt. Carbon cathodes can also reduce O$_2$, but carbon is rarely used for O$_2$ sensing. Often the O$_2$ sensitivity and best polarization voltage depend on how the carbon tip was prepared. Beran et al.[11] evaluated 4 different O$_2$ electrode cathode/anode combinations, including Au or Pt cathodes with Ag/AgCl anodes, a Ag cathode and a Pb/PbO$_2$ anode, and a Au cathode with a Cd anode. They tested 10 electrodes with each cathode/anode combination continuously for 80 h in flowing bicarbonate buffers in a tonometer. Cathodes were coated with 400- to 600-μm-thick polyhydroxyethyl methacrylate (HEMA, or Hydron) membranes. Anodes were placed externally in the buffer and not integrated into the sensor. O$_2$ electrodes with Au cathodes and Ag/AgCl anodes stabilized more quickly and had the least drift compared to the other cathode/anode combinations. The Pt cathode and Ag/AgCl anode O$_2$ electrodes had only slightly more drift, but the stability was much worse for the other two designs.

[8] J. Grote, O. Laue, P. Eiring, and M. Wehler, J. Autonom. Nerv. Sys. **57**, 168 (1996).
[9] D. W. Crawford and M. A. Cole, J. Appl. Physiol. **58**, 1400 (1985).
[10] D. B. Cater and I. A. Silver, in "Reference Electrodes" (J. G. Ives and J. G. Janz, eds.), p. 464. Academic Press, New York, 1961.
[11] A. V. Beran, G. Y. Shigezawa, D. A. Whiteside, H. N. Yeung, and R. F. Huxtable, J. Appl. Physiol. **44**, 969 (1978).

Recessed Metal Alloy Microelectrodes

Dowben and Rose[12] filled a glass microelectrode using an alloy of 50% In and 50% Sn with a relatively low temperature melting point (110°). The molten alloy was drawn into a small open tip glass pipette by suction and then allowed to cool. The exposed metal alloy tip was electroplated with a layer of Au, followed by a layer of Pt. However, this was a bare tip design and therefore subject to measurement errors. Whalen et al.[7] refined this method to produce recessed PO_2 microelectrodes with very small tips (<2 μm) and long tapers to minimize tissue damage during the penetration of vascular tissues. Earlier, Davies and Brink[13] used a glass-insulated recessed 25-μm-diameter Pt wire electrode with a relatively large (200 μm) tip for tissue PO_2 measurements in brain and other tissues. While accurate PO_2 measurements could be made with this electrode, the time response was very slow, the spatial resolution was poor, and the potential for damage to the microcirculation was much greater than for true microelectrodes.

A schematic drawing for the Whalen-type recessed PO_2 microelectrode design is shown in Fig. 1A (left) along with a photomicrograph of a microelectrode that has a tip diameter ~3 μm (right). The glass micropipette is filled with Wood's metal (50% Bi, 26.7% Pb, 13.3% Sn, 10% Cd), which has a low-temperature melting point (73–75°). Another metal alloy (44.7% Bi, 22.6% Pb, 19.1% In, 8.3% Sn, 5.3% Cd) with a lower melting point (47°) was used by Linsenmeier and Yancy.[14] This alloy has a lower thermal expansion coefficient than Wood's metal, and they observed that it caused fewer microfractures in the glass tips after cooling. Many alloys with other combinations of metals and different melting points are available commercially. Several different methods can be used to fill glass micropipettes with metal alloy. Whalen et al.[7] accomplished this by filling glass capillary tubes with molten Wood's alloy, which were then pulled out on a standard micropipette puller. This requires higher temperatures than empty glass capillary tubes. Usually, the glass micropipette is filled completely to the tip with alloy. Occasionally, gaps can form in the metal alloy during pulling, which can be removed by local heating. The tip can be beveled on a grinding wheel at this point to expose the metal. Some of the metal alloy must then be removed from the tip by electrochemical etching in cyanide solution using AC or DC voltage to form a recess approximately 20 to 50 μm deep. Good results have also been reported using slightly acidic 1 to 3 M $ZnCl_2$ solution for etching out metal alloy to form

[12] R. M. Dowben and J. E. Rose, *Science* **118,** 22 (1953).
[13] P. W. Davies and F. Brink, Jr., *Rev. Sci. Instrum.* **13,** 524 (1942).
[14] R. A. Linsenmeier and C. M. Yancey, *J. Appl. Physiol.* **63,** 2554 (1987).

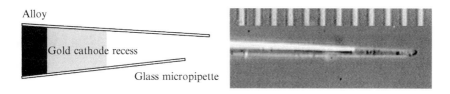

A. Whalen-type recessed tip PO$_2$ microelectrode

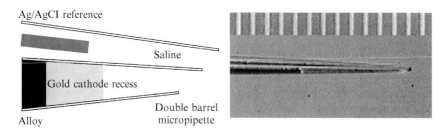

B. Double-barrel cell membrane potential/recessed tip PO$_2$ microelectrode

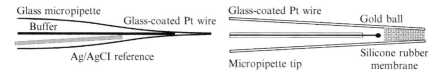

C. Miniaturized clark-type PO$_2$ microelectrode

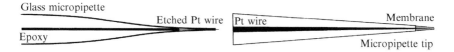

D. Etched needle-type PO$_2$ microelectrode

FIG. 1. Schematic drawings for different types of PO$_2$ microelectrodes (calibration bars with 10-μm spacing in photomicrographs).

deep recesses.[15] If small fragments of alloy are left in the recess after etching, they can be dissolved by immersing the tip in an acidic solution.

The author uses an alternate method to fill micropipettes. Molten alloy is vacuumed into small-bore (24 or 30 gauge) Teflon tubing, which is

[15] A. N. Bezbaruah and T. C. Zhang, *Water Res.* **36,** 4428 (2002).

stripped off with a razor blade after cooling. This creates alloy wires that can be inserted into the back of empty glass micropipettes with open, beveled tips. Applying force to the back of the wire and locally heating the wire at the leading end allows the molten alloy to be pushed upward into the tip. By removing the heat source quickly, it is possible to leave part of the tip empty to form a ~50-μm deep recess.

After a recess is formed, the alloy is then electroplated with either Au or Pt. The tip is submerged into the plating solution, allowing sufficient time to completely fill the recess. A vacuum can be applied to speed up this step. A 10- to 20-μm-thick layer of Au is plated onto the metal alloy at a negative potential of 0.7 to 1 V relative to a Au reference electrode. After terminating the electroplating, the final recess length should be at least 10 times the tip diameter. The tip is then soaked in distilled water to remove all plating solution. A hydrophobic membrane can be introduced into the recess by soaking in a dilute aqueous solution of Rhoplex polymer (Rohm and Haas) and then drying. Other membrane materials, such as collodion or polystyrene, have also been used. The author uses liquid Nafion polymer (5% wt) dissolved in aliphatic alcohols. Dry microelectrodes can be stored indefinitely, but the membrane must be rehydrated in saline or distilled water before use. As a matter of practice it is advisable to polarize the microelectrodes for several days prior to experimental use to ensure that they are stable and do not have excessive drift. Polarographic currents for recessed PO_2 microelectrodes with tip dimensions ~3 μm typically range between 10^{-11} and 10^{-10} A in room air-equilibrated saline. Zero currents in the absence of O_2 (background currents) are usually less than 1% of the room air calibration current. Response times for 90% completion of a step change in PO_2 are <100 ms.

Techniques for improved electromagnetic shielding have been investigated by physically surrounding the microelectrode with the anode, for example, by coating the outside with conductive Ag paint. Nair et al.[16] coated the outer glass surface of recessed PO_2 microelectrodes with Ag using metal vapor sputtering in a vacuum. The external Ag/AgCl coating could be used as a reference anode. They also electroplated iron on the outside of the tip, which could be electrodeposited into tissue for later histological identification of the measurement site using an appropriate stain. This technique was used to confirm which type of cells the microelectrode had taken measurements from in the cat carotid body.

[16] P. Nair, J. I. Spande, and W. J. Whalen, *J. Appl. Physiol.* **49**, 916 (1980).

Double-Barrel Microelectrodes

A double-barrel O_2 microelectrode was designed by Tsacopoulos and Lehmenkühler[17] by sealing an etched Pt wire in one glass barrel with wax. The second barrel was filled with physiological saline and connected to a Ag/AgCl wire. This double-barrel microelectrode was used for simultaneous measurements of PO_2 and bioelectrical activity in rat brain and in the honeybee retina. The metal alloy-recessed cathode PO_2 microelectrode has also been fabricated in a double-barrel configuration shown schematically in Fig. 1B (left) along with a photomicrograph of a double-barrel microelectrode that has a combined tip diameter <5 μm (right). Whalen *et al.*[18] used a spring-mounted, "floating" double-barrel microelectrode to measure simultaneous cardiac action potentials and tissue PO_2 in the beating cat heart. Linsenmeier and Yancey[14] used a similar double-barrel microelectrode for PO_2 and electrophysiological measurements in the cat eye. These types of double-barrel microelectrodes must be stored wet in the same solution that the second barrel is filled with. This can cause current leakage problems between the two channels if salt bridges form on the electrical connections at the top of the sensor.

Buerk and Riva[19,20] designed double-barrel-recessed electrodes for simultaneous PO_2 and NO measurements in the cat eye. Because the tip size was relatively large, our measurements were limited to the surface of the optic nerve head. Other combinations, such as pH or ion-sensitive barrels, could also be fabricated in combination with PO_2 microelectrodes, although this has not been pursued by many investigators.

Miniaturized Clark-Type PO_2 Microelectrodes

Revsbech[21] developed a miniaturized version of the Clark O_2 electrode to measure O_2 in marine sediments. Schematic drawings for this design are shown in Fig. 1C for the upper body (left) and tip (right). A 0.1-mm-diameter Pt wire was etched electrically in KCN solution to a tip diameter of 1–4 μm using 2–7 V AC voltage. The etched Pt wire was coated with soda-lime capillary glass that was melted in an electrically heated nichrome wire loop. The tip was ground on a rotating disc covered with 0.25-μm grain

[17] M. Tsacopoulos and A. Lehmenkühler, *Experientia* **33**, 1337 (1977).
[18] W. J. Whalen, P. Nair, and D. G. Buerk, *in* "Oxygen Supply: Theoretical and Practical Aspects of Oxygen Supply and Microcirculation of Tissue" (M. Kessler, D. F. Bruley, L. C. Clark, Jr., D. W. Lübbers, I. A. Silver, and J. Strauss, eds.), p. 199. Urban and Schwarzenberg, Berlin, Germany, 1973.
[19] D. G. Buerk and C. E. Riva, *Microvasc. Res.* **55**, 103 (1998).
[20] D. G. Buerk and C. E. Riva, *Microvasc. Res.* **64**, 254 (2002).
[21] N. P. Revsbech, *Limnol. Oceanogr.* **34**, 474 (1989).

diamond paste to remove excess glass and expose the metal. After grinding, the tip of the Pt wire was etched further to form a 5- to 10-μm deep recess. Au was electroplated into the recess using a 5% KAu(CN)$_2$ solution. A larger Au ball is sometimes plated onto the Pt tip to increase the cathode size. Finally, the micropipette tip was filled with a double membrane of collodion and polystyrene. Alternately, silicone rubber can be used as the membrane at the end of the micropipette. In some designs, the Ag/AgA1 anode is placed in the electrolyte in the upper body and a finer Ag wire "guard cathode" is positioned closer to the working Pt/Au cathode. Additional fabrication details for miniaturized Clark-type microelectrodes and a review of some applications in the environmental sciences are given by Lu and Yu.[22] Liss et al.[23] used this microelectrode design for tissue PO$_2$ measurements in the kidney. Tips are generally larger than the metal alloy-recessed microelectrode design, on the order of 5 to 10 μm, and the response times are a little slower, generally <2 s.

Etched Metal Needle-Type Microelectrodes

Glass-coated etched metal wire microelectrodes, shown schematically in Fig. 1D, can be more durable than metal alloy-filled glass micropipettes and have more rapid time responses, depending on the type of membrane and its thickness. Wolbarsht et al.[24] appear to be the first to use an electrochemical etching technique to produce metal microelectrodes with very fine points, which were used for electrophysiology measurements. Cater and Silver[10] etched stainless-steel needles, which were plated with Au, insulated with glass, and then covered with an O$_2$-permeable cellulose membrane for tissue PO$_2$ measurements. Wolbarsht et al.[24] etched Pt–Ir (70:30) wire using an AC voltage in a solution of 50% NaCN with 30% NaOH added to prevent the formation of cyanide gas. The bath was agitated during the initial etching stage at 6 to 10 V amplitude relative to a carbon rod reference electrode. The taper was controlled by adjusting the length of wire submersed in the bath. The final electropolishing step used a lower AC voltage amplitude of 0.8 V, without agitating the bath, to produce tip diameters <1 μm. A mechanical dipping process in combination with electrolysis can also be used to produce finely tapered metal microelectrodes. Grote et al.[8] modified the etched needle design using electrochemical etching after the tip was coated with glass to create a recessed cathode. The recess was then filled with a membrane.

[22] R. Lu and T. Yu, *J. Environ. Eng. Sci.* **1,** 225 (2002).
[23] P. Liss, A. Nygren, N. P. Revsbech, and H. Ulfendahl, *Pflüg. Arch.* **434,** 705 (1997).
[24] M. L. Wolbarsht, E. F. MacNichol, and H. G. Wagner, *Science* **132,** 1309 (1960).

Baumgärtl and Lübbers[25,26] used similar fabrication methods to fabricate coaxial needle-type PO$_2$ microelectrodes with an external Ag/AgCl coating. After etching, a layer of Au was electroplated onto the Pt tip. The Au layer enlarges the cathode surface and improves the current signal compared to the Pt surface. The shaft of the glass-coated Pt wire was then coated sequentially with thin layers of Ta, Pt, and then Ag. The Ta layer adheres to the glass surface and improves the coating of the intermediate Pt layer, which ensures that there is good electrical conductivity. The final thin layer of Ag/AgCl on the outside of the glass-coated Pt wire shields the microelectrode from electrical noise and serves as the reference anode. Problems with the etched needle microelectrode design include hydration and weakening of the glass insulation at the tip and loss of the membrane when penetrating tissues. If the membrane does not adequately limit diffusion to the tip, there may be significant calibration errors, which can lead to an underestimation of the absolute tissue PO$_2$ or overestimation due to stirring artifacts if the tip is subjected to convective mixing.

Hypodermic Needle Electrodes

Whalen and Spande[27] developed an O$_2$ electrode with either single or multiple glass-coated 5- to 10-μm-diameter Au wire cathodes sealed with epoxy in a 23-gauge stainless-steel hypodermic needle, which could be inserted into tissue. The Au cathodes were recessed back into the glass coating by electrochemical etching. The stainless-steel needle could be used as a reference anode without excessive drift or an external Ag/AgCl reference could be used. Similar needle-type electrodes with a single, unrecessed cathode were used clinically in several German hospitals and led to the commercial Eppendorf histograph system licensed to Helzel Medical Systems in Kaltenkirchen, Germany. The system has a new user interface and is available commercially as the Phoenix tissue hypoximeter. The hypodermic needle probe is programmed to advance a specified distance and then retreat, partially before measuring the local PO$_2$. This maneuver is thought to relieve any increase in interstitial pressure that may build up during penetration, which might compress the microcirculation. A series of profiles are obtained through the tissue, with the information outputted as PO$_2$ frequency histogram in 2.5-torr increments. The system includes a

[25] H. Baumgärtl and D. W. Lübbers, *in* "Oxygen Supply: Theoretical and Practical Aspects of Oxygen Supply and Microcirculation of Tissue," p. 130. University Park Press, Baltimore, MD, 1973.

[26] H. Baumgärtl and D. W. Lübbers, *in* "Polarographic Oxygen Sensors" (E. Gnaiger and H. Forstner, eds.), p. 37. Springer-Verlag, New York, 1983.

[27] W. J. Whalen and J. I. Spande, *J. Appl. Physiol.* **48**, 186 (1980).

tonometer chamber for calibrating the O_2 electrode. The Eppendorf electrode has been used in radiooncology studies at more than 70 sites worldwide since 1989 and has been referred to as the "gold standard" for measuring the oxygenation of cancer tumors.[28] Although the cathode is small (10–15 μm diameter), the needle housing the cathode is ~300 μm in diameter so it is not a true microelectrode. Mathematical modeling of the needle sensor[29,30] suggests that it does not give the true tissue PO_2 distribution primarily due to spatial averaging effects, which lead to an underestimation of the expected distributions that were used in the model. Although damage to the microcirculation was not considered in these models, this is likely to be another significant source of error.

Surface PO_2 Electrodes

The Clark O_2 electrode has been modified for transcutaneous and subcutaneous measurements by placing it on skin or on the surface of exposed tissues. The extensive literature on these measurements will not be reviewed here. In the case of heated transcutaneous sensors or pulsed oximeters, the measurement is intended for the continuous monitoring of arterial blood PO_2. However, there has been interest in a relatively noninvasive clinical measurement of subcutaneous PO_2 in wound healing and monitoring trauma or burn victims. Usually this is a single value reflecting complex geometry, flow, O_2 delivery, and usage underneath the sensor. The time trend is of clinical interest rather than an accurate depiction of the tissue PO_2 distribution. There has been interest in using arrays of cathodes for surface O_2 electrodes to sample more than one location. Kessler and Grunwald[31] developed a multiwire surface O_2 electrode, which was subsequently used in numerous animal experiments as well as for clinical studies. Both bare cathode and membrane-covered arrays have been used. Individual currents, usually from eight or more symmetrically spaced, 15-μm-diameter Pt cathodes were monitored. Frequency distribution histograms were generated for the tissue PO_2 values measured on the various organ surfaces that were studied. However, spatial averaging will also be a factor with these types of sensors and care must be taken that the sensor does not compress the tissue and impede blood flow. Also, attention must be paid to cathode spacing so that the "catchment volumes" do not significantly overlap and influence the measurements.

[28] H. B. Stone, J. M. Brown, T. L. Philips, and R. M. Sutherland, *Radiat. Res.* **136,** 422 (1993).
[29] I. Toma-Dasu, A. Waites, A. Dasu, and J. Denekamp, *Physiol. Meas.* **22,** 713 (2001).
[30] I. Toma-Dasu, A. Dasu, A. Waites, J. Denekamp, and J. F. Fowler, *Radiother. Oncol.* **64,** 109 (2002).
[31] M. Kessler and W. Grunwald, *in* "International Symposium on Oxygen Pressure Recording," p. 147. Karger, New York, 1969.

Fiber-Optic Probes

A number of different optical technologies have been developed to measure blood and tissue oxygenation,[32,33] which will not be reviewed here. However, it should be noted that the pioneering work of Opitz and Lübbers[34] laid the scientific foundation for fiber-optic systems that are now available commercially for clinical applications with single-use, disposable sensors. A combination fiber-optic probe for PO$_2$, PCO$_2$, and pH with a thermocouple for temperature was approved by the FDA in March 2000 (Neuro Trend, Diametrics Medical, Inc., St. Paul, MN) for monitoring brain tissue parameters. A brain tissue PO$_2$ and temperature monitor (LICOX, Integra LifeSciences, Plainsboro, NJ) was approved by the FDA in January 2001. The technology was developed by GMSmbH in Germany. Another system (OxyLite, Oxford Optronix Ltd., Oxford, UK) has been approved for clinical studies in Europe. The company is seeking FDA approval this year or next for a clinical instrument combining the O$_2$ sensor with fiber optics for laser Doppler flowmetry to evaluate blood flow in addition to tissue oxygenation. At the present time, the fiber-optic probe tip diameters are ~500 μm for the NeuroTrend, ~460 μm for the LICOX, and ~230 μm for the OxyLite (~500 μm when combined with laser Doppler fiber optics) and therefore cannot be classified as microelectrodes. However, because there is very little O$_2$ used by the optical methods, O$_2$ depletion by the sensor is not a factor. Also, the stability for optical sensors is excellent so that drift over long periods of time is superior to electrochemical sensors. The optical sensors may replace the conventional Clark electrode in many applications. However, sensor size is still an issue with regard to damage when inserted into tissue, and the sensor signal represents a spatial average over a relatively large area.

Applications

In Vivo Tissue PO$_2$ Distributions

Measurements in different organs in animals and human tissues from numerous laboratories using microelectrodes or small O$_2$ sensors clearly demonstrate that tissue PO$_2$ is heterogeneous. Consequently, it is difficult to compare findings using different types of sensors. One comprehensive

[32] D. G. Buerk, "Biosensors: Theory and Applications." Technomic Publishing Co., Inc., Lancaster, PA, 1993.
[33] D. G. Buerk, in "Handbook of Chemical and Biological Sensors" (R. F. Taylor and J. S. Schultz, eds.), p. 83. Institute of Physics Publishing, Bristol, UK, 1996.
[34] N. Opitz and D. W. Lübbers, Int. Anesthesiol. Clin. 25, 177 (1987).

study compared the distributions of murine fibrosarcoma and normal subcutis tissue PO_2 measured by the Eppendorf histograph method in six different laboratories.[35] They concluded that these tissue PO_2 measurements were useful, despite notable differences in frequency distributions and median PO_2 between laboratories, but recommended caution in interpreting PO_2 data.

We reported very different results for Eppendorf and recessed PO_2 microelectrode measurements in untreated and vascular endothelial grow factor (VEGF)-transfected human melanoma (NIH1286) xenografts grown in athymic nude mice.[36] Tissue PO_2 frequency distributions had significantly higher values when measured with recessed PO_2 microelectrodes. Mean and median tissue PO_2 were 5.6 and 4.0 torr, respectively, in untreated tumors and 19.4 and 13.7 torr, respectively, in the hypervascular VEGF-transfected tumors. Eppendorf measurements were markedly lower, with 90% of PO_2 values <2.5 torr for the untreated tumors and a median PO_2 of 2.5 torr for the VEGF-treated group. Because the Eppendorf needle is much larger ($\sim$300 μm diameter) compared to the recessed PO_2 microelectrode tip (<5 μm), we suspected that it caused much more damage to the tumor microcirculation.

Braun et al.[37] compared recessed PO_2 microelectrode and optical sensor (OxyLite) measurements in tumors and normal tissues. They demonstrated with a simple model that the larger "measurement volume" of the optical sensor ($\sim$230 μm diameter) truncates upper and lower tissue PO_2 frequency distribution values compared to PO_2 measurements with microelectrodes. The larger fiber-optic probe consistently measured a significantly higher fraction of hypoxic PO_2 values compared to microelectrode data. This may reflect damage to the microcirculation rather than spatial averaging errors. Another laboratory compared OxyLite and Eppendorf PO_2 measurements in murine tumors[38] and found that the fraction of hypoxic values <2.5 torr were 50 and 69%, respectively, with the Eppendorf sensor reporting significantly more hypoxic values. Another comparison of OxyLite and Eppendorf PO_2 measurements in P22 rat carcinosarcoma reported no statistically significant differences in median PO_2

[35] M. Nozue, I. Lee, F. Yuan, B. A. Teicher, D. M. Brizel, M. W. Dewhirst, C. G. Milross, L. Milas, C. W. Song, C. D. Thomas, M. Guichard, S. M. Evans, C. J. Koch, E. M. Lord, R. K. Jain, and H. D. Suit, *J. Surg. Oncol.* **66**, 30 (1997).

[36] G. M. Polin, T. W. Bauer, D. G. Buerk, D. A. Hsi, I. Prabakaran, C. Menon, and D. L. Fraker, *Proc. Wangensteen Surg. Forum* **55**, 296 (2000).

[37] R. D. Braun, J. L. Lanzen, S. A. Snyder, and M. W. Dewhirst, *Am. J. Physiol. Heart Circ. Physiol.* **280**, H2533 (2001).

[38] D. R. Collingridge, M. K. Young, B. Vojnovic, P. Wardman, E. M. Lynch, S. A. Hill, and D. J. Chaplin, *Radiat. Res.* **147**, 329 (1997).

or hypoxic fractions for the two sensors.[39] A discrepancy between LICOX and Eppendorf measurements in tumors was reported by another laboratory, who concluded that the Eppendorf histograph method was more suitable for making tissue PO_2 measurements.[40]

It should be recognized that tissue PO_2 is not the whole story when trying to characterize the metabolic status and adequacy of O_2 transport to tissue. Blood flow, capillary density, capillary hematocrit, tissue O_2 diffusion (D) and solubility (S) coefficients, O_2 consumption (Q), and Michaelis–Menten kinetics (K_m for O_2 dependent metabolism) are also key parameters that are usually not available to fully interpret experimental or clinical tissue PO_2 measurements. Also, there is not a good understanding of what adequate O_2 levels are in different organs and tissue, as there are complex molecular events with transcriptional, posttranscriptional, and posttranslational genetic changes that are associated with lack of O_2. There may not be a clearly defined "critical tissue PO_2" to identify when the tissue is hypoxic, and many other factors may need to be taken into account.

Analysis of Steady-State PO_2 Gradients

When a one-dimensional tissue PO_2 gradient can be measured experimentally, an analysis of the PO_2 profile may be possible using simple O_2 diffusion and consumption models. Differential equations for mass transport, general solutions for one-dimensional gradients, and mass fluxes (J) for these simple cases are summarized in Table I based on the assumption that Q, D, and S are homogeneous and constant in tissue. Constants a and b for the general solutions can be determined from boundary conditions for the experiments. Typically, these include continuity of mass flux and partial pressure at a specified boundary. Measured tissue PO_2 gradients can be fit directly to the predicted gradients (shown in Fig. 2 for the three basic geometries) using least-squares minimization methods or plotted in transformed coordinates where a linear regression can be performed. Only the lumped parameter ratio (Q/DS) can be determined from the measured PO_2 gradients, and other methods are needed to directly or indirectly determine individual values for Q, D, or S in tissue, as described later.

As an example, *in vitro* tissue PO_2 measurements from brain and liver slices[41] were fit to the planar model. As illustrated in Fig. 2 (left), only a

[39] B. M. Seddon, D. J. Honess, B. Vojnovic, G. M. Tozer, and P. Workman, *Radiat. Res.* **155**, 837 (2001).

[40] W. Kehrl, C. Sagowski, S. Wenzel, and F. Zywietz, *Laryngorhinootologie* **80**, 318 (2001).

[41] D. G. Buerk and I. S. Longmuir, *Microvasc. Res.* **13**, 345 (1977).

TABLE I
SIMPLE ONE-DIMENSIONAL DIFFUSION MODELS WITH CONSTANT O_2 CONSUMPTION

Differential equation	PO_2 gradient	Flux
Planar model $DS\frac{d^2P}{dx^2} - Q = 0$	$P(x) = a + bx - \frac{Q}{2DS}x^2$	$J(x) = -(bDS - Qx)$
Cylindrical model $\frac{DS}{r}\frac{d}{dr}\left(r\frac{dP}{dr}\right) - Q = 0$	$P(x) = a + b\log(r) - \frac{Q}{4DS}r^2$	$J(r) = -\left(\frac{bDS}{r} - \frac{Q}{2}r\right)$
Spherical model $\frac{DS}{r^2}\frac{d}{dr}\left(r^2\frac{dP}{dr}\right) - Q = 0$	$P(x) = a + \frac{b}{r} - \frac{Q}{6DS}r^2$	$J(r) = -\left(-\frac{bDS}{r^2} - \frac{Q}{3}r\right)$

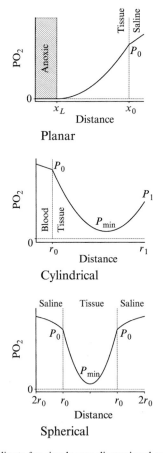

FIG. 2. Predicted PO_2 gradients for simple one-dimensional geometries used in Table I.

finite amount of tissue can be oxygenated and deeper locations would be anoxic for a thick slice of tissue. The maximum depth of tissue that can be supplied with O$_2$ (x_L) can be calculated from

$$x_L = \sqrt{\frac{P_0 DS}{Q}} \qquad (1)$$

where P_0 is the PO$_2$ at the surface of the slice ($x_0 = 0$). The extent of oxygenation can also be determined for the other geometries. We found that our data were not described adequately by the planar model, as experimental PO$_2$ profiles[41] extended deeper into the slice than expected for constant Q (zero-order reaction). A Michaelis–Menten model for O$_2$ consumption provided a better fit, and the maximum Q and K_m were determined from the optimum curve fit.[42] *In vivo* examples of simple diffusion models include PO$_2$ microelectrode measurements across avascular regions of the vascular wall[43,44] and the eye,[45,46] which have been fit to analytical solutions for one or two layer planar models with constant Q/DS in each layer. Models for the vascular wall were also developed for cylindrical geometry[43] (with gradients similar to the middle panel of Fig. 2). A model with spherical geometry (with gradients similar to the right panel of Fig. 2) was used by Mueller-Klieser[47] to analyze PO$_2$ measurements in multicellular, spherical tumors.

Information can also be derived from PO$_2$ gradients measured in stagnant saline above a tissue slice, above a confluent monolayer of cultured cells, near an individual cell, or around blood vessels. Differential equations, analytical solutions, and fluxes for all three one-dimensional cases can be defined by setting $Q = 0$ in the equations summarized in Table I. For the planar model, the relationship between the Krogh constant in tissue $(DS)_t$ relative to $(DS)_s$ in saline can be estimated from the PO$_2$ gradient $(\Delta P/\Delta x)$ in the stagnant boundary layer above the slice from

$$(DS)_t/(DS)_s = \frac{\Delta P}{\Delta x}\frac{2x_L}{P_0} \qquad (2)$$

Mueller-Klieser[47] measured PO$_2$ gradients in unstirred boundary layers of growth media next to multicellular, spherical tumors and also obtained

[42] D. G. Buerk and G. M. Saidel, *Microvas. Res.* **16,** 391 (1978).
[43] D. G. Buerk and T. K. Goldstick, *Am. J. Physiol. Heart Circ. Physiol.* **243,** H948 (1982).
[44] S. M. Santilli, A. S. Tretinyak, and E. S. Lee, *Am. J. Physiol. Heart Circ. Physiol.* **279,** H1518 (2000).
[45] R. A. Linsenmeier and L. Padnick-Silver, *Invest. Ophthalmol. Vis. Sci.* **41,** 3117 (2000).
[46] S. J. Cringle and D.-Y. Yu, *Comp. Biochem. Physiol. A* **132,** 61 (2002).
[47] W. Mueller-Klieser, *Biophys. J.* **46,** 343 (1984).

(Q/DS) from PO_2 gradients measured within the tumor. These two pieces of information allowed separate calculations for Q and the Krogh constant $(DS)_t$ in the tumor.

A method for accurately measuring O_2 fluxes near individual cells is described by Land et al.[48] They designed a mechanical stepper motor system to repetitively move a recessed PO_2 microelectrode back and forth at a programmed step size. Inward O_2 fluxes were determined in the external culture medium near 50- to 100-μm-diameter neurons that had been dissected from the abdominal ganglion of *Aplysia californica*. The microelectrode position was changed back and forth every 3 s with a step size of 5 μm, and 50 data points were signal averaged to determine the flux at different positions relative to the neuron surface. Variations in O_2 flux were quantified after altering Q with different drugs. Outward O_2 fluxes were also measured from photosynthetic chloroplasts during changes from light to dark. This "self-referencing" method can be considered a version of scanning electrochemical microscopy (SECM), where a sensor is programmed to move systematically either horizontally or vertically over a surface. This technique has been used with nanometer scale Pt microelectrodes to obtain a topographical map of O_2 generated by photosynthesis over a 120×120-μm section of a leaf.[49] Shiku et al.[50] used a SECM technique with a PO_2 microelectrode to characterize O_2 consumption of single bovine embryos during different stages of growth after fertilization. Their experimental PO_2 measurements were found to be in good agreement with the theoretical gradients predicted from spherical diffusion in the culture medium. An *in vivo* example is the determination of O_2 fluxes out of retinal arterioles and into some venules due to countercurrent exchange in the cat eye from PO_2 gradients measured by recessed PO_2 microelectrodes in the vitreous humor immediately adjacent to small blood vessels using an analysis based on cylindrical geometry.[51]

Determining Q from Transient Measurements

Following World War I, Otto Heinrich began investigating how O_2 is consumed by living cells. His research led to the identification of the role of respiratory cytochromes. In 1931, Warburg was awarded the Nobel Prize for Physiology or Medicine for his research on respiratory enzymes. He also investigated photosynthesis and was the first to observe that the growth of

[48] S. C. Land, D. M. Porterfield, R. H. Sanger, and P. J. S. Smith, *J. Exp. Biol.* **202**, 211 (1999).
[49] S. Mancuso, G. Papeshi, and A. M. Marras, *Planta* **211**, 384 (2000).
[50] H. Shiku, T. Shiraishi, H. Ohya, T. Matsue, H. Abe, H. Hoshi, and M. Kobayashi, *Anal. Chem.* **73**, 3751 (2001).
[51] D. G. Buerk, R. D. Shonat, and C. E. Riva, *Microvasc. Res.* **45**, 134 (1993).

malignant cells requires markedly smaller amounts of O$_2$ than normal cells. He introduced the use of manometry as a means of studying O$_2$ consumption by slices of tissue or suspensions of cells. The manometric technique measures the amount of gas being consumed or evolved by determining the change in the pressure in a closed system while maintaining a constant temperature and volume based on ideal gas laws. This type of apparatus is known as the Warburg respirometer and has since been modified with O$_2$ electrodes to measure the fall in PO$_2$ after closing the system.

Several methods have been developed for measuring O$_2$ disappearance rates to determine Q. The simplest analysis, which can be used to determine Q for stirred suspensions of cells or mitochondria in a known volume of aqueous medium, is given by

$$Q = S\frac{dP}{dt} \tag{3}$$

where S is the O$_2$ solubility of the medium. The mass or volume of O$_2$-consuming cells or mitochondria must also be known precisely. The Michaelis–Menten constant (K_m) can also be characterized by determining the PO$_2$ where dP/dt decreases to 50% of the initial rate. There is some uncertainty in this analysis, as there could be stagnant boundary layers around the particles or very steep PO$_2$ gradients between the medium and the center of the cells or mitochondria.

We have used recessed PO$_2$ microelectrodes to measure *in vivo* O$_2$ disappearance rates (dP/dt) after stopping blood flow in gerbil brain[52] and cat carotid body[53,54] to determine Q and K_m. An equivalent solubility based on a blood and tissue compartment model was developed to calculate Q, taking into account the additional amount of O$_2$ carried in the bloodstream. In a simplified form, the model is given by

$$Q = (S + \beta)\frac{dP}{dt} \tag{4}$$

where S is the O$_2$ solubility in tissue and β is a nonlinear function involving the relative sizes of the two compartments, hematocrit in the microcirculation, and the slope of the oxyhemogobin equilibrium curve.[55] The blood supply to the carotid body was isolated via a silicon rubber tubing loop, which could be clamped. A bilateral carotid artery occlusion method can

[52] D. G. Buerk and P. Nair, *J. Appl. Physiol.* **74**, 1723 (1993).
[53] D. G. Buerk, P. K. Nair, and W. J. Whalen, *J. Appl. Physiol.* **67**, 60 (1989).
[54] D. G. Buerk, P. K. Nair, and W. J. Whalen, *J. Appl. Physiol.* **67**, 1578 (1989).
[55] D. G. Buerk, P. K. Nair, E. W. Bridges, and T. R. Hanley, *Adv. Exp. Med. Biol.* **200**, 151 (1986).

be used to stop blood flow in the gerbil brain, as there are no collateral blood vessels from the basilar arteries. In our study of the cortex and hippocampus of the aging gerbil brain,[52] we found minimal changes in brain tissue PO_2 distributions with age, which was consistent with the reduction in Q with age that we quantified. We have also conducted numerous *in vitro* studies of neural activity, tissue PO_2, and Q using the O_2 disappearance technique in saline-perfused cat and rat carotid bodies,[56–58] which do not require the correction for O_2 carried by blood, and recently confirmed that nitric oxide inhibits carotid body Q.[59]

Determining D from Transient Measurements

Takahashi *et al.*[60] used a conventional Clark electrode to determine D from studies with a flat tissue (rabbit cornea) analyzing the time course of PO_2 change at the bottom of the cornea after covering the top of the O_2 electrode with the tissue sample. When the steady state was reached, the top surface of the cornea was then covered with an impermeable barrier (glass coverslip), and Q was determined from the rate of O_2 disappearance [Eq. (3)] based on an assumed value for S in corneal tissue. Mahler[61,62] and Evans *et al.*[63] used this method to determine D for tissue slices by measuring the change in PO_2 on one side following a step change in PO_2 on the other side. Dowse *et al.*[64] described the kinetics in terms of resistance and capacitance in an electrical circuit analogy for diffusion.

The model for a step change at one boundary is shown schematically in Fig. 3. The initial steady state gradient across the slice (bottom curve) is given by

$$P(x) = P_1 - \frac{Q}{2DS}x^2 \tag{5}$$

In order for this solution to be valid, the slice must be thin enough that none of the tissue is anoxic so that $P_3 > 0$ is on the opposite side for the initial PO_2 (P_1) before the step change. An analytical solution for the

[56] D. G. Buerk, R. Itturiaga, and S. Lahiri, *J. Appl. Physiol.* **76**, 1317 (1994).

[57] S. Lahiri, D. G. Buerk, D. K. Chugh, S. Osanai, and A. Mokashi, *Brain Res.* **684**, 194 (1995).

[58] S. Lahiri, D. G. Buerk, S. Osanai, A. Mokashi, and D. K. Chugh, *Brain Res.* **66**, 1 (1997).

[59] D. G. Buerk and S. Lahiri, *in* "Oxygen Sensing: Molecule to Man" (S. Lahiri, N. Prabhakar, and R. E. Forster, II, eds.), p. 337. Plenum Press, New York, 2000.

[60] G. Takahashi, I. Fatt, and T. K. Goldstick, *J. Gen. Physiol.* **50**, 317 (1966).

[61] M. Mahler, *J. Gen. Physiol.* **71**, 533 (1978).

[62] M. Mahler, C. Louy, E. Homsher, and A. Peskof, *J. Gen. Physiol.* **86**, 105 (1985).

[63] N. T. S. Evans, P. F. D. Naylor, and T. H. Quinon, *Respir. Physiol.* **43**, 179 (1981).

[64] H. B. Dowse, S. Norton, and B. D. Sidell, *J. Theor. Biol.* **207**, 531 (2000).

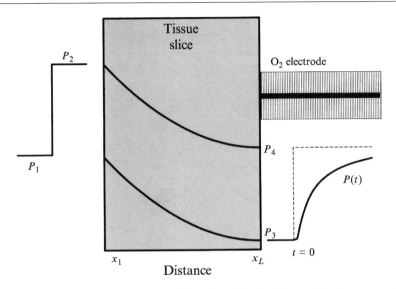

FIG. 3. Schematic drawing of an experimental method for determining tissue properties from transient PO$_2$ changes measured after a step change in PO$_2$ at the opposite surface.

PO$_2$ transient $P(t)$ at the bottom of the slice (thickness x_L) is given by an infinite series of exponentials

$$P(t) = P_1 + (P_2 - P_1)\left[1 - \frac{2}{\pi}\sum_{n=0}^{\infty}\frac{(-1)^n}{n+\frac{1}{2}}e^{-(n+\frac{1}{2})^2\left(\frac{\pi}{x_L}\right)^2 Dt}\right] - \frac{Q}{2DS}x_L^2 \quad (6)$$

where P_1 is the initial and P_2 is the final PO$_2$. D can be calculated from the PO$_2$ transient using a few terms in the series or estimated from a single exponential ($n = 0$) for data at longer time. The final steady-state gradient (top curve in Fig. 3) is given by

$$P(x) = P_2 - \frac{Q}{2DS}x^2 \quad (7)$$

Weind et al.[65] used this technique to determine Q and D for aortic valve cusps. However, their results suggested that Q and D were probably not uniform across the tissue sample. It is not possible to evaluate spatial variations in Q and D with these methods because the model assumes that the tissue properties are uniform and homogeneous. It might be possible

[65] K. L. Weind, D. R. Boughner, L. Rigutto, and C. G. Ellis, Am. J. Physiol. Heart Circ. Physiol. 281, H2604 (2001).

to approach this problem by measuring D with PO_2 microelectrodes positioned at known depths in a slice of tissue by analyzing the time course of tissue PO_2 change following a step change in PO_2 at one surface. An average value for D in the tissue between the boundary and the microelectrode tip can be determined from the PO_2 transient if the depth is known accurately.

An alternative method, exploiting the properties of bare or very shallowly recessed microelectrodes, can be used to measure D with a detailed spatial resolution in tissue. We analyzed transient polarographic currents measured after turning the microelectrode on (step change in voltage from 0 to -0.7 V) to determine D in the vascular wall.[66] "Turn-on" transients were measured at different locations in the vascular wall and compared with transient measurements in stagnant saline above the vascular strip, as illustrated in Fig. 4. The principle is based on the time course for the O_2 diffusion field to spread outward as O_2 is depleted from tissue (or saline) around the tip after the electrode is polarized. At longer times, the transient approaches the normalized current for a bare spherical cathode

$$\frac{I(t)}{I_{ss}} = 1 + \sqrt{\frac{t_c}{t}} \qquad (8)$$

The steady-state current is

$$I_{ss} = \pi n F D S r_0 PO_2 \qquad (9)$$

where F is the Faraday constant and $n = 4$ electrons transferred with the electrochemical reduction of one O_2 molecule. The time constant is

$$t_c = \frac{r_0^2}{\pi D} \qquad (10)$$

for a bare spherical cathode with radius r_0.[67] Analytical solutions for planar electrode geometry are discussed by Buerk.[32]

Differences in D between tissue and saline were determined by comparing the time constants. The normalized transient currents can be plotted with respect to $1/t^{0.5}$ to determine t_c at different locations (x_1, x_2) from linear regression analysis, as illustrated by the bottom graph in Fig. 4. Because the current transient reaches steady state rapidly, a series of repeated measurements can be obtained in a relatively short time period to improve the accuracy of the method by signal averaging. We found that D was significantly lower ($\sim 24\%$ of D in water) on the intimal side compared to the adventitial side (37%).[66] This "turn-on transient" method is not valid in

[66] D. G. Buerk and T. K. Goldstick, *Ann. Biomed. Eng.* **20,** 629 (1992).
[67] H. A. Laitinen and I. M. Kolthoff, *J. Am. Chem. Soc.* **61,** 3344 (1939).

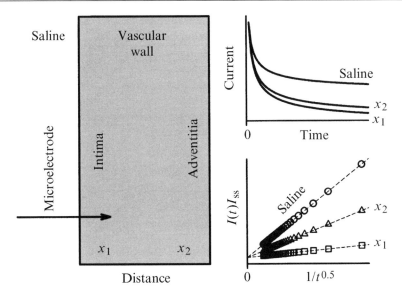

FIG. 4. Schematic drawing of an experimental method for determining O$_2$ diffusion coefficient D from transient currents (top graph) after turning the microelectrode on in saline and at different locations in the vascular wall based on a comparison of time constants estimated from transformed data (bottom graph).

flowing solutions or in blood perfused tissues, as the current transient would be influenced by convection. Also, the response during the early part of the transient departed from Eq. (8). More recently, Zoski and Mirkin[68] used numerical methods to solve the general transient diffusion problem for a cone-shaped tip and reported that their computer simulations were very close to an earlier empirical model for a disk-shaped tip by Shoup and Szabo.[69] The empirical equation adds an exponential term to Eq. (8), which would improve the description of the current during the initial part of the transient and might yield a better estimate for t_c to calculate D.

Determining S from Transient Measurements

S can be determined from a tissue sample if respiration is completely arrested, for example, by freezing the tissue. After thawing, the tissue can be completely deoxygenated or equilibrated with a known partial pressure of O$_2$ and then immersed in a closed system with a precisely known volume

[68] C. G. Zoski and M. V. Mirkin, *Anal. Chem.* **74**, 1986 (2002).
[69] D. Shoup and A. Szabo, *J. Electroanal. Chem.* **140**, 237 (1982).

of water or saline that is equilibrated at a higher PO_2. The PO_2 in the external medium will decline until a new equilibrium is reached. This can be measured by an O_2 electrode in the closed chamber, allowing calculation of S from the amount of O_2 entering the sample. Errors can result from O_2 consumed by the electrode or tissue edema associated with cell damage after freezing.

Validation of In Vivo Optical Measurements

The noninvasive optical O_2 phosphorescence quenching technique provides information about intravascular PO_2, as the dye is bound to albumin and usually remains in the bloodstream until excreted by the kidney.[70] Intravascular PO_2 distributions should be right shifted compared to tissue PO_2 distributions measured by recessed PO_2 microelectrodes, but a comparison of the two techniques has not been done. In tumors, which are known to have leaky blood vessels, it appears that the dye diffuses out into tissue and allows optical measurements of tissue PO_2 to be made.[71] This has also been observed to occur in avascular regions of hamster skinfold measured in a closed dorsal chamber.[72] We have confirmed the accuracy of phosphorescence quenching measurements in tissue from recessed PO_2 measurements obtained in a narrow optical window around the tip.[73] The dorsal chamber was opened and avascular sites were chosen for tissue PO_2 measurements, with no visible blood vessels in the center of the region. No significant difference was found between the mean PO_2 measured with recessed PO_2 microelectrodes and the mean tissue PO_2 determined from the time course of phosphorescence decay. Paired measurements were made over a wide range, from 2 to 46 torr, with good correlation between the two methods ($r = 0.93$).

Clinical Measurements

There have been relatively limited clinical applications with true PO_2 microelectrodes, mostly with etched metal needle-type microelectrodes. Ehrly[74] reviewed tissue PO_2 measurements of human skeletal muscle in the lower limbs of patients with intermittent claudication, where tissue

[70] W. L. Rumsey, J. M. Vanderkooi, and D. F. Wilson, *Science* **241,** 1649 (1989).
[71] I. P. Torres-Filho, M. Leunig, F. Yuan, M. Intaglietta, and R. K. Jain, *Proc. Natl. Acad. Sci. USA* **91,** 2081 (1994).
[72] I. P. Torres-Filho, H. Kerger, and M. Intaglietta, *Microvasc. Res.* **51,** 202 (1996).
[73] D. G. Buerk, A. G. Tsai, M. Intaglietta, and P. C. Johnson, *Microcirculation* **5,** 219 (1998).
[74] A. M. Ehrly, *in* "Drugs and the Delivery of Oxygen to Tissue" (J. S. Fleming, ed.), p. 1. C. R. C. Press, Boca Raton, FL, 1990.

PO_2 was found to be lower than in muscles of a control group of healthy volunteers. Histograms were determined by tabulating over 400 values while withdrawing the microelectrode over a 10-mm distance. The effectiveness of various drug treatments to increase muscle tissue PO_2 in intermittent claudication patients was also investigated. Fleckenstein et al.[75] used a hypodermic needle type electrode to measure skeletal muscle PO_2 in healthy volunteers and in critically ill patients and demonstrated that dopamine infusions could increase the tissue PO_2. In another clinical application, Sakaue et al.[76] used a small polarographic O_2 electrode during vitreous surgery to measure aqueous humor O_2 tension in the human eye. Austin et al.[77] used a 25-μm-diameter, Teflon-coated platinum wire "microelectrode" to measure brain tissue PO_2 at a depth of approximately 2 mm in the human cortex. This size was too large to be considered as a true microelectrode, and they did not attempt to calculate absolute PO_2 values. They did show that brain tissue PO_2 levels were improved after microsurgery to create anastomoses from the superficial temporal artery in patients with symptoms of transient ischemic episodes or with incomplete strokes. Many other clinical studies used larger surface PO_2 electrodes for a wide range of problems.

More recent clinical applications have shifted toward commercially available sensors. Many clinical studies have been conducted with the Eppendorf system, especially for evaluating tumor oxygenation. For example, Collingridge et al.[78] used the Eppendorf histograph system to measure tissue PO_2 in human gliomas and surrounding brain tissue while patients were under local sedation or deeper general anesthesia. Tumor tissue PO_2 values tended to be higher in patients with local sedation. Using the NeuroTrend system, Charbel et al.[79] were able to monitor brain tissue PO_2, PCO_2, and pH for 7 to 10 days in a group of 10 intensive care patients diagnosed with subarachnoid hemorrhage. In this group, 3 patients developed vasospasm during monitoring. Although tissue PO_2 did not change in these patients, PCO_2 was elevated and pH decreased significantly. Although this is a small sample, perhaps it demonstrates that there may be some clinical situations where tissue PO_2 is not the most useful indicator.

[75] W. Fleckenstein, K. Reinhart, T. Kersting, R. Dennhardt, A. Jasper, C. Weiss, and K. Eyrich, Adv. Exp. Med. Biol. 180, 609 (1984).

[76] H. Sakaue, Y. Tsukahara, A. Negi, N. Ogino, and Y. Honda, Jpn. J. Ophthalmol. 33, 199 (1989).

[77] G. Austin, G. Haugen, and J. LaManna, in "Oxygen and Physiological Function" (F. F. Jöbsis, ed.), p. 531. Prof. Inform. Library, Dallas, TX, 1977.

[78] D. R. Collingridge, J. M. Piepmeier, S. Rockwell, and J. P. S. Knisely, Radiother. Oncol. 53, 127 (1999).

[79] F. T. Charbel, X. Du, W. E. Hoffman, and J. I. Ausman, Surg. Neurol. 54, 432 (2000).

The combined laser Doppler blood flow and LICOX PO_2 probe appears to be well suited for monitoring the viability of reconstructive surgery flaps, according to a 3-year study by Kamolz et al.[80] They did not find any false positives or negatives in evaluating vascular patency in 60 tissue transfers to head and neck, trunk, and upper and lower extremities. The combination laser Doppler blood flow and OxyLite system has been used in animals to investigate blood O_2 level-dependent (BOLD) contrast-based functional magnetic resonance imaging (MRI)[81] of tumors and might be a way to help in quantifying the analysis of BOLD fMRI signals for future clinical applications. However, the time response of fiber-optic O_2 sensors might be a limitation for resolving temporal changes in tissue PO_2. Using laser Doppler flowmetry with rapid-responding recessed PO_2 microelectrodes, Ances et al.[82] were able to resolve the elusive "initial dip" in rat somatosensory cortex tissue PO_2 that occurs in the ~1-s time interval prior to the increase in blood flow after functional activation of this brain region following the onset of electrical stimulation to the forepaw. An "initial dip" in blood O_2 has been observed in animal experiments using optical methods by some, but not all, investigators. There will continue to be a need to validate various sensor technologies in animal models to establish the potential for clinical use, and there is no question that studies with invasive PO_2 microelectrodes will continue to contribute toward efforts to better characterize how well the spatially averaged signals from larger probes reflect more detailed heterogeneous tissue PO_2 distributions.

Future Directions

There are promising new technologies under development in micromachining, electrochemical microscopy, biomaterials, and biosensors, as well as a movement toward chemical sensing at nanometer scales. These research efforts could lead to further advancements in physiological monitoring, not only for O_2, but other substrates of interest. There was some initial development of sensors using semiconductor photolithographic techniques by Prohaska et al.,[83] and a multiple PO_2 and temperature probe fabricated on a glass needle was used experimentally in rat brain by LaManna et al.[84]

[80] L. P. Kamolz, P. Giovanoli, W. Haslik, R. Koller, and M. Frey, *J. Reconstr. Microsurg.* **18,** 487 (2002).
[81] C. Baudelet and B. Gallez, *Magn. Reson. Med.* **48,** 980 (2002).
[82] B. M. Ances, D. G. Buerk, J. H. Greenberg, and J. A. Detre, *Neurosci. Lett.* **306,** 106 (2001).
[83] O. J. Prohaska, F. Olcaytug, P. Pfunder, and H. Dragaun, *IEEE Trans. Biomed. Eng.* **33,** 223 (1986).
[84] J. C. LaManna, K. A. McCracken, M. Patil, and O. J. Prohaska, *Metab. Brain Dis.* **4,** 225 (1989).

to study the effects of direct electrical stimulation. The glass substrate was 100 μm thick and 500 μm wide, with a knife-like shape and sharp point for penetrating tissue. O$_2$ electrodes were fabricated with a 20-μm-diameter Au cathode and a larger Ag/AgCl anode in a small enclosed structure on the glass substrate with a 20-μm-diameter opening. The multisensor probe was initially recognized as a potentially useful research tool in the neurosciences, but did not become a viable commercial venture. Similar technology is being applied toward biosensor development by several other laboratories so the potential exists today for further development of this concept. Also, ceramic substrates are being used to fabricate smaller and more robust probes, and arrays of chemical sensors can be fabricated on small electronic chips.

There has been strong interest in the development of electrochemical sensors with nanometer-sized tips for SECM and other electrochemical techniques. New materials for electrode surfaces or that coat conventional metal or carbon surfaces, such as diamond nanoparticles or carbon nanotubes, are currently under investigation. For these applications, bare surfaces are preferred so that the "catchment volume" extends out into the sample that is being scanned. The extremely small tip allows studies of ultrafast electrochemical kinetics as well as superb spatial resolution. New insulating materials are being investigated as an alternative to glass, such as dip coating with insulating polymers or electrochemical coating with poly(acrylic acid) anodic paint. A new type of combination optical and electrochemical probe with an extremely fine tip size has been described by Lee and Bard.[85] This probe is made by stripping the polymer coating off a quartz glass optical fiber (3.7 μm diameter core, with 125-μm cladding) and then pulling the quartz glass out to form a fine tip using a micropipette puller with a high-temperature laser heat source (Sutter Instruments, Novato, CA). The tapered fiber is then coated with Au by vacuum evaporation and is tested for gaps in the coating by shining laser light through the probe. Anodic electrophoretic paint is applied for insulation and is then heat cured. The curing process pulls a small amount of the insulation away, exposing only the extreme end of the tip. The exposed Au near the tip is used as a ring-type microelectrode. Lee and Bard[85] demonstrated that this new design allows simultaneous optical and electrochemical measurements. As these new ways to fabricate micro- and nanosensors are developed, there will undoubtably be new tools available for biomedical and clinical research.

It is reasonable to expect that existing commercial systems will seek methods to further miniaturize their invasive devices or to incorporate

[85] Y. Lee and A. J. Bard, *Anal. Chem.* **74,** 3626 (2002).

more types of optical or electrochemical sensors into the same probe size. Continued advancements in noninvasive techniques are also expected, and clever experiments to validate these methods using more refined invasive methods will be needed to drive the process forward.

Dedicated to the memory of William J. Whalen (1915–2003).

[44] Identifying Oxygen Sensors by Their Photochemical Action Spectra

By DAVID F. WILSON

Introduction

Cellular oxygen sensing takes place through molecules that interact with oxygen and generate a signal that can be used as a measure of the oxygen concentration. In response to that signal, the cells alter their properties and/or generate extracellular signals that can be "read" by other cells and tissue. The overall process is referred to as oxygen sensing if the signal is a highly selective measure of oxygen pressure and a response elicited by the signal that is physiologically appropriate to the sensed level of oxygen, for example, increased or decreased erythropoetin production, breathing rate, and vascular resistance. Thus, oxygen sensors are defined not by their interaction with oxygen *per se,* but by their regulatory role and the "downstream" messenger system that transmits the oxygen measurement into a physiological response. Cells contain many metabolic systems that use oxygen but only a few of these "measure" changes in oxygen pressure and provide signals used to regulate tissue responses to these changes.

One of the most important roles for heme in cells is in the active site of proteins that bind and/or metabolize oxygen. These include cytochrome oxidase (mitochondrial oxidative phosphorylation), cytochrome P450 (drug metabolism), nitric oxide synthase (NO synthesis), guanylate synthase (c-GMP synthesis), etc. In heme oxidases and heme oxygenases, where oxygen is consumed in the reaction, the CO compounds are inhibitory because they decrease the effective concentration of the enzyme available for reaction with oxygen. For heme proteins that do not catalyze an oxygen-consuming reaction (such as hemoglobin, myoglobin, guanylate cyclase), carbon monoxide may still bind to iron of the heme. For hemoglobin and myoglobin, this decreases the ability to store/transport oxygen, whereas for guanylate cyclase and other regulatory sites it may act as either an agonist or an antagonist of oxygen. Depending on the properties of the reaction site, binding CO may either induce a conformation change similar to that for oxygen (therefore acting as an agonist) or it may bind without inducing such a change and, while it is present in the site, prevent oxygen binding (antagonist). A generic representation for the reaction of CO with a heme protein (Fe^{2+}), including the photodissociation by light (L), is

$$Fe^{2+} + O_2 \underset{k_{-1}}{\overset{k_1}{\rightleftharpoons}} Fe^{2+} - O_2 + \text{substrate} \xrightarrow{k_3} Fe^{3+} + \text{oxidized substrate} \quad (1)$$

$$Fe^{2+} + CO \underset{k_{-2}}{\overset{k_1}{\rightleftharpoons}} Fe^{2+} - CO \quad (2)$$

$$Fe^{2+} - CO + L \xrightarrow{k_{-2}*} Fe^{2+} + CO \quad (3)$$

Further treatment has been simplified by assuming k_3, the rate constant for the formation of product from the Fe^{2+}–O_2 complex, is much smaller than k_{-1}.

Photochemical Action Spectra

When CO is bound to many heme proteins, absorption of a photon results in energy being transmitted to the CO–iron bond and dissociation of the CO[1-5] as shown in Eq. (4). The light introduces a second mechanism for dissociation of CO, a second-order reaction between the CO compound and a photon with a rate constant $k_{-2}*$. The rate constant is a function of the absorption of the CO compound at that wavelength times the probability that the excited state will decay by dissociation of the CO. The overall dissociation of CO has a rate constant that is the sum of the rate constants $(k_{-2} + Lk_{-2}*)$. Thus, the dissociation constant (K_d) for dissociation of CO from the Fe^{2+}–CO compound can be written as

$$\frac{(k_{-2} + Lk_{-2}*)}{k_2} = \frac{[Fe^{2+}][CO]}{[Fe^{2+} - CO]} = K_d \quad (4)$$

Binding of CO can be quite high affinity, and for reduced cytochrome c oxidase, the value is approximately 0.4 μM.[6] When light of a wavelength absorbed by the CO compound is applied, the apparent dissociation constant

[1] J. Haldane and J. L. Smith, *J. Physiol.* **20**, 497 (1896).
[2] O. Warburg, *Biochem. Z.* **177**, 471 (1926).
[3] O. Warburg and E. Negelein, *Biochem. Z.* **193**, 339 (1928).
[4] O. Warburg and E. Negelein, *Biochem. Z.* **214**, 64 (1929).
[5] L. N. Castor and B. Chance, *J. Biol. Chem.* **217**, 453 (1955).
[6] J. G. Lindsay and D. F. Wilson, *FEBS Lett.* **48**, 45 (1974).

increases with increasing light intensity (photon flux) as indicated in Eq. (4). At sufficiently high intensity, CO binding is effectively eliminated.

Light as a Chemical Reactant

There are several very experimentally useful properties of light as a reactant.

1. Only photons absorbed by the Fe^{2+}–CO compound can contribute to the experimentally measured response to light. This makes the method highly selective and independent of any other pigments and/or reactants, such as the other cellular pigments, that may be present in the sample. Although the other pigments absorb light, the photons absorbed do not affect the experimentally measured response.

2. The effective light flux can be changed from zero to saturation within milliseconds. This permits change very rapidly in the amount of bound CO at any fraction saturation of the site. The result is an effective mechanism for modifying the CO effect, in the limiting case providing an on/off switch that is completely under experimental control.

3. The intensity of the light can be varied and when this is done the result is similar to varying the concentration of other reactants. Increasing the photon flux gives saturation behavior (Michaelis–Menten kinetics) with a response that is nearly linear below about 50% saturation.

4. The effective K_M for photon flux measured for the light of different wavelengths is directly proportional to the probability of photon absorption (absorption coefficient) by the Fe^{2+}–CO compound. Thus, when illuminating with light of different wavelengths, as long as the photon fluxes less than half saturate the response, the response is proportional to the absorption coefficient of the CO compound at each wavelength. Conversely, if the photon flux necessary for the same response at each wavelength is determined, then a plot of the photon flux vs wavelength is the inverse of the absorption spectrum of the Fe^{2+}–CO compound.

The CO:O_2 Partition Coefficient

Consistent with competitive binding of the oxygen and CO, the relative efficacy of CO as a competitive inhibitor of oxygen can be characterized by a "partition coefficient"[7] (K).

$$K = [a/(1 - a)]*(pCO/pO_2) \qquad (5)$$

[7] O. Warburg, *Biochem. Z.* **189**, 354 (1927).

The activity in the absence of CO and the activity in the presence of CO are both divided by the activity in the absence of CO, giving values of 1 and a, respectively. The pressures of CO and O_2 are represented as pCO and pO_2, respectively. As indicated by Equation (5), the effect of CO is dependent on the ratio of pCO to pO_2. The same degree of the CO effect, such as a selected decrease (such as 50%), can be attained at a variety of pCO values, as long as the pO_2 is varied to keep the ratio constant. This indicates that O_2 and CO compete for the same site on the enzyme. As long as both inhibitor and substrate concentrations are well above the K_M, the inhibition is a function of the concentration ratio. In this case, the "partition coefficient" is the ratio of the inhibitor constant to the substrate K_M, and the partition coefficient (pCO/pO_2) is a measure of the relative efficacy of CO and O_2 in competing for a reaction site. Because CO binds to the iron of the heme only when it is reduced and the K_M for oxygen can be dependent on enzyme "turnover," the partition coefficient can vary widely with the experimental conditions. Warburg[7] reported a value of 9–11 for oxygen consumption in yeast mitochondrial cytochrome c oxidase but noted that this was dependent on the substrate being oxidized. Values of 1–9 have been reported for microsomal P450, with the value increasing with increasing saturation of the enzyme system with reducing enzyme.[8,9]

Relationship of Absorption Spectra and Photochemical
    Action Spectra

The binding of CO to the iron of hemoproteins results in the absorption spectrum characteristic of that hemoprotein changing to that of the CO compound. The photochemical action spectrum is identical to the absorption spectrum, and the question arises as to why it is important to measure the action spectrum. The answer is in the specificity conferred by the function used to measure action spectra. Biological samples typically contain more than one hemoprotein that can bind CO. Each of these has a characteristic absorption spectrum and many of the CO compounds can be dissociated by light. The aspects of photochemical action spectra that make them an extraordinary powerful research tool include the following.

1. Photochemical action spectra are measured through the effect of the light on a reaction, whether it be oxygen consumption, drug hydroxylation, neural activity, and so on, that is a specific biochemical/physiological response. Thus, only the CO compound directly involved in that specific

[8] R. Estabrook, *Biochem. Z.* **338,** 741 (1963).
[9] D. Y. Cooper, H. Schleyer, and O. Rosenthal, *Ann. N.Y. Acad. Sci.* **174,** 205 (1970).

response contributes to the action spectrum. In contrast, absorption spectra are measured changes in the total number of photons absorbed. In biological samples containing both hemoglobin and mitochondria, for example, CO-induced absorption changes are the sum of those for CO compounds of both hemoglobin and cytochrome a_3, as well as any other hemoproteins that are present. The photochemical action spectrum, however, is measured by the light-induced reversal of inhibition of the rate of oxygen consumption by CO. Because dissociation of the hemoglobin–CO compound does not affect the rate of oxygen consumption, the action spectrum is the absorption spectrum of the cytochrome a_3–CO complex, the one that influences the rate of oxygen consumption.

2. The action spectrum requires only dissociation of sufficient CO complex to allow the measured reaction to proceed, and this may be only a small fraction of the total CO compound. The photochemical action spectrum Warburg attributed to "atmungsferment" was recognized by Keilin and Hartree to be similar to the absorption spectrum of the cytochrome a_3–CO complex.[10,11] However, light intensities that fully reversed the inhibition of oxygen consumption by CO did not significantly diminish the strength of the absorption band. As a result, many workers were hesitant to conclude that they were the same entity. This apparent discrepancy arose because the unliganded heme produced by dissociation of the CO by light can react with either O_2 or CO. The "on" rate constant for the reaction with O_2 is $4 \times 10^8\ M^{-1}\ s^{-1}$, whereas that for CO is $8 \times 10^4\ M^{-1}\ s^{-1}$.[12] As a result, even when the oxygen concentration is 10-fold less than that of CO, 99.5% of the cytochrome a_3 from which CO is photodissociated reacts with O_2 and only 0.5% reacts with CO. In addition, there is a very high activity of the cytochrome oxidase (turnover number per cytochrome a_3 of about $5000\ s^{-1}$) relative to the overall respiratory chain activity (less than $20\ s^{-1}$). These factors combine to restore the respiratory capacity of mitochondria by photodissociation of only a very small fraction of the total cytochrome a_3–CO complex.

Experimental Protocols

The first application of the photochemical action spectrum was for identifying the enzyme responsible for oxygen consumption (respiration) by cells.[2–4,7] During the period prior to 1960, oxygen consumption was

[10] D. Keilin, in "The History of Cell Respiration and Cytochrome," Chap. 7, p. 117. Cambridge Univ. Press, 1966.

[11] D. Keilin and E. F. Hartree, Proc. Roy. Soc. B. **127,** 167 (1939).

[12] M. Erecinska and D. F. Wilson, Pharmacol. Ther. **8,** 1 (1980).

typically measured manometrically, that is, by the volume of gas lost from
the flask. A small amount (2–3 ml) of a suspension of the cells or of tissue
homogenates was placed in a glass reaction flask (15 ml) with a glass stopper. A small volume of alkaline solution, usually KOH, was placed in a central well where it was exposed to the atmosphere within the flask. The
remaining volume in the flask was filled with an appropriate gas mixture
and the flask was sealed with a glass stopper attached to a mercury manometer. The vials were then incubated in constant temperature water baths
with shaking to facilitate the exchange of CO_2 and oxygen between the
sample suspension and the atmosphere.

Carbon dioxide produced by metabolism was absorbed by the alkaline
solution in the central well so the change in gas volume was entirely due to
oxygen consumption. The mercury manometer was used to measure the
change in gas volume at constant pressure, providing very accurate measurements of the amount of oxygen removed from the gas. In this experimental system, photochemical action spectra were obtained by including
CO in the gas and illuminating the sample through a glass window in the
bottom of the water bath (Figs. 1 and 2).

Subsequent to the work of Warburg and co-workers, similar experimental methods were used by several workers to identify the primary

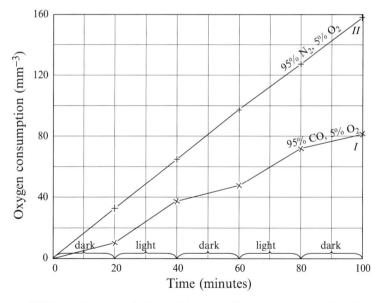

FIG. 1. Inhibition of yeast respiration and its reversal by light as measured by the rate of
oxygen consumption (taken with revision from Warburg[2]).

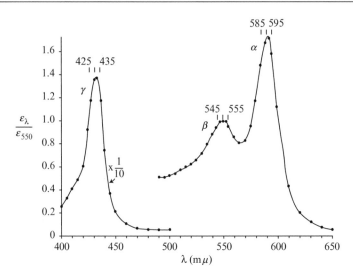

FIG. 2. Photochemical action spectrum for yeast respiration taken from Castor and Chance[5] with the light response presented relative to that at 550 nm. This spectrum is similar to those reported by Kubowitz and Haas[13] and others.

oxygen-consuming oxidases of various tissues[14–16] and many different organisms.[10,17] In mammalian tissues, photochemical action spectra for the primary oxidase are consistent with those of the mitochondrial cytochrome a_3–CO complex. In microorganisms, photochemical action spectra have been used to identify a number of oxidases dependent on the organism and growth conditions. Inhibition of drug hydroxylation by CO and its reversal by light showed a maximal sensitivity at 450 nm, identifying cytochrome P450 as the mixed function oxidase responsible for hydroxylation of a wide range of drugs.[8] The presence of a heme protein that reacted with CO to give an absorption spectrum with a maximum at 450 nm was first observed when extracts of liver (microsomes) were treated with reducing agent and CO.[18,19] It was the spectrum of the light-induced reversal of CO inhibition, however, that identified cytochrome P450 as the primary hydroxylase of drug metabolism.

[13] F. Kubowitz and E. Haas, *Biochem. Z.* **255,** 247 (1932).
[14] J. L. Melnick, *J. Biol. Chem.* **141,** 269 (1941).
[15] J. L. Melnick, *J. Biol. Chem.* **146,** 385 (1942).
[16] H. Laser, *Biochem. J.* **31,** 1677 (1937).
[17] L. N. Castor and B. Chance, *J. Biol. Chem.* **234,** 1587 (1959).
[18] D. Garfinkel, *Arch. Biochem.* **77,** 493 (1958).
[19] M. Klingenberg, *Arch. Biochem.* **75,** 376 (1958).

Experimental Design and Equipment

1. A high-intensity light source. A light source with a relatively uniform light output over the visible range (400 to 650 nm) provides the most versatile source. Tungsten-halide lamps are very good because their spectral output rises steadily from shorter to longer wavelengths and there are no significant peaks or troughs in the profile. The luminosity (photons per mm^2 surface) is lower than for electrical arc lamps, however, and must be used if intensities are required that are above those available from tungsten-iodide lamps. Arc lamps with medium to high internal gas pressure are preferred because the higher plasma pressure substantially broadens the emission lines. Whatever the light source, the variation in intensity with wavelength should be measured and appropriate compensation made. Whatever the light source, care must be taken to remove the near-infrared light (heat) to the point that there is no change in sample temperature when the light is applied. This can be accomplished by a variety of methods either singly or in combination as required for the lamp being used. Heat-absorbing filters include infrared-absorbing glass, gold mirrors, and copper sulfate solution.

2. A mechanism for obtaining high-intensity monochromatic light with bandwidths less than about 20 nm. This application can tolerate higher stray light than normal photometric measurements so there are a number of suitable choices, ranging from high-efficiency grating monochromators to specialized interference filters such as the Schott Veril linear variable interference filter (Edmund Industrial Optics, Barrington, NJ). Sets of interference of appropriate wavelengths may also be used[8] but this is less versatile than continuously variable monochromators. In this application the absolute intensity of the light is not important as long as it is sufficient to provide a substantial reversal of the CO effect. Typically, however, higher-intensity lamps are better to have more intensity than necessary and then to attenuate it as desired.

3. An accurate method for measuring light intensity at different wavelengths. There are a number of accurate commercial devices for measuring light and these are preferred. A silicon photodiode and a good millivoltmeter will, however, suffice for this application. The photocurrent generated by photodiodes as a function of wavelength is highly reproducible and most suppliers provide detailed information on the spectral responsivity for their particular photodiode.

4. A mechanism for precise attenuation of the intensity of the illumination light. This is necessary to compensate for the large differences in absorption of the CO compounds at different wavelengths. Light absorption, and therefore the functional response, ranges from zero at

wavelengths where there is no absorption by the CO compound to very high near the absorption maxima. This is best done in two stages, a diaphragm or rotating "venetian blind" louver[5] for continuously varying the intensity followed by, or preceded by, a holder for neutral density filters. The former is continuously variable, but cannot be used readily to make precise and reproducible intensity changes, whereas the latter gives precise and highly repeatable steps of attenuation.

5. A mechanism for conducting the monochromatic light to the sample. The wide variety of possible samples and sample chambers precludes recommending a single method for illuminating the sample. The wavelengths of light used (400 to 650 nm) are very efficiently carried through liquid-filled light guides, which can readily be equipped with lenses for illumination of the sample. The liquid-filled light guides have significantly higher transmission than the glass fiber bundles (0.7 vs 0.5). This can be of significant advantage when the light is limiting, as is often the case.

Identifying the Oxygen Sensor of the Carotid Body

The carotid body (CB) is a small organ of neural origin that lies at the bifurcation of the carotid artery and is supplied with blood from the carotid artery. It has the physiological role of measuring O_2 and CO_2 pressure in the blood in the carotid artery. Changes in the CO_2 and/or oxygen pressure in the arterial blood are detected, and the resulting signal alters the electrical activity of the afferent sinus nerves for the control of breathing and other functions. Sensing of CO_2 is independent of that for oxygen, although both sensory systems modulate the afferent neural activity. The afferent neural activity increases with decreasing oxygen pressure and increasing carbon dioxide. The time for changes in oxygen pressure pO_2 or pCO_2 resulting in altered breathing rate is short (seconds) (Fig. 3).

The carotid body has been studied extensively, and the use of inhibitors and other methods has resulted in a wide range of proposals for the identity of the oxygen sensor. These have included mitochondrial cytochrome oxidase,[20,21] an oxygen-sensitive K^+ channel,[22,23] and NADH oxidase,[24]

[20] E. Mulligan, S. Lahiri, and B. T. Storey, J. Appl. Physiol. **51**(2), 438 (1981).

[21] D. F. Wilson, A. Mokashi, D. Chugh, S. A. Vinogradov, S. Osanai, and S. Lahiri, FEBS Lett. **351**, 370 (1994).

[22] J. López-Barneo, R. Montoro, P. Ortega-Sáaenz, and J. Ureña, in "Oxygen Regulation of Ion Channels and Gene Expression" (J. López-Barneo and E. K. Weir, eds.), p. 127. Futura, Amonk, NY, 1998.

[23] J. López-Barneo, R. Pardal, and P. Ortega-Sáenz, Annu. Rev. Physiol. **63**, 259 (2001).

[24] L. He, J. Chen, B. Dinger, K. Sanders, K. Sundar, J. Hoidal, and S. Fidone, Am. J. Physiol. Cell Physiol. **282**(1), C27 (2002).

Fig. 3. A flow chart for oxygen sensing by the carotid body.

as well as other entities.[25–28] The development of an isolated perfused carotid body preparation opened the way for a number of experiments that were not possible in animals *in vivo*. In this isolated preparation, the carotid body is removed still attached to the carotid artery and with a length of sinus nerves attached. Electrodes can be placed against the sinus nerve bundle to measure the afferent neural activity. Thus, the physiologically relevant afferent nerve activity can be measured without many of the experimental restrictions encountered in the intact animal (Fig. 4).

In intact animals, the addition of even low levels of CO results in its binding to hemoglobin. This removes the oxygen-carrying ability of hemoglobin and profoundly depresses oxygen delivery to all tissues, including the carotid body. The isolated organ, however, can be perfused with hemoglobin-free physiological saline and the addition of CO does not affect delivery of oxygen to the tissue. Although the oxygen-carrying capacity of saline is low relative to blood when both are equilibrated with the same oxygen pressure, oxygen delivery can be increased by equilibration with gases containing greater than 21% oxygen and/or by an increased perfusion rate. Figure 5 shows simultaneous measurements of the afferent nerve activity of an isolated perfused/superfused cat carotid body and the oxygen pressure in the carotid microcirculation measured by imaging the phosphorescence of a phosphor dissolved in the perfusate.[29] The perfusate flow was stopped at time zero and remained off until the afferent activity reached maximum (about 65 s). Perfusion was then restored, and the neural activity decreased again rapidly to control values. During perfusion with the air-saturated saline solution, oxygen levels in the microcirculation of the carotid body were maintained at near physiological levels as indicated by both direct measurement (right ordinate) and low afferent neural

[25] C. Eyzaguirre and H. Koyano, *J. Physiol.* **178**(3), 410 (1965).

[26] C. Eyzaguirre and H. Koyano, *J. Physiol.* **178**(3), 385 (1965).

[27] C. L. Eyzaguirre and H. Koyano, *J. Physiol.* **178**, 485 (1965).

[28] T. Streller, C. Huckstorf, C. Pfeiffer, and H. Acker, *FASEB J.* **16**(10), 1277 (2002).

[29] W. L. Rumsey, R. Iturriaga, D. Spergel, S. Lahiri, and D. F. Wilson, *Am. J. Physiol.* **261**, C614 (1991).

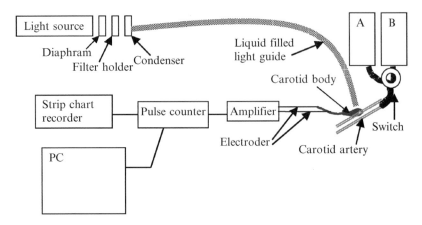

Fig. 4. A schematic of the experimental apparatus used for measurements in the isolated perfused/superfused carotid body. The carotid body is placed in a small chamber in a temperature-controlled water bath and is superfused with Tyrode solution equilibrated with a gas phase containing nitrogen (estimated O_2, 10–20 mm Hg). The carotid artery is perfused with Tyrode solution equilibrated with either a gas mixture containing % argon (A) or one containing the same oxygen pressure but consisting of air and CO (B). The medium perfusing the carotid artery (which supplies the carotid body) can be switched between two solutions. The two electrodes measure afferent electrical pulses by differential amplification of the measured potential, resulting in an electrical "spike" as each impulse passes down the fiber. Afferent activity is measured by counting the number of impulses per second.

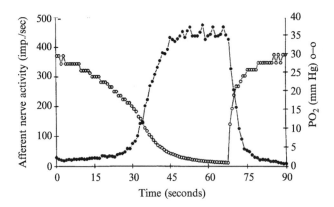

Fig. 5. The dependence of afferent neural activity of the isolated, perfused, superfused rat carotid body on the oxygen pressure in the microvasculature. The experimental conditions are given in Rumsey et al.[29] The carotid body was superfused with a physiological Tyrode solution equilibrated with a low oxygen gas (argon) and perfused with the same solution equilibrated with air containing 1% bovine serum albumin and 3 μM Pd-meso-tetra-(4-carboxyphenyl) porphyrin. At time zero the perfusate flow was stopped and remained off for 65 s. It was then restarted and the flow returned to normal. Oxygen was measured by the oxygen-dependent quenching of phosphorescence using an intensified CCD camera.

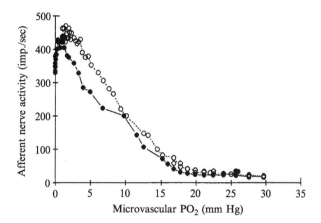

FIG. 6. Afferent neural activity plotted against microvascular PO_2 is plotted for two independent experiments. Conditions are the same as in Fig. 5.

activity. The afferent activity increases severalfold as the oxygen pressure in the equilibrating gas is lowered or the rate of perfusion is decreased, consistent with oxygen-sensing responses observed *in vivo*.

Afferent neural activity is plotted in Fig. 6 against the microvascular oxygen pressure. The neural activity begins to rise as the microvascular oxygen pressure falls below about 20 mm Hg and reaches maximal values many times normoxia by 1–2 mm Hg (these are averaged values so there are regions the tissue are anoxic). With prolonged time at very low oxygen pressure, the activity again falls. The decrease in afferent activity with approach to anoxia is progressive with increasing time of exposure and is reversed by again increasing the oxygen pressure. Thus it is likely due to the requirement for oxidative energy metabolism in order to maintain the neural activity.

The measured responses to light can best be measured using steady-state changes, that is, leaving the light on until a new steady state is obtained. In the case of the isolated perfused carotid body, however, this would mean a relatively long period for each measurement (several seconds to reach the steady state followed by many seconds for recovery). Because there is significant degradation of the response over periods of many minutes, it is important to minimize the time necessary per measurement if substantial fractions of the spectrum are to be measured in a single experiment. An illumination period of about 6 s was found to give a near maximal response, indicating that the mechanism controlling the oxygen response has a response time of 2–3 s. When the light was then switched

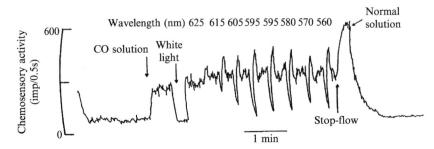

Fig. 7. Reversal of the CO-induced increase in the oxygen-dependent sinus nerve activity of the carotid cat body effect by high-intensity white light and moderate-intensity monochromatic light of different wavelengths. After an initial period of perfusion (38°) with medium equilibrated with air, the perfusate was changed to one equilibrated with a gas mixture containing O_2:CO at 130 and 560 mm Hg, respectively (CO solution). The CO-induced increase in neural activity was completely eliminated by illuminating the carotid body with a bright white light but only while the light was on. The carotid body was then illuminated for 6-s intervals with monochromatic light separated by a recovery period. The wavelength of the monochromatic light was varied from 625 to 560 nm to show the wavelength dependence of the reversal. From Wilson et al.,[21] with modification.

off, the control produced an "overshoot" in which the afferent neural activity rises transiently to well above the dark value and then returns. As shown in Fig. 7, the response is the same whether measured by the light-induced decrease in neural activity or by the total light/dark excursion. The latter is larger and less affected by normal fluctuations in the activity, properties that can be used to advantage when the measured changes are small. Using a short illumination period maximized the number of different wavelengths of light that could be measured in an experiment to the point that it was possible to do a complete spectrum with about 10-nm resolution in a single experiment.

The wavelength dependence of light-induced alterations in neural activity for a typical experiment are plotted in Fig. 8. The responses are uncorrected for the differences in light intensity at the different wavelengths, but clearly show maxima near 440 and 590 nm. The intensity at each wavelength was measured using a 1-cm-diameter silicon photodiode and these were used to correct the responses to all at the same light quantum flux (same number of light quanta per second) and the corrected values plotted in Fig. 9. The peak positions were determined in separate experiments in which a series of measurements were made near the peak position using 2-nm intervals in wavelength. The peak positions of 432 ± 2 and 590 ± 2 nm, as well as the fact that the 432-nm peak is six to nine times as strong as the 590-nm peak, are typical of the CO compound of mitochondrial cytochrome c oxidase (cytochrome a_3). This is the only known pigment

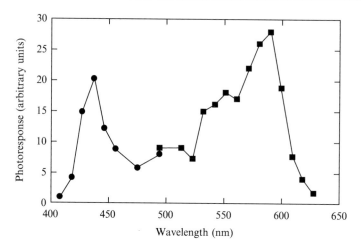

FIG. 8. Wavelength dependence of the light-induced reversal of the CO effect before correcting for the difference in light flux at the different wavelengths. From Wilson et al.,[21] with modification.

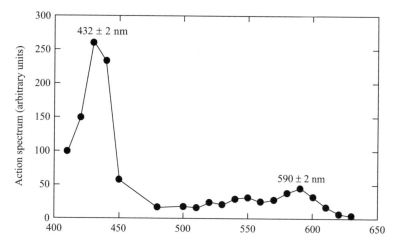

FIG. 9. Wavelength dependence of the light-induced reversal of the CO effect after correction to the same light flux for all wavelengths. From Wilson et al.,[21] with modification.

with these spectral properties, and the response time of 2–3 s is typical of that for cellular energy metabolism. Thus oxygen sensing by the carotid body is a function of mitochondrial oxidative phosphorylation with cytochrome a_3 acting as the oxygen sensor.

Conclusion

Photochemical action spectra can be used to identify heme proteins responsible for specific metabolic/physiological functions. This method is limited to a selected class of heme proteins that bind CO as an inhibitor/agonist competitive with oxygen and for which the CO compound is light sensitive. Within this class, however, the specificity conferred by measuring the light-induced reversal of the effect of CO allows unambiguous identification of the heme protein responsible for the measured function.

Acknowledgment

Supported in part by Grants NS-31465 and HL-43413 from the U.S. National Institutes of Health.

[45] Fabrication and *In Vivo* Evaluation of Nitric Oxide-Releasing Electrochemical Oxygen-Sensing Catheters

By Megan C. Frost and Mark E. Meyerhoff

Introduction

Considerable research effort has focused on developing implantable chemical sensors capable of continuous, real-time monitoring of physiologically important analytes such as O_2, CO_2, and glucose *in vivo*. A major challenge to obtaining reliable analytical results from such indwelling sensors is the biological response elicited by the sensor itself.[1] Catheter-type sensors implanted within veins and arteries will encounter several processes that can lead to errant analytical results. Upon exposure to blood, proteins such as fibrinogen and von Willebrand's factor can adsorb on onto the surface of the sensor.[2] These proteins provide sites for platelet adhesion. Platelet adhesion and subsequent activation can accelerate the coagulation cascade, leading to a network of fibrin and entrapped blood cells covering the sensor. The sensor then monitors the analyte of interest within the microenvironment of the clot rather than its true level within bulk blood. Even a monolayer of metabolically active cells on the surface of the sensor can cause errant analytical results due to local consumption/production of

[1] M. C. Frost and M. E. Meyerhoff, *Curr. Opin. Chem. Biol.* **6**, 633 (2002).
[2] R. Banerjee, K. Nageswari, and R. R. Puniyani, *J. Biomater. Appl.* **12**, 57 (1997).

analytes such as O_2 and CO_2 by cellular respiration within the thrombus. Further, vasoconstriction around the indwelling sensor can lead to reduced blood flow at the implant site.[3] Pooling of blood will enhance bulk metabolic changes resulting from cellular metabolism, producing the same errant analytical results seen from thrombus formation. Similarly, the specific location of the sensor within the blood vessel can also result in local changes in analyte concentration. If the sensing element of the device is placed so that it touches the blood vessel wall, endothelial cell metabolism will also affect the flux of analyte to the sensor (the wall effect).[4]

Nitric oxide (NO) has been shown to have a wide range of physiological roles *in vivo*.[5] Nitric oxide is released endogenously on a continuous basis by endothelial cells that line all blood vessels at an estimated surface flux of $1.0–4.0 \times 10^{-10}$ mol/cm$^2 \cdot$ min.[6] Endothelial cell-derived NO diffuses into the lumen of the blood vessel and inhibits platelet adhesion and activation on healthy vessel walls.[7,8] Nitric oxide also diffuses into the smooth muscle cells that surround blood vessels, causing vasodilatation, and maintains normal vascular tone and blood pressure.[9] If intravascular sensors can be fabricated from materials that mimic the endogenous NO release of endothelial cells without interfering with the analytical performance of the devices, thrombus formation and vasoconstriction may be inhibited, allowing for a more reliable real-time, continuous *in vivo* monitoring of physiologically important analytes. The design of the sensor and specific placement of the sensing elements (e.g., sensing windows placed in multiple locations, distributed radially around the sensor) can minimize the wall effect.[10] Studies have shown that catheter-type Clarke-style amperometric oxygen sensors fabricated with NO-release coatings exhibit greatly enhanced hemocompatibility and improved analytical performance when tested in canine and porcine carotid and femoral arteries for extended periods of time (16 h).[11,12]

[3] M. E. Meyerhoff, *Trends Anal. Chem.* **12**, 257 (1993).
[4] C. K. Mahutte, C. S. H. Sassoon, J. R. Muro, D. R. Hansmann, T. P. Maxwell, W. W. Miller, and M. Yafuso, *J. Clin. Monit.* **6**, 147 (1990).
[5] P. L. Feldman, O. W. Griffith, and D. J. Stuehr, *Chem. Eng. News Dec.* 20, 26 (1993).
[6] M. W. Vaughn, L. Kuo, and J. C. Liao, *Am. J. Physiol. (Heart Circ. Physiol.)* **274**, H2163 (1998).
[7] M. W. Radomski, R. M. J. Palmer, and S. Moncada, *Br. J. Pharmacol.* **92**, 639 (1987).
[8] B. T. Mellion, L. J. Ignarro, E. H. Ohlstein, E. G. Pontecorvo, A. L. Hyman, and P. H. Kadowitz, *Blood* **57**, 946 (1981).
[9] M. T. Gewaltig and G. Kojda, *Cardiovasc. Res.* **55**, 250 (2002).
[10] C. K. Mahutte, *Clin. Biochem.* **31**, 119 (1998).
[11] M. C. Frost, S. M. Rudich, H. Zhang, M. A. Maraschio, and M. E. Meyerhoff, *Anal. Chem.* **74**, 5942 (2002).
[12] M. H. Schoenfisch, K. A. Mowery, M. V. Rader, N. Baliga, J. A. Wahr, and M. E. Meyerhoff, *Anal. Chem.* **72**, 1119 (2000).

This article describes the procedure employed to construct functional NO-releasing catheter-type amperometric oxygen sensors, including coating the sensor with NO-release materials, assembly of the sensor itself, and *in vivo* evaluation of the analytical performance and hemocompatibility of the device (details of the surgical protocol used for animal studies are described elsewhere[11]). The following procedure describes the specific fabrication of an intravascular catheter-type sensor (for use in animal studies only) that is introduced into an artery via a 14-gauge, 1.16-in. angiocath. The sensor (see Fig. 1) is fixed into a four-way stopcock to allow a means to attach the sensor to the catheter securely and to introduce a saline drip to prevent blood from pooling in the catheter. The length and diameter of the sensor can be adjusted by selecting tubing of the appropriate diameters and adjusting lengths of wires and the sensor body to accommodate a wide variety of sizes and specific methods of introducing the sensor

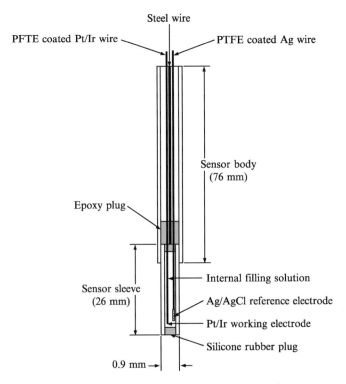

FIG. 1. Schematic of the catheter-type Clarke-style amperometric oxygen sensor.

into the blood vessel or tubing to be monitored (e.g., sensors could also be used for on-line PO_2 measurement in perfusion fluids in extracorporeal circuitry, etc.).

NO Release Material: DACA-6/N_2O_2-SR

Materials

   Poly(dimethylsiloxane), viscosity 2000 cSt (PDMS) (Sigma-Aldrich, Milwaukee, WI)
   N-(6-Aminohexyl)-3-aminopropyltrimethoxysilane (DACA-6) (Gelsdt, Tullytown, PA)
   Dibutyltin dilaurate (Sigma-Aldrich)
   Silicone dioxide, amorphous hexamethyldisilazane treated (fumed silica) (Gelsdt)
   Toluene
   Tetrahydrofuran (THF)
   Nitric oxide gas, 99% pure (Matheson Tri-Gas, Montgomeryville, PA)

Procedure

   Several NO-release materials have been developed that could be potentially used to fabricate more biocompatible sensors.[13–15] Modified silicone rubber (SR) materials have been used as the NO-release coating of choice for the catheter-type oxygen sensors because oxygen diffuses readily through SR. One of these SR materials is made with diamine containing cross-links that can be functionalized to diazeniumdiolate adducts *in situ* to release NO after the SR has cured (this material is referred to as DACA-6/N_2O_2-SR).[13] Details of the synthesis and characterization of DACA-6/N_2O_2-SR are outside the scope of this article; however, to make DACA-6-SR, combine 1.6 g of PDMS, 0.3 g of DACA-6, and 3.2 mg of dibutyltin dilaurate in 8 ml toluene and vortex to mix thoroughly. Allow the solution to stand at room temperature for 24 h to partially cross-link the SR. Suspend 0.3 g of trimethylated-fumed silica in 4 ml of toluene. Four milliliters of the PDMS solution and 3 ml of the fumed silica solution are then combined and used to dip coat onto the sensor sleeves (see later).

[13] H. Zhang, G. M. Annich, J. Miskulin, K. Osterholzer, S. I. Merz, R. H. Bartlett, and M. E. Meyerhoff, *Biomaterials* **23**, 1485 (2002).
[14] H. Zhang, G. M. Annich, J. Miskulin, K. Stankiewicz, K. Osterholzer, S. I. Merz, R. H. Bartlett, and M. E. Meyerhoff, *J. Am. Chem. Soc.* **125**, 5015–5024 (2003).
[15] M. M. Batchelor, S. L. Reoma, P. S. Fleser, V. K. Nuthakki, R. E. Callahan, C. J. Shanley, J. K. Politis, J. Elmore, S. I. Merz, and M. E. Meyerhoff, *J. Med. Chem.* **46**, 5153–5161 (2003).

Coating Sensor Sleeves

Materials

> Silastic medical grade tubing (0.55 mm i.d. × 0.94 mm o.d.) (Helix
>   Medical, Inc., Carpinteria, CA)
> Dow Corning RTV-3140 silicone rubber (World Precision Instruments,
>   Inc., Sarasota, FL)
> 26-gauge steel wire (hangers for dip coating sensor sleeves)

Procedure

1. Cut 26-mm lengths of Silastic medical grade tubing (0.51 mm i.d. × 0.94 mm o.d.) to make sensor sleeves (see Fig. 1). One end of the tubing is filled with a SR plug by putting a small bead of Dow Corning 3140 RTV-SR in a weighing dish and gently dipping one end of the tubing into the SR several times (see Fig. 2A). Continue dipping until about 2 mm of the end of the tubing is filled with SR. It is important to remove excess SR from the outside of the tubing so that the sensor sleeves will be smooth. Allow SR plug to cure overnight under ambient conditions.

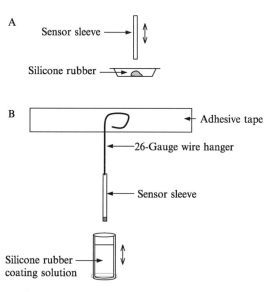

FIG. 2. Illustration of process for (A) forming a sensor sleeve by plugging one end of silicone rubber tubing with silicone rubber and (B) arrangement for dip coating sensor sleeves with NO-release material.

2. Sensor sleeves are placed on 26-gauge wire hangers such that they are hanging vertically and can be dip coated with the SR coating (see Fig. 2B). DACA-6 SR in toluene is dip coated at 10-min intervals until the desired coating thickness is achieved ($\sim$100-μm-thick coating with the described polymer solution requires eight coats). Care should be taken to ensure that the solvent has had sufficient time to evaporate between coats. If the SR solution is sagging and leaving drops at the lower end of the sensor sleeve as the solvent dries, dilute the coating solution with additional solvent so that each coat is slightly thinner. It is extremely important that the coatings on the sensor sleeves are as smooth and as even as possible. A single top coat of plain SR is applied (dissolve 0.6 g Dow corning RTV-3140 in 4 ml THF) as the final coat. Allow the SR to cure overnight under ambient temperature and moisture. Cured sensor sleeves are then placed in a high-pressure (80 psi) NO reactor for 24–36 h to form the diazeniumdiolate NO donor within the polymer matrix.[13]

Sensor Fabrication

Materials

Teflon-coated platinum/iridium wire (0.076 mm o.d.) (Medwire, Mount Vernon, NY)
Teflon-coated silver wire (0.076 mm o.d.) (Medwire)
Ferric chloride ($FeCl_3$)
Hydrochloric acid (HCl)
Silastic medical grade tubing (0.91 mm i.d. $\times$ 1.2 mm o.d.) (Medtronic Corp., Minneapolis, MN)
30-gauge, 1-in. disposable syringe needles (Becton-Dickinson, Franklin Lakes, NJ)
1-ml disposable syringes
Methocel 90 HG (Fluka, Milwaukee, WI)
Potassium chloride (KCl)
Quick-setting two-part epoxy (Super Glue Corp., Rancho Cucamonga, CA)
30-gauge steel wire (steel support wire)
Ultra four-way stopcocks (Medex, Hillard, OH)

Procedure

3. Electrodes are made by cutting 150-mm lengths of Teflon-coated Ag and Pt/Ir wire. Approximately 12 mm of Teflon is gently scraped off both ends of the Ag wire. One end of the Ag is then dipped into 1 M $FeCl_3$

dissolved in 1 M HCl for approximately 20 s to deposit AgCl for the reference elctrode. The remaining exposed end provides a site for electrical contact for the sensor. One end of the Pt/Ir wire is also stripped of Teflon, again to provide a site for electrical contact, and the other end is cut with a sharp razor blade, perpendicular to the length of the wire to form a round disk electrode to serve as the working cathode electrode. Care should be taken to carefully cut the Pt/Ir wire with a clean, square cut so that the surface area of the working electrode is as reproducible as possible from sensor to sensor.

4. The reference electrode and working electrodes are then threaded through the sensor body, which is a 75-mm length of Silastic medical grade tubing (0.91 mm i.d. × 1.2 mm o.d.). The electrode ends are aligned and pushed through the tubing until about 3 cm of wire extends past the end of the tube. After the electrodes are in place, retract the Ag/AgCl electrode slightly so that it is offset and shorter than the Pt/Ir working electrode (see Fig. 3). This avoids the potential problem of the electrodes directly touching one another. The length of the wires employed to make the

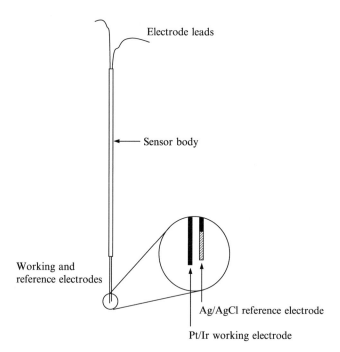

FIG. 3. Diagram of position of electrodes in the sensor body; the end of the Ag/AgCl reference electrode is slightly offset from the end of the Pt/Ir working electrode, as shown in the inset.

electrodes and the sensor body can be adjusted depending on the size of the catheter used to implant the sensor. The dimensions suggested here are appropriate for use with 1.16-in.-long catheters.

5. Using a 30-gauge needle and a disposable syringe, fill the sensor sleeve with 0.15 M KCl/1.5%(wt) Methocel solution. Gentle heating is required to fully dissolve the Methocel to prepare the solution, but avoid overheating and boiling the solution. To ensure that no air bubbles are trapped in the sensor sleeve, fill the needle with filling solution prior to inserting it into the sleeve and gently push the needle all the way to the SR plug at the distal tip of the sleeve. Begin filling the sensor sleeve from the distal tip of the sleeve and slowly remove the needle as solution is injected into the sleeve.

6. Thread the electrodes into the filled sensor sleeve. Gently pull the wires from the back of the sensor body until the sensor sleeve is within approximately 3 mm of the sensor body. Fill ~3–4 mm of the end of the sensor body with 5-min epoxy and slide the sensor sleeve into the sensor body, sealing the internal filling solution into the sensor sleeve and joining the sensor sleeve to the sensor body. Gently pull the electrodes from the back of the sensor body to ensure that the working electrode is not pushed against the SR plug in the distal tip of the sensor sleeve. Insert a ~90-mm steel support wire through the sensor body and into the epoxy plug used to join the sensor sleeve and sensor body. Wipe any excess epoxy from the outside of the sensor and allow the epoxy to completely cure.

7. Fix the sensor into a four-way stopcock to allow attachment to the catheter used to introduce the sensor into the vessel. Figure 4 shows positioning of the sensor within the stopcock. Prior to placing the sensor in the stopcock, the valve should be turned to the completely open position because the valve cannot be turned once the sensor transects the stopcock. The end that the electrode leads protrude from is then sealed with 5-min epoxy, and the end of the sensor body and the steel wire are also fixed into place with the epoxy. After the epoxy has dried completely, trim the steel support wire to the sensor body.

Sensor Calibration and Use

Materials

Chemical Microsensor potentiostat (Diamond Electro-Tech, Ann Arbor, MI)
DATAQ-700 USB data acquisition card (DATAQ Instruments, Akron, OH)

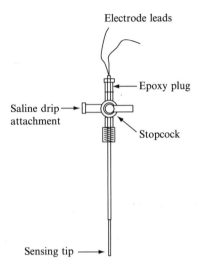

Electrode leads

Epoxy plug

Saline drip
attachment

Stopcock

Sensing tip

Fɪɢ. 4. Diagram of the fully assembled oxygen sensor positioned into a stopcock.

Procedure

8. The cathodic current ($\sim$0–300 nA, depending on the exact surface area of the Pt/Ir working electrode and O_2 present) is monitored at an applied potential of -0.7 V relative to the Ag/AgCl reference. The potentiostat leads are attached to the bare metal ends of the Pt/Ir and Ag wires. The Pt/Ir electrode should be allowed to polarize for $\sim$1 h while the sensor is soaking in phosphate-buffered saline (PBS) prior to making analytical measurements. (*Note.* The sensor should be left in PBS to avoid evaporation of the internal filling solution.)

9. For bench-top experiments, the sensor can be calibrated by tonometering PBS solutions with 0, 5, 10, 21, and 30% $O_{2(g)}$ balanced with $N_{2(g)}$. The steady-state current response is directly proportional to the PO_2 (calibration curves for these sensors should have excellent linearity). For *in vivo* experiments, a one-point, *in situ* calibration is made by noting the current from the sensor and measuring PO_2 of a discrete blood sample *in vitro* (using conventional blood gas analysis) and assuming that 0 oxygen corresponds to 0 current and that the current of the sensor corresponds to the oxygen level determined for the discrete sample.

10. Calibrated continuous outputs from an NO-release sensor and a control sensor implanted in the same animal are recorded. The *in vivo* analytical performance of the sensors is assessed by comparing the PO_2 determined from the sensors to the PO_2 level measured *in vitro* from

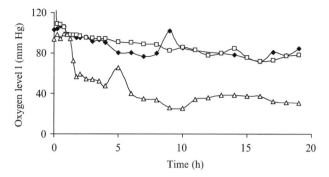

FIG. 5. Oxygen levels determined by an NO-releasing oxygen sensor (□) and a non-NO-releasing sensor (△) implanted in porcine carotid arteries for 19 h compared to oxygen levels determined by conventional blood gas analysis of discrete blood samples (◆).

periodically drawn discrete blood samples. Figure 5 shows PO_2 levels determined from two implanted catheter-type sensors in porcine carotid arteries over a 19-h period with an NO-release sensor and a non-NO-release sensor (control) compared to PO_2 levels measured *in vitro*.

Assessing Sensor Hemocompatiblity

Materials

    Phosphate-buffered saline, pH 7.4
    Glutaraldehyde
    Ethanol (200 proof)
    Hexamethyldisilazane

Procedure

    11. Sensors should be removed from the *in vivo* environment without disrupting the biolayer that has accumulated (do *not* pull the sensor out of the catheter). The entire segment of the vessel surrounding the sensor should be excised. A surgical snap can be placed slightly distal to the barrel of the catheter to hold the vessel and sensor in place. Using a pair of wire cutters, cut the stopcock and barrel of the catheter off, leaving the sensor tip and catheter tip in the artery (see Fig. 6). The artery is then carefully dissected open, allowing the sensor surface to be examined. After removal of the surrounding vessel, the cut end of the sensor is pushed completely through the remaining portion of the catheter so that the sensing tip that was exposed to blood is never dragged through the catheter, preserving any biolayer or thrombus formation present.

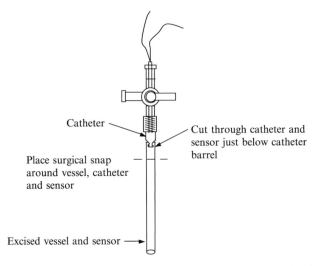

FIG. 6. Schematic of the explanted vessel containing the catheter sensor showing where to position the surgical snap prior to excising the vessel and the location for cutting the catheter and sensor, allowing the biolayer on the surface of the sensor to be preserved.

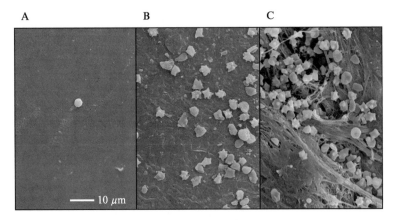

FIG. 7. SEM images illustrating the range of biological response to implanted sensors: nonactivated platelet adhering to an NO-releasing sensor surface (A) and activated platelets and fibrin network with entrapped platelets on non-NO-releasing surfaces (B and C, respectively).

12. Gently rinse the sensor tip in PBS to remove loosely adsorbed blood elements and place in a 1% (v/v) glutaraldehyde/PBS solution for 8 to 24 h to fix cells and proteins deposited on the surface of the sensor.

13. Before dehydrating the sensor for scanning electron microscopy (SEM) examination, the sensor sleeve should be trimmed below the epoxy

plug that joined the sensor sleeve and sensor body, sealing the internal filling solution. This will allow the internal filling solution to be removed, eventually leading to complete dehydration of the sensor sleeve. The sensor sleeve is dehydrated in a series of ethanol/water solutions of increasing alcohol concentrations [35, 50, 70, 90, 95, 99, 100, 100% ethanol/water (v/v)] for 10-min intervals. The final step in dehydration is to soak the sensor in neat hexamethyldisilazane for 20 min. The sensor sleeve is allowed to dry overnight under ambient conditions. The hexamethyldisilazane causes the SR tubing to swell slightly; upon overnight evaporation the sensor sleeve returns to its original size.

14. The distal-most 1 cm of the sensor tip exposed to flowing blood is sputter coated with gold (50 torr pressure, 40 mA current for 200 s). The gold-coated sensor tip is then examined using SEM (20 kV acceleration voltage and 1.5 nA beam current). Figure 7 shows examples of SEM images of a nonactivated platelet on a DACA-6/N_2O_2-SR sensor tip (A) and (B) activated platelets and activated platelets entangled in fibrin on non-NO-releasing sensor tips (B and C, respectively).

[46] Methods in Optical Oxygen Sensing: Protocols and Critical Analyses

By DMITRI B. PAPKOVSKY

Introduction

The detection of molecular oxygen by luminescence quenching has a long history,[1,2] but only recently has this technology become available in a convenient and cost-effective form suitable for routine use in research laboratories and industrial applications. The last decade has been particularly fruitful, with major technological developments in advanced sensing materials, detection principles, instrumentation based on low-cost components and signal processing electronics, assay formats, and applications.[3,4] A number of commercial systems have appeared, thus opening new opportunities and niches in oxygen sensing and competing with established

[1] H. Kautsky and A. Hirsch, *Ber. Dtsch. Chem. Ges.* **64,** 2677 (1931).
[2] D. W. Lubbers and N. Opitz, *Z. Naturforsch. Teil C.* **30,** 532 (1975).
[3] J. N. Demas, B. A. DeGraff, and P. B. Coleman, *Anal. Chem.* **71,** 793A (1999).
[4] J. P. Kerry and D. B. Papkovsky, *in* "Research Advances in Food Science" (R. M. Mohan, ed.), Vol. 3, p. 121. Global Research Network, Kerala, India, 2002.

techniques such as electrochemical Clark-type oxygen electrodes and paramagnetic sensors.

The inherent advantages of luminescence-based oxygen sensors are a nonchemical and reversible mechanism of sensing in which both the probe and the analyte are not consumed, high sensitivity and selectivity,[4,5] potential for miniaturization,[6] rapid response times down to milliseconds,[6–8] and the possibility of contactless measurements through a (semi)transparent vessel or material.[9,10] Oxygen is one of the principal metabolites, which diffuses rapidly across cell membranes, tissues, and biological fluids, so it can be measured outside the cells and *in situ*. Optical oxygen sensing allows noninvasive measurements with biological objects, including living cells, tissues and whole organisms, parallel monitoring of multiple samples,[9,11,12] and imaging,[13] thus providing high utility and multiple applications for life sciences, general cell biology, drug discovery, and food, medicine, biotechnology, environmental, and biomedical areas. The optical oxygen sensor can serve as a transducer for bioanalytical systems based on biological recognition.[5,14] In particular, coupling with enzymes enables the development of biosensors for metabolites such as glucose,[15,16] with microorganisms and cellular components–BOD biosensors,[17] and with antibodies and enzyme label-sensitive immunosensors.[18] Optical oxygen sensors can be coupled with sensors with other analytes to allow multiplexed assays.[2,19]

[5] O. S. Wolfbeis (ed.), *Fiber Optic Chemical Sensors and Biosensors*. CRC Press, Boca Raton, FL, 1991.

[6] I. Klimant, F. Ruckruh, G. Liebsch, C. Stangelmayer, and O. S. Wolfbeis, *Mikrochim. Acta* **131,** 35 (1999).

[7] J. Kavandi, J. Callis, M. Gouterman, G. Khalil, D. Wright, E. Green, D. Burns, and B. McLachlan, *Rev. Sci. Instrum.* **61,** 3340 (1990).

[8] C. Kolle, W. Gruber, W. Trettnak, K. Biebernik, C. Dolezal, F. Reininger, and P. O'Leary, *Sens. Actuat. B* **38,** 141 (1997).

[9] S. Zanetti, F. Ardito, L. Sechi, M. Sanguinetti, P. Molicotti, G. Delogu, M. P. Pinna, A. Nacci, and G. Fadda, *J. Clin. Microbiol.* **35**(8), 2072 (1997).

[10] D. B. Papkovsky, N. Papkovskaia, A. Smyth, and V. I. Ogurtsov, *Anal. Let.* **33,** 1755 (2000).

[11] T. C. O'Riordan, D. Buckley, V. I. Ogurtsov, R. O'Connor, and D. B. Papkovsky, *Anal. Biochem.* **278,** 221 (2000).

[12] M. Wodnicka, R. D. Guarino, J. J. Hemperly, M. R. Timmins, D. Stitt, and J. B. Pitner, *J. Biomol. Screen.* **5**(3), 141 (2000).

[13] W. L. Rumsey, J. M. Vanderkooi, and D. F. Wilson, *Science* **241,** 1649 (1988).

[14] W. Trettnak, *in* "Fluorescence Spectroscopy: New Methods and Applications" (O. S. Wolfbeis, ed.), p. 79. Springer-Verlag, Berlin, 1993.

[15] W. Trettnak, M. J. P. Leiner, and O. S. Wolfbeis, *Analyst* **113,** 1519 (1988).

[16] D. B. Papkovsky, A. N. Ovchinnikov, V. I. Ogurtsov, and T. Korpela, *Sens. Actuat. B* **51,** 137 (1998).

[17] C. Preininger, I. Klimant, and O. S. Wolfbeis, *Anal. Chem.* **66,** 1841 (1994).

[18] D. B. Papkovsky, T. C. O'Riordan, and G. G. Guilbault, *Anal. Chem.* **71,** 1568 (1999).

[19] R. Fraatz, C. Sachs, J. F. Mollard, P. Finetti, and J. Tusa, *Clin. Chem.* **42,** 768 (1996).

Numerous optical oxygen-sensing systems have been described, including proof of concept, sophisticated prototypes, laboratory studies, integrated systems, and commercial devices. They employ different sensing materials, detection principles, and measurement formats and, therefore, differ very much in their general design, performance characteristics, and range of applications. For example, oxygen sensors for blood gas analysis,[19] oxygen imaging in tissues,[13,20] microsensors,[6] packaging,[10,21] pressure imaging,[7,22] environmental,[23] and bioreactor[24] applications have demonstrated high utility and performance characteristics, but many of them are too specialized to suit other applications and research tasks. Biological applications, particularly studies with living cells and *in vivo* measurements, impose additional requirements on the sensor, which often cannot be met by the existing chemical or medical sensors, but require their redesign and tuning by the user.

The objectives of this article are to familiarize researchers entering this area with basic principles of quenched luminescence oxygen sensing and to provide general strategies for the selection of the appropriate oxygen probe, detection principle, assay format, and experimental setup to suit the particular application they intend to explore. This is illustrated with several typical applications and assay protocols, which have high practical utility and employ commercial instrumentation, materials, and accessory tools, namely (1) noninvasive measurement of the oxygen concentration in a sealed vessel; (2) monitoring of oxygen uptake rates by living organisms and cells; and (3) microplate-based screening assays for cell viability and effector action on cells. These systems are then analyzed critically to outline their characteristics features, performance characteristics, advantages and limitations, and potential bioanalytical uses.

Principles of Quenched-Luminescence Oxygen Sensing

The process of luminescence quenching by molecular oxygen is described by the following scheme:

[20] D. F. Wilson, S. A. Vinogradov, B. W. Dugan, D. Biruski, L. Waldron, and S. A. Evans, *Comp. Biochem. Physiol. A* **132**(1), 153 (2002).
[21] D. B. Papkovsky, M. Smiddy, N. Yu. Papkovskaia, and J. P. Kerry, *J. Food Sci.* **67**(8), 3164 (2002).
[22] E. Puklin, B. Carlson, S. Gouin, C. Costin, E. Green, S. Ponomarev, H. Tanji, and M. Gouterman, *J. Appl. Pol. Sci.* **77**, 2795 (2000).
[23] M. Hill, V. I. Ogurtsov, D. B. Papkovsky, and V. Yershov, *in* "Proceedings 1st SENSPOL Workshop," p. 153. Alcala, Spain, 2001.
[24] H. Trubel and W. K. Barnikol, *Biomed. Tech. (Berl.)* **43**, 302 (1998).

$$D + h\nu_1 \rightarrow D^* - \text{excitation} \tag{1}$$

$$D^* \xrightarrow{k_L} D + h\nu_2 - \text{emission} \tag{2}$$

$$D^* + O_2 \xrightarrow{k_Q} D + O_2^* = \text{collisional quenching, no emission} \tag{3}$$

where D and D^* and O_2 and O_2^* are ground and excited states of the dye and oxygen molecules and k_L and k_Q are luminescence and quenching rate constants, respectively.

Because the last two processes of luminescence emission (radiative, monomolecular) and quenching (nonradiative, bimolecular) are competing with each other, at a constant concentration of a fluorophore and excitation intensity, its emission intensity and lifetime are inversely proportional to oxygen concentration, as described by the Stern–Volmer equation:[25]

$$I_o/I = \tau_o/\tau = 1 + k_Q * \tau_o * [O_2] \tag{4}$$

where I_o and I and τ_o and τ are emission intensities and lifetimes of the dye in the absence and in the presence of the oxygen concentration $[O_2]$, respectively. By measuring the luminescence intensity or lifetime, one can quantify oxygen concentration in the sample using a predetermined calibration function:

$$[O_2] = (I_o - I)/(I * k_Q * \tau_o) = (\tau_o - \tau)/(\tau * \tau_o * k_Q) \tag{5}$$

The oxygen probe comprises a combination of an oxygen-sensitive photoluminescent dye in the appropriate quenching medium. Such oxygen-sensitive material produces a distinct luminescence response to oxygen present in the sample at certain concentrations, that is, analytical range. Oxygen diffuses from the sample (macrophase) to the dye microenvironment and quenches it by a dynamic (collisional) mechanism.

The dye defines main spectral characteristics of such an oxygen probe, luminescence efficiency (quantum yield ϕ and I_o), kinetic characteristics (τ_o), and quenchability by oxygen (the range of k_Q and τ_o). However, quenching medium is as important as the dye, as it defines the partitioning and diffusion of oxygen and other sample components in the dye microenvironment. Hence, the sensor main working characteristics, including calibration function, k_Q, τ, and I at a given $[O_2]$, temperature behavior, cross-sensitivity, and sensor response time, are determined by the dye and its quenching medium, and also by the detection system used.

[25] J. R. Lakowicz, "Principles of Fluorescence Spectroscopy," 2nd Ed. Plenum Press, New York, 1999.

Many oxygen sensors display heterogeneity of their luminescence and quenching properties, which results in more complex analytical dependences and nonlinearity of Stern–Volmer plots.[26] Such sensor behavior requires a more thorough optimization of sensor material, its modeling, and characterization.[27,28] This, however, is not a major technical obstacle, and commercial sensors usually have predictable behavior and work reliably and reproducibly.

Although present at relatively high concentrations in the environment (about 21 kPa in air or 200 μM in air-saturated aqueous solutions), molecular oxygen is a relatively weak dynamic quencher.[25] Therefore, dyes with relatively long emission lifetimes, typically within the range of 1 to 100 μs, are required for making oxygen probes. Conventional short-lived fluorophores are rather insensitive to oxygen, whereas phosphorescent dyes with very long lifetimes are almost completely quenched by ambient oxygen and their emission is hardly measurable. The dye spectral characteristics are important to minimize interference by scattering, autofluorescence and ensure high selectivity of its detection in complex biological samples. Longwave oxygen-sensitive dyes with well-resolved excitation and emission bands, compatible with LED/diode laser excitation, enable one to develop efficient oxygen sensors and simple and inexpensive instrumentation.[4,5,29]

Not many dyes fulfill these requirements and therefore are suitable for optical oxygen sensing. Initially, polycyclic aromatics such as decacyclene were used in oxygen sensors,[2,5,14] but their drawbacks are the need of UV excitation and very short lifetimes (50–100 ns). These dyes were superseded by the fluorescent complexes of ruthenium(II)[30,31] and phosphorescent Pt(II)- and Pd(II)-porphyrins.[7,13,29,32] These two classes of oxygen-sensitive dyes are now dominating in oxygen sensing. Few other dyes have been described, including long-wave Os(II) complexes[33] and some others[34]; however, their advantages in comparison with ruthenium complexes and metalloporphyrins are marginal. The most common probes developed for general oxygen sensing are listed in Table I.

[26] P. Hartmann, M. J. P. Leiner, and M. E. Lippitsch, *Sens. Actuat. B* **B29,** 251 (1995).

[27] J. N. Demas, B. A. Degraff, and W. Y. Xu, *Anal. Chem.* **67,** 1377 (1995).

[28] V. I. Ogurtsov, D. B. Papkovsky, and N. Yu. Papkovskaia, *Sens. Actuat. B* **B81,** (2001).

[29] D. B. Papkovsky, G. V. Ponomarev, W. Trettnak, and P. O'Leary, *Anal. Chem.* **67,** 4112 (1995).

[30] J. R. Bacon and J. N. Demas, *Anal. Chem.* **59,** 2780 (1987).

[31] O. S. Wolfbeis, M. J. P. Leiner, and H. E. Posch, *Mikrochim. Acta (Vienna)* **III,** 359 (1986).

[32] S. A. Vinogradov and D. F. Wilson, *J. Chem. Soc. Perkin Trans.* **2,** 103 (1995).

[33] W. Xu, K. A. Kneas, J. N. Demas, and B. A. DeGraff, *Anal. Chem.* **68,** 2605 (1996).

[34] Y. Kostov, K. A. van Houten, P. Harms, R. S. Pilato, and G. Rao, *Appl. Spectrosc.* **54**(6), 864 (2000).

TABLE I
LUMINESCENT OXYGEN PROBES AND THEIR PROPERTIES

Oxygen-sensitive material[a]	Excitation (nm)	Emission (nm)	Lifetime (τ_o, μs)	Reference
Solid-state oxygen probes				
Ru(bipy)$_3$Cl$_2$ in silicagel/silicone rubber, ormosil glass, PS	460	605	0.8 μs	12, 19, 30, 31
Ru (dpp)$_3$Cl$_2$ in PDMS	460	610	4 μs	35
Os-bis(terpy) in PDMS-acrylate	650	710	270 ns	33
PtPFPP in fluoroacrylic polymer	520	670	40 μs	22
PtOEP in polystyrene	535	650	100 μs	36
PtOEPK in polystyrene	590	760	64 μs	16, 21, 29
Water-soluble oxygen probes				
(Ru[dpp(SO$_3$Na)$_2$]$_2$Cl$_2$	480	615	3.7 μs	37
PtCP–BSA conjugate	380, 535	650	~80 μs	38
PdCP free	390	670	990 μs	39
PdTPCPP free/albumin complex	415, 425	700	~1 ms	13, 40
Pd–TBP conjugate	620	780	450 μs	41

[a] PtTFPP, platinum(II)-tetrakis(pentafluorophenyl)porphine; PtTPCPP, platinum(II)-(p-carboxyphenyl) porphine; PtOEP, platinum(II)-octaethylporphine; PtOEPK, platinum(II)-octaethylporphine-ketone; PdTBP, substituted palladium(II)-tetrabenzo-porphine; PtCP, PdCP, platinum(II) and palladium(II)-coproporphyrin-I; Ru(bipy)$_3$, ruthenium(II)-trisbipyridine; Ru(dpp)$_3$, ruthenium(II)-diphenylphenantroline; Os-bis(-terpy), osmium(II)-bisterpyridine; PS, polystyrene; PDMS, poly(dimethylsyloxane).

The two main types of oxygen probes (see Table I) are

i. solid-state oxygen sensors comprising the luminescent dye embedded in a polymer medium, usually a film coating applied on a solid support, which is permeable to oxygen and exposed to the sample
ii. water-soluble oxygen probes added directly to the sample, which acts as a quenching medium for the oxygen-sensitive dye.

[35] W. Xu, R. C. McDonough, B. Langsdorf, J. N. Demas, and B. A. DeGraff, *Anal. Chem.* **66**, 4133 (1994).
[36] D. B. Papkovsky, J. Olah, I. V. Troyanovsky, N. A. Sadovsky, V. D. Rumyantseva, A. F. Mironov, A. I. Yaropolov, and A. P. Savitsky, *Biosens. Bioelectron.* **7**, 199 (1991).
[37] F. N. Castellano and J. R. Lakowicz, *Photochem. Photobiol.* **67**(2), 179 (1998).
[38] J. Hynes, S. Floyd, A. Soini, R. O'Connor, and D. B. Papkovsky, *J. Biomol. Screen.* **8**, 264 (2003).
[39] D. B. Papkovsky, A. P. Savitsky, and A. I. Yaropolov, *J. Anal. Chem.* **45**, 1441 (1990).
[40] J. M. Vanderkooi, G. Maniara, T. J. Green, and D. F. Wilson, *J. Biol. Chem.* **262**(12), 5476 (1987).
[41] I. Dunphy, S. A. Vinogradov, and D. F. Wilson, *Anal. Biochem.* **310**(2), 191 (2002).

Advantages of the first type of probes are better defined spectral and quenching properties, which are usually unaffected by the sample due to the dye immobilization and shielding effect of the solid-state material (gas-diffusion barrier). Dyes with a very different sensitivity to oxygen can be used in solid-state sensors by selecting a proper polymer material (with high solubility for oxygen for short-decay dyes and with low solubility for long-decay dyes), and a large amount of the dye can be loaded to give high and reproducible signals and no self-quenching.[4,5,35] Sensor support material can also play an active role in achieving optimal performance and special features[16,18]; an additional covering layer can be applied for optical insulation, selectivity filter, and mechanical protection.[5,14] Solid-state sensors are robust, relatively inexpensive, do not contaminate the sample, withstand biodegradation, and allow long-term operation. They are used in industrial and medical applications.

With respect to biological applications, solid-state sensors have some limitations. These are rather complex manufacturing procedures (casting, spin coating, dip coating[5]), fixed shape and permanent attachment of the sensor to its support (membrane, optical fiber, microplate, or flow device), and there are problems with sensor sterility and reuse, hydrophobicity, poor compatibility with cells, and phototoxic effects.[12] Instrument-specific or assay-specific optimization of such sensors by the user and their use in some applications is difficult. In screening applications, when large numbers of samples are to be analyzed, significant costs and waste of sensing materials become a serious issue.[38]

Water-soluble oxygen probes can overcome many of these drawbacks of the solid-state sensors, but at the same time they have their own limitations. In particular, sample contamination by the probe, effects of sample components on the probe quenching characteristics, self-quenching, and so on have to be considered.[40] Cross-sensitivity to other compounds present in the sample, interference by sample scattering and autofluorescence, and photobleaching, especially when used at low concentrations, can be significant for the soluble probes (solid-state sensors are susceptible to these factors as well[12,38,42]).

In general, solid-state sensors are more appropriate for measuring absolute oxygen concentrations, long-term use, and continuous monitoring of oxygen in relatively large samples. Water-soluble probes are better suited to kinetic and screening assays, when relative changes or rates of oxygen uptake/release are monitored, and when large numbers of samples are analyzed in parallel and on a microscale. They allow adjustment of the probe working concentration by the user for a particular sample, assay, or instrument,

[42] BD Biosciences, *Technical Bulletin #449*, http://www.bdbiosciences.com (2002).

miniaturization, cost saving, automation (robotic dispensing), and high sample throughput. These two types of oxygen probes complement each other, thus facilitating the development of various assays and applications.

Ruthenium- and porphyrin-based probes are used most commonly in oxygen sensors. Photophysical characteristics of these dyes, including bright long-wave emission well resolved with excitation bands, good photochemical stability, long lifetimes, and a moderate response to oxygen, provide satisfactory performance in many oxygen-sensing systems and applications. Ruthenium dyes have significantly shorter emission lifetimes than porphyrins (see Table I) and therefore can be used with a limited number of polymers. Their quenching by oxygen is not always strong enough (few tenths of a percent[30,31,35,37]) and, due to blue excitation, they are more prone to optical interference by biological samples.[42] Porphyrins allow a broader choice of polymers and provide a more optimal response (two- to five-fold quenching at ambient oxygen), both in polymers and in water-soluble forms.[22,29,36,38,40] Long-wave absorption bands (500–650 nm) and far-red emission (650–850 nm) make them less susceptible to optical interferences. Lifetime-based oxygen sensing and time-resolved detection are technically more simple for porphyrin probes (see later).

Molecular oxygen alters the luminescence intensity and lifetime of the probe [see Eq. (4)] so these parameters can be used for the quantitation of oxygen in a sample.[3,4,26] Intensity-based oxygen sensors are relatively simple and straightforward. They can provide satisfactory performances,[8,19] especially with fixed measurement geometry (permanently attached sensors), optical insulation of the sensor, and/or clear samples (gases, water). However, intensity-based sensors operating with arbitrary luminescent signals (relative fluorescence units, RFU) are dependent on many factors, such as optical geometry, sensor positioning (detached sensors), dye photobleaching, light source fluctuations, detector and electronics fluctuations, optical properties of sample, and sensor aging.[4,26] These may result in instability of the calibration function, significant measurement errors, and a need for regular calibrations.

Luminescence lifetime-based oxygen sensing overcomes many of the limitations of intensity-based sensors. Lifetime is a self-referencing parameter of a luminescent material [also related to oxygen concentration; Eq. (5)], which is essentially independent on the dye concentration and measurement geometry. Luminescence lifetime-based sensing provides more stable calibration and improved performance.[29,36,40,43,44] Lifetime-based oxygen sensing can be performed by direct lifetime measurement using pulsed excitation or by phase measurements.

In phase-fluorimetric oxygen sensors, the active element is illuminated with modulated light (sine or square-wave excitation) while monitoring the luminescence signal with respect to excitation. The delay or emission (phase shift ϕ, measured in degrees angle) relates to the lifetime of the dye as follows:

$$\tan(\phi) = 2 * \pi * \nu * \tau \qquad (6)$$

where ν is the frequency of excitation (i.e., device working frequency) and π is 3.14 ....[25] Combining Eqs. (5) and (6), one can work out the relationship between phase shift and oxygen concentration. More complex mechanisms of quenching produce more complex relationships.[28]

The technical realization of phase measurements is relatively simple. For long-decay oxygen probes requiring modulation frequencies of excitation in the kilohertz range, this can be achieved using inexpensive LEDs and photodiodes. Phase measurements enable discrimination of a specific (modulated) signal from the sensor from ambient light (nonmodulated) by frequency filtering. Such systems can operate under ambient light, with detached sensors and in real time, including remote and contactless measurements through a semitransparent material,[4] and this methodology is becoming increasingly popular in oxygen sensing.[6,20,29,44,45] Sample autofluorescence and scattering, if comparable with the luminescence from the sensor, are potential sources of interference in phase measurements. However, using long-wave dyes with large Stokes' shift and solid-state sensors with high luminescent signals, phase-fluorimetric sensors are usually very reliable and accurate.

Direct lifetime measurements are more complex and yet uncommon in oxygen sensing, although such systems have been described.[3,43] They usually employ short excitation pulses (shorter than luminophore lifetime), a fast and sensitive detector, a time scaler with multiple bins, and fast signal processing electronics to measure the luminescence decay curve, reconstruct, and determine its kinetic characteristics (lifetime).[25] Being more costly and cumbersome, this approach has the potential of discriminating probe luminescence from a short-lived background (scattering and autofluorescence) and getting more information about sensor heterogeneity and mechanisms of quenching by oxygen.[3,25]

Time-resolved fluorescence detection can eliminate optical interferences and achieve high sensitivity of the probe detection in oxygen-sensing systems using existing time-resolved fluorescence plate readers. Some of

[43] M. E. Lippitsch, J. Pusterhofer, M. J. P. Leiner, and O. S. Wolfbeis, *Anal. Chim. Acta* **205,** 1 (1988).

[44] R. B. Thompson and J. R. Lakowicz, *Anal. Chem.* **65,** 835 (1993).

[45] W. Trettnak, C. Kolle, F. Reininger, C. Dolezal, and P. O'Leary, *Sens. Actuat. B* **36,** 506 (1996).

these devices, such as Victor family (PerkinElmer Life Sciences), enable simultaneous measurements of luminescence intensity at two or more delay times, thus allowing lifetime-based sensing and/or internal referencing.[38] Time-resolved detection, combined with phase measurements, has been used to measure oxygen transcutaneously[46,47] in conditions of high scattering and autofluorescence and a relatively weak probe signal.

Several examples of typical sensing materials, detection systems, and assay formats, which we consider of high utility for biological applications, are discussed.

Standard Protocols and Applications

Protocol 1. Contactless Monitoring of Oxygen Concentration in a Sealed Vessel Using the Phase-Fluorimetric Oxygen Sensor System

Experimental Setup

   Assay substrate: Plastic hypoxia chamber (Billups-Rothenberg) with shelves for samples and valves for flushing with gas mixtures.

   Oxygen probe: PtOEPK-based oxygen sensor stickers (12 × 26 mm, Type KPS-P, Luxcel Biosciences), batch calibrated. The sensor is attached to the inner wall of the chamber.

   Detector: Benchtop phosphorescence phase detector (Luxcel) with a 6-mm fiber-optic probe, T probe, PC communication port (RS-232), and software.

   Detection mode: Phase measurements (lifetime-based oxygen sensing).

   Temperature: $4°–40°$, normally $37°$ (in an incubator).

   Samples: Headspace oxygen (gas phase). Measurement in liquid is also possible.

   Measurement Procedure. The fiber-optic probe of the phase detector is brought into the vicinity of the sensor so that the instrument can detect its luminescent signal. Without the sensor, measured intensity (primary signal) is negligible and therefore phase readings are chaotic. When the luminescence from the sensor is detected (the probe can be 0–15 mm away from the sensor), the instrument phase readings become stable and correspond to a particular oxygen concentration in the sample. The exact distance between the sensor and fiber-optic probe and their orientation is

[46] H. M. Rowe, S. P. Chan, J. N. Demas, and B. A. DeGraff, *Anal. Chem.* **74**(18), 4821 (2002).
[47] S. B. Bambot, G. Rao, G. M. Carter, J. Sipior, E. Terpetchnig, and J. R. Lakowicz, *Biosens. Bioelectron.* **10**, 643 (1995).

not critical. Once the phase shift signal does not fluctuate much, it can be used for calculating the oxygen concentration. The oxygen concentration is calculated by the user or software based on current phase readings, sample temperature, and sensor calibration, which is provided by the manufacturer for each batch of sensors. Measurement setup and sensor calibration are shown in Figs. 1 and 2.

Results of a typical experiment are shown in Fig. 3. When the chamber is filled with air (20.5 kPa oxygen), phase readings are about 12° (at a given temperature, see Fig. 2). Purging it with a gas mixture containing 5% CO_2, 1.0% oxygen balanced with nitrogen (used in hypoxic experiments with cells) brings the phase signal up to about 37°. When the phase signal becomes stable and has no drift ($\pm 0.02-0.05°$), purging is terminated and the inlet and outlet valves are closed. After that, the phase signal is checked periodically to ensure that the chamber has proper gas composition inside and no leaks. Over the next 24 h the phase signal decreases by about 1°, which corresponds to the increase of oxygen concentration in the chamber by about 0.16 kPa (or by 0.16%). This indicates that the diffusion of

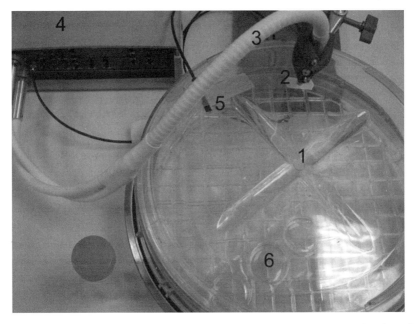

FIG. 1. Measurement setup for contactless oxygen sensing in a sealed vessel: 1, hypoxia chamber; 2, sensor sticker; 3, fiber-optic probe; 4, phase detector; 5, T probe; and 6, petri dishes with biological samples.

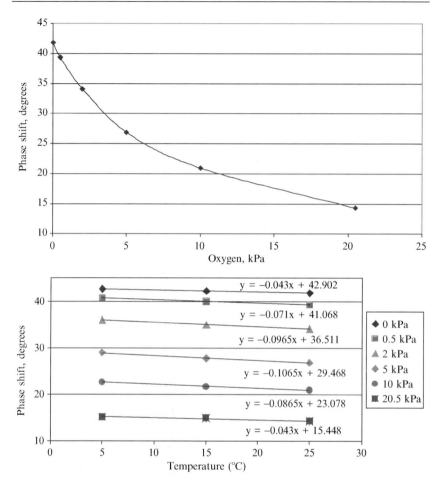

Fig. 2. Calibration of the phase-fluorimetric oxygen sensor system described in Protocol 1 (top) and its temperature dependence (bottom).

ambient air into the chamber is considerable. For this particular experiment the changes in actual gas composition are quite modest, but for anaerobic experiments they would be very significant.

Critical Analysis. The bench-top phase detector made of low-cost optoelectronic components allows stand-alone operation and external control from a PC. The instrument can operate with different types of PtOEPK-based sensors and at different working frequencies to achieve optimal performance and sensitivity within the desired oxygen range.[4]

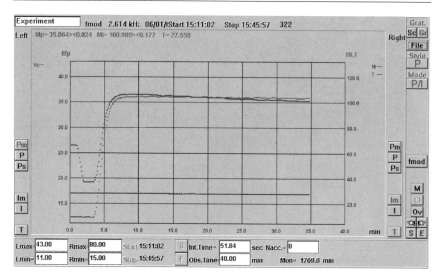

FIG. 3. Signal profiles of a typical oxygen-sensing experiment with a hypoxia chamber at 23°: 0 min, chamber filled with air, start of monitoring; ~2 min, fiber-optic probe was moved to the edge of the sensor, causing a significant drop in intensity but no change in the phase signal; ~4 min, chamber was flushed with a standard gas mixture containing 0.5% oxygen; and ~10 min, gas supply was discontinued, valves were closed, and sealed chamber was monitored (minor drift diffusion of air).

Using batch-calibrated disposable sensors, the system provides accuracy better than 0.1 kPa of oxygen (0.1% in air). In the continuous monitoring mode, phase changes of 0.02° are reliably detectable, which corresponds to changes in oxygen concentration of 0.004 kPa at zero oxygen or 0.03 kPa in the ambient oxygen range (i.e., around 20.5 kPa). For accurate quantitation of oxygen, measurements must be compensated for variation of sample temperature. The sensors described in this example have relatively flat temperature dependence of calibration described by a set of linear functions (see Fig. 2). Temperature correction can be carried out using standard software packages such as MathLab or incorporated into instrument software. Sensors can operate over a broad temperature range of $-15-+40°$.[10] The phase detector has a T probe for parallel recording of the sample temperature. The sensor response is fast (~20 s) and reversible; the system allows long-term continuous operation for many days under ambient light without recalibration. Sensors costing a few dollars each can be stored for years and used on a disposable basis.

The system provides flexibility and multiple applications. One instrument can operate with multiple detached sensors and monitor sample oxygen in a contactless fashion. It has been applied to the nondestructive measurement of oxygen in sealed food packs[21] and bioreactors.[23] The sensor can be attached to the end of a fiber-optic probe and used to measure oxygen concentrations in gas, aqueous samples, and tissue. In the liquid phase, sensor calibration remains practically the same.[10]

Protocol 2. Monitoring of Oxygen Uptake by Small Organisms in Microwell Strips Using Oxygen Sensor Inserts and Phase Measurements

Experimental Setup

Assay substrate: Standard 8- or 12-well strips (Nunc) in a T-controlled holder.

Oxygen probe: PtOEPK-based microporous sensor membranes (6 mm in diameter, Type KPS-D, Luxcel), disposable, and batch calibrated.

Detector: Bench-top phosphorescence phase detector (Luxcel) with a 3-mm fiber-optic probe, T probe, PC communication port (RS-232), and software.

Detection mode: Phase measurements through the bottom of the wells.

Temperature range: $5°-40°$.

Samples: Small marine organisms, *Tigriopus brevicornis*, in seawater. Measurement of aerobic bacteria, yeast, mammalian cells, and other microorganisms in appropriate growth media is possible.

Measurement Procedure. Sensor inserts are placed in wells of an 8- or 12-well flat-bottom clear strip (Nunc). The strip is aligned on a special metal holder to which the fiber-optic probe of the phosphorescent phase detector is connected, which interrogates with the sensors through the bottom of microwells. The phase detector is connected to a PC, which provides instrument control, graphic presentation of data in real time, and data storage by means of dedicated Windows-based software supplied with the instrument. The holder is temperature controlled by an external water bath. Measurement setup is shown in Fig. 4.

To perform an assay, 0.2 ml of seawater is added to each well, and the strip is equilibrated at the desired temperature for 15–20 min. Animals are brought to the wells (5 per well), and 0.1 ml of mineral oil (M5904, Sigma) is applied on top of each sample (barrier for ambient air oxygen). Measurements are started from the first sample, which is monitored for 1–5 min, and then the strip is moved to the next well and so on. Time profiles of the phase shift signal for each sample are generated by the software on the screen. Once the oxygen gradients are established (typically 20–200 min), monitoring is terminated and readings are saved as spreadsheet files. Measured data are analyzed to determine the initial slopes of the phase signal, which reflect the rates of oxygen uptake. Changes in animal respiration as a function of temperature and salinity are shown in Fig. 5.

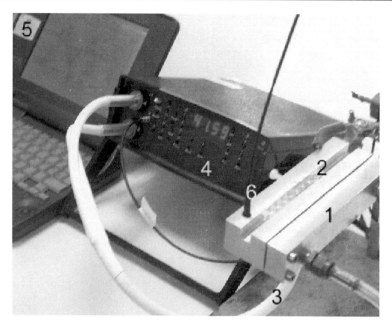

FIG. 4. Experimental setup for Protocol 2: 1 metal holder; 2, microwell strip with sensor inserts in each microwell; 3, fiber-optic probe; 4, phosphorescence phase detector; 5, PC; and 6, T probe.

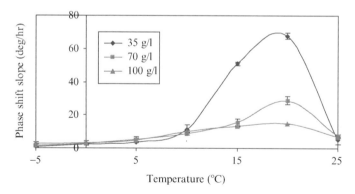

FIG. 5. Typical profiles of respiratory activity of *Tigriopus brevicornis* at different temperatures and water salinities measured with the optical oxygen sensor system.

Critical Analysis. This format of respirometric measurements employs the same instrument and sensors as in the previous example so its performance characteristics remain practically the same. In such kinetic experiments, absolute oxygen concentrations and rates of oxygen uptake are not very relevant, as the samples are not fully sealed from ambient air oxygen. Instead, apparent oxygen gradients and rates of oxygen uptake are usually analyzed with respect to some control sample(s) and/or are presented as raw data (slopes of the phase shift) in arbitrary scales or normalized.[11,12]

A partial seal from ambient oxygen, which still diffuses into the sample at considerable rates through mineral oil and the polystyrene body of the strip, is the only main limitation of this assay format, which may result in an inability to detect low rates of oxygen uptake by the samples and/or necessitate the use of relatively large numbers of animals/cells per well. Temperature effects may also result in baseline drift, inconsistency, and variation of results. Due to manual processing, this assay format is limited to a small number of samples. The use of 96- and 384-well plates is problematic, but the measurement of larger samples (e.g., 6- and 24-well plates) is possible. By using completely sealed chambers filled with samples, it is possible to enhance assay sensitivity and measure absolute rates of oxygen uptake and small oxygen gradients.

This approach was initially developed for monitoring cellular respiration and was applied to monitoring *Schizosaccharomyces pombe* yeast, including cell numbers, cell viability, and drug/effector action.[11] High signals from the sensors and bottom reading allow one to work with highly absorbing samples, such as soil leachate.[23] Similarly, responses of test organisms to toxicants, stress, environmental factors, and so on can be assessed. The sensors described in this application can be coupled with various oxygen-dependent systems of biological recognition of important analytes to produce corresponding biosensors with the optical oxygen transducer. A sponge-like structure of the oxygen sensor membranes provides high capacity for biomaterial, simple immobilization means, and the possibility of contactless measurements. Oxygen sensors coupled with a glucose oxidase enzyme were used in a reagentless glucose biosensor,[16] flow injection analysis of glucose,[48] in immune sensors, and for sensitive detection of enzymatic activity.[18]

Protocol 3. Cell Viability Assays Using the Water-Soluble Oxygen Probe, Standard Microplates, and Fluorescence Plate Reader

Experimental Setup

Assay substrate: 96-well plates, clear or black; 24-384-well plates can also be used.

---

[48] A. N. Ovchinnikov, V. I. Ogurtsov, W. Trettnak, and D. B. Papkovsky, *Anal. Lett.* **32,** 701 (1999).

Oxygen probe: Water-soluble Pt–porphyrin-based type A65N-1 (Luxcel).

Detector: A prompt fluorescence plate reader such as Spectramax Gemini (Molecular Devices) with an excitation wavelength of 380 nm and an emission of 650 nm or a time-resolved fluorescence plate reader, such as Victor[2] (PerkinElmer Life Sciences) a set of 340/642-nm filters, a delay time of 30 μs, a gate time of 90 μs, and an integration time of 1 s per well ($\sim$200 flashes).

Detection mode: Kinetic intensity measurements by scanning the wells periodically.

Temperature: Typically 37° or ambient.

Samples: Various suspension cells including bacteria, yeast, and mammalian cells in appropriate growth media, with or without serum and pH indicator.

Assay Procedure. A stock solution of the oxygen probe is prepared according to the manufacturer's instructions by dissolving the vial content in the assay buffer. To determine the probe optimal working dilution, three-fold serial dilutions of the stock are made in assay medium and measured on the reader. Typically, a probe dilution giving a signal of at least 3 times the blank signal (i.e., without the probe) for prompt fluorescence readers and at least 10 times the blank signal for time-resolved readers is selected for subsequent experiments with cells.

Test cells are cultured, harvested, diluted with growth medium, and dispensed in 150-μl aliquots in wells of a 96-well plate at a final concentration of 10^5–10^7 cells/ml. An oxygen probe is added to each sample at the optimal working concentration (20 nM for the Gemini reader used). The plate is equilibrated at 37° for 10 minutes, and then 100 μl of mineral oil (M5904, Sigma) is applied on top of each sample to exclude atmospheric oxygen. Phosphorescence intensity is measured at regular time intervals (1- to 5-min. cycle) over a period of 30–200 min. Time profiles thus obtained are used to determine the initial slopes of the phosphorescent signal for each sample, which correlate with the rates of oxygen uptake, metabolic activity, and cell numbers. The relationship between the probe signal (slope of the phosphorescence intensity) and cell numbers is shown in Fig. 6.

To investigate the effects of drugs, toxicants, effectors, and hormones, cells are taken at particular cell numbers producing easily measurable respiration profiles in assay conditions, treated with these compounds, and then analyzed at certain time intervals in comparison with untreated cells. Typical profiles of phosphorescence are shown in Fig. 7.

Critical Analysis. This application is similar to the previous example, but the water-soluble oxygen probe is used instead. Spectral characteristics of the probe make it compatible with standard fluorescent plate readers

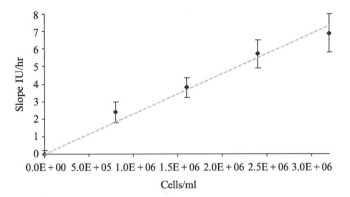

FIG. 6. Graph demonstrating the dependence of the slope of luminescence intensity *vs* cell numbers measured on a steady-state fluorescence plate reader (Gemini) ($n = 6$).

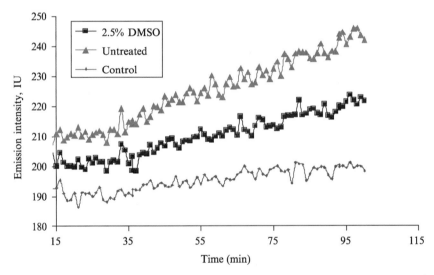

FIG. 7. The effect of dimethyl sulfoxide (DMSO) on the respiration of B cells (1.5×10^5 cells/well): ▲ untreated cells; ■, cells treated with 2.5% DMSO (v/v) and analyzed directly after treatment; and control sample without cells.

available in many research laboratories, and assays can be carried out in standard 96-well plates. This assay format is simple and cost effective, it requires no specialized equipment, and allows flexibility, high sample throughput, automation, and miniaturization (e.g., 384-well plates). It is particularly suited for cell-based screening and drug discovery applications.[38] Assays can be used to measure cell viability, cell proliferation, cell numbers, efficacy of drug/effector, cytobolic and metabolic toxicity, resistance to antibiotics, antimicrobial agents, subcellular systems, for example, mitochondria, oxidative enzymes (CyP450, cytochrome oxidase),[12,38,42] and monitoring of oxygen respiration of small animals.

The lifetime range of the probe (about 20 μs in air-saturated solution and about 80 μs in deoxygenated solutions) is also within the resolution of standard time-resolved plate readers, which can provide additional enhancement of the assay performance. Time-resolved detection allows for higher selectivity, less background and interference by sample components, reduced probe consumption, and the possibility of lifetime-based oxygen sensing.[38] Work with surface-attached cells can be problematic due to relatively small cell numbers per well.

A similar system commercialized by Becton-Dickinson (BD Biosensor) operates with 96- and 384-well plates coated with a Ru-based oxygen probe at the bottom of each well.[12] Such coated plates are used on conventional fluorescence plate readers to measure cell viability, activity of cytochrome P450 enzymes, and ADME toxicity studies of potential drug candidates[42] in a relatively simple and convenient way. Relevant technical information and protocols can be found on the company Web site. However, this system is not suitable for time-resolved detection and lifetime-based sensing due to the short lifetime of the Ru probe. It was reported to have some problems with growing cells in such coated plates: interference by fluorescent compounds.[42] The waste of materials and assay costs are significantly higher than for the water-soluble oxygen probe.

All these assay formats have the same limitations as the ones described in Protocol 2, particularly the inability to monitor low rates of oxygen uptake and, hence, the need of relatively high cell numbers for such assays. A low-volume platform with a more efficient sealing of samples from ambient air and suitable for monitoring small cell numbers and surface-attached cells is under development by Luxcel. The microfluidic LabCD system (Tecan) developed for toxicity screening[49] in which measured samples are isolated from ambient oxygen provides better sensitivity.

Other Oxygen-Sensing Systems

Oxygen sensors by Ocean Optics (USA) operate with a Ru-based coating applied to the tip of an optical fiber linked to a compact diode array detector and a PC. It employs blue LED excitation and intensity

[49] P. Bansal, "LabCD Assay Platform Targets ADME-Tox Bottlenecks," p. 12. Pharmaceutical Laboratory, August (2001).

measurements covering the range of 0–100% (mole percent) of oxygen with a resolution of 0.01% at zero oxygen and 0.04% at ambient oxygen. It was developed for monitoring oxygen in seawater, oxygen respiration, and some other applications. However, sample optical properties, sensor photobleaching, may affect the calibration and system performance. Oxygen microsensors by PreSens GmbH (Germany) employ a similar design, but phase measurements, which provide them improved performance and less interference, do not. Using micromanipulators, these sensors were applied to measure oxygen gradients in sediments with micrometer spatial resolution,[50] tissue oxygenation,[51] oxygen concentration in liquid samples and soil, and so on. Disadvantages of these two systems are they are fragile, not detachable, and have rather expensive sensing tips.

Near-infrared water-soluble oxygen probes based on Pd-tetraphenyl-porphyrins and Pd-benzoporphyrins[40] developed by Professor Wilson's group for oxygen imaging in tissues are administered intravenously or applied locally to animal tissue.[13,20] These systems are described elsewhere.[52] A dedicated phase phosphorimeter was developed for these probes by Oxygen Enterprises (USA), which uses Xe flashlamp excitation and a 6-mm fiber-optic probe. These probes were also suggested for the monitoring of cellular respiration and screening applications.[41] They have more long-wave spectral characteristics than Pt-coproporphyrins (see Table I) and therefore are less compatible with conventional plate readers and spectrometers, which are usually quite insensitive at above 650 nm.

Conclusions

Optical oxygen sensor technology and a number of systems described earlier provide a powerful and versatile tool for biochemical research and allow multiple applications, including noninvasive sensing of oxygen in a closed vessel, measurement of the rates of oxygen uptake by living cells, organisms, oxygen-dependent enzymes, and metabolites, and miniaturized and automated screening assays of cell viability and drug/effector action on living cells. Corresponding products and assay tools are already on the market, making this technology available for routine use in many research laboratories and on a large scale.

[50] I. Klimant, G. Holst, and M. Kuhl, *Limnol. Oceanogr.* **42,** 1638 (1997).
[51] K. Kellner, G. Liebsch, I. Klimant, O. S. Wolfbeis, T. Blunk, M. B. Schulz, and A. Gopferich, *Biotechnol. Bioeng.* **80,** 73 (2002).
[52] D. F. Wilson, *Methods Enzymol.* **381**[44], 2003 (this volume).

Acknowledgments

Financial support of parts of this work by the Irish Health Research Board, Grant RP/106/2002, is gratefully acknowledged. I thank my co-workers and collaborators for their enthusiastic work.

[47] Concurrent ^{31}P Nuclear Magnetic Resonance Spectroscopy and Fiber-Optic Oxygen Consumption Measurements in Perfused Rat Hearts

By PIYU ZHAO, YADONG ZHAO, A. DEAN SHERRY, and PAUL PANTANO

Introduction

High-resolution nuclear magnetic resonance (NMR) spectroscopy is a powerful tool for probing the intermediary metabolism of intact organs and whole animals. Relative substrate utilization by the citric acid cycle, for example, is easily monitored in isolated perfused hearts by ^{13}C NMR spectroscopy, but conversion of such data into quantitative flux values requires an additional measure of tissue oxygen consumption.[1] The calculation of myocardial oxygen consumption requires the measurement of three critical variables (arterial O_2 content, venous O_2 content, and coronary flow rate).[2] Traditionally, arterial and venous O_2 content have been measured by withdrawing one sample of blood or perfusate from the pulmonary artery and another from arterial circulation. However, sampling oxygen from small rodent hearts is problematic because a significant amount of perfusate is required cumulatively for *ex situ* blood gas analyzer determinations. In addition, despite advances in O_2-sampling devices (e.g., in gas-tight syringe systems,[3–5] tube-based systems,[6,7] and sealed glass

---

[1] C. R. Malloy, J. G. Jones, F. M. H. Jeffrey, M. E. Jesson, and A. D. Sherry, *MAGMA* **4,** 35 (1996).

[2] N. B. Radford, B. Wan, C. Storey, A. Richman, and R. W. Moreadith, *Circulation* **98,** I-3303 (1998).

[3] F. M. H. Jeffrey, L. Alvarez, V. Diczku, A. D. Sherry, and C. R. Malloy, *J. Cardiovasc. Pharmacol.* **25,** 469 (1995).

[4] J. R. Neely, H. Liebermeister, E. J. Battersby, and H. E. Morgan, *Am. J. Physiol.* **212,** 804 (1967).

[5] X. Yu, L. T. White, C. Doumen, L. A. Damico, K. F. LaNoue, N. M. Alpert, and E. D. Lewandowski, *Biophys. J.* **69,** 2090 (1995).

[6] Y. Hata, T. Sakamoto, S. Hosogi, T. Ohe, H. Suga, and M. Takaki, *J. Mol. Cell Cardiol.* **30,** 2137 (1998).

chambers[8,9]), obtaining samples without room air contamination from the very small (<1 mm i.d.) pulmonary artery of a mouse heart is technically challenging.

The primary advantages of using *in situ* fiber-optic probes in biomedical NMR applications is that they are miniature and flexible, they alleviate the need to remove circulating medium from perfused organs, they are immune to electrical interferences, and they do not affect the static magnetic and pulsed electromagnetic fields of NMR spectrometers.[10–16] Herein we describe the fabrication of a small (350-μm-diameter) imaging-fiber oxygen sensor (IFOS) that employs a luminescent ruthenium complex entrapped inside a polysiloxane membrane immobilized on the distal face of a coherent fiber-optic bundle (i.e., an imaging fiber). In addition, we report the simultaneous monitoring of phosphorus metabolism (by ^{31}P NMR) and oxygen consumption in isolated rat hearts perfused in a standard narrow-bore 11.75-T magnet.

Optical Oxygen Sensors

The advantages/disadvantages of electrochemical and optical transducers in clinical oxygen partial pressure (PO_2) determinations have been reviewed.[17–21] In brief, the major limitation of a Clark-type electrode occurs at low PO_2s when the quantity of oxygen consumed by electrochemical detection is significant. Conversely, the reversible quenching of a luminescent indicator by oxygen does not consume oxygen.

[7] Y. Hata, T. Sakamoto, S. Hosogi, T. Ohe, H. Suga, and M. Takaki, *Jpn. J. Physiol.* **48,** 197 (1998).

[8] P. M. Matthews, J. L. Bland, D. G. Gadian, and G. K. Radda, *Biochem. Biophys. Acta* **721,** 312 (1982).

[9] K. Ugurbil, M. Petein, R. Maidan, S. Michurski, and A. H. From, *Biochemistry* **25,** 100 (1986).

[10] J. P. Legendre, R. Misner, G. V. Forester, and Y. Geoffrion, *Magn. Reson. Med.* **3,** 953 (1986).

[11] F. M. H. Jeffrey and C. R. Malloy, *Biochem. J.* **287,** 117 (1992).

[12] A. Mayevsky, S. Nioka, and B. Chance, *Adv. Exp. Med. Biol.* **222,** 365 (1988).

[13] F. G. Shellock, S. M. Myers, and K. J. Kimble, *Am. J. Roentgenol.* **158,** 663 (1992).

[14] A. C. Brau, C. T. Wheeler, L. W. Hedlund, and G. A. Johnson, *Magn. Reson. Med.* **47,** 314 (2002).

[15] R. Jayasundar, L. D. Hall, and N. M. Bleehen, *NMR Biomed.* **5,** 360 (1992).

[16] S. Sade and A. Katzir, *Magn. Reson. Imaging* **19,** 287 (2001).

[17] M. A. Arnold, *Anal. Chem.* **64,** 1015A (1992).

[18] M. E. Collison and M. E. Meyerhoff, *Anal. Chem.* **62,** 425A (1990).

[19] C. E. Hahn, *Analyst* **123,** 57R (1998).

[20] B. R. Soller, *IEEE Eng. Med. Biol.* **13,** 327 (1994).

[21] O. S. Wolfbeis, *in* "Molecular Luminescence Spectroscopy, Methods and Applications: Part 2" (S. G. Schulman, ed.), p. 129. Wiley, New York, 1988.

A variety of luminescence-based, fiber-optic oxygen sensors have been constructed using imaging fibers.[22–25] A typical high-resolution 350-μm-diameter imaging fiber comprises ~6000 individually cladded ~3-μm-diameter optical fibers. The use of a bundle is advantageous in that it provides more light-gathering power than a similarly sized individual optical fiber.[26] IFOSs were fabricated using a technically expedient, photoinitiated polymerization reaction. Ultraviolet radiation was directed through the imaging fiber to the distal face submerged in a polymerization solution (i.e., monomer, initiator, and indicator). The central area (r $\approx$ 100 μm) was illuminated for a predetermined length of time during which polymer growth occurred only across the illuminated region (Fig. 1). This immobilization method was chosen because the sensing layer diameter and thickness could be controlled in a straightforward manner. In addition, planar-sensing layers (required for imaging through the bundle and sensing layer[27]) can be fabricated by spin coating the polymerization solution onto the fiber before illumination.

The oxygen-sensing layer used in this work was composed of a transition metal complex, Ru(Ph$_2$phen)$_3^{2+}$, entrapped in a gas-permeable photopolymerizable siloxane membrane (PS802). The transduction mechanism was based on the oxygen collisional quenching of the ruthenium complex luminescence. In brief, appropriate excitation radiation was coupled to the proximal face of the imaging fiber and was propagated to the distal face of the fiber. The indicator–analyte interaction modulated the luminescence intensity, and a portion of the isotropically emitted light returning through the same imaging fiber was directed to a suitable detector [e.g., photomultiplier tube, photodiode array, or charge-coupled device (CCD) camera]. Figure 2 shows a CCD image of luminescence collected from an IFOS that was immersed in buffers with different oxygen tensions. The mean luminescence intensities from IFOS images were used for calibration purposes (Fig. 3). All oxygen consumption values presented were calculated using the postcalibration curves, as these curves were acquired from IFOSs that were exposed to the biological matrix.

[22] B. G. Healey and D. R. Walt, *Anal. Chem.* **69,** 2213 (1997).
[23] J. A. Ferguson, B. G. Healey, K. S. Bronk, S. M. Barnard, and D. R. Walt, *Anal. Chim. Acta* **340,** 123 (1997).
[24] L. Li and D. R. Walt, *Anal. Chem.* **67,** 3746 (1995).
[25] Y. Zhao, A. Richman, C. Storey, N. B. Radford, and P. Pantano, *Anal. Chem.* **71,** 3887 (1999).
[26] J. S. Schultz, in "Biosensors: Fundamentals and Applications" (A. P. F. Turner, I. Karube, and G. S. Wilson, eds.), p. 638. Oxford University Press, Oxford, 1987.
[27] P. Pantano and D. R. Walt, *Anal. Chem.* **67,** 481A (1995).

Fɪɢ. 1. Representative scanning electron micrograph of an IFOS distal tip with a hemispherical sensing layer; the white bar denotes 50 μm. Scanning electron microscopy was performed with a Philips XL 30 LaB$_6$ scanning electron microscope (Mahwah, NJ) operated at $\leq$20 keV. IFOSs were sputter coated with a ~50-Å-thick gold/palladium layer using a Denton Vacuum Desk II sputter coater (Moorestown, NJ).

Concurrent Phosphate Metabolite and Oxygen Consumption Measurements

The determination of myocardial oxygen consumption provides a measure of the total metabolism of a heart, as cardiac tissue relies almost exclusively on the oxidation of substrates for energy generation. In addition, myocardial oxygen consumption measurements can be used as a physiological tool to monitor the hemodynamic status of the heart because it is influenced by a number of parameters such as wall stress, contractility, and heart rate.[28] Previously, we have used IFOSs to acquire *in situ* oxygen

[28] E. Braunwald, "Heart Disease: A Textbook of Cardiovascular Medicine." Saunders, Philadelphia, 1992.

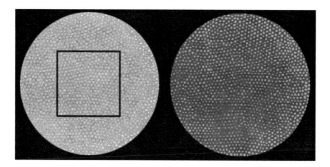

FIG. 2. IFOS luminescence images acquired with a 3.0-s CCD exposure time. The same IFOS was placed in a N_2-purged *(left)* and an O_2-purged MKH buffer *(right)*. White represents high luminescence intensities, and the black box represents a typical region of interest (ROI).

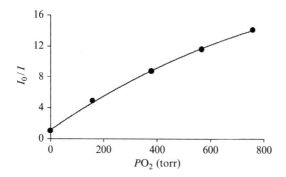

FIG. 3. Representative Stern–Volmer plot for an IFOS.

consumption measurements from intact beating mouse hearts (under basal and pharmacologically enhanced conditions), and we have validated the IFOS method through concurrent blind, *ex situ* blood gas analyzer measurements.[25] The present goal is to combine IFOS monitoring with ^{31}P NMR spectroscopy to acquire concurrent metabolic and oxygen consumption data from isolated perfused rat hearts.

Continuous ^{31}P NMR and IFOSs measurements were made in single hearts during four physiological states: normoxic perfusion, KCl arrest, after KCl washout, and during stimulation of heart function by isoproterenol. The metabolites inorganic phosphate (P_i), phosphocreatine (PCr), and ATP from Langendorff-perfused rat hearts were measured by ^{31}P NMR while two IFOSs [one positioned in the pulmonary artery and the other just above the aorta (retrograde perfusion)] compared oxygen levels (Fig. 4).

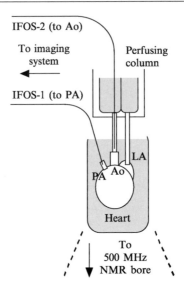

FIG. 4. Schematic diagram of the Langendorff rat heart perfusion system. One IFOS was placed in the perfusion column where 95% O_2-saturated MKH buffer was delivered to the aorta (Ao). The other IFOS was inserted in the pulmonary artery (PA). Both IFOSs were interfaced to the CCD/imaging system.

Table I shows O_2 consumption data and other physiological parameters measured at each stage. The mean O_2 consumption in control hearts was 19.6 ± 1.4 μmol/min/gdw. While coronary flow did not decrease significantly after the addition of 20 mM KCl, the mean O_2 consumption did drop to 3.0 ± 2.5 μmol/min/gdw as anticipated.[6] After KCl washout, left ventricle pressure fully recovered, but the heart rate typically recovered to 80–85% and oxygen consumption returned to 60–65% of the pre-KCl value (12.5 ± 5.8 μmol/min/gdw). After the KCl-washed hearts were exposed to 7.0 μmol isoproterenol, the heart rate, coronary flow, left ventricle pressure, and oxygen consumption (43.5 ± 2.1 μmol/min/gdw) all increased significantly.

Figure 5 shows representative ^{31}P NMR spectra and Table II summarizes the combined phosphorus metabolite results collected from the same heart during each perfusion protocol. While the mean ^{31}P peak areas for ATP tended to be lower in hearts treated with isoproterenol, the measured ATP values did not differ significantly from control, KCl-arrested, or reperfusion conditions. While the mean ^{31}P peak areas for P_i, PCr, and ATP did not change during arrest with KCl or after reperfusion, there was a significant increase in P_i and a significant decrease in PCr during stimulation by isoproterenol.

TABLE I
PHYSIOLOGICAL PARAMETERS OF PERFUSED RAT HEARTS[a]

Perfusion condition	Coronary flow (ml/min)	Heart rate (beats/min)	Left ventricle pressure (mm Hg)	O₂ consumption (μmol/min/gdw)
A (control)	7.0 ± 0.9	310 ± 30	77 ± 3.0	19.6 ± 1.4
B (+KCl)	5.5 ± 0.3	—	—	3.0 ± 2.5
C (recovery after removal of KCl)	5.7 ± 0.3	290 ± 10	75 ± 3.0	12.5 ± 5.8
D (+isoproterenol)	14.0 ± 1.0^b	455 ± 22^b	98 ± 10^b	43.5 ± 2.1^b

[a] Mean $\pm$ standard deviation ($n = 4$ hearts).
[b] $p < 0.01$.

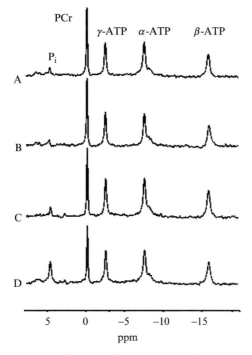

FIG. 5. ³¹P NMR spectra at 500 MHz for a Langendorff rat heart under the following perfusion conditions: (A) MKH buffer; (B) MKH buffer + 20 mmol KCl; (C) MKH buffer; and (D) MKH buffer + 7.0 μmol isoproterenol. Resonances are assigned as follows: inorganic phosphate (P$_i$), phosphocreatine (PCr), γ-phosphate of ATP (γ-ATP), α-phosphate of ATP (α-ATP), and β-phosphate of ATP (β-ATP).

TABLE II
Phosphate Metabolites in Rat Hearts Perfused with 2 mM Acetate[a]

Perfusion condition	(P_i) (μmol/gdw)	(PCr) (μmol/gdw)	(β-ATP) (μmol/gdw)
A (control)	4.0 ± 0.2	23.9 ± 1.6	17.8 ± 0.5
B (+KCl)	3.9 ± 1.6	25.8 ± 1.8	17.7 ± 0.4
C (recovery after removal of KCl)	5.3 ± 0.7	24.7 ± 3.0	17.8 ± 1.5
D (+isoproterenol)	14.3 ± 1.9^b	19.3 ± 2.2^c	16.6 ± 0.8

[a] Mean $\pm$ standard deviation ($n = 4$ hearts); the fully relaxed ^{31}P spectrum was acquired with an 11-s interpulse delay using a 45° pulse. Subsequent ^{31}P spectra of the same heart were acquired with a 1.3-s interpulse delay. Therefore, the saturation factor of each peak was calculated and used for saturation correction in partially saturated spectra with a pulse delay of 1.3 s.
[b] $p < 0.01$.
[c] $p < 0.05$.

In summary, an *in situ* fiber-optic sensor was demonstrated to monitor O_2 consumption in isolated, perfused rat hearts within the 51-mm-wide bore of a high-field magnet. We believe the combined NMR and IFOS methodology could prove important for the sequential measurement of metabolism and O_2 consumption in the organs of transgenic mice,[29] and thus find utility in future investigations of the physiological and biological consequences of genetic engineering in mouse models.

Experimental Methods

Materials and Solutions

Chemicals. Benzoin ethyl ether, isoproterenol, and 3-(trimethoxysilyl)-propyl methacrylate (TPM) are from Aldrich Chemical Co. (Milwaukee, WI). (Acryloxy)propylmethyl siloxane ($\sim$18%) + dimethylsiloxane ($\sim$82%) copolymer (PS802) is from Gelest Inc. (Tullytown, PA). Tris(4,7-diphenyl-1,10-phenanthroline) ruthenium (II) chloride hydrate [Ru(Ph$_2$phen)$_3^{2+}$] is synthesized according to the procedure of Klimant *et al.*[30]; it is now available commercially from Fluka Chemical (Milwaukee, WI) and GFS Chemical Co. (Powell, OH). Deionized water (18 MΩ-cm) is from a Nanopure Infinity water purification system (Barnstead, Dubuque,

[29] R. W. Moreadith and N. B. Radford, *J. Mol. Med.* **75**, 208 (1997).
[30] I. Klimant and O. S. Wolfbeis, *Anal. Chem.* **67**, 3160 (1995).

IA). All other chemicals are from commercially available sources and are of the highest quality available.

PS802 Prepolymer Solution. The PS802 prepolymer solution is prepared by mixing 200 μl PS802 and 15 mg benzoin ethyl ether with a 400-μl CH_2Cl_2 solution of 2.0 mg/ml $Ru(Ph_2phen)_3^{2+}$.

Modified Krebs–Henseleit (MKH) Buffer. MKH buffer contains 118 mM NaCl, 5.0 mM KCl, 1.25 mM $CaCl_2$, 1.0 mM $MgSO_4$, 25 mM $NaHCO_3$, and 5 mM acetate as the only oxidizable substrate.

Imaging Fibers. Silica imaging fibers with a $\sim$350-μm-total diameter comprising $\sim$6000 individually cladded $\sim$3-μm-diameter optical fibers are from Sumitomo Electric Industries (Part No. IGN-035/06; Torrance, CA). The active imaging diameter of the coherent bundle is $\sim$270 μm (i.e., the individual optical fibers are fused within a $\sim$15-μm-thick silica tube, and this tube is coated with a $\sim$25-μm-thick silicone jacket).

Imaging Fiber Oxygen Sensor Fabrication

IFOS fabrication is optimized by careful selection of the oxygen-sensing dye ($Ru(Ph_2phen)_3^{2+}$), the gas-permeable polymer matrix (PS802), and the immobilization method (photopolymerization) to produce sensitive and accurate IFOSs for use across the entire 0- to 760-torr PO_2 range.[25] IFOS fabrication begins by polishing both ends of a 20-foot-long imaging fiber successively with 12-, 3-, 1-, and 0.3-μm lapping films (Mark V Laboratory, East Granby, CT) using a fiber-polishing chuck (General Fiber Optics Model No. 30–21; Fairfield, NJ). Polished fiber tips are sonicated in deionized water for 30 s after each step to remove any lapping film residuals and fiber particulates. Polished imaging fibers are silanized in a 10% (v/v) TPM/acetone solution for 1 h and are cured at room temperature for at least 1 h. The TPM-modified imaging fiber tip is placed in the PS802 prepolymer solution, and the proximal face is mounted in the epifluorescence microscope stage (see later discussion). A $\sim$25-s radiation pulse is delivered to initiate the photopolymerization reaction. All IFOSs are rinsed with ethanol and are eventually stored in deionized water.

Photopolymerization/Imaging System

A modified epifluorescence microscope system (Labophot 1A; Nikon, Irving, TX) similar to that described in Bronk *et al.*[31] is used for photopolymerization and for luminescence imaging. The collimated radiation from a 100-W mercury arc lamp is reflected 90° by a 410-nm-long pass filter

[31] K. S. Bronk, K. L. Michael, P. Pantano, and D. R. Walt, *Anal. Chem.* **67,** 2750 (1995).

(Chroma Technology, Battleboro, VT) and focused onto the proximal face of the imaging fiber by a 20× E-Plan microscope objective. Ultraviolet radiation is transmitted through the imaging fiber to the distal face where it initiates PS802 polymerization. A manual shutter is used to control the photopolymerization time.

In luminescence detection mode, the proximal face of the imaging fiber is mounted to the microscope stage, and the distal tip of the IFOS (Fig. 1) is immersed in MKH buffer. The collimated excitation radiation is passed through two neutral density filters (ND2 + ND4) to control the intensity. The light is passed through a 500 ± 40-nm excitation filter (Omega Optical, Battleboro, VT) reflected by a 560 DRLP dichroic mirror (Omega Optical) and is focused on the proximal face of the imaging fiber by a 10× E-Plan microscope objective. Light is transmitted through the imaging fiber to the distal face where it excites the immobilized luminescent-sensing dye. The luminescence is collected by the same fiber and the same microscope objective. The luminescence is transmitted through the dichroic mirror and is filtered by a 580 ± 15-nm emission filter (Omega Optical). The filtered luminescence image is captured by a scientific-grade CCD camera. The 768 × 512 pixel, 100-kHz, 16-bit, −30° cooled Teleris 2 CCD camera is from SpectraSource Instruments (Westlake Village, CA). CTXK imaging software and a Pentium-II PC are used to control CCD functions and to process all images. In general, a 200 × 200 region of interest (ROI) is selected from the image of a sensing layer of an imaging fiber (Fig. 2) for quantitation purposes. All luminescence images are background subtracted to account for scattered light; specifically, the mean intensity from a ROI of the silica jacket is subtracted from the mean luminescence intensity of the ROI of the sensing layer.

Imaging Fiber Oxygen Sensor Calibration

IFOSs are calibrated at 37° in MKH buffers that have different oxygen tensions (Fig. 3). The different oxygen tensions are achieved by bubbling MKH buffers for ≥15 min through fritted stones connected to the appropriate oxygen/nitrogen-compressed gas tanks. O_2 (99.99 + %), N_2 (99.99 + %), and O_2/N_2 mixtures (74.8 and 49.9% O_2, balance high purity N_2) are from Air Liquide (Grand Prairie, TX). All gases are humidified at 37°, and all oxygen partial pressures (PO_2) are corrected using the daily barometric pressure and partial pressure of water (PH_2O) at 37°.[32] All IFOS measurements are fitted using the two-site nonlinear Stern–Volmer equation:

[32] A. C. Guyton, *in* "Textbook of Medical Physiology," 5th Ed. Saunders, Philadelphia, 530–542 (1976).

$$I_0/I = 1/\{f_{01}/(1 + K_{sv1} \bullet PO_2) + f_{02}/(K_{sv2} \bullet PO_2)\}$$

where f_{0i} values are the fraction of the total emission from each component under unquenched conditions and K_{svi} values are the associated Stern–Volmer quenching constants for each component.[33] IFOSs are calibrated before and after each rat heart perfusion. There was no significant difference between PO_2 values calculated from pre- and postcalibration curves ($p \leq 0.05$; students t test, $n = 4$ experiments, data not shown), indicating that IFOSs are not fouled by exposure to wet tissue.

Langendorff Heart Perfusion and Oxygen Consumption Measurements

Male Sprague–Dawley rats, 250–300 g, are from Sasco Inc. (Houston, TX). The care and treatment of all rats in this study are in accord with the recommendations of the National Institutes of Health and the U.S. Department of Health and Human Services. The Langendorff rat heart perfusion method has been detailed previously.[34,35] Briefly, rats are anesthetized in an ether atmosphere before hearts are removed surgically and placed in an ice-cold MKH perfusion medium (protocol approved by the University of Texas Animal Care Committee). Aortas are cannulated within 15–30 s, and hearts are perfused using standard Langendorff methods at a column height of 70 cm H_2O (Fig. 4). A short, open-ended cannula is inserted into the pulmonary artery (PA) and secured with a suture. The MKH buffer is oxygenated with 95% O_2/5% CO_2 (Air Liquide), and the entire recirculation system is water jacketed and maintained at 37°. One end of small-diameter polyethylene tubing (~3 m in length, filled with buffer) is held in the left ventricle by a suture while the other end is connected to an OHMEDA pressure transducer (Model P23XL; Madison, WI). Spontaneous heart rate and left ventricle pressure are measured on a Colbourne Instruments polygraph (Lehigh Valley, PA). Coronary flow is measured sequentially over 5-min intervals throughout the experimental protocol.

Luminescence measurements are acquired concurrently from two IFOSs to determine differential PO_2 across the functioning heart. The distal end of one IFOS is placed in the distal opening of the PA cannula while the distal end of a second IFOS is placed a few millimeters above the aortic

[33] J. N. Demas and B. A. Degraff, *in* "Topics of Fluorescence Spectroscopy," p. 71. Plenum Press, New York, 1994.
[34] C. R. Malloy, D. C. Buster, M. M. Castro, C. F. Geraldes, F. M. H. Jeffrey, and A. D. Sherry, *Magn. Reson. Med.* **15**, 33 (1990).
[35] A. D. Sherry, P. Zhao, A. J. Wiethoff, F. M. H. Jeffrey, and C. R. Malloy, *Am. J. Physiol.* **274**, H591 (1998).

cannula. The perfusion apparatus (including the pressure transducer line and two IFOSs) is placed inside the narrow-bore magnet. The proximal face of each IFOS is interfaced sequentially to the CCD/imaging system, and three luminescence measurements from each IFOS are acquired and averaged during a 7-min NMR spectral accumulation.

A Varian INOVA 500 (11.75T) spectrometer operating at 202 MHz for ^{31}P is equipped with an 18-mm probe that can be tuned to either ^{23}Na or ^{31}P. The magnetic field homogeneity is adjusted by using the ^{23}Na-free induction decay to a linewidth of $\sim$14 Hz. ^{31}P spectra are acquired using a 45° pulse and a 1.3-s repetition rate over 7 min (360 scans).

^{31}P spectra are acquired sequentially during perfusion with MKH buffer (control), MKH buffer + 20 mmol KCl (arrest), MKH buffer (recovery), and MKH buffer + 7.0 μmol isoproterenol (Fig. 5). Specifically, rat hearts are perfused first with MKH buffer under normoxic conditions for $\sim$20 min while ^{31}P spectra and oxygen consumption measurements are acquired. Next, 20 mmol KCl is added to the perfusate, and a second set of spectra and oxygen measurements are collected. The buffer is switched back to normal MKH buffer (nonrecirculating for $\sim$15 min) to remove excess KCl, and a third series of spectra and oxygen measurements are collected. Finally, 7.0 μmol isoproterenol is added to the perfusate to stimulate oxygen consumption while a fourth data set is acquired.

At the conclusion of the experimental protocol, each heart is removed from the cannula, blotted dry, and placed in a $\sim$60° oven to constant dry weight. Oxygen consumption is calculated as

$$\text{Oxygen consumption } (\mu\text{mol/min/gdw}) = (\Delta AV_{PO_2} \times \text{CFR} \times 1.18 \times 10^{-3})/\text{gdw}$$

where ΔAV_{PO_2} (mm Hg) is the oxygen partial pressure difference between input and output MKH perfusates; CFR is the coronary flow rate in milliliters per minute; gdw is the dry weight of heart tissue in grams: and 1.18×10^{-3} (units = μmol/ml•mm Hg) is a constant derived from the Bunsen solubility coefficient for O_2 (0.03 μl O_2/ml•mm Hg) and the volume of 1.0 μmol O_2 (25.4 μl) at 37°.

Acknowledgments

We gratefully acknowledge the Robert A. Welch Foundation (P.P) and the National Institutes of Health (A.D.S.) for financial support of this work.

[48]   Microximetry: Simultaneous Determination
of Oxygen Consumption and Free Radical
Production Using Electron Paramagnetic
Resonance Spectroscopy

By GOVINDASAMY ILANGOVAN,
JAY L. ZWEIER, and PERIANNAN KUPPUSAMY

Introduction

The respiration of cells and cellular components such as mitochondria is conventionally studied using a respirometer, which uses the Clark electrode to measure oxygen concentration. The Clark electrode is also used to measure oxygen consumption in enzymatic reactions. The amount of sample required to use this technique is typically about 2–5 ml.[1] However, on many occasions the sample amount limits the use of this technique. Moreover, although this technique is suitable in the concentration range of 10–100 mm Hg pO_2, it imposes problems at higher and lower ranges. At lower ranges, the readouts are not accurate, as during the measurement it depletes the oxygen, perturbing the local oxygen concentration. At higher concentration ranges, it loses the linearity in the response (the current) vs concentration plot. Thus it is essential to develop an alternative technique with equal or even higher accuracy and resolution, but with a lower sample requirement. In this regard we envision that electron paramagnetic resonance (EPR) spectroscopy-based oximetry with a highly sensitive probe such as lithium phthalocyanine (LiPc) microcrystals could be very useful. The sample requirement for this technique would be less than 10 μl ("microximetry"). Because the microximetry is based on EPR, an added advantage of this technique is that it is also possible to simultaneously study the free radicals generated, if any, in the enzymatic reactions or during the cellular respirations using a spin-trapping technique with an appropriate spin trap added to the medium. Thus both the oxygen consumption rate and the quantitation of free radicals can be obtained in a single experiment. This requirement is very important, as in almost all enzymes and cellular systems, oxygen consumption is always associated with the generation of reactive oxygen species. In the existing methods, oxygen consumption and free radical generation should be done independently. This article illustrates the use of this technique in two examples: the

[1] E. Gnaiger, *Respir. Physiol.* **128,** 277 (2001).

xanthine/xanthine oxidase and stimulated human leukocyte systems. The simultaneous measurement of oxygen consumption and free radical production should yield a correlation and interpretation of oxygen/free radical flux in many complex enzymatic reactions and cellular respirations.

Principle of Electron Paramagnetic Resonance Oximetry

Molecular oxygen is paramagnetic and has two unpaired electrons in the ground state. Nevertheless, it cannot be detected directly by EPR at ambient conditions due to fast relaxation of spins from the excited state, resulting in broad and low-amplitude EPR spectrum. However, it can be measured indirectly from the oxygen-induced changes in the EPR spectrum of other suitable paramagnetic probes (with a relatively lower relaxation rate) in the system. The EPR spectrum of the spin probe is broadened due to the O_2. From oxygen-induced EPR line broadening, one can quantitate the oxygen concentration.[2] EPR line broadening occurs due to the Heisenberg spin–spin exchange between molecular oxygen and the spin probe. This exchange shortens the relaxation time of the spin probe, causing an increase in the peak-to-peak width in the first-derivative EPR spectrum. This spin exchange occurs when a collision between molecular oxygen and spin probe occurs. As the number of such a collision increases, the spin exchange increases proportionately, resulting in proportionately increased EPR line widths. In systems wherein the spin probe and molecular oxygen are freely tumbling, the effect of oxygen on the spin probe can be quantified in terms of bimolecular collision frequency ω, which is related to the oxygen concentration as defined by the classic Smoluchowski equation,[3]

$$\omega = 4\pi R \{D_{O2} + D_{SL}\} [O_2] \tag{1}$$

where R is the interaction distance, D_{O2} and D_{SL} are the diffusion coefficients of oxygen and the spin probe, respectively, and $[O_2]$ is the concentration of the oxygen. The peak-to-peak line width (LW) increase, caused by the molecular oxygen, is directly proportional to the collision rate and hence proportional to the O_2 concentration as per Eq. (1). Thus a plot of LW *vs* $[O_2]$ is expected to be linear and hence the relationship can be used to obtain any unknown oxygen concentration. This relationship has been proved experimentally and, using this principle, many magnetic resonance

[2] H. M. Swartz and J. F. Goyner, *in* "EPR Imaging and In Vivo EPR" (G. R. Eaton, S. S. Eaton, and K. Ohno, eds.). CRC Press Boca Raton, FL, 1991.

[3] J. S. Hyde and W. K. Subczynski, *in* "Biological Magnetic Resonance" (L. R. Berliner, ed.), Vol. 18. Plenum Press, New York, 1988.

imaging techniques, including nuclear magnetic resonance imaging,[4] Overhauser magnetic resonance imaging,[5] and EPR imaging,[6] have been developed to enable the spatially resolved mapping of oxygen concentrations.

Particulate EPR Oximetry Probes

The EPR oximetry is unique in its potential for high-sensitivity local oxygen measurements. The EPR technique provides real-time and nonperturbing measurements of oxygen. However, its application has been limited by the difficulties in preparing suitable spin probes capable of undergoing effective spin exchange with O_2 and providing high-resolution oxygen data over a wide range of experimental conditions. Although there are a variety of soluble EPR spin probes, such as nitroxides, suitable for oximetry, they lack in sensitivity, especially in the most desired pathophysiological oxygen range (0–100 mm Hg). Moreover, they are susceptible to bioreduction/oxidation by the cells *in vivo*, resulting in instability of the probe. Thus, many particulate probes, with a relatively high spin density, such as charcoal,[7] fusinite,[8] and India ink,[9,10] have been developed. Similarly, many synthetic materials, such as lithium phthalocyanine (LiPc, Fig. 2) and sugar chars, have also been recognized as highly sensitive probes.[11–17] The particulate probe-based EPR oximetry has been demonstrated as a viable technique and is accurate and reproducible for repeated measurements of local

[4] M. K. Stehling, F. Schmitt, and R. Ladebeck, *J. Magn. Reson. Imag.* **3**, 471 (1993).

[5] J. H. Ardenkjaer-Larsen, I. Laursen, I. Leunbach, G. Ehnholm, L. G. Wistrand, J. S. Petersson, and K. Golman, *J. Magn. Reson.* **133**, 1 (1998).

[6] S. S. Velan, R. G. Spencer, J. L. Zweier, and P. Kuppusamy, *Magn. Reson. Med.* **43**, 804 (2000).

[7] G. He, R. A. Shankar, M. Chzhan, A. Samouilov, P. Kuppusamy, and J. L. Zweier, *Proc. Natl. Acad. Sci. USA* **96**, 4586 (1999).

[8] N. Vahidi, R. B. Clarkson, K. J. Liu, S. W. Norby, M. Wu, and H. M. Swartz, *Magn. Reson. Med.* **31**, 139 (1994).

[9] F. Goda, K. J. Liu, T. Walczak, J. A. O'Hara, J. Jiang, and H. M. Swartz, *Magn. Reson. Med.* **33**, 237 (1995).

[10] H. M. Swartz, K. J. Liu, F. Goda, and T. Walczak, *Magn. Reson. Med.* **31**, 229 (1994).

[11] G. Ilangovan, J. L. Zweier, and P. Kuppusamy, *J. Phys. Chem. B* **104**, 9404 (2000).

[12] G. Ilangovan, H. Li, J. L. Zweier, and P. Kuppusamy, *J. Phys. Chem. B* **105**, 5325 (2001).

[13] G. Ilangovan, J. L. Zweier, and P. Kuppusamy, *J. Phys. Chem. B* **104**, 4047 (2000).

[14] G. Ilangovan, H. Li, J. L. Zweier, M. C. Krishna, J. B. Mitchell, and P. Kuppusamy, *Magn. Reson. Med.* **48**, 723 (2002).

[15] G. Ilangovan, H. Li, J. L. Zweier, and P. Kuppusamy, *Mol. Cell Biochem.* **234**, 393 (2002).

[16] J. L. Zweier, M. Chzhan, U. Ewert, G. Schneider, and P. Kuppusamy, *J. Magn. Reson. B* **105**, 52 (1994).

[17] P. Kuppusamy, H. Li, G. Ilangovan, J. L. Zweier, and J. B. Mitchell, *Free Radic. Biol. Med.* **31**, 415 (2001).

concentrations of molecular oxygen in various *in vivo* environments such as cerebral interstitial regions of brain,[18] tumor,[19] and the ischemic heart.[16,20] The LiPc, in particular, has several other advantages, including an extremely sharp EPR absorption line (26 mG) and a very short response time (i.e., when exposed to O_2 the line width changes within a few milliseconds), etc.

Lithium Phthalocyanine Microcrystals as Oxygen-Measuring Probe

LiPc in the form of microcrystals (Fig. 2) has been found to be a useful EPR oximetry probe for *in vivo* biological applications.[12,14,15,17,21,22] However, the synthesis of LiPc in the pure and desirable form and usable in EPR oximetry is hitherto considered to be a critical issue.[11] It shows polymorphism and, depending on the preparation condition, it can crystallize in three different structures, namely the oxygen-sensitive (i.e., capable of undergoing spin exchange with O_2) x form, the commonly obtained α form, and the relatively rare β form. Critical changes in their magnetic behavior are observed for these polymorphs due to their very different crystal structures. Detailed EPR studies[23–31] of these three polymorphs have shown very different magnetic properties strongly correlating to the nature of molecular packing. Among the three known polymorphs of LiPc, Heisenberg spin exchange with O_2 is reported to be effective only in the x form, and thus line broadening in the presence of O_2 is observed only for the x form (oxygen sensitive[30]). The line width of the x form is observed to be

[18] K. J. Liu, G. Bacic, P. J. Hoopes, J. Jiang, H. Du, L. C. Ou, J. F. Dunn, and H. M. Swartz, *Brain Res.* **685,** 91 (1995).

[19] J. F. Dunn, S. Ding, J. A. O'Hara, K. J. Liu, E. Rhodes, J. B. Weaver, and H. M. Swartz, *Magn. Reson. Med.* **34,** 515 (1995).

[20] B. J. Friedman, O. Y. Grinberg, K. A. Isaacs, T. M. Walczak, and H. M. Swartz, *J. Mol. Cell Cardiol.* **27,** 2551 (1995).

[21] B. J. Friedman, O. Y. Grinberg, S. A. Grinberg, and H. M. Swartz, *J. Mol. Cell Cardiol.* **29,** 2855 (1997).

[22] K. J. Liu, P. Gast, M. Moussavi, S. W. Norby, N. Vahidi, T. Walczak, M. Wu, and H. M. Swartz, *Proc. Natl. Acad. Sci. USA* **90,** 5438 (1993).

[23] H. Wachtel, J.-J. Andre, W. Bietsch, and J. U. von Shulz, *J. Chem. Phys.* **102,** 5088 (1995).

[24] F. Bensebaa and J.-J. Andre, *J. Phys. Chem.* **92,** 5739 (1992).

[25] M. Brinkmann and J.-J. Andre, *J. Mater. Chem.* **9,** 1511 (1999).

[26] M. Brinkmann, S. Graff, C. Chaumont, and J.-J. Andre, *J. Mater. Res.* **14,** 2162 (1999).

[27] M. Brinkmann, P. Turek, and J.-J. Andre, *J. Mater. Chem.* **8,** 675 (1998).

[28] M. Brinkmann, P. Turek, and J.-J. Andre, *Thin Solid Films* **303,** 107 (1997).

[29] H. Wachtel, J. C. Wittmann, B. Lotz, M. A. Petit, and J.-J. Andre, *Thin Solid Films* **250,** 219 (1994).

[30] M. Brinkmann, C. Chaumont, H. Wachtel, and J.-J. Andre, *Thin Solid Films* **283,** 97 (1996).

[31] H. Wachtel, J. C. Wittmann, B. Lotz, and J.-J. Andre, *Synth. Met.* **61,** 139 (1993).

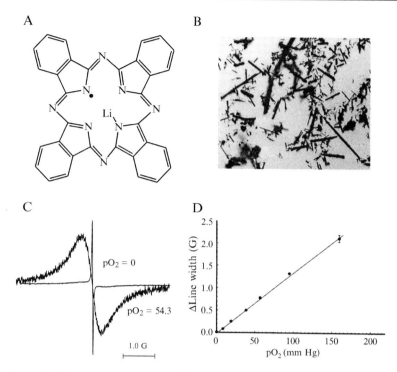

FIG. 1. (A) Molecular structure of a lithium phthalocyanine (LiPc)-free radical oximetry probe. (B) Micrograph of LiPc microcrystals. (C) Effect of oxygen on the EPR spectrum of LiPc. (D) Calibration plot: a linear relationship between pO_2 and LW (peak to peak) is observed for LiPc microcrystalline powder.

extremely narrow, about 26 mG in high vacuum, whereas the other two forms exhibit line widths of about 200 and 400 mG, respectively.[25,26] Figure 1 shows micrographs of the needle-shaped microcrystals of the x form.

Spin Trap for Free Radicals: 5, 5-(Diethoxyphosphoryl)-5-methyl-1-pyrroline-N-oxide (DEPMPO)

Direct detection of free radicals such as superoxide ($O_2^{\bullet-}$) and hydroxyl ($^{\bullet}OH$) radicals is extremely difficult in biological or physiological experimental conditions because these species are short lived. Thus consequences of their generation are measured as the experimental observable. For example, $O_2^{\bullet-}$ species react with hydroethidine to produce ethidine, which is fluorescent, and measurement of the fluorescence is taken as the indirect

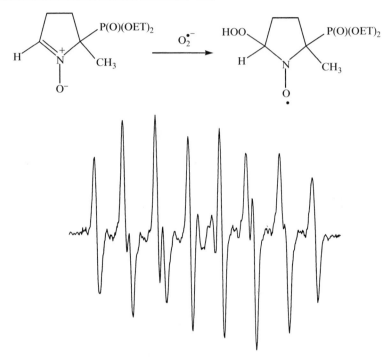

Fɪɢ. 2. (Top) The EPR silent nitrone (left) traps the superoxide radical species generated in the biochemical and physiological reaction and becomes a stable nitroxide radical with a half-life time of 14 min. (Bottom) EPR spectrum of the DEPMPO-OOH adduct. Following the intensity of one of the peaks, one can measure the formation and decay kinetics of the spin adduct.

measure of free radical generation. Similarly, another easy and viable technique is EPR-based "spin trapping," meaning that the generated $O_2^{\bullet-}$ can be trapped rapidly using a spin trap (generally the nitrones), which results in a stable-free radical and can be subsequently detected using EPR spectroscopy (Fig. 2). The resulting spin adduct spectrum is the fingerprint of the nature of the free radicals generated. From the position of spectral lines and amplitude ratios within the obtained EPR spectrum, the free radicals can be identified. There are several issues to be considered before choosing a trap and to reliably use this technique to measure $O_2^{\bullet-}$. However, the key issue is the rate of the reaction between the $O_2^{\bullet-}$ and the trap (trapping efficiency) and the stability of the resulting spin adduct. The DEPMPO has many advantages over the other traps in use. We have reported on the estimation of $O_2^{\bullet-}$ using DEPMPO as the spin trap.[32,33] The DEPMPO reacts very fast with superoxide and other radicals and is stable for longer time, approximately 14 min as its half-life time.

A typical EPR spectrum of the DEPMPO superoxide adduct (DEPMPO-OOH) has multiple lines due to hyperfine splitting caused by the magnetic N and P nuclei coupling shown in Fig. 2. One important advantage of the splitting in the EPR spectrum is that the positions of the split lines are different for the $O_2^{\bullet-}$ and $OH^{\bullet}$ adducts and thus one can evaluate each fraction from the magnitudes of the split lines of a single measurement. Although there are multiple peaks, they can be analyzed conveniently and the components can be quantified using simulation and fitting procedures.

Simultaneous Measurements of Both Oxygen and Free
   Radical Concentrations

Spin trapping and EPR oximetry have been used as independent techniques for some time. However, the possibility of combining these two techniques in a single experiment was developed in our laboratory after the advent of LiPc microcrystals as an oximetry probe. Because LiPc microcrystals are small enough to accommodate in the microreaction vessel (and also being chemically inert), it can be used conveniently along with the experimental solution. More importantly, the EPR characteristics of the LiPc microcrystals and spin traps are different. We used the difference in EPR properties of LiPc microcrystals and DEPMPO-OOH spin adducts to measure the oxygen and the free radicals simultaneously as follows. (i) The microwave power saturation properties of the two species are widely different such that in the extremely low microwave power (about 50 μW) and the small field sweep window (1–5 G) used for the detection of the LiPc signal, the EPR spectrum of the DEPMPO-OOH spin adducts is not observed. However, with the power levels (about 20 mW) used for recording the EPR spectrum of the DEPMPO spin adducts, the LiPc is completely saturated and hence its EPR spectrum appears in the middle of the spin adduct spectrum overlapping with only two middle peaks. Quantitative analysis can still be carried out accurately. (ii) The solid-state nature of the LiPc has the advantage that it does not interfere chemically with the spin trap or the oxygen radicals. Thus, we can measure both

[32] V. Roubaud, S. Sankarapandi, P. Kuppusamy, P. Tordo, and J. L. Zweier, *Anal. Biochem.* **247**, 404 (1997).
[33] V. Roubaud, S. Sankarapandi, P. Kuppusamy, P. Tordo, and J. L. Zweier, *Anal. Biochem.* **257**, 210 (1998).

quantities of oxygen and oxygen radicals simultaneously by the switching of microwave power settings during the reaction and the associated kinetics can be evaluated in real time.

Experimental Methods

Electrochemical Synthesis and Calibration of LiPc Probes

Because there is no commerical source of this oximetry probe, it needs to be synthesized in the laboratory. We have established the synthesis procedure. Electrochemical synthesis of the LiPc microcrystals can be carried out using the published procedure.[13] Electrosynthesized LiPc microcrystals are dark green-colored, needle-shaped crystals as illustrated in Fig. 1. The calibration curve (the EPR line width vs pO_2) is constructed by measuring the EPR line width at different known pO_2, as described elsewhere.[12] The slope of the calibration curve measured in the range of pO_2 10–100 mm Hg is 6.3 mG/mm Hg.

Microximetry Experimental Setup

EPR measurements of the two illustrated examples in this article were carried out using a Bruker ER-300 EPR spectrometer operating at the X band (9.78 GHz) fitted with a TM_{110} microwave cavity. Because EPR-based oximetry is highly sensitive, a few microliters of reaction mixture are required to carry out these measurements in enzymatic reactions or cellular respirations. The EPR line width sensitivity to oxygen does not depend on the solution volume. Two kinds of experiments can be carried out using the setup described in Fig. 3. A custom-designed, gas-permeable Teflon tube (Zeus Industrial Products, Orangeburg, SC) with an inner diameter of 0.8 mm can be used as a microreaction vessel. The advantage of such a setup is that because the Teflon tube wall is gas permeable, the pO_2 of the reaction mixture or cell suspensions inside the Teflon tube can be varied by simply passing the required O_2 through the inlet in the 3-mm EPR tube (Fig. 3). Alternatively, microtubes with a gas-impermeable quartz wall and with a typical volume of 25 μl or less can also be used whenever an oxygen concentration change needs to be followed.

Experimental Procedure

The reaction mixture is saturated with the required amounts of oxygen (such as room air or any desired percentage of oxygen) and about 20 μg LiPc microcrystals, and the DEPMPO (10 mM) is added and sampled into

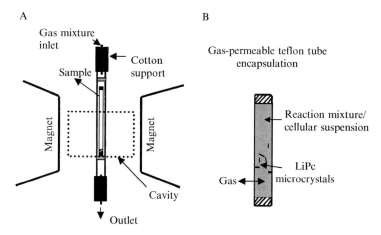

FIG. 3. (A) Schematic illustration of the details of EPR measurements of O_2 and free radicals from reaction mixture of an enzyme or cellular respiration. (B) Expanded view of the microencapsulation of cells and the enzymatic reaction mixture.

10-μl quartz microtubes. The tube is sealed at both ends using special clay (Critoseal, Oxford Labware, St. Louis, MO). While sealing, care needs to be taken to ensure that there is no air gap present inside the tube, as such a gap may act as an additional oxygen source. The sealed tube is placed inside the microwave cavity, and data acquisition is started immediately. Additional care needs to be taken to avoid settling down of both LiPc particulates and cells at the bottom. Such a situation may result in excess consumption and a vibrant change in pO_2 at the bottom of the microtube, and the oxygen consumption pattern may change, leading to the wrong conclusion that cellular oxygen consumption follows biphasic pathways. To avoid settling down of the LiPc particles and cells in the tube, the EPR cavity should be rotated 90° with an appropriate wave guide so that the tube is kept in the horizontal position. Data acquisition for the examples presented in this article is carried out using EPR 2000, a custom-developed PC software, which is capable of doing automated data acquisition and processing. Because the sweep width and microwave power required are widely different for DEPMPO spin adducts and LiPc, the software is programmed to perform an automated acquisition of spectrum with two sets of predefined parameters. The EPR spectra of DEPMPO adducts is measured using a 200-G sweep width, 20 mW power, whereas the LiPc spectrum is obtained using a 2.0-G sweep width and 20 μW power.

Procedure for Simultaneous Measurement of O_2 Consumption and Superoxide Generation

The formation and decay reaction sequence of DEPMPO-OOH adduct is analyzed quantitatively by considering the following reactions:

$$O_2 \xrightarrow[k_0]{\text{Enzyme}} O_2^{\bullet-} \xrightarrow[k_f]{\text{DEPMPO}} \text{DEPMPO-OOH} \xrightarrow{k_d} \text{Decomposed product}$$

In the aforementioned reactions, the reaction between $O_2^{\bullet-}$ and DEPMPO is very fast and thus is not the rate-limiting step. The kinetic expressions for the consecutive reaction can be deduced to estimate the oxygen consumption rate and the actual concentration of $O_2^{\bullet-}$ that is formed in the reaction. The following rate equations can be derived:

$$[O_2] = [O_2]_0 \exp(-k_0 t) \tag{2}$$

$$[\text{DEPMPO-OOH}] = \{k_f[\text{DEPMPO-OOH}]_{max}/(k_d - k_f)\}$$
$$[\exp(-k_f t) - \exp(-k_d t)] \tag{3}$$

where $[O_2]_0$ is the initial concentration of O_2 (220 μM if the solution is room air saturated), $[\text{DEPMPO-OOH}]_{max}$ is the maximum DEPMPO-OOH concentration possible in the event of no decomposition, and k_0, k_f, and k_d are pseudo-first-order rate constants of oxygen consumption, DEPMPO-OOH formation, and decay, respectively. Formation and decay constants are expressed as overall rate constants, which contain various individual steps embedded. Oxygen concentration and spin adduct data are fit to Eqs. (2) and (3) to obtain the rate constant of the overall oxygen consumption in the medium and superoxide formation and decay.

Illustrated Examples

Quantitation of Oxygen Consumption and Superoxide Generation by Xanthine/Xanthine Oxidase (X/XO)

Application of the EPR microximetry for the simultaneous measurements of O_2 and free radicals was demonstrated in the X/XO system and stimulated human neutrophils as model systems, as these systems are known to produce relatively large amounts of free radicals. The enzymatic conversion of xanthine (X) to uric acid by xanthine oxidase (XO) is an important metabolic process that utilizes oxygen, and a fraction of consumed oxygen is leaked as oxygen radicals (Fig. 4), otherwise an energetically tight and well-conserved sequence of coupled redox processes. Because

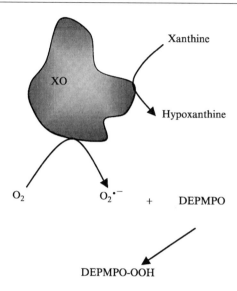

F IG . 4. Schematic illustration of superoxide trapping by DEPMPO in a X/XO reaction. In this method, one can simultaneously follow oxygen decay as well as the superoxide produced.

oxygen radicals are short lived compared to the EPR detection time scale, they are normally stabilized by reacting with the spin traps.

To follow the kinetics of oxygen consumption and $O_2^{\bullet-}$ production, EPR data are acquired at regular time intervals using two separate sweep widths and other settings corresponding to the LiPc and DEPMPO-OOH, as described earlier, for a reaction mixture containing X (0.1 mM), XO (0.005 U/ml), DTPA (0.5 mM), DEPMPO (50 mM), and LiPc particles ($\sim$25 μg) in phosphate-buffered saline equilibrated with ambient air, filled in a 25-μl microcapillary tube, and sealed both ends (Fig. 2). Serial EPR data acquisition is initiated to measure both the spin adduct and LiPc spectra as separate data files. Typical EPR spectra of LiPc obtained at various times of X/XO reaction are shown in Fig. 5. The continuous utilization of oxygen is revealed in the progressive decrease in the line width of the LiPc spectrum. The measured peak-to-peak line widths are used in combination with calibration data to compute the oxygen concentrations in the medium. Because very low microwave power (20 μW) and low modulation amplitude (5 mG) are used in a narrow sweep window (2.0 G), there is no influence of the other peaks from the DEPMPO-OOH. The O_2 consumption is evaluated by monitoring the line width change of LiPc spectrum with time, whereas the superoxide is measured as the DEPMPO-OOH adduct.

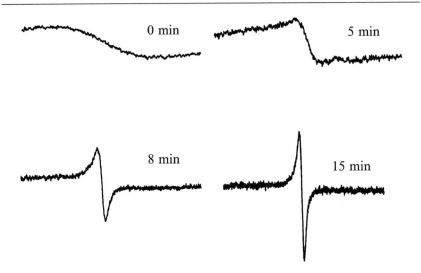

FIG. 5. EPR of LiPc in saline containing a stimulated X/XO reaction. The EPR line width narrows down due to the continuous consumption of oxygen by PMNs.

Figure 6 shows two typical EPR spectra of the DEPMPO-OOH adduct that were obtained simultaneously from the same X/XO reaction mixture. Spectra show essentially two major components: a multipeaked component from DEPMPO-OOH, whose intensity can be used to quantify the superoxide adduct, and a strong singlet peak of LiPc in the center of the spectrum, whose line width is overmodulated due to the higher modulation used to acquire the DEPMPO-OOH spectrum. The DEPMPO-OOH adduct has 12 EPR hyperfine lines, as expected, which can be split into two identical sets such that each set has six lines.[34] Two of these six lines overlap very closely with two other peaks so that the set would appear as a 1:2:2:1 quartet.[32,33] The DEPMPO-OOH spectrum obtained after 20 min, as shown in Fig. 6, also has some small contributions from DEPMPO-OH, which is seen as shoulders inside the DEPMPO-OOH spectrum. These DEPMPO-OH peaks are known to result from either the breakdown of the DEPMPO-OOH adduct or directly from the Fenton-type reaction.[34] The time course of DEPMPO spin adduct spectra is analyzed and quantified to obtain the individual concentrations of the different spin adducts and their decay into EPR inactive forms. The evaluated parameters, namely the coupling constants, are the same as reported earlier.[34]

[34] C. Frejaville, H. Karoui, B. Tuccio, F. Le Moigne, M. Culcasi, S. Pietri, R. Lauricella, and P. Tordo, *J. Med. Chem.* **38,** 258 (1995).

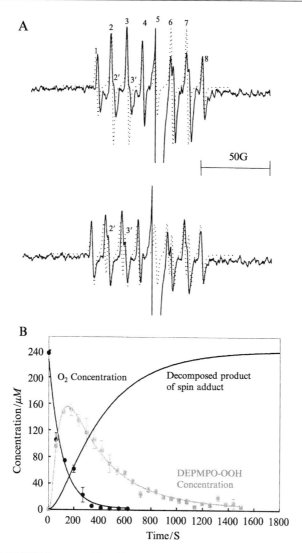

FIG. 6. (A) DEPMPO–superoxide adduct EPR spectrum after 5 (top) and 30 (bottom) min (represented by solid lines). The DEPMPO-OOH spin adduct yields 12 line EPR spectrum. Overlap of the LiPc spectrum occurs in the middle peaks. However, the fitting of experimental data can be optimized using the other peaks as shown to quantitate the concentration of the spin adduct. (B) Kinetic analysis and concentration profiles of O_2 and DEPMPO-OOH and its decomposed product. Exponential decay was analyzed using Eq. (2), and the rise and fall of the DEPMPO-OOH were analyzed by fitting with Eq. (3) (see text). In both cases, a good fit was observed illustrating the validity of the model.

The total DEPMPO-OOH concentration peaked to 24 μM at about 3 min and then decayed with time (Fig. 5). It should be noted that the paramagnetic adduct DEPMPO-OOH is not stable in solution and is known to undergo slow decomposition to give DEPMPO-OH and nonparamagnetic (EPR inactive) products.[34] EPR measurements give the total accumulated concentration of the adduct at any given time. The observed variation in the adduct concentration clearly reflects the formation-decay characteristic of the adduct. The DEPMPO-OH concentration remained small and almost constant at approximately 3 μM throughout the time of the study. The oxygen concentration was evaluated from the EPR line width of the LiPc spectra. Both the oxygen and DEPMPO-OOH concentrations were plotted as a function of time in Fig. 6. As can be seen, the O_2 concentration decreases exponentially and approaches zero within about 7 min. The oxygen depletion was analyzed using Eq. (2). The formation and decay reaction sequence of the DEPMPO-OOH adduct can be analyzed quantitatively by considering the consecutive reaction model described earlier and in Eq. (3). Oxygen concentration data were analyzed to obtain the rate constant of the overall oxygen consumption in the medium. We evaluated the decay rate constant of DEPMPO-OOH to be $4.1 \pm 0.2 \times 10^{-3}$ s^{-1}. We evaluated the values of k_f and $[\text{DEPMPO-OOH}]_{max}$ as $7.9 \pm 0.2 \times 10^{-3}$ s^{-1} and 42.5 ± 5.0 μM, respectively. The total amount of superoxide generated in the system can be computed based on the trapping efficiency of the DEPMPO. The efficiency for the concentration of the trap used is 65%.[32,33] This gives an estimate of 65.4 ± 7.7 μM for the total quantity of $O_2^{\bullet -}$. This value suggests that about 30% of the consumed oxygen is converted to $O_2^{\bullet -}$.

Measurement of Oxygen Consumption and Superoxide Generation in Human Polymorphonuclear Neutrophils

This study demonstrated that the present technique can be used for the measurement of oxygen and superoxide during cellular respiration by taking human leukocytes as an example. Oxygen is a ubiquitous electron acceptor in aerobic biological systems, including the "respiratory burst" of human leukocytes (polymorphonuclear neutrophils, PMNs). PMNs, during the immune defense against bacterial infection, reach the site of infection, are activated, and increase their O_2 uptake from the surrounding medium with a concomitant generation of large amounts of $O_2^{\bullet -}$ that lead to the formation of toxic O_2 metabolites. It is known that a major portion of extracellular or cytoplasmic free radicals observed are produced by membrane-bound NADPH oxidase, whereas the free radicals generated by mitochondria are scavenged by the antioxidants present in the mitochondria. The NADPH oxidase is a complex enzyme, which requires translocation and association in the plasma membrane of several proteins.[35] Activation of the complex NADPH oxidase leads to the reduction of molecular oxygen to $O_2^{\bullet-}$. Despite these understandings, questions remain regarding the quantitative relation between the amount of O_2 consumed during the respiratory burst and the quantity of toxic metabolites, including $O_2^{\bullet-}$ produced. We illustrate here how both O_2 and free radicals can be quantitated simultaneously using the proposed method.

In order to understand the kinetics of oxygen consumption and superoxide production, we simultaneously performed spin trapping in conjunction with LiPc oximetry. In the reaction mixture containing 5×10^6 cells/ml of PMNs stimulated by PMA, the EPR spectrum of LiPc sharpened as illustrated in Fig. 5 for X/XO. The EPR line width sharpens gradually due to the continued consumption of O_2. The O_2 concentrations obtained from line width data were observed to decrease exponentially. $O_2^{\bullet-}$ radicals generated by stimulated PMNs were trapped as DEPMPO-OOH as in the case of X/XO described earlier. EPR spectra obtained at various times consisted of both DEPMPO-OOH and DEPMPO-OH adduct signals. The superoxide adduct is the direct product of the cellular consumption of oxygen, and the hydroxyl adduct is formed by the decomposition of the superoxide adduct.[32,33] The kinetics of O_2 consumption and $O_2^{\bullet-}$ production by PMNs were analyzed using the sequential reaction model as in the case of X/XO, and the corresponding parameters were obtained using the same kinetic pattern and equations. An exponential fit to data showed a pseudo-first-order rate constant [k_o of Eq. (2)] of $7.0 \pm 0.2 \times 10^{-3}$ s^{-1} for oxygen consumption in this system. The DEPMPO-OOH generation profile was similar to that of the X/XO system (data not shown). A rate constant of $3.1 \pm 0.2 \times 10^{-3}$ s^{-1} was obtained from data for the DEPMPO-OOH adduct formation [k_f, Eq. (3)] using the decay rate constant (k_d) of $1.7 \pm 0.2 \times 10^3$ s^{-1} determined as in the case of the X/XO system. The corresponding [DEPMPO-OOH]$_{max}$ concentration was 50.0 ± 6.0 μM, which gives an estimate of 76.0 ± 9.0 μM for the total amount of the superoxide generated.

Summary and Conclusions

The results presented in this article demonstrate that EPR oximetry based on the LiPc microcrystalline powder as a probe in combination with the DEPMPO spin trap can be used for the simultaneous measurement of

[35] I. Maridonneau-Parini, S. E. Malawista, H. Stubbe, F. Russo-Marie, and B. S. Polla, *J. Cell Physiol.* **156**, 204 (1993).

oxygen consumption and reactive oxygen-free radical species in enzymatic reactions and cellular respirations. Because the sample requirement is less than 20 μl, the present technique is ideally suited for cases where very little sample is available. From the illustrated examples, it is clear that such an experiment is not restricted only to enzymatic reactions, but can also be applied for cellular respirations. Especially in the lower oxygen concentration range, the EPR oximetry method is very accurate, unlike the established methods. The DEPMPO used in the present work is also a reasonably good spin trap with a relatively higher spin adduct half-life time of the spin adduct. The quantitative analysis of the concentration is a little complex, as there is also simultaneous decomposition. However, with appropriate computations, the absolute concentration of superoxide and OH radicals could be determined. From the results obtained in these studies, it is estimated that about 30% of the consumed oxygen is leaked as free radicals. Although we have illustrated simple uses of this technique as examples, this technique can be used to study the effects of specifically blocked or induced individual steps on the overall oxygen consumption and superoxide productions in different cell lines.

Acknowledgments

We acknowledge the financial support from National Cancer Institute Grant CA 78886 and NIH Grant RR 12190. GI was supported by American Heart Association Postdoctoral Research Grant 0120509U during the tenure of this work.

Author Index

Numbers in parentheses are footnote reference numbers and indicate that an author's work is referred to although the name is not cited in the text.

A

Aaronson, P. I., 3, 71, 72, 77(2; 3), 78(2; 9), 79(3; 9), 81(9), 82, 82(9), 251, 252, 258
Abbott, B. M., 91, 387
Abboud, F. M., 108
Abe, H., 680
Abe, K., 399, 401(2)
Abee, Y., 593
Abernathy, K., 460
Abman, S. H., 92, 93, 99, 100(48)
Abo, A., 185
Abraham, N. G., 141, 142
Abramovitch, R., 346
Achebach, S., 628
Acinni, L., 228
Acker, H., 44, 47(13), 133, 377, 403, 488, 489, 490, 490(17; 27), 491, 492, 492(31), 494, 495, 495(17), 496(29), 498(29; 46), 504, 509(55), 510(55), 589, 699
Acker, H. T., 508
Acker, T., 93, 344
Acworth, I. N., 601
Adamietz, P., 570
Adamson, T., 216(7), 217
Adelman, J. P., 258
Adnot, S., 90, 91
Adriaensen, D., 4, 14(8), 15(8), 16, 267(29), 268, 270(29), 271(29)
Afanas'ev, I. B., 173, 174(19)
Afzal, M., 47
Agah, A., 511, 519(5)
Agani, F., 122, 218, 219(2), 345, 346, 347, 383, 387, 520, 589
Agapito, M. T., 40, 41, 44, 44(3), 46(3), 47(10), 48(3; 10), 50, 52(3; 10), 67(51), 69(51)

Agarwal, R., 170
Ahn, B.-W., 594
Ahrens, M., 399
Ailhaud, G., 388
Akaike, T., 347
Akakura, N., 347
Akça, O., 552, 580
Aki, M., 625
Akita, S., 471
al-Achi, A., 352
Alam, M., 618, 619(1)
Alani, E., 661
Alarcon, R. M., 336
Alberts, D., 104
Albina, J. E., 527, 528(1), 529, 529(1), 530(7), 532(1), 533(1), 565, 566(9)
Albrecht, J. A., 640
Alder, N. P., 607, 615(36), 616(36)
Aldinger, A. M., 248, 252(53), 256(53)
Alexander, D. L., 391
Alexander, S., 502
Alger, L., 103, 106(51)
Alger, L. A., 91, 92, 97(30), 99(30), 101(30), 102(30)
Ali, M. A., 570
Allard, J. D., 91, 96
Allen, C. B., 451
Allen, D. G., 275
Allen, L. A., 645, 646, 646(9), 647(9), 651(12)
Almaraz, L., 11, 41, 43(5), 49, 58, 63, 64
Alonso-Galicia, M., 142, 143(23; 25), 144(25), 145(20), 149(20), 154(20; 22), 155(25), 160(23), 164(23), 166(23)
Alpert, N. M., 735
Alt, F. W., 218, 219(3), 223(3)
Althaus, F. R., 572
Alvarez, L., 735

Krishna, U. M., 141, 143(17), 144, 149(30), 154(30), 155
Krishnamachary, B., 347
Krishnamurti, L., 209
Kroetz, D. L., 142
Kropat, J., 605, 607
Kruuv, J., 335
Kubota, S., 351
Kubowitz, F., 696
Kucho, K., 605, 610(21)
Kuge, Y., 417
Kuhl, M., 734
Kuipers, J. G., 104
Kukreja, R. C., 168
Kulisz, A., 45
Kull, F. C., Jr., 568
Kumar, G. K., 107
Kummer, W., 46, 238, 255(26), 403, 495, 504
Kunert, M. P., 142, 143(25), 144, 144(22; 25), 149(22), 154(22), 155(25)
Kunst, F., 643
Kuo, L., 705
Kupper, H., 133
Kupper, J. H., 569
Kuppusamy, P., 579, 747, 749, 750(11; 12; 14–17), 753, 754(12; 13), 761(32; 33)
Kurachi, Y., 255
Kurz, A., 552, 580
Kusumi, A., 635
Kuwano, M., 377
Kwast, K. E., 589, 591, 601, 644, 645, 646(10), 648, 648(10), 653(5), 654(5), 656(5), 660(5), 661(3)
Kwiatkowski, A. V., 223
Kwon, M., 12
Kwon, O., 629
Kyle, J. W., 277

L

Labarre, J., 602
Labialle, S., 383
Lad, C., 72, 77(8)
Ladanyi, M., 104
Ladd, C., 104
Ladebeck, R., 749
Laderoute, K. R., 348
Ladhoff, A., 135
Lafferty, K., 556, 557(24)
La Fontaine, S., 605

Lagadic-Gossmann, D., 234, 240(17)
Lagniel, G., 602
Lahiri, S., 9, 70, 488, 489, 490(26), 491, 492(31), 494, 504, 682, 698, 699, 700(29), 701(29), 702(21)
Lai, L.-C., 645, 653(5), 654(5), 656(5), 660(5)
Laitinen, H. A., 684
Lakatta, E. G., 241, 275, 289(1)
Lakowicz, J. R., 718, 719(25), 720, 722(37; 44), 723, 723(25), 724
Lal, A., 345
Lalonde, B., 661
Lam, K. P., 220
LaManna, J. C., 589, 687, 688
Lambeth, J. D., 49, 495
Lamers, W. H., 360
Lamouroux, A., 438
Land, S. C., 680
Lander, E. S., 102, 104
Lando, D., 133, 305, 467, 468(5), 475(5), 477(5), 479(4), 480(4), 481(4), 490, 508(28)
Landzberg, M., 91
Lane, P., 527
Lane, W. S., 305, 306(4), 308(4), 309(4), 467
Lanfumey, L., 90
Lange, A. R., 141, 142(9; 16), 143(9; 16), 144(9), 149(9), 160(16), 166(16)
Lange, B., 46
Langleben, D., 90, 91
Langness, U., 567
Langsdorf, B., 720, 721(35), 722(35)
Laniado-Schwartzman, M., 141, 142, 143(17)
LaNoue, K. F., 735
Lansford, R., 218, 219(3), 223(3)
Lanzen, J. L., 676
Lapadat, R., 106
Larsen, R. W., 618, 619(1)
LaRusch, J., 347
Laser, H., 696
Lassègue, B., 167, 169(5)
Latorre, R., 258
Lau, Y.-F., 452
Laue, O., 667, 672(8)
Laughner, E., 122, 218, 219(2), 345, 383, 387
Launay, J. M., 90
Lauricella, R., 758
Laurindo, F. R., 47
Laursen, I., 749

Mowery, K. A., 705
Mueller-Klieser, W., 342, 495, 679
Mukai, T., 133
Mukherji, M., 305, 313(9), 467, 482(2)
Mukhopadhyay, D., 574
Mukkerji, M., 305, 306(5), 309(5)
Mulligan, E., 9, 698
Mulshine, J. L., 15
Mulvany, M. J., 72, 75(14)
Münck, E., 629, 636(13)
Mungai, P. T., 168
Münzel, T., 173, 174(19)
Murakami, A., 593
Murakami, M., 225
Murali, K. U., 149, 151(38), 154(38)
Murali, S., 90
Muramatsu, M., 393
Murdock, P. R., 262, 265(24)
Muro, J. R., 705
Murphy, E. D., 228
Muschen, M., 377, 495
Musson, D. G., 467, 469(9), 470(9), 483(9)
Mustacich, D., 55
Myers, E. N., 256
Myers, S. M., 736
Myssiorek, D., 256

N

Nacci, A., 716
Naczki, C., 488
Nagai, K., 619
Nagao, M., 321, 400, 405(23)
Nageswari, K., 704
Nair, P., 42, 43(7), 494, 666, 668(7),
    670, 671, 681, 682(52)
Nakajima, H., 618, 619(9)
Nakamasu, K., 393
Nakamoto, S. S., 605, 607,
    615(28), 617(38)
Nakamura, M., 225
Nakamura, Y., 104, 605, 610(22)
Nakashima, T., 342
Nakata, T., 546
Nakazawa, T., 416, 417(3)
Nana-Sinkam, P., 90, 106(9)
Nanbu-Wakao, R., 393
Narasimhan, L. S., 275
Narayanan, J., 141, 142(9; 14; 16), 143(9; 16),
    144(9), 149(9), 160(16), 166(16)

Narayanan, L., 250(68), 349
Narravula, S., 513
Nasjletti, A., 141, 142, 143(17)
National Center for Health Statistics, 121
Nauck, M., 361, 366(19)
Nawashiro, H., 399(12), 400
Naylor, P. F. D., 682
Near, J. A., 8
Nebigil, C. G., 90
Neckers, L., 348
Nedelec, A. S., 399, 400(5), 401(5), 403(5),
    406(5), 407(5), 411(5), 412(5)
Neely, J. R., 735
Neeman, M., 346
Negelein, E., 691, 694(3; 4)
Negi, A., 687
Negrel, R., 388
Neher, E., 242, 282
Neidhardt, F. C., 635
Nelin, L. D., 93
Nelson, D. P., 241, 252, 256
Nemery, B., 93
Neubauer, M. G., 520
Neverova, I., 192
Newby, A. C., 91
Newman, A. B., 107
Newman, C., 26, 29(5), 31(5), 34(5), 37(5)
Nguyen, K., 495
Nguyen, T. B., 475
Nguyen, X., 143, 145, 155(31)
Nguyen-Huu, L., 248, 249(54)
Nichols, C. G., 241
Nicolaisen, E. M., 468
Nicolaus, B., 569
Nies, A. S., 91, 96(15)
Nieto, F. J., 107
Niinikoski, J., 532, 548
Niinobe, M., 399
Niitsu, Y., 457
Nilzeki, H., 347
Ninomiya, Y., 5
Nioka, S., 736
Nisbet, J. W., 10, 243, 249(43),
    253(43), 254(43)
Nishida, T., 351
Nishimura, T., 91
Nishino, T., 9
Nishio, S., 399
Nithipatikom, K., 141, 142(16), 143(16),
    160(16), 166(16)

Subject Index

A

Absorption spectrophotometry
heme proteins in afferent carotid sinus
nerve activity
cytochrome a592
characterization, 492–494
cytochrome b558 characterization, 495
overview, 490
photometry, 490–491
spectral deconvolution, 492
spectra relationship with photochemical
action spectra, 693–694
Adipogenesis, adipocyte differentiation and
oxygen regulation
cell culture systems, 388–389
Northern blot analysis of transcription
factors, 391–393
Oil Red O staining, 389, 391
transcription factors, 388
Western blot analysis of transcription
factors, 393–394
Aerotaxis, *see* HemAT
Amperometry
hypoxia-evoked exocytosis in PC12
cells, 6–11
neuroepithelial body serotonin release
in hypoxia
advantages, 35, 37
calcium flux role, 40
data analysis, 39–40
equipment, 37, 39
solutions, 37
Anemia, fetal sheep model
adult catheterization studies, 213–215
anesthesia, 212
capillary density determination, 215–217
delivery, 213
hemodynamic changes, 211
maximal coronary conductance
response, 211
persistence of effects, 211
surgery, 212

Apoptosis, hypoxia induction, 335–336
Autoantibodies, measurement in
hypoxia-inducible factor-1α-
deficient chimeric mouse
antidouble-stranded DNA
antibodies, 225–226
rheumatoid factors, 226

B

Bacillus subtilis oxygen sensing, *see* HemAT
B-cell, hypoxia-inducible factor-1α-deficient
chimeric mouse studies of autoimmunity
and lymphocyte development
autoantibody production measurement
antidouble-stranded DNA
antibodies, 225–226
rheumatoid factors, 226
B-cell development, 227–228
flow cytometry analysis of lymphoid cell
phenotypes, 224–225
RAG-2-deficient complementation assay,
218–221, 223
Brain stem catecholaminergic cells
carotid chemodenervation for oxygen
sensing studies
anesthesia, 426
materials, 425–426
surgery, 426–427
hypoxia evaluation
biochemical analysis of catecholamines
and metabolites, 447–449
hypoxia condition induction, 433–434
immunohistochemistry of tyrosine
hydroxylase and hypoxia-inducible
factor-1α expression, 442–443
norepinephrine turnover
assay, 443–444
oxygen tension, intracerebral
measurements, 437
respiratory motor output
recording, 436–437
respiratory neuronal activity, 437

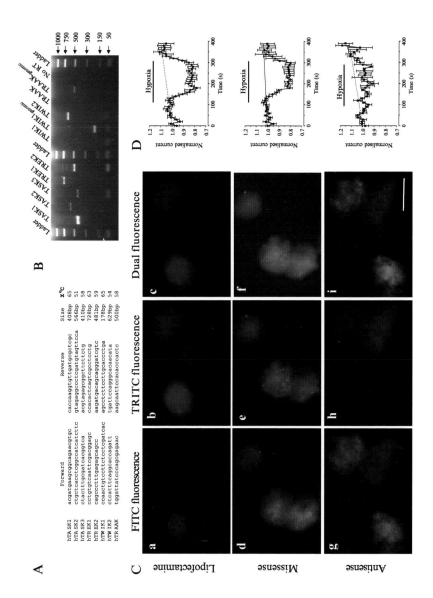

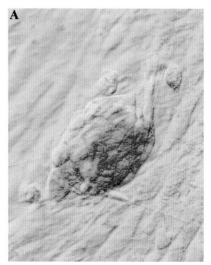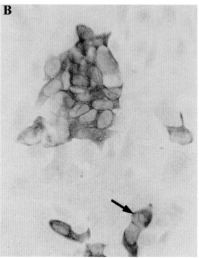

Cutz *et al.*, Chapter 2, Fig. 4. (A) Vital staining of a 3-day culture of NEB isolated from rabbit fetal lung (E26) with neutral red visualized under Hoffman modulating contrast optics. The culture was incubated with 0.02 mg/ml neutral red for 30 min at $37°$. Positive selective staining of the NEB cell cytoplasm (red) contrasts with other lung cells, which are unstained (neutral red, $\times$550). (B) NEB in a 3-day culture fixed in neutral-buffered formalin and immunostained for 5-HT. NEB forms a closely packed cluster of 5-HT-immunoreactive cells (dark cytoplasm) and negative nuclei. Few isolated 5-HT-immunoreactive NEB cells are seen outside of the cell cluster (arrow) (immunoperoxidase method for 5-HT, $\times$550).

Kemp and Peers, Chapter 1, Fig. 6. RT-PCR screening for tandem P-domain K^+ channel mRNA and loss-of-function experiments in H146 cells. (A) Primer pair, optimal annealing temperatures, and predicted amplicon size employed for RT-PCR screening. (B) A 2% agarose gel showing products amplified from reverse-transcribed H146 mRNA employing the specific primer pairs complementary to the K^+ channels indicated above each lane. Also shown is a reaction in which mRNA was not reverse transcribed (No RT) and three DNA ladders running at the base pair number indicated to the right. For TWIK1 and TRAAK, reactions employing genomic DNA are also shown as a positive control for the efficiency of those reactions. The forward and reverse primer sequences employed to amplify each specific channel mRNA, the optimized annealing temperature (in $°$C), and the expected product size for each reaction are shown in A. (C) Exemplar high-power images from lipofectamine-only (a–c), missense (d–f), or antisense oligodeoxynucleotide (g–i)-treated H146 cells. Column 1 (a, d, and g) shows FITC localization of the probe, where applicable. The second column (b, e, and h) shows TRITC localization of the anti-hTASK-1 antibody. The third column (c, f, and i) shows dual-fluorescence images and demonstrates that only antisense transfection resulted in specific hTASK-1 protein knock down. The scale bar represents 10 μm and applies to all panels. (D) Mean time courses of the effect of acute hypoxia (15–25 mm Hg, applied for the period indicated by the horizontal bar) on cells treated only with lipofectamine (A; $n = 11$), cells treated with missense oligodeoxynucleotide (B; $n = 5$), and cells treated with antisense oligodeoxynucleotide (C; $n = 12$). Adapted from Hartness *et al.*

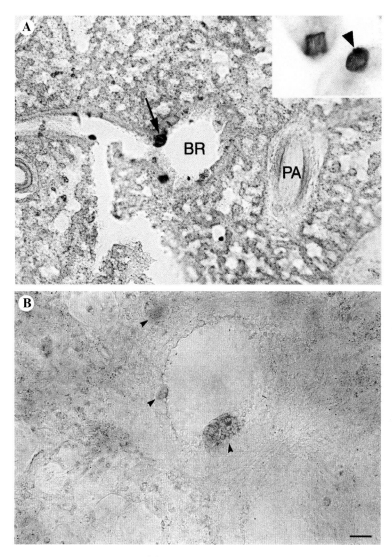

CUTZ *ET AL.*, CHAPTER 2, FIG. 5. (A) Low magnification view of a neonatal rabbit lung slice (~200 μm thick) fixed in neutral-buffered formalin and immunostained for 5-HT. A small bronchiole (BR) with a NEB cell cluster strongly positive for 5-HT is located at airway bifurcation (arrow). (Immunoperoxidase method for 5-HT; bar: 100 μm.) PA, pulmonary artery. (Inset) Higher magnification of airway epithelium with two 5-HT-immunoreactive NEBs facing each other with their apical surfaces exposed to airway lumen (arrowhead) (immunoperoxidase method for 5-HT; bar: 5 μm). From Fu *et al.* (B) Staining of NEB cells. Arrowheads show neutral red staining of NEB cells in a fresh slice (~200 μm thick) of neonatal rabbit lung (2 days old) (bar: 5 μm). From Fu *et al.*

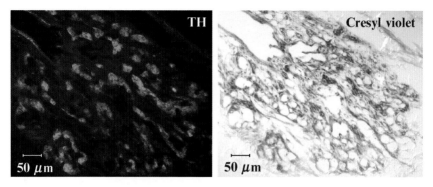

GONZALEZ *ET AL.*, CHAPTER 3, FIG. 1. Histological sections near the equator of a rat carotid body. The section was first immunostained using a monoclonal antibody against tyrosine hydroxylase (TH) and a secondary antibody labeled with fluorescein isothiocyanate to label the TH-containing chemoreceptor cells that are grouped in clusters. The section was later counterstained with cresyl violet to show the great density of capillaries in the CB tissue.

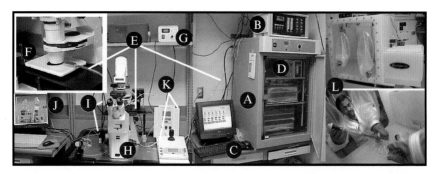

ROY *ET AL.*, CHAPTER 8, FIG. 2. Cell culture and live cell imaging under a controlled O_2 environment. Cells are grown in an incubator (A) where the gaseous environment can be regulated variably as a function of time using the OxyCycler (Biospherix, Redfield, NY) (B) controlled by PC software (C) capable of cycling (in minutes) and maintaining ambient O_2 and CO_2 concentrations based on feedback information from O_2 sensors installed in the incubator (A). Mixed gas is delivered by pump (D) via tubing (E) to a cell culture enclosure (F, zoomed inset), which is heated by G. Cells in F are imaged by an Axiovert 200 M (Zeiss) fully motorized fluorescence microscope (H) supported by dual (color and B/W) AxioCam digital cameras (I) and rested on an air table. Time-lapse images are collected and analyzed using Axiovision software (Zeiss) installed in a PC (J). The microscope contains necessary hardware/software to image cells grown on standard plastic culture plates (growing on glass coverslips is not necessary). A twin-tip micromanipulator (InjectMan NI2, Eppendorf) and microinjector (Femtojet, Eppendorf) system (K) is attached to the microscope to perform microinjections/manipulations as well as the collection of nuclear materials from single cells. For splitting/seeding of cells under a controlled gas environment, a specialized glove box (L) fitted with an O_2 controller (PRO-OX, Biospherix) is available. Six gas-controlled incubators are available for parallel experimentation.

Roy *ET AL.*, Chapter 8, Fig. 3. Live cell imaging using fluorescence microscopy. Cotransfection of actin-EGFP (green fluorescence) and nuclear-targeted DS2-red (golden fluorescence) plasmids are used to visualize cytoskeletal and nuclear changes in live cells. Three images are shown after the indicated time from a 2-h time-lapse series where images were collected in a digital video format. Images were collected using the live cell microscopy system as described (Fig. 2).

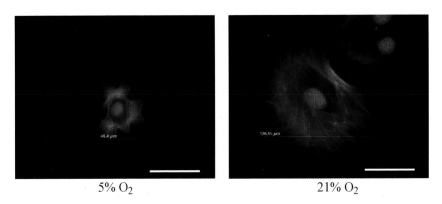

Roy *ET AL.*, Chapter 8, Fig. 4. Freshly isolated CF are phenotypically closer to CF cultured at 5% O_2 compared to CF cultured at higher O_2 tensions. After isolation, CF were cultured at 5 or 21% O_2 for 8 days. For morphologic comparisons, CF cultured at various O_2 tensions for 8 days were stained with phalloidin (actin, red) and nuclei were stained with DAPI (blue). Imaging was performed using a Zeiss microscope (Fig. 2). Scale bar: 50 μm.

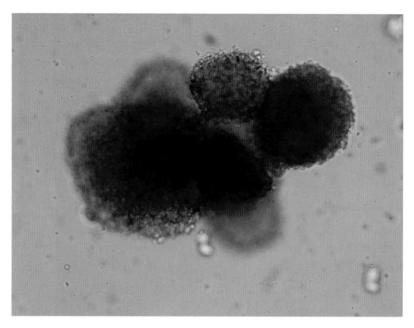

JELINEK AND PRCHAL, CHAPTER 14, FIG. 2. Multicentric erythroid colony derived from a BFU-E progenitor cell.

KEMP *ET AL.*, CHAPTER 18, FIG. 1. Using a gene chip microarray to screen for expression cation channel mRNA. (A) Principles of microarraying. Total RNA is extracted from cells or tissues before, during, and after a biological or environmental change, such as adaptation to hypoxia (I). Messenger (m)RNA is used as a template for the synthesis of first-strand cDNA by reverse transcriptase (RT) extension of annealed oligo(dT)primers (II). The cDNA can be labeled either directly through the enzymatic incorporation of fluorochrome-modified nucleotides during the RT reaction or indirectly through the incorporation of aminoallyl-dUTP in the RT reaction, followed by chemical coupling of the fluorochrome label to the aminoallyl-cDNA. Messenger RNAs isolated from cells or tissues under different experimental conditions are labeled with different fluorochromes, such as Cy3 and Cy5, and pooled together in a single hybridization reaction (III). After hybridization, the fluorescence intensity at each spot is measured at the appropriate emission wavelength for both fluorochromes and the results are processed and analyzed using commercially available image analysis software (IV). The intensity of fluorescence at each spot is, to a large extent, related to the relative abundance of transcripts complementary to the arrayed DNA probes. Consequently, if gene "A" is being expressed under "normal" conditions, but not under "experimental" conditions, then the probe for gene "A" will fluoresce only at the emission wavelength of the fluorochrome used to label "normal" cDNA. Conversely, if gene "B" is activated in response to environmental or biological change, then fluorescence will be detected only at the wavelength of the fluorochrome used to label "experimental" cDNA. Fluorescence will be detected with approximately equal intensity at both wavelengths when the expression levels of the genes represented on the array do not change in response to the environmental or biological changes under study. (B) Basal expression levels of ion channels in native

A

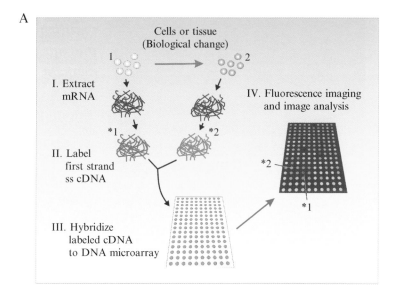

B

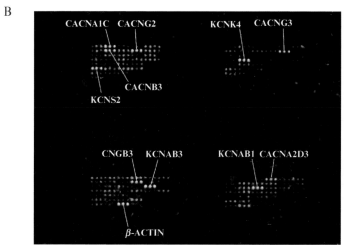

untransfected HEK293 cells. An experimental array of 149 DNA sequences (spotted in triplicate), consisting of expressed sequence tags (ESTs) and PCR-amplified, gene-specific sequences, was designed to measure the expression levels of 90 different human cation channel genes (some of which are amplified from more than one EST of the same gene). Total RNA was extracted from HEK293 cells growing under conditions of normoxia, and equal amounts of RNA were used to generate Cy3- and Cy5-labeled cDNA as described in A. Labeled cDNAs were pooled and hybridized to the array, and fluorescence intensities at each spot were measured. Signal intensities were normalized to the β-actin-positive control and superimposed using ArrayPro4 software. Results suggest that only a minimal array of cation channels are expressed in native untransfected HEK293 cells.

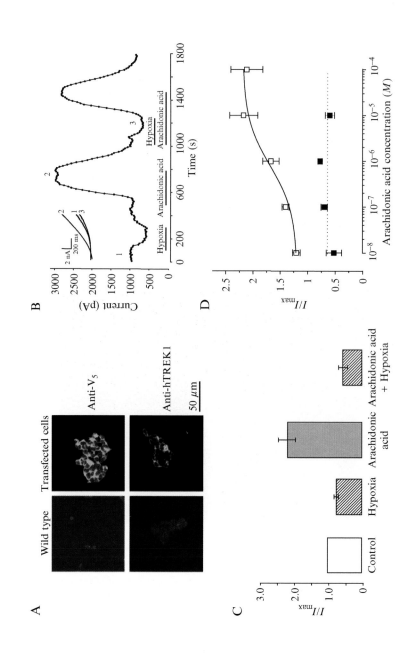

K<small>EMP</small> *ET AL.*, C<small>HAPTER</small> 18, F<small>IG</small>. 4. Occlusion of hTREK1 regulation by hypoxia. (A) Immunocytochemical verification of hTREK1 stable expression using anti-V_5 (top) and anti-hTREK1 (bottom) antibodies in transfected (right) and untransfected (left) HEK293 cells. (B) Exemplar time series plot of hTREK1 whole cell K^+ currents during hypoxic challenge (*ca.* 40 mm Hg) and application of 10 μM arachidonic acid or both as indicated by the horizontal bars. Currents were recorded during the final 10 ms of the 200-ms voltage step from −70 to 60 mV. (Inset) Numbered ramp currents corresponding to the points indicated on the time series plot. (C) Mean, normalized whole cell hTREK1 K^+ currents recorded at 60 mV during hypoxic challenge (*ca.* 40 mm Hg) and application of 10 μM arachidonic acid or both as indicated at the bottom of each bar. (D) Mean concentration–response data for arachidonic acid in normoxia (□) and hypoxia (■). Taken from Kemp *et al.*

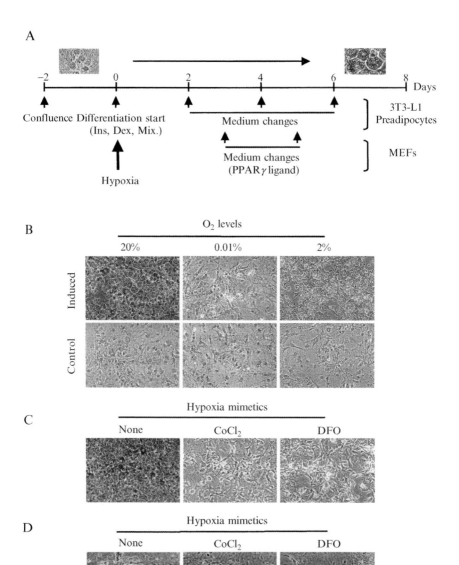

A

−2   0   2   4   6   8   Days

Confluence Differentiation start
(Ins, Dex, Mix.)

Medium changes   3T3-L1 Preadipocytes

Medium changes
(PPARγ ligand)   MEFs

Hypoxia

B

O$_2$ levels

20%   0.01%   2%

Induced

Control

C

Hypoxia mimetics

None   CoCl$_2$ DFO

D

Hypoxia mimetics

None   CoCl$_2$ DFO

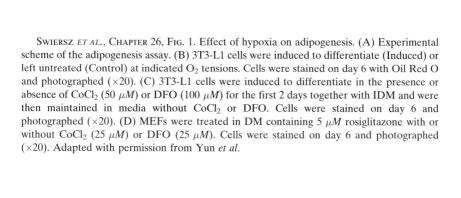

SWIERSZ *ET AL.*, CHAPTER 26, FIG. 1. Effect of hypoxia on adipogenesis. (A) Experimental scheme of the adipogenesis assay. (B) 3T3-L1 cells were induced to differentiate (Induced) or left untreated (Control) at indicated O_2 tensions. Cells were stained on day 6 with Oil Red O and photographed (×20). (C) 3T3-L1 cells were induced to differentiate in the presence or absence of $CoCl_2$ (50 μM) or DFO (100 μM) for the first 2 days together with IDM and were then maintained in media without $CoCl_2$ or DFO. Cells were stained on day 6 and photographed (×20). (D) MEFs were treated in DM containing 5 μM rosiglitazone with or without $CoCl_2$ (25 μM) or DFO (25 μM). Cells were stained on day 6 and photographed (×20). Adapted with permission from Yun *et al.*